CALCULUS
and Analytic Geometry

CALCULUS
and Analytic Geometry

Philip S. Clarke, Jr.
Los Angeles Valley College

D.C. HEATH AND COMPANY
Lexington, Massachusetts Toronto London

Preface

The philosophy underlying this work is one that has gradually developed from the author's experience in teaching calculus. During this time calculus books have come and gone, and the mathematical community's views on what a first course in calculus should be have undergone substantial change. My experience in teaching from books of widely divergent natures has convinced me that while the beginning calculus student stands to gain significantly from considerable exposure to rigor and the theoretical foundations of calculus, pedagogy very definitely does impose limits to the quantity of the latter that can profitably be presented. In short, while there can be little defense for excluding all rigor from a first course in calculus, there can also be little defense for trying to make such a course into an advanced calculus course in which computational applications appear as asides. Accordingly, I have attempted to state precisely and display conspicuously all definitions and theorems used, and to prove as many theorems as experience leads me to believe can be grasped by a significant number of beginning students. I have not attempted to treat in detail the completeness property of the real numbers or topological notions such as limit points, nor have I attempted to present proofs of deeper theorems in analysis, such as the attainment of extreme values by a continuous function on a closed interval, integrability of continuous functions, and so on. I have tried to convey some of the ideas behing the proofs presented whenever this seems feasible, in order to avoid the impression many beginners come away with that all mathematical proofs consist of ad hoc unmotivated cleverness beyond the ability of all but an elite corps of geniuses to produce.

It should be noted that the major concern in this text is not merely with how much and what kind of theory to present, but also with presenting a considerable array of applications of calculus to other fields including physics, biology, and economics. In these chapters, for example, one finds in Section 7.2 a discussion of applications of antidifferentiation. Chapters 5, 6, and 8 contain a great deal more in the way of applications, as can be inferred from the section titles, and in general I have tried to include applications where they arise naturally in the development of the techniques of calculus.

In conclusion, I want to thank the reviewers of the manuscript, Professor James Hurley of the University of Connecticut, Professor David Sanchez of the University of California at Los Angeles, and Professor Charles Seekins of Occidental College for their many constructive suggestions, and to express my appreciation to Professor Hurley and his wife, Cecile, for their fine work in preparing the student guide that accompanies this text. Also, I must acknowledge the worthwhile assistance and counsel of my agent, Paul E. Harris, Jr. of Academic Authors. Finally, I am especially indebted to my wife, Louise, who typed the major portion of the manuscript and was a constant source of encouragement during this entire project.

Philip S. Clarke, Jr.
Van Nuys, California

Contents

9. Exponential and Logarithmic Functions

10. Trigonometric and Inverse Trigonometric Functions: Hyperbolic Functions

11. Methods of Integration

12. Conics

13. Numerical Methods; Indeterminate Forms; Improper Integrals

CALCULUS
and Analytic Geometry

1. Real Numbers; Inequalities

Since this text is concerned with the study of analytic geometry and calculus, it is appropriate at the outset to comment briefly on the nature of these two branches of mathematics. Analytic geometry, which was discovered about 1637 by René Descartes (1596–1650), a French mathematician and philosopher, deals with the use of methods of algebra to solve geometric problems. The use of analytic geometry paved the way for the development of calculus in the latter part of the 17th century. The invention of calculus is usually credited jointly to Sir Isaac Newton (1642–1727), the English physicist and mathematician, and Gottfried Wilhelm Leibniz (1646–1716), a German philosopher and mathematician, since their discoveries occurred independently and at about the same time. However, it was not until the end of the 19th century that calculus finally attained its present refined form. Among the eminent mathematicians who participated in the refining of calculus were Augustin Louis Cauchy (1789–1857), a Frenchman, and Karl Weierstrass (1815–1897), a German.

Calculus has two main subdivisions, *differential calculus* and *integral calculus,* each dealing with a different form of the concept of *limit.* Differential calculus was originally devised to solve problems dealing with the tangent to a curve and the velocity of a particle at an instant in time. Integral calculus arose originally from attempts to calculate areas of regions with curved boundaries. However, as we shall discover, the scope of calculus is much broader than the solution of the problems mentioned above.

1.1   Sets; Real Numbers

In the study of any branch of mathematics one is always concerned with the mathematical behavior of collections of objects or entities. We commonly call each such collection a *set,* and the objects comprising the collection the *elements* or *members* of the set. For example, one might consider the set of all people living in Los Angeles, California, the set of all molecules in a given volume of water, or, as better serves our purpose here, various sets of numbers.

To indicate that a set S has an element a, we write

$$a \in S.$$

The expression "$a \in S$" is read "a is an element of S." On the other hand, if b is not a member of S one writes

$$b \notin S,$$

read "b is not an element of S." When speaking of elements a and b in a set, the statement "$a = b$" means that a and b represent the same element of the set.

A set may be defined by listing its elements, enclosed in braces. For example,

$$\{1, 3, 4\}$$

is the set consisting of just the numbers 1, 3, and 4, while

$$\left\{ -\tfrac{7}{2} \right\}$$

is the set consisting of the single number $-\tfrac{7}{2}$. If a set contains too many elements for convenient listing, a characterization of its elements may be indicated in some understandable abbreviated fashion. For example, the set of all *natural numbers* is denoted by the symbol

$$\{1, 2, 3, \ldots\}$$

where the dots written after the 3 mean "and so on."

As a convenience we also introduce the *null set*, a set which has no elements. The null set will be denoted by the symbol $\varnothing$.

The terms *constant* and *variable* are often used in discussing sets. A variable is a symbol that may represent an arbitrary member of a particular set that has at least one element while a symbol that can denote only one element of a set is called a constant. A set is sometimes defined by using a variable to express a property common to every element in the set. For example,

$$\{x : x^2 > 25\}$$

is the set of all numbers x such that $x^2 > 25$. The colon used here is translated as "such that."

If each element of a set S is also an element of a set T, we say S is a *subset* of T and denote this relationship between the sets by writing $S \subset T$. The symbol $\subset$ is called an *inclusion symbol*. When S is a subset of T, the relation between the sets may be visualized by a schematic diagram like Figure 1-1.

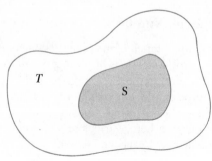

Figure 1-1

As an illustration of the notion of a subset,

$$\{1, 3, 5\} \subset \{1, 2, 3, 4, 5\}.$$

Also, note that any set is a subset of itself. If

$$S \subset T \qquad \text{and} \qquad T \subset S,$$

we say S and T are *equal sets* and write

$$S = T.$$

The *union* of sets E and F, denoted by $E \cup F$, is the set defined by

$$E \cup F = \{x : x \in E \text{ or } x \in F\}. \qquad \qquad \text{(Figure 1–2(a))}$$

The symbol $E \cup F$ is frequently read "E union F." The union of two sets consists of all objects that are elements of either of the sets (or both). If, for example, $E = \{1, 4, 6, 10\}$ and $F = \{0, 4, 7, 9, 10\}$, then $E \cup F = \{0, 1, 4, 6, 7, 9, 10\}$.

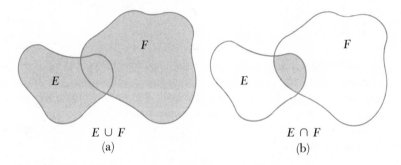

$$
\begin{array}{cc}
E \cup F & E \cap F \\
\text{(a)} & \text{(b)}
\end{array}
$$

Figure 1–2

The *intersection* of the sets E and F (shaded in Figure 1–2(b)) is the set defined by

$$E \cap F = \{x : x \in E \text{ and } x \in F\}.$$

We read $E \cap F$ as "E intersection F." The intersection of E and F consists of all objects that are elements of *both* E and F. If, for example, $E = \{1, 4, 6, 10\}$ and $F = \{0, 4, 7, 9, 10\}$, then $E \cap F = \{4, 10\}$. However, if $E = \{1, 3, 5\}$ and $F = \{2, 4, 6\}$, then $E \cap F = \varnothing$.

Throughout this text we will be concerned with the set of all *real numbers*. This set, which we denote by R, includes among other numbers, the *integers* . . . -3, -2, -1, 0, 1, 2, 3, 4, ... It also includes quotients of integers, which are called *rational numbers*. For example, $\frac{2}{3}$ and $-\frac{11}{5}$ are rational numbers, as are the integers themselves. Those real numbers that are not rational numbers are called *irrational numbers* and include, for example, $\sqrt{2}$, $\sqrt[3]{5}$, π, and $-1 + \sqrt{3}$.

In this section and in the remaining sections of this chapter some of the important properties of real numbers will be reviewed in preparation for their later use in the text. We begin by noting that the addition and multiplication of real numbers satisfy the following properties, which we will accept as axioms. In the statement of these axioms a, b, and c denote arbitrary real numbers.

Closure and uniqueness	a and b have a unique sum $a + b$. a and b have a unique product ab.
Associativity	$(a + b) + c = a + (b + c)$; $(ab)c = a(bc)$
Commutativity	$a + b = b + a \qquad ab = ba$
Identity elements	There are real numbers 0 and 1 where $0 \neq 1$ such that

$$a + 0 = 0 + a = a \qquad a \cdot 1 = 1 \cdot a = a.$$

Inverse elements	For every real number a there is a real number $-a$ such that

$$a + (-a) = (-a) + a = 0.$$

For every real number $a \neq 0$ there is a real number $1/a$ such that

$$a\left(\frac{1}{a}\right) = \left(\frac{1}{a}\right)a = 1.$$

Distributivity	$a(b + c) = ab + ac$

The rational numbers also satisfy these axioms. Each of the axioms except for the statement about the multiplicative inverse $1/a$ is also true for the integers. The natural numbers fail to satisfy the inverse element axioms and the axiom introducing the existence of zero, but satisfy all of the others.

Subtraction and division may be defined in terms of addition and multiplication, respectively:

$$a - b = a + (-b). \qquad \text{If } b \neq 0, \frac{a}{b} = a \cdot \frac{1}{b}.$$

We will assume the truth of the usual properties of real numbers that may be derived from the above axioms. For example,

$$a(-b) = -ab \qquad (-a)(-b) = ab$$

and

$$ab = 0 \quad \text{if and only if} \quad a = 0 \text{ or } b = 0. \tag{1.1}$$

Statement (1.1), which contains the first occurrence in the text of the phrase "if and only if," is equivalent to the condition that

$$ab = 0 \text{ if } a = 0 \text{ or } b = 0$$

and

$$\text{if } ab = 0, \text{ then } a = 0 \text{ or } b = 0. \tag{1.2}$$

The phrase "only if" in (1.1) gives rise to (1.2), which is equivalent to the statement

$$ab = 0 \text{ only if } a = 0 \text{ or } b = 0.$$

NOTE: The phrase "if and only if" is often abbreviated "iff."

The positive numbers form a subset of R, which by definition, has the following properties:

(*Trichotomy law*) For any real number a, one, and only
one, of the following statements is true: (1.3)

$$a \text{ is positive} \qquad a = 0 \qquad -a \text{ is positive}$$

The sum of any two positive numbers is positive.   (1.4)
The product of any two positive numbers is positive.   (1.5)

If $a \neq 0$, then from (1.3) either a is positive or $-a$ is positive. If a is
positive, so is $a^2 = a \cdot a$ by (1.5). If $-a$ is positive, so is $a^2 = (-a)(-a)$. Hence,

$$\text{If } a \neq 0, \text{ then } a^2 \text{ is positive.} \tag{1.6}$$

In particular from (1.6), $1^2 = 1$ is positive. Since $2 = 1 + 1$, 2 is positive by (1.4);
since $3 = 2 + 1$, 3 is positive, and so on. Thus the natural numbers are also called
the *positive integers*.

Negative numbers are defined by requiring that

$$a \text{ is negative} \quad \text{iff} \quad -a \text{ is positive.} \tag{1.7}$$

In particular, -1 is negative, since $-(-1) = 1$ is positive.

Because of the trichotomy law a real number that is zero or a positive
number is called a *non-negative number*. Also, two nonzero real numbers are said
to have the *same sign* if both are positive or both are negative. However, two
nonzero real numbers are of *opposite signs* if one of them is positive and the other
is negative.

If n is an *odd* natural number and a is a real number, there is a unique
real number x such that $x^n = a$. Also, if n is an *even* natural number, and a is
non-negative, there is a unique non-negative number x such that $x^n = a$. In either
of these cases x is called the *principal nth root* of a and is denoted by $\sqrt[n]{a}$ or $a^{1/n}$.
If, however, n is even and a is negative, there is no real number x such that $x^n = a$
and hence $\sqrt[n]{a}$ is not defined (as a real number). For example,

$$\sqrt[3]{64} = 4 \qquad \sqrt[3]{-64} = -4 \qquad \sqrt[4]{16} = 2$$

but $\sqrt[4]{-16}$ is not defined. It should be noted that the principal square root of a
non-negative number a, $\sqrt{a}$, is a non-negative number. Thus, for example,

$$\sqrt{9} = 3 \quad \text{(but not } -3 \text{ or } \pm 3\text{)} \qquad \sqrt{(-4)^2} = \sqrt{16} = 4.$$

If $\sqrt[n]{a}$ and $\sqrt[n]{b}$ are real numbers,

$$\sqrt[n]{ab} = \sqrt[n]{a}\sqrt[n]{b} \qquad \sqrt[n]{\frac{a}{b}} = \frac{\sqrt[n]{a}}{\sqrt[n]{b}}$$

where $b \neq 0$ in the equation on the right.

It is possible to represent any real number by a point on a line if a
point called the *origin* is chosen for the number 0 and a unit of distance selected
for the line. If the line is horizontal (Figure 1–3), a positive number a is customarily
associated with the point a units to the right of the origin and a negative number
a is associated with the point $-a$ units to the left of the origin. In this way there

exists a *one-to-one correspondence* between the real numbers and the points on the line. That is, for each real number there corresponds a unique point on the line, and each point on the line is the correspondent of a unique real number. A line thus associated with the set of all real numbers is called a *number line* and the number corresponding to a point on the line is termed the *coordinate* of the point. If a point P on a number line has coordinate a, P is often called "the point a." In Figure 1-3 the points -3, -2, $-\frac{4}{3}$, -1, 0, $\frac{1}{2}$, 1, $\sqrt{2}$, 2, 3, and π have been located.

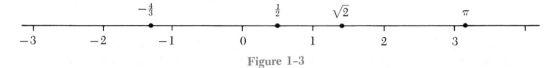

Figure 1-3

Exercise Set 1.1

1. If $A = \{1, 2, 3, 4, 5, 6\}$, $B = \{1, 3, 5, 7, 9\}$, and $C = \{2, 3, 4\}$, find:

   (a) $A \cup B$ (b) $A \cap C$

   (c) $B \cup (A \cup C)$ (d) $C \cup (A \cap B)$

   (e) $A \cap (B \cap C)$

2. Answer the following questions and justify your answers.

   (a) Is $\{2\} \subset \{x : x^2 = 2x\}$?

   (b) Is $\{x : x^2 = 2x\} \subset \{2\}$?

   (c) Is $R \subset \left\{ x : \dfrac{x^2 - 9}{x - 3} = x + 3 \right\}$?

   (d) Is $\{3\} \subset \{x : \sqrt{2x + 3} + \sqrt{x + 1} = 1\}$?

   (e) Is $\{x : \sqrt{2x + 3} + \sqrt{x + 1} = 1\} \subset \{3, -1\}$?

   (f) Is $\{x : \sqrt{2x + 3} + \sqrt{x + 1} = 1\} = \{3, -1\}$?

3. Simplify:

   (a) $\dfrac{(x + h)^3 - x^3}{h}$ (b) $\dfrac{\dfrac{x + h}{x - 2 + h} - \dfrac{x}{x - 2}}{h}$

4. Let $S = \{x : x = m + n\sqrt{3}$ where m and n are any integers$\}$.

   (a) Which of the following numbers, if any, are elements of S?

   $$3 - 5\sqrt{3},\ 1,\ \frac{1}{\sqrt{3}},\ 4 + \sqrt{75},\ -2 - \sqrt{6},\ \frac{3 - 2\sqrt{3}}{3},\ \sqrt{3}$$

   (b) If $x \in S$ and $y \in S$, show that $x + y \in S$.
   HINT: Let $x = m_1 + n_1\sqrt{3}$ and $y = m_2 + n_2\sqrt{3}$ where m_1, n_1, m_2, and n_2 are integers.

   (c) If $x \in S$ and $y \in S$, show that $xy \in S$.

5. An integer m is *odd* if there is an integer n such that $m = 2n + 1$. Show that m^2 is odd if and only if m is odd.

1.2 Mathematical Induction

In mathematics one frequently encounters statements that are expressed in terms of n where n is an arbitrary positive integer. As examples, consider the following statements:

For every positive integer n,

$$1^2 + 2^2 + 3^2 + \cdots + n^2 = \frac{n(n + 1)(2n + 1)}{6}.$$

and

For every positive integer n, $3^n - 1$ is divisible by 2.

To prove such statements, the following property of natural numbers is useful.

Mathematical Induction Axiom

Let S be a subset of N, the set of all natural numbers. If

(i) $1 \in S$

and

(ii) $k + 1 \in S$ *whenever $k \in S$,*

then $S = N$.

The method of proof using this axiom is called *mathematical induction* and is illustrated in the next three examples.

Example 1 Prove that for any natural number n,

$$1^2 + 2^2 + 3^2 + \cdots + n^2 = \frac{n(n + 1)(2n + 1)}{6}.$$

PROOF Let $S = \{n \in N : 1^2 + 2^2 + 3^3 + \cdots + n^2 = [(n(n + 1)(2n + 1)]/6\}$. $1 \in S$ since $1^2 = [1(1 + 1)(2 \cdot 1 + 1)]/6 = 1$. Thus S satisfies (i) of the mathematical induction axiom. If $k \in S$, then from the definition of S,

$$1^2 + 2^2 + \cdots + k^2 = \frac{k(k + 1)(2k + 1)}{6}. \tag{1.8}$$

To show that $k + 1 \in S$ it must be proved that

$$1^2 + 2^2 + \cdots + (k + 1)^2 = \frac{(k + 1)[(k + 1) + 1][2(k + 1) + 1]}{6}. \tag{1.9}$$

The left member of (1.9) is obtained by adding $(k + 1)^2$ to the left member of (1.8). Then if $(k + 1)^2$ is added to both members of (1.8)

$$1^2 + 2^2 + \cdots + k^2 + (k + 1)^2 = \frac{k(k + 1)(2k + 1)}{6} + (k + 1)^2$$

$$= \frac{k(k + 1)(2k + 1) + 6(k + 1)^2}{6}$$

$$= \frac{(k + 1)[k(2k + 1) + 6(k + 1)]}{6}$$

$$= \frac{(k + 1)(2k^2 + 7k + 6)}{6}$$

$$= \frac{(k + 1)(k + 2)(2k + 3)}{6}$$

$$= \frac{(k + 1)[(k + 1) + 1][2(k + 1) + 1]}{6}.$$

Thus (1.9) is obtained, and S satisfies (ii) in the mathematical induction axiom. Hence $S = N$; that is, the equation given in the conclusion is true for every natural number n.

Example 2 Prove that for every positive integer n,

$$3 + 3^2 + 3^3 + \cdots + 3^n = \tfrac{3}{2}(3^n - 1).$$

PROOF Let $S = \{n \in N: 3 + 3^2 + \cdots + 3^n = \tfrac{3}{2}(3^n - 1)\}$. $1 \in S$ since $3 = \tfrac{3}{2}(3^1 - 1)$. If $k \in S$, then

$$3 + 3^2 + \cdots + 3^k = \tfrac{3}{2}(3^k - 1). \tag{1.10}$$

To show that $k + 1 \in S$, it suffices to prove that

$$3 + 3^2 + \cdots + 3^{k+1} = \tfrac{3}{2}(3^{k+1} - 1). \tag{1.11}$$

To derive (1.11) from (1.10), add 3^{k+1} to each side of (1.10) to obtain

$$3 + 3^2 + \cdots + 3^k + 3^{k+1} = \tfrac{3}{2}(3^k - 1) + 3^{k+1}$$

$$= \frac{3^{k+1}}{2} - \frac{3}{2} + 3^{k+1}$$

$$= \tfrac{3}{2} \cdot 3^{k+1} - \tfrac{3}{2}$$

$$= \tfrac{3}{2}(3^{k+1} - 1).$$

Hence $k + 1 \in S$. Then $S = N$, and the conclusion is true.

Example 3 Prove that for every positive integer n, $3^n - 1$ is divisible by 2.

PROOF Recall that if p and q are natural numbers, then p is divisible by q if and only if there is a natural number r such that $p = qr$. Letting $S = \{n : 3^n - 1 \text{ is divisible by } 2\}$, $1 \in S$, as $3^1 - 1 = 2$ is divisible by itself. If $k \in S$, there is a natural number r such that $3^k - 1 = 2r$. To show that $k + 1 \in S$, we write

$$
\begin{aligned}
3^{k+1} - 1 &= 3^{k+1} - 3^k + 3^k - 1 \\
&= 3^k(3 - 1) + (3^k - 1) \\
&= 2 \cdot 3^k + 2r \\
&= 2(3^k + r).
\end{aligned}
$$

Thus $3^{k+1} - 1$ is divisible by 2 and $k + 1 \in S$. Then $S = N$ and the conclusion follows.

The proof by mathematical induction that

$$
a^n - b^n = (a - b)(a^{n-1} + a^{n-2}b + a^{n-3}b^2 + \cdots + b^{n-1}) \tag{1.12}
$$

for every natural number n is left as an exercise.

Exercise Set 1.2

Use mathematical induction to prove each of the following statements for every natural number n.

1. $2 + 6 + 10 + \cdots + (4n - 2) = 2n^2$

2. $4 + 7 + 10 + \cdots + (3n + 1) = \dfrac{n}{2}(3n + 5)$

3. $1 \cdot 2 + 2 \cdot 3 + 3 \cdot 4 + \cdots + n(n + 1) = \dfrac{n(n + 1)(n + 2)}{3}$

4. $1^3 + 2^3 + 3^3 + \cdots + n^3 = \dfrac{n^2(n + 1)^2}{4}$

5. $\dfrac{1}{1 \cdot 2} + \dfrac{1}{2 \cdot 3} + \dfrac{1}{3 \cdot 4} + \cdots + \dfrac{1}{n(n + 1)} = \dfrac{n}{n + 1}$

6. $\dfrac{1}{1 \cdot 3} + \dfrac{1}{3 \cdot 5} + \dfrac{1}{5 \cdot 7} + \cdots + \dfrac{1}{(2n - 1)(2n + 1)} = \dfrac{n}{2n + 1}$

7. $2 + 2^2 + 2^3 + \cdots + 2^n = 2^{n+1} - 2$

8. $2 \cdot 5 + 2 \cdot 5^2 + 2 \cdot 5^3 + \cdots + 2 \cdot 5^n = \frac{5}{2}(5^n - 1)$

9. $5^n - 1$ is divisible by 4.

10. $n^3 + 2n$ is divisible by 3.

11. $n^4 + 5n$ is divisible by 2.

12. $3^{2n} - 2^{2n}$ is divisible by 5.

13. Equation (1.12) in this section

1.3   Order Relations for Real Numbers

In this section we shall discuss the following order relations for real numbers:

"is greater than"
"is less than"
"is greater than or equal to"
"is less than or equal to"

These types of inequality are defined in the following statement.

1.3.1   Definition

Let a and b be any real numbers. Then:

$$a > b \quad \text{iff} \quad a - b \text{ is positive}$$
$$a < b \quad \text{iff} \quad b > a$$
$$a \geq b \quad \text{iff} \quad a > b \text{ or } a = b$$
$$a \leq b \quad \text{iff} \quad a < b \text{ or } a = b$$

As an illustration of the definitions of $>$ and $<$,

$$16 > 11 \quad \text{and} \quad 11 < 16$$

since $16 - 11 = 5$, a positive number.

A number of well-known properties of inequalities will now be derived using Definition 1.3.1 and the properties (1.3) through (1.7), which were given in Section 1.1. In the statement of these properties it is assumed that a, b, and c are arbitrary real numbers unless otherwise noted.

The first property is sometimes used to define a positive number.

1.3.2   Theorem

a is positive   if and only if $a > 0$.

PROOF   Suppose a is positive. Since $a - 0 = a$, a positive number, from the definition of $>$, $a > 0$. To prove the converse, we note that if $a > 0$, then $a - 0 = a$ is positive.

The proof of the following companion property of Theorem 1.3.2 is left as an exercise.

1.3.3   Theorem

a is negative   if and only if $a < 0$.

Because of Theorem 1.3.2 the statements "$a > 0$" and "a is positive" may be used interchangeably; also, from Theorem 1.3.3 the same is true of the statements "a is negative" and "$a < 0$."

The proofs of some of the following properties are also left as exercises.

1.3.4   Theorem

If a and b are positive, a/b is positive.

PROOF   We first show $1/b > 0$ using an indirect argument. From (1.3), either $1/b > 0$, $1/b = 0$, or $-1/b > 0$. If $1/b = 0$, then $1 = b \cdot 1/b = b \cdot 0 = 0$, which is impossible. If $-1/b > 0$, then $b(-1/b)$ should be positive by (1.5). However, $b(-1/b) = -(b \cdot 1/b) = -1$, which is negative. Thus, the only remaining possibility is $1/b > 0$. Hence $a/b = a \cdot 1/b$ is positive by (1.5).

By the next theorem *the direction of the inequality is unchanged if two inequalities in the same direction are added term by term.*

1.3.5   Theorem

If $a > b$ and $c \geq d$, then $a + c > b + d$.

PROOF   It can be proved that $(a + c) - (b + d)$ is positive, using Definition 1.3.1. Since $a > b$ and $c \geq d$, $a - b$ is positive and $c - d$ is non-negative. Thus $(a + c) - (b + d) = (a - b) + (c - d)$ is positive, being either the sum of two positive numbers or the sum of a positive number and zero.

As an illustration of Theorem 1.3.5, since $2 > -3$ and $5 > 1$, we have $7 > -2$.

From the next theorem *the direction of an inequality is unchanged if each member of the inequality is multiplied by a positive number.*

1.3.6   Theorem

If $a > b$ and c is positive, then $ac > bc$.

PROOF   Here we need to show that $ac - bc$ is positive. Now $ac - bc = c(a - b)$. Since $a > b$, $a - b$ is positive. Thus $ac - bc$ is the product of two positive numbers and is therefore positive.

For example, since $10 > 4$, multiplying each member by 3 gives $30 > 12$.

The following theorem states that *if each member of an inequality is multiplied by a negative number, the direction of the inequality is reversed.*

1.3.7 Theorem

If $a > b$ and c is negative, then $ac < bc$.

For example, since $2 > -5$, multiplying each member by -2 gives $-4 < 10$.

1.3.8 Theorem (Transitive Law)

If $a > b$ and $b > c$, then $a > c$.

If $a > b$ and $b > c$, we may combine these inequalities and write $a > b > c$ or $c < b < a$. However, two inequalities in the opposite direction are never combined in this manner. For example, one never writes $a > b < c$ or $a < b > c$.

Theorems 1.3.5–1.3.8 remain true if the directions of the inequalities are reversed. From Theorem 1.3.7, for example, we have the statement: "If $a < b$ and c is negative, then $ac > bc$."

1.3.9 Theorem

Let n be a positive integer. If $a > b \geq 0$, then:

(a) $a^n > b^n$

(b) $\sqrt[n]{a} > \sqrt[n]{b}$

1.3.10 Theorem

If $a > b > 0$, then $\dfrac{1}{a} < \dfrac{1}{b}$.

PROOF Note that

$$\frac{1}{b} - \frac{1}{a} = \frac{a - b}{ab}.$$

Since $a > b$, $a - b$ is positive. Also ab is positive since it is the product of two positive numbers. Hence $(1/b) - (1/a)$ is positive by Theorem 1.3.4.

The following two examples are also included to further illustrate the technique used in proving inequalities.

Example 1 Prove: If $a > 0$, $b > 0$, and $a \neq b$, then $\dfrac{a}{b} + \dfrac{b}{a} > 2$.

PROOF　$\dfrac{a}{b} + \dfrac{b}{a} - 2 = \dfrac{a^2 - 2ab + b^2}{ab} = \dfrac{(a - b)^2}{ab}.$

By (1.6) and (1.5), respectively, $(a - b)^2$ and ab are positive. Thus $(a/b) + (b/a) - 2$ is positive by Theorem 1.3.4.

Example 2　Prove: $3x^2 - 7x + 9 > 0$ for every x.

PROOF　$3x^2 - 7x + 9 = 3(x^2 - \tfrac{7}{3}x) + 9$

$$= 3(x^2 - \tfrac{7}{3}x + \tfrac{49}{36}) + 9 - \tfrac{49}{12}$$

After completing the square in x,

$$3x^2 - 7x + 9 = 3(x - \tfrac{7}{6})^2 + \tfrac{59}{12}.$$

Now $(x - \tfrac{7}{6})^2 \geq 0$ by (1.1) and (1.6). Hence by properties (1.4) and (1.5) for positive numbers, $3x^2 - 7x + 9$ is positive.

　　　　Since we shall be almost exclusively concerned with real numbers, in the interest of brevity the term "number" will customarily be used henceforth to mean "real number."

Exercise Set 1.3

1. Prove: If $a = b$ and $c < d$, then $a - c > b - d$.

2. Prove: Theorem 1.3.3

3. Prove: Theorem 1.3.7

4. Prove: Theorem 1.3.8

5. Prove: Theorem 1.3.9(a). HINT: Use (1.12).

6. Prove: Theorem 1.3.9(b). HINT: Use (1.12) with $a - b$ in place of $a^n - b^n$.

7. Prove: If a, b, and c are positive, then $\dfrac{a}{b + c} < \dfrac{a}{b}.$

8. If a and b are unequal positive numbers, prove that:

   (a) $\dfrac{a + b}{2} > \sqrt{ab}$

   (b) $\sqrt{ab} > \dfrac{2ab}{a + b}$

   (c) $\dfrac{b}{a^2} + \dfrac{a}{b^2} > \dfrac{1}{a} + \dfrac{1}{b}$

9. Prove: If $a > b > 0$ and $c > d > 0$, then $ac > bd$.

10. If $a > b > 0$, prove $a^3 - b^3 > (a - b)^3$.

11. If x is an arbitrary number, prove

   (a) $x^2 + 5 > 2x$

   (b) $3x^2 > 6x - 5$

   (c) $2x^2 - 3x + 5 > 0$

12. Prove that for every positive integer n, $n! \geq 2^{n-1}$.

13. Let $x > -1$. Prove that for every positive integer n, $(1 + x)^n \geq 1 + nx$. (Bernoulli's inequality)

14. Prove that the sum of two negative numbers is negative.

1.4   Intervals; Solutions of Inequalities

Sets of real numbers are often defined using inequalities. For example, the set of all numbers between 1 and 3 can be denoted by

$$\{x: 1 < x < 3\}. \tag{1.13}$$

Among the sets of numbers that are defined using inequalities are the *intervals*. If a and $(b > a)$ are arbitrary numbers, there are four *finite* intervals associated with these numbers, which are defined and illustrated in Figure 1–4.

$[a, b]$, the *closed interval* from a to b

$$[a, b] = \{x: a \leq x \leq b\} \tag{1.14}$$

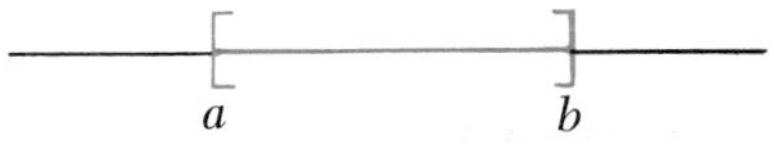

(a, b), the *open interval* from a to b

$$(a, b) = \{x: a < x < b\} \tag{1.15}$$

$[a, b)$, the *left-closed interval* from a to b

$$[a, b) = \{x: a \leq x < b\} \tag{1.16}$$

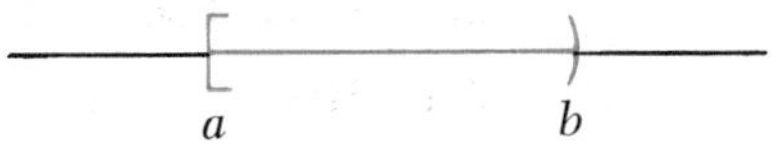

$(a, b]$, the *right-closed interval* from a to b

$$(a, b] = \{x: a < x \leq b\} \tag{1.17}$$

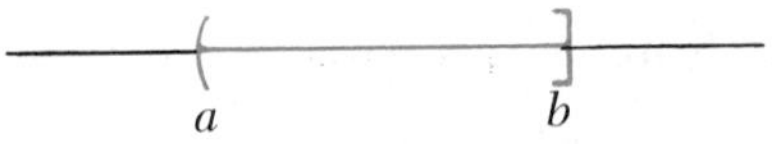

Figure 1-4

The numbers a and b are called the *endpoints* of these intervals. Observe in the notation for these intervals and in the corresponding sketches shown in Figure 1–4 that a bracket is associated with an endpoint that is in the interval and a parenthesis with an endpoint that is not in the interval. Note also that the set (1.13) above is the open interval from 1 to 3 and can be denoted by the symbol $(1, 3)$.

If a is an arbitrary number, we also have four *infinite intervals* that are associated with the endpoint a.

$$[a, +\infty) = \{x : x \geq a\}$$ —————————————————→ (1.18)

$$(a, +\infty) = \{x : x > a\}$$ —————————————————→ (1.19)

$$\quad\quad\quad\quad\quad\quad\quad\quad\quad\quad\quad a$$

$$(-\infty, a] = \{x : x \leq a\}$$ ←————————————————— (1.20)

$$\quad\quad\quad\quad\quad\quad\quad\quad\quad\quad a$$

$$(-\infty, a) = \{x : x < a\}$$ ←————————————————— (1.21)

$$\quad\quad\quad\quad\quad\quad\quad\quad\quad\quad a$$

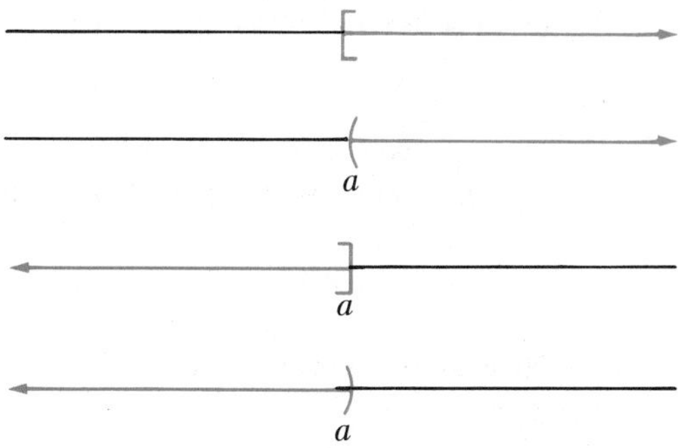

Figure 1-5

The set R of all real numbers is also considered an infinite interval and we can write

$$R = (-\infty, +\infty). \tag{1.22}$$

Note that the symbol ∞ (*infinity*) used in defining the intervals of Statements (1.18) through (1.22) is not a number, but serves to indicate that the interval extends indefinitely in a given direction.

An interval that contains its endpoints is said to be *closed*. The interval (1.14) is a finite closed interval while (1.18) and (1.20) are infinite closed intervals. An interval that contains none of its endpoints is said to be *open*. The intervals (1.15), (1.19), and (1.21) are examples of open intervals. The set R, which has no endpoints, is considered to be both open and closed. The *interior* of an interval is the set consisting of all numbers in the interval that are not endpoints of the interval. A number in the interior of an interval is called an *interior point* of the interval.

Intervals are important because they are encountered in the solution of inequalities in one unknown, as will be noted in the examples below. To solve such an inequality, we must obtain the set of all numbers that satisfy the inequality, the *solution set* of the inequality. The inequality properties expressed by Theorems 1.3.5, 1.3.6, and 1.3.7 are useful in obtaining these solution sets.

Example 1 Solve the inequality $\frac{4}{3}x + 10 > \frac{5}{2}x - 4$.

SOLUTION If x is a number such that

$$\tfrac{4}{3}x + 10 > \tfrac{5}{2}x - 4,$$

then multiplying each member by 6, using Theorem 1.3.6, we have

$$8x + 60 > 15x - 24.$$

Adding $-15x - 60$ to each member of this inequality, using Theorem 1.3.5, we obtain

$$-7x > -84.$$

Then if each side here is multiplied by $-\frac{1}{7}$, from Theorem 1.3.7

$$x < 12. \tag{1.23}$$

We have proved that if $\frac{4}{3}x + 10 > \frac{5}{2}x - 4$, then $x < 12$. Now each of the foregoing steps is reversible. If we start with (1.23) and multiply each side by -7 then

$$-7x > -84.$$

Then if $15x + 60$ is added to each side

$$8x + 60 > 15x - 24,$$

and multiplying each side by $\frac{1}{6}$ gives

$$\tfrac{4}{3}x + 10 > \tfrac{5}{2}x - 4.$$

Thus

$$\tfrac{4}{3}x + 10 > \tfrac{5}{2}x - 4 \quad \text{iff} \quad x < 12$$

and the solution set is the interval $(-\infty, 12)$, which is illustrated in Figure 1–6.

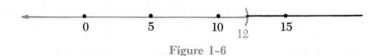

Figure 1–6

Notice the similarity between the processes used in solving equations and inequalities. From Theorem 1.3.5, the terms of an inequality can be "transposed" to give an equivalent inequality—that is, one having the same solutions as the given inequality. The same is true if each member of an inequality is multiplied by the same nonzero number, although, from Theorem 1.3.7, multiplying by a negative number reverses the direction of the given inequality.

Example 2 Solve $-1 \le 5 - 3x < 4$.

SOLUTION If $-1 \le 5 - 3x < 4$, adding -5 to each member of the inequality gives

$$-6 \le -3x < -1.$$

Then if each member is multiplied by $-\frac{1}{3}$

$$2 \ge x > \tfrac{1}{3}.$$

Since the above steps are reversible, the solution set of the given inequality is the interval $(\frac{1}{3}, 2]$ (Figure 1–7).

Figure 1–7

Example 3   Solve $\dfrac{x+3}{3x+1} \geq 1$.

SOLUTION   It will be convenient to obtain an equivalent inequality with a right member of 0 and the left member as a simple fraction. Transposing 1 to the left,

$$\frac{x+3}{3x+1} - 1 \geq 0;$$

and after combining terms on the left,

$$\frac{-2x+2}{3x+1} \geq 0.$$

The latter inequality is satisfied if and only if $-2x+2 = 0$ and $x \neq -\frac{1}{3}$ or else $-2x+2$ and $3x+1$ have the same sign. Note that $-2x+2 = 0$ iff $x = 1$. To obtain those values of x for which $-2x+2$ and $3x+1$ have the same sign, we need to know where these expressions are positive and where they are negative. By solving the appropriate inequalities, we find that:

$$-2x+2 > 0 \quad \text{iff} \quad x < 1 \qquad -2x+2 < 0 \quad \text{iff} \quad x > 1$$
$$3x+1 > 0 \quad \text{iff} \quad x > -\tfrac{1}{3} \qquad 3x+1 < 0 \quad \text{iff} \quad x < -\tfrac{1}{3}$$

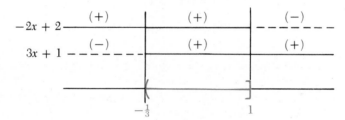

Figure 1–8

In Figure 1–8, the algebraic signs of the expressions $-2x+2$ and $3x+1$ are represented for every number x. Where the expression is positive it has been represented by a continuous line, and where the expression is negative it is shown by a dashed line. An inspection of the above statements or Figure 1–8 shows that $-2x+2$ and $3x+1$ have the same sign if $-\frac{1}{3} < x < 1$. However, since 1 is a solution of the given inequality, the solution set for the given inequality is the interval $(-\frac{1}{3}, 1]$.

NOTE: The student is cautioned against blindly multiplying each side of the given inequality by $3x+1$ and obtaining

$$x + 3 \geq 3x + 1. \tag{1.24}$$

This inequality does follow from the given inequality if $3x+1 > 0$, but if $3x+1 < 0$, then, instead of (1.24), we have

$$x + 3 \leq 3x + 1.$$

Example 4   Solve $(2x + 5)(1 - 3x)(4x - 5) < 0$.

SOLUTION   Note that the product on the left is negative iff

(a) any two of the factors are positive and the remaining one is negative, or

(b) all three of the factors are negative.

By solving appropriate inequalities between each of the factors and zero, we obtain a sketch similar to Figure 1–8 representing the algebraic signs of each of the factors.

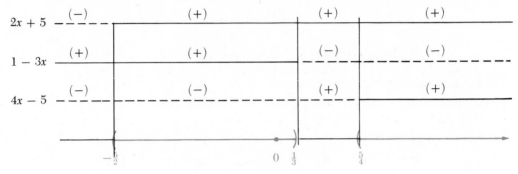

Figure 1-9

The solution set of the inequality is $(-\frac{5}{2}, \frac{1}{3}) \cup (\frac{5}{4}, +\infty)$.

A typical expression in x for which the sketches are made in Figures 1–8 and 1–9 is of the form $ax + b$, where $a \neq 0$. Note that $ax + b = 0$ if $x = -(b/a)$. If $a > 0$, the sketch showing the algebraic sign of $ax + b$ resembles that in Figure 1–10 where the positive portion extends to the right from $-(b/a)$ and

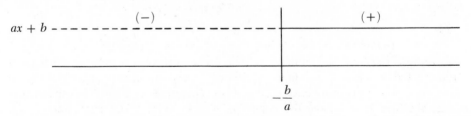

Figure 1-10

the negative portion extends to the left from $-(b/a)$. If however, $a < 0$, the positive and negative portions are reversed and we have a representation for $ax + b$ like that in Figure 1–11.

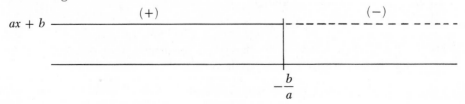

Figure 1-11

Exercise Set 1.4

Solve the inequalities in Exercises 1–20.

1. $3x - 9 < 6$

2. $12 - 4x > 3$

3. $2x + \frac{3}{5} > \frac{3}{4}x - 1$

4. $2 - 4x < \frac{2}{3}x + 1$

5. $0 \leq 2x + 7 < 3$

6. $10 > 1 - 4x > 5$

7. $(3x + 7)(x - 2) < 0$

8. $(3x - 2)(1 - x)(2x + 1) \geq 0$

9. $x^2 - x - 2 < 0$

10. $4x^2 + 3 \geq 7x$

11. $x^2 \leq 3$

12. $3x^2 + x - 1 < 0$

13. $x - 4 < x^2$

14. $\dfrac{3}{x} > \dfrac{5}{4}$

15. $\dfrac{x - 1}{x} \leq \dfrac{1}{3}$

16. $\dfrac{2x - 1}{2 - x} \leq -1$

17. $\dfrac{x + 2}{x - 3} > \dfrac{x}{x - 2}$

18. $2 < \dfrac{1}{x - 1} \leq 5$

19. $-2 < \dfrac{5 - 3x}{x + 1} < 4$

20. $x^3 + 3x^2 < x + 3$

In Exercises 21–26 determine the set of numbers for which each of the given numbers is real.

21. $\sqrt{4x^2 - 25}$ HINT: For what numbers x is $4x^2 - 25 \geq 0$?

22. $\sqrt{9 - x^2}$

23. $\sqrt{15 + x - 2x^2}$

24. $\sqrt{2x + x^2 - x^3}$

25. $\sqrt{\dfrac{3 - 4x}{x + 2}}$

26. $\sqrt{\dfrac{3x + 5}{3 - x}}$

1.5 Absolute Value

As we mentioned in the introduction to Section 1.1, a fundamental notion in the study of calculus is the concept of a limit. Beginning in Chapter 3 we will discuss several different types of limits, most of which are defined using the notion of the *absolute value* of a number. We recall from pre-calculus mathematics that the absolute value of a number a, which is denoted by $|a|$, can be defined as follows:

1.5.1 Definition

$$|a| = \begin{cases} a & \text{if } a \geq 0, \\ -a & \text{if } a < 0. \end{cases}$$

For example:

$$|5| = 5 \qquad |-5| = -(-5) = 5 \qquad |0| = 0$$

It will be noted from Definition 1.5.1 that the absolute value of a number is always *non-negative*. Recalling the definition of the principal square root, we also have

$$\sqrt{a^2} = \begin{cases} a & \text{if } a \geq 0, \\ -a & \text{if } a < 0. \end{cases} \tag{1.25}$$

Thus, from Definition 1.5.1 and (1.25),

$$|a| = \sqrt{a^2} \tag{1.26}$$

and hence

$$|a|^2 = a^2. \tag{1.27}$$

The notion of absolute value is used to define the distance between two points on a number line.

1.5.2   Definition

Suppose P_1 and P_2 are any two points on a number line with coordinates x_1 and x_2, respectively. The distance between P_1 and P_2, denoted by $|P_1P_2|$, is given by

$$|P_1P_2| = |x_2 - x_1|. \tag{1.28}$$

If $x_2 > x_1$ (Figure 1–12(a)), then from (1.28)

$$|P_1P_2| = x_2 - x_1,$$

but if $x_2 < x_1$ (Figure 1–12(b)), then

$$|P_1P_2| = -(x_2 - x_1) = x_1 - x_2.$$

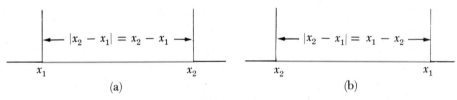

(a)   (b)

Figure 1-12

In particular, from Definition 1.5.2, the number $|a|$ is the distance on the number line between the origin and the point with coordinate a.

In the statements of the following theorems about the absolute-value of a number, a and b are arbitrary numbers unless otherwise noted.

1.5.3   Theorem

Let $b > 0$. Then:

(a) $|a| < b$ iff $-b < a < b$
(b) $|a| = b$ iff $a = b$ or $a = -b$
(c) $|a| > b$ iff $a > b$ or $a < -b$

PROOF OF (a) We first show that if $|a| < b$, then $-b < a < b$. Consider the separate cases, $a \geq 0$ and $a < 0$.

If $a \geq 0$, then $a = |a| < b$ and hence $-b < 0 \leq a < b$.

If $a < 0$, then $-a = |a| < b$. Multiplying by -1 gives $a > -b$, and hence

$$-b < a < 0 < b.$$

Next we prove the converse: if $-b < a < b$, then $|a| < b$.

If $a \geq 0$, then $|a| = a < b$.

If $a < 0$, then $|a| = -a$, and thus $-|a| = a > -b$. Multiplying the left and right members by -1 then yields $|a| < b$.

The proofs of (b) and (c) are left as exercises.

Theorem 1.5.3 is utilized in Examples 1–3 below.

Example 1 Solve $|3x - 2| = 4$.

SOLUTION From Theorem 1.5.3(b):

$$|3x - 2| = 4 \quad \text{iff} \quad 3x - 2 = 4 \text{ or } 3x - 2 = -4$$
$$\text{iff} \quad 3x = 6 \text{ or} \quad 3x = -2$$
$$\text{iff} \quad x = 2 \text{ or} \quad x = -\tfrac{2}{3}$$

Thus 2 and $-\tfrac{2}{3}$ are the solutions of the given equation.

Example 2 Solve $|3x - 2| < 4$.

SOLUTION From Theorem 1.5.3(a):

$$|3x - 2| < 4 \quad \text{iff} \quad -4 < 3x - 2 < 4$$
$$\text{iff} \quad -2 < 3x < 6$$
$$\text{iff} \quad -\tfrac{2}{3} < x < 2$$

Hence the solution set is the interval $(-\tfrac{2}{3}, 2)$.

Example 3 Solve $|3x - 2| > 4$.

SOLUTION From Theorem 1.5.3(c):

$$|3x - 2| > 4 \quad \text{iff} \quad 3x - 2 > 4 \text{ or } 3x - 2 < -4$$
$$\text{iff} \quad 3x > 6 \text{ or} \quad 3x < -2$$
$$\text{iff} \quad x > 2 \text{ or} \quad x < -\tfrac{2}{3}$$

The solution set is $(-\infty, -\tfrac{2}{3}) \cup (2, +\infty)$.

Example 4 Solve $|2x + 1| = x - 3$.

SOLUTION   Although the equation could be solved using Theorem 1.5.3(b), it will be easier to square both members of the equation, using (1.27), and obtain:

$$(2x + 1)^2 = (x - 3)^2$$
$$4x^2 + 4x + 1 = x^2 - 6x + 9$$
$$3x^2 + 10x - 8 = 0$$
$$(3x - 2)(x + 4) = 0$$
$$x = \tfrac{2}{3} \text{ or } -4$$

However, neither of the numbers $\tfrac{2}{3}$ and -4 satisfies the given equation, and hence the equation has no solutions.

Example 5   Solve $|1 - 2x| \leq |4x - 3|$.

SOLUTION   By Theorem 1.3.9(a) and (b):

$$
\begin{aligned}
|1 - 2x| \leq |4x - 3| \quad &\text{iff} & (1 - 2x)^2 &\leq (4x - 3)^2 \\
&\text{iff} & 1 - 4x + 4x^2 &\leq 16x^2 - 24x + 9 \\
&\text{iff} & 0 &\leq 12x^2 - 20x + 8 \\
&\text{iff} & 0 &\leq 3x^2 - 5x + 2 \\
&\text{iff} & 0 &\leq (3x - 2)(x - 1) \\
&\text{iff} & 3x - 2 & \text{ and } x - 1 \text{ are of the same sign} \\
& & & \text{or } 3x - 2 = 0 \text{ or } x - 1 = 0
\end{aligned}
$$

Using the graphical scheme introduced in Section 1.4 to analyze the signs of $3x - 2$ and $x - 1$ (Figure 1–13), the solution set is found to be $(-\infty, \tfrac{2}{3}] \cup [1, +\infty)$.

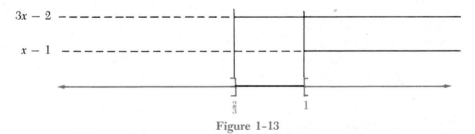

Figure 1-13

NOTE: The above examples could be solved by considering the separate cases which arise when the expressions inside the absolute value bars are non-negative and when they are negative. However, this consideration of cases is usually more tedious than the methods outlined above when solving inequalities with absolute values.

Example 6   Find an $M > 0$ such that $|x^2 - 3x + 5| < M$ if $|x - 2| < 1$.

SOLUTION   If $|x - 2| < 1$ then by Theorem 1.5.3(a)

$$-1 < x - 2 < 1. \tag{1.29}$$

Adding 2 to each member of (1.29) using Theorem 1.3.5 gives

$$1 < x < 3. \tag{1.30}$$

Multiplying each member of (1.30) by -3 yields

$$-9 < -3x < -3, \tag{1.31}$$

and upon applying Theorem 1.3.9(a) to (1.30), we have

$$1 < x^2 < 9. \tag{1.32}$$

From (1.31) and (1.32) using Theorem 1.3.5

$$-8 < x^2 - 3x < 6,$$

and if 5 is added to each member of this inequality, then

$$-3 < x^2 - 3x + 5 < 11.$$

Hence

$$|x^2 - 3x + 5| < 11$$

and 11 is a suitable number for M.

Theorems 1.5.4, 1.5.5, and 1.5.7 below will be useful in our work with limits in Chapter 3.

1.5.4   Theorem

$$|ab| = |a||b|.$$

PROOF   From (1.26)

$$|ab| = \sqrt{(ab)^2} = \sqrt{a^2 b^2} = \sqrt{a^2}\sqrt{b^2} = |a||b|.$$

Theorem 1.5.4 states that the *absolute value of the product of two numbers is the product of their absolute values*.

As an illustration of Theorem 1.5.4

$$|x^3 - 3x - 10| = |(x - 5)(x + 2)| = |x - 5||x + 2|.$$

Also

$$|-a| = |(-1)a| = |-1||a| = 1|a|$$
$$= |a|. \tag{1.33}$$

1.5.5   Theorem

$$\left| \frac{a}{b} \right| = \frac{|a|}{|b|} \qquad if \ b \neq 0.$$

The proof of Theorem 1.5.5, which is analogous to that of Theorem 1.5.4, is left as an exercise.

By Theorem 1.5.5 the *absolute value of the quotient of two numbers is the quotient of their absolute values*.

1.5.6 Theorem

$$-|a| \le a \le |a|.$$

PROOF If $b \ge 0$, from Theorem 1.5.3(a) and (b),

$$|a| \le b \quad \text{iff} \quad -b \le a \le b. \tag{1.34}$$

Since it is certainly true that $|a| \le |a|$, we can replace b by $|a|$ in (1.34) and obtain the conclusion.

1.5.7 Theorem

$$|a \pm b| \le |a| + |b|.$$

PROOF From Theorem 1.5.6

$$-|a| \le a \le |a| \quad \text{and} \quad -|b| \le b \le |b|.$$

Then from Theorem 1.3.5

$$-(|a| + |b|) \le a + b \le |a| + |b|, \tag{1.35}$$

and from (1.34) and (1.35) we have

$$|a + b| \le |a| + |b|. \tag{1.36}$$

From (1.33) and (1.36)

$$\begin{aligned} |a - b| = |a + (-b)| &\le |a| + |-b| \\ &\le |a| + |b|. \end{aligned} \tag{1.37}$$

Inequalities (1.36) and (1.37) can be combined to give the conclusion.

Exercise Set 1.5

In Exercises 1–10 solve the given equation.

1. $|2x + 5| = 9$

2. $|2 - 3x| = 4$

3. $\left| \dfrac{3x + 7}{2x - 3} \right| = 3$

4. $\left| \dfrac{x - 4}{3x - 1} \right| = 5$

5. $|x - 3| = |2 - 3x|$

6. $|3x + 1| = |x - 2|$

7. $|2x - 5| = 1 - 4x$

8. $|4 - 3x| = 2x - 1$

9. $|x| + |2x - 1| = 5$

10. $|2x - 3| - |x + 2| = 7$

In Exercises 11–20 solve the given inequality.

11. $|2x + 3| < 5$ 12. $|4 - 7x| > 2$

13. $|5x - 4| \geq 8$ 14. $|3x - 5| \leq 4$

15. $\left| \dfrac{1 - 4x}{x - 2} \right| \leq 5$ 16. $\left| \dfrac{3x - 1}{4x - 3} \right| < \dfrac{1}{4}$

17. $\left| \dfrac{2x}{4x + 3} \right| > \dfrac{2}{3}$ 18. $\left| \dfrac{3 - x}{5 - 3x} \right| \geq 3$

19. $\left| \dfrac{2}{2x - 1} \right| \geq \left| \dfrac{5}{x - 1} \right|$ 20. $\left| \dfrac{3}{3x - 2} \right| < \left| \dfrac{1}{x - 2} \right|$

In Exercises 21–24 find an $M > 0$ for which the given statement is true.

21. $|3x + 5| < M$ if $|x - 3| < 1$

22. $|4x - 1| < M$ if $|x + 2| < \frac{1}{2}$

23. $|x^2 + 4x - 9| < M$ if $|x - 3| < \frac{1}{3}$

24. $|(x + 1)(x^2 + 1)| < M$ if $|x - 1| < 1$

25. Prove Theorem 1.5.3(b).

26. Prove Theorem 1.5.3(c).

27. Prove Theorem 1.5.5.

In Exercises 28 and 29, $a_1, a_2, \ldots, a_n$ are any numbers.

28. Prove that $|a_1||a_2| \cdots |a_n| = |a_1 a_2 \cdots a_n|$ for every positive integer n.

29. Prove that $|a_1 + a_2 + \cdots + a_n| \leq |a_1| + |a_2| + \cdots + |a_n|$ for every positive integer n.

30. Prove that for any numbers a and b, $|a| - |b| \leq |a \pm b|$.

2. Fundamentals of Analytic Geometry

2.1 Rectangular Coordinate System

We recall from Section 1.1 that the points on a line can be put into a one-to-one correspondence with the real numbers. Similarly, it will be shown in this section that the points in a plane can be put into a one-to-one correspondence with *pairs* of real numbers that are called *coordinates* of the points. This association between points in a plane and pairs of real numbers will permit the solution of geometric problems concerned with the plane by algebraic methods. The study of geometry in the plane utilizing the coordinates of points in the plane is called (*plane*) *analytic geometry*.

In order to develop the basic ideas in analytic geometry, the words "plane," "point," "line," and "distance" will be left as undefined terms just as in Euclidean geometry. The usual definitions and postulates from Euclidean geometry will be used, and also we will assume the truth of the common theorems from that branch of mathematics. It might be more satisfactory mathematically to develop the study of analytic geometry directly from the real number system rather than from a basis in Euclidean geometry. However, such a procedure would be overly time-consuming and therefore will not be followed in this text.

We begin by considering two perpendicular lines, which we call the *x axis* and the *y axis,* and their intersection point O. Usually the x axis is visualized as a horizontal line and the y axis as a vertical line (Figure 2–1). Following the procedure discussed at the end of Section 1.1, a coordinate system is introduced on the x and y axes in which the origin for both number lines is the point O. The positive direction of the x axis will be to the right from O and the negative direction to the left. For the y axis the positive direction is upward and the negative direction downward. Collectively, these number lines are termed the *coordinate axes* and their point of intersection is called the *origin*. The coordinate axes divide the plane into four regions called *quadrants*, which are numbered counterclockwise as shown in Figure 2–1.

In discussions about points in the plane the notion of an *ordered pair* of numbers will be used. By an ordered pair of numbers is meant a set of two numbers in which there is a specified first element and a specified second element. The ordered pair in which p is the *first element* and q the *second element* will be

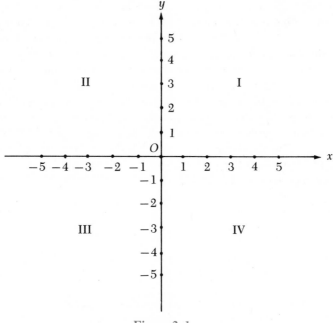

Figure 2-1

denoted, using parentheses, by (p, q). It is important to observe that (p, q) does not have the same meaning as the set $\{p, q\}$, which has p and q as elements but specifies no ordering for these elements. Thus, $(p, q) = (q, p)$ if and only if $p = q$.

For every point P in the plane a unique ordered pair of numbers (Figure 2–2) can be associated in the following way: we construct perpendiculars from P to the coordinate axes. If a and b are the coordinates of the intersection points of these perpendiculars with the x and y axes respectively, a is called the x *coordinate*

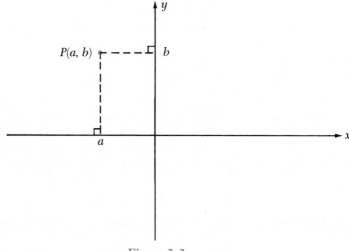

Figure 2-2

or *abscissa* of P, b the *y coordinate* or *ordinate* of P, and (a, b) is termed the *rectangular coordinate* representation of P. By the foregoing construction there cannot be any point except P with the representation (a, b) since there is exactly one perpendicular to a given line at a point on the line. Conversely, if the ordered pair (a, b) is given, by constructing perpendiculars to the x and y axes at the points with coordinates a and b, respectively, we obtain a unique point in the plane having the representation (a, b). Hence *there is a one-to-one correspondence between the set of points in the coordinate plane and the set of ordered pairs of real numbers.* Because of this one-to-one correspondence, the notation P(a, b) is often used to denote the point P in the plane associated with the ordered pair of numbers (a, b).

In Figure 2–3 the points $(-4, 0)$, $(-4, 1)$, $(3, 3)$, $(5, \frac{1}{3})$, $(2, -\sqrt{2})$, and $(0, -3)$ have been located in a sketch of the plane.

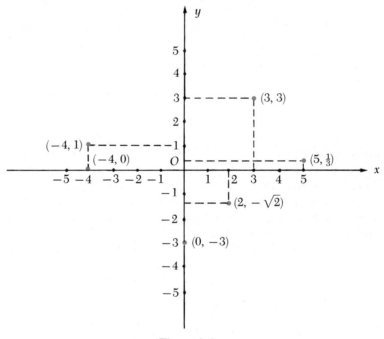

Figure 2–3

A point is on the x axis if and only if its rectangular coordinate representation is of the form (a, 0), while a point is on the y axis if and only if it has rectangular coordinates (0, b).

In view of our remark about the one-to-one correspondence between the points in the plane and the ordered pairs of real numbers, the set

$$R^2 = \{(x, y): x \in R \text{ and } y \in R\}$$

will be termed the *Cartesian plane,* and the elements (x, y) of R^2 can be referred to as the "points" of R^2.

Our immediate need here is for a formula for the distance between any two points in the plane. If P_1 and P_2 are two such points, the distance between P_1 and P_2 will be denoted by $|P_1P_2|$. In the next theorem we will obtain a formula, called the *two-point distance formula*, for calculating this distance.

2.1.1 Theorem

The distance between the points $P_1(x_1, y_1)$ and $P_2(x_2, y_2)$ in the plane is given by

$$|P_1P_2| = \sqrt{(x_2 - x_1)^2 + (y_2 - y_1)^2}. \tag{2.1}$$

PROOF We first construct perpendiculars from P_1 and P_2 to the coordinate axes (Figure 2–4). The perpendiculars from P_1 will intersect the x axis and the y axis at points $A_1(x_1, 0)$ and $B_1(0, y_1)$, respectively. Also, the perpendiculars from P_2 will intersect the x axis and y axis at $A_2(x_2, 0)$ and $B_2(0, y_2)$, respectively. The line segments P_1B_1 and P_2A_2, or their extensions, will then intersect at the point $Q(x_2, y_1)$. Hence P_1P_2Q is a right triangle with a right angle at Q. By (1.28),

$$|A_1A_2| = |x_2 - x_1| \quad \text{and} \quad |B_1B_2| = |y_2 - y_1|.$$

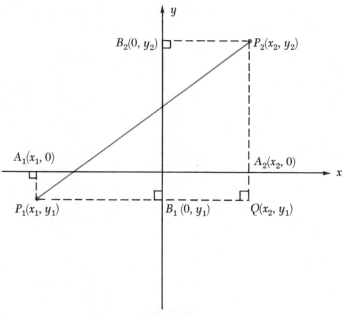

Figure 2–4

Since A_1A_2 and P_1Q are opposite sides of a rectangle,

$$|P_1Q| = |A_1A_2| = |x_2 - x_1|. \tag{2.2}$$

For the same reason

$$|QP_2| = |B_1B_2| = |y_2 - y_1|. \tag{2.3}$$

By the Pythagorean theorem

$$|P_1P_2|^2 = |P_1Q|^2 + |QP_2|^2$$

and from (2.2) and (2.3)

$$|P_1P_2|^2 = |x_2 - x_1|^2 + |y_2 - y_1|^2$$
$$= (x_2 - x_1)^2 + (y_2 - y_1)^2. \tag{2.4}$$

Hence from (2.4), Equation (2.1) is obtained.

We note from (2.1) that

$$|P_2P_1| = \sqrt{(x_1 - x_2)^2 + (y_1 - y_2)^2} = |P_1P_2|,$$

so that the distance between P_1 and P_2 can be denoted by either $|P_1P_2|$ or $|P_2P_1|$. Sometimes the terminology *Euclidean plane* is used to denote the Cartesian plane with distance defined by Formula (2.1).

Example 1   Find the distance between the points $A(3, -2)$ and $B(-1, 0)$ (Figure 2–5).

    SOLUTION   In (2.1) we let $(x_1, y_1) = (3, -2)$ and $(x_2, y_2) = (-1, 0)$. Then from Equation (2.1)

$$|AB| = \sqrt{(-1 - 3)^2 + (0 + 2)^2} = \sqrt{20} = 2\sqrt{5}.$$

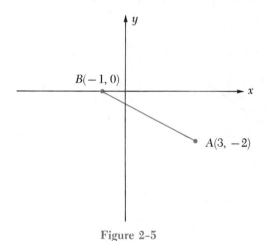

Figure 2–5

Example 2   Are the points $A(-1, 2)$, $B(5, 1)$, and $C(1, 14)$ vertices of a right triangle (Figure 2–6)?

    SOLUTION   From elementary geometry a triangle is a right triangle if the sum of the squares of the lengths of two of its sides equals the square of the length of the third side. From (2.1) the lengths of the sides of triangle ABC are

$$|AB| = \sqrt{37} \qquad |BC| = \sqrt{185} \qquad |AC| = \sqrt{148}$$

Since $|AB|^2 + |AC|^2 = 185 = |BC|^2$, ABC is a right triangle with A as the right angle.

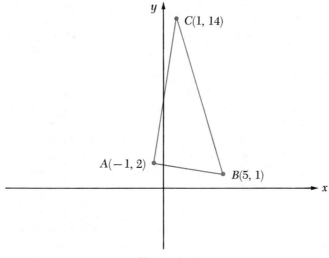

Figure 2-6

Example 3   Using (2.1) determine if the points $A(1, -2)$, $B(6, 8)$, and $C(3, 2)$ are *collinear* (that is, lie on the same line) (Figure 2–7).

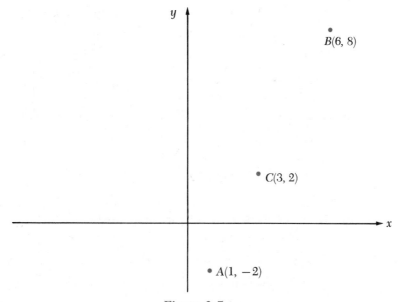

Figure 2-7

SOLUTION   The points A, B, and C are collinear if and only if the sum of any two of the distances $|AC|$, $|BC|$, and $|AB|$ is equal to the third distance. From (2.1):

$$|AC| = \sqrt{20} = 2\sqrt{5}$$
$$|BC| = \sqrt{45} = 3\sqrt{5}$$
$$|AB| = \sqrt{125} = 5\sqrt{5}$$

Since $|AC| + |BC| = 5\sqrt{5} = |AB|$, the points are collinear.

It should be emphasized that the solutions of the illustrative examples in this section were based on the application of the two-point distance formula and in no way depended on the drawings provided (Figures 2–5, 2–6, and 2–7). Nevertheless, the student is strongly urged to supply a graphical sketch for his solutions of problems in analytic geometry. Such a sketch is useful in detecting possible errors and also in reducing the number of possibilities in a problem which must be investigated. For instance, from Figure 2–6 it is seen that BC is the only candidate for the hypotenuse of a right triangle in Example 2.

Example 4   Find an equation in simplest form satisfied by the coordinates of any point $P(x, y)$ equidistant from the points $A(-1, 3)$ and $B(6, 7)$ (Figure 2–8).

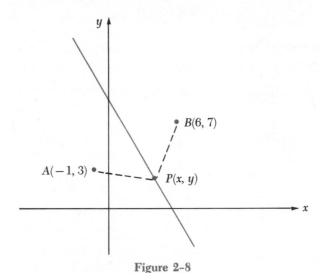

Figure 2-8

SOLUTION   If P is equidistant from A and B, we can write successively

$$|PA| = |PB|$$
$$\sqrt{(x + 1)^2 + (y - 3)^2} = \sqrt{(x - 6)^2 + (y - 7)^2}.$$

After squaring each member of the latter equation we obtain

$$(x + 1)^2 + (y - 3)^2 = (x - 6)^2 + (y - 7)^2,$$

$$x^2 + 2x + 1 + y^2 - 6y + 9 = x^2 - 12x + 36 + y^2 - 14y + 49,$$

and combining terms gives the equation

$$14x + 8y = 75.$$

The student will recall that the set of all points P equidistant from A and B is the perpendicular bisector of line segment AB. Thus P is in this perpendicular bisector if and only if the coordinates of the point satisfy the equation $14x + 8y = 75$.

Exercise Set 2.1

1. In each of the following exercises find the length of the line segments having the given pair of points as endpoints..

   (a) $(4, -5)$, $(0, -2)$ (b) $(-2, 4)$, $(4, 3)$

   (c) $(\frac{2}{3}, \frac{1}{6})$, $(-\frac{11}{3}, -\frac{7}{6})$ (d) $(-6\sqrt{3}, 3\sqrt{5})$, $(\sqrt{3}, -3\sqrt{5})$

2. In each of the following exercises determine if the given points are collinear.

   (a) $(-1, 4)$, $(3, 7)$, $(10, 12)$ (b) $(3, -1)$, $(-1, 2)$, $(11, -7)$

3. In the following examples the given points are vertices of a triangle. Which of the triangles are isosceles triangles? right triangles? neither?

   (a) $(1, 2)$, $(4, 0)$, $(8, 6)$ (b) $(-3, 4)$, $(1, 2)$, $(-4, -3)$

   (c) $(1, -1)$, $(-3, 0)$, $(2, 4)$ (d) $(2, 5)$, $(3, 1)$, $(6, 10)$

   (e) $(-1, 2)$, $(4, 0)$, $(-3, -3)$ (f) $(3, 7)$, $(0, 2)$, $(10, -4)$

4. In each of the following exercises name the quadrilateral having the given points as vertices.

   (a) $(1, 1)$, $(5, -2)$, $(2, -3)$, $(-2, 0)$

   (b) $(1, -4)$, $(5, -3)$, $(6, -7)$, $(2, -8)$

   (c) $(-5, -5)$, $(-1, 2)$, $(7, 3)$, $(3, -4)$

5. Find the point that is equidistant from the points $(7, 3)$, $(6, -4)$, and $(3, -5)$. HINT: Let (x, y) be the desired point. Set up and solve simultaneously two equations in x and y.

6. Find the point that is equidistant from the points $(3, 7)$, $(0, 2)$, and $(10, -4)$.

2.2   Line Segments; Lines

In this section we will derive equations that give the coordinates of any point P on a given line segment. Then it will be shown that equations of the same form also express the coordinates of any point on the line containing the line segment. To begin our discussion, let $P_1(x_1, y_1)$ and $P_2(x_2, y_2)$ be distinct points in the plane and let $P(x, y)$ be a point on the line segment P_1P_2 (Figure 2–9) such that

$$\frac{|P_1P|}{|P_1P_2|} = t \quad \text{where } 0 \le t \le 1. \tag{2.5}$$

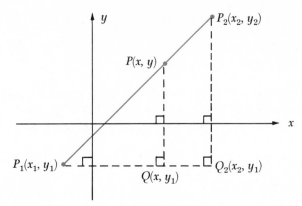

Figure 2-9

Then suppose that lines are constructed perpendicular to the coordinate axes from P, P_1, and P_2 as shown in Figure 2-9.

If $x_1 \neq x_2$, $y_1 \neq y_2$ and P is distinct from P_1, the triangles $P_1P_2Q_2$ and P_1QP that are formed are similar. Since the corresponding sides of these triangles are proportional

$$\frac{|P_1P|}{|P_1P_2|} = \frac{|P_1Q|}{|P_1Q_2|}$$

$$= \frac{|x - x_1|}{|x_2 - x_1|} = t \qquad (2.6)$$

where $0 < t \leq 1$.

Equation (2.6) still holds even if $y_1 = y_2$ or if P coincides with P_1 provided $x_1 \neq x_2$ as we will now show. If P coincides with P_1, then $|P_1P| = |x - x_1| = 0 = t$ and Equation (2.6) holds. If $y_1 = y_2$ (Figure 2–10), P_1P_2 lies on a line that is perpendicular to the y axis and hence $y = y_1 = y_2$. Then from Theorem 2.1.1, $|P_1P| = |x - x_1|$ and $|P_1P_2| = |x_2 - x_1|$ and again (2.6) holds.

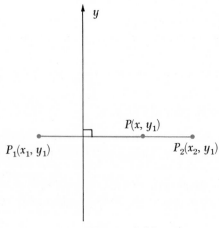

Figure 2-10

Since P is in P_1P_2, either $x - x_1$ and $x_2 - x_1$ agree in sign or $x = x_1$. Thus the absolute value bars enclosing $x - x_1$ and $x_2 - x_1$ in (2.6) can be dropped and we can write

$$\frac{x - x_1}{x_2 - x_1} = t \quad \text{where } 0 \leq t \leq 1.$$

This equation can be solved for x to yield

$$x = x_1 + t(x_2 - x_1) \quad \text{where } 0 \leq t \leq 1. \tag{2.7}$$

It should be also noted that (2.7) is satisfied by x if $x_1 = x_2$ because then P_1P_2 would lie in a line perpendicular to the x axis and we would have $x = x_1 = x_2$.

Hence, regardless of the placement of the endpoints P_1 and P_2 and the location of P in P_1P_2, x is given by Equation (2.7). Using a similar proof, we can also show that the y coordinate of the point P is given by

$$y = y_1 + t(y_2 - y_1) \quad \text{where } 0 \leq t \leq 1. \tag{2.8}$$

The filling in of the details, which is analogous to the above derivation, is left as an exercise.

We also leave as an exercise the proof of the converse: If $P_1(x_1, y_1)$ and $P_2(x_2, y_2)$ are distinct points and $P(x, y)$ is an arbitrary point whose coordinates satisfy (2.7) and (2.8), then P is on the line segment P_1P_2; that is, $|P_1P| + |PP_2| = |P_1P_2|$, and also (2.5) is true.

Thus we have the following theorem.

2.2.1   Theorem

The point $P(x, y)$ is on the line segment with endpoints $P_1(x_1, y_1)$ and $P_2(x_2, y_2)$ and (2.5) is satisfied if and only if

$$\left.\begin{aligned} x &= x_1 + t(x_2 - x_1) \\ y &= y_1 + t(y_2 - y_1) \end{aligned}\right\} \; \text{where } 0 \leq t \leq 1 \tag{2.9}$$

The formulas in (2.9) are called the *point of division* formulas for a line segment.

Example 1   Find the point $P(x, y)$ on the line segment with endpoints $P_1(-2, 8)$ and $P_2(4, 3)$ which is three times as far from P_1 as from P_2 (Figure 2–11).

SOLUTION   From the statement of the problem

$$|P_1P| = 3|PP_2|, \tag{2.10}$$

and since $|P_1P_2| = |P_1P| + |PP_2|$,

$$|P_1P_2| = 4|PP_2|. \tag{2.11}$$

From (2.10) and (2.11)

$$\frac{|P_1P|}{|P_1P_2|} = \frac{3}{4}.$$

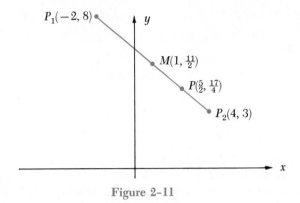

Figure 2-11

Then letting $t = \frac{3}{4}$, $(x_1, y_1) = (-2, 8)$, and $(x_2, y_2) = (4, 3)$ in (2.9), we obtain:

$$x = -2 + \tfrac{3}{4}(4 - (-2)) = \tfrac{5}{2}$$
$$y = 8 + \tfrac{3}{4}(3 - 8) = \tfrac{17}{4}$$

The desired point P is therefore $(\frac{5}{2}, \frac{17}{4})$.

By letting $t = \frac{1}{2}$ in (2.9), we obtain the midpoint of the line segment P_1P_2. The coordinates of this midpoint are:

$$x = \frac{x_1 + x_2}{2} \qquad y = \frac{y_1 + y_2}{2} \tag{2.12}$$

The formulas in (2.12) are called the *midpoint formulas*. From (2.12) it is seen that the x and y coordinates of the midpoint of a line segment are, respectively, the average of the x coordinates and the average of the y coordinates of the endpoints of the line segment.

Example 2   Find the midpoint of the line segment with endpoints $P_1(-2, 8)$ and $P_2(4, 3)$ (Figure 2–11).

SOLUTION   From (2.12) the coordinates of the midpoint are:

$$x = \frac{-2 + 4}{2} = 1 \qquad y = \frac{8 + 3}{2} = \frac{11}{2}$$

so the midpoint of the line segment is $(1, \frac{11}{2})$. This midpoint is denoted by M in Figure 2–11.

In the next example the formulas (2.12) are used to give an analytic proof of a theorem from elementary geometry.

Example 3   Prove that the midpoint of the hypotenuse of a right triangle is equidistant from the vertices of the triangle.

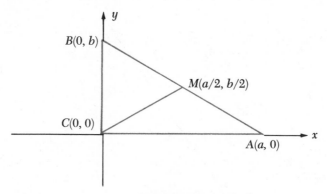

Figure 2-12

SOLUTION   It should first be noted that coordinate axes do not come with the plane. These number lines must be provided in order to give the required analytic proof. We are free, however, to locate the coordinate axes for this proof in some convenient manner with respect to the triangle. Thus the vertices of the triangle can be chosen to be the points $A(a, 0)$, $B(0, b)$, and $C(0, 0)$, where a and b are positive, thereby making the angle at C the required right angle (Figure 2–12). Then from (2.12) the midpoint of the hypotenuse is $M(a/2, b/2)$ and from Theorem 2.1.1

$$|MC| = |AM| = |MB| = \tfrac{1}{2}\sqrt{a^2 + b^2}.$$

Thus the theorem is proved.

By the next theorem the equations (2.9) can be used to express the coordinates of any point on the line through $P_1(x_1, y_1)$ and $P_2(x_2, y_2)$ even if the restriction that $t \in [0, 1]$ is removed.

2.2.2   Theorem

The point $P(x, y)$ is on the line L passing through the distinct points $P_1(x_1, y_1)$ and $P_2(x_2, y_2)$ if and only if for some t the coordinates of P satisfy the equations

$$x = x_1 + t(x_2 - x_1) \qquad and \qquad y = y_1 + t(y_2 - y_1). \tag{2.13}$$

PROOF   We first note that P is on line L if and only if P satisfies one of the following conditions:

(i) P is on P_1P_2 (Figure 2–13(a))
(ii) P_2 is on P_1P (Figure 2–13(b))
(iii) P_1 is on PP_2 (Figure 2–13(c))

From (2.9) condition (i) holds if and only if for some $t \in [0, 1]$ Equations (2.13) are valid.

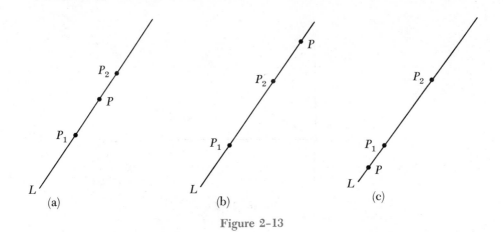

Figure 2-13

Also from (2.9), (ii) holds if and only if for some $s \in (0, 1]$

$$x_2 = x_1 + s(x - x_1) \qquad \text{and} \qquad y_2 = y_1 + s(y - y_1). \qquad (2.14)$$

(We note here that $s \neq 0$ for otherwise P_1 and P_2 would coincide.) From Equations (2.14)

$$x = \frac{1}{s}x_2 + \frac{s-1}{s}x_1 \qquad \text{and} \qquad y = \frac{1}{s}y_2 + \frac{s-1}{s}y_1,$$

and if we regroup terms

$$x = x_1 + \frac{1}{s}(x_2 - x_1) \qquad \text{and} \qquad y = y_1 + \frac{1}{s}(y_2 - y_1). \qquad (2.15)$$

Hence Equations (2.15) are equivalent to Equations (2.13) if $t = 1/s$. Here $t \geq 1$ since $s \in (0, 1]$.

Finally, from (2.9), (iii) holds if and only if for some $u \in [0, 1)$

$$x_1 = x + u(x_2 - x) \qquad \text{and} \qquad y_1 = y + u(y_2 - y) \qquad (2.16)$$

(Here $u \neq 1$ for otherwise P_1 and P_2 would coincide.) From Equations (2.16),

$$x = \frac{u}{u-1}x_2 - \frac{1}{u-1}x_1 \qquad \text{and} \qquad y = \frac{u}{u-1}y_2 - \frac{1}{u-1}y_1,$$

and after regrouping terms

$$x = x_1 + \frac{u}{u-1}(x_2 - x_1) \qquad \text{and} \qquad y = y_1 + \frac{u}{u-1}(y_2 - y_1). \qquad (2.17)$$

Hence Equations (2.17) are equivalent to Equations (2.13) if $t = u/(u - 1)$. Here $t < 0$ since $u \in [0, 1)$.

Since one of the conditions (i), (ii), and (iii) will hold if and only if Equations (2.13) are satisfied by the coordinates of P for some t, the theorem is proved.

Example 4 Using Equations (2.13) show that the point $(5, -1)$ is on the line L through the points $(2, 1)$ and $(-4, 5)$.

SOLUTION   If we let $(x_1, y_1) = (2, 1)$ and $(x_2, y_2) = (-4, 5)$, then from (2.13) the point $(5, -1)$ is on L if there is some number t which satisfies both

$$5 = 2 + t(-4 - 2) \qquad \text{and} \qquad -1 = 1 + t(5 - 1).$$

Since each of these equations is satisfied by $t = -\frac{1}{2}$, the point $(5, -1)$ is on L.

Equations (2.13) will be utilized in Section 2.4 to obtain a definition of the slope of a line. Then in Section 2.5 the equations will again be used to derive the first of various forms for an equation of a line.

Exercise Set 2.2

1. If $P_1(1, -5)$ and $P_2(-3, -2)$ are the endpoints of a line segment, find the point P on the line segment satisfying the following conditions:

   (a) $\dfrac{|P_1P|}{|P_1P_2|} = \dfrac{2}{5}$ (b) $\dfrac{|P_1P|}{|P_1P_2|} = \dfrac{2}{3}$ (c) $\dfrac{|PP_2|}{|P_1P_2|} = \dfrac{1}{4}$

   (d) P is twice as far from P_2 as from P_1.

   (e) P is $1\frac{1}{3}$ times as far from P_1 as from P_2.

2. Find the midpoint of the line segment having endpoints

   (a) $(2, -4)$, and $(-3, -6)$;   (b) $(-3, 4)$ and $(5, 9)$.

3. The point $P(3, -7)$ lies on a line segment AB. If A is the point $(5, 6)$, find B if

   (a) P is the midpoint of AB;

   (b) P is three times as far from A as from B.

4. Using Equations (2.13), determine if the point $(9, 3)$ is on the line through $(2, -3)$ and $(6, -1)$.

5. Using Equations (2.13), determine if the point $(18, -4)$ is on the line through $(8, 0)$ and $(3, 2)$.

6. Find x if the point $(x, 2)$ is on the line segment with endpoints $(3, -4)$ and $(6, 1)$.

7. Find y if the point $(-1, y)$ is on the line segment with endpoints $(-3, 1)$ and $(5, 4)$.

8. Show that the diagonals of the quadrilateral with vertices $(-1, -2)$, $(1, 3)$, $(-3, -3)$, and $(-5, -8)$ bisect each other.

9. Find the lengths of the medians of the triangle with vertices $A(0, 3)$, $B(5, 6)$, and $C(-3, -8)$.

10. Prove, without using similar triangles, that if $P(x, y)$ is on the line segment with endpoints $P_1(x_1, y_1)$ and $P_2(x_2, y_2)$ and

$$\frac{|P_1P|}{|PP_2|} = \frac{r_1}{r_2}$$

where $r_1 > 0$ and $r_2 > 0$, then

$$x = \frac{r_2 x_1 + r_1 x_2}{r_1 + r_2} \quad \text{and} \quad y = \frac{r_2 y_1 + r_1 y_2}{r_1 + r_2}.$$

11. Prove analytically that the quadrilateral formed by joining consecutive midpoints of the sides of a rectangle is a rhombus.

12. Prove analytically that the medians of a triangle pass through a point that is two-thirds of the distance from any vertex of the triangle to the midpoint of the opposite side.

13. Derive Equation (2.8).

14. Prove: If $P_1(x_1, y_1)$ and $P_2(x_2, y_2)$ are distinct points and $P(x, y)$ is an arbitrary point whose coordinates satisfy (2.7) and (2.8), then P is on the line segment $P_1 P_2$.

2.3 Graphs of Equations

We shall begin this section with a brief discussion of an equation in the variables x and y. An ordered pair (a, b) is a *solution* of such an equation, or saying the same thing, *satisfies* the equation, if and only if the same number is obtained from each side of the equation when a is substituted for x and b for y. For example, the ordered pair $(3, -2)$ is a solution of the equation $3x + 2y = 5$. Also, two equations in x and y are said to be *equivalent* if and only if every solution of one of the equations is also a solution of the other equation.

We shall define the *graph of an equation* in x and y.

2.3.1 Definition

The *graph of a set* of ordered pairs of real numbers is the set of all points in the plane that are associated with the ordered pairs. In particular, the *graph of an equation* in x and y is the graph of the set of all solutions of the equation.

In this section we will be concerned with the process of sketching the graph of an equation in x and y—that is, *graphing* the equation. In order to graph such an equation, it is necessary to plot a sufficient number of points whose coordinates satisfy the equation. The exact number of points to be plotted would depend upon the nature of the equation to be plotted, the accuracy desired in the sketch, and the familiarity of the person doing the graphing with certain information from analytic geometry and calculus that facilitates the graphing process.

In the first example below, the sketch will be made by plotting some points and then drawing a curve through them.

Example 1 Graph the equation $y = x^2 - 4$ (Figure 2–14).

SOLUTION   We first complete the following table giving the corresponding x and y coordinates for certain points in the graph. In each case an x

coordinate is assigned, and the corresponding y coordinate is calculated using the given equation. Hence each pair (x, y) is a solution of the equation.

x	0	1	-1	2	-2	3	-3	4	-4
y	-4	-3	-3	0	0	5	5	12	12

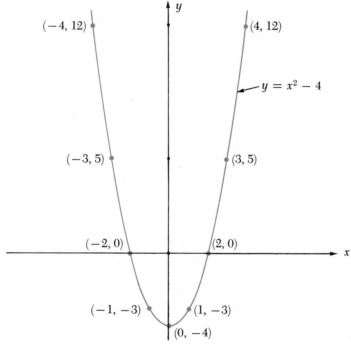

Figure 2-14

It should be noted that we are justified in connecting the plotted points by drawing a curve since there is a point in the graph having any number as its x coordinate. As in Figure 2–14, the sketch of the graph should, if possible, show its intersections with the coordinate axes and other distinctive features—for example, any "high" and "low" points of the graph.

The graphing of an equation is facilitated by a prior discussion of the graph with respect to the following analytic considerations: intercepts, symmetry to a coordinate axis or the origin, and extent. We shall first define the x and y *intercepts* of a graph.

2.3.2 Definition

Let S be a set of points in the plane and let a and b be any numbers. Then a is said to be an x *intercept* of S if and only if the point $(a, 0)$ is in S. Also, b is a y *intercept* of S if and only if the point $(0, b)$ is in S.

For example, 2 and -2 are x intercepts of the graph of the equation $y = x^2 - 4$ since $(2, 0)$ and $(-2, 0)$ are points in the graph. Also, -4 is a y intercept of the graph since the point $(0, -4)$ is in the graph.

We next define the symmetry of a graph to a line.

2.3.3 Definition

A set S of points is *symmetric to a line* L if and only if for every point P_1 in S there is a point P_2 in S such that L is the perpendicular bisector of P_1P_2 (Figure 2–15). The line L is called an *axis of symmetry* of S.

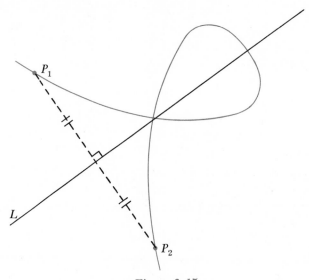

Figure 2-15

Thus, if a graph is symmetric to a line L, the portion of the graph on one side of L is a mirror image through L of the portion on the other side of L.

Since the y axis is the perpendicular bisector of the line segment joining any points $(-x, y)$ and (x, y), from Definition 2.3.3

$$S \text{ is symmetric to the } y \text{ axis} \quad \text{iff} \quad \begin{cases} \text{the point } (-x, y) \in S \\ \quad \text{when } (x, y) \in S. \end{cases} \quad (2.18)$$

Also, since the x axis is the perpendicular bisector of the line segment joining any points $(x, -y)$ and (x, y),

$$S \text{ is symmetric to the } x \text{ axis} \quad \text{iff} \quad \begin{cases} \text{the point } (x, -y) \in S \\ \quad \text{when } (x, y) \in S. \end{cases} \quad (2.19)$$

From (2.18), *the graph of an equation in x and y is symmetric to the y axis if and only if an equivalent equation is obtained by replacing x by $-x$ in the equation of the graph.* For example, the graph of $y = x^2 - 4$ (Figure 2–14) is symmetric to the y axis since the equation $y = (-x)^2 - 4$ is equivalent to $y = x^2 - 4$. Also, from (2.19) the *graph of an equation in x and y is symmetric*

to the *x* axis if and only if an equivalent equation is obtained by replacing *y* by
−*y* in the equation of the graph. The graph of $y = x^2 - 4$ is not symmetric to
the *x* axis since the equation $-y = x^2 - 4$ is not equivalent to $y = x^2 - 4$. How-
ever, the graph of $x = |y|$ (Figure 2–16) is symmetric to the *x* axis since the equation
$x = |-y|$ is equivalent to $x = |y|$.

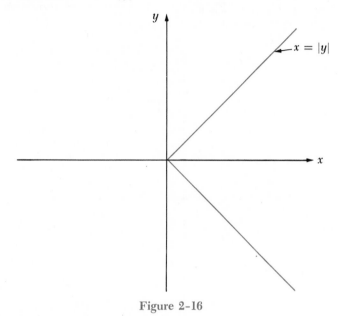

Figure 2-16

We next define the symmetry of a graph to a point.

2.3.4   Definition

A set *S* of points is *symmetric to a point M* if and only if for every point P_1 in
S there is also a point P_2 in *S* such that *M* is the midpoint of P_1P_2 (Figure 2–17).

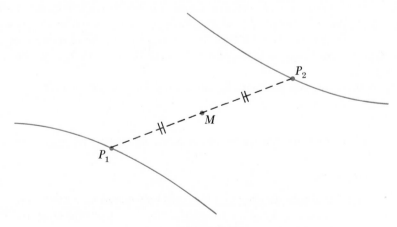

Figure 2-17

In particular from Definition 2.3.4 and the midpoint formulas,

S is symmetric to the origin   iff   the point $(-x, -y) \in S$ when $(x, y) \in S$.

Thus, the graph of an equation in x and y is symmetric to the origin if and only if an equivalent equation is obtained by replacing x by $-x$ and y by $-y$. For example, the graph of $y = x^3$ (Figure 2–18) is symmetric to the origin since the equation $-y = (-x)^3$ is equivalent to $y = x^3$.

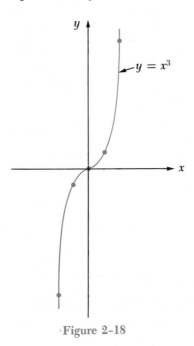

·Figure 2–18

To determine the *extent* of a graph, one must obtain the set of all numbers x for which there are points (x, y) in the graph and also the set of all numbers y for which there are points (x, y) in the graph. If the given equation in x and y can be solved for y in terms of x, then for every number x for which we can obtain a value of y there will be a point (x, y) in the graph of the equation. Similarly, if the equation can be solved for x in terms of y, then for every number y for which we can obtain a value of x there will be a point (x, y) in the graph of the equation.

Two examples will now be considered in which intercepts, symmetry, and extent are discussed.

Example 1   Graph the equation $4x^2 + y^2 = 16$.

SOLUTION

(i) *Intercepts*   Letting $y = 0$ in the given equation, we obtain $x = \pm 2$, which are the x intercepts. Letting $x = 0$ in the equation, the y intercepts, $y = \pm 4$, are obtained.

(ii) *Symmetry*   Since the equation $4x^2 + (-y)^2 = 16$ is equivalent to $4x^2 + y^2 = 16$, the graph is symmetric to the x axis. Also, since the equation $4(-x)^2 + y^2 = 16$ is equivalent to $4x^2 + y^2 = 16$, the graph is symmetric to the y axis. Finally, since $4(-x)^2 + (-y)^2 = 16$ is equivalent to $4x^2 + y^2 = 16$, the graph is symmetric to the origin.

(iii) *Extent*   Solving the equation $4x^2 + y^2 = 16$ for y in terms of x, we have

$$y = \pm\sqrt{16 - 4x^2} = \pm 2\sqrt{4 - x^2}. \tag{2.20}$$

Thus y is defined if and only if $x \in [-2, 2]$. Also, if the given equation is solved for x in terms of y, one obtains

$$x = \pm\tfrac{1}{2}\sqrt{16 - y^2},$$

and hence x is defined if and only if $y \in [-4, 4]$. Therefore, from our discussion above, there will be points (x, y) in the graph for every $x \in [-2, 2]$ and for every $y \in [-4, 4]$.

From (2.20) the following table of corresponding x and y values is obtained.

x	0	$\pm\frac{1}{2}$	± 1	$\pm\frac{3}{2}$	± 2
y	± 4	$\pm\sqrt{15}$	$\pm 2\sqrt{3}$	$\pm\sqrt{7}$	0

The sketch in Figure 2–19 can be drawn from this table. It will be noted that the graph is an ellipse. It should also be noted that the portion of the graph above the x axis has the equation $y = 2\sqrt{4 - x^2}$ and the portion below the x axis has the equation $y = -2\sqrt{4 - x^2}$.

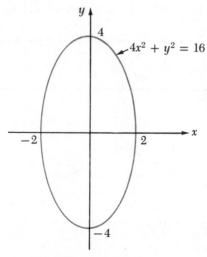

Figure 2–19

Example 2 Graph the equation $|x| + 2y = 4$.

SOLUTION

(i) *Intercepts* Letting $y = 0$ in this equation, we obtain $|x| = 4$ and hence the x intercepts are ± 4. If we let $x = 0$ in the equation, we have $2y = 4$ and the y intercept is 2.

(ii) *Symmetry* Since $|x| + 2(-y) = 4$ is not equivalent to $|x| + 2y = 4$, the graph is not symmetric to the x axis. The graph is, however, symmetric to the y axis since the equation $|-x| + 2y = 4$ is equivalent to $|x| + 2y = 4$. Finally, the graph is not symmetric to the origin since the equation $|-x| + 2(-y) = 4$ is not equivalent to $|x| + 2y = 4$.

(iii) *Extent* Solving the equation $|x| + 2y = 4$ for y in terms of x, we have

$$y = \frac{4 - |x|}{2}.$$

Thus y is defined for every x. If the given equation is solved for x in terms of y, one obtains

$$|x| = 4 - 2y.$$

Since $|x|$ is never negative, x is defined only if $4 - 2y \geq 0$. Conversely, if $4 - 2y \geq 0$, then from the definition of absolute value:

$$x = \begin{cases} 4 - 2y & \text{if } x \geq 0 \\ -(4 - 2y) & \text{if } x < 0 \end{cases} \tag{2.21}$$

Thus x is defined if and only if $4 - 2y \geq 0$—that is, if and only if $y \leq 2$.

From (2.21) the following table of corresponding x and y values is obtained.

x	0	± 1	± 2	± 3	± 4	± 5	± 6
y	2	$\frac{3}{2}$	1	$\frac{1}{2}$	0	$-\frac{1}{2}$	-1

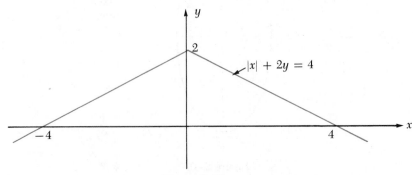

Figure 2-20

The sketch in Figure 2–20 can be drawn from this table. It will be noted that the graph consists of two half lines drawn from the point $(0, 2)$. Since $|x| = x$ when $x \geq 0$ and $|x| = -x$ when $x < 0$, the right half line has the equation $x + 2y = 4$ and the left half line has the equation $-x + 2y = 4$.

Exercise Set 2.3

In Exercises 1–22 discuss the graph of each equation with regard to intercepts, symmetry, and extent. Graph the equation.

1. $2x + 3y = 5$ 2. $4x - 5y = 10$

3. $x^2 = 9$ 4. $y^2 = 5$

5. $y = 4x - x^2$ 6. $x = 8 - 2y - y^2$

7. $y = x^3 - 4x^2 - x + 4$ 8. $x^2 + y^2 = 16$

9. $3x^2 + 4y^2 = 48$ 10. $3x^2 - 4y^2 = 12$

11. $y = \sqrt{36 - x^2}$ 12. $x = \sqrt{4y^2 + 9}$

13. $y^2 = 4x^2$ 14. $x^2 + 4y^2 = 0$

15. $x = |y + 2|$ 16. $y = |2x - 5|$

17. $y^2 = x^3 + 2x$ 18. $|x| + |y| = 4$

19. $|x + 2| - |y - 3| = 4$ 20. $|x - 1| + 2|y - 2| = 6$

21. $x^{1/2} + y^{1/2} = 4$ 22. $x^{2/3} + y^{2/3} = 4$

23. Graph the equation $(x + y)(2x - 3y + 5) = 0$. HINT: $(x + y)(2x - 3y + 5) = 0$ if and only if $x + y = 0$ or $2x - 3y + 5 = 0$.

24. Graph the equation $(3x + 4y - 10)(y^2 - x) = 0$.

2.4   Slope of a Line

Suppose L is a line passing through the distinct points (x_1, y_1) and (x_2, y_2), where $x_2 \neq x_1$, and suppose (x', y') and (x'', y'') are any two points on the line (Figure 2–21). From Equations (2.13) there are numbers t' and t'' such that:

$$y'' = y_1 + t''(y_2 - y_1) \qquad x'' = x_1 + t''(x_2 - x_1) \qquad (2.22)$$
$$y' = y_1 + t'(y_2 - y_1) \qquad x' = x_1 + t'(x_2 - x_1) \qquad (2.23)$$

Then from (2.22) and (2.23)

$$\frac{y'' - y'}{x'' - x'} = \frac{(t'' - t')(y_2 - y_1)}{(t'' - t')(x_2 - x_1)}$$

$$= \frac{y_2 - y_1}{x_2 - x_1}. \qquad (2.24)$$

Equation (2.24) says that as a point moves along line L from (x', y') to (x'', y''), the *ratio of the change in the y coordinate to the change in the x coordinate is constant*, the constant being $(y_2 - y_1)/(x_2 - x_1)$. This ratio, intuitively, seems like a good measure of the steepness or *slope* of line L.

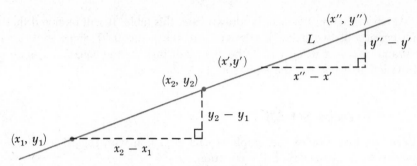

Figure 2-21

2.4.1 Definition

Let $P_1(x_1, y_1)$ and $P_2(x_2, y_2)$, where $x_1 \neq x_2$, be any two distinct points on a line. The *slope, m*, of the line is given by

$$m = \frac{y_2 - y_1}{x_2 - x_1}. \tag{2.25}$$

If $y_2 > y_1$ when $x_2 > x_1$ as in Figure 2–21, the line slopes up to the right, and from (2.25) the slope is positive. If $y_2 < y_1$ when $x_2 > x_1$ (Figure 2–22(a)) the line slopes up to the left, and by (2.25) its slope is negative. If $y_1 = y_2$ (Figure 2–22(b)) the line is perpendicular to the y axis, and by (2.25) its slope is zero. If $x_1 = x_2$, the line is perpendicular to the x axis, and its slope is undefined.

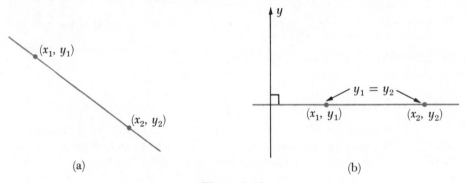

Figure 2-22

Example 1 Find the slope of the line passing through the points $(-2, -3)$ and $(-6, 0)$ (Figure 2–23).

SOLUTION From Definition 2.4.1, if we let $(x_1, y_1) = (-2, -3)$ and $(x_2, y_2) = (-6, 0)$,

$$m = \frac{0 - (-3)}{-6 - (-2)} = -\frac{3}{4}.$$

Geometrically, the slope $-\frac{3}{4}$ for the line in Example 1 means that in traversing the line, we move down 3 units for every 4 we move to the right.

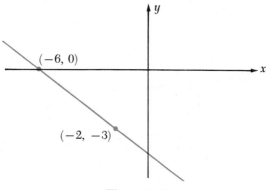

Figure 2-23

It should be noted that *if the slope of a line L exists, then any number is the x coordinate of some point on L.* To show this, we let *a* be any number and consider the line *L'* that consists of all points having an *x* coordinate of *a*. It is apparent that *L'* is perpendicular to the *x* axis at $(a, 0)$. Now, if there were no point in *L* having *a* as its *x* coordinate, *L* and *L'* would not intersect—that is, *L* and *L'* would be parallel. Then *L* would be perpendicular to the *x* axis, which is impossible since the slope of *L* exists.

The preceding italicized statement is used in proving a property of the slopes of parallel lines:

2.4.2   Theorem

Let L_1 and L_2 be distinct lines with respective slopes m_1 and m_2. L_1 and L_2 are parallel if and only if

$$m_1 = m_2.$$

PROOF   If $m_1 = m_2$, suppose L_1 and L_2 intersect at (a, b). If (x_1, y_1) and (x_1, y_2) are points distinct from (a, b) on L_1 and L_2, respectively, then from (2.25)

$$m_1 = \frac{b - y_1}{a - x_1} = \frac{b - y_2}{a - x_1} = m_2. \tag{2.26}$$

Hence from (2.26), $y_1 = y_2$ and (x_1, y_1) and (x_1, y_2) would be the same point. Thus L_1 and L_2 would intersect in two points, which is impossible. Hence L_1 and L_2 are parallel.

In proving the converse, we assume that L_1 and L_2 are parallel. Then if *P* is any point on L_1, we construct a line *L* through *P* with slope m_2. From the "if" part of this theorem, which we have just proved, *L* is parallel to L_2 (since these lines have the same slope). Thus we have two lines, L_1 and *L*, through *P* that are each parallel to L_2, and so *L* and L_1 must be the same line. Hence, the slopes of *L* and L_1 are equal; that is, $m_2 = m_1$.

Example 2   Show using slopes that the quadrilateral with vertices $A(-3, -1)$, $B(1, -4)$, $C(10, -5)$, and $D(6, -2)$ is a parallelogram (Figure 2–24).

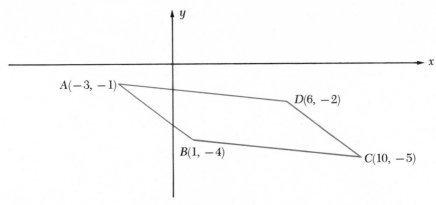

Figure 2-24

SOLUTION From (2.25), the slopes of AB, BC, CD, and DA are:

$$m_{AB} = \frac{-4-(-1)}{1-(-3)} = -\frac{3}{4} \qquad m_{BC} = \frac{-5-(-4)}{10-1} = -\frac{1}{9}$$

$$m_{DC} = \frac{-5-(-2)}{10-6} = -\frac{3}{4} \qquad m_{AD} = \frac{-2-(-1)}{6-(-3)} = -\frac{1}{9}$$

Since $m_{AB} = m_{DC}$ and $m_{BC} = m_{AD}$, by Theorem 2.4.2

$$AB \parallel DC \qquad \text{and} \qquad BC \parallel AD$$

and hence $ABCD$ is a parallelogram.

Example 3 Prove analytically that the opposite sides of a parallelogram are equal.

SOLUTION We shall locate the coordinate axes so that the vertices of the parallelogram can be denoted consecutively by $A(0, 0)$, $B(a, 0)$, $C(b, c)$, and $D(d, e)$ (Figure 2-25). By definition

$$AB \parallel CD \qquad \text{and} \qquad AD \parallel BC.$$

Then from Theorem 2.4.2

$$m_{AB} = m_{CD} \qquad \text{and} \qquad m_{AD} = m_{BC},$$

and from (2.25)

$$0 = \frac{c-e}{b-d} \qquad \text{and} \qquad \frac{e}{d} = \frac{c}{b-a}. \qquad (2.27)$$

From (2.27)

$$c = e \qquad \text{and} \qquad d = b - a;$$

hence we can rewrite the coordinates of the vertices as $A(0, 0)$, $B(a, 0)$, $C(a + d, c)$, and $D(d, c)$. Then, calculating the lengths of the opposite sides, we have

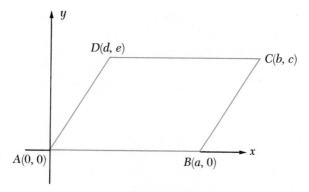

Figure 2-25

$$|AB| = a = |CD| \qquad \text{and} \qquad |AD| = \sqrt{d^2 + c^2} = |BC|,$$

and the conclusion is proved.

The following theorem gives the relation between the slopes of perpendicular lines.

2.4.3   Theorem

Let L_1 and L_2 be distinct lines with respective slopes m_1 and m_2 where $m_1 \neq 0 \neq m_2$. Then L_1 and L_2 are perpendicular if and only if

$$m_1 m_2 = -1, \qquad\qquad (2.28)$$

that is, the slopes are the negative reciprocals of each other.

PROOF   Suppose L_1 and L_2 intersect at the point $P(a, b)$ and $P_1(a + h_1, b + k_1)$ and $P_2(a + h_2, b + k_2)$ are points distinct from P on L_1 and L_2, respectively (Figure 2–26).

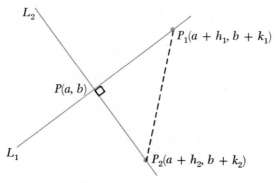

Figure 2-26

In proving the "only if" part we assume that L_1 and L_2 are perpendicular lines. Then by the Pythagorean theorem

$$|PP_1|^2 + |PP_2|^2 = |P_1P_2|^2$$

and if the lengths $|PP_1|$, $|PP_2|$, and $|P_1P_2|$ are calculated from the two-point distance formula,

$$(h_1^2 + k_1^2) + (h_2^2 + k_2^2) = (h_2 - h_1)^2 + (k_2 - k_1)^2.$$

Then after expanding the squares in each member of this equation and simplifying,

$$0 = -2h_1h_2 - 2k_1k_2$$
$$2k_1k_2 = -2h_1h_2.$$

Upon dividing each member by $2h_1h_2$ we obtain

$$\frac{k_1k_2}{h_1h_2} = -1. \tag{2.29}$$

From (2.25) the slopes of L_1 and L_2, which exist by hypothesis, are respectively

$$m_1 = \frac{k_1}{h_1} \quad \text{and} \quad m_2 = \frac{k_2}{h_2}.$$

Hence from (2.29)

$$m_1m_2 = -1,$$

which proves the "only if" part.

To prove the "if" part, one starts with $m_1m_2 = -1$. Because of the coordinates assigned to the points P_1 and P_2 we have $m_1 = k_1/h_1$ and $m_2 = k_2/h_2$, and therefore (2.29) holds. Then we can reverse the steps in the "only if" proof back to $|PP_1|^2 + |PP_2|^2 = |P_1P_2|^2$, thus proving that L_1 and L_2 are perpendicular.

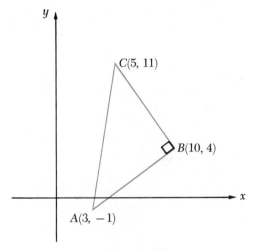

Figure 2-27

Example 4   Prove using slopes that the triangle with vertices $A(3, -1)$, $B(10, 4)$, and $C(5, 11)$ is a right triangle (Figure 2–27).

SOLUTION   From (2.25)

$$m_{AB} = \tfrac{5}{7} \quad \text{and} \quad m_{BC} = -\tfrac{7}{5}.$$

Hence by Theorem 2.4.3, AB and BC are perpendicular and ABC is a right triangle.

Exercise Set 2.4

1. Find the slopes of the lines determined by the following pairs of points.
   (a) $(-4, 2)$, $(-1, 8)$ (b) $(2, -5)$, $(6, 1)$
   (c) $(-\tfrac{3}{2}, \tfrac{1}{9})$, $(-\tfrac{15}{4}, \tfrac{11}{6})$ (d) $(\tfrac{13}{3}, -\tfrac{7}{2})$, $(-\tfrac{21}{4}, \tfrac{13}{6})$

2. In each of the following exercises determine, using slopes, if the given points are collinear.
   (a) $(2, 0)$, $(7, -4)$, $(22, -16)$ (b) $(1, 5)$, $(9, 8)$, $(-15, -1)$
   (c) $(4, -1)$, $(7, 1)$, $(11, 4)$ (d) $(0, -3)$, $(4, -5)$, $(12, -9)$

3. Identify the quadrilaterals having the following sets of points as vertices.
   (a) $(4, -6)$, $(8, -5)$, $(5, 2)$, $(1, 1)$
   (b) $(5, 0)$, $(1, 5)$, $(-1, 2)$, $(3, -3)$
   (c) $(-10, 1)$, $(-6, 7)$, $(-3, 5)$, $(-7, -1)$
   (d) $(-2, 3)$, $(2, 8)$, $(7, 4)$, $(3, -1)$
   (e) $(4, -1)$, $(0, 2)$, $(-3, -4)$, $(5, -10)$
   (f) $(5, -1)$, $(-5, -5)$, $(-7, 0)$, $(3, 4)$
   (g) $(-3, -4)$, $(0, 2)$, $(6, -1)$, $(3, -7)$
   (h) $(2, 6)$, $(-3, 4)$, $(-4, -2)$, $(6, 2)$

4. Find y if the line through $(3, y)$ and $(-4, 5)$ is parallel to the line through $(8, 2)$ and $(-2, 6)$.

5. Find x if the line through $(x, 4)$ and $(-2, -9)$ is perpendicular to the line through $(4, 1)$ and $(0, 3)$.

6. A line with slope $\tfrac{3}{4}$ passes through the point $(1, -2)$. Find two points on the line each 10 units from $(1, -2)$.

7. A line with slope $-\tfrac{5}{12}$ passes through the point $(-4, 1)$. Find two points on the line each 13 units from $(-4, 1)$.

8. Find the fourth vertex of a parallelogram having vertices $(4, 1)$, $(7, -2)$, and $(1, -5)$. (Three solutions)

9. Prove that if the consecutive midpoints of the sides of a quadrilateral are joined by line segments, the figure formed is a parallelogram.

10. Prove that if the diagonals of a parallelogram are equal, the parallelogram is a rectangle.

11. Prove that if the diagonals of a parallelogram are perpendicular, the parallelogram is a rhombus.

12. Prove that if the diagonals of a quadrilateral bisect each other, the quadrilateral is a parallelogram.

2.5   Equations of a Line

We begin this section by stating a definition.

2.5.1   Definition

Let S be a set of points in the plane and suppose that the point $P(x, y) \in S$ if and only if the ordered pair (x, y) satisfies an equation in x and y. We then say this equation is *an equation of S*.

For instance, a point $P(x, y)$ is 4 units distant from the point $(0, 0)$ if and only if

$$\sqrt{x^2 + y^2} = 4. \tag{2.30}$$

Since the set of all such points $P(x, y)$ is a circle with center at the origin and radius 4, from Definition 2.5.1 Equation (2.30) is an equation of this circle.

Recalling Definition 2.3.1, it follows that an equation in x and y is an equation of a set S of points in the plane if and only if S is the graph of the equation.

In this section various forms for an equation of a line will be derived. The first form is the *two-point form* for an equation of a line.

2.5.2   Theorem

An equation for the line L passing through the distinct points $P_1(x_1, y_1)$ and $P_2(x_2, y_2)$ is

$$(x - x_1)(y_2 - y_1) = (x_2 - x_1)(y - y_1). \tag{2.31}$$

PROOF   From Theorem 2.2.2 a point (x, y) is on L if and only if for some t

$$x - x_1 = t(x_2 - x_1) \quad \text{and} \quad y - y_1 = t(y_2 - y_1). \tag{2.32}$$

From (2.32)

$$(x - x_1)(y_2 - y_1) = t(x_2 - x_1)(y_2 - y_1)$$
$$= (x_2 - x_1)(y - y_1).$$

Hence a point (x, y) is on L only if (x, y) satisfies (2.31).

Conversely, suppose (x, y) satisfies (2.31). We note that since P_1 and P_2 are distinct points, either $x_1 \neq x_2$ or $y_1 \neq y_2$. If $x_1 \neq x_2$, we divide each member of (2.31) by $x_2 - x_1$, obtaining

$$y - y_1 = \frac{x - x_1}{x_2 - x_1}(y_2 - y_1). \tag{2.33}$$

If we let

$$t = \frac{x - x_1}{x_2 - x_1} \tag{2.34}$$

in (2.33), the equation for y in (2.32) is obtained. The equation for x in (2.32) is obtained by solving (2.34) for x.

If $x_1 = x_2$, then $y_1 \neq y_2$ and this time we divide each member of (2.31) by $y_2 - y_1$. The equations in (2.32) can then be obtained after letting $t = (y - y_1)/(y_2 - y_1)$. Thus the theorem is true.

Example 1   Write an equation for the line passing through the points $(-1, 4)$ and $(3, 7)$.

SOLUTION   If in (2.31) we let $x_1 = -1$, $y_1 = 4$, $x_2 = 3$, and $y_2 = 7$, we have

$$(x + 1) \cdot 3 = 4(y - 4),$$

or in simpler form

$$3x - 4y = -19.$$

For economy of expression the line in Example 1 could be called "the line $3x - 4y = -19$" instead of using the more lengthy description "the line having an equation $3x - 4y = -19$." We will follow a similar usage in describing sets in the plane having known equations.

If for the line L in Theorem 2.5.2, it is true that $x_1 = x_2$, then L is the *vertical line through* $(x_1, 0)$. Since $y_1 \neq y_2$ when $x_1 = x_2$, we obtain from (2.31) the equation

$$x = x_1 \tag{2.35}$$

for this vertical line.

If $x_1 \neq x_2$ for the line L in Theorem 2.5.2, Equation (2.31) is equivalent to

$$y - y_1 = \frac{y_2 - y_1}{x_2 - x_1}(x - x_1),$$

and hence

$$y - y_1 = m(x - x_1). \tag{2.36}$$

Equation (2.36) is called the *point-slope form* of an equation of a line. As an illustration of (2.31), a line with slope $\frac{2}{3}$ that passes through the point $(2, -5)$ has an equation of the form

$$y + 5 = \tfrac{2}{3}(x - 2)$$

or

$$-2x + 3y = -19.$$

Suppose a line has slope m and y-intercept b—that is, $(0, b)$ is a point in the line. From (2.36) the line has an equation

$$y - b = m(x - 0)$$

or

$$y = mx + b. \tag{2.37}$$

Equation (2.37) is termed the *slope-intercept* form of an equation of a line. It is a useful form, particularly in calculating y for a given value of x since the equation expresses y explicitly in terms of x.

If a, b, and c are numbers such that a and b are not both zero, the equation

$$ax + by = c \tag{2.38}$$

is called a *linear equation in x and y*.

The next theorem justifies the use of the term linear for equations of the form of (2.38).

2.5.3   Theorem

Every line in the plane has a linear equation in x and y. Conversely, every linear equation in x and y is an equation of some line in the plane.

PROOF   If L is a line in the plane passing through the distinct points (x_1, y_1) and (x_2, y_2), it has an equation of the form

$$(x - x_1)(y_2 - y_1) = (x_2 - x_1)(y - y_1),$$

which is equivalent to

$$(y_2 - y_1)x - (x_2 - x_1)y = x_1(y_2 - y_1) - y_1(x_2 - x_1)$$
$$= x_1 y_2 - x_2 y_1. \tag{2.39}$$

Since (x_1, y_1) and (x_2, y_2) are distinct points, $x_2 - x_1$ and $y_2 - y_1$ cannot both equal zero. Equation (2.39) is a linear equation in x and y since by comparison with (2.38) $a = y_2 - y_1$, $b = -(x_2 - x_1)$, and $c = x_1 y_2 - x_2 y_1$.

To prove the converse, we consider the separate cases for the linear equation (2.38) where $b = 0$ and $b \neq 0$. If $b = 0$, then $a \neq 0$ since a and b are not both zero, and hence from (2.38)

$$ax = c$$

which is equivalent to

$$x = \frac{c}{a}. \tag{2.40}$$

Equation (2.40) is the equation of the vertical line passing through $(c/a, 0)$.

If $b \neq 0$, Equation (2.38) can be solved for y to yield

$$y = -\frac{a}{b}x + \frac{c}{b}. \tag{2.41}$$

By comparison with (2.37) it is seen that (2.41) is an equation of the line with slope $-a/b$ and y intercept c/b.

We leave as an exercise the proof of the next theorem.

2.5.4 Theorem

Let A and B be any numbers that are not both zero.

(a) *The lines $Ax + By = C$ and $Ax + By = D$ are parallel if $C \neq D$.*
(b) *The lines $Ax + By = C$ and $Bx - Ay = D$ are perpendicular.*

Exercise Set 2.5

In Exercises 1–13 find an equation of the line satisfying the given condition.

1. Passes through $(2, 3)$ with slope $\frac{5}{4}$
2. Passes through $(3, -2)$ and $(8, 4)$
3. Passes through $(2, 5)$ and $(7, -1)$
4. Has slope $-\frac{3}{2}$ and y intercept 4
5. Has slope $\frac{3}{2}$ and x intercept -2
6. Is the median of the triangle with vertices $A(-4, -2)$, $B(-1, 5)$, and $C(10, -4)$ drawn from the vertex A
7. Passes through $(2, -1)$ and is parallel to the line $3x + 2y = 10$
8. Passes through $(2, -1)$ and is perpendicular to the line $3x + 2y = 10$
9. Passes through $(-3, 5)$ and is perpendicular to the line $2x - 7y = -12$
10. Passes through $(-3, 5)$ and is parallel to the line $2x - 7y = -12$
11. Is the perpendicular bisector of the line segment having endpoints $(3, -2)$ and $(-5, 1)$.
12. Passes through $(10, 0)$ and is at a distance of 5 units from $(0, 0)$
13. Passes through $(-3, 4)$ and the ratio of the x-intercept to the y-intercept is $2:5$
14. Obtain the slope and y-intercept of the following lines.

    (a) $y = x - 6$ (b) $3x - 4y + 5 = 0$

    (c) $x = 3y - 11$ (d) $\dfrac{x}{3} + \dfrac{y}{5} = 1$

    (e) $\dfrac{3x - y}{4} - \dfrac{x + 3y}{2} = 2$

15. Prove Theorem 2.5.4.

16. Show that an equation for a line with *x-intercept a* and *y-intercept b*, where $a \neq 0 \neq b$, is

$$\frac{x}{a} + \frac{y}{b} = 1.$$

The equation here is called the intercept form of an equation of a line.

2.6   The Circle

We begin by recalling the definition of a circle and then obtain an equation for a circle having a given center and radius.

2.6.1   Definition

Let r be a positive number and C be a point in the plane. The circle with center C and radius r is the set of all points P in the plane such that $|PC| = r$.

We can easily derive an equation for this circle using the two-point distance formula.

2.6.2   Theorem

The circle with center $C(h, k)$ and radius r has the equation

$$(x - h)^2 + (y - k)^2 = r^2. \tag{2.42}$$

PROOF   By Definition 2.6.1, if $P(x, y)$ is on the circle, then

$$|PC| = r, \tag{2.43}$$

that is,

$$\sqrt{(x - h)^2 + (y - k)^2} = r, \tag{2.44}$$

and hence (2.42) is satisfied. Conversely, if the coordinates of $P(x, y)$ satisfy (2.42), then (2.44) and hence (2.43) are satisfied. Thus from Definition 2.6.1, (2.42) is an equation of the circle.

Equation (2.42) is called the *center-radius form* for an equation of a circle. If the center of a circle of radius r is at the origin, then $h = k = 0$ and the circle has the equation

$$x^2 + y^2 = r^2.$$

Example 1   Find an equation of the circle with center $(-2, 4)$ and radius 5.

SOLUTION   If $(h, k) = (-2, 4)$ and $r = 5$ in (2.42), then the equation

$$(x + 2)^2 + (y - 4)^2 = 25$$

is obtained.

An expansion of the squared terms in (2.42) gives the equation

$$x^2 + y^2 - 2hx - 2ky + h^2 + k^2 - r^2 = 0,$$

which is of the form

$$x^2 + y^2 + Mx + Ny + P = 0 \qquad (2.45)$$

since we can let $M = -2h$, $N = -2k$, and $P = h^2 + k^2 - r^2$. Thus every circle in the plane has an equation of the form (2.45).

It is also reasonable to ask the converse question: "Is every equation of the form (2.45) an equation of some circle?" By completing the square in x and y in (2.45) we obtain

$$\left(x^2 + Mx + \frac{M^2}{4}\right) + \left(y^2 + Ny + \frac{N^2}{4}\right) = \frac{M^2}{4} + \frac{N^2}{4} - P,$$

or equivalently,

$$\left(x + \frac{M}{2}\right)^2 + \left(y + \frac{N}{2}\right)^2 = \frac{M^2 + N^2 - 4P}{4}. \qquad (2.46)$$

If $M^2 + N^2 - 4P > 0$, then by comparison with Equation (2.42), the relation (2.46) and hence (2.45) are equations of a circle with center $(-M/2, -N/2)$ and radius $\frac{1}{2}\sqrt{M^2 + N^2 - 4P}$. However, if $M^2 + N^2 - 4P < 0$, the graph of (2.45) is the empty set since the sum of two squares $(x + (M/2))^2 + (y + (N/2))^2$ is never negative. If $M^2 + N^2 - 4P = 0$, Equation (2.46) and hence (2.45) are satisfied if and only if $(x, y) = (-M/2, -N/2)$, that is, the graph of (2.45) consists of the single point $(-M/2, -N/2)$.

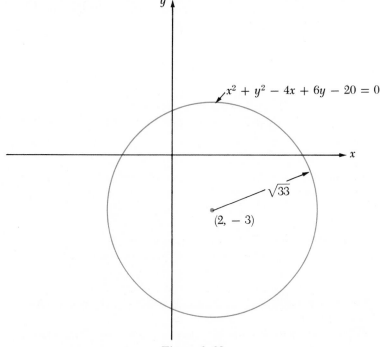

Figure 2-28

Example 2   Discuss the graph of the equation

$$x^2 + y^2 - 4x + 6y - 20 = 0 \text{ (Figure 2-28).}$$

SOLUTION   Completing the square in x and y in the given equation, we have

$$(x^2 - 4x + 4) + (y^2 + 6y + 9) = 20 + 4 + 9,$$

or

$$(x - 2)^2 + (y + 3)^2 = 33.$$

From (2.42) the graph of this equation is a circle with center $(2, -3)$ and radius $\sqrt{33}$.

Example 3   Write an equation for the circle passing through the three points $A(8, 1)$, $B(-1, 4)$, and $C(6, -3)$.

SOLUTION   From the discussion following Example 1, a circle has an equation of the form

$$x^2 + y^2 + Mx + Ny + P = 0. \tag{2.45}$$

Since the given points are on the circle, their coordinates satisfy this equation. Hence from (2.45)

$$8^2 + 1^2 + M \cdot 8 + N \cdot 1 + P = 0$$
$$(-1)^2 + 4^2 + M(-1) + N \cdot 4 + P = 0$$
$$6^2 + (-3)^2 + M \cdot 6 + N \cdot (-3) + P = 0.$$

and if these equations are simplified,

$$8M + N + P = -65$$
$$-M + 4N + P = -17$$
$$6M - 3N + P = -45.$$

These equations can be solved simultaneously to yield the solutions:

$$M = -6 \qquad N = -2 \qquad P = -15 \tag{2.47}$$

After a substitution from (2.47) into (2.45), one obtains for an equation of the circle

$$x^2 + y^2 - 6x - 2y - 15 = 0 \tag{2.48}$$

ALTERNATE SOLUTION   In this solution we first find the perpendicular bisectors of two of the three chords having the given points as endpoints. From a theorem in elementary geometry these perpendicular bisectors intersect in the center of the circle. Suppose we write equations for the perpendicular bisectors of the chord connecting $A(8, 1)$ and $B(-1, 4)$ and the chord connecting $A(8, 1)$ and $C(6, -3)$ (Figure 2-29). The midpoint of AB is $(\frac{7}{2}, \frac{5}{2})$ and its slope is $-\frac{1}{3}$. Since the slope of the

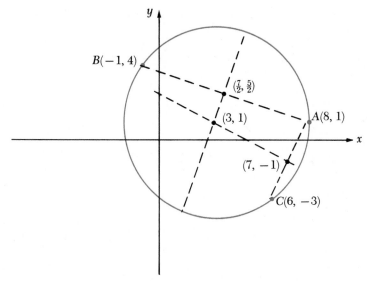

Figure 2-29

perpendicular bisector of *AB* is 3, from (2.36) an equation of this perpendicular bisector is

$$y - \tfrac{5}{2} = 3(x - \tfrac{7}{2}),$$

or in simplest form

$$-3x + y = -8. \tag{2.49}$$

Also, the midpoint of *AC* is $(7, -1)$ and its slope is 2. Since the slope of the perpendicular bisector of *AC* is $-\tfrac{1}{2}$, an equation of this perpendicular bisector is

$$y + 1 = -\tfrac{1}{2}(x - 7)$$

or

$$x + 2y = 5. \tag{2.50}$$

Solving simultaneously (2.49) and (2.50), the solution

$$x = 3 \qquad y = 1$$

is obtained for the coordinates of the center of the circle. The radius of the circle can be found by calculating the distance from the center to any of the points *A*, *B*, and *C*, and is 5. The equation of the circle, in center-radius form, is therefore

$$(x - 3)^2 + (y - 1)^2 = 25,$$

which is equivalent to Equation (2.48), as can be seen by completing the square in (2.48).

Example 4   Find the intersection points of the circles

$$x^2 + y^2 - 2x + 5y + 7 = 0$$

and

$$x^2 + y^2 - 4x + 2y = 0.$$

SOLUTION Substituting the second equation from the first gives

$$2x + 3y + 7 = 0,$$

from which we obtain

$$x = \frac{-3y - 7}{2}. \tag{2.51}$$

If we substitute for x in terms of y in, say, the second equation,

$$\left(\frac{-3y - 7}{2}\right)^2 + y^2 - 4\left(\frac{-3y - 7}{2}\right) + 2y = 0,$$

and hence:

$$\frac{9y^2 + 42y + 49}{4} + y^2 + 6y + 14 + 2y = 0$$

$$13y^2 + 74y + 105 = 0$$

$$(13y + 35)(y + 3) = 0$$

$$y = -\tfrac{35}{13} \quad \text{or} \quad y = -3$$

After substituting these y values in (2.51) we obtain the intersection points $(1, -3)$ and $(\tfrac{7}{13}, -\tfrac{35}{13})$.

Exercise Set 2.6

In Exercises 1–12 write an equation for the circle satisfying the given conditions.

1. Center $(-1, 1)$, radius 4

2. Center $(2, -3)$, radius 6

3. Center $(4, 5)$, passing through $(7, -2)$

4. Center $(0, -4)$, passing through $(4, 3)$

5. Having a diameter with endpoints $(-2, -5)$ and $(6, -1)$

6. Passing through $(2, 4)$, $(1, 5)$, and $(2, -2)$

7. Passing through $(-3, 0)$, $(2, 3)$, and $(5, -2)$

8. Tangent to the line $3x - 4y = 10$ at $(2, -1)$ and having its center on the line $x - 2y = 15$

9. Tangent to the line $2x + 3y = -7$ at $(1, -3)$ and passing through $(11, -1)$

10. Tangent to the line $y = -3x$ at $(-2, 6)$ and passing through $(6, 2)$

11. Tangent to the line $3x + 4y = -6$ at $(-2, 0)$ and having a radius of 5 (2 solutions)

12. Tangent to the line $2x - y = 8$ and having the center $(-1, 5)$

In Exercises 13–19 describe the graph of the given equation.

13. $x^2 + y^2 + 8x - 2y + 8 = 0$

14. $x^2 + y^2 - 12x + 14y + 4 = 0$

15. $2x^2 + 2y^2 + x + 3y - 10 = 0$

16. $3x^2 + 3y^2 + 2x - y - 12 = 0$

17. $x^2 + y^2 - 4x - 10y + 29 = 0$

18. $x^2 + y^2 + 4x + 12y + 50 = 0$

19. $x^2 + y^2 - 6x + 2y + 20 = 0$

In Exercises 20 and 21 find the intersection points of the circles having the given equations.

20. $x^2 + y^2 + 3x - y - 10 = 0$, $x^2 + y^2 - 2x - 3y + 2 = 0$

21. $x^2 + y^2 + 2x + y - 2 = 0$, $x^2 + y^2 + 3x - 2y - 6 = 0$

22. Show that the circles $x^2 + y^2 - 10x + 4y = 35$ and $x^2 + y^2 - 2x - 2y = 7$ are tangent to each other.

23. Find an equation of the tangent to the circle $x^2 + y^2 + 2x - 2y - 39 = 0$ at the point $(3, -4)$.

24. Find equations for the tangents to the circle $(x - 1)^2 + (y - 3)^2 = 13$ passing through the point $(-7, 2)$.

25. Find equations for the lines with slope m which are tangent to the circle $x^2 + y^2 = r^2$.

3. Functions;
Limit of a Function

3.1 Functions

In calculus one frequently encounters situations in which two variables, say x and y, are given and for every value that x may assume there is a unique corresponding value of y. We then say that y is *a function of* x. Consider the following examples:

1. y is a function of x when x and y are related by the equation $y = 3x + 1$.
2. The area A of a circle is a function of its radius r, since $A = \pi r^2$.
3. The cost C in dollars of a telephone call from Van Nuys, California to Long Beach, exclusive of taxes levied, is a function of t, the duration of the call in minutes. C can be calculated from the following rule:

$$\text{If } 0 < t \le 3, \quad C = .40$$
$$3 < t \le 4, \quad C = .50$$
$$4 < t \le 5, \quad C = .60$$
$$5 < t \le 6, \quad C = .70$$

4. The rate of decay, y, of radioactive material at any time is a function of x, the weight of the material present.
5. y is *not* a function of x if x and y are related by the equation $x^2 + y^2 = 25$. If for example $x = 3$, there are the *two* corresponding values, $y = \pm 4$.

In a situation in which y is a function of x, we can form ordered pairs of values by taking an x value as the first component of an ordered pair and the corresponding y value as the second component of the ordered pair.

Furthermore, since for each value of x there is a unique corresponding value of y, no two different ordered pairs thus formed can have the same first component. In illustration 3, above, since C is a function of t and, for example,

$$C = .40 \quad \text{when } t = 1$$
$$C = .40 \quad \text{when } t = 3$$
$$C = .50 \quad \text{when } t = 4$$
$$C = .70 \quad \text{when } t = 5.3$$

we can form the ordered pairs $(1, .40)$, $(3, .40)$, $(4, .50)$, and $(5.3, .70)$.

This brings us to the definition of a *function*.

3.1.1 Definition

A *function* is a set of ordered pairs of elements in which no distinct ordered pairs have the same first element.

For example, the set

$$\{(-1, 2), (0, -\tfrac{4}{3}), (2, 0), (3, 2)\} \tag{3.1}$$

is a function. However, $\{(1, -1), (2, 4), (1, 2), (4, 3)\}$ is not a function since the ordered pairs $(1, -1)$ and $(1, 2)$ have the same first element, 1.

Associated with any function are two important sets, the *domain* and the *range*. The domain of a function is the set of all first elements of the ordered pairs of the function, and the range of a function is the set of all second elements of the ordered pairs of the function. In particular, the domain and range of the function (3.1) are the sets $\{-1, 0, 2, 3\}$ and $\{2, -\tfrac{4}{3}, 0\}$, respectively. A function is said to be a *real-valued function of a real variable* if and only if its domain and range are subsets of the set R of all real numbers. In Chapters 3 through 14 of this text we will be concerned exclusively with such functions. Later, we will study *vector-valued functions* (Chapter 15) and functions of more than one variable (Chapter 18).

Functions are frequently denoted by letters such as f, F, g, G, or Greek letters: ϕ, ψ, and so on. The domain and range of a function f will be represented by $\mathfrak{D}_f$ and $\mathfrak{R}_f$, respectively, in this text. If (x, y) is an arbitrary ordered pair of a function f, we could write

$$f = \{(x, y) \mid x \in \mathfrak{D}_f\}.$$

The variable x, here, which denotes an arbitrary element in the domain of f, is called the *independent variable* for the function. Also, since y, here, stands for the unique element in the range of f which depends upon a corresponding element in the domain of f, y is termed the *dependent variable* of the function.

If x is an element in the domain $\mathfrak{D}_f$, the corresponding element in the range $\mathfrak{R}_f$ is denoted by

$$f(x),$$

which is read "f at x" or "f of x." When $f(x)$ exists (as a real number), we say f

is *defined* at x and call $f(x)$ *the value of f at x*. For example, if the function (3.1) is denoted by f, one may write

$$f(-1) = 2 \qquad f(0) = -\tfrac{4}{3} \qquad f(2) = 0 \qquad f(3) = 2.$$

The symbol $f(x)$ is an example of what is called *functional notation*.†

A function is often defined using functional notation. For example, the function $F = \{(x, y): y = \sqrt{5 - 2x}\}$, can be defined by writing

$$F(x) = \sqrt{5 - 2x}. \tag{3.2}$$

In the absence of any explicit restriction on $F(x)$ in (3.2), it is understood that $\mathfrak{D}_F$ is the set of all numbers x for which $F(x)$ exists. Thus

$$x \in \mathfrak{D}_F \qquad \text{if and only if} \qquad 5 - 2x \geq 0,$$

and hence $\mathfrak{D}_F = (-\infty, \tfrac{5}{2}]$. In particular from (3.2),

$$F(-2) = 3 \qquad F(0) = \sqrt{5} \qquad F(\tfrac{1}{2}) = 2 \qquad F(\tfrac{5}{2}) = 0.$$

By Definition 2.3.1 the graph of f, a real-valued function of a real variable, is the set of all points in the plane associated with ordered pairs of f. For every number x in $\mathfrak{D}_f$ there is a *unique* point (x, y) in the graph of f. Also, for every number y in the range of f, there is *at least one* point (x, y) in its graph. Thus the graph of f is the graph of the equation $y = f(x)$.

A sketch of the graph of the function (3.1) is shown in Figure 3–1.

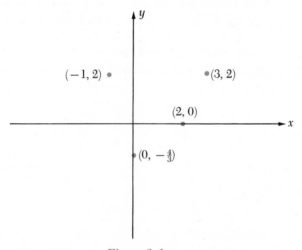

Figure 3–1

Example 1 Graph the function F where $F(x) = \sqrt{5 - 2x}$.

SOLUTION   The graph of F is the graph of the equation

$$y = \sqrt{5 - 2x}. \tag{3.3}$$

---

† The first recorded appearance of functional notation was in 1734 in a paper by Leonard Euler, a Swiss mathematician who made outstanding contributions to the development of calculus.

We note that $\frac{5}{2}$ is the x intercept and $\sqrt{5}$ is the y intercept of the graph. From our prior discussion there is a unique point (x, y) in the graph for every x in $\mathfrak{D}_F = (-\infty, \frac{5}{2}]$. If (3.3) is solved for x in terms of y, we obtain successively

$$y^2 = 5 - 2x \quad \text{where } y \geq 0,$$

$$x = \frac{5 - y^2}{2} \quad \text{where } y \geq 0. \tag{3.4}$$

Since x is defined in (3.4) for every $y \geq 0$, there is at least one point (x, y) in the graph for every such number y. In other words, $\mathfrak{R}_F = [0, +\infty)$.

The graph of F can be sketched (Figure 3–2) from the following table, which is obtained by assigning certain values to x and then calculating the corresponding values for y, using (3.3).

x	$\frac{5}{2}$	2	1	0	-1	-2
$y = F(x)$	0	1	$\sqrt{3}$	$\sqrt{5}$	$\sqrt{7}$	3

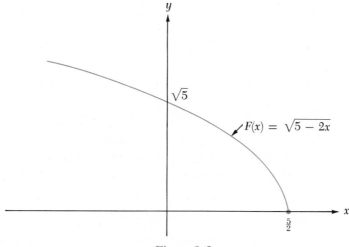

Figure 3-2

In contrast to the function F given by (3.2) we consider the function G where

$$G = \{(x, y) : y = \sqrt{5 - 2x} \text{ and } x > -1\}.$$

G can be defined, using functional notation, by writing

$$G(x) = \sqrt{5 - 2x} \quad \text{if } x > -1. \tag{3.5}$$

Example 2 Sketch the graph of the function G given by (3.5).

SOLUTION The domain of G is the set of all numbers x for which $x \leq \frac{5}{2}$

and $x > -1$. Therefore $\mathfrak{D}_G = (-1, \frac{5}{2}]$. Solving $y = \sqrt{5 - 2x}$ where $x > -1$ for x yields

$$x = \frac{5 - y^2}{2} > -1 \quad \text{where } y \geq 0. \tag{3.6}$$

From (3.6) x is defined for every y satisfying $0 \leq y < \sqrt{7}$. A sketch of the graph of G is given in Figure 3–3. The point $(-1, \sqrt{7})$ is indicated by an open circle in the sketch to show that it is not in the graph.

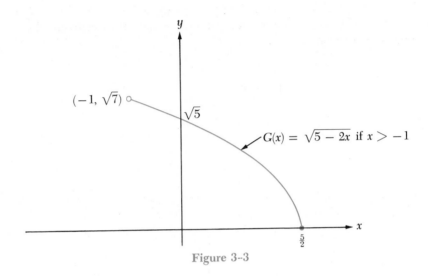

Figure 3–3

Unlike the function F in (3.2) and G in (3.5), the next function is defined using more than one equation. Suppose f is given by:

$$f(x) = \begin{cases} x + 1 & \text{if } x < 0 \\ |x - 2| & \text{if } x > 0 \end{cases} \tag{3.7}$$

If one wished to calculate, for example, $f(-1)$ or $f(-2)$, the rule expressed in (3.7) for forming $f(x)$ when $x < 0$ would be used. Thus:

$$f(-1) = -1 + 1 = 0 \qquad f(-2) = -2 + 1 = -1$$

If, however, we wanted to obtain $f(\frac{2}{3})$ or $f(4)$, the rule given in (3.7) for finding $f(x)$ when $x > 0$ would be employed. Thus:

$$f(\tfrac{2}{3}) = |\tfrac{2}{3} - 2| = \tfrac{4}{3} \qquad f(4) = |4 - 2| = 2$$

Example 3 Sketch the graph of the function f given by (3.7).

SOLUTION We note that since $f(x)$ is defined for every number $x \neq 0$, there is a point (x, y) in the graph of f for every $x \neq 0$. Also, from the definition of the absolute value of a number, f could be redefined by writing

$$f(x) = \begin{array}{ll} x + 1 & \text{if } x < 0 \\ 2 - x & \text{if } 0 < x < 2 \\ x - 2 & \text{if } x \geq 2 \end{array} \qquad (3.8)$$

By graphing the equations $y = x + 1$, $y = 2 - x$, and $y = x - 2$ subject to the restrictions expressed in (3.8), the required sketch (Figure 3-4) is obtained.

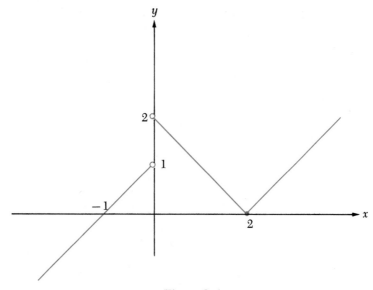

Figure 3-4

Before discussing the next example, we define $[a]$, the *greatest integer in a*. Let a be any number, and n be an integer such that $n \leq a < n + 1$. Then $[a] = n$. In particular:

$$[3] = 3 \qquad [3.6] = 3 \qquad [-1.2] = -2 \qquad [-0.6] = -1$$

Example 4 Sketch the graph of the function g where

$$g(x) = [2x - 1]. \qquad (3.9)$$

SOLUTION   It should be noted that

$$n = [2x - 1] \quad \text{if and only if} \quad n \leq 2x - 1 < n + 1$$
$$\text{if and only if} \quad n + 1 \leq 2x < n + 2$$
$$\text{if and only if} \quad \frac{n + 1}{2} \leq x < \frac{n + 2}{2} \qquad (3.10)$$

By assigning integral values to n in (3.10), the table below is obtained.

$y = n$	-2	-1	0	1	2
x	$-\frac{1}{2} \leq x < 0$	$0 \leq x < \frac{1}{2}$	$\frac{1}{2} \leq x < 1$	$1 \leq x < \frac{3}{2}$	$\frac{3}{2} \leq x < 2$

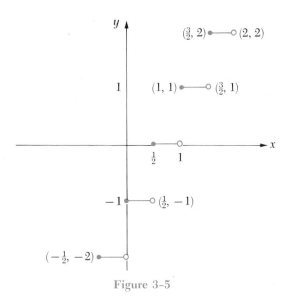

Figure 3-5

The sketch of the graph of g (Figure 3–5) can then be made from this table.

An important function in the study of calculus is a *polynomial function.* If $a_0, a_1, \ldots, a_n$ are any numbers and n is a non-negative integer, the function P, where

$$P(x) = a_0x^n + a_1x^{n-1} + \cdots + a_{n-1}x + a_n,$$

is called a polynomial function. Note that the domain of P is the set R. If $n = 0$, P is called a *constant function.* If $a_0 \neq 0$, then P is termed a polynomial function of *degree n.* For example, if

$$P(x) = x^4 - 2x^2 + 5x - 3,$$

then P is of degree 4. A polynomial function of degree 1 is called a *linear function,* a polynomial function of degree 2 is a *quadratic function,* and a polynomial function of degree 3 is a *cubic function.* If P and Q are polynomial functions, the function f defined by

$$f(x) = \frac{P(x)}{Q(x)} \tag{3.11}$$

is called a *rational function.* The domain of the function given by (3.11) is the set $\{x : Q(x) \neq 0\}$. The function g where

$$g(x) = \frac{3x + 2}{x^2 - 2x - 3}$$

is an example of a rational function.

It should be observed that the graph of a function f intersects a line perpendicular to the x axis in at most one point (Figure 3–6(a)) since for every number x in $\mathcal{D}_f$ there is a unique ordered pair $(x, f(x))$ in f. In Figure 3–6(b), however, a line perpendicular to the x axis intersects the set of points C in two points (x, y_1) and (x, y_2). Thus C cannot be the graph of a function since the ordered pairs (x, y_1) and (x, y_2) have the same first element x.

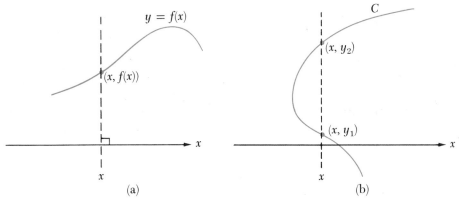

Figure 3-6

Recalling the definition of equal sets, the functions f and g are equal if and only if each function is a subset of the other. Accordingly, we have the following definition.

3.1.2   Definition

Let f and g be functions.

$$f = g \quad \text{iff} \quad \mathcal{D}_f = \mathcal{D}_g \text{ and } f(x) = g(x)$$

whenever x is in the common domain of the functions.

Example   Does $f = g$ when $f(x) = (x^2 - 9)/(x - 3)$ and $g(x) = x + 3$? (See Figure 3–7(a), (b) on page 72.)

SOLUTION g is a polynomial function and hence $\mathcal{D}_g = R$. f is a rational function and is defined for every x except where the denominator $x - 3 = 0$. Thus $\mathcal{D}_f = \{x : x \neq 3\}$. By Definition 3.1.2, the functions are unequal since $\mathcal{D}_f \neq \mathcal{D}_g$.

Exercise Set 3.1

1. State which of the following sets of ordered pairs are functions.
   (a) $\{(1, 2), (0, -4), (2, 5), (-1, -2)\}$
   (b) $\{(3, 4), (1, 0), (2, 1), (3, -2)\}$
   (c) $\{(0, 3)\}$
   (d) $\{(-1, 0), (0, 0), (1, 1), (2, -4)\}$

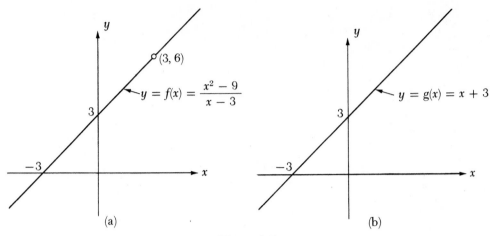

Figure 3-7

(e) $\{(x, y): y^3 = x\}$

(f) $\{(x, y): x^2 + 2y^2 = 4\}$

(g) $\{(x, y): x^2 = y^2\}$

(h) $\{(x, y): x^2 - y^2 = 4\} \cap \{(x, y): y = 2\}$

(i) $\{(x, y): y = \sqrt{2 - x}\} \cup \{(x, y): y = \sqrt{x - 2}\}$

In each of Exercises 2–17 find the domain of the function defined in the exercise and sketch the graph of the function.

2. $f(x) = x^3 - 5x$

3. $f(x) = 4 - 3x - x^2$

4. $g(x) = \begin{cases} x + 5 & \text{if } x > 0 \\ -2 & \text{if } x = 0 \\ -x + 2 & \text{if } x < 0 \end{cases}$

5. $h(x) = \begin{cases} x & \text{if } x > 1 \\ 2 - x^2 & \text{if } x < 1 \end{cases}$

6. $\phi(x) = \sqrt{3x + 4}$

7. $F(x) = \sqrt{1 - x^2}$

8. $G(x) = \sqrt{x^2 - 2x - 15}$

9. $F(x) = \sqrt{\dfrac{6 - x}{x + 3}}$

10. $g(x) = |x + 2|$

11. $f(x) = |2x - 5|$

12. $\psi(x) = \begin{cases} \dfrac{4x}{x - 1} & \text{if } x < 1 \\ -1 & \text{if } x \geq 1 \end{cases}$

13. $\phi(x) = \dfrac{x + 1}{|x + 1|}$

14. $g(x) = |2x + 1| - |2 - x|$

15. $f(x) = \dfrac{|x - 1|}{[x - 1]}$

16. $g(x) = |x| - [x]$

17. $h(x) = 2[x] - [2x]$

18. Which of the functions defined below are equal? Why?

(a) $f(x) = |x - 1|$

(b) $g(x) = \sqrt{1 - 2x + x^2}$

(c) $\phi(x) = \sqrt{\dfrac{(x-1)^3}{x-1}}$

(d) $\psi(x) = \left| \dfrac{x^2 - 3x + 2}{x - 2} \right|$

(e) $F(x) = \begin{cases} \left| \dfrac{x^2 + 2x - 3}{x + 3} \right| & \text{if } x \neq -3 \\ 4 & \text{if } x = -3 \end{cases}$

19. In each of the following parts write the appropriate equation that describes the given functional relationship.

   (a) The area A of a circle is a function of the circumference C of the circle. HINT: From the equations $A = \pi r^2$ and $C = 2\pi r$ obtain A in terms of C.

   (b) The area A of an isosceles triangle with base of length 4 is a function of x, the length of one of the equal sides of the triangle.

   (c) The total cost C in dollars of a chartered train is a function of x, the number of passengers on the train, if the cost per passenger on the train is \$15 when there are 200 passengers and is reduced 10¢ for each additional passenger over 200.

   (d) The distance s in miles between the ships is a function of the time t in hours after 12:00 noon in the following situation: Ship A is steaming north at 20 miles per hour at the same time as ship B is steaming west at 10 miles per hour. At noon A is 100 miles south of B.

3.2   Functional Notation; Operations on Functions

In the last section the notation $f(x)$ was introduced to denote the second component of the ordered pair of the function f having x as its first component. If f is defined by specifying $f(x)$ for an arbitrary number x in $\mathfrak{D}_f$, we can obtain the value of f at some other number a in $\mathfrak{D}_f$ by replacing x in the expression for $f(x)$ by a. This procedure would be followed regardless of whether a is a constant or is itself a variable.

Example 1   If $f(x) = (2x + 1)/(x + 2)$, find:

   (a) $f(3)$ (b) $f(-2)$ (c) $f(x - 4)$ (d) $f(3x)$

   (e) $f\left(\dfrac{1}{x}\right)$ (f) $f\left(\dfrac{2x + 1}{x + 2}\right)$ (g) $\dfrac{f(x + h) - f(x)}{h}$

SOLUTION

   (a) $f(3) = \dfrac{2 \cdot 3 + 1}{3 + 2} = \dfrac{7}{5}$

   (b) $f(-2)$ is undefined since division by 0 is not defined.

   (c) $f(x - 4) = \dfrac{2(x - 4) + 1}{(x - 4) + 2} = \dfrac{2x - 7}{x - 2}$

(d) $f(3x) = \dfrac{2 \cdot 3x + 1}{3x + 2} = \dfrac{6x + 1}{3x + 2}$

(e) $f\left(\dfrac{1}{x}\right) = \dfrac{2(1/x) + 1}{(1/x) + 2} = \dfrac{2 + x}{1 + 2x}$

(f) $f\left(\dfrac{2x + 1}{x + 2}\right) = \dfrac{2\left(\dfrac{2x + 1}{x + 2}\right) + 1}{\dfrac{2x + 1}{x + 2} + 2} = \dfrac{5x + 4}{4x + 5}$

(g) $\dfrac{f(x + h) - f(x)}{h} = \dfrac{\dfrac{2(x + h) + 1}{(x + h) + 2} - \dfrac{2x + 1}{x + 2}}{h}$

$$= \dfrac{(x + 2)(2x + 2h + 1) - (2x + 1)(x + h + 2)}{h(x + 2)(x + h + 2)}$$

$$= \dfrac{3h}{h(x + 2)(x + h + 2)} = \dfrac{3}{(x + 2)(x + h + 2)}$$

Example 2   If $F(x) = \sqrt{5 - 2x}$, find:

(a) $F(-2)$ (b) $F(-3x)$ (c) $F(5 - 2x)$

(d) $\dfrac{F(x + h) - F(x)}{h}$

SOLUTION

(a) $F(-2) = \sqrt{5 - 2(-2)} = \sqrt{9} = 3$

(b) $F(-3x) = \sqrt{5 - 2(-3x)} = \sqrt{5 + 6x}$

(c) $F(5 - 2x) = \sqrt{5 - 2(5 - 2x)} = \sqrt{4x - 5}$

(d) $\dfrac{F(x + h) - F(x)}{h} = \dfrac{\sqrt{5 - 2x - 2h} - \sqrt{5 - 2x}}{h}$ and if we ration-

alize the numerator of the right member by multiplying numerator and denominator by $\sqrt{5 - 2x - 2h} + \sqrt{5 - 2x}$, we have

$$\dfrac{F(x + h) - F(x)}{h}$$

$$= \dfrac{(\sqrt{5 - 2x - 2h} - \sqrt{5 - 2x})(\sqrt{5 - 2x - 2h} + \sqrt{5 - 2x})}{h(\sqrt{5 - 2x - 2h} + \sqrt{5 - 2x})}$$

$$= \dfrac{(5 - 2x - 2h) - (5 - 2x)}{h(\sqrt{5 - 2x - 2h} + \sqrt{5 - 2x})}$$

$$= \dfrac{-2}{\sqrt{5 - 2x - 2h} + \sqrt{5 - 2x}}$$

It is convenient to define the *sum, difference, product,* and *quotient* of two functions.

3.2.1   Definition

If f and g are functions with respective domains $\mathfrak{D}_f$ and $\mathfrak{D}_g$, then:

(a) $f + g = \{(x, f(x) + g(x)): x \in \mathfrak{D}_f \cap \mathfrak{D}_g\}$

(b) $f - g = \{(x, f(x) - g(x)): x \in \mathfrak{D}_f \cap \mathfrak{D}_g\}$

(c) $fg = \{(x, f(x)g(x)): x \in \mathfrak{D}_f \cap \mathfrak{D}_g\}$

(d) $\dfrac{f}{g} = \left\{\left(x, \dfrac{f(x)}{g(x)}\right): x \in \mathfrak{D}_f \cap \mathfrak{D}_g \text{ and } g(x) \neq 0\right\}$

It is clear from these definitions that $\mathfrak{D}_{f+g} = \mathfrak{D}_{f-g} = \mathfrak{D}_{fg} = \mathfrak{D}_f \cap \mathfrak{D}_g$ and that $\mathfrak{D}_{f/g} = \{x : x \in \mathfrak{D}_f \cap \mathfrak{D}_g \text{ and } g(x) \neq 0\}$. Thus if $x \in \mathfrak{D}_f \cap \mathfrak{D}_g$, then:

$$(f + g)(x) = f(x) + g(x)$$
$$(f - g)(x) = f(x) - g(x)$$
$$(fg)(x) = f(x)g(x)$$
$$\left(\frac{f}{g}\right)(x) = \frac{f(x)}{g(x)} \quad \text{if } g(x) \neq 0$$

We also define the sum and product of an arbitrary number of functions as a generalization of Definitions 3.2.1(a) and (c).

3.2.2   Definition

Let $f_1, f_2, \ldots, f_n$ be functions with domains $\mathfrak{D}_{f_1}, \mathfrak{D}_{f_2}, \ldots, \mathfrak{D}_{f_n}$, respectively. Then:

(a) $f_1 + f_2 + \cdots + f_n = \{(x, f_1(x) + f_2(x) + \cdots + f_n(x))\}$

(b) $f_1 f_2 \cdots f_n = \{(x, f_1(x)f_2(x) \cdots f_n(x))\}$

x here is an arbitrary number in the intersection of $\mathfrak{D}_{f_1}, \mathfrak{D}_{f_2}, \ldots, \mathfrak{D}_{f_n}$, the domain of each of the functions defined in (a) and (b). For such numbers x we see that

$$(f_1 + f_2 + \cdots + f_n)(x) = f_1(x) + f_2(x) + \cdots + f_n(x)$$
$$(f_1 f_2 \cdots f_n)(x) = f_1(x)f_2(x) \cdots f_n(x)$$

3.2.3   Definition

Let f and g be any functions. The *composite function of f and g*, which is denoted by $f \circ g$ ("read f circle g"), is defined by

$$(f \circ g)(x) = f(g(x)).$$

The domain of $f \circ g$ is $\{x : x \in \mathfrak{D}_g \text{ and } g(x) \in \mathfrak{D}_f\}$.

Example 1   Let $f = \{(1, 2), (2, -3), (-1, 4), (3, 5)\}$ and $g = \{(1, 0), (2, 2), (3, 1), (4, -2)\}$. Find:

(a) $f + g$ (b) $f - g$ (c) fg (d) f/g (e) $f \circ g$ (f) $g \circ f$.

SOLUTION In defining each of the following functions it is desirable to first obtain their domains. Since

$$\mathcal{D}_f = \{1, 2, -1, 3\} \qquad \text{and} \qquad \mathcal{D}_g = \{1, 2, 3, 4\} \tag{3.12}$$

Then:

$$\mathcal{D}_f \cap \mathcal{D}_g = \{1, 2, 3\} = \mathcal{D}_{f+g} = \mathcal{D}_{f-g} = \mathcal{D}_{fg}$$
$$\mathcal{D}_{f/g} = \{x \in \mathcal{D}_f \cap \mathcal{D}_g : g(x) \neq 0\} = \{2, 3\}$$

Since

$$f(1) = 2 \qquad f(2) = -3 \qquad f(-1) = 4 \qquad f(3) = 5$$
$$g(1) = 0 \qquad g(2) = 2 \qquad g(3) = 1 \qquad g(4) = -2$$

we therefore have:

$$\left. \begin{aligned} (f+g)(1) &= f(1) + g(1) = 2 + 0 = 2 \\ (f+g)(2) &= f(2) + g(2) = -1 \\ (f+g)(3) &= f(3) + g(3) = 6 \end{aligned} \right\} \text{ and } f + g = \\ \{(1, 2), (2, -1), (3, 6)\}$$

$$\left. \begin{aligned} (f-g)(1) &= f(1) - g(1) = 2 - 0 = 2 \\ (f-g)(2) &= f(2) - g(2) = -5 \\ (f-g)(3) &= f(3) - g(3) = 4 \end{aligned} \right\} \text{ and } f - g = \\ \{(1, 2), (2, -5), (3, 4)\}$$

$$\left. \begin{aligned} (fg)(1) &= f(1)g(1) = 2 \cdot 0 = 0 \\ (fg)(2) &= f(2)g(2) = (-3)2 = -6 \\ (fg)(3) &= f(3)g(3) = 5 \cdot 1 = 5 \end{aligned} \right\} \text{ and } fg = \{(1, 0), (2, -6), (3, 5)\}$$

$$\left. \begin{aligned} (f/g)(2) &= \frac{f(2)}{g(2)} = \frac{-3}{2} \\ (f/g)(3) &= \frac{f(3)}{g(3)} = \frac{5}{1} = 5 \end{aligned} \right\} \text{ and } f/g = \{(2, -\tfrac{3}{2}), (3, 5)\}$$

We note from (3.12) that $g(1) = 0 \notin \mathcal{D}_f$, $g(2) = 2 \in \mathcal{D}_f$, $g(3) = 1 \in \mathcal{D}_f$, and $g(4) = -2 \notin \mathcal{D}_f$, so

$$\mathcal{D}_{f \circ g} = \{x : x \in \mathcal{D}_g \text{ and } g(x) \in \mathcal{D}_f\} = \{2, 3\}.$$

Then

$$\left. \begin{aligned} (f \circ g)(2) &= f(g(2)) = f(2) = -3 \\ (f \circ g)(3) &= f(g(3)) = f(1) = 2 \end{aligned} \right\} \text{ and } f \circ g = \{(2, -3), (3, 2)\}$$

Also, from (3.12), $f(1) = 2 \in \mathcal{D}_g$, $f(-1) = 4 \in \mathcal{D}_g$, $f(2) = -3 \notin \mathcal{D}_g$, and $f(3) = 5 \notin \mathcal{D}_g$. Thus

$$\mathcal{D}_{g \circ f} = \{x : x \in \mathcal{D}_f \text{ and } f(x) \in \mathcal{D}_g\} = \{-1, 1\}.$$

Then

$$\left. \begin{aligned} (g \circ f)(-1) &= g(f(-1)) = g(4) = -2 \\ (g \circ f)(1) &= g(f(1)) = g(2) = 2 \end{aligned} \right\} \text{ and } (g \circ f) = \{(-1, -2), (1, 2)\}.$$

Example 2   Let $f(x) = x^2 - 1$ if $x < \sqrt{5}$ and $g(x) = \sqrt{3x - 1}$. Define:

(a) $f + g$ (b) $f - g$ (c) fg (d) f/g (e) $f \circ g$ (f) $g \circ f$

SOLUTION   We note that

$$\mathfrak{D}_f = (-\infty, \sqrt{5}) \quad \text{and} \quad \mathfrak{D}_g = [\tfrac{1}{3}, +\infty), \tag{3.13}$$

so

$$\mathfrak{D}_f \cap \mathfrak{D}_g = [\tfrac{1}{3}, \sqrt{5}) = \mathfrak{D}_{f+g} = \mathfrak{D}_{f-g} = \mathfrak{D}_{fg}.$$
$$\mathfrak{D}_{f/g} = \{x \in \mathfrak{D}_f \cap \mathfrak{D}_g : g(x) \neq 0\} = (\tfrac{1}{3}, \sqrt{5}).$$

Thus:

$$(f + g)(x) = f(x) + g(x) = x^2 - 1 + \sqrt{3x - 1} \quad \text{if } x \in [\tfrac{1}{3}, \sqrt{5})$$
$$(f - g)(x) = f(x) - g(x) = x^2 - 1 - \sqrt{3x - 1} \quad \text{if } x \in [\tfrac{1}{3}, \sqrt{5})$$
$$(fg)(x) = f(x)g(x) = (x^2 - 1)\sqrt{3x - 1} \quad \text{if } x \in [\tfrac{1}{3}, \sqrt{5})$$
$$\left(\frac{f}{g}\right)(x) = \frac{f(x)}{g(x)} = \frac{x^2 - 1}{\sqrt{3x - 1}} \quad \text{if } x \in \left(\frac{1}{3}, \sqrt{5}\right)$$

From (3.13)

$$\mathfrak{D}_{f \circ g} = \{x : x \in \mathfrak{D}_g \text{ and } g(x) \in \mathfrak{D}_f\} = \{x : x \geq \tfrac{1}{3} \text{ and } g(x) < \sqrt{5}\}.$$

In solving simultaneously the inequalities $x \geq \tfrac{1}{3}$ and $g(x) < \sqrt{5}$, we obtain successively

$$x \geq \tfrac{1}{3} \text{ and } \sqrt{3x - 1} < \sqrt{5}$$
$$x \geq \tfrac{1}{3} \text{ and } 3x - 1 < 5$$
$$x \geq \tfrac{1}{3} \text{ and } x < 2 \tag{3.14}$$

From (3.14) we have $\mathfrak{D}_{f \circ g} = [\tfrac{1}{3}, 2)$ and thus

$$(f \circ g)(x) = f(g(x)) = (g(x))^2 - 1 = (3x - 1) - 1 \quad \text{if } \tfrac{1}{3} \leq x < 2$$
$$= 3x - 2 \quad \text{if } \tfrac{1}{3} \leq x < 2$$

From (3.13)

$$\mathfrak{D}_{g \circ f} = \{x : x \in \mathfrak{D}_f \text{ and } f(x) \in \mathfrak{D}_g\} = \{x : x < \sqrt{5} \text{ and } f(x) \geq \tfrac{1}{3}\}.$$

Solving simultaneously the inequalities $x < \sqrt{5}$ and $f(x) \geq \tfrac{1}{3}$, we have

$$x < \sqrt{5} \text{ and } x^2 - 1 \geq \tfrac{1}{3}$$
$$x < \sqrt{5} \text{ and } x^2 \geq \tfrac{4}{3}$$
$$x < \sqrt{5} \text{ and } \left(x \geq \frac{2}{\sqrt{3}} \text{ or } x \leq -\frac{2}{\sqrt{3}}\right)$$
$$x \leq -\frac{2}{\sqrt{3}} \text{ or } \frac{2}{\sqrt{3}} \leq x < \sqrt{5} \tag{3.15}$$

From (3.15), $\mathfrak{D}_{g \circ f} = \left(-\infty, -\dfrac{2}{\sqrt{3}}\right] \cup \left[\dfrac{2}{\sqrt{3}}, \sqrt{5}\right)$ and thus

$$(g \circ f)(x) = g(f(x)) = \sqrt{3f(x) - 1}$$

$$= \sqrt{3(x^2 - 1) - 1}$$

$$= \sqrt{3x^2 - 4} \quad \text{if } x \leq -\frac{2}{\sqrt{3}} \text{ or } \frac{2}{\sqrt{3}} \leq x < \sqrt{5}$$

Exercise Set 3.2

In Exercises 1–8, find the following:

(a) $f(-1)$ (b) $f(\frac{2}{5})$ (c) $f(2x)$

(d) $f(3x - 2)$ (e) $f\left(\dfrac{1}{x}\right)$ (f) $\dfrac{f(x + h) - f(x)}{h}$

1. $f(x) = 2x - 3$ 2. $f(x) = x^2 + 4$

3. $f(x) = 4 - 3x - x^2$ 4. $f(x) = x^3$

5. $f(x) = \dfrac{1}{x + 2}$ 6. $f(x) = \dfrac{2x + 1}{x}$

7. $f(x) = \sqrt{3x - 1}$ 8. $f(x) = |2x + 3|$

In Exercises 9–12, find   (a) $f(\frac{4}{3})$ (b) $f(-\frac{5}{2})$.

9. $f(x) = \begin{cases} \dfrac{x + 5}{x} & \text{if } x > 0 \\ -2 & \text{if } x = 0 \\ 1 - x & \text{if } x < 0 \end{cases}$ 10. $f(x) = \begin{cases} \dfrac{4x}{x - 1} & \text{if } x < 1 \\ -1 & \text{if } x > 1 \end{cases}$

11. $f(x) = \sqrt{\dfrac{x + 4}{x + 5}}$ 12. $f(x) = |[x]|$

In Exercises 13–19 define   (a) $f + g$ (b) fg (c) f/g (d) g/f (e) $f \circ g$ (f) $f \circ f$ (g) $g \circ f$. Give the domain of each set.

13. $f = \{(1, 3), (2, 4), (3, 1), (4, 3)\}$ $g = \{(1, 2), (2, 2), (3, 0), (4, 3)\}$

14. $f(x) = \dfrac{x^2 + 1}{x - 1}$ $g(x) = x - 3$

15. $f(x) = \dfrac{x - 1}{x - 2}$ $g(x) = \dfrac{x - 2}{x - 1}$

16. $f(x) = \sqrt{x - 4}$ $g(x) = \dfrac{1}{x^2}$

17. $f(x) = \sqrt{x + 2}$ $g(x) = x - 2$

18. $f(x) = \sqrt{x - 1}$ $g(x) = |x - 1|$

19. $f(x) = \sqrt{2x - 1}$ $g(x) = \sqrt{x + 2}$

20. If $f(x) = x^2$, find a function g such that $f(g(x)) = x$ for every $x \in \mathcal{D}_g$. Is $g(f(x)) = x$ for every $x \in \mathcal{D}_f$?

21. If $f(x) = (x + 1)/(x + 2)$, find a function g such that $f(g(x)) = x$ for every $x \in \mathcal{D}_g$. Is $g(f(x)) = x$ for every $x \in \mathcal{D}_f$?

22. If $f(x) = (1 - x^2)^{1/2}$, find $f(f(x))$ and $f(f(f(x)))$.

23. If $f(x + h) = f(x) + f(h)$ for any numbers h and x, show that
    (a) $f(0) = 0$ (b) $f(-x) = -f(x)$

24. A function f is said to be an *even function* if and only if $f(-x) = f(x)$ for every x in $\mathcal{D}_f$. A function f is said to be an *odd function* if and only if $f(-x) = -f(x)$ for every x in $\mathcal{D}_f$. If f is even and g is odd determine whether the following functions are even or odd.
    (a) fg (b) gg (c) $f \circ g$ (d) $g \circ g$

3.3   Limit of a Function

In order to introduce the notion of the limit of a function, a basic concept in the study of calculus, we shall study the function f, graphed in Figure 3-8, where

$$f(x) = \begin{cases} 4x - 3 & \text{if } x \neq 2, \\ 1 & \text{if } x = 2. \end{cases} \qquad (3.16)$$

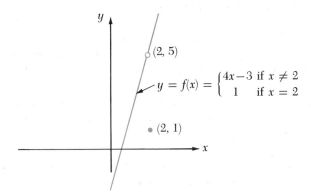

Figure 3-8

We consider the values, $f(x)$, for f when x is close to but unequal to 2. For certain of these numbers x, the corresponding values $f(x)$ are included in the table below.

x	1.75	1.9	1.95	1.99	1.995	1.9995	2.0005	2.005	2.01	2.05	2.1	2.25
$f(x)$	4	4.6	4.8	4.96	4.98	4.998	5.002	5.02	5.04	5.2	5.4	6

It is intuitively clear that $f(x)$ is close to 5 when x is close to but unequal to 2. Now suppose we ask the question:

"How close to 2 must $x \neq 2$ be chosen so that $f(x)$ differs from 5 by less than $\frac{1}{10}$?"

Since the closeness of x to 2 is given by $|x - 2|$ and the closeness of $f(x)$ to 5 is expressed by $|f(x) - 5|$, our problem here is to find a $\delta > 0$ such that

$$|f(x) - 5| < \tfrac{1}{10} \quad \text{if } 0 < |x - 2| < \delta.$$

To obtain this number δ, we can write

$$|f(x) - 5| = |(4x - 3) - 5| = |4(x - 2)| = 4|x - 2|. \tag{3.17}$$

If $0 < |x - 2| < \tfrac{1}{4 \cdot 10} = \tfrac{1}{40}$, from (3.17)

$$|f(x) - 5| = 4|x - 2| < 4 \cdot \tfrac{1}{40} = \tfrac{1}{10}.$$

In this problem we therefore have $\delta = \tfrac{1}{40}$, which means that $f(x)$ will differ from 5 by less than $\tfrac{1}{10}$ if $x \neq 2$ differs from 2 by less than $\tfrac{1}{40}$.

If we were to repeat the question just posed with $\tfrac{1}{500}$ in place of $\tfrac{1}{10}$, we would find using the same technique as before that $x \neq 2$ must differ from 2 by less than $\tfrac{1}{2000}$. Similarly, if we had replaced $\tfrac{1}{10}$ by ϵ, an arbitrary positive number, $x \neq 2$ must differ from 2 by less than $\epsilon/4$.

Suppose that a function f is defined at least throughout some open interval having the number c as an endpoint. We say that f has the *limit L at c* if and only if

    (i) $f(x)$ can be made to differ from L by as little as we please for every x sufficiently close to, but unequal to c. (It is tacitly assumed here that x must be an element of $\mathcal{D}_f$.)

When f has the limit L at c, one expresses this idea by writing $\lim_{x \to c} f(x) = L$. We can rewrite statement (i) in an equivalent form without the unprecise phrases "as little as we please" and "sufficiently close to." As a first refinement:

    (ii) $f(x)$ can be made to differ from L by an amount less than ϵ, where ϵ is an arbitrary positive number, for every x sufficiently close to, but unequal to c.

Now x will be sufficiently close to c when x differs from c by an amount less than some positive number δ, which usually depends upon ϵ. Hence (ii) can be further refined to read

    (iii) For every $\epsilon > 0$ there is a $\delta > 0$ such that $f(x)$ can be made to differ from L by an amount less than ϵ if $x \neq c$ differs from c by an amount less than δ.

Since the distance between $f(x)$ and L is $|f(x) - L|$ and the distance between x and c is $|x - c|$, we shall state the definition below:

3.3.1 Definition

Let f be defined, at least, throughout some open interval having the number c as an endpoint. Then $\lim_{x \to c} f(x) = L$ if and only if for every number $\epsilon > 0$ there is a number $\delta > 0$ such that

$$|f(x) - L| < \epsilon \quad \text{if } 0 < |x - c| < \delta \text{ (and } x \in \mathcal{D}_f) \qquad (3.18)$$

The student should make a serious effort to understand Definition 3.3.1 because it is one of the foundations of calculus. Probably the best plan would be to memorize the definition in the beginning and then try to gain an understanding of it through following the illustrative examples.

Example 1   For the function f given by (3.16) we showed that $\lim_{x \to 2} f(x) = 5$ since for every $\epsilon > 0$ there is a $\delta = \epsilon/4$ such that $|f(x) - 5| < \epsilon$ if $0 < |x - 2| < \epsilon/4$.

The student should note that in using Definition 3.3.1:

(a) It is not necessary that $\lim_{x \to c} f(x) = f(c)$. In fact, $f(c)$ may not even be defined. (For the function f given by Equation (3.16) we saw that $\lim_{x \to 2} f(x) = 5$, but $f(2) = 1$.)

(b) The number ϵ is an arbitrary positive number and may be chosen as small as we wish. For a given function f and numbers c and L, the δ obtained generally depends upon ϵ. Usually the smaller we select ϵ, the smaller is the corresponding δ. (Consider the function f defined by (3.16) with $c = 2$ and $L = 5$. When $\epsilon = \frac{1}{10}$, $\delta = \frac{1}{40}$ will suffice; when $\epsilon = \frac{1}{500}$, $\delta = \frac{1}{2000}$ will suffice.)

(c) The choice of the δ for a given ϵ is not unique. (For f given by (3.16) with $c = 2$ and $L = 5$, $\frac{1}{40}$ is actually the *maximum* permissible δ corresponding to $\epsilon = \frac{1}{10}$. Any positive number less than $\frac{1}{40}$, for example $\frac{1}{55}$, would suffice for δ since $0 < |x - 2| < \frac{1}{55}$ implies $0 < |x - 2| < \frac{1}{40}$ and hence $|f(x) - 5| < \frac{1}{10}$.)

(d) Statement (3.18) in the definition is equivalent to saying

$$L - \epsilon < f(x) < L + \epsilon \quad \text{if } x \neq c \text{ and } c - \delta < x < c + \delta.$$

(Recall Theorems 1.5.3(a) and 1.3.5.)

Note that (d) is illustrated in Figure 3–9. If $\lim_{x \to c} f(x) = L$, then corresponding to an open interval $(L - \epsilon, L + \epsilon)$ on the y axis, there is some open interval $(c - \delta, c + \delta)$ on the x axis such that if $x \neq c$ is in $(c - \delta, c + \delta)$, then $f(x)$ will be in $(L - \epsilon, L + \epsilon)$. Saying the same thing in another way, if $x \neq c$ is chosen in the open interval $(c - \delta, c + \delta)$, the corresponding point on the graph of f, $(x, f(x))$, will be inside the rectangle bounded by the lines $y = L - \epsilon$, $y = L + \epsilon$, $x = c - \delta$, and $x = c + \delta$. This rectangle is shaded in Figure 3–9.

(e) If there is no number L such that $\lim_{x \to c} f(x) = L$, we say that $\lim_{x \to c} f(x)$ fails to exist (in the sense of Definition 3.3.1). In particular, this limit fails to exist if f is not defined for every $x \neq c$ in some open interval containing c.

(f) Let ϵ be an arbitrary positive number. If there is an $M > 0$ such that

$$|f(x) - L| \leq M|x - c| \quad \text{for every } x \qquad (3.19)$$

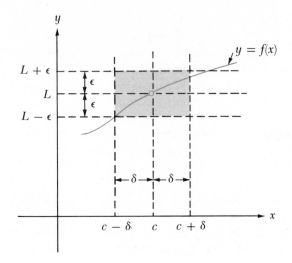

Figure 3-9

and if $0 < |x - c| < \epsilon/M$, from (3.19) we have

$$|f(x) - L| < M \cdot \frac{\epsilon}{M} = \epsilon.$$

Hence $\lim_{x \to c} f(x) = L$. This approach is used in Example 2 below.
(g) Let ϵ be an arbitrary positive number. Suppose there are positive numbers M and γ such that

$$|f(x) - L| \le M|x - c| \quad \text{if } |x - c| < \gamma. \qquad (3.20)$$

If $0 < |x - c| < \delta = smaller$ of γ and ϵ/M, from (3.20) we have

$$|f(x) - L| < M \cdot \frac{\epsilon}{M} = \epsilon$$

and hence $\lim_{x \to c} f(x) = L$. This approach is used in Examples 3, 4, and 5 below.

Example 2 Prove: $\lim_{x \to 5} (2x + 4) = 14$.

PROOF Let ϵ be an arbitrary positive number. We note that

$$|(2x + 4) - 14| = |2x - 10| = |2(x - 5)| = 2|x - 5|, \qquad (3.21)$$

and thus we have a condition of the type (3.19). (Here $M = 2$.) By choosing $0 < |x - 5| < \epsilon/2 = \delta$, from (3.21)

$$|(2x + 4) - 14| = 2|x - 5| < 2 \cdot \frac{\epsilon}{2} = \epsilon.$$

The conclusion then follows from (3.18).

Example 3   Prove: $\lim_{x \to 2} x^2 = 4$.

PROOF   Let ϵ be an arbitrary positive number. We note that

$$|x^2 - 4| = |(x + 2)(x - 2)| = |x + 2||x - 2|. \qquad (3.22)$$

If we can find a positive number M such that $|x + 2| \le M$ when x is sufficiently close to 2, we obtain a condition of the type (3.20). To find this M, we could consider numbers x such that $|x - 2| < 1$. Then

$$-1 < x - 2 < 1,$$

and if 4 is added to each member of this inequality, we have

$$3 < x + 2 < 5.$$

From this inequality

$$|x + 2| < 5. \qquad (3.23)$$

Hence from (3.22) and (3.23)

$$|x^2 - 4| < 5|x - 2| \quad \text{if } |x - 2| < 1, \qquad (3.24)$$

which is a condition of the type (3.20). If $0 < |x - 2| < \delta = $ smaller of 1 and $\epsilon/5$, from (3.24), $|x^2 - 4| < 5 \cdot \epsilon/5 = \epsilon$ and the conclusion follows from (3.18).

The choice to restrict x here so that $|x - 2| < 1$ was purely arbitrary. If initially we had restricted x so that $|x - 2| < \frac{1}{2}$ instead of 1, we could obtain the conclusion by showing that $|x^2 - 4| < \epsilon$ if $0 < |x - 2| < \delta = $ smaller of $\frac{1}{2}$ and $2\epsilon/9$.

Example 4   Prove: $\lim_{x \to -2} (1/x) = -1/2$.

PROOF   Let ϵ be an arbitrary positive number. We note that

$$\left| \frac{1}{x} - \left(-\frac{1}{2} \right) \right| = \left| \frac{2 + x}{2x} \right| = \frac{|x + 2|}{|2x|}. \qquad (3.25)$$

To get a condition of the type (3.20), we seek a positive number M such that $1/|2x| \le M$ when x is sufficiently close to -2. To find this M, we consider these numbers x such that

$$|x + 2| < 1.$$

Then

$$-1 < x + 2 < 1$$

and

$$-3 < x < -1.$$

Hence

$$|x| > 1$$
$$|2x| > 2$$

and by Theorem 1.3.10

$$\frac{1}{|2x|} < \frac{1}{2} \qquad (3.26)$$

Hence from (3.25) and (3.26)

$$\left| \frac{1}{x} - \left(-\frac{1}{2} \right) \right| = \frac{|x+2|}{|2x|} < \frac{1}{2}|x+2| \quad \text{if } |x+2| < 1 \qquad (3.27)$$

a condition of the type (3.20). If $0 < |x+2| < \delta = $ smaller of 1 and 2ϵ, from (3.27), $|1/x - (-\frac{1}{2})| < \frac{1}{2} \cdot 2\epsilon = \epsilon$ and the conclusion follows.

In using the approach suggested in note (g) above for the proof, we must select a $\gamma > 0$ small enough so that there is, in fact, an $M > 0$ such that

$$\frac{1}{|2x|} \leq M \qquad (3.28)$$

when $|x+2| < \gamma$. Suppose, for example, we had initially restricted x so that $|x+2| < 3$. Although this inequality is satisfied by $x = 0$, there is no $M > 0$ such that $x = 0$ also satisfies (3.28). Thus the number $\gamma = 3$ is too large for our purposes here.

Example 5 Prove: $\lim_{x \to 8} x^{1/3} = 2$.

PROOF Let $\epsilon > 0$ be given. Since we can write $x - 8 = (x^{1/3} - 2)(x^{2/3} + 2x^{1/3} + 4)$,

$$|x^{1/3} - 2| = \frac{|x-8|}{|x^{2/3} + 2x^{1/3} + 4|}. \qquad (3.29)$$

To find an $M > 0$ such that $1/|x^{2/3} + 2x^{1/3} + 4| \leq M$ when x is sufficiently close to 8, we need only choose x close enough to 8 so that $x > 0$, which is guaranteed if $|x - 8| < 8$. For if $|x - 8| < 8$, then $x - 8 > -8$, and $x > 0$. Hence $|x^{2/3} + 2x^{1/3} + 4| > 4$ and

$$\frac{1}{|x^{2/3} + 2x^{1/3} + 4|} < \frac{1}{4} \quad \text{if } |x-8| < 8. \qquad (3.30)$$

Substituting from (3.30) and (3.29), we have

$$|x^{1/3} - 2| < \tfrac{1}{4}|x - 8| \quad \text{if } |x - 8| < 8. \qquad (3.31)$$

If $0 < |x - 8| < \delta = $ smaller of 8 and 4ϵ, from (3.31), $|x^{1/3} - 2| < \frac{1}{4} \cdot 4\epsilon = \epsilon$ and the conclusion follows.

We next prove that the limit of a function is unique when it exists.

3.3.2 Theorem

If $\lim_{x \to c} f(x) = L_1$ and $\lim_{x \to c} f(x) = L_2$, then $L_1 = L_2$.

PROOF Suppose $L_1 \neq L_2$. Then there is an $\epsilon > 0$ such that

$$|L_1 - L_2| \geq \epsilon. \qquad (3.32)$$

Since $\lim_{x\to c} f(x) = L_1$, there is a $\delta_1 > 0$ such that

$$|f(x) - L_1| < \frac{\epsilon}{2} \tag{3.33}$$

if $0 < |x - c| < \delta_1$. Also, since $\lim_{x\to c} f(x) = L_2$, there is a $\delta_2 > 0$ such that

$$|f(x) - L_2| < \frac{\epsilon}{2} \tag{3.34}$$

if $0 < |x - c| < \delta_2$. If $\delta = $ smaller of δ_1 and δ_2 and we choose x so that $0 < |x - c| < \delta$, then (3.33) and (3.34) are true and by Theorem 1.5.7

$$|L_1 - L_2| = |[f(x) - L_2] - [f(x) - L_1]|$$
$$\leq |f(x) - L_2| + |f(x) - L_1| < \frac{\epsilon}{2} + \frac{\epsilon}{2} = \epsilon,$$

which is a contradiction of (3.32). Since the supposition that $L_1 \neq L_2$ leads to a contradiction, it must be true that $L_1 = L_2$.

We conclude this section by stating a theorem which will be useful in developing the theory of calculus.

3.3.3   Theorem

If $\lim_{x\to c} f(x) = L > 0$ (or $L < 0$), there is an open interval containing c such that for every $x \neq c$ in the interval and in $\mathcal{D}_f$, $f(x) > a > 0$ (or $f(x) < a < 0$) for some number a.

PROOF IF $L > 0$ Since $\lim_{x\to c} f(x) = L$ there is a $\delta > 0$ such that

$$|f(x) - L| < \frac{L}{2} \tag{3.35}$$

if $0 < |x - c| < \delta$. From (3.35) if $0 < |x - c| < \delta$,

$$-\frac{L}{2} < f(x) - L < \frac{L}{2}, \tag{3.36}$$

and if we add L to the left and middle members of (3.36), we have

$$f(x) > \frac{L}{2} > 0 \quad \text{if } 0 < |x - c| < \delta.$$

We note that $L/2$ is the number a mentioned in the conclusion.
The proof if $L < 0$ is similar and is left as an exercise.

Exercise Set 3.3

1. If $f(x) = 5 - 4x$, find a $\delta > 0$ such that $|f(x) - (-7)| < \epsilon$ if $0 < |x - 3| < \delta$ when:

   (a) $\epsilon = \frac{1}{5}$ (b) $\epsilon = .001$ (c) $\epsilon > 0$ is arbitrary

2. If $f(x) = 3x + 2$, find a $\delta > 0$ such that $|f(x) - 5| < \epsilon$ if $0 < |x - 1| < \delta$ when:

(a) $\epsilon = \frac{1}{5}$ (b) $\epsilon = .001$ (c) $\epsilon > 0$ is arbitrary

In Exercises 3–13 verify the given limits using (3.18).

3. $\lim_{x \to 3} (2x + 1) = 7$

4. $\lim_{x \to 1} f(x) = 2$ where $f(x) = \begin{cases} 4x - 2 & \text{if } x \neq 1 \\ 3 & \text{if } x = 1 \end{cases}$

5. $\lim_{x \to 4} (x^2 + 2x) = 24$

6. $\lim_{x \to 2} x^3 = 8$

7. $\lim_{x \to 9} \sqrt{x} = 3$

8. $\lim_{x \to -1} (x^2 - 3x + 4) = 8$

9. $\lim_{x \to 3} \dfrac{2}{x} = \dfrac{2}{3}$

10. $\lim_{x \to 5} \dfrac{x}{x + 2} = \dfrac{5}{7}$

11. $\lim_{x \to 0} \dfrac{5x + 2}{x + 1} = 2$

12. $\lim_{x \to 0} \sqrt[3]{x} = 0$

13. $\lim_{x \to 1} \sqrt{x - 1} = 0$

14. $\lim_{x \to 3} \sqrt{2x + 3} = 3$

15. Prove Theorem 3.3.3 for the case $L < 0$.

16. It can be proved that $\lim_{x \to 2} (x^3 - x - 5) = 1$. Then why is there a $\delta > 0$ such that $x^3 - x - 5 > 0$ when x is in the open interval $(2 - \delta, 2 + \delta)$?

17. Prove: If $\lim_{x \to c} |f(x)| = 0$, then $\lim_{x \to c} f(x) = 0$.

18. Prove: $\lim_{x \to c} f(x) = L$ if and only if $\lim_{x \to c} [f(x) - L] = 0$.

19. Prove: If $\lim_{x \to c} f(x) = L$, then $\lim_{x \to c} |f(x)| = |L|$. HINT: Use the inequality in Exercise 30, Exercise Set 1.5.

20. Prove: If $\lim_{x \to c} f(x) = L$, there is an $M > 0$ such that $|f(x)| \leq M$ for every $x \neq c$ in $\mathcal{D}_f$ and in some open interval containing c.

21. Suppose F is defined for every $x \neq c$ in some open interval with endpoint c. $\mathrm{Lim}_{x \to c} f(x) \neq L$ if L does not satisfy the condition expressed in Definition 3.3.1—that is, there is some $\epsilon > 0$ such that for every $\delta > 0$ there is an x (in $\mathcal{D}_f$) such that $0 < |x - c| < \delta$ and $|f(x) - L| \geq \epsilon$. Prove that $\lim_{x \to 2} 3x \neq 9$.

3.4   Limit Theorems

The student has probably noticed that the verification of limits using the ϵ-δ technique is generally a tedious procedure. It will therefore be convenient to derive certain theorems from Definition 3.3.1 which enable us to easily calculate limits by by-passing the ϵ-δ process. These theorems, which are called *limit theorems*, will also aid in the development of the theory.

3.4.1   Theorem

Let c, m, and b be any numbers. Then

$$\lim_{x \to c} (mx + b) = mc + b.$$

PROOF We consider separately the cases where $m = 0$ and $m \neq 0$. If $m = 0$, let $\epsilon > 0$ be arbitrary. Then for *any* $\delta > 0$ if $0 < |x - c| < \delta$, we have $|(mx + b) - (mc + b)| = |b - b| = 0 < \epsilon$, and hence the conclusion follows from Definition 3.3.1.

The proof when $m \neq 0$ is left as an exercise.

As an illustration of Theorem 3.4.1,

$$\lim_{x \to -2} (3x + 10) = 3(-2) + 10 = 4.$$

Also, from Theorem 3.4.1 if $f(x) = b$ for every x,

$$\lim_{x \to c} f(x) = \lim_{x \to c} b = b. \tag{3.37}$$

3.4.2 Theorem

Let $\sqrt[q]{c}$ be a real number. Then

$$\lim_{x \to c} \sqrt[q]{x} = \sqrt[q]{c}.$$

PROOF If $c \neq 0$, a proof can be written which is analogous to that for Example 5 of Section 3.3. The proof, which is left as an exercise, is based on the equation

$$x - c = (x^{1/q} - c^{1/q})(x^{(q-1)/q} + c^{1/q}x^{(q-2)/q} + c^{2/q}x^{(q-3)/q} + \cdots$$
$$+ c^{(q-1)/q}), \tag{3.38}$$

which can be derived from (1.12) by letting $a = x^{1/q}$ and $b = c^{1/q}$. Next, suppose $c = 0$. We note that when $\sqrt[q]{x}$ exists, $|\sqrt[q]{x}| = \sqrt[q]{|x|}$. For a given $\epsilon > 0$ if $0 < |x| < \epsilon^q$, then $|\sqrt[q]{x}| = \sqrt[q]{|x|} < \sqrt[q]{\epsilon^q} = \epsilon$, using Theorem 1.3.9(b). The conclusion is obtained from Definition 3.3.1.

For example,

$$\lim_{x \to 4} \sqrt[3]{x} = \sqrt[3]{4}.$$

By the next theorem the limit of the sum, difference, product, and quotient of two functions is equal, respectively, to the sum, difference, product, and quotient of the limits of the functions, provided these limits exist. In this theorem and in the remaining theorems of this chapter *it will be tacitly assumed, whenever the limit of a function at a number is to be computed, that the function is actually defined on some open interval having that number as an endpoint* (recall Definition 3.3.1).

3.4.3 Theorem

If $\lim_{x \to c} f(x) = L_1$ and $\lim_{x \to c} g(x) = L_2$, then:

(a) $\lim_{x \to c} (f + g)(x) = \lim_{x \to c} [f(x) + g(x)] = L_1 + L_2$

(b) $\lim_{x \to c} (f - g)(x) = \lim_{x \to c} [f(x) - g(x)] = L_1 - L_2$

(c) $\lim_{x \to c} (fg)(x) = \lim_{x \to c} f(x)g(x) = L_1 L_2$

(d) $\lim_{x \to c} \left(\dfrac{f}{g}\right)(x) = \lim_{x \to c} \dfrac{f(x)}{g(x)} = \dfrac{L_1}{L_2}$ when $L_2 \neq 0$

PROOF OF THEOREM 3.4.3(a)   In this proof and in the proofs of the other parts of this theorem $\epsilon > 0$ is given. The conclusion of (a) can be proved by invoking Definition 3.3.1—that is, by showing there is a $\delta > 0$ such that

$$|[f(x) + g(x)] - (L_1 + L_2)| < \epsilon \quad \text{if } 0 < |x - c| < \delta.$$

First note that by Theorem 1.5.7

$$|[f(x) + g(x)] - (L_1 + L_2)| = |[f(x) - L_1] + [g(x) - L_2]|$$
$$\leq |f(x) - L_1| + |g(x) - L_2|. \tag{3.39}$$

Since $\lim_{x \to c} f(x) = L_1$, there is a $\delta_1 > 0$ such that

$$|f(x) - L_1| < \frac{\epsilon}{2} \tag{3.40}$$

if $0 < |x - c| < \delta_1$. Also, since $\lim_{x \to c} g(x) = L_2$, there is a $\delta_2 > 0$ such that

$$|g(x) - L_2| < \frac{\epsilon}{2} \tag{3.41}$$

if $0 < |x - c| < \delta_2$. Then, letting $\delta = $ smaller of δ_1 and δ_2 and choosing x so that $0 < |x - c| < \delta$, *both* (3.40) and (3.41) are satisfied and hence from (3.39)

$$|[f(x) + g(x)] - (L_1 + L_2)| < \frac{\epsilon}{2} + \frac{\epsilon}{2} = \epsilon.$$

Thus the conclusion is obtained.

The proof of Theorem 3.4.3(b) is analogous to the foregoing proof and is left as an exercise.

PROOF OF THEOREM 3.4.3(c)   Here we shall prove the conclusion by showing there is a $\delta > 0$ such that

$$|f(x)g(x) - L_1 L_2| < \epsilon \quad \text{if } 0 < |x - c| < \delta.$$

From subtracting and adding $L_1 g(x)$ we obtain

$$|f(x)g(x) - L_1 L_2| = |f(x)g(x) - L_1 g(x) + L_1 g(x) - L_1 L_2|$$
$$= |g(x)[f(x) - L_1] + L_1[g(x) - L_2]|.$$

Hence by Theorems 1.5.7 and 1.5.4,

$$|f(x)g(x) - L_1 L_2| \leq |g(x)||f(x) - L_1| + |L_1||g(x) - L_2|. \tag{3.42}$$

We first seek an $M > 0$ such that $|g(x)| \leq M$ when $x \neq c$ is sufficiently close to c. Since $\lim_{x \to c} g(x) = L_2$, there is a $\delta_1 > 0$ such that

$$|g(x) - L_2| < 1 \qquad (3.43)$$

if $0 < |x - c| < \delta_1$, although in (3.43) any positive number could have been used in place of 1. Using Theorem 1.5.7,

$$|g(x)| = |[g(x) - L_2] + L_2| \le |g(x) - L_2| + |L_2|,$$

and from this statement and that obtained by adding $|L_2|$ to each member of (3.43), we have

$$|g(x)| < 1 + |L_2| \qquad (3.44)$$

if $0 < |x - c| < \delta_1$. Since $\lim_{x \to c} f(x) = L_1$, there is a $\delta_2 > 0$ such that

$$|f(x) - L_1| < \frac{\epsilon}{2(1 + |L_2|)} \qquad (3.45)$$

if $0 < |x - c| < \delta_2$. Also, since $\lim_{x \to c} g(x) = L_2$, there is a $\delta_3 > 0$ such that

$$|g(x) - L_2| < \frac{\epsilon}{2(1 + |L_1|)} \qquad (3.46)$$

if $0 < |x - c| < \delta_3$. We then let $\delta =$ smallest of δ_1, δ_2, and δ_3 and choose x so that $0 < |x - c| < \delta$. Then (3.44), (3.45), and (3.46) are true, and the remainder of the proof, which is left as an exercise, can be readily shown.

It should be noted that in writing (3.46) we chose $|g(x) - L_2| < \epsilon/2(1 + |L_1|)$ instead of $\epsilon/2|L_1|$ since the latter number is undefined when $L_1 = 0$.

As a corollary of Theorem 3.4.3(c), if k is any number, then from Theorem 3.4.3(c) and (3.37)

$$\lim_{x \to c} kg(x) = (\lim_{x \to c} k)(\lim_{x \to c} g(x))$$

$$= k \lim_{x \to c} g(x) \quad \text{if } \lim_{x \to c} g(x) \text{ exists.} \qquad (3.47)$$

PROOF OF THEOREM 3.4.3(d)   We shall first prove that $\lim_{x \to c} [1/g(x)] = 1/L_2$. From Theorem 3.3.3 since $\lim_{x \to c} g(x) = L_2 \ne 0$, there is some $\delta_1 > 0$ such that

$$|g(x)| > a > 0 \qquad (3.48)$$

if $0 < |x - c| < \delta_1$. We can write, using Theorems 1.5.4 and 1.5.5

$$\left| \frac{1}{g(x)} - \frac{1}{L_2} \right| = \left| \frac{L_2 - g(x)}{L_2 g(x)} \right| = \frac{|g(x) - L_2|}{|L_2||g(x)|}. \qquad (3.49)$$

Now since $\lim_{x \to c} g(x) = L_2$, there is a $\delta_2 > 0$ such that

$$|g(x) - L_2| < a|L_2|\epsilon \qquad (3.50)$$

if $0 < |x - c| < \delta_2$. We leave as an exercise, using (3.48), (3.49), and (3.50), the proof that there is a small enough positive number δ such that

$$\left| \frac{1}{g(x)} - \frac{1}{L_2} \right| < \epsilon \qquad (3.51)$$

if $0 < |x - c| < \delta$. Thus we have

$$\lim_{x \to c} \frac{1}{g(x)} = \frac{1}{L_2}.$$

Since $(f/g)(x) = f(x)/g(x) = f(x) \cdot 1/g(x)$, from Theorem 3.4.3(c)

$$\lim_{x \to c} \left(\frac{f}{g}\right)(x) = \lim_{x \to c} \left[f(x) \cdot \frac{1}{g(x)}\right] = \lim_{x \to c} f(x) \cdot \lim_{x \to c} \frac{1}{g(x)}$$

$$= L_1 \cdot \frac{1}{L_2} = \frac{L_1}{L_2}.$$

Example 1   Find $\lim_{x \to 2} (4x - 9 + \sqrt{x})$.

SOLUTION

$$\lim_{x \to 2} (4x - 9 + \sqrt{x}) = \lim_{x \to 2} (4x - 9) + \lim_{x \to 2} \sqrt{x} \qquad \text{by Theorem 3.4.3(a)}$$

$$= -1 + \sqrt{2} \qquad \qquad \text{by Theorems 3.4.1 and 3.4.2}$$

Example 2   If c is any number, find $\lim_{x \to c} x^2$.

SOLUTION $\lim_{x \to c} x^2 = (\lim_{x \to c} x)(\lim_{x \to c} x)$ $\qquad\qquad$ by Theorem 3.4.3(c)

$$= c \cdot c = c^2. \qquad\qquad\qquad\qquad \text{by Theorem 3.4.1}$$

Example 3   Find $\lim_{x \to -1} \dfrac{-3x + 4}{x + 3}$.

SOLUTION $\lim_{x \to -1} \dfrac{-3x + 4}{x + 3} = \dfrac{\lim_{x \to -1} (-3x + 4)}{\lim_{x \to -1} (x + 3)}$ $\qquad$ by Theorem 3.4.3(d)

$$= \tfrac{7}{2}. \qquad\qquad\qquad\qquad \text{by Theorem 3.4.1}$$

Theorem 3.4.3(d) could not be used to evaluate $\lim_{x \to 2} (x^2 - 4)/(x - 2)$, for example, since $\lim_{x \to 2} (x - 2) = 0$. The existence of the limit of the quotient of two functions when the limit of the denominator function is zero will be considered in the next section.

From Theorems 3.4.3(a) and (c) respectively the following corollaries can be proved by mathematical induction.

3.4.4   Theorem

If $\lim_{x \to c} f_1(x) = L_1$, $\lim_{x \to c} f_2(x) = L_2$, ..., *and* $\lim_{x \to c} f_n(x) = L_n$, *then:*

(a) $\lim_{x \to c} [f_1(x) + f_2(x) + \cdots + f_n(x)] = L_1 + L_2 + \cdots + L_n$

(b) $\lim\limits_{x\to c} f_1(x)f_2(x) \cdots f_n(x) = L_1L_2 \cdots L_n$

We next prove

3.4.5   Theorem

If c is any number and P is a polynomial function,

$$\lim_{x\to c} P(x) = P(c).$$

PROOF   It will be recalled that $P(x)$ can be represented in the form

$$P(x) = a_0x^n + a_1x^{n-1} + \cdots + a_n$$

where n is a non-negative integer and $a_0, a_1, \ldots, a_n$ are any numbers. From Theorem 3.4.4(b) if $i = 1, 2, \ldots, n$,

$$\lim_{x\to c} a_ix^{n-i} = (\lim_{x\to c} a_i)(\underbrace{\lim_{x\to c} x)(\lim_{x\to c} x) \cdots (\lim_{x\to c} x)}_{n-i \text{ factors}}$$

$$= a_i c^{n-i}$$

and from Theorem 3.4.4(a)

$$\lim_{x\to c} P(x) = \lim_{x\to c} a_0x^n + \lim_{x\to c} a_1x^{n-1} + \cdots + \lim_{x\to c} a_n$$

$$= a_0c^n + a_1c^{n-1} + \cdots + a_n = P(c).$$

Example 4   Find $\lim_{x\to 3} (-2x^4 + \frac{1}{3}x^3 + \sqrt{3}x - \pi)$.

SOLUTION   From Theorem 3.4.5

$$\lim_{x\to 3} (-2x^4 + \frac{1}{3}x^3 + \sqrt{3}x - \pi) = -2 \cdot 3^4 + \frac{1}{3} \cdot 3^3 + \sqrt{3} \cdot 3 - \pi$$

$$= -153 + 3\sqrt{3} - \pi.$$

The proofs of the next two theorems are left as exercises.

3.4.6   Theorem

If P and Q are polynomial functions and c is any number, the limit of the rational function P/Q at c is given by

$$\lim_{x\to c} \left(\frac{P}{Q}\right)(x) = \lim_{x\to c} \frac{P(x)}{Q(x)} = \frac{P(c)}{Q(c)} \quad \text{if } Q(c) \neq 0.$$

3.4.7   Theorem

Let q > 0 and p be integers. Then

$$\lim_{x\to c} x^{p/q} = c^{p/q}.$$

The student is reminded again that if $\lim_{x \to c} f(x)$ exists, the limit is not necessarily equal to $f(c)$, Theorems 3.4.1, 3.4.2, 3.4.5, 3.4.6, and 3.4.7 notwithstanding. (Remember Example 1 of Section 3.3.) In the next section we will consider examples where $\lim_{x \to c} f(x)/g(x)$ exists, but $f(c)/g(c)$ is undefined since $g(c) = 0$.

Exercise Set 3.4

In Exercises 1–7 evaluate the required limits and mention the limit theorems used in each calculation.

1. $\lim_{x \to -1} (x^2 - 4x + 6)$

2. $\lim_{x \to 2} (2x - 3)(2 - 5x - x^2)$

3. $\lim_{x \to -2} (2x^2 - 6x + 5)^2$

4. $\lim_{x \to 3} \sqrt[3]{2x^2 + x + 6}$

5. $\lim_{x \to 16} 5x^{-5/4}$

6. $\lim_{y \to 4} \dfrac{3(3y - 2)}{2y(y^2 + 4)}$

7. $\lim_{t \to 2} \dfrac{t^2 - 3t - 6}{2t^2 + 5t}$

8. Complete the proof of Theorem 3.4.1 by showing that $\lim_{x \to c} (mx + b) = mc + b$ if $m \neq 0$.

9. Complete the proof of Theorem 3.4.2 for $c \neq 0$. HINT: If $c \neq 0$, it can be shown that if $|x - c| < |c|$, then c and x are of the same sign, and hence $c^{(i-1)/q} x^{(q-i)/q}$ is positive for $i = 1, 2, \ldots, q$.

10. Prove Theorem 3.4.3(b).

11. Complete the proof of Theorem 3.4.3(c).

12. Complete the proof of Theorem 3.4.3(d).

13. Prove Theorem 3.4.4(a).

14. Prove Theorem 3.4.4(b).

15. Prove Theorem 3.4.6.

16. Prove Theorem 3.4.7.

17. Suppose $\lim_{x \to c} f(x)$ and $\lim_{x \to c} g(x)$ exist. Prove that $\lim_{x \to c} f(x) = \lim_{x \to c} g(x)$ if and only if $\lim_{x \to c} [f(x) - g(x)] = 0$.

3.5   Limit Theorems (continued)

Suppose one wished to evaluate, if possible,

$$\lim_{x \to 1} \frac{x - 1}{x^2 - 1}. \tag{3.52}$$

We note that since $\lim_{x \to 1} (x^2 - 1) = 0$, one cannot use Theorem 3.4.3(d) or 3.4.6 directly. (If one did attempt to employ either of these theorems here, the meaningless result $0/0$ would be obtained.)

To facilitate the evaluation of the limit (3.52), the following theorem is useful.

3.5.1   Theorem

Let $F(x) = G(x)$ for every $x \neq c$ in some open interval containing c. Then:

(a) $\lim_{x \to c} F(x) = L$ *if* $\lim_{x \to c} G(x) = L$
(b) $\lim_{x \to c} F(x)$ *fails to exist if* $\lim_{x \to c} G(x)$ *fails to exist*

The proofs of (a) and (b) are left as exercises. The proof of (b) follows from the proof of (a) using an indirect argument.

Example 1   Find $\lim_{x \to 1} \dfrac{x - 1}{x^2 - 1}$.

SOLUTION   We remarked in the beginning that since the limit of the denominator at 1 is zero neither Theorem 3.4.3(d) or 3.4.6 applies directly. However, we note that for every $x \neq 1$ (and thus for every $x \neq 1$ in some open interval containing 1),

$$\frac{x - 1}{x^2 - 1} = \frac{x - 1}{(x - 1)(x + 1)} = \frac{1}{x + 1}.$$

Hence by Theorem 3.5.1

$$\lim_{x \to 1} \frac{x - 1}{x^2 - 1} = \lim_{x \to 1} \frac{1}{x + 1} = \frac{1}{2}.$$

Example 2   If $a > 0$, find $\lim_{x \to a} (\sqrt{x} - \sqrt{a})/(x - a)$.

SOLUTION   Since $\lim_{x \to a} (x - a) = 0$, Theorem 3.4.3(d) cannot be used directly. Now

$$\frac{\sqrt{x} - \sqrt{a}}{x - a} = \frac{\sqrt{x} - \sqrt{a}}{(\sqrt{x} - \sqrt{a})(\sqrt{x} + \sqrt{a})} = \frac{1}{\sqrt{x} + \sqrt{a}} \qquad \text{if } x \neq a.$$

By Theorem 3.4.3(d), 3.4.3(a), and 3.4.2

$$\lim_{x \to a} \frac{1}{\sqrt{x} + \sqrt{a}} = \frac{\lim_{x \to a} 1}{\lim_{x \to a} (\sqrt{x} + \sqrt{a})} = \frac{1}{2\sqrt{a}}$$

and hence by Theorem 3.5.1

$$\lim_{x \to a} \frac{\sqrt{x} - \sqrt{a}}{x - a} = \lim_{x \to a} \frac{1}{\sqrt{x} + \sqrt{a}} = \frac{1}{2\sqrt{a}}.$$

Example 3   Find $\lim_{h \to 0} \dfrac{(h + 2)^2 - 4}{h}$.

SOLUTION   If $h \neq 0$, then

$$\frac{(h + 2)^2 - 4}{h} = \frac{h^2 + 4h}{h} = h + 4.$$

Hence by Theorem 3.5.1

$$\lim_{h \to 0} \frac{(h + 2)^2 - 4}{h} = \lim_{h \to 0} (h + 4) = 4.$$

According to the next theorem, if $\lim_{x \to c} g(x) = 0$ and $\lim_{x \to c} f(x)$ exists, then $\lim_{x \to c} f(x)/g(x)$ *fails to exist* unless $\lim_{x \to c} f(x) = 0$ also.

3.5.2 Theorem

If $\lim_{x \to c} g(x) = 0$ and $\lim_{x \to c} f(x) = L \neq 0$, then f/g has no limit at c.

PROOF Suppose $\lim_{x \to c} f(x)/g(x) = L'$. Then for every $x \neq c$ in some open interval with endpoint c, $f(x)/g(x)$ is defined and $f(x) = g(x) \cdot f(x)/g(x)$. Hence by Theorems 3.5.1 and 3.4.3(c), $\lim_{x \to c} f(x) = \lim_{x \to c} g(x) \cdot f(x)/g(x) = \lim_{x \to c} g(x) \cdot \lim_{x \to c} f(x)/g(x) = 0 \cdot L' = 0$, which is contrary to our second hypothesis. Thus the conclusion is obtained.

Example 4 Show that $\lim_{x \to 2} (2x + 3)/(x - 2)$ does not exist.

SOLUTION   Since $\lim_{x \to 2} (x - 2) = 0$ and $\lim_{x \to 2} (2x + 3) = 7$, by Theorem 3.5.2 the conclusion is obtained.

The following limit theorems will also be utilized in our later work.

3.5.3 Theorem

If for every $x \neq c$ in some open interval containing c, $f(x) \leq g(x) \leq h(x)$ and if $\lim_{x \to c} f(x) = L = \lim_{x \to c} h(x)$, then $\lim_{x \to c} g(x) = L$ also.

PROOF Let I be the open interval mentioned in the hypothesis. Then

$$f(x) - L \leq g(x) - L \leq h(x) - L \quad \text{if } x \neq c \text{ and } x \in I. \tag{3.53}$$

We next let ϵ be an arbitrary positive number. Since $\lim_{x \to c} f(x) = L$, there is a $\delta_1 > 0$ such that

$$|f(x) - L| < \epsilon \quad \text{if } x \in I \text{ and } 0 < |x - c| < \delta_1.$$

Also there is a $\delta_2 > 0$ such that

$$|h(x) - L| < \epsilon \quad \text{if } x \in I \text{ and } 0 < |x - c| < \delta_2.$$

Now suppose $x \in I$ and $0 < |x - c| < \delta =$ smaller of δ_1 and δ_2. Then by Theorem 1.5.3(a)

$$-\epsilon < f(x) - L \quad \text{and} \quad h(x) - L < \epsilon. \tag{3.54}$$

From (3.53) and (3.54)

$$-\epsilon < g(x) - L < \epsilon$$

and hence by Theorem 1.5.3(a)

$$|g(x) - L| < \epsilon \quad \text{if } x \in I \text{ and } 0 < |x - c| < \delta.$$

The conclusion follows from the latter statement.

3.5.4   Theorem

$\lim_{x \to c} f(x) = L$ *if and only if* $\lim_{h \to 0} f(c + h) = L$.

The proof of Theorem 3.5.4 is left as an exercise.

Exercise Set 3.5

Evaluate the indicated limits in Exercises 1–14 or show they fail to exist.

1. $\lim\limits_{x \to 2} \dfrac{3x - 6}{x^2 - x - 2}$

2. $\lim\limits_{x \to 4} \dfrac{x^2 - 3x - 4}{2x^2 - 9x + 4}$

3. $\lim\limits_{x \to 3} \dfrac{x^2 - 4x}{x^2 - 5x + 6}$

4. $\lim\limits_{x \to 4} \dfrac{3x^2 - 7x - 20}{x^2 - 8x + 16}$

5. $\lim\limits_{t \to 1} \dfrac{t^2 - 1}{t^3 - 1}$

6. $\lim\limits_{y \to -2} \dfrac{3y^2 + 2y - 8}{2y^3 + y^2 - 8y - 4}$

7. $\lim\limits_{u \to 4} \dfrac{\dfrac{u + 3}{u - 1} - \dfrac{7}{3}}{u - 4}$

8. $\lim\limits_{h \to 0} \dfrac{(a + h)^3 - a^3}{h}$

9. $\lim\limits_{h \to 0} \dfrac{\dfrac{2}{x + h} - \dfrac{2}{x}}{h}$

10. $\lim\limits_{h \to 0} \dfrac{\dfrac{x + h + 1}{x + h} - \dfrac{x + 1}{x}}{h}$

11. $\lim\limits_{x \to a} \dfrac{\dfrac{1}{\sqrt[3]{x}} - \dfrac{1}{\sqrt[3]{a}}}{x - a}$

12. $\lim\limits_{y \to a} \dfrac{\dfrac{1}{y^2} - \dfrac{1}{a^2}}{y - a}$

13. $\lim\limits_{x \to 1} \left(\dfrac{x}{2x - 2} - \dfrac{1}{x^2 - 1} \right)$

14. $\lim\limits_{x \to -1} \dfrac{\dfrac{x + 3}{x^2 + x} - \dfrac{4}{x + 1}}{\dfrac{2x}{x^2 - 1}}$

15. Prove Theorem 3.5.1(a).

16. Prove Theorem 3.5.4.

3.6 One-Sided Limits

Consider the function F given by

$$F(x) = \begin{cases} 1 - x & \text{if } x \geq 1 \\ -x & \text{if } x < 1 \end{cases} \tag{3.55}$$

(Figure 3–10). Intuition suggests that $F(x)$ can be made to differ from 0 by as little as desired if we choose $x > 1$ but sufficiently close to 1. Also, it would appear that $F(x)$ can be made to differ from -1 by as little as we please if $x < 1$ but sufficiently close to 1.

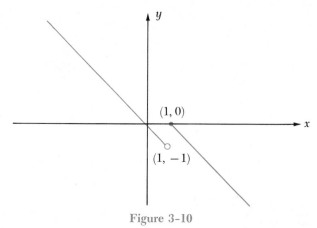

Figure 3–10

It can be shown that the function F defined by (3.55) has no limit at 1; however, it is still meaningful to consider the *one-sided limits* of F at 1.

The notation used for the one-sided limits of a function f at a number c is as follows:

$\lim\limits_{x \to c^+} f(x)$ denotes the "right-hand limit of f at c"

$\lim\limits_{x \to c^-} f(x)$ denotes the "left-hand limit of f at c"

We shall define these limits.

3.6.1 Definition

Let f be defined on some open interval (c, b). $\lim\limits_{x \to c^+} f(x) = L$ if and only if for every $\epsilon > 0$ there is a $\delta > 0$ such that

$$|f(x) - L| < \epsilon \quad \text{if } c < x < c + \delta. \tag{3.56}$$

3.6.2 Definition

Let f be defined on some open interval (a, c). $\lim\limits_{x \to c^-} f(x) = L$ if and only if for every $\epsilon > 0$ there is a $\delta > 0$ such that

$$|f(x) - L| < \epsilon \quad \text{if } c - \delta < x < c. \tag{3.57}$$

To distinguish it from the one-sided limits of f at c, $\lim_{x \to c} f(x)$ is sometimes called the *two-sided* limit of f at c. The following theorems are helpful in evaluating the one-sided limits of the function F given by (3.55).

3.6.3   Theorem

Let $f(x) = g(x)$ for every x in some open interval (c, b). If $\lim_{x \to c} g(x) = L$, then $\lim_{x \to c+} f(x) = L$.

PROOF   Let $\epsilon > 0$ be arbitrary. By hypothesis there is a δ satisfying $0 < \delta < b - c$ such that

$$|g(x) - L| < \epsilon \quad \text{if } 0 < |x - c| < \delta. \tag{3.58}$$

If x is chosen so that $c < x < c + \delta$, then $0 < x - c < \delta$, and hence $0 < |x - c| < \delta$. For such a number x, $f(x) = g(x)$, and from (3.58) $|f(x) - L| < \epsilon$. From this assertion the conclusion follows.

Example 1   For the function F in (3.55), find $\lim_{x \to 1+} F(x)$.

SOLUTION   Since $F(x) = 1 - x$ for $x > 1$ and $\lim_{x \to 1} (1 - x) = 0$, by Theorem 3.6.3 $\lim_{x \to 1+} F(x) = 0$.

In particular from Theorem 3.6.3 if $f(x)$ is defined for every x in some open interval with c as its left endpoint, and if $\lim_{x \to c} f(x) = L$, then $\lim_{x \to c+} f(x) = L$ also. For example:

$$\lim_{x \to 2^+} x^2 = \lim_{x \to 2} x^2 = 4$$

$$\lim_{x \to 0^+} \sqrt{x} = \lim_{x \to 0} \sqrt{x} = 0$$

3.6.4   Theorem

Let $f(x) = g(x)$ for every x in some open interval (a, c). If $\lim_{x \to c} g(x) = L$, then $\lim_{x \to c-} f(x) = L$.

The proof of Theorem 3.6.4 is left as an exercise.

Example 2   For the function F in (3.55), find $\lim_{x \to 1-} F(x)$.

SOLUTION   Since $F(x) = -x$ for $x < 1$ and $\lim_{x \to 1} (-x) = -1$, by Theorem 3.6.4, $\lim_{x \to 1-} F(x) = -1$.

From Theorem 3.6.4 if $f(x)$ is defined for every x in some open interval with c as its right endpoint, and if $\lim_{x \to c} f(x) = L$, then $\lim_{x \to c^-} f(x) = L$ also. Thus

$$\lim_{x \to 2^-} x^2 = \lim_{x \to 2} x^2 = 4.$$

The next theorem relates a two-sided limit of a function and its one-sided limits.

3.6.5   Theorem

Let f be defined at every number except c in some open interval containing c. $\lim_{x \to c} f(x) = L$ if and only if $\lim_{x \to c^+} f(x) = L$ and $\lim_{x \to c^-} f(x) = L$.

PROOF OF "ONLY IF" PART   If $\lim_{x \to c} f(x) = L$, then from our remarks following Example 1

$$\lim_{x \to c^+} f(x) = \lim_{x \to c} f(x) = L.$$

Similarly, from the remarks following Example 2

$$\lim_{x \to c^-} f(x) = \lim_{x \to c} f(x) = L.$$

The proof of the "if" part of the theorem is left as an exercise.

As an illustration of Theorem 3.6.5, since $\lim_{x \to 2} x^2 = 4$, we obtain

$$\lim_{x \to 2^+} x^2 = 4 \qquad \text{and} \qquad \lim_{x \to 2^-} x^2 = 4.$$

By Theorem 3.6.5 $\lim_{x \to c} f(x)$ fails to exist if the one-sided limits of f at c exist but are unequal. Thus from the results of Examples 1 and 2 the function F given by (3.55) has no limit at 1.

We consider some further examples of Theorems 3.6.3, 3.6.4, and 3.6.5.

Example 3   If $f(x) = \begin{cases} 2x + x^2 & \text{if } x > 0 \\ 2 & \text{if } x = 0 \\ [x + 1] & \text{if } x < 0 \end{cases}$ find

(a) $\lim_{x \to 0^+} f(x)$ (b) $\lim_{x \to 0^-} f(x)$ (c) $\lim_{x \to 0} f(x)$ (Figure 3–11).

SOLUTION   To evaluate the one-sided limits of f at 0, we must examine $f(x)$ for numbers x on each side of 0. If $x > 0$, $f(x) = 2x + x^2$ and by Theorem 3.6.3

$$\lim_{x \to 0^+} f(x) = \lim_{x \to 0} (2x + x^2) = 0.$$

If $-1 < x < 0$, then $0 < x + 1 < 1$, and $f(x) = [x + 1] = 0$. Hence by Theorem 3.6.4

$$\lim_{x \to 0^-} f(x) = \lim_{x \to 0} 0 = 0.$$

From Theorem 3.6.5

$$\lim_{x \to 0} f(x) = \lim_{x \to 0^+} f(x) = \lim_{x \to 0^-} f(x) = 0.$$

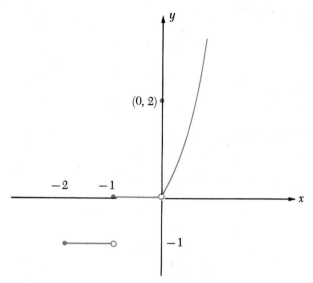

Figure 3-11

Example 4   If $G(x) = (x^2 - 1)/|x - 1|$ find:
(a) $\lim_{x \to 1^+} G(x)$ (b) $\lim_{x \to 1^-} G(x)$ (c) $\lim_{x \to 1} G(x)$ (Figure 3-12).

SOLUTION   We examine $G(x)$ for numbers x on each side of 1. If $x > 1$, $|x - 1| = x - 1$ and

$$G(x) = \frac{(x - 1)(x + 1)}{x - 1} = x + 1.$$

By Theorem 3.6.3

$$\lim_{x \to 1^+} G(x) = \lim_{x \to 1} (x + 1) = 2.$$

If $x < 1$, $|x - 1| = -(x - 1) = 1 - x$ and

$$G(x) = \frac{(x - 1)(x + 1)}{1 - x} = -x - 1.$$

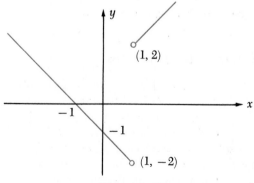

Figure 3-12

By Theorem 3.6.4

$$\lim_{x \to 1^-} G(x) = \lim_{x \to 1} (-x - 1) = -2.$$

Since the one-sided limits of G at 1 exist and are unequal, by Theorem 3.6.5 $\lim_{x \to 1} G(x)$ fails to exist.

Exercise Set 3.6

In Exercises 1–10, obtain the one-sided limits of the given functions at the given numbers or show that the one-sided limits fail to exist. In each case evaluate the two-sided limit of the function or show that it fails to exist.

1. $f(x) = \begin{cases} 2x^2 & \text{if } x < 0 \\ 3x + 4 & \text{if } x > 0 \end{cases}$ at 0

2. $F(x) = \begin{cases} \sqrt{x} & \text{if } x \leq 1 \\ 2x - 1 & \text{if } x > 1 \end{cases}$ at 1

3. $F(x) = \begin{cases} 4 & \text{if } x < 2 \\ 1 & \text{if } x = 2 \\ (x - 4)^2 & \text{if } x > 2 \end{cases}$ at 2

4. $f(t) = \begin{cases} 3t - 2 & \text{if } t < -1 \\ 4 & \text{if } t = -1 \\ 2 - t & \text{if } t > -1 \end{cases}$ at -1

5. $f(x) = |2x + 3|$ at $-\frac{3}{2}$

6. $g(x) = -2 - |6 - 3x|$ at 2

7. $f(x) = \dfrac{|x - 2|}{x - 2}$ at 2

8. $G(y) = \dfrac{1 + 2y}{1 - 2|y|}$ at 0

9. $f(x) = x - [x]$ at -2

10. $f(x) = [x + [x]]$ at 0

11. Prove Theorem 3.6.4.

12. Prove the "if" part of Theorem 3.6.5.

3.7 Continuous Functions

We think of a function as being *continuous* at a number c in its domain if its graph is "connected" at $x = c$. Thus the function having the graph sketched in Figure 3–13 would be continuous at every number in its domain except a_1, a_2, and a_3. At $x = a_1$ the graph has a "hole," while at $x = a_2$ and $x = a_3$ the graph has a "finite jump" and an "infinite jump," respectively. At a_1 the function is undefined even though the limit of the function exists there. At a_2 the function is defined but fails to have a limit. Finally, at a_3 the function both is undefined

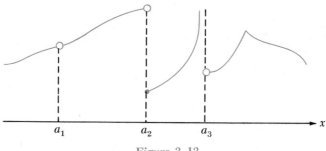

Figure 3–13

and has no limit. Continuity of a function at a number in its domain appears to require that (i) the function be defined at the number, (ii) the function have a limit at the number, and (iii) that the value of the function from (i) and the limit from (ii) be equal.

3.7.1   Definition

A function f is *continuous* at a number c in its domain if and only if the following conditions hold:

(i) $f(c)$ is defined

(ii) $\lim\limits_{x \to c} f(x)$ exists

(iii) $\lim\limits_{x \to c} f(x) = f(c)$

Otherwise, f is said to be *discontinuous* at c.

The notion of continuity can be extended to a set.

3.7.2   Definition

A function is *continuous on a set* contained in its domain if and only if the function is continuous at every number in the set.

We now consider some examples which illustrate continuity at a number.

Example 1   If $F(x) = \begin{cases} 1 - x & \text{if } x \geq 1 \\ -x & \text{if } x < 1 \end{cases}$, is the function F continuous at 1 (Figure 3–10)?

SOLUTION   We shall check the function F against the conditions (i), (ii), and (iii) of Definition 3.7.1. Since $F(1) = 1 - 1 = 0$, (i) is satisfied. From Examples 1 and 2 of Section 3.6

$$\lim_{x \to 1^-} F(x) = -1 \qquad \text{and} \qquad \lim_{x \to 1^+} F(x) = 0.$$

Hence by Theorem 3.6.5, $\lim_{x \to 1} F(x)$ fails to exist and (ii) is not satisfied. The function F is therefore discontinuous at 1.

We note that the graph of F (Figure 3–10) has a "jump" at 1.

Example 2   If $f(x) = (x^2 - 9)/(x - 3)$, is the function f continuous at 3? (See Figure 3–7(a).)

SOLUTION   Since $f(3)$ is undefined, condition (i) of Definition 3.7.1 is not fulfilled, and hence f is discontinuous at 3.

Example 3   If $F(x) = \begin{cases} \dfrac{x^2 - 9}{x - 3} & \text{if } x \neq 3 \\ 1 & \text{if } x = 3 \end{cases}$,   is the function F continuous at 3 (Figure 3–14)?

SOLUTION   Since $F(3) = 1$, (i) is satisfied at 3. Also, since

$$\lim_{x \to 3} F(x) = \lim_{x \to 3} \frac{(x + 3)(x - 3)}{x - 3} = \lim_{x \to 3} (x + 3) = 6,$$

(ii) is satisfied at 3. However, since $\lim_{x \to 3} F(x) \neq F(3)$, (iii) is not satisfied and F is discontinuous at 3.

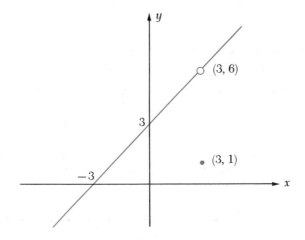

Figure 3–14

Example 4   If $f(x) = \begin{cases} x^2 + 2 & \text{if } x < 1 \\ 4 - x & \text{if } x \geq 1 \end{cases}$,   is f continuous at 1 (Figure 3–15)?

SOLUTION   Since $f(1) = 4 - 1 = 3$, (i) is satisfied at 1. Also

$$\lim_{x \to 1^-} f(x) = \lim_{x \to 1} (x^2 + 2) = 3$$

and

$$\lim_{x \to 1^+} f(x) = \lim_{x \to 1} (4 - x) = 3.$$

Hence by Theorem 3.6.5, $\lim_{x \to 1} f(x) = 3$ and thus (ii) is satisfied at 1. Since $\lim_{x \to 1} f(x) = 3 = f(1)$, (iii) is satisfied at 1. Since f satisfies (i), (ii), and (iii) at 1, f is continuous at 1.

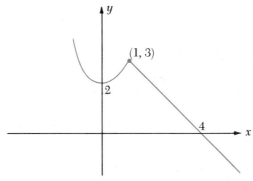

Figure 3–15

The usual functions that we study in this text are continuous at most, if not at all, numbers in their domains. From Theorem 3.4.5 a polynomial function is everywhere continuous; from Theorem 3.4.6 a rational function is continuous on its domain; and by Theorem 3.4.7, the function f defined by $f(x) = x^{p/q}$, where $q > 0$ and p are integers, is continuous on its domain.

The following theorems are useful in proving functions continuous. Their proofs are left as exercises.

3.7.3   Theorem

If $f_1, f_2, \ldots, f_n$ are continuous at c, then $f_1 + f_2 + \cdots + f_n$ and $f_1 f_2 \cdots f_n$ are continuous at c.

3.7.4   Theorem

If f and g are continuous at c and $g(c) \neq 0$, f/g is continuous at c.

The definition of continuity at a number can be restated in ϵ-δ language.

3.7.5   Definition

Let f be defined on some open interval having the number c as an endpoint. f is continuous at c if and only if for every $\epsilon > 0$ there is a $\delta > 0$ such that

$$|f(x) - f(c)| < \epsilon \quad \text{if } 0 < |x - c| < \delta \text{ (and } x \in \mathcal{D}_f).$$

This alternate form of Definition 3.7.1 will be used to prove the main theorem in Section 3.8.

Exercise Set 3.7

1. For what numbers x is f continuous if $f(x) = (x^2 - x - 2)/(x^2 - x - 6)$?

2. If $f(x) = (x^3 - 8)/(x - 2)$ for $x \neq 2$, how can $f(2)$ be defined so that f is continuous at 2?

3. If $f(x) = (x^2 + 2x - 1)/(2x + 1)$ for $x \neq -\frac{1}{2}$, can $f(-\frac{1}{2})$ be defined so that f is continuous at $-\frac{1}{2}$? Why?

4. If $f(x) = (P(x))^n$ where P is a polynomial function and n is a positive integer, prove that f is everywhere continuous.

In Exercises 5–11 determine if the given function is continuous at the given number. Justify your answers.

5. $f(x) = \begin{cases} \dfrac{x^2 - x - 6}{x - 3} & \text{if } x \neq 3 \\ 2 & \text{if } x = 3 \end{cases}$ at 3

6. $g(x) = \begin{cases} \dfrac{x^2 - x - 6}{x - 3} & \text{if } x \neq 3 \\ 5 & \text{if } x = 3 \end{cases}$ at 3

7. $f(x) = \begin{cases} x + 1 & \text{if } x \geq 0 \\ -2x & \text{if } x < 0 \end{cases}$ at 0

8. $G(x) = \begin{cases} \dfrac{2(c + x)^3 - 2c^3}{x} & \text{if } x \neq 0 \\ 6c^2 & \text{if } x = 0 \end{cases}$ at 0

9. $h(x) = \begin{cases} [x + 3] & \text{if } x < -1 \\ |x| & \text{if } x \geq -1 \end{cases}$ at -1

10. $f(x) = |[x - 2]|$ at 2

11. $f(x) = \begin{cases} \dfrac{2x - 2}{|x - 1|} & \text{if } x \neq 1 \\ 0 & \text{if } x = 1 \end{cases}$ at 1

12. Prove Theorem 3.7.3.

13. Prove Theorem 3.7.4.

3.8   Composite Function Limit Theorem

We now consider

$$\lim_{x \to -2} \sqrt{9 - x^2}. \tag{3.59}$$

This limit cannot be evaluated using theorems that have already been discussed;

however, it should be recognized that the function F defined by

$$F(x) = \sqrt{9 - x^2}$$

is the composite function $f \circ g$ (see Section 3.2) where f and g are defined by

$$f(x) = \sqrt{x} \quad \text{and} \quad g(x) = 9 - x^2.$$

The limit (3.59) and limits of composite functions, in general, can be evaluated by the following theorem.

3.8.1   Theorem   (Composite Function Limit Theorem)

If $\lim_{x \to c} g(x) = L$, *and* f *is continuous at* L, *then*

$$\lim_{x \to c} (f \circ g)(x) = \lim_{x \to c} f(g(x)) = f(\lim_{x \to c} g(x)) = f(L).$$

PROOF   Let ϵ be an arbitrary positive number. If f is continuous at L, then by Definition 3.7.5

$$|f(s) - f(L)| < \epsilon \tag{3.60}$$

when $|s - L| < \delta$. Also if $\lim_{x \to c} g(x) = L$, there is a $\gamma > 0$ such that

$$|g(x) - L| < \delta \tag{3.61}$$

when $0 < |x - c| < \gamma$. For every x such that $0 < |x - c| < \gamma$, (3.61) follows and s can be replaced by $g(x)$ in (3.60) to obtain

$$|f(g(x)) - f(L)| < \epsilon.$$

The conclusion then follows from this assertion.

The remaining theorems in this section are corollaries of Theorem 3.8.1.

3.8.2   Theorem

Suppose $\lim_{x \to c} g(x) = L$ *and* $L^{p/q}$ *is a real number where* $q > 0$ *and* p *are integers. Then*

$$\lim_{x \to c} (g(x))^{p/q} = L^{p/q}.$$

PROOF   Letting $f(s) = s^{p/q}$, it will be recalled from the comments preceding Theorem 3.7.3 that f is continuous at L. The conclusion then follows from Theorem 3.8.1.

Example 1   Find $\lim_{x \to -2} \sqrt{9 - x^2}$.

SOLUTION   Since $\lim_{x \to -2} (9 - x^2) = 5$,   by   Theorem   3.8.2 $\lim_{x \to -2} \sqrt{9 - x^2} = \sqrt{5}$.

Example 2   Find $\lim\limits_{x \to 0} \left(\dfrac{x + 2}{1 - x} \right)^{4/3}$.

SOLUTION   Since $\lim\limits_{x \to 0} \left(\dfrac{x + 2}{1 - x} \right) = 2$, by Theorem 3.8.2

$$\lim_{x \to 0} \left(\frac{x + 2}{1 - x} \right)^{4/3} = 2^{4/3}.$$

Example 3   If $f(x) = \sqrt{(3 - x)/x}$, at what numbers c is f continuous?

SOLUTION   Since the only numbers for which f can be continuous are those in $\mathcal{D}_f$, we shall first determine this set. Now $x \in \mathcal{D}_f$ if and only if $(3 - x)/x \geq 0$. Upon solving this inequality, it is seen that $\mathcal{D}_f = (0, 3]$. If $c \in (0, 3]$, then

$$\lim_{x \to c} \frac{3 - x}{x} = \frac{3 - c}{c},$$

and by Theorem 3.8.2,

$$\lim_{x \to c} \sqrt{\frac{3 - x}{x}} = \sqrt{\frac{3 - c}{c}}.$$

Thus f is continuous on the interval $(0, 3]$.

3.8.3   Theorem

If $\lim_{x \to c} g(x) = L$, *then* $\lim_{x \to c} |g(x)| = |L|$.

The proof of Theorem 3.8.3 is left as an exercise.

Example 4   Find $\lim_{x \to -1} |x^2 + 3x - 6|$.

SOLUTION   Since $\lim_{x \to -1} (x^2 + 3x - 6) = -8$, by Theorem 3.8.3 $\lim_{x \to -1} |x^2 + 3x - 6| = |-8| = 8$.

The next theorem is useful in discussing the continuity of a composite function. Its proof is also left as an exercise.

3.8.4   Theorem

If g *is continuous at* c, *and* f *is continuous at* $g(c)$, *then* $f \circ g$ *is continuous at* c.

Exercise Set 3.8

1. Evaluate the following limits, or show that they fail to exist.

   (a) $\lim\limits_{x \to -1} \sqrt{4 - x^2}$

   (b) $\lim\limits_{y \to 3} \sqrt[3]{y^2 + 2y + 3}$

   (c) $\lim\limits_{x \to 4} \sqrt[3]{\dfrac{2x^2 - 3x}{2x + 4}}$

   (d) $\lim\limits_{x \to -3} \sqrt{\dfrac{x^2 - 4x + 5}{x^2 - 5}}$

   (e) $\lim\limits_{x \to 2} \dfrac{\sqrt{2 - x}}{2 - x}$

   (f) $\lim\limits_{x \to 3} \sqrt[3]{\dfrac{x^2 - 9}{x^2 - 4x + 3}}$

   (g) $\lim\limits_{t \to -2} \sqrt{\left| \dfrac{2t^2 + 4t}{2t^2 + 11t + 7} \right|}$

   (h) $\lim\limits_{h \to 0} \dfrac{\sqrt{2x + 2h + 1} - \sqrt{2x + 1}}{h}$ HINT: Rationalize the numerator.

   (i) $\lim\limits_{h \to 0} \dfrac{\sqrt[3]{1 + h} - \sqrt[3]{1 - h}}{2h}$

2. For what numbers is each of the following functions continuous?

   (a) f where $f(x) = \dfrac{1}{\sqrt[3]{8 - 2x - x^2}}$

   (b) g where $g(x) = \sqrt{\dfrac{x^2 - x}{x^2 - 1}}$

   (c) F where $F(x) = \sqrt[4]{\dfrac{8 + 10x - 3x^2}{x^2 - x - 12}}$

3. Prove Theorem 3.8.3.

4. Prove Theorem 3.8.4.

5. Show by a counter-example that the following statement is false: If $\lim_{s \to L} f(s) = M$, and $\lim_{x \to c} g(x) = L$, then $\lim_{x \to c} f(g(x)) = M$.

3.9 Infinite Limits

Suppose g is the function defined by

$$g(x) = \frac{2}{x - 3} \tag{3.62}$$

From Figure 3–16 and the following table it is clear that if $x > 3$ but sufficiently close to 3, $g(x)$ can be made as large as desired and hence $1/g(x)$ can be made to differ from 0 by as little as desired.

x	5	4	3.5	3.1	3.01	3.001
$g(x)$	1	2	4	20	200	2000
$\frac{1}{g(x)}$	1	$\frac{1}{2}$	$\frac{1}{4}$	$\frac{1}{20}$	$\frac{1}{200}$	$\frac{1}{2000}$

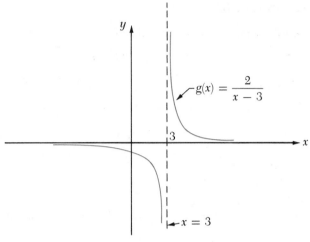

Figure 3–16

We then say the function g defined by (3.62) has a right-hand limit of $+\infty$ at 3 and express this situation by writing

$$\lim_{x\to3^+} \frac{2}{x-3} = +\infty.$$

This right-hand limit of g, of course, does not exist in the sense of Definition 3.6.1.

The one-sided *infinite limits* of a function f at a number c are denoted by the equations below. To the right of each equation there is a sentence which expresses the infinite limit in ordinary language.

$\lim_{x\to c^+} f(x) = +\infty$ "The right-hand limit of f at c is positive infinity."

$\lim_{x\to c^-} f(x) = +\infty$ "The left-hand limit of f at c is positive infinity."

$\lim_{x\to c^+} f(x) = -\infty$ "The right-hand limit of f at c is negative infinity."

$\lim_{x\to c^-} f(x) = -\infty$ "The left-hand limit of f at c is negative infinity."

3.9.1   Definition

(a) $\lim_{x\to c^+} f(x) = +\infty$ iff $\lim_{x\to c^+} 1/f(x) = 0$ and $f(x) > 0$ for every x in some open interval (c, b).

(b) $\lim_{x\to c^+} f(x) = -\infty$ iff $\lim_{x\to c^+} 1/f(x) = 0$ and $f(x) < 0$ for every x in some open interval (c, b).

3.9.2   Definition

(a) $\lim_{x\to c^-} f(x) = +\infty$ iff $\lim_{x\to c^-} 1/f(x) = 0$ and $f(x) > 0$ for every x in some open interval (a, c).

(b) $\lim_{x \to c^-} f(x) = -\infty$ iff $\lim_{x \to c^-} 1/f(x) = 0$ and $f(x) < 0$ for every x in some open interval (a, c).

Example 1   Prove that $\lim\limits_{x \to 3^+} \dfrac{2}{(x - 3)} = +\infty$ (Figure 3–16).

   PROOF   Since $\lim_{x \to 3^+} [(x - 3)/2] = 0$ and $2/(x - 3) > 0$ for every $x > 3$, by Definition 3.9.1(a) $\lim_{x \to 3^+} [2/(x - 3)] = +\infty$.

Example 2   Prove that $\lim\limits_{x \to 3^-} \dfrac{2}{(x - 3)} = -\infty$.

   PROOF   Since $\lim_{x \to 3^-} [(x - 3)/2] = 0$ and $2/(x - 3) < 0$ for every $x < 3$, by Definition 3.9.2(b) $\lim_{x \to 3^-} [2/(x - 3)] = -\infty$.

   The line $x = 3$ is called a *vertical asymptote* of the graph of the function g given by (3.62). The subject of asymptotes will be studied in detail in Section 5.8, which deals with the sketching of graphs of equations using concepts from calculus.

   If $\lim_{x \to c^+} f(x) = L \neq 0$ and $\lim_{x \to c^+} g(x) = 0$, the function f/g *may have* a right-hand infinite limit at c. Similarly, if in the preceding sentence we replace $x \to c^+$ by $x \to c^-$, f/g may have a left-hand infinite limit at c.

Example 3   Find $\lim\limits_{x \to -2^+} \dfrac{1 - x}{x^2 - 4}$.

   SOLUTION   Since $\lim_{x \to -2^+} (1 - x) = 3$ and $\lim_{x \to -2^+} (x^2 - 4) = 0$, we suspect from the preceding discussion that $\lim_{x \to -2^+} [(1 - x)/(x^2 - 4)]$ might be $+\infty$ or $-\infty$. Since

$$\lim_{x \to -2^+} \frac{x^2 - 4}{1 - x} = 0$$

and also

$$\frac{1 - x}{x^2 - 4} = \frac{1 - x}{(x - 2)(x + 2)} < 0$$

if $x \in (-2, 1)$, by Definition 3.9.1(b)

$$\lim_{x \to -2^+} \frac{1 - x}{x^2 - 4} = -\infty.$$

   It is also useful to consider the *two-sided infinite limits* of a function f at a number c, which are denoted as follows:

$\lim\limits_{x \to c} f(x) = +\infty$, which is read "The limit of f at c is positive infinity."

$\lim\limits_{x \to c} f(x) = -\infty$, which is read "The limit of f at c is negative infinity."

3.9.3 Definition

Suppose f is defined, at least, throughout some open interval having the number c as an endpoint.

(a) $\lim_{x \to c} f(x) = +\infty$ iff $\lim_{x \to c} 1/f(x) = 0$, and there is some open interval I containing c such that $f(x) > 0$ for every $x \neq c$ in $I \cap \mathfrak{D}_f$.

(b) $\lim_{x \to c} f(x) = -\infty$ iff $\lim_{x \to c} 1/f(x) = 0$, and there is some open interval I containing c such that $f(x) < 0$ for every $x \neq c$ in $I \cap \mathfrak{D}_f$.

Example 4 Find $\lim_{x \to 2} \dfrac{1}{(x-2)^2}$ (Figure 3–17).

SOLUTION Since $\lim_{x \to 2} (x-2)^2 = 0$ and $1/(x-2)^2 > 0$ for every $x \neq 2$, by Definition 3.9.3(a) we have $\lim_{x \to 2} [1/(x-2)^2] = +\infty$.

The last theorem of this section relates the one-sided and two-sided infinite limits of a function when they occur.

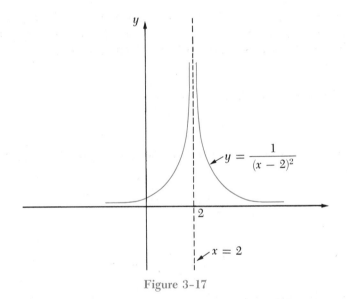

$$y = \frac{1}{(x-2)^2}$$

$x = 2$

Figure 3–17

3.9.4 Theorem

Let $f(x)$ be defined for every $x \neq c$ in some open interval containing c.

(a) $\lim_{x \to c} f(x) = +\infty$ *iff* $\lim_{x \to c+} f(x) = +\infty$ *and* $\lim_{x \to c-} f(x) = +\infty$.

(b) $\lim_{x \to c} f(x) = -\infty$ *iff* $\lim_{x \to c+} f(x) = -\infty$ *and* $\lim_{x \to c-} f(x) = -\infty$.

PROOF OF (a) From Definition 3.9.3 $\lim_{x \to c} f(x) = +\infty$ if and only if (i) $\lim_{x \to c} 1/f(x) = 0$ and (ii) $f(x) > 0$ for every $x \neq c$ in some open interval containing c. Since f is defined at every number except c in some open interval containing c, (i) is equivalent to saying that $\lim_{x \to c+} 1/f(x) = 0 = \lim_{x \to c-} 1/f(x)$ and (ii) is equivalent to saying that f assumes positive values on some open intervals (a, c) and (c, b). Thus (i) and (ii) occur if and only if $\lim_{x \to c+} f(x) = +\infty$ and $\lim_{x \to c-} f(x) = +\infty$, which proves the theorem.

The proof of (b) is analogous and is therefore omitted.

Exercise Set 3.9

In Exercises 1–4 obtain (a) $\lim_{x \to c+} f(x)$ (b) $\lim_{x \to c-} f(x)$ (c) $\lim_{x \to c} f(x)$ if they exist as infinite limits.

1. $f(x) = \dfrac{x + 1}{x - 3}$ $c = 3$
   $\qquad\qquad\qquad$
   2. $f(x) = \dfrac{2x + 1}{x^2 - x - 2}$ $c = 2$

3. $f(x) = \dfrac{2x - 1}{(x - 1)^4}$ $c = 1$
   $\qquad\qquad\qquad$
   4. $f(x) = \dfrac{2}{x} + \dfrac{1}{x^2}$ $c = 0$

5. (a) Find $\displaystyle\lim_{x \to -2+} \dfrac{\sqrt{4 - x^2}}{x + 2}$.

   (b) What can you say about $\displaystyle\lim_{x \to -2-} \dfrac{\sqrt{4 - x^2}}{x + 2}$?

   (c) What can you say about $\displaystyle\lim_{x \to -2} \dfrac{\sqrt{4 - x^2}}{x + 2}$?

6. Suppose f is defined at every number in some open interval (c, b). Prove that $\lim_{x \to c+} f(x) = +\infty$ if and only if for every $B > 0$ there is a $\delta > 0$ such that $f(x) > B$ if $c < x < c + \delta$.

7. Suppose f is defined at every number in some open interval (a, c). State and prove a theorem suggested by Exercise 6 that begins: $\lim_{x \to c-} f(x) = -\infty$ if and only if . . . .

8. If $\lim_{x \to c} f(x) = +\infty$, prove that f cannot have a limit at c of the type given by Definition 3.3.1. HINT: Use an indirect argument.

4. Derivative of a Function

4.1 Tangent to a Graph

In this section we will define the *tangent* (*line*) to a graph of a function at a point on the graph. The reader will subsequently note that the definition of this line involves a limit that also appears in Section 4.2 in a discussion of velocity and throughout the text in the notion of the *derivative* of a function. Therefore, the topic of the tangent to a graph has been selected to introduce this chapter on the derivative.

It will be recalled from elementary geometry that a tangent to a circle is a line that intersects the circle in exactly one point. However, for our purposes, such a definition would be unsatisfactory since a line that seems perfectly plausible as the tangent to a graph of a function at a point might also intersect the graph in another point. For example, the line that intuitively seems like the tangent to the graph of $y = x^3$ at the point $(1, 1)$ (Figure 4–1) also appears to intersect the graph in a second point in the third quadrant. Also, we note that the y axis, which intersects the parabola $y = x^2$ only at the origin, would hardly be considered a tangent to the parabola at that point.

Suppose f is a function that is continuous at a number c in its domain and is defined on some interval containing c. Intuition suggests that if the tangent T to the graph of f at the point $(c, f(c))$ exists, then the slope of T can be approximated by the slope of a secant L through $(c, f(c))$ and a nearby point on the graph, $(c + h, f(c + h))$ (Figure 4–2). Here, we could have $h > 0$, as indicated in Figure 4–2, or $h < 0$.

Intuition also suggests that the slope of L, given by

$$\frac{f(c - h) - f(c)}{h}$$

can be made to differ from the slope of T by as little as we please if $h \neq 0$ is taken sufficiently close to 0.

It is therefore reasonable to *define* the slope of the tangent to the graph of f at the point $(c, f(c))$ as

$$\lim_{h \to 0} \frac{f(c + h) - f(c)}{h} \tag{4.1}$$

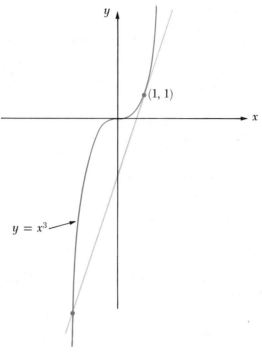

Figure 4-1

if this limit exists. This limit is also called the *slope of the graph* of f at $(c, f(c))$.

If, however, the limit (4.1) is $+\infty$ or $-\infty$, the secant L approaches a vertical position as h approaches 0, and we say the line $x = c$ is a *vertical tangent* to the graph of f at $(c, f(c))$. Vertical tangents will be discussed in more detail in Section 4.9.

The preceding discussion can be summarized with the following definition of a tangent to a graph.

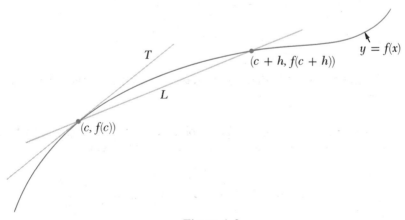

Figure 4-2

4.1.1   Definition

Let f be continuous at c and be defined on some interval containing c. The *tangent* to the graph of f at the point $(c, f(c))$ is

(a) the line through $(c, f(c))$ with slope $\displaystyle\lim_{h \to 0} \frac{f(c + h) - f(c)}{h}$

or

(b) the line $x = c$ if $\displaystyle\lim_{h \to 0} \frac{f(c + h) - f(c)}{h}$ is $+\infty$ or $-\infty$.

The point $(c, f(c))$, here, is called the *point of tangency.*

If $\lim_{h \to 0} (f(c + h) - f(c))/h$ fails to exist, but is not $+\infty$ or $-\infty$, the graph of f has no tangent at $(c, f(c))$.

Example 1   Find an equation for the tangent to the graph of the equation $y = x^3$ at $(1, 1)$ (Figure 4-1).

SOLUTION   Letting $f(x) = x^3$,

$$\frac{f(1 + h) - f(1)}{h} = \frac{(1 + h)^3 - 1}{h} = \frac{3h + 3h^2 + h^3}{h}$$

$$= 3 + 3h + h^2$$

Then by Definition 4.1.1 the slope of the tangent is

$$\lim_{h \to 0} \frac{f(1 + h) - f(1)}{h} = \lim_{h \to 0} (3 + 3h + h^2)$$

$$= 3.$$

An equation of the tangent using the point-slope form is

$$y - 1 = 3(x - 1),$$

or

$$3x - y = 2.$$

Example 2   Find the slope of the graph of the equation $y = 1/(x - 2)$ at the point $(3, 1)$ (Figure 4-3).

SOLUTION   The slope of the graph at a point is the slope of the tangent to the graph at the point. Letting $f(x) = 1/(x - 2)$, we obtain

$$\frac{f(3 + h) - f(3)}{h} = \frac{\dfrac{1}{1 + h} - 1}{h} = -\frac{h}{h(1 + h)}$$

$$= -\frac{1}{1 + h}.$$

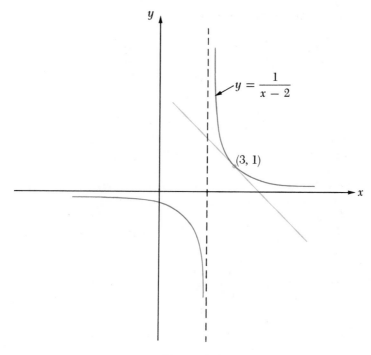

Figure 4–3

Then by Definition 4.1.1 the slope of the tangent is

$$\lim_{h \to 0} \frac{f(3 + h) - f(3)}{h} = \lim_{h \to 0} - \frac{1}{1 + h},$$
$$= -1,$$

which is the slope of the graph at $(3, 1)$.

Exercise Set 4.1

In Exercises 1–9 obtain equations for the tangent to the graphs at the given points. Sketch the graphs of the given equation and the required tangent.

1. $y = x^2 + 1$ at $(-1, 2)$

2. $y = \dfrac{2}{x}$ at $\left(\dfrac{1}{2}, 4\right)$

3. $y = \sqrt{x}$ at $\left(\frac{9}{4}, \frac{3}{2}\right)$

4. $y = (x - 2)^2$ at $(3, 1)$

5. $y = \dfrac{x^2}{x^2 + 1}$ at $\left(-2, \dfrac{4}{5}\right)$

6. $y = -x^2 + 3x - 1$ at $(1, 1)$

7. $y = \sqrt{x + 1}$ at $(3, 2)$

8. $y = \sqrt[3]{x}$ at $(1, 1)$

9. $y = \begin{cases} x^2 - x & \text{if } x \le 2 \\ 3x - 4 & \text{if } x > 2 \end{cases}$ at $(2, 2)$

10. At what point(s) does the graph of $y = -x^2 + 3x - 1$ have a horizontal tangent?

11. At what point(s) does the graph of $y = x^3 - 6x^2$ have a horizontal tangent?

4.2 Instantaneous Velocity

In this section the limit $\lim_{h \to 0} (f(c + h) - f(c)/h$ is encountered in a different setting. From physics we know that if a body falls from rest, the distance s in feet it travels in t seconds can be expressed fairly accurately by the equation

$$s = 16t^2. \tag{4.2}$$

Suppose the following question were to be asked: "How fast is the body traveling at the *instant* $t = 3$ seconds?" Before such a question can be answered, one must know how to define the velocity of the falling body at an instant of time, a task which is beyond the realm of pre-calculus mathematics.

We will attempt to answer the question using our intuitive notion of what is meant by velocity at an instant of time, employing an analogous approach to that used to define the slope of a tangent in Section 4.1. First note that the *average velocity* of the falling body in the time interval between $t = 3$ and $t = 3 + h$ is given by

$$\text{Average velocity} = \frac{\text{Distance fallen}}{\text{Elapsed time}} = \frac{16(3 + h)^2 - 16 \cdot 3^2}{h}$$

$$= 96 + 16h \text{ ft/sec.} \tag{4.3}$$

Here h may be positive or negative. Intuition suggests that (4.3) can be made to differ from the velocity at $t = 3$ by as little as we please if $h \neq 0$ is taken sufficiently small. Hence from (4.3), we actually *define* the velocity of the falling body at $t = 3$ as $\lim_{h \to 0} (96 + 16h) = 96$ ft/sec and the velocity of a body falling in accordance with (4.2) at any time $t = c$ as

$$\lim_{h \to 0} \frac{16(c + h)^2 - 16c^2}{h} = \lim_{h \to 0} (32c + 16h) = 32c \text{ ft/sec.}$$

A body falling from rest is one illustration of straight line motion. Straight line motion of a particle is often described by an equation of the form

$$s = f(t) \tag{4.4}$$

where s is the coordinate of the location of the particle on a number line at time t. Equation (4.4) would then be termed an *equation of motion* of the particle. The foregoing discussion of a falling body suggests the following general definition for the velocity associated with straight line motion.

4.2.1 Definition

If a particle moves in a straight line in accordance with the equation of motion (4.4), its *velocity* v at time $t = c$ is given by

$$v = \lim_{h \to 0} \frac{f(c + h) - f(c)}{h}$$

if this limit exists. The velocity v defined here is often termed the *instantaneous velocity* at $t = c$.

From Definition 4.1.1 the velocity v at time $t = c$ given in Definition 4.2.1 is the slope of the tangent to the graph of Equation (4.4) at the point $(c, f(c))$.

Example 1 A particle moves in a straight line in accordance with the equation of motion

$$s = t^2 - 4t - 3.$$

Find (a) the velocity of the particle at $t = 5$, and (b) the time when the velocity of the particle is 0.

SOLUTION From Definition 4.2.1 the velocity v at any time t is given by

$$v = \lim_{h \to 0} \frac{f(t + h) - f(t)}{h}$$

$$= \lim_{h \to 0} \frac{[(t + h)^2 - 4(t + h) - 3] - [t^2 - 4t - 3]}{h}$$

$$= \lim_{h \to 0} \frac{2th + h^2 - 4h}{h} = \lim_{h \to 0} (2t + h - 4)$$

$$= 2t - 4. \tag{4.5}$$

Letting $t = 5$ in (4.5), the velocity $v = 6$ in part (a) is obtained. To solve (b), we solve the equation $2t - 4 = 0$ thereby obtaining $t = 2$.

A more complete discussion of straight line motion and a wider assortment of exercises is contained in Section 6.4. By that time the student will have learned certain formulas that facilitate the calculation of limits of the type $\lim_{h \to 0} (f(t + h) - f(t))/h$ when they exist.

Exercise Set 4.2

In Exercises 1–4 a particle moves along a number line in accordance with the given equation of motion. Find the velocity at the given instant of time.

1. $s = 3t - 2$, $t = 4$

2. $s = \dfrac{1}{t + 2}$, $t = 2$

3. $s = \sqrt{t - 3}$, $t = 5$

4. $s = t^3 - 3t^2 - 30t + 5$, $t = -2$

5. Suppose the equation of motion of a body falling from rest is given by $s = 16t^2$ where s and t have the meanings given at the beginning of Section 4.2. Find, using Definition 4.2.1:

   (a) the velocity of the body at time $t = 4$ sec

   (b) the time t when the velocity of the body is 80 ft/sec

   (c) the velocity of the body after it has fallen 108 ft

   (d) the distance the body falls during the time interval when its velocity increases from $v = 80$ ft/sec to $v = 96$ ft/sec

6. A particle moves in a straight line in such a way that its distance s (ft) from the origin at time t (sec) is given by the following table:

t	0	1	2	3	4	5	6	7
s	52.0	87.1	124.8	145.7	155.0	152.9	137.3	104.6

Plot the points corresponding to the ordered pairs (t, s) given by this table and sketch a smooth curve through these points. From this sketch estimate the velocity of the particle at the following times: (a) $t = 0$, (b) $t = 4$, (c) $t = 6.5$. At what time t does the particle have velocity 0?

4.3   Definition of the Derivative

We have seen the usefulness of the limit

$$\lim_{h \to 0} \frac{f(c + h) - f(c)}{h}$$

in defining the slope of a tangent and the instantaneous velocity of a particle in straight line motion (Definitions 4.1.1 and 4.2.1). Because of its importance in these and other applications, the limit is formalized in the definition of the derivative, which is the basic notion in the study of differential calculus.

4.3.1   Definition

The *derivative* of a function f is the function f' defined by writing

$$f'(x) = \lim_{h \to 0} \frac{f(x + h) - f(x)}{h}. \tag{4.6}$$

The domain of f', therefore, is the set of all numbers $x \in \mathcal{D}_f$ for which the limit (4.6) exists.

If x is in $\mathcal{D}_{f'}$, we say f is *differentiable at x.* $f'(x)$ (which is read "f prime at x") is called the *derivative of f at x.*

In view of Definition 4.3.1, Definitions 4.1.1 and 4.2.1 could be rephrased by replacing $\lim_{h \to 0} (f(c + h) - f(c))/h$ by $f'(c)$.

The notation $D_x(f(x))$ is frequently used in place of $f'(x)$, and when x and y are related by the equation $y = f(x)$, the symbols y' and $D_x y$ are often used instead of $f'(x)$. We then sometimes speak of y' or $D_x y$ as the *derivative of y with respect to x,* or the *rate of change of y with respect to x.* The latter terminology is appropriate since $D_x y$ is the limiting value of $(f(x + h) - f(x))/h$, the ratio of the change in y to the change in x.

The process of obtaining $f'(x)$ from a given $f(x)$ is termed *differentiation,* or more specifically, *differentiation with respect to x.* Some examples in which differentiation is performed using Definition 4.3.1 follow.

Example 1   Find $f'(x)$ if $f(x) = 2x^2 - 3x + 1$. Also find $f'(2)$.

SOLUTION   We note that $f(x + h) = 2(x + h)^2 - 3(x + h) + 1$. Hence

$$\frac{f(x + h) - f(x)}{h} = \frac{[2(x + h)^2 - 3(x + h) + 1] - (2x^2 - 3x + 1)}{h}$$

$$= \frac{4hx + 2h^2 - 3h}{h}$$

$$= 4x + 2h - 3 \quad \text{if } h \neq 0.$$

Then by Theorem 3.5.1

$$f'(x) = \lim_{h \to 0} \frac{f(x + h) - f(x)}{h} = \lim_{h \to 0} (4x + 2h - 3)$$

$$= 4x - 3. \tag{4.7}$$

In particular from (4.7), $f'(2) = 4 \cdot 2 - 3 = 5$.

Example 2   Find $g'(x)$ if $g(x) = x/(x + 2)$. Also, find $g'(-1)$.

SOLUTION   Note that $\mathfrak{D}_g = \{x : x \neq -2\}$. Since $g(x + h) = \dfrac{x + h}{x + h + 2}$,

$$\frac{g(x + h) - g(x)}{h} = \frac{\dfrac{x + h}{x + h + 2} - \dfrac{x}{x + 2}}{h}$$

$$= \frac{(x + h)(x + 2) - x(x + h + 2)}{h(x + 2)(x + h + 2)}$$

$$= \frac{2h}{h(x + 2)(x + h + 2)}$$

$$= \frac{2}{(x + 2)(x + h + 2)} \quad \text{if } h \neq 0.$$

Again using Theorem 3.5.1,

$$g'(x) = \lim_{h \to 0} \frac{g(x + h) - g(x)}{h} = \lim_{h \to 0} \frac{2}{(x + 2)(x + h + 2)}$$

$$= \frac{2}{(x + 2)^2} \quad \text{if } x \neq -2. \tag{4.8}$$

Also, from (4.8)

$$g'(-1) = \frac{2}{(-1 + 2)^2} = 2.$$

Example 3   Find $\phi'(x)$ if $\phi(x) = \sqrt[q]{x}$. Does $\phi'(0)$ exist?

SOLUTION   Here $\mathfrak{D}_\phi = \{x : x \in R\}$ if q is odd and $\{x : x \geq 0\}$ if q is even.

Note that if $x^{1/q}$ exists, $x \neq 0$, and i is any positive integer, then

$$x^{(q-i)/q} = \frac{x}{(x^{1/q})^i}$$

also, exists. We first write

$$\frac{\phi(x+h) - \phi(x)}{h} = \frac{\sqrt[q]{x+h} - \sqrt[q]{x}}{h}. \tag{4.9}$$

We cannot directly obtain $\phi'(x) = \lim_{h \to 0} (\phi(x+h) - \phi(x))/h$ from (4.9) since the limits at $h = 0$ of both the numerator and denominator of the right member of (4.9) are zero. To express $(\phi(x+h) - \phi(x))/h$ in a form from which we can readily obtain $\phi'(x)$, consider the identity (compare (3.38))

$$a - b = (a^{1/q} - b^{1/q})(a^{(q-1)/q} + a^{(q-2)/q}b^{1/q} + a^{(q-3)/q}b^{2/q} + \cdots + b^{(q-1)/q}),$$

from which is obtained

$$a^{1/q} - b^{1/q} = \frac{a - b}{a^{(q-1)/q} + a^{(q-2)/q}b^{1/q} + a^{(q-3)/q}b^{2/q} + \cdots + b^{(q-1)/q}}.$$

Then letting $a = x + h$ and $b = x$, we have

$$\sqrt[q]{x+h} - \sqrt[q]{x}$$

$$= \frac{h}{(x+h)^{(q-1)/q} + (x+h)^{(q-2)/q}x^{1/q} + (x+h)^{(q-3)/q}x^{2/q} + \cdots + x^{(q-1)/q}}, \tag{4.10}$$

and substituting from (4.10) into (4.9)

$$\frac{\phi(x+h) - \phi(x)}{h} = \frac{1}{(x+h)^{(q-1)/q} + (x+h)^{(q-2)/q}x^{1/q} + \cdots + x^{(q-1)/q}} \quad \text{if } h \neq 0.$$

Thus

$$\phi'(x) = \lim_{h \to 0} \frac{\phi(x+h) - \phi(x)}{h}$$

$$= \lim_{h \to 0} \frac{1}{(x+h)^{(q-1)/q} + (x+h)^{(q-2)/q}x^{1/q} + \cdots + x^{(q-1)/q}}. \tag{4.11}$$

Each of the q terms in the denominator of the right member of (4.11) is of the form $(x+h)^{(q-i)/q}x^{(i-1)/q}$ where $i = 1, 2, \ldots, q$. By Theorems 3.4.3(c) and 3.8.2

$$\lim_{h \to 0} (x+h)^{(q-i)/q}x^{(i-1)/q} = x^{(q-i)/q}x^{(i-1)/q} = x^{(q-1)/q}, \tag{4.12}$$

and from (4.11) and (4.12)

$$\phi'(x) = \frac{1}{qx^{(q-1)/q}} \quad \text{if } x \in \mathcal{D}_\phi \text{ and } x \neq 0.$$

From Definition 4.3.1, even though $\phi(0)$ is defined,

$$\phi'(0) = \lim_{h \to 0} \frac{\sqrt[q]{h} - 0}{h} = \lim_{h \to 0} \frac{1}{h^{(q-1)/q}},$$

which does not exist.

Example 4 Find $f'(x)$ if

$$f(x) = \begin{cases} 2x & \text{if } x \geq 1, \\ 3x^2 - 4x + 3 & \text{if } x < 1. \end{cases}$$

Does $f'(1)$ exist?

SOLUTION If $x > 1$, then for all $h \neq 0$ such that $x + h > 1$

$$\frac{f(x + h) - f(x)}{h} = \frac{2(x + h) - 2x}{h} = 2.$$

Hence

$$f'(x) = \lim_{h \to 0} \frac{f(x + h) - f(x)}{h} = \lim_{h \to 0} 2$$

$$= 2 \quad \text{if } x > 1.$$

If, however, $x < 1$, then for all $h \neq 0$ such that $x + h < 1$,

$$\frac{f(x + h) - f(x)}{h} = \frac{[3(x + h)^2 - 4(x + h) + 3] - (3x^2 - 4x + 3)}{h}$$

$$= \frac{6hx + 3h^2 - 4h}{h}$$

$$= 6x + 3h - 4.$$

Therefore

$$f'(x) = \lim_{h \to 0} \frac{f(x + h) - f(x)}{h} = \lim_{h \to 0} (6x + 3h - 4)$$

$$= 6x - 4 \quad \text{if } x < 1.$$

So far we have calculated $f'(x)$ for every $x \neq 1$. To investigate the existence of $f'(1)$, we calculate the one-sided limits of $(f(1 + h) - f(1))/h$ at $h = 0$. Note that

$$\lim_{h \to 0^+} \frac{f(1 + h) - f(1)}{h} = \lim_{h \to 0} \frac{2(1 + h) - 2}{h} = \lim_{h \to 0} 2 = 2,$$

and also that

$$\lim_{h \to 0^-} \frac{f(1 + h) - f(1)}{h} = \lim_{h \to 0} \frac{3(1 + h)^2 - 4(1 + h) + 3 - 2}{h}$$

$$= \lim_{h \to 0} \frac{2h + 3h^2}{h}$$

$$= \lim_{h \to 0} (2 + 3h) = 2.$$

Hence by Theorem 3.6.5

$$f'(1) = \lim_{h \to 0} \frac{f(1 + h) - f(1)}{h} = 2.$$

Thus

$$f'(x) = \begin{cases} 2 & \text{if } x \geq 1, \\ 6x - 4 & \text{if } x < 1. \end{cases}$$

From the next theorem we have a relationship between differentiability and continuity of a function which will be useful in our work.

4.3.2 Theorem

If $f'(c)$ exists, then f is continuous at c.

PROOF If $f'(c)$ exists, by definition

$$f'(c) = \lim_{h \to 0} \frac{f(c + h) - f(c)}{h}.$$

Now

$$\lim_{h \to 0} [f(c + h) - f(c)] = \lim_{h \to 0} \left[\frac{f(c + h) - f(c)}{h} \cdot h \right]$$

$$= \lim_{h \to 0} \frac{f(c + h) - f(c)}{h} \cdot \lim_{h \to 0} h = f'(c) \cdot 0$$

$$= 0. \tag{4.13}$$

From (4.13)

$$\lim_{h \to 0} f(c + h) = f(c).$$

Then by Theorem 3.5.4

$$\lim_{x \to c} f(x) = f(c)$$

and the conclusion is obtained.

Stated another way, Theorem 4.3.2 says that *if a function is discontinuous at a number, it is not differentiable there.*
The converse of Theorem 4.3.2 is false. As a counter-example the function ϕ where $\phi(x) = \sqrt[9]{x}$, which is continuous at 0, does not have a derivative at 0 (compare Example 3).
Finally, we mention the following theorem, which is analogous to Theorem 3.5.1. Its proof is left as an exercise.

4.3.3 Theorem

If $F(x) = G(x)$ for every $x \neq c$ in some open interval containing c, and $G'(c)$ exists, then $F'(c) = G'(c)$.

Exercise Set 4.3

In Exercises 1–13 obtain $f'(x)$ from Definition 4.3.1 and find $f'(c)$ for the given value of c.

1. $f(x) = 2x, c = 2$ 　　　　　　　　2. $f(x) = 5 - 3x, c = -1$
3. $f(x) = x^2 + 2x + 4, c = 1$ 　　　　4. $f(x) = 3x^2 - x - 5, c = 2$

5. $f(x) = 4 - x^2$, $c = -2$

6. $f(x) = x^3$, $c = -2$

7. $f(x) = x^4$, $c = \frac{1}{2}$

8. $f(x) = 1/x^2$, $c = -3$

9. $f(x) = \dfrac{1}{x+1}$, $c = -4$

10. $f(x) = \dfrac{x-1}{x-2}$, $c = 4$

11. $f(x) = \sqrt{2x}$, $c = 8$

12. $f(x) = \sqrt[3]{x}$, $c = -1$

13. $f(x) = (x+2)^{-2/3}$, $c = 6$

In each of the Exercises 14–20 a number c is given. For each exercise: (a) find $f'(x)$ if $x \neq c$, (b) determine if f is continuous at c, (c) find $f'(c)$ if it exists, or state why it fails to exist.

14. $f(x) = \begin{cases} x^2 & \text{if } x \leq 2 \\ 3x - 2 & \text{if } x > 2 \end{cases}$, $c = 2$

15. $f(x) = |x|$, $c = 0$

16. $f(x) = x|x|$, $c = 0$

17. $f(x) = \begin{cases} 2 & \text{if } x < 1 \\ 3x - 4 & \text{if } x \geq 1 \end{cases}$, $c = 1$

18. $f(x) = \begin{cases} -2x^2 & \text{if } x < 0 \\ x^3 & \text{if } x \geq 0 \end{cases}$, $c = 0$

19. $f(x) = \begin{cases} |x - 2| & \text{if } x \geq -1 \\ \frac{1}{2}x^2 + \frac{5}{2} & \text{if } x < -1 \end{cases}$, $c = -1$

20. $f(x) = \begin{cases} 2x + 1 & \text{if } x \leq 3 \\ x^2 & \text{if } x > 3 \end{cases}$, $c = 3$

21. State the definition of the derivative of a function in ϵ-δ language.

22. Prove Theorem 4.3.3.

4.4 Differentiation Formulas

At this point in our discussion of the derivative of a function, the student is doubtlessly well aware of the laboriousness of the differentiation process using Definition 4.3.1 directly. We are in somewhat the same situation here as in Section 3.4 when we needed the limit theorems to bypass the tediousness resulting from the use of the ϵ-δ technique to establish limits. The following differentiation formulas, which are derived from Theorem 4.3.1, will facilitate the process of differentiation.

The first formula says that the derivative of a constant function has the value zero at each number in its domain.

4.4.1 Theorem

If k is any number and $f(x) = k$ for every number x, then $f'(x) = 0$, that is,

$$D_x k = 0. \tag{4.14}$$

PROOF   By hypothesis $f(x + h) = k$ for every x and h, so

$$\frac{f(x + h) - f(x)}{h} = \frac{k - k}{h} = 0 \quad \text{if } h \neq 0.$$

Hence

$$f'(x) = \lim_{h \to 0} \frac{f(x + h) - f(x)}{h} = \lim_{h \to 0} 0 = 0.$$

As an illustration of (4.14), $D_x(-3) = 0$.

4.4.2   Theorem

If $f(x) = x^n$ where n is a positive integer, then $f'(x) = nx^{n-1}$, that is,

$$D_x x^n = nx^{n-1}. \tag{4.15}$$

PROOF $\dfrac{f(x + h) - f(x)}{h} = \dfrac{(x + h)^n - x^n}{h},$

and after expanding $(x + h)^n$ by the binomial formula,

$$\frac{f(x + h) - f(x)}{h} = \frac{x^n + nx^{n-1}h + \dfrac{n(n - 1)}{2!}x^{n-2}h^2 + \cdots + h^n - x^n}{h}$$

$$= nx^{n-1} + \frac{n(n - 1)}{2!}x^{n-2}h + \cdots + h^{n-1} \quad \text{if } h \neq 0.$$

Since each term after the first on the right has h as a factor, its limit at $h = 0$ is zero. Hence

$$f'(x) = \lim_{h \to 0} \left[nx^{n-1} + \frac{n(n - 1)}{2!}x^{n-2}h + \cdots + h^{n-1} \right] = nx^{n-1}.$$

As an illustration of (4.15)

$$D_x x^4 = 4x^3.$$

Note that (4.15) is valid for every x when $n > 1$. But when $n = 1$, the formula holds just for every $x \neq 0$. (If $n = 1$ and $x = 0$, the formula fails since 0^0 is undefined.) However, for $n = 1$ one easily obtains

$$D_x x = 1 \tag{4.16}$$

from Definition 4.3.1.

4.4.3   Theorem

Let k be any number and let $f(x) = kg(x)$. Then if $g'(x)$ exists, $f'(x) = kg'(x)$, that is,

$$D_x[kg(x)] = kD_x g(x). \tag{4.17}$$

The proof of Theorem 4.4.3 is left as an exercise.

Example 1 Find $D_x(\frac{4}{3}x^{10})$.

SOLUTION $D_x(\frac{4}{3}x^{10}) = \frac{4}{3}D_x x^{10}$ from **(4.17)**

$$= \frac{4}{3} \cdot 10x^{10-1}$$ from **(4.15)**

$$= \frac{40}{3}x^9$$

We next prove that the derivative of the sum or difference of two differentiable functions is the sum or difference of their derivatives. In this theorem and in the remaining theorems of this chapter it will be tacitly assumed, whenever the derivative of a function at some number x is to be computed, that the function is defined on some open interval having x as an endpoint. Otherwise, the derivative at x could not exist (recall Definition 4.3.1).

4.4.4 Theorem

If $f'(x)$ and $g'(x)$ exist, then $(f \pm g)'(x) = f'(x) \pm g'(x)$, that is,

$$D_x[f(x) \pm g(x)] = D_x f(x) \pm D_x g(x). \qquad (4.18)$$

PROOF Letting $\phi(x) = (f + g)(x) = f(x) + g(x)$,

$$\frac{\phi(x + h) - \phi(x)}{h} = \frac{[f(x + h) + g(x + h)] - [f(x) + g(x)]}{h}$$

$$= \frac{f(x + h) - f(x)}{h} + \frac{g(x + h) - g(x)}{h}.$$

Then

$$\phi'(x) = \lim_{h \to 0} \frac{\phi(x + h) - \phi(x)}{h}$$

$$= \lim_{h \to 0} \frac{f(x + h) - f(x)}{h} + \lim_{h \to 0} \frac{g(x + h) - g(x)}{h}$$

$$= f'(x) + g'(x).$$

The proof that $(f - g)'(x) = f'(x) - g'(x)$ is analogous and is therefore left as an exercise.

The results of Theorem 4.4.4 can be extended to the sum of an arbitrary number of functions.

4.4.5 Theorem

If $f'_1(x), f'_2(x), \ldots, f'_n(x)$ exist, then

$$(f_1 + f_2 + \cdots + f_n)'(x) = f'_1(x) + f'_2(x) + \cdots + f'_n(x),$$

that is,

$$D_x[f_1(x) + f_2(x) + \cdots + f_n(x)] = D_x f_1(x) + D_x f_2(x) + \cdots + D_x f_n(x). \quad (4.19)$$

We omit the proof of Theorem 4.4.5, which follows from Theorem 4.4.4 using mathematical induction.

Example 2 Find y' if $y = x^3 - \frac{5}{4}x^2 + \sqrt{2}x - \frac{3}{7}$.

SOLUTION $y' = D_x x^3 + D_x(-\frac{5}{4}x^2) + D_x(\sqrt{2}x) + D_x(-\frac{3}{7})$ from **(4.19)**
$$= 3x^2 - \frac{5}{4}D_x x^2 + \sqrt{2}D_x x$$
$$= 3x^2 - \frac{5}{2}x + \sqrt{2}$$

The following formula for the derivative of the product of two functions is probably unexpected!

4.4.6   Theorem

If $f'(x)$ and $g'(x)$ exist, then

$$(fg)'(x) = f(x)g'(x) + g(x)f'(x),$$

that is,

$$D_x[f(x)g(x)] = f(x)D_x g(x) + g(x)D_x f(x). \quad (4.20)$$

PROOF   Letting $\phi(x) = (fg)(x) = f(x)g(x)$,

$$\frac{\phi(x + h) - \phi(x)}{h} = \frac{f(x + h)g(x + h) - f(x)g(x)}{h}.$$

To obtain a more convenient form for $(\phi(x + h) - \phi(x))/h$, we subtract and add $f(x + h)g(x)$ in the numerator of the right member of this equation. Then

$$\frac{\phi(x + h) - \phi(x)}{h}$$

$$= \frac{f(x + h)g(x + h) - f(x + h)g(x) + f(x + h)g(x) - f(x)g(x)}{h}$$

$$= f(x + h)\frac{g(x + h) - g(x)}{h} + g(x)\frac{f(x + h) - f(x)}{h}. \quad (4.21)$$

By hypothesis $f'(x)$ exists, and hence f is continuous at x by Theorem 4.3.2. From Theorem 3.5.4 we then have

$$\lim_{h \to 0} f(x + h) = f(x). \quad (4.22)$$

If we take limits at $h = 0$ using (4.22), we obtain from (4.21)

$$\phi'(x) = \lim_{h \to 0} \left[f(x + h)\frac{g(x + h) - g(x)}{h} + g(x)\frac{f(x + h) - f(x)}{h} \right]$$

$$= f(x)g'(x) + g(x)f'(x).$$

Thus the derivative of the product of two differentiable functions is the first factor times the derivative of the second factor plus the second factor times the derivative of the first factor.

Example 3 Find $g'(x)$ if $g(x) = (3x - 2)(2x^2 - 5x + 1)$.

SOLUTION

$$g'(x) = (3x - 2)D_x(2x^2 - 5x + 1) + (2x^2 - 5x + 1)D_x(3x - 2)$$
$$= (3x - 2)(4x - 5) + (2x^2 - 5x + 1)\cdot 3 \qquad \text{from (4.20)}$$
$$= 18x^2 - 38x + 13$$

ALTERNATIVE SOLUTION   Another solution could be obtained by first performing the indicated multiplication in the expression for $g(x)$ and then differentiating:

$$g(x) = 6x^3 - 19x^2 + 13x - 2$$
$$g'(x) = 18x^2 - 38x + 13$$

We next obtain a formula for the derivative of the quotient of two functions.

4.4.7   Theorem

If $f'(x)$ and $g'(x)$ exist, and $g(x) \neq 0$, then

$$\left(\frac{f}{g}\right)'(x) = \frac{g(x)f'(x) - f(x)g'(x)}{(g(x))^2},$$

that is,

$$D_x\left[\frac{f(x)}{g(x)}\right] = \frac{g(x)D_xf(x) - f(x)D_xg(x)}{(g(x))^2}. \qquad (4.23)$$

PROOF   Letting $\phi(x) = f(x)/g(x)$,

$$\frac{\phi(x + h) - \phi(x)}{h} = \frac{\dfrac{f(x + h)}{g(x + h)} - \dfrac{f(x)}{g(x)}}{h}$$

$$= \frac{g(x)f(x + h) - f(x)g(x + h)}{hg(x)g(x + h)}.$$

If we subtract and add $f(x)g(x)$ in the numerator of the right member of the preceding equation, then

$$\frac{\phi(x + h) - \phi(x)}{h} = \frac{\dfrac{g(x)f(x + h) - f(x)g(x) - f(x)g(x + h) + f(x)g(x)}{h}}{g(x)g(x + h)}.$$

The remainder of the proof is left as an exercise. Before supplying the details it would be instructive for the student to reread the proof of Theorem 4.4.6.

The derivative of the quotient of two differentiable functions is equal to the denominator times the derivative of the numerator minus the numerator times the derivative of the denominator all divided by the square of the denominator.

Example 4 Find $D_x\left(\dfrac{3x}{x^2 + 4}\right)$.

SOLUTION   By (4.23)

$$D_x\left(\frac{3x}{x^2 + 4}\right) = \frac{(x^2 + 4)D_x 3x - 3x D_x(x^2 + 4)}{(x^2 + 4)^2}$$

$$= \frac{(x^2 + 4)\cdot 3 - 3x \cdot 2x}{(x^2 + 4)^2}$$

$$= \frac{12 - 3x^2}{(x^2 + 4)^2}.$$

Example 5 Find $f'(x)$ if $f(x) = \begin{cases} x/(x - 1) & \text{if } x < 1 \\ x^3 - 4x + 2 & \text{if } x \geq 1 \end{cases}$.

SOLUTION   If $x > 1$, by Theorem 4.3.3

$$f'(x) = D_x(x^3 - 4x + 2) = 3x^2 - 4.$$

If $x < 1$, by the same theorem and (4.23)

$$f'(x) = D_x\left(\frac{x}{x - 1}\right) = -\frac{1}{(x - 1)^2}.$$

$f'(1)$ does not exist as f is discontinuous at 1. Hence:

$$f'(x) = \begin{cases} -\dfrac{1}{(x - 1)^2} & \text{if } x < 1 \\ 3x^2 - 4 & \text{if } x > 1 \end{cases}$$

As a corollary of Theorem 4.4.7 we have the following extension of Theorem 4.4.2.

4.4.8   Theorem

If n is any integer, and $f(x) = x^n$, then

$$f'(x) = nx^{n-1} \quad \text{except if } x = 0 \text{ when } n \leq 1.$$

PROOF   As the theorem has already been proved for $n > 0$, we give a proof when $n \leq 0$. Since $D_x x^0 = D_x 1 = 0 = 0 \cdot x^{0-1}$, the theorem is true when $n = 0$. If $n < 0$, then $n = -m$, where $m > 0$, and from (4.23)

$$D_x x^n = D_x(x^{-m}) = D_x\left(\frac{1}{x^m}\right)$$

$$= \frac{x^m \cdot 0 - 1 \cdot mx^{m-1}}{x^{2m}} = -mx^{-m-1}$$

$$= nx^{n-1}.$$

Thus the conclusion is obtained.

Example 6 Find $D_x(3/x^6)$.

SOLUTION From Theorem 4.4.8

$$D_x\left(\frac{3}{x^6}\right) = D_x(3x^{-6}) = -18x^{-7} = -\frac{18}{x^7}.$$

Exercise Set 4.4

In Exercises 1–14 differentiate with respect to x using the appropriate differentiation formulas.

1. $f(x) = -2x^3$

2. $g(x) = 4/x^7$

3. $f(x) = 2x^2 + 5x - 7$

4. $f(x) = 1 + 4x - 2x^2 - x^3$

5. $y = 100\left(x^2 - \frac{x^4}{1000}\right)$

6. $y = (3x - 4)^2$

7. $f(x) = \frac{-4x^2 + x - 3}{2x^2}$

8. $f(x) = \frac{3x + 5}{2x - 1}$

9. $F(x) = \frac{3}{x^2 + 2x}$

10. $\phi(x) = \frac{x^2 + 2x + 7}{2x^2 - 3x - 5}$

11. $g(x) = \frac{x^3 - 4x + 2}{(x - 2)^2}$

12. $f(x) = |4 - x^2|$

13. $f(x) = \begin{cases} -5/x^4 & \text{if } x > 0 \\ x^2 - 2x + 5 & \text{if } x < 0 \end{cases}$

14. $g(x) = |x| - |x + 2|$

In Exercises 15–18 differentiate with respect to the independent variable.

15. $\phi(h) = \frac{2}{3}\pi h^2(2a - h)$

16. $F(t) = (2t - 3)(3t + 4)$

17. $f(s) = (2s + 1)^3$

18. $g(t) = 10/(t - 4)$

19. Obtain $D_x(2x + 1)(x^2 - 3x + 4)$ by two methods: (a) by first performing the indicated multiplication and then differentiating (b) by differentiating using Theorem 4.4.6.

20. Prove Theorem 4.4.3.

21. Complete the proof of Theorem 4.4.4 by showing that
$$D_x[f(x) - g(x)] = D_x f(x) - D_x g(x)$$
if $f'(x)$ and $g'(x)$ exist.

22. Prove Theorem 4.4.5.

23. Complete the proof of Theorem 4.4.7.

24. Prove: If the functions f, g, and h have derivatives at x, then $(fgh)'(x) = f(x)g(x)h'(x) + f(x)h(x)g'(x) + g(x)h(x)f'(x)$.

25. If a hemispherical bowl of radius r is being filled with water, find the rate of change of the area A of the water surface with respect to the depth x.

4.5   Composite Function Differentiation Formula

Suppose we wish to differentiate the function F where
$$F(x) = (x^3 - 4)^{10}. \tag{4.24}$$

Using the differentiation formulas developed so far, we would be obliged to expand the right member of (4.24) and then differentiate term-by-term to obtain $F'(x)$, a tedious and time-consuming procedure.

Note, however, that the function F defined by (4.24) is actually the composite function $f \circ g$ where
$$f(s) = s^{10} \qquad \text{and} \qquad s = g(x) = x^3 - 4. \tag{4.25}$$

Thus, our purpose here is to obtain a formula for differentiating a composite function. In the formula that will be derived, $(f \circ g)'(x)$ is given in terms of $f'(g(x))$, the derivative of f at $g(x)$, and also $g'(x)$. The notion of $f'(g(x))$ can be troublesome and therefore an illustration should be helpful. Suppose f and g are given by (4.25). Then
$$f'(s) = 10s^9. \tag{4.26}$$

To obtain $f'(g(x))$, we merely replace s by $g(x)$ in (4.26). Thus
$$f'(g(x)) = f'(x^3 - 4) = 10(x^3 - 4)^9.$$

We now give the composite function differentiation formula.

4.5.1   Theorem (Chain Rule)

Suppose f and g are functions such that $f'(g(x))$ and $g'(x)$ exist. Then $f \circ g$ is differentiable at x and
$$(f \circ g)'(x) = D_x[f(g(x))] = f'(g(x)) \cdot g'(x). \tag{4.27}$$

PROOF   We note from Definition 4.3.1 that $f \circ g$ has a derivative at x if and only if
$$(f \circ g)'(x) = \lim_{h \to 0} \frac{f(g(x + h)) - f(g(x))}{h}$$

exists. To compute this limit, we shall first rewrite $[f(g(x + h)) - f(g(x))]/h$, and therefore we consider the function ϕ where

$$\phi(k) = \begin{cases} \dfrac{f(s + k) - f(s)}{k} - f'(s) & \text{if } k \neq 0, \\ 0 & \text{if } k = 0. \end{cases} \tag{4.28}$$

By Theorem 3.5.1

$$\begin{aligned} \lim_{k \to 0} \phi(k) &= \lim_{k \to 0} \left[\frac{f(s + k) - f(s)}{k} - f'(s) \right] \\ &= f'(s) - f'(s) = 0 \\ &= \phi(0). \end{aligned} \tag{4.29}$$

Thus ϕ is continuous at 0. If $k \neq 0$, then from (4.28)

$$f(s + k) - f(s) = f'(s) \cdot k + \phi(k) \cdot k. \tag{4.30}$$

Note that (4.30) is also true when $k = 0$. Let

$$s = g(x) \quad \text{and} \quad k = g(x + h) - g(x) \tag{4.31}$$

where $h \neq 0$. Then

$$s + k = g(x + h), \tag{4.32}$$

and if we substitute from (4.31) and (4.32) into (4.30) and divide by h,

$$\begin{aligned} \frac{f(g(x + h)) - f(g(x))}{h} &= f'(g(x)) \cdot \frac{g(x + h) - g(x)}{h} \\ &\quad + \phi(g(x + h) - g(x)) \cdot \frac{g(x + h) - g(x)}{h}. \end{aligned} \tag{4.33}$$

Equation (4.33) gives the promised rewriting of $[f(g(x + h)) - f(g(x))]/h$. Since $g'(x)$ exists,

$$g'(x) = \lim_{h \to 0} \frac{g(x - h) - g(x)}{h}. \tag{4.34}$$

To obtain $\lim_{h \to 0} \phi(g(x + h) - g(x))$, note that g is continuous at x since $g'(x)$ exists. Hence by Theorem 3.5.4

$$\lim_{h \to 0} g(x + h) = g(x), \tag{4.35}$$

and therefore from Theorem 3.8.1, using (4.35),

$$\begin{aligned} \lim_{h \to 0} \phi(g(x + h) - g(x)) &= \phi(\lim_{h \to 0} [g(x + h) - g(x)]) \\ &= \phi(0) = 0. \end{aligned} \tag{4.36}$$

Taking the limit at $h = 0$ in (4.33) using (4.34) and (4.36),

$$\frac{f(g(x + h)) - f(g(x))}{h} = f'(g(x)) \cdot g'(x) + 0 \cdot g'(x)$$

or

$$D_x[f(g(x))] = f'(g(x))g'(x).$$

We next prove a corollary of Theorem 4.5.1 that can be used to differentiate the function F defined in (4.24).

4.5.2 Theorem

If n is any integer, and g is differentiable at x, then

$$D_x[(g(x))^n] = n(g(x))^{n-1} \cdot g'(x) \quad \text{except if } g(x) = 0 \text{ when } n \leq 1.$$

PROOF If $f(s) = s^n$, then by Theorem 4.4.8

$$f'(s) = ns^{n-1} \quad \text{except if } s = 0 \text{ when } n \leq 1.$$

Hence if g is differentiable at x, and $g(x) \neq 0$ when $n \leq 1$, by Theorem 4.5.1

$$D_x[f(g(x))] = f'(g(x))g'(x),$$

or saying the same thing,

$$D_x[(g(x))^n] = n(g(x))^{n-1} \cdot g'(x).$$

Example 1 Find $F'(x)$ if $F(x) = (x^3 - 4)^{10}$.

SOLUTION From Theorem 4.5.2 letting $g(x) = x^3 - 4$

$$F'(x) = 10(x^3 - 4)^9 D_x(x^3 - 4)$$
$$= 30x^2(x^3 - 4)^9.$$

Example 2 Find $D_x\left[\dfrac{5}{(x^2 - 3x - 5)^2}\right]$.

SOLUTION $D_x\left[\dfrac{5}{(x^2 - 3x - 5)^2}\right] = D_x[5(x^2 - 3x - 5)^{-2}]$

and by Theorem 4.5.2

$$D_x\left[\frac{5}{(x^2 - 3x - 5)^2}\right] = 5(-2)(x^2 - 3x - 5)^{-3} D_x(x^2 - 3x - 5)$$

$$= \frac{-10(2x - 3)}{(x^2 - 3x - 5)^3}.$$

Example 3 Find $D_x y$ if $y = (3x - 2)^4/(2x + 5)^3$.

SOLUTION From Theorem 4.4.7

$$D_x y = \frac{(2x + 5)^3 D_x(3x - 2)^4 - (3x - 2)^4 D_x(2x + 5)^3}{(2x + 5)^6}$$

and from Theorem 4.5.2

$$D_x y = \frac{(2x + 5)^3 \cdot 4(3x - 2)^3 \cdot 3 - (3x - 2)^4 \cdot 3(2x + 5)^2 \cdot 2}{(2x + 5)^6}$$

$$= \frac{6(2x + 5)^2(3x - 2)^3[2(2x + 5) - (3x - 2)]}{(2x + 5)^6}$$

$$= \frac{6(3x - 2)^3(x + 12)}{(2x + 5)^4}.$$

Example 4 Find $f'(x)$ if $f(x) = 3x(x^2 + 1)^{-4}$.

SOLUTION From Theorem 4.4.6

$$f'(x) = 3xD_x(x^2 + 1)^{-4} + (x^2 + 1)^{-4}D_x 3x$$

and from Theorem 4.5.2

$$f'(x) = 3x(-4)(x^2 + 1)^{-5}D_x(x^2 + 1) + (x^2 + 1)^{-4} \cdot 3$$
$$= -12x(x^2 + 1)^{-5} \cdot 2x + (x^2 + 1)^{-4} \cdot 3$$
$$= -3(x^2 + 1)^{-5}[8x^2 - (x^2 + 1)]$$
$$= -3(x^2 + 1)^{-5}(7x^2 - 1).$$

Exercise Set 4.5

In Exercises 1–15 find $f'(x)$.

1. $f(x) = (x^3 - 4x + 5)^5$

2. $f(x) = 1/(5 - 3x)^4$

3. $f(x) = 2/3(x^2 - x + 1)^2$

4. $f(x) = (4 - 9x^2)^{-3}$

5. $f(x) = 3x(3x^2 - 2x + 7)^{10}$

6. $f(x) = x(4x - 1)^7$

7. $f(x) = (x^2 - 3x)^3(x^2 + 4)^5$

8. $f(x) = (2 - x)^2(2x^2 - x - 5)^9$

9. $f(x) = (x^2 + 3x)^2(x^2 + 4)^{-3}$

10. $f(x) = (3x - 1)^{-2}(2x + 7)^{-1}$

11. $f(x) = \dfrac{(3x + 4)^{12}}{x^3}$

12. $f(x) = \dfrac{5x^2}{(2x^2 - 3x - 4)^2}$

13. $f(x) = \left(\dfrac{2x - 1}{2x + 1}\right)^5$

14. $f(x) = \left(\dfrac{x^2 + 1}{x^2}\right)^{-3}$

15. $f(x) = (x + 2)^2(2x - 3)^3(x^2 + 1)^4$

16. If $F(x) = f\left(\dfrac{x - 1}{x + 1}\right)$ and $f'(x) = x^3$, find $F'(x)$.

17. If $F(x) = f(x^3)$ and $f'(x) = \dfrac{x - 1}{x + 1}$, find $F'(x)$.

18. Two cars begin a drag race from the same point on a straight track at time $t = 0$. Car A's elapsed time at any point on the track is $\frac{9}{10}$ of car B's time at that point. Using the chain rule, compare their velocities at any point on the track.

19. Prove using the chain rule: The derivative of an even function is an odd function and the derivative of an odd function is an even function (see Exercise 24, Exercise Set 3.2).

20. Give an example where two functions f and g are each differentiable at some number c and yet $(f \circ g)'(c)$ fails to exist. Does this example contradict the chain rule?

4.6   General Power Formula

Our primary purpose here is to obtain a formula for the differentiation of $(g(x))^r$ where r is a rational number. As a first step toward obtaining this formula, we mention a differentiation formula that was derived in Example 3 of Section 4.3.

4.6.1   Theorem

If q is a positive integer, then

$$D_x(x^{1/q}) = \frac{1}{q}x^{(1/q)-1}. \qquad (4.37)$$

(It should be understood that (4.37) holds if $x^{1/q}$ exists and $x \neq 0$.)

As illustrations of Theorem 4.6.1:

$$D_x x^{1/4} = \frac{1}{4}x^{-3/4}$$

$$D_x \sqrt{3x} = D_x(3^{1/2}x^{1/2}) = 3^{1/2}D_x x^{1/2} = 3^{1/2} \cdot \frac{1}{2}x^{-1/2} = \frac{1}{2}\sqrt{\frac{3}{x}}$$

In the next theorem we extend the result obtained in Theorem 4.6.1.

4.6.2   Theorem

If r is any rational number, and $F(x) = x^r$,

$$F'(x) = rx^{r-1}.$$

($x \in \mathfrak{D}_{F'}$, if x^r exists and $x \neq 0$ when $r \leq 1$.)

PROOF   By hypothesis, $r = p/q$ where $q > 0$ and p are integers. Suppose $x \neq 0$ and $x^{1/q}$ exists. Then

$$F(x) = x^{p/q} = (x^{1/q})^p.$$

If $f(s) = s^p$, then by Theorem 4.4.8 $f'(s) = ps^{p-1}$ and

$$f'(x^{1/q}) = p(x^{1/q})^{p-1}.$$

Hence from (4.37), using the chain rule,

$$F'(x) = D_x f(x^{1/q}) = f'(x^{1/q}) D_x x^{1/q}$$

$$= p(x^{1/q})^{p-1} \cdot \frac{1}{q} x^{(1/q)-1}$$

$$= \frac{p}{q} x^{(p/q)-1}. \tag{4.38}$$

For the case where $r > 1$ and $x = 0$

$$F'(0) = \lim_{h \to 0} \frac{F(h) - F(0)}{h} = \lim_{h \to 0} \frac{h^r}{h} = \lim_{h \to 0} h^{r-1}$$

$$= 0 = r \cdot 0^{r-1}. \tag{4.39}$$

The conclusion follows from (4.38) and (4.39).

As illustrations of Theorem 4.6.2:

$$D_x(x^{2/3}) = \tfrac{2}{3} x^{-1/3}$$

$$D_x(x^{-3/5}) = -\tfrac{3}{5} x^{-8/5}$$

4.6.3   Theorem (General Power Formula)

If r is any rational number and g is differentiable at x, then if $(g(x))^r$ exists and $g(x) \neq 0$ when $r \leq 1$,

$$D_x((g(x))^r) = r(g(x))^{r-1} \cdot g'(x) \tag{4.40}$$

The proof of this important formula, which depends upon the chain rule, is left as an exercise. We recognize that this theorem is an extension of Theorem 4.5.2.

Example 1   Find $f'(x)$ if $f(x) = (4 - x^2)^{-1/2}$.

SOLUTION   From (4.40)

$$f'(x) = -\tfrac{1}{2}(4 - x^2)^{-3/2} D_x(4 - x^2)$$

$$= -\tfrac{1}{2}(4 - x^2)^{-3/2}(-2x)$$

$$= x(4 - x^2)^{-3/2}.$$

Example 2   Find $D_x[x^2(3x + 4)^{5/3}]$.

SOLUTION   By the product rule for differentiation

$$D_x[x^2(3x + 4)^{5/3}] = x^2 D_x(3x + 4)^{5/3} + (3x + 4)^{5/3} D_x x^2,$$

and from (4.40)

$$D_x[x^2(3x + 4)^{5/3}] = x^2 \cdot \tfrac{5}{3}(3x + 4)^{2/3} \cdot 3 + (3x + 4)^{5/3} \cdot 2x$$
$$= x(3x + 4)^{2/3}[5x + 2(3x + 4)]$$
$$= x(3x + 4)^{2/3}(11x + 8).$$

Example 3 Find y' if $y = \sqrt[4]{\dfrac{1 - x}{1 + x}}$.

SOLUTION From (4.40)

$$y' = \frac{1}{4}\left(\frac{1 - x}{1 + x}\right)^{-3/4} D_x\left(\frac{1 - x}{1 + x}\right)$$

$$= \frac{1}{4}\left(\frac{1 - x}{1 + x}\right)^{-3/4} \frac{(1 + x)(-1) - (1 - x)(1)}{(1 + x)^2}$$

$$= \frac{1}{4}\left(\frac{1 + x}{1 - x}\right)^{3/4} \frac{-2}{(1 + x)^2}$$

$$= \frac{-1}{2(1 + x)^{5/4}(1 - x)^{3/4}}.$$

The proof of the next theorem, a corollary of Theorem 4.6.3, is left as an exercise.

4.6.4 Theorem

If $g'(x)$ exists, and $g(x) \neq 0$, then

$$D_x|g(x)| = \frac{g(x) \cdot g'(x)}{|g(x)|}.$$

Exercise Set 4.6

In Exercises 1–13 find the derivative of the function defined by the given equation.

1. $f(x) = 5x^{3/2} - 3x^{1/2} - x^{-1/2}$

2. $F(x) = (3x^{1/3} - 2x^{-1/3})^2$

3. $f(x) = \sqrt[3]{x^2 + 1}$

4. $g(x) = (2x + x^2)^{3/2}$

5. $\phi(x) = (2x^2 - x - 5)^{-4/3}$

6. $F(t) = \dfrac{(3t^2 - 2t + 7)^2}{t^{3/4}}$

7. $G(y) = \dfrac{y^{2/5}}{(3y + 1)^2}$

8. $f(x) = \dfrac{x^2}{\sqrt{x - 3}}$

9. $f(x) = \sqrt{\dfrac{x^2 + a^2}{x^2 - a^2}}$

10. $f(t) = \sqrt{2t - 1}\sqrt[3]{2t + 1}$

11. $g(t) = \sqrt[3]{3 - t}\,(t^2 - 1)^3$

12. $G(x) = \dfrac{(x^2 - 3x - 2)^{2/3}}{(x^2 - x - 4)^{1/2}}$

13. $y = \dfrac{(1 + \sqrt{x})^2}{\sqrt{1 + x^2}}$

14. Prove Theorem 4.6.4. HINT: $|g(x)| = \sqrt{(g(x))^2}$.

In Exercises 15–18 differentiate using Theorem 4.6.4.

15. $y = |x^3 - 4x + 2|$

16. $y = |3x^2 - 2x - 5|$

17. $f(x) = \left|\dfrac{x + 2}{x - 3}\right|$

18. $f(x) = \sqrt{|3x - 2|}$

19. For what numbers x is the rate of change with respect to x of the distance from the point $(0, -1)$ to the point $(x, 2 - x^2)$ equal to zero?

20. The volume and surface area of a sphere are given by the formulas $V = \frac{4}{3}\pi r^3$ and $S = 4\pi r^2$, respectively, where r is the radius of the sphere. Find the rate of change of the volume of the sphere with respect to the surface area.

4.7 Higher Derivatives

Let f' be the derivative of a function f. If f' itself has a derivative, we denote this derivative by f'' (read "f double prime"). The function f'' is called the *second derivative* of f. Similarly the derivative of f'', if it exists, is denoted by f'''.† f''', if it exists, is called the *third derivative* of f. For higher order derivatives, we shall use the symbols $f^{(4)}, f^{(5)}, \ldots$. In this hierarchy of derivatives, f' itself is termed the *first derivative* of f.

We may denote the nth *derivative* of f, where n is any positive integer, by $f^{(n)}$.

4.7.1 Definition

If $n > 1$ is an integer,

$$f^{(n)}(x) = D_x[f^{(n-1)}(x)].$$

If $y = f(x)$, we shall use the notation:

$$y'' = D_x^2 y = f''(x)$$
$$y''' = D_x^3 y = f'''(x), \ldots$$
$$y^{(n)} = D_x^{\,n} y = f^{(n)}(x)$$

† The notation f', f'', f''', $\ldots$ for the first and higher derivatives of the function f was introduced by the eminent French mathematician Joseph Louis Lagrange (1736–1813) in a text in 1797. Lagrange is also remembered in mathematics for his work with convergent series and for contributions in theory of numbers, spherical trigonometry, and calculus of variations. He wrote the first text in which the study of mechanics was developed using calculus. Such was the fame of Lagrange that he was invited in 1766 by Frederick the Great to become his court mathematician in a letter which stated that the "greatest king in Europe" wanted the "greatest mathematician in Europe."

Example 1 If $f(x) = 2x^4 - 5x^2 + 4x - 1$, find the first five derivatives
of f.

SOLUTION $f'(x) = 8x^3 - 10x + 4$
$f''(x) = 24x^2 - 10$
$f'''(x) = 48x$
$f^{(4)}(x) = 48$
$f^{(5)}(x) = 0$

Example 2 If $y = \sqrt{x^2 + a^2}$, find y', y'', and y'''.

SOLUTION $y' = \dfrac{2x}{2\sqrt{x^2 + a^2}} = \dfrac{x}{\sqrt{x^2 + a^2}}$

$y'' = \dfrac{\sqrt{x^2 + a^2} \cdot 1 - x \cdot \frac{1}{2}(x^2 + a^2)^{-1/2}(2x)}{x^2 + a^2}$

and if we multiply the numerator and denominator by $(x^2 + a^2)^{1/2}$

$$y'' = \frac{(x^2 + a^2) - x^2}{(x^2 + a^2)^{3/2}}$$

$$= \frac{a^2}{(x^2 + a^2)^{3/2}},$$

$$y''' = -3a^2 x(x^2 + a^2)^{-5/2}.$$

In Section 6.4 we will see that if the equation of motion of a particle
in straight line motion is given by $s = f(t)$, its *instantaneous acceleration* at any
time t is $D_t^2 s = f''(t)$. We will also observe in Chapter 5 the usefulness of the second
derivative of a function in sketching the graph of the function.

Exercise Set 4.7

In Exercises 1–4 find $f'(x)$, $f''(x)$, and $f'''(x)$. Also, evaluate f', f'', and f''' at the
given number c.

1. $f(x) = 3x^4 - 5x^3 + 6x - 1$, $c = -2$
2. $f(x) = -2x^3 + 4x^2 - 3x + 2$, $c = 3$
3. $f(x) = (2x - 3)^4$, $c = \frac{5}{2}$
4. $f(x) = \sqrt[3]{2x + 1}$, $c = -1$

In Exercises 5–8 obtain y' and y''.

5. $y = \sqrt[4]{x^2 + 2}$

6. $y = \dfrac{1}{\sqrt{3x - x^2}}$

7. $y = x(3x - 4)^2$

8. $y = \dfrac{x^2}{(2x + 3)^3}$

9. Prove by mathematical induction: If n is a positive integer, and $f(x) = (ax + b)^n$, then $f^{(n)}(x) = n!a^n$.

10. Prove by mathematical induction: If n is a positive integer, and $f(x) = 1/x^2$, then

$$f^{(n)}(x) = \frac{(-1)^n(n + 1)!}{x^{n+2}}.$$

4.8 Implicit Differentiation

The equation $y = f(x)$ is said to define the function f *explicitly* and to define y as an *explicit function* of x. For example, the function $\{(x, \sqrt{9 - x^2})\}$ is defined explicitly by the equation

$$y = \sqrt{9 - x^2}. \tag{4.41}$$

One can differentiate in (4.41) and obtain

$$y' = -\frac{x}{\sqrt{9 - x^2}}. \tag{4.42}$$

An equation in x and y in which y is not isolated on one side of the equation may also define y as a function of x. The equation is then said to define y as an *implicit function* of x. For example, the equation

$$x^2 + y^2 = 9 \tag{4.43}$$

defines implicitly the functions given by $y = \sqrt{9 - x^2}$ and $y = -\sqrt{9 - x^2}$. It can also be shown that

$$x^2y + 2x = 9y + 4 \tag{4.44}$$

defines implicitly the function given by

$$y = \frac{4 - 2x}{x^2 - 9}.$$

It will be noted here that the functions that were implicitly defined by Equations (4.43) and (4.44) could readily be obtained by solving the equations for y in terms of x.

Sometimes, however, it is difficult to solve an equation in x and y for y in order to determine if an implicit function is actually defined—for example, the equation

$$xy^3 = (x + y)^2 - 5.$$

Also, it may happen that an equation in x and y does not define y as a function of x. For example,

$$x^4 + y^4 = -5$$

does not define a function since $x^4 + y^4$ cannot be negative.

If an equation in x and y defines y as a differentiable function of x, we can obtain $D_x y$ by a process called *implicit differentiation*. In this process y is assumed equal to some unstated $f(x)$, and hence each member of the equation is considered to be a function of x. We then differentiate each side with respect to x, equate the expressions obtained, and solve for $D_x y$, which generally will be expressed in terms of x and y.

The theoretical justification for implicit differentiation more properly belongs in an advanced calculus course and is omitted here.

In each of the examples below it will be tacitly assumed that the derivative of y with respect to x exists, and that implicit differentiation may therefore be performed.

Example 1 If $x^2 + y^2 = 9$ (Equation (4.43)), find $D_x y$ by implicit differentiation.

SOLUTION From the given equation we form

$$D_x(x^2 + y^2) = D_x 9,$$

and using the sum formula

$$D_x x^2 + D_x y^2 = 0. \tag{4.45}$$

Now $D_x x^2 = 2x$, but $D_x y^2$ *must be obtained using the general power formula* (*Theorem* 4.6.3). From this formula

$$D_x y^2 = 2y D_x y.$$

Thus from (4.45) we obtain successively:

$$2x + 2y D_x y = 0$$

$$D_x y = -\frac{x}{y}. \tag{4.46}$$

The expression obtained for the derivative in (4.46) can be reconciled with (4.42) if we let $y = \sqrt{9 - x^2}$.

Example 2 Using implicit differentiation, find $D_x y$ if $x^2 y + 2x = 9y + 4$ (see (4.44)).

SOLUTION Again we assume y equals some unstated differentiable $f(x)$ and form

$$D_x(x^2 y + 2x) = D_x(9y + 4).$$

From the sum formula for differentiation

$$D_x(x^2 y) + D_x 2x = D_x(9y + 4). \tag{4.47}$$

Since y is considered a differentiable function of x, the first differentiation on the left in (4.47) must be performed using the product rule. Thus

$$D_x(x^2 y) = x^2 D_x y + y D_x x^2$$
$$= x^2 D_x y + 2xy.$$

Hence from (4.47):

$$x^2 D_x y + 2xy + 2 = 9 D_x y$$

$$D_x y = \frac{-2xy - 2}{x^2 - 9}.$$

It is instructive to consider an example in which the second derivative of y with respect to x is obtained by implicit differentiation.

Example 3   Find y'' by implicit differentiation if $x^2 + y^2 = 9$ (see Example 1).

SOLUTION   In Example 1 we obtained

$$y' = -\frac{x}{y}. \tag{4.48}$$

Then after differentiation with respect to x in (4.48), using the quotient rule,

$$y'' = -\frac{y \cdot 1 - xy'}{y^2}.$$

A substitution from (4.48) yields

$$y'' = -\frac{y - x(-x/y)}{y^2}$$

$$= -\frac{y^2 + x^2}{y^3}. \tag{4.49}$$

Since $x^2 + y^2 = 9$, (4.49) simplifies to

$$y'' = -\frac{9}{y^3}.$$

Exercise Set 4.8

In Exercises 1–4 obtain y' by implicit differentiation.

1. $3x^2 - 4y^2 = 12$
2. $x^{1/3} y^{-2/3} + x^{-2/3} y^{1/3} = 1$
3. $\sqrt[3]{x^2 y} = (x + y)^2$
4. $x^2 y^2 = 2(x^2 + y^2)$

In Exercises 5–10 obtain y' and y'' by implicit differentiation.

5. $2x^2 - 3y^2 = 6$
6. $y^2 = x^2 + 3y$
7. $x^{1/2} + y^{1/2} = a^{1/2}$
8. $x^{2/3} + y^{2/3} = 1$
9. $x^2 - xy + y^2 = 3$
10. $2xy^2 + y^3 = 5$

11. For the equation in Exercise 9 obtain y' and y'' at the point $(1, -1)$.
12. For the equation in Exercise 10 obtain y' and y'' at the point $(2, 1)$.

4.9 Tangents and Normals

We recall from Definition 4.1.1 that the slope of the tangent to the graph of the function f at the point $(c, f(c))$ is

$$\lim_{h \to 0} \frac{f(c + h) - f(c)}{h}$$

provided this graph has a nonvertical tangent at the point. Since this limit is $f'(c)$, Definition 4.1.1 can be restated using the notion of the derivative.

4.9.1 Definition

Let f be continuous at c and be defined on some interval containing c. The tangent to the graph of f at the point $(c, f(c))$ is

(a) the line with equation $y - f(c) = f'(c) \cdot (x - c)$ if $f'(c)$ exists

or

(b) the line $x = c$ if $f'(c) = +\infty$ or $-\infty$.

If $f'(c)$ does not exist, but is not $+\infty$ or $-\infty$, the graph of f has no tangent at $(c, f(c))$.

The calculation of the slope of the tangent is facilitated by the use of appropriate differentiation formulas, as in the next example. The formulas, of course, were not available to us in Section 4.1.

Example 1 Find an equation of the tangent T to the graph of $y = \frac{1}{4}x^4 - x^3 + 2$ at the point $(2, -2)$ (Figure 4-4).

SOLUTION Here $y' = f'(x) = x^3 - 3x^2$ and $f'(2) = -4$. From part (a) of Definition 4.9.1, the tangent has an equation

$$y + 2 = -4(x - 2)$$

or

$$y = -4x + 6.$$

The notions of the tangent and *normal* to a graph are inseparable as noted in the following statement.

4.9.2 Definition

Suppose the graph of a function f has a tangent at the point $(c, f(c))$. The *normal* to the graph of f at $(c, f(c))$ is the line that is perpendicular to the tangent at this point.

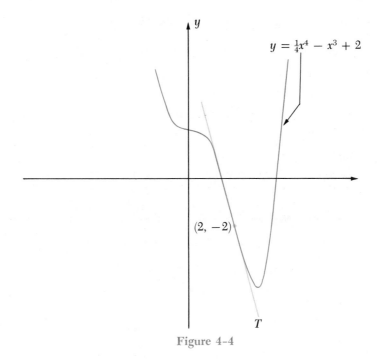

$$y = \tfrac{1}{4}x^4 - x^3 + 2$$

$(2, -2)$

T

Figure 4-4

In Figure 4–5 the tangent and normal to the graph of f at $(c, f(c))$ are denoted by T and N, respectively.

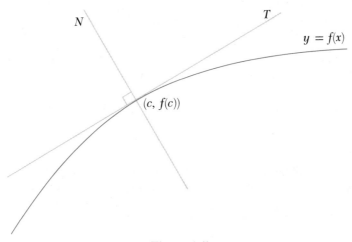

N T

$y = f(x)$

$(c, f(c))$

Figure 4-5

Example 2   Find an equation of the normal to the graph of $y = \tfrac{1}{4}x^4 - x^3 + 2$ at the point $(2, -2)$ (see Example 1).

SOLUTION   We note from the equation obtained for the tangent to the graph in Example 1, that the slope of the tangent is -4. Hence the slope of the normal at the same point is $\frac{1}{4}$ and an equation of the normal can be written

$$y + 2 = \tfrac{1}{4}(x - 2)$$

or

$$y = \tfrac{1}{4}x - \tfrac{5}{2}.$$

In the next example we obtain the tangent to a graph of a function that is defined implicitly.

Example 3 Find the tangent of the graph of $x^2 + xy + y^2 = 7$ at the point $(-1, 3)$.

SOLUTION   We obtain y' at $(-1, 3)$ by implicit differentiation:

$$2x + xy' + y + 2yy' = 0$$

$$y' = -\frac{2x + y}{x + 2y}.$$

At $(-1, 3)$,

$$y' = -\frac{2(-1) + 3}{-1 + 2(3)} = -\frac{1}{5}$$

and an equation for the tangent is

$$y - 3 = -\tfrac{1}{5}(x + 1)$$

or

$$y = -\tfrac{1}{5}x + \tfrac{14}{5}.$$

Example 4 If a tangent to the graph of $y^2 = x$ passes through the point $(-2, 0)$, find the point(s) of tangency.

SOLUTION   Let (x_1, y_1) be a point of tangency. Then from the equation of the graph

$$y_1^2 = x_1. \tag{4.50}$$

We shall obtain another equation in x_1 and y_1 to solve simultaneously with (4.50). Such an equation can be obtained since there are two different ways of expressing the slope m of the tangent at (x_1, y_1). First, by the two-point formula from analytic geometry,

$$m = \frac{y_1 - 0}{x_1 + 2}. \tag{4.51}$$

From implicit differentiation in the equation of the graph

$$2yy' = 1$$

$$y' = \frac{1}{2y},$$

and therefore

$$m = \frac{1}{2y_1}. \tag{4.52}$$

After equating the expressions for m from (4.51) and (4.52) and cross-multiplying, the equation

$$2y_1{}^2 = x_1 + 2 \tag{4.53}$$

is obtained. Solving Equations (4.50) and (4.53) for (x_1, y_1) we obtain the points of tangency $(2, \sqrt{2})$ and $(2, -\sqrt{2})$.

In the next example a vertical tangent to a graph is found.

Example 5 Show that the graph of $y = (x - 1)^{1/3}$ has a tangent at $(1, 0)$ (Figure 4–6).

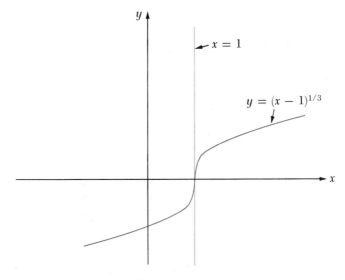

Figure 4–6

SOLUTION Here $y' = f'(x) = \frac{1}{3}(x - 1)^{-2/3}$, but $f'(1)$ is undefined. However,

$$\lim_{h \to 0} \frac{f(1 + h) - f(1)}{h} = \lim_{h \to 0} \frac{h^{1/3}}{h} = \lim_{h \to 0} \frac{1}{h^{2/3}}.$$
$$= +\infty.$$

Thus the line $x = 1$ is a vertical tangent to the graph at $(1, 0)$.

Example 6 Show that the graph of $f(x) = x^{2/3}$ has no tangent at $(0, 0)$ (Figure 4–7).

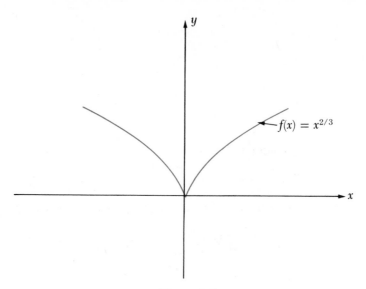

Figure 4-7

SOLUTION   Note that $f'(x) = \frac{2}{3}x^{-1/3}$ if $x \neq 0$, but $f'(0)$ is undefined. Also

$$\frac{f(h) - f(0)}{h} = \lim_{h \to 0^+} \frac{h^{2/3}}{h} = \lim_{h \to 0^+} \frac{1}{h^{1/3}}$$

$$= +\infty$$

but

$$\lim_{h \to 0^-} \frac{f(h) - f(0)}{h} = \lim_{h \to 0^-} \frac{1}{h^{1/3}} = -\infty.$$

Hence $\lim_{h \to 0} (f(h) - f(0))/h$ does not exist and is not $+\infty$ or $-\infty$. Therefore by Definition 4.9.1 the graph of f does not have a tangent at $(0, 0)$.

Exercise Set 4.9

In Exercises 1–10 obtain equations for the tangents and normals to the graphs of the given equations at the indicated points.

1. $f(x) = x^2 + 2x$, $(1, 3)$

2. $g(x) = \dfrac{2x - 1}{x + 3}$, $(3, \frac{5}{6})$

3. $y = \sqrt{2x + 5}$, $(2, 3)$

4. $y = (3x - 2)^{-1/3}$, $(-2, -\frac{1}{2})$

5. $x^2 - y^2 = 5$, $(-3, 2)$

6. $y^3 - xy^2 = 9$, $(2, 3)$

7. $x^{1/2} + y^{1/2} = a^{1/2}$, $(a/4, a/4)$

8. $x^2 + xy + y^2 = 3$, $(2, -1)$

9. $y = \sqrt{4 - x^2}$, $(-2, 0)$

10. $y = (x - 2)^{3/5} + 4$, $(2, 4)$

11. Show that the graph of $y = |x|$ has no tangent at $(0, 0)$.

12. Show that the graph of f where $f(x) = \begin{cases} 9 - x^2 & \text{if } x > 2 \\ 2x + 1 & \text{if } x \leq 2 \end{cases}$ has no tangent at $(2, 5)$.

13. Find the constants a, b, and c if the graph of $y = ax^2 + bx + c$ passes through the points $(0, -3)$ and $(2, 5)$ and has the slope 1 at $(2, 5)$.

14. At what point on the graph of $y = x^2$ is the tangent parallel to the line $2x - y = 5$?

15. If a tangent to the graph of $y = x^2 - 2x - 3$ passes through the point $(5, 3)$, find the point(s) of tangency.

16. Find equations of the tangents to the circle $x^2 + y^2 = a^2$ having slope m.

17. Find equations of the tangents to the ellipse $3x^2 + 4y^2 = 16$ passing through the point $(6, -5)$.

18. Show that a tangent to the graph of the circle $x^2 + y^2 = a^2$ at an arbitrary point (x_1, y_1) on the circle intersects the circle in exactly one point.

19. At what points on the graph of $y = x^2$ do the normal(s) pass through the point $(5, 4)$?

4.10   Angle of Intersection

We digress from our discussion of tangents to a graph to consider the notion of the angle between two lines and to obtain a formula for calculating this angle. Then, when two graphs intersect, the angle between the graphs at an intersection point is defined as the angle between their tangents at the point if this angle exists.

Suppose L_1 and L_2 are intersecting lines (Figure 4–8). From elementary

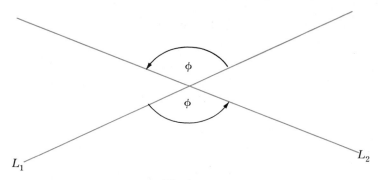

Figure 4–8

geometry the pairs of vertical angles formed are equal and an angle in one pair is supplementary to an angle in the other pair. It is desirable to define the angle between L_1 and L_2 in such a way as to avoid ambiguity, as is done in the next definition.

4.10.1   Definition

Let L_1 and L_2 be distinct lines in the plane.

    (a) If L_1 and L_2 intersect, the *angle between L_1 and L_2 is the smallest angle measured counterclockwise from L_1 to L_2.* Note that the angle here is measured counterclockwise from the *first* line mentioned in the phrase "between L_1 and L_2," namely, L_1, to the *second* line mentioned in the phrase, which is L_2. In Figure 4–8 either of the angles denoted by ϕ is the angle between L_1 and L_2.

(b) If L_1 and L_2 are parallel, the angle between L_1 and L_2 is defined to be $0°$.

If ϕ is the angle between L_1 and L_2, then

$$0° \leq \phi < 180°$$

and also $180° - \phi$ is the angle between L_2 and L_1.

4.10.2 Definition

The *inclination* θ of a line L is the angle between the x axis and L (Figure 4–9).

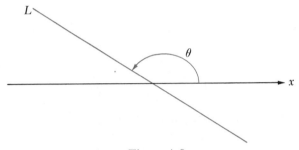

Figure 4-9

There is a relation between the inclination of a line and its slope. Suppose that L is any line in the plane and θ is its inclination. If L does not pass through the origin, then there is a line L_0 through this point which is parallel to L (Figure 4–10). Hence by elementary geometry θ is also the inclination of L_0. From basic

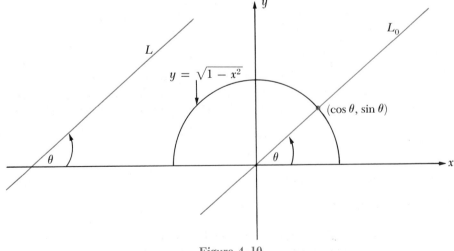

Figure 4-10

trigonometry we know that L_0 intersects the semicircle $y = \sqrt{1 - x^2}$ at the point $(\cos\theta, \sin\theta)$. The slope of L_0 is given by

$$\frac{\sin\theta - 0}{\cos\theta - 0} = \tan\theta.$$

Since L is parallel to L_0, the slope of L is also $\tan\theta$. Therefore, we have proved the following theorem.

4.10.3   Theorem

The slope m of a line L with inclination θ is given by

$$m = \tan\theta. \tag{4.54}$$

In particular, a line with inclination $45°$ has slope $m = \tan 45° = 1$ and a line with inclination $120°$ has slope $\tan 120° = -\sqrt{3}$, but the slope of a line with inclination $90°$ is undefined.

Example 1   Find the inclination of the tangent to the graph of $y = x^2$ at the point $(\frac{3}{2}, \frac{9}{4})$ (Figure 4–11).

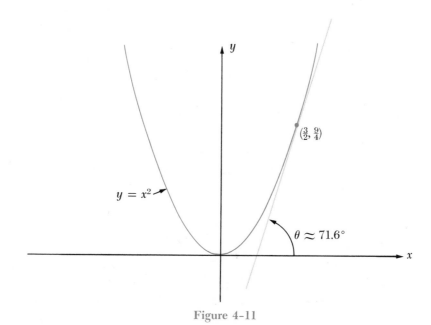

Figure 4-11

SOLUTION   Since $D_x y = 2x$, the slope of the tangent at $(\frac{3}{2}, \frac{9}{4})$ is $2 \cdot \frac{3}{2} = 3$. From (4.54) the inclination θ of the tangent satisfies

$$\tan\theta = 3.$$

Since $0° \le \theta < 180°$, $\theta \approx 71.6°$ using Table I in the appendix of this text.

We next derive a relation between the angle between two lines and the slopes of the lines.

4.10.4   Theorem

L_1 and L_2 are distinct lines with respective slopes m_1 and m_2 and respective inclinations θ_1 and θ_2. If the angle between L_1 and L_2 has measure $\phi \neq 90°$, then

$$\tan \phi = \frac{m_2 - m_1}{1 + m_1 m_2}. \qquad (4.55)$$

PROOF   If L_1 and L_2 intersect, either (i) $\theta_2 > \theta_1$ (Figure 4–12(a)) or (ii) $\theta_2 < \theta_1$ (Figure 4–12(b)).

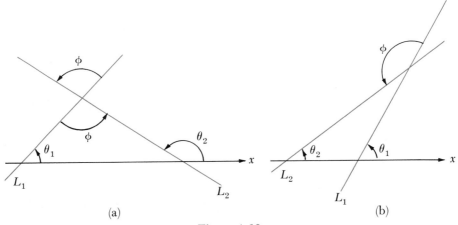

(a)   (b)

Figure 4-12

In case (i) $\theta_2 = \theta_1 + \phi$, and hence $\phi = \theta_2 - \theta_1$. Then

$$\tan \phi = \tan (\theta_2 - \theta_1)$$

$$= \frac{\tan \theta_2 - \tan \theta_1}{1 + \tan \theta_1 \tan \theta_2}$$

and from (4.54)

$$\tan \phi = \frac{m_2 - m_1}{1 + m_1 m_2}.$$

In case (ii) $\theta_1 = (180° - \phi) + \theta_2$, and hence $\phi = 180° + (\theta_2 - \theta_1)$. Then $\tan \phi = \tan [180° + (\theta_2 - \theta_1)] = \tan (\theta_2 - \theta_1)$ and (4.55) is obtained as in case (i).

Tan ϕ would not exist if $1 + m_1 m_2 = 0$. In that case $m_1 m_2 = -1$ and L_1 and L_2 would be perpendicular.

Example 2   Find the angle between the lines $x - 4y = 4$ and $2x + 3y = 6$ (Figure 4–13).

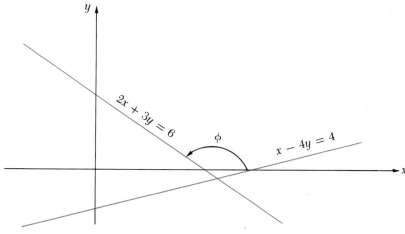

Figure 4-13

SOLUTION   Rewriting the equations of the lines in slope-intercept form, we obtain $y = -\frac{2}{3}x + 2$ and $y = \frac{1}{4}x - 1$, and hence the slopes of the lines are respectively $-\frac{2}{3}$ and $\frac{1}{4}$. Since the angle is measured counterclockwise from the line $x - 4y = 4$ to the line $2x + 3y = 6$, we let $m_2 = -\frac{2}{3}$ and $m_1 = \frac{1}{4}$ in (4.55). Then from (4.55)

$$\tan \phi = \frac{-\frac{2}{3} - \frac{1}{4}}{1 + \frac{1}{4}(-\frac{2}{3})} = \frac{-\frac{11}{12}}{\frac{5}{6}} = -\frac{11}{10}$$

and correct to the nearest tenth of a degree

$$\phi \approx 180° - 47.7° = 132.3°.$$

In the next statement the angle of intersection between two graphs is defined.

4.10.5   Definition

Suppose the graphs C_1 and C_2 intersect at a point P and T_1 and T_2 are their respective tangents at P. The *angle of intersection* between C_1 and C_2 at P is defined as the angle between T_1 and T_2. C_1 and C_2 are tangent to each other at P if and only if T_1 and T_2 are the same line. C_1 and C_2 are *orthogonal* at P if and only if T_1 and T_2 are perpendicular.

Example 3   Find the angle of intersection between the graphs of $y^2 = x$ and $x^2 + 2y^2 = 8$ at their intersection points (Figure 4–14).

SOLUTION   To find the intersection points of the graphs, the equations are solved simultaneously. Replacing y^2 by x in the equation $x^2 + 2y^2 = 8$, we obtain

$$x^2 + 2x = 8$$
$$x^2 + 2x - 8 = 0$$
$$(x + 4)(x - 2) = 0$$
$$x = -4 \quad \text{or} \quad x = 2.$$

-4 cannot be the x coordinate of an intersection point, but from $x = 2$ the intersection points $(2, \sqrt{2})$ and $(2, -\sqrt{2})$ are obtained.

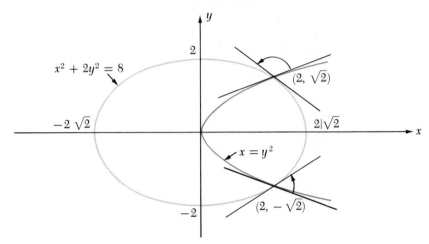

Figure 4–14

Differentiating with respect to x,

$$\text{from} \qquad y^2 = x, \quad \text{we obtain} \quad y' = \frac{1}{2y};$$

$$\text{from} \quad x^2 + 2y^2 = 8, \quad \text{we obtain} \quad y' = -\frac{x}{2y}.$$

At $(2, \sqrt{2})$, the respective slopes are $1/2\sqrt{2}$ and $-1/\sqrt{2}$, and in (4.55), we let $m_2 = -1/\sqrt{2}$ and $m_1 = 1/2\sqrt{2}$ since the angle is measured counterclockwise from the tangent to $y^2 = x$ to the tangent to $x^2 + 2y^2 = 8$. Then from (4.55)

$$\tan \phi = \frac{-(1/\sqrt{2}) - (1/2\sqrt{2})}{1 + (1/2\sqrt{2})(-1/\sqrt{2})} = \frac{-(3/2\sqrt{2})}{\frac{3}{4}} = -\sqrt{2} \approx -1.414.$$

Hence from Table I

$$\phi \approx 180° - 54.7° = 125.3°.$$

At $(2, -2^{1/2})$, the slopes are $m_2 = 1/\sqrt{2}$ and $m_1 = -1/2\sqrt{2}$. Then from (4.55)

$$\tan \phi \approx 1.414$$

and

$$\phi \approx 54.7°.$$

Exercise Set 4.10

1. Find the slope of the line with inclination:

   (a) $30°$ (b) $135°$ (c) $90°$ (d) $60°$ (e) $150°$

2. Find to the nearest tenth of a degree the inclination of a line with slope:

   (a) $\frac{2}{3}$ (b) $\frac{3}{4}$ (c) $-\frac{5}{2}$ (d) $-\frac{3}{5}$ (e) 2

3. Find to the nearest tenth of a degree the angle between the lines:

    (a) $5x + 2y = 4$ and $5x - 2y = 10$

    (b) $3x + 4y = 12$ and $x + 2y = -8$

    (c) $2x - 3y = 0$ and $x + 3y = 7$

    (d) $4x - 5y = -10$ and $2x - y = 4$

4. Find the angles of the triangle having vertices:

    (a) $(4, 1)$, $(0, 2)$, and $(-2, -1)$

    (b) $(1, -3)$, $(2, 3)$, and $(-4, 5)$

5. Find the lines passing through $(-2, 5)$ that make an angle of $45°$ with the line $3x - y = 7$.

6. What is the inclination of the tangent to the graph of the equation $y = \sqrt{x}$ at the point $(4, 2)$?

7. What is the inclination of the tangent to the ellipse $2x^2 + 3y^2 = 13$ at the point $(\sqrt{2}, \sqrt{3})$?

In Exercises 8–11 find the angles of intersection between the graphs of the equations in the order given.

8. $y = 3x$

    $y = 3x^2 - x + 1$

9. $y = \dfrac{2x^2}{x^2 + 1}$

    $y = -x$

10. $y^2 - x^2 = 9$

    $y = 2x$

11. $y^2 = 4x$

    $x^2 + 2y^2 = 48$

12. Show that the graphs of the equations $x^2 + y^2 = 25$ and $3x + 4y = 0$ intersect orthogonally.

13. Let a and b be any numbers. Show that the graphs of $x^2 - y^2 = a^2$ and $xy = b$ intersect orthogonally.

14. Prove that the circle $(x - \frac{5}{2})^2 + (y + 1)^2 = 8$ and the parabola $y^2 = 2x$ are tangent at the point $(\frac{1}{2}, 1)$.

5. Curve Sketching Using Differential Calculus

5.1 Extreme Values on an Interval

In application of calculus to practical problems as well as in the graphing of equations it will be observed that a knowledge of the maximum and minimum values of a particular function is often very useful. Therefore, we will begin this chapter with the following definition.

5.1.1 Definition

Let S be a set contained in the domain of a function f and let c be an element of S.

(a) $f(c)$ is a *maximum (value)* of f on S if and only if $f(x) \leq f(c)$ for every x in S.

(b) $f(c)$ is a *minimum (value)* of f on S if and only if $f(x) \geq f(c)$ for every x in S.

A maximum or minimum of a function on a set is often called an *extreme value* or *extremum* (plural *extrema*) of the function on the set.

Example 1 Find the extrema of the function f where $f(x) = x^2$ on the interval $[-1, 2)$ and on the closed interval $[-1, 2]$ (Figure 5–1).

SOLUTION Since $f(0) = 0$ and $f(x) = x^2 \geq 0$ *for every x*, $f(0) = 0$ is the minimum of f on both the intervals $[-1, 2)$ and $[-1, 2]$.

Note that $f(x) < 4$ if $x \in [-1, 2)$. Thus if f has a maximum M on this interval, $M < 4$. But since $\lim_{x \to 2} f(x) = 4$, it is possible to choose $x \in [-1, 2)$ so that $M < f(x) < 4$, which contradicts the defining property of M. Hence f has no maximum on $[-1, 2)$. Nevertheless, $f(2) = 4$ is a maximum of f on $[-1, 2]$.

A function may not have any extrema on an interval contained in its

154

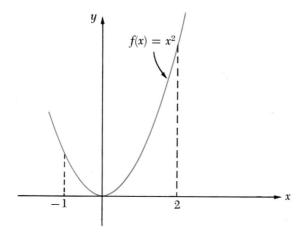

Figure 5-1

domain. The function g where $g(x) = x + 1$, for example, has no maximum or minimum on the open interval $(0, 2)$. On the other hand, a constant function has both a maximum and a minimum at every number in any interval contained in its domain.

We come now to an important theorem that is intuitively believable but has a proof beyond the scope of this text.

5.1.2   Theorem

If f is continuous on the closed interval $[a, b]$, then f has both a maximum and a minimum on $[a, b]$.

Theorem 5.1.2 does not say where on the interval $[a, b]$ these extreme values are assumed. The theorem only guarantees their existence. The next theorem, however, is helpful in determining where extreme values are assumed on the closed interval.

5.1.3   Theorem (Fermat's Theorem)†

If the following conditions are satisfied:

   (i) *$f(c)$ is an extremum of f on an interval I,*
   (ii) *c is in the interior of I,*
   (iii) *$f'(c)$ exists,*

then $f'(c) = 0$.

---

† Pierre de Fermat (1601–1665) was a French mathematician who made outstanding contributions to the theory of numbers and analytic geometry. Although Fermat lived before the invention of calculus, he used methods in his studies of tangents to graphs which resembled the limit process.

PROOF Suppose $f(c)$ is a minimum of f on I. If $c + h \in I$, from Definition 5.1.1

$$f(c + h) \geq f(c)$$

and

$$f(c + h) - f(c) \geq 0. \qquad (5.1)$$

If $h > 0$, dividing each side of (5.1) by h gives

$$\frac{f(c + h) - f(c)}{h} \geq 0, \qquad (5.2)$$

but if $h < 0$, we obtain

$$\frac{f(c + h) - f(c)}{h} \leq 0. \qquad (5.3)$$

Since $f'(c)$ exists, either $f'(c) > 0$, $f'(c) < 0$, or $f'(c) = 0$. If $f'(c) > 0$, by Theorem 3.3.3 for every $h \neq 0$ sufficiently close to 0, we have

$$\frac{f(c + h) - f(c)}{h} > 0,$$

which is a contradiction of (5.3). Hence we must have either $f'(c) < 0$ or $f'(c) = 0$. If $f'(c) < 0$, for every $h \neq 0$ sufficiently close to 0 we have

$$\frac{f(c + h) - f(c)}{h} < 0,$$

which is a contradiction of (5.2). Thus $f'(c) = 0$ is the only remaining possibility. The proof when $f(c)$ is a maximum is analogous.

An extremum of a continuous function on a closed interval will obviously occur either at an endpoint of the interval or in the interior of the interval. From Theorem 5.1.3, if the extremum is assumed at a number c in the interior of the interval, either $f'(c) = 0$ or $f'(c)$ fails to exist.

5.1.4   Definition

A number c in the domain of a function f is a *critical number* of f if and only if $f'(c) = 0$ or $f'(c)$ fails to exist.

In summary, to locate the extreme values of a continuous function on a closed interval, we evaluate the function at the endpoints and at its critical numbers in the interval. The largest value thus obtained is the maximum of the function on the interval, and the least value is the minimum of the function on the interval.

Example 2 Find the extrema of the function f where $f(x) = x^{5/3} - 5x^{2/3}$ on the interval $[-1, 4]$ (Figure 5-2).

SOLUTION We first note that f is continuous on $[-1, 4]$ and hence by Theorem 5.1.2 has a maximum and a minimum on the interval. To obtain the critical numbers for f, we compute

$$f'(x) = \tfrac{5}{3}x^{2/3} - \tfrac{10}{3}x^{-1/3}$$
$$= \tfrac{5}{3}x^{-1/3}(x - 2). \tag{5.4}$$

Note from (5.4) that $f'(x) = 0$ if and only if $x = 2$, and $f'(x)$ fails to exist but $x \in \mathcal{D}_f$ if and only if $x = 0$. Thus 0 and 2 are the only critical numbers of f. We then calculate:

$$f(-1) = -6$$
$$f(0) = 0$$
$$f(2) = 2^{5/3} - 5 \cdot 2^{2/3} = -3 \cdot 2^{2/3}$$
$$f(4) = 4^{5/3} - 5 \cdot 4^{2/3} = -4^{2/3}$$

From our immediately preceding discussion

$$f(0) = 0 \text{ is the maximum of } f \text{ on } [-1, 4]$$

and

$$f(-1) = -6 \text{ is the minimum of } f \text{ on } [-1, 4].$$

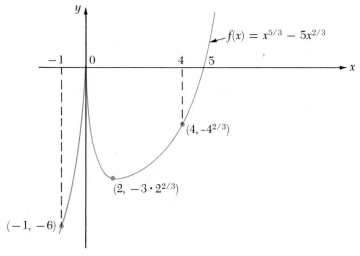

Figure 5-2

Example 3 Find the extrema of the function g where $g(x) = (4x + 4)/(x^2 + 3)$ on the interval $[-4, 3]$ (Figure 5-3).

SOLUTION We note that the function g is continuous on $[-4, 3]$ and hence has a maximum and a minimum on the interval.

To obtain the critical numbers for g, we form

$$g'(x) = \frac{(x^2 + 3) \cdot 4 - (4x + 4)(2x)}{(x^2 + 3)^2}$$

$$= \frac{-4x^2 - 8x + 12}{(x^2 + 3)^2}$$

$$= -\frac{4(x + 3)(x - 1)}{(x^2 + 3)^2}.$$

Since $g'(x) = 0$ if and only if $x = 1$ or $x = -3$, 1 and -3 are the only critical numbers of g. We then calculate:

$$g(-4) = -\tfrac{12}{19}$$
$$g(-3) = -\tfrac{2}{3}$$
$$g(1) = 2$$
$$g(3) = \tfrac{4}{3}$$

Hence

$$g(1) = 2 \text{ is the maximum of } g \text{ on } [-4, 3]$$

and

$$g(-3) = -\tfrac{2}{3} \text{ is the minimum of } g \text{ on } [-4, 3].$$

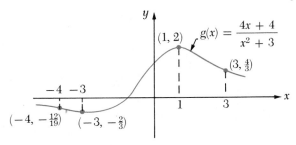

Figure 5-3

Example 4   Find the extrema of the function ϕ where

$$\phi(x) = \begin{cases} \dfrac{1}{x - 2} & \text{if } x \neq 2 \\ 4 & \text{if } x = 2 \end{cases} \qquad \text{on the interval } [2, 5] \text{ (Figure 5-4).}$$

SOLUTION   Since $\lim_{x \to 2} \phi(x)$ fails to exist, ϕ is discontinuous at 2 and hence Theorem 5.1.2 does not apply on the interval $[2, 5]$. It is intuitively apparent that $\phi(x)$ decreases as x increases when $x > 2$ and therefore that $\phi(5) = \tfrac{1}{3}$ is the minimum of ϕ on $[2, 5]$. To prove the latter contention rigorously, we first note that if $x \in (2, 5]$, then $5 - x \geq 0$ and $x - 2 > 0$. Hence

$$\phi(x) - \frac{1}{3} = \frac{1}{x - 2} - \frac{1}{3}$$

$$= \frac{5 - x}{3(x - 2)} \geq 0.$$

Thus $\phi(x) \geq \tfrac{1}{3}$ if $x \in (2, 5]$ and since $\phi(2) = 4$, $\phi(5) = \tfrac{1}{3}$ is the minimum of ϕ on $[2, 5]$. However, since

$$\lim_{x \to 2^+} \phi(x) = \lim_{x \to 2^+} \frac{1}{x - 2} = +\infty,$$

we can obtain $\phi(x)$ as large as we please by selecting $x > 2$ but sufficiently close to 2. Hence ϕ has no maximum on $[2, 5]$.

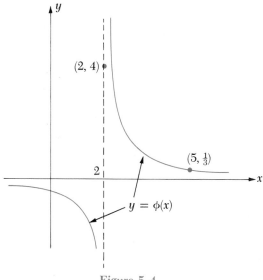

Figure 5-4

Exercise Set 5.1

In Exercises 1–14 obtain the maximum and minimum of the given function on the closed interval $[-1, 3]$.

1. $f(x) = x^3 - 2x^2 + x - 5$

2. $f(x) = \frac{1}{2}x^4 - 2x^3 + 2$

3. $f(x) = x^4 - 4x^2 + 4$

4. $f(x) = \dfrac{x}{x + 2}$

5. $f(x) = (x^2 - 3x + 2)^{1/3}$

6. $f(x) = (x^2 - 1)^6$

7. $f(x) = |x - 1| - 2$

8. $f(x) = \frac{1}{6}x(3 - x)^{5/2}$

9. $f(x) = x\sqrt[3]{1 - x}$

10. $f(x) = x^{1/3}(x - 3)^3$

11. $f(x) = \dfrac{x^2}{\sqrt{x^2 + 1}}$

12. $f(x) = (x + 1)^{1/2}(4 - x)^{3/2}$

13. $f(x) = \begin{cases} x^2 - 4x & \text{if } x \geq 1 \\ 3x - 6 & \text{if } x < 1 \end{cases}$

14. $f(x) = \begin{cases} 9 - x^2 & \text{if } x > 2 \\ 2x + 1 & \text{if } x \leq 2 \end{cases}$

In Exercises 15–17 find the extrema of the functions on the indicated intervals if these extrema exist.

15. $f(x) = \dfrac{x}{x^2 - 4}$, $[0, 2]$

16. $f(x) = \sqrt{9 - x^2}$, $(-3, 3]$

17. $f(x) = x + [x - 1]$, $(0, 3]$

18. Prove Theorem 5.1.3 if $f(c)$ is a maximum of f on I.

5.2   The Mean Value Theorem

Suppose f is a function that satisfies the following conditions:

(i) f is continuous on the closed interval $[a, b]$,
(ii) f is differentiable on the open interval (a, b),
(iii) $f(a) = f(b) = 0$.

Here f may have the value zero everywhere on the interval $[a, b]$ (Figure 5–5(a)) or f may assume values different from zero on the interval (Figure 5–5(b)). From (ii) above, for every number c in the interval (a, b), the graph has a nonvertical tangent at the point $(c, f(c))$. Hence, as is suggested by Figure 5–5(a) and (b), there is at least one number c between a and b such that the tangent to the graph of f at $(c, f(c))$ is horizontal.

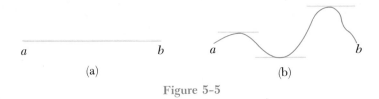

| a | | b | a | | b |

(a)   (b)

Figure 5-5

5.2.1   Rolle's Theorem

If f satisfies the conditions (i), (ii), and (iii) mentioned above on the interval $[a, b]$, then there is a number c in the open interval (a, b) such that $f'(c) = 0$.

PROOF   *Case 1:* ($f(x) = 0$ for every $x \in [a, b]$). If we select *any number* c in the open interval (a, b), we obtain $f'(c) = 0$.

Case 2: ($f(x) > 0$ for some number $x \in [a, b]$). Using (i) and Theorem 5.1.2, f has a maximum $f(c)$ on $[a, b]$ and $f(c) \geq f(x) > 0$. We cannot have $c = a$ or $c = b$, for then by (iii) we would have $f(c) = 0$. Since $c \in [a, b]$ but $a \neq c \neq b$, we must have $c \in (a, b)$. From (ii), $f'(c)$ exists, and hence by Theorem 5.1.3, $f'(c) = 0$.

Case 3: ($f(x) < 0$ for some number $x \in [a, b]$). The proof is left as an exercise.

Example 1   Show that the function f where $f(x) = x^2 + x - 2$ satisfies the hypotheses of Rolle's Theorem on $[-2, 1]$. Find a number c in $(-2, 1)$ such that $f'(c) = 0$.

SOLUTION   We note that f is everywhere continuous, being a polynomial function. Also $f'(x) = 2x + 1$ for every x, and finally $f(-2) = f(1) = 0$. Hence by Rolle's Theorem, there is a c in $(-2, 1)$ such that $f'(c) = 2c + 1 = 0$. Thus $c = -\frac{1}{2}$.

Example 2   Does f where $f(x) = x^{2/3} - 1$ satisfy the hypotheses of Rolle's theorem on $[-1, 1]$? Is there a c in $(-1, 1)$ such that $f'(c) = 0$?

SOLUTION   Even though f is continuous everywhere, $f'(0)$ is undefined, so f is not differentiable on $(-1, 1)$. Thus hypothesis (ii) of Rolle's Theorem is not satisfied. There is no c in $(-1, 1)$ such that $f'(c) = \frac{2}{3}c^{-1/3} = 0$.

The main reason for mentioning Rolle's Theorem is that it is used to prove the mean value theorem, one of the most important theorems in calculus.

5.2.2   Theorem (Mean Value Theorem)

If f is continuous on the closed interval $[a, b]$, and differentiable on the open interval (a, b), there is a number c in the interval (a, b) such that

$$\frac{f(b) - f(a)}{b - a} = f'(c).$$

PROOF   We consider the function ϕ where

$$\phi(x) = f(x) - f(a) - \frac{f(b) - f(a)}{b - a}(x - a). \qquad (5.5)$$

It can be readily shown that for any number x in $[a, b]$, $\phi(x)$ expresses the vertical "signed" distance from the secant through the points $(a, f(a))$ and $(b, f(b))$ to the graph of f (Figure 5–6) (Exercise 8, Exercise Set 5.2).

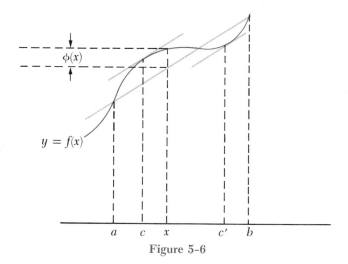

Figure 5-6

Since f is continuous on $[a, b]$ so is ϕ, and since f is differentiable on (a, b), for each x in (a, b)

$$\phi'(x) = f'(x) - \frac{f(b) - f(a)}{b - a}.$$

From (5.5) we obtain $\phi(a) = 0$ and $\phi(b) = 0$, so the function ϕ satisfies the hypotheses of Rolle's Theorem on $[a, b]$. Hence by Rolle's Theorem there is a c in (a, b) such that

$$\phi'(c) = f'(c) - \frac{f(b) - f(a)}{b - a} = 0,$$

from which we obtain the conclusion

$$f'(c) = \frac{f(b) - f(a)}{b - a}.$$

The geometric significance of the mean value theorem is that if f satisfies the hypotheses of the theorem, then for some number c (not necessarily unique) between a and b, the tangent to the graph of f at the point $(c, f(c))$ is parallel to the secant through the points $(a, f(a))$ and $(b, f(b))$ (Figure 5–6).

Example 3 If $f(x) = x^3$ on the interval $[1, 4]$, find a number c in $(1, 4)$ such that

$$\frac{f(4) - f(1)}{4 - 1} = f'(c).$$

SOLUTION Since f satisfies the hypotheses of the mean value theorem on $[1, 4]$, by this theorem there is a c in $(1, 4)$ such that $f'(c) = (f(4) - f(1))/(4 - 1)$ or, saying the same thing,

$$3c^2 = \frac{64 - 1}{4 - 1} = 21.$$

Then

$$c^2 = 7$$
$$c = \pm\sqrt{7}.$$

However, since $-\sqrt{7} \notin (1, 4)$, $c = \sqrt{7}$ is the required number.

Exercise Set 5.2

In Exercises 1–5 show that the given function satisfies the hypotheses of the mean value theorem on the given interval. Then find a number c in this interval which results from the application of this theorem.

1. $f(x) = \sqrt{x}$, $[4, 9]$

2. $g(x) = 1/x$, $[\frac{1}{2}, 3]$

3. $f(x) = x^3 - 4x^2 + 5$, $[0, 2]$

4. $f(x) = x^4 - 4x$, $[-1, 3]$

5. $f(x) = (2x + 1)^3$, $[1, 2]$

6. (a) If $f(x) = 1/x^2$, tell why the mean value theorem cannot be invoked to say there is some number c in the interval $(-1, 1)$ such that

$$\frac{f(1) - f(-1)}{1 - (-1)} = 0 = f'(c).$$

(b) Answer the same question as in (a) if $f(x) = |x|$.

(c) Answer the same question as in (a) if

$$f(x) = \begin{cases} x + 1 & \text{if } -1 \le x < 1 \\ 0 & \text{if } x = 1 \end{cases}.$$

7. Complete the proof of Rolle's Theorem by supplying the details in the proof of Case 3.

8. Verify the remark made in the proof of the mean value theorem about the geometric significance of $\phi(x)$.

9. For a particular function f, f' is continuous on the interval $[a, b]$. Prove that there are numbers u and v in the interval for which $f(a) + (b - a)f'(u) \le f(b) \le f(a) + (b - a)f'(v)$.

5.3   Monotonic Functions;
First Derivative Test

We first define an increasing function and a decreasing function on an interval.

5.3.1   Definition

Let the function f be defined on an interval I.

(a) f is *increasing* on I if and only if for every pair of numbers x_1 and x_2 in I where $x_2 > x_1$ we have $f(x_2) > f(x_1)$ (Figure 5–7(a)).

(b) f is *decreasing* on I if and only if for every pair of numbers x_1 and x_2 in I where $x_2 > x_1$, we have $f(x_2) < f(x_1)$ (Figure 5–7(b)).

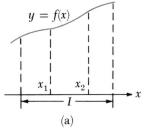   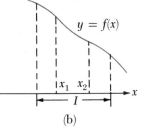

(a)   (b)

Figure 5–7

A function which is either increasing or decreasing on an interval is said to be *monotonic* on the interval.

The function f where $f(x) = x^2$ is increasing on the interval $[0, +\infty)$ since whenever $x_2 > x_1 \ge 0$, we have $x_2{}^2 > x_1{}^2$. The function g where $g(x) = 3 - x$ is decreasing on the interval $(-\infty, +\infty)$ since whenever $x_2 > x_1$ we have $3 - x_2 < 3 - x_1$.

Rather than attempting to prove that a function is monotonic on an interval using Definition 5.3.1 directly, it will generally be easier to accomplish this using theorems derived from Definition 5.3.1.

5.3.2   Theorem

Let f be continuous on an interval I and differentiable in the interior of the interval.

(a) *f is increasing on I if $f'(x) > 0$ for every x in the interior of I.*
(b) *f is decreasing on I if $f'(x) < 0$ for every x in the interior of I.*

PROOF OF (a)   Let x_1 and x_2 be any two numbers in I such that $x_2 > x_1$. f satisfies the hypotheses of the mean value theorem on the interval $[x_1, x_2]$ and therefore by this theorem there is a $c \in (x_1, x_2)$ such that

$$\frac{f(x_2) - f(x_1)}{x_2 - x_1} = f'(c) > 0. \qquad (5.6)$$

Since $x_2 > x_1$, $x_2 - x_1 > 0$, and if we multiply the left and right members of (5.6) by $x_2 - x_1$, we obtain $f(x_2) - f(x_1) > 0$. Hence

$$f(x_2) > f(x_1).$$

The conclusion then follows from Definition 5.3.1(a).
    The proof of Theorem 5.3.2(b) is similar and is therefore omitted.

Example 1   Let $f(x) = (x + 2)^{2/3}(4 - x)$. In what intervals is f increasing? decreasing? (Figure 5–9)

SOLUTION   We shall examine the sign of $f'(x)$ using Theorem 5.3.2.

$$f'(x) = (x + 2)^{2/3}(-1) + \tfrac{2}{3}(x + 2)^{-1/3}(4 - x)$$
$$= \tfrac{1}{3}(x + 2)^{-1/3}[-3(x + 2) + 2(4 - x)]$$
$$= \tfrac{1}{3}(x + 2)^{-1/3}(-5x + 2).$$

We graph the factors of $f'(x)$ as to algebraic sign (Figure 5–8). From Figure 5–8, we note that $f'(x) < 0$ on $(-\infty, -2)$ and $(\tfrac{2}{5}, +\infty)$. Hence f is decreasing on $(-\infty, -2]$ and $[\tfrac{2}{5}, +\infty)$ by Theorem 5.3.2(b).

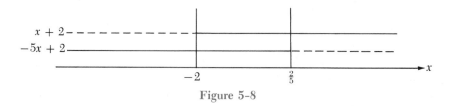

Figure 5–8

Also, since $f'(x) > 0$ on $(-2, \tfrac{2}{5})$, f is increasing on $[-2, \tfrac{2}{5}]$ by Theorem 5.3.2(a).

The graph of the function f considered in Example 1 (Figure 5–9) has the point $(\tfrac{2}{5}, \tfrac{18}{5}(\tfrac{12}{5})^{2/3})$ which appears to be above all other points in the vicinity

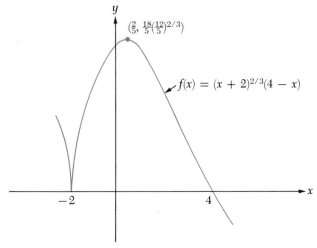

Figure 5-9

of $x = \frac{2}{5}$ and also has the point $(-2, 0)$ which seems below all other points in the vicinity of $x = -2$. We will prove that $f(\frac{2}{5}) = \frac{18}{5}(\frac{12}{5})^{2/3}$ is a *relative maximum* and $f(-2) = 0$ is a *relative minimum* of the function f.

5.3.3 Definition

Let f be defined at c.

    (a) $f(c)$ is a *relative maximum* (*value*) of the function f if and only if there is an open interval (a, b) containing c such that $f(c)$ is a maximum of f on (a, b).

    (b) $f(c)$ is a *relative minimum* (*value*) of the function f if and only if there is an open interval (a, b) containing c such that $f(c)$ is a minimum of f on (a, b).

    A relative maximum or minimum value of a function is often called a *relative extremum* of the function.

    Suppose f is continuous on an interval $[a, b]$ and $c \in (a, b)$. If f is increasing on $[a, c]$ and decreasing on $[c, b]$, then intuition suggests that $f(c)$ must be a relative maximum of f (note the sketch of the graph of f in Figure 5–10(a)).

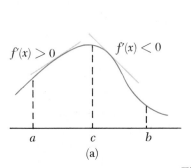

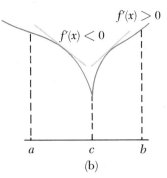

Figure 5-10

By Theorem 5.3.2 f is increasing on $[a, c]$ and decreasing on $[c, b]$ if its derivative, f', is positive-valued on (a, c) and negative-valued on (c, b). Thus, part (a) of the following theorem is plausible. A similar discussion suggests the reasonableness of part (b) of the theorem.

5.3.4 Theorem (First Derivative Test)

Let f be continuous on the closed interval $[a, b]$ and let c, a critical number of f, be in the open interval (a, b).

> (a) *If $f'(x) > 0$ for all $x \in (a, c)$ and $f'(x) < 0$ for all $x \in (c, b)$, then $f(c)$ is a relative maximum of f (Figure 5–10(a)).*
> (b) *If $f'(x) < 0$ for all $x \in (a, c)$ and $f'(x) > 0$ for all $x \in (c, b)$, then $f(c)$ is a relative minimum of f (Figure 5–10(b)).*
> (c) *If $f'(x) > 0$ or $f'(x) < 0$ for all $x \in (a, c) \cup (c, b)$, then $f(c)$ is not a relative extremum of f.*

PROOF OF THEOREM 5.3.4(a) The proof is based upon our preliminary remarks made before stating this theorem. Note from the hypotheses given for part (a) that f satisfies the hypotheses of Theorem 5.3.2(a) on $[a, c]$ and of Theorem 5.3.2(b) on $[c, b]$. Hence by these theorems f is increasing on $[a, c]$ and decreasing on $[c, b]$. Therefore by Definition 5.3.1(a)

$$f(x) < f(c) \quad \text{whenever } x \in (a, c),$$

and by Definition 5.3.1(b)

$$f(x) < f(c) \quad \text{whenever } x \in (c, b).$$

Thus $f(c)$ is a maximum of f on (a, b) and by Definition 5.3.3(a) $f(c)$, is a relative maximum of f.

The proofs for Theorem 5.3.4(b) and (c) are left as exercises.

When the hypotheses given in the first sentence of Theorem 5.3.4 are satisfied, $f(c)$ is a relative extremum if and only if there is a *change in sign* of $f'(x)$ as x increases past c. If this sign change in $f'(x)$ is from $+$ to $-$, then $f(c)$ is a relative maximum by Theorem 5.3.4(a). If the sign change is from $-$ to $+$, $f(c)$ is a relative minimum by Theorem 5.3.4(b). When there is no change in sign of $f'(x)$ as x increases past c, $f(c)$ is not a relative extremum by Theorem 5.3.4(c).

In determining relative extrema using the First Derivative Test, the following steps should be followed.

> (i) Obtain the critical numbers $c_1, c_2, \ldots, c_k$ of the function f.
> (ii) Examine the sign of $f'(x)$ on each of the intervals $(-\infty, c_1), (c_1, c_2),$ $\ldots, (c_k, +\infty)$ determined by the critical numbers.
> (iii) Apply the First Derivative Test to each critical number.

Example 2 If $f(x) = (x + 2)^{2/3}(4 - x)$, find the relative extrema of f.

SOLUTION   This function, which was discussed in Example 1 (Figure 5–9), has -2 and $\frac{2}{5}$ as its only critical numbers. From Example 1

$$f'(x) < 0 \quad \text{if } x \in (-\infty, -2)$$
$$f'(x) > 0 \quad \text{if } x \in (-2, \tfrac{2}{5})$$
$$f'(x) < 0 \quad \text{if } x \in (\tfrac{2}{5}, +\infty)$$

Therefore by Theorem 5.3.4(b) $f(-2) = 0$ is a relative minimum and by Theorem 5.3.4(a) $f(\tfrac{2}{5}) = \tfrac{18}{5}(\tfrac{12}{5})^{2/3}$ is a relative maximum.

Example 3   If $f(x) = \frac{1}{4}x^4 - x^3 + 2$, find the relative extrema of f (Figure 5–12).

SOLUTION   To find the critical numbers of f, we take

$$f'(x) = x^3 - 3x^2 = x^2(x - 3)$$

Since $f'(x) = 0$ if and only if $x = 0$ or 3, 0 and 3 are the only critical numbers of f.

From the sketch of the signs of the factors of $f'(x)$ (Figure 5–11), we

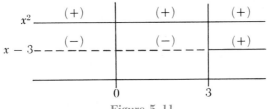

Figure 5-11

find that

$$f'(x) < 0 \quad \text{if } x \in (-\infty, 0)$$
$$f'(x) < 0 \quad \text{if } x \in (0, 3)$$
$$f'(x) > 0 \quad \text{if } x \in (3, +\infty)$$

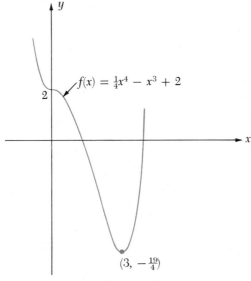

Figure 5-12

By Theorem 5.3.4(c), $f(0) = 2$ is not a relative extremum, and by Theorem 5.3.4(b) $f(3) = -\frac{19}{4}$ is a relative minimum.

Exercise Set 5.3

In Exercises 1–13 determine (a) the intervals where the function is increasing, (b) the intervals where the function is decreasing, and (c) the relative extrema of the function. Sketch the graph of the function.

1. $f(x) = x^3 + \frac{3}{2}x^2 - 6x + 1$ 2. $f(x) = -\frac{5}{3}x^3 + \frac{13}{2}x^2 + 6x - 1$

3. $f(x) = x^5 - 15x^3$ 4. $f(x) = x^6 - 12x^4$

5. $f(x) = \dfrac{4x}{x^2 + 1}$ 6. $f(x) = x^{1/3} + 3x^{-1/3}$

7. $f(x) = (3x - x^2)^{1/3}$ 8. $f(x) = x^{1/2}(x + 2)^2$

9. $f(x) = x\sqrt{2 - x}$ 10. $f(x) = \dfrac{2x - 3}{(x + 1)^2}$

11. $f(x) = (x + 1)^{1/2}(2x - 1)^{3/2}$ 12. $f(x) = (x^2 - 1)^6$ if $\mathcal{D}_f = [-2, 1]$

13. $f(x) = \begin{cases} |1 - x^2| & \text{if } x \geq -1 \\ 0 & \text{if } x < -1 \end{cases}$

In Exercises 14–19 answer the question "T" if always true or "F" if not always true. Justify your answer.

14. A function f could not have a relative extremum at c if $f'(c)$ exists but is $\neq 0$.

15. If $f(c)$ is a relative minimum, then $f'(c) = 0$.

16. A function f may have a relative extremum at an endpoint of its domain.

17. If $f'(x) = 0$ for every x in some open interval containing c, $f(c)$ is a relative minimum of the function f.

18. A relative maximum of a function is at least as large as any relative minimum of the function.

19. $f(c)$ may be a relative extremum of f even though f is discontinuous at c.

20. Prove Theorem 5.3.4(b).

21. Prove Theorem 5.3.4(c).

5.4   Concavity of a Graph

       In sketching the graph of a function, it is desirable to know where the graph of the function is *concave upward* and where it is *concave downward*. In ordinary language, the graph of a function f is concave upward at the point $(c, f(c))$ if the points on the graph of f near $(c, f(c))$ are above the tangent to the graph of f at $(c, f(c))$ (Figure 5–13(a)). Also the graph of f is concave downward at $(c, f(c))$ if the points on the graph of f near $(c, f(c))$ are below the tangent to the graph of f at $(c, f(c))$ (Figure 5–13(b)).

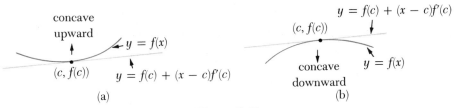

Figure 5-13

Since the tangent to the graph of f at $(c, f(c))$ has an equation $y = f(c) + (x - c)f'(c)$, we can give the following more precise definition of concavity of a graph at a point.

5.4.1   Definition

Let f be differentiable at c.

(a) The graph of f is *concave upward* at $(c, f(c))$ if and only if there is some open interval I containing c and contained in $\mathcal{D}_f$ such that

$$f(x) > f(c) + (x - c)f'(c) \quad \text{if } x \neq c \text{ is in } I.$$

(b) The graph of f is *concave downward* at $(c, f(c))$ if and only if there is some open interval I containing c and contained in $\mathcal{D}_f$ such that

$$f(x) < f(c) + (x - c)f'(c) \quad \text{if } x \neq c \text{ is in } I.$$

For example, the graph of $f(x) = x^2$ is concave upward at $(0, 0)$ since $x^2 > 0 + (x - 0) \cdot 0$ for every $x \neq 0$.

The following theorem, utilizing the second derivative of a function, is useful in determining the concavity of the graph of the function at a point.

5.4.2   Theorem

Let f be differentiable on an open interval containing c.

(a) *If $f''(c) > 0$, the graph of f is concave upward at $(c, f(c))$.*
(b) *If $f''(c) < 0$, the graph of f is concave downward at $(c, f(c))$.*

PROOF OF (a)   If $f''(c) > 0$, by Theorem 3.3.3, there is a $\delta > 0$ such that

$$\frac{f'(c + h) - f'(c)}{h} > 0 \quad \text{if } h \neq 0 \text{ and } h \in (-\delta, \delta). \tag{5.7}$$

Letting $x = c + h$, (5.7) can be restated to read

$$\frac{f'(x) - f'(c)}{x - c} > 0 \quad \text{if } x \neq c \text{ and } x \in (c - \delta, c + \delta). \tag{5.8}$$

Now consider

$$\phi(x) = f(x) - f(c) - (x - c)f'(c) \quad \text{if } x \neq c \text{ and } x \in (c - \delta, c + \delta). \quad (5.9)$$

$\phi(x)$ represents the vertical "signed" distance to the graph of f from a point (x, y) on the tangent to the graph at $(c, f(c))$ (Figure 5–14). ϕ satisfies the hypotheses of

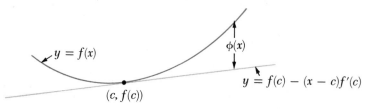

Figure 5-14

the mean value theorem on the closed interval with endpoints c and x. Then by the mean value theorem

$$\phi(x) = (x - c)f'(d) - (x - c)f'(c) \quad \text{where } d \text{ is between } c \text{ and } x$$
$$= (x - c)(f'(d) - f'(c)). \quad (5.10)$$

From (5.8) $f'(x) - f'(c)$ and $x - c$ are of the same sign. Also, since d is between c and x, $d \in (c - \delta, c + \delta)$, and therefore from (5.8) $f'(d) - f'(c)$ and $d - c$ are of the same sign. Since d is between c and x, $d - c$ and $x - c$ are of the same sign. Hence

$$f'(d) - f'(c) \text{ and } x - c \text{ are of the same sign.}$$

From (5.10) $\phi(x) > 0$ and hence from (5.9)

$$f(x) > f(c) + (x - c)f'(c) \quad \text{if } x \neq c \text{ and } x \in (c - \delta, c + \delta).$$

Therefore the graph of f is concave upward at $(c, f(c))$ by Definition 5.4.1(a).

The proof of (b) is left as an exercise.

Example 1 If $f(x) = x^4 - 6x^3 + 12x^2 - 8x + 1$, at what points $(x, f(x))$ is the graph of f concave upward? concave downward? (Figure 5–15)

SOLUTION From the given equation we obtain:

$$f'(x) = 4x^3 - 18x^2 + 24x - 8$$
$$f''(x) = 12x^2 - 36x + 24$$
$$= 12(x^2 - 3x + 2)$$
$$= 12(x - 1)(x - 2)$$

From considering the signs of the factors $x - 1$ and $x - 2$, we note that

$$f''(x) > 0 \quad \text{if } x < 1 \text{ or } x > 2$$
$$f''(x) < 0 \quad \text{if } 1 < x < 2.$$

From Theorem 5.4.2(a) the graph is concave upward where $x < 1$ or $x > 2$, and from Theorem 5.4.2(b) the graph is concave downward where $1 < x < 2$.

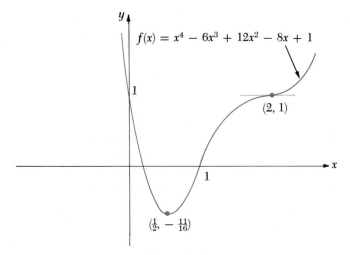

Figure 5-15

Example 2   If $f(x) = 4x^{1/3} - x^{4/3}$, at what points $(x, f(x))$ is the graph of f concave upward? concave downward? (Figure 5-16)

SOLUTION   From the given equation we obtain

$$f'(x) = \tfrac{4}{3}x^{-2/3} - \tfrac{4}{3}x^{1/3}$$
$$f''(x) = -\tfrac{8}{9}x^{-5/3} - \tfrac{4}{9}x^{-2/3}$$
$$= -\tfrac{4}{9}x^{-5/3}(2 + x).$$

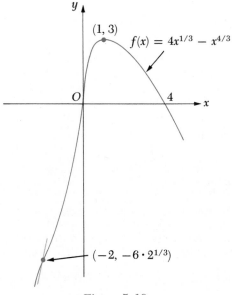

Figure 5-16

From considering the signs of the factors $-\frac{4}{9}x^{-5/3}$ and $2 + x$, we note

$$f''(x) > 0 \quad \text{if } -2 < x < 0$$
$$f''(x) < 0 \quad \text{if } x < -2 \text{ or } x > 0.$$

From Theorem 5.4.2(a) the graph is concave upward where $-2 < x < 0$ and from Theorem 5.4.2(b) the graph is concave downward where $x < -2$ or $x > 0$.

We note that the direction of concavity of the graph of $f(x) = 4x^{1/3} - x^{4/3}$ changes at $x = -2$ and $x = 0$. Points on a graph of a function where the direction of concavity changes are named in Definition 5.4.3 below.

It should be emphasized that the converses of Theorem 5.4.2(a) and (b) are not true. For example, the graph of $f(x) = x^4$ is concave upward at $(0, 0)$ yet $f''(0) = 0$. The student is asked to supply another counter-example in Exercise 12 of Exercise Set 5.4.

5.4.3 Definition

Let f be a function that is differentiable for every $x \neq c$ in some open interval (a, b) containing c and let the graph of f have a tangent at the point $(c, f(c))$. Then $(c, f(c))$ is an *inflection point* of the graph if and only if

(a) the graph is concave upward at $(x, f(x))$ if $x \in (a, c)$ and concave downward at $(x, f(x))$ if $x \in (c, b)$

or

(b) the graph is concave downward at $(x, f(x))$ if $x \in (a, c)$ and concave upward at $(x, f(x))$ if $x \in (c, b)$.

For example, the graph of $f(x) = x^4 - 6x^3 + 12x^2 - 8x + 1$ (Figure 5–15) has inflection points at $(1, 0)$ and $(2, 1)$ and the graph of $f(x) = 4x^{1/3} - x^{4/3}$ (Figure 5–16) has inflection points at $(0, 0)$ and $(-2, -6 \cdot 2^{1/3})$. An inflection point can only occur where there is a change in the direction of concavity of the graph. Thus a point $(c, f(c))$ in the graph of f *may be* an inflection point of the graph if either (i) $f''(c) = 0$ or (ii) f is continuous at c and $f''(c)$ fails to exist. In connection with (i) the student is asked to prove that if $(c, f(c))$ is an inflection point of the graph of f, and f'' is continuous on an open interval containing c, then $f''(c) = 0$.

Exercise Set 5.4

In Exercises 1–10 find (a) the points $(x, f(x))$ at which the graph of f is concave upward, (b) the points $(x, f(x))$ at which the graph is concave downward, (c) the inflection points of the graph, and (d) the slope of the tangent to the graph at the inflectional point. Sketch the graph of the function.

1. $f(x) = x^2 - 4x + 6$ 2. $f(x) = (x - 3)^3$

3. $f(x) = -5 + 15x + 6x^2 - x^3$ 4. $f(x) = x^3 - 2x^2 - 4x + 1$

5. $f(x) = 1 - 18x^2 + 8x^3 - x^4$ 6. $f(x) = x^4 - 4x + 2$

7. $f(x) = x^{5/3} - 5x^{2/3}$ 8. $f(x) = x^{2/3} + 3x^{1/3} + 2$

9. $f(x) = \dfrac{x}{\sqrt{2x+1}}$ 10. $f(x) = x\sqrt{25 - 4x^2}$

11. Prove Theorem 5.4.2(b).

12. State a value for n such that the graph of a function $f(x) = x^n$ is concave upward at $(0, 0)$ yet $f''(0)$ fails to exist.

13. Prove: If the graph of f is concave upward at $(c, f(c))$ and $f''(c)$ exists but is not zero, then $f''(c) > 0$.

14. Prove: If the graph of f is concave downward at $(c, f(c))$ and $f''(c)$ exists but is not zero, then $f''(c) < 0$.

15. If $f''(x)$ exists on some open interval containing c, $f''(c) = 0$, and $f'''(c) \neq 0$, then $(c, f(c))$ is an inflection point of the graph of f.

16. Prove: If $(c, f(c))$ is an inflection point of the graph of f and f'' is continuous on an open interval containing c, then $f''(c) = 0$.

5.5   Second Derivative Test

Suppose that at a number c in the domain of a function f, $f'(c) = 0$. Geometrically, this means that the graph of f has a horizontal tangent at the point $(c, f(c))$. If f is also differentiable on an open interval containing c, and $f''(c) > 0$, then the graph of f is concave upward at $(c, f(c))$ (Figure 5–17(a)), and intuitively it would seem that $f(c)$ is a relative minimum of f. If instead of having $f''(c) > 0$, we had $f''(c) < 0$, the graph of f would be concave downward at $(c, f(c))$ (Figure 5–17(b)), and $f(c)$ would appear to be a relative maximum of f.

These conclusions are stated in the next theorem.

$f''(c) > 0$
$f'(c) = 0$ $y = f(x)$
$(c, f(c))$
(a)

$(c, f(c))$
$f''(c) < 0$ $y = f(x)$
$f'(c) = 0$
(b)

Figure 5–17

5.5.1   Theorem (Second Derivative Test)

Let f be differentiable on some open interval containing c.

(a) *If $f'(c) = 0$, and $f''(c) > 0$, then $f(c)$ is a relative minimum of f.*
(b) *If $f'(c) = 0$, and $f''(c) < 0$, then $f(c)$ is a relative maximum of f.*

PROOF OF (a)   Since $f''(c) > 0$, by Theorem 5.4.2, the graph of f is concave upward at $(c, f(c))$. Then from Definition 5.4.1 there is some open interval I containing c and contained in $\mathcal{D}_f$ such that

$$f(x) > f(c) + (x - c)f'(c) \quad \text{if } x \neq c \text{ is in } I. \tag{5.11}$$

Since $f'(c) = 0$ by hypothesis, from (5.11) we have

$$f(x) \geq f(c) \quad \text{if } x \in I$$

and thus $f(c)$ is a relative minimum of f.

The proof of (b) is analogous and is left as an exercise.

The second derivative test is frequently useful in determining relative extrema in situations where the hypotheses of the test are satisfied. If, however, c is a critical number of the function f, but $f''(c) = 0$ or $f''(c)$ fails to exist, the second derivative test gives no information as to whether $f(c)$ is a relative extremum, and in such cases the first derivative test should be used, if possible. The first derivative test should also be used if $f''(c)$ is particularly difficult to calculate.

Example 1 Find the relative extrema of the function f where $f(x) = x^3 - 3x^2 + 4$ (Figure 5–18).

PROOF   We first obtain the critical numbers of f. Since

$$f'(x) = 3x^2 - 6x$$
$$= 3x(x - 2)$$

the critical numbers are $x = 0$ and 2, and

$$f'(0) = 0$$

and

$$f'(2) = 0.$$

The second derivative of f is given by

$$f''(x) = 6x - 6.$$

Since

$$f''(0) = -6 < 0$$
$$f''(2) = 6 > 0,$$

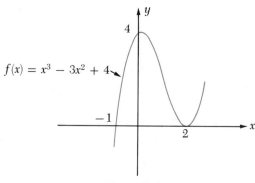

$$f(x) = x^3 - 3x^2 + 4$$

Figure 5–18

by Theorem 5.5.1(a), $f(2) = 0$ is a relative minimum, and by Theorem 5.5.1(b), $f(0) = 4$ is a relative maximum.

Example 2   Find the relative extrema of the function f where $f(x) = x(3 - x)^{5/3}$ (Figure 5–19).

PROOF   We can write

$$f'(x) = (3 - x)^{5/3} - x\tfrac{5}{3}(3 - x)^{2/3}$$
$$= \tfrac{1}{3}(3 - x)^{2/3}[3(3 - x) - 5x]$$
$$= \tfrac{1}{3}(3 - x)^{2/3}(9 - 8x)$$

so the critical numbers are $x = 3$ and $\tfrac{9}{8}$. At each of these numbers f' assumes the value zero. We then obtain

$$f''(x) = \tfrac{1}{3}[(3 - x)^{2/3}(-8) + (9 - 8x) \cdot \tfrac{2}{3}(3 - x)^{-1/3}(-1)]$$
$$= -\tfrac{2}{9}(3 - x)^{-1/3}[12(3 - x) + (9 - 8x)]$$
$$= -\tfrac{2}{9}(3 - x)^{-1/3}(45 - 20x). \tag{5.12}$$

From (5.12), $f''(\tfrac{9}{8}) = -5(\tfrac{15}{8})^{-1/3} < 0$ and hence by Theorem 5.5.1(b), $f(\tfrac{9}{8}) = \tfrac{9}{8}(\tfrac{15}{8})^{5/3}$ is a relative maximum. Since $f''(3)$ does not exist, the second derivative test gives no information as to whether $f(3)$ is a relative extremum. However, by the first derivative test, $f(3) = 0$ is not a relative extremum. It can be shown that $(3, 0)$ is an inflection point of the graph of f in this example.

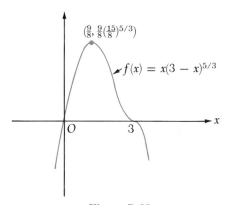

$$\left(\tfrac{9}{8}, \tfrac{9}{8}(\tfrac{15}{8})^{5/3}\right)$$

$$f(x) = x(3 - x)^{5/3}$$

Figure 5–19

Exercise Set 5.5

In Exercises 1–14 find the relative extrema of f. Use Theorem 5.5.1 if possible.

1. $f(x) = 3x^2 - 4x + 2$

2. $f(x) = (x - 2)^3$

3. $f(x) = (x + 1)^4$

4. $f(x) = x^3 - 2x^2 - 4x + 1$

5. $f(x) = x^3 + 2x^2 - 7x - 4$

6. $f(x) = 3x^5 - 5x^4$

7. $f(x) = 3x^4 + 4x^3 - 12x^2 + 12$

8. $f(x) = \dfrac{x + 2}{x^2 + 2x + 9}$

9. $f(x) = \dfrac{10x - 10}{x^2 + 8}$ 10. $f(x) = \dfrac{x^2}{\sqrt{x + 1}}$

11. $f(x) = x^2 \sqrt[3]{x + 4}$ 12. $f(x) = (x - 1)^{2/3}(x + 2)^{1/3}$

13. $f(x) = x^{3/2}(x - 4)^{1/2}$

14. $f(x) = (a_1 - x)^2 + (a_2 - x)^2 + \cdots + (a_n - x)^2$

15. Prove Theorem 5.5.1(b).

16. Determine a, b, and c so that the function f where $f(x) = x^3 + ax^2 + bx + c$ has a critical number at 2 and $f(2) = -3$ is not a relative extremum of f.

17. Prove that if the function f where $f(x) = x^3 + ax^2 + bx + c$ and $a^2 > 3b$ has a critical number, then f has both a relative maximum and a relative minimum.

5.6   Other Limits Involving Infinity

Suppose we consider the graph (Figure 5–20) of the function f where

$$f(x) = \frac{2x^2}{x^2 + 3}. \tag{5.13}$$

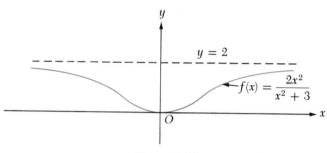

Figure 5–20

We note that if the numerator and denominator of the right member of (5.13) are each divided by x^2, then $f(x)$ can be expressed in the form

$$f(x) = \begin{cases} \dfrac{2}{1 + (3/x^2)} & \text{if } x \neq 0, \\ 0 & \text{if } x = 0. \end{cases}$$

We can readily show that $\dfrac{2}{1 + (3/x^2)} < 2$ if $x \neq 0$. Intuition suggests that as x increases without bound, $3/x^2$ decreases toward 0, and hence $\dfrac{2}{1 + (3/x^2)}$ increases toward 2. It would also appear that we could have $f(x)$ differ from 2 by as little as we please if x is chosen sufficiently large. For example if $x = 10$, then $f(10) = \frac{200}{103}$; if $x = 1000$, then $f(1000) = \frac{2,000,000}{1,000,103}$.

The foregoing discussion suggests a new kind of limit, $\lim_{x \to +\infty} f(x) = L$, which is read "the limit of f at $+\infty$ is L." This limit is defined in the next statement.

5.6.1 Definition

Suppose $(c, +\infty) \subset \mathcal{D}_f$ for some number c. $\lim_{x \to +\infty} f(x) = L$ if and only if for every $\epsilon > 0$ there is an $M > 0$ such that $|f(x) - L| < \epsilon$ if $x > M$.

Example 1 Prove that $\lim\limits_{x \to +\infty} \dfrac{2x^2}{x^2 + 3} = 2$.

PROOF Let $\epsilon > 0$ be given. We can write

$$\left| \frac{2x^2}{x^2 + 3} - 2 \right| = \left| \frac{-6}{x^2 + 3} \right| = \frac{6}{x^2 + 3} < \frac{6}{x^2}.$$

Now $6/x^2 < \epsilon$ if $6 < \epsilon x^2$ and hence if $x > \sqrt{6/\epsilon}$. Since we have shown that

$$\left| \frac{2x^2}{x^2 + 3} - 2 \right| < \epsilon$$

if $x > M = \sqrt{6/\epsilon}$, the conclusion follows from Definition 5.6.1.

We also note that as x decreases without bound, the numbers $f(x)$ defined by (5.13) also increase toward 2. For example, if $x = -10$, $f(-10) = \frac{200}{103}$; if $x = -1,000, f(-1000) = \frac{2,000,000}{1,000,003}$. Thus, we are led to the limit, $\lim_{x \to -\infty} f(x) = L$, which is read "the limit of f at $-\infty$ is L."

5.6.2 Definition

Suppose $(-\infty, c) \subset \mathcal{D}_f$ for some number c. $\lim_{x \to -\infty} f(x) = L$ if and only if for every $\epsilon > 0$ there is an $M > 0$ such that $|f(x) - L| < \epsilon$ if $x < -M$.

We could show by an argument similar to that used in Example 1 that for every $\epsilon > 0$,

$$\left| \frac{2x^2}{x^2 + 3} - 2 \right| < \epsilon \quad \text{if } x < -\sqrt{\frac{6}{\epsilon}}$$

and hence

$$\lim_{x \to -\infty} \frac{2x^2}{x^2 + 2} = 2.$$

We also consider another example.

Example 2 Prove that $\lim_{x \to -\infty} \dfrac{3x^3 + 2}{4x^3} = \dfrac{3}{4}$.

PROOF Let $\epsilon > 0$ be given; we can write

$$\left| \frac{3x^3 + 2}{4x^3} - \frac{3}{4} \right| = \left| \frac{1}{2x^3} \right| = \frac{1}{2|x|^3}.$$

Now $\dfrac{1}{2|x|^3} < \epsilon$ if $1 < 2\epsilon|x|^3$

$\qquad\qquad\qquad\qquad\qquad$ if $\dfrac{1}{2\epsilon} < |x|^3$

$\qquad\qquad\qquad\qquad\qquad$ if $\dfrac{1}{(2\epsilon)^{1/3}} < |x|$ $\qquad$ by **Theorem 1.3.9(a)**

and hence $\qquad\qquad\qquad\qquad$ if $x < -\dfrac{1}{(2\epsilon)^{1/3}}.$ $\qquad$ by **Theorem 1.5.3(c)**

We could show by a similar argument that $\lim\limits_{x \to +\infty} \dfrac{3x^3 + 2}{4x^3} = \dfrac{3}{4}.$
The proof of the next theorem is left as an exercise.

5.6.3   Theorem

If n is any positive rational number, and k is any number

(a) $\lim\limits_{x \to +\infty} \dfrac{k}{x^n} = 0;$

(b) $\lim\limits_{x \to -\infty} \dfrac{k}{x^n} = 0.$

As an illustration of Theorem 5.6.3(a)

$$\lim\limits_{x \to +\infty} 3x^{-4/5} = \lim\limits_{x \to +\infty} \dfrac{3}{x^{4/5}} = 0.$$

The following limit theorems can be proved. In Theorems 5.6.4 to 5.6.10 the proofs, which are analogous to those for corresponding limit theorems in Chapter 3, are omitted. These theorems are also true if we replace "$x \to +\infty$" by "$x \to -\infty$", "$(c, +\infty)$" by "$(-\infty, c)$", and "$x > c$" by "$x < c$".

5.6.4   Theorem

If k is any number, $\lim_{x \to +\infty} k = k.$

5.6.5   Theorem

If $\lim_{x \to +\infty} f(x) = L_1$ and $\lim_{x \to +\infty} g(x) = L_2$, and $(c, +\infty) \subset \mathcal{D}_f \cap \mathcal{D}_g$ for some number c,

(a) $\lim\limits_{x \to +\infty} [f(x) + g(x)] = L_1 + L_2;$

(b) $\lim\limits_{x \to +\infty} [f(x) - g(x)] = L_1 - L_2;$

(c) $\displaystyle\lim_{x\to+\infty} f(x)g(x) = L_1 L_2$;

(d) $\displaystyle\lim_{x\to+\infty} \frac{f(x)}{g(x)} = \frac{L_1}{L_2}$ *when* $L_2 \neq 0$.

5.6.6   Theorem

If $\lim_{x\to+\infty} f_1(x) = L_1, \lim_{x\to+\infty} f_2(x) = L_2, \dots$, *and* $\lim_{x\to+\infty} f_n(x) = L_n$ *and* $(c, +\infty)$ *is a subset of the intersection of the domains of* $f_1, f_2, \dots, f_n$ *for some number c, then*

(a) $\displaystyle\lim_{x\to+\infty} [f_1(x) + f_2(x) + \cdots + f_n(x)] = L_1 + L_2 + \cdots + L_n$;

(b) $\displaystyle\lim_{x\to+\infty} f_1(x) f_2(x) \cdots f_n(x) = L_1 L_2 \cdots L_n$.

5.6.7   Theorem

Let $F(x) = G(x)$ *for every* $x > c$ *where c is some number. Then*

(a) $\displaystyle\lim_{x\to+\infty} F(x) = L$ *if* $\displaystyle\lim_{x\to+\infty} G(x) = L$;

(b) $\displaystyle\lim_{x\to+\infty} F(x)$ *fails to exist if* $\displaystyle\lim_{x\to+\infty} G(x)$ *fails to exist.*

5.6.8   Theorem

If $f(x) \leq g(x) \leq h(x)$ *for every* $x > c$ *where c is some number, and* $\lim_{x\to+\infty} f(x) = \lim_{x\to+\infty} h(x) = L$, *then* $\lim_{x\to+\infty} g(x) = L$.

5.6.9   Theorem (Composite Function Limit Theorem)

If $\lim_{x\to+\infty} g(x) = L$, *f is continuous at L, and* $(c, +\infty) \subset \mathcal{D}_{f\circ g}$ *for some number c, then*

$$\lim_{x\to+\infty} (f \circ g)(x) = \lim_{x\to+\infty} f(g(x)) = f(\lim_{x\to+\infty} g(x)) = f(L).$$

5.6.10   Theorem

If $\lim_{x\to+\infty} g(x) = L$ *and* $L^{p/q}$ *is a real number where* $q > 0$ *and p are integers, then*

$$\lim_{x\to+\infty} (g(x))^{p/q} = L^{p/q}.$$

5.6.11   Theorem

If P is a polynomial function other than a constant function, $\lim_{x\to+\infty} P(x)$ *fails to exist.*

The proof of Theorem 5.6.11 is left as an exercise.

Example 1   Find $\lim\limits_{x \to +\infty} \dfrac{3x^2 - 5x - 1}{7 + 2x - 4x^2}$.

SOLUTION   From Theorem 5.6.11 the limits of the numerator and denominator at $+\infty$ fail to exist. To express $(3x^2 - 5x - 1)/(7 + 2x - 4x^2)$ in a more manageable form, we divide the numerator and denominator by the highest power of x in either the numerator or denominator, namely x^2. Then

$$\lim_{x \to +\infty} \frac{3x^2 - 5x - 1}{7 + 2x - 4x^2} = \lim_{x \to +\infty} \frac{3 - \dfrac{5}{x} - \dfrac{1}{x^2}}{\dfrac{7}{x^2} + \dfrac{2}{x} - 4} \qquad \text{by Theorem 5.6.7(a)}$$

$$= \frac{\lim\limits_{x \to +\infty} \left(3 - \dfrac{5}{x} - \dfrac{1}{x^2}\right)}{\lim\limits_{x \to +\infty} \left(\dfrac{7}{x^2} + \dfrac{2}{x} - 4\right)} \qquad \text{by Theorem 5.6.5(d)}$$

$$= \frac{\lim\limits_{x \to +\infty} 3 + \lim\limits_{x \to +\infty} \left(-\dfrac{5}{x}\right) + \lim\limits_{x \to +\infty} \left(-\dfrac{1}{x^2}\right)}{\lim\limits_{x \to +\infty} \dfrac{7}{x^2} + \lim\limits_{x \to +\infty} \dfrac{2}{x} + \lim\limits_{x \to +\infty} (-4)} \qquad \text{by Theorem 5.6.6(a)}$$

$$= \frac{3 + 0 + 0}{0 + 0 - 4} = -\frac{3}{4}. \qquad \text{by Theorems 5.6.3(a), 5.6.4}$$

Example 2   Find $\lim\limits_{x \to -\infty} \dfrac{2x + 5}{x^3 - 4x - 7}$.

SOLUTION   Since the limits of the numerator and denominator at $-\infty$ fail to exist, we divide numerator and denominator by x^3, the highest power of x in either the numerator or denominator and obtain

$$\lim_{x \to -\infty} \frac{2x + 5}{x^3 - 4x - 7} = \lim_{x \to -\infty} \frac{\dfrac{2}{x^2} + \dfrac{5}{x^3}}{1 - \dfrac{4}{x^2} - \dfrac{7}{x^3}}$$

$$= \frac{\lim\limits_{x \to -\infty} \left(\dfrac{2}{x^2} + \dfrac{5}{x^3}\right)}{\lim\limits_{x \to -\infty} \left(1 - \dfrac{4}{x^2} - \dfrac{7}{x^3}\right)}$$

$$= \frac{0 + 0}{1 - 0 - 0} = 0.$$

Example 3   Find $\lim\limits_{x \to -\infty} \dfrac{\sqrt{2x^2 + 1}}{x + 2}$.

SOLUTION   If we divide numerator and denominator by $\sqrt{x^2} = -x$ when $x < 0$,

$$\lim_{x \to -\infty} \frac{\sqrt{2x^2 + 1}}{x + 2} = \lim_{x \to -\infty} \frac{\sqrt{2 + \dfrac{1}{x^2}}}{-1 - \dfrac{2}{x}} \qquad \textbf{by Theorem 5.6.7(a)}$$

$$= \frac{\displaystyle\lim_{x \to -\infty} \sqrt{2 + \dfrac{1}{x^2}}}{\displaystyle\lim_{x \to -\infty} \left(-1 - \dfrac{2}{x}\right)}$$

$$= \frac{\sqrt{\displaystyle\lim_{x \to -\infty} \left(2 + \dfrac{1}{x^2}\right)}}{-1} \qquad \textbf{by Theorem 5.6.10, etc.}$$

$$= \frac{\sqrt{2}}{-1} = -\sqrt{2}.$$

Exercise Set 5.6

In Exercises 1–17 evaluate the given limits if they exist.

1. $\displaystyle\lim_{x \to +\infty} 3/x^5$

2. $\displaystyle\lim_{y \to -\infty} 2y^{-2/3}$

3. $\displaystyle\lim_{x \to +\infty} (x^3 - 3x + 6)$

4. $\displaystyle\lim_{s \to -\infty} (-s^2 + 4s - 3)$

5. $\displaystyle\lim_{x \to +\infty} \frac{3 - 2x - x^2}{3x^2 + x + 4}$

6. $\displaystyle\lim_{y \to -\infty} \frac{4y^2 - y - 1}{5y^2 + 2y - 6}$

7. $\displaystyle\lim_{x \to -\infty} \frac{2x^2 + 5x - 6}{x^3 - 2x^2 - 2}$

8. $\displaystyle\lim_{x \to +\infty} \frac{x^2 - 3x - 1}{4x^3 + 5x}$

9. $\displaystyle\lim_{u \to +\infty} \frac{3u}{\sqrt{u^2 + 1}}$

10. $\displaystyle\lim_{t \to -\infty} \frac{3t - 1}{\sqrt{2t^2 + t - 2}}$

11. $\displaystyle\lim_{x \to -\infty} \frac{\sqrt{x^2 - 4}}{2x + 3}$

12. $\displaystyle\lim_{x \to +\infty} \frac{x^2 + x + 1}{x + 2}$

13. $\displaystyle\lim_{x \to +\infty} \frac{\sqrt[3]{x^2 - 1}}{\sqrt{x^2 + 1}}$

14. $\displaystyle\lim_{x \to -\infty} \frac{x^{1/3} - 2x^{-1/3} - 5}{2x^{1/3} - x^{-1/3} + 1}$

15. $\displaystyle\lim_{x \to +\infty} (\sqrt{x^2 + 4} - x)$

16. $\displaystyle\lim_{x \to +\infty} (\sqrt[3]{x + 1} - \sqrt[3]{x})$

17. $\displaystyle\lim_{x \to +\infty} \left(2x - \frac{2x^2 + 3}{\sqrt{x^2 + 1}}\right)$

18. Prove Theorem 5.6.3.

19. Using Definition 5.6.1 prove

(a) $\displaystyle\lim_{x \to +\infty} \frac{2x - 1}{5x + 7} = \frac{2}{5}$

(b) $\displaystyle\lim_{x \to +\infty} \frac{x}{x - 3} = 1$

(c) $\displaystyle\lim_{x\to+\infty}\frac{4}{(x+3)^2}=0$ (d) $\displaystyle\lim_{x\to+\infty}\frac{2x}{x^2-1}=0$

20. Prove Theorem 5.6.11. HINT: Suppose $P(x)$ is of degree $n\geq 1$ and $\lim_{x\to+\infty}P(x)=L$. Then consider $\lim_{x\to+\infty}P(x)/x^n$ and use an indirect proof.

21. Prove that $\lim_{x\to+\infty}F(x)=L$ if and only if $\lim_{t\to 0^+}F(1/t)=L$.

5.7   Infinite Limits at Infinity

We consider the function ϕ where

$$\phi(x)=\sqrt{x+4}.\qquad\qquad\text{(Figure 5–21)}$$

When $x>0$, $\phi(x)>\sqrt{x}$, and intuitively it seems that $\phi(x)$ can be made as large as desired simply by choosing x sufficiently large. In fact, if B is an arbitrary positive number, we can insure that $\phi(x)>B$ by choosing $x>B^2$.

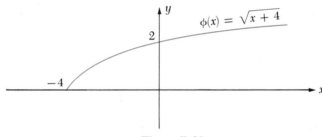

Figure 5-21

The situation when $f(x)$ increases without bound as x increases without bound is dealt with in the next definition.

5.7.1   Definition

Suppose $(c,+\infty)\subset \mathcal{D}_f$ for some number c. $\lim_{x\to+\infty}f(x)=+\infty$ if and only if for every $B>0$ there is an $M>0$ such that $f(x)>B$ when $x>M$.

From the immediately preceding discussion

$$\lim_{x\to+\infty}\sqrt{x+4}=+\infty$$

since for every $B>0$

$$\sqrt{x+4}>B\quad\text{when }x>M=B^2.$$

For a given function f it may happen that $f(x)$ decreases without bound as x increases without bound.

5.7.2   Definition

Suppose $(c,+\infty)\subset \mathcal{D}_f$ for some number c. $\lim_{x\to+\infty}f(x)=-\infty$ if and only if for every $B>0$ there is an $M>0$ such that $f(x)<-B$ when $x>M$.

The following theorem is useful in proving that a function has an infinite limit at $+\infty$.

5.7.3   Theorem

Suppose $\lim_{x\to+\infty} f(x)/x^r = L \neq 0$ where r is a positive rational number.

(a) *If $L > 0$, then $\lim_{x\to+\infty} f(x) = +\infty$.*

(b) *If $L < 0$, then $\lim_{x\to+\infty} f(x) = -\infty$.*

PROOF OF (a)   Let $B > 0$ be arbitrary. By hypothesis, there is an $M > 0$ such that

$$\left| \frac{f(x)}{x^r} - L \right| < \frac{L}{2} \quad \text{if } x > M. \tag{5.14}$$

From (5.14) for $x > M$

$$\frac{f(x)}{x^r} - L > -\frac{L}{2}$$

and hence

$$\frac{f(x)}{x^r} > \frac{L}{2}$$

$$f(x) > \frac{Lx^r}{2} > \frac{LM^r}{2}. \tag{5.15}$$

In stating (5.14) it should be emphasized that the number M used is not unique. In fact, if (5.14) is true for a certain M, the statement is also true if M is replaced by any number greater than M. Thus we could require that any M for which (5.14) is true also satisfy $M > (2B/L)^{1/r}$. Then from (5.15) we would have

$$f(x) > \frac{LM^r}{2} > \frac{L}{2} \cdot \frac{2B}{L} = B \quad \text{if } x > M > (2B/L)^{1/r}.$$

The conclusion then follows from Definition 5.7.1.

The proof of (b) is analogous and is left as an exercise.

Example 1   Prove that $\lim_{x\to+\infty} (2x^3 - 3x^2 + 5) = +\infty$.

PROOF   Since the highest power of x in any term of the polynomial $2x^3 - 3x^2 + 5$ is x^3, we consider

$$\lim_{x\to+\infty} \frac{2x^3 - 3x^2 + 5}{x^3} = \lim_{x\to+\infty} \left(2 - \frac{3}{x} + \frac{5}{x^3} \right) = 2.$$

Hence from Theorem 5.7.3(a), $\lim_{x\to+\infty} (2x^3 - 3x^2 + 5) = +\infty$.

Example 2 Find $\lim_{x \to +\infty} \sqrt{x^3 + 1}$.

SOLUTION   Since $\sqrt{x^3 + 1} \approx x^{3/2}$ when x is large, we evaluate

$$\lim_{x \to +\infty} \frac{\sqrt{x^3 + 1}}{x^{3/2}} = \lim_{x \to +\infty} \sqrt{\frac{x^3 + 1}{x^3}}$$

$$= \sqrt{\lim_{x \to +\infty} \left(1 + \frac{1}{x^3}\right)} = 1$$

using Theorem 5.6.10. Hence from Theorem 5.7.3(a), $\lim_{x \to +\infty} \sqrt{x^3 + 1} = +\infty$.

We also define the infinite limits of a function at $-\infty$.

5.7.4   Definition

Let $(-\infty, c) \subset \mathcal{D}_f$ for some number c.

(a) $\lim_{x \to -\infty} f(x) = +\infty$ if and only if for every $B > 0$ there is an $M > 0$ such that $f(x) > B$ when $x < -M$.

(b) $\lim_{x \to -\infty} f(x) = -\infty$ if and only if for every $B > 0$ there is an $M > 0$ such that $f(x) < -B$ when $x < -M$.

The following theorems are also useful. Their proofs are left for the reader.

5.7.5   Theorem

Suppose $\lim_{x \to -\infty} f(x)/x^n = L \neq 0$ where n is a positive integer.

(a) *If $L > 0$, then* $\displaystyle \lim_{x \to -\infty} f(x) = \begin{cases} +\infty & \text{if } n \text{ is even,} \\ -\infty & \text{if } n \text{ is odd.} \end{cases}$

(b) *If $L < 0$, then* $\displaystyle \lim_{x \to -\infty} f(x) = \begin{cases} -\infty & \text{if } n \text{ is even,} \\ +\infty & \text{if } n \text{ is odd.} \end{cases}$

Example 3 Prove that $\lim_{x \to -\infty} (-x^2 + 4x - 1) = -\infty$.

PROOF   Since the highest power of x in the given polynomial is x^2, we consider

$$\lim_{x \to -\infty} \frac{-x^2 + 4x - 1}{x^2} = \lim_{x \to -\infty} \left(-1 + \frac{4}{x} - \frac{1}{x^2}\right) = -1.$$

Hence by Theorem 5.7.5(b) the conclusion is obtained.

Exercise Set 5.7

In Exercises 1–7 obtain the given infinite limits. Justify your answers by citing the appropriate limit theorems used.

1. $\lim\limits_{x \to +\infty} (1 - x^4)$

2. $\lim\limits_{x \to +\infty} (x^{1/3} - 2x^{-1/3})^2$

3. $\lim\limits_{x \to -\infty} (1 - x)^3$

4. $\lim\limits_{x \to +\infty} \dfrac{1 + x}{\sqrt{x}}$

5. $\lim\limits_{x \to +\infty} \sqrt{x^2 + 3x + 6}$

6. $\lim\limits_{x \to -\infty} \sqrt[3]{x^2 + 3x + 6}$

7. $\lim\limits_{x \to +\infty} \dfrac{x^{1/2} + 2x^{-1/2}}{x^{1/3}}$

8. Prove: $\lim_{x \to +\infty} f(x) = +\infty$ if and only if $\lim_{x \to +\infty} (-f(x)) = -\infty$.

9. Prove Theorem 5.7.3(b).

10. Prove: If $a_0 \neq 0$, $b_0 \neq 0$, and $m > n$ where m and n are positive integers,

$$\lim_{x \to +\infty} \frac{a_0 x^m + a_1 x^{m-1} + \cdots + a_m}{b_0 x^n + b_1 x^{n-1} + \cdots + b_n} = \begin{cases} +\infty & \text{if } \dfrac{a_0}{b_0} > 0, \\[2ex] -\infty & \text{if } \dfrac{a_0}{b_0} < 0. \end{cases}$$

11. Using the results of Exercise 10, find $\lim\limits_{x \to +\infty} \dfrac{x^4 + 4x + 5}{4x^2 - 3x + 1}$.

12. Using the results of Exercise 10, find $\lim\limits_{x \to +\infty} \dfrac{(1 - x)^3}{(1 + x)^2}$.

5.8   Curve Sketching

In graphing the equation in x and y in Section 2.3, we found that the required sketches could be drawn with more confidence and usually with greater ease if we first investigated the equation with regard to extent, symmetries, and intercepts. If the graph of the equation to be sketched can also be expressed in the form $y = f(x)$ where the first and second derivatives of f exist at all but a few isolated numbers in the domain of f, then our preliminary investigation can be expanded to include consideration of where the graph is rising or falling, is concave upward or concave downward, and where extrema are assumed or inflection points are located.

In addition to the above properties of a graph having an equation of the form $y = f(x)$, we are also interested in whether the graph has *asymptotes*. In ordinary terms, a line is an asymptote to a graph if the distance between the graph and the line can be made less than any positive number for all points on the line past a certain point on the line.

We shall define the different kinds of asymptotes.

5.8.1   Definition

The line $x = c$ is a *vertical asymptote* of the graph of $y = f(x)$ if and only if one of the following conditions holds: $\lim_{x \to c+} f(x) = +\infty$, $\lim_{x \to c+} f(x) = -\infty$, $\lim_{x \to c-} f(x) = +\infty$, or $\lim_{x \to c-} f(x) = -\infty$.

The graph of $y = 3/\sqrt{x + 1}$ (Figure 5–22) has a vertical asymptote at

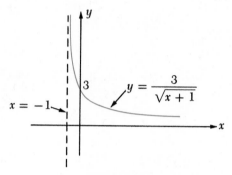

Figure 5-22

$x = -1$ since $\lim_{x \to -1+} 3/\sqrt{x + 1} = +\infty$. The graph of $y = (x^2 - 4)/x^2$ (Figure 5–23) has a vertical asymptote at $x = 0$ since

$$\lim_{x \to 0+} \frac{x^2 - 4}{x^2} = -\infty \qquad \text{and} \qquad \lim_{x \to 0-} \frac{x^2 - 4}{x^2} = -\infty.$$

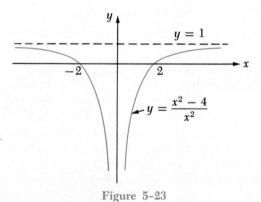

Figure 5-23

The line $x = c$ should be suspected as a possible vertical asymptote of the graph of an equation of the form $y = F(x)/G(x)$ if $G(c) = 0$, but $F(c) \neq 0$.

5.8.2   Definition

The line $y = mx + b$ is an asymptote of the graph of $y = f(x)$ if and only if $\lim_{x \to +\infty} [f(x) - (mx + b)] = 0$ or $\lim_{x \to -\infty} [f(x) - (mx + b)] = 0$. If $m = 0$, the line

$y = b$ is called a *horizontal asymptote* of the graph; if $m \neq 0$ the line $y = mx + b$ is termed an *oblique asymptote* of the graph.

The x axis is a horizontal asymptote of the graph of $y = 3/\sqrt{x + 1}$ (Figure 5–22) since $\lim_{x \to +\infty} 3/\sqrt{x + 1} = 0$. The line $y = 1$ is a horizontal asymptote of the graph of $y = (x^2 - 4)/x^2$ since $\lim_{x \to +\infty} (x^2 - 4)/x^2 = 1$ and $\lim_{x \to -\infty} (x^2 - 4)/x^2 = 1$.

A line $y = mx + b$ where $m \neq 0$ would seem to be an oblique asymptote to the graph of a function f if

$$f(x) \approx mx + b$$

when x is large "positively" or large "negatively." For example, since

$$\frac{x^2 + 3x + 2}{x} = x + 3 + \frac{2}{x} \approx x + 3$$

when x is large "positively" or large "negatively," the line $y = x + 3$ would seem to be an oblique asymptote to the graph of $y = (x^2 + 3x + 2)/2$ (Figure 5–24). This conjecture is easily verified using Definition 5.8.2 since

$$\frac{x^2 + 3x + 2}{x} - (x + 3) = \frac{2}{x}$$

and therefore

$$\lim_{x \to +\infty} \frac{2}{x} = 0 = \lim_{x \to -\infty} \frac{2}{x}.$$

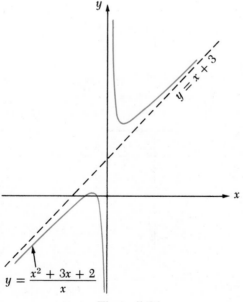

Figure 5–24

Example 1   Sketch the graph of the equation $xy + 3y - 2x + 4 = 0$ (Figure 5–25) using an analysis based upon the considerations mentioned in this section.

SOLUTION

(i) *Extent*   If we solve the given equation for y in terms of x, we have

$$y = \frac{2x - 4}{x + 3},$$

and hence y is defined for all $x \neq -3$. Solving the given equation for x in terms of y gives

$$x = \frac{-3y - 4}{y - 2}.$$

Hence x is defined for every $y \neq 2$.

(ii) *Symmetry*   The graph is not symmetric to the x axis, y axis, or origin by the methods introduced in Section 2.3.

(iii) *Intercepts*   The intercepts are $x = 2$ and $y = -\frac{4}{3}$.

(iv) *Discussion using $f'(x)$*   Differentiating in the equation $f(x) = (2x - 4)/(x + 3)$ we obtain

$$f'(x) = \frac{2(x + 3) - (2x - 4) \cdot 1}{(x + 3)^2} = \frac{10}{(x + 3)^2}.$$

Since $f'(x) > 0$ for every $x \neq -3$, f is increasing on the intervals $(-\infty, -3)$ and $(-3, +\infty)$. By the first derivative test, the function

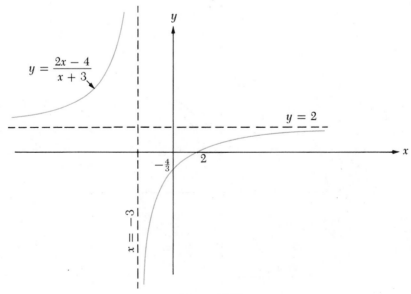

Figure 5–25

f has no relative extrema.

(v) *Discussion using $f''(x)$* We next obtain

$$f''(x) = \frac{-20}{(x+3)^3}.$$

Since $f''(x) > 0$ if $x < -3$, the graph is concave upward to the left of $x = -3$ and since $f''(x) < 0$ if $x > -3$, the graph is concave downward to the right of $x = -3$. There is no inflection point at $x = -3$ since y does not exist when $x = -3$.

(vi) *Vertical asymptotes* The line $x = -3$ is a vertical asymptote as

$$\lim_{x \to -3^+} \frac{2x-4}{x+3} = -\infty \qquad \lim_{x \to -3^-} \frac{2x-4}{x+3} = +\infty.$$

(vii) *Horizontal or oblique asymptotes* The line $y = 2$ is a horizontal asymptote as

$$\lim_{x \to +\infty} \frac{2x-4}{x+3} = 2 = \lim_{x \to -\infty} \frac{2x-4}{x+3}.$$

Example 2 Sketch the graph of the equation $y = 8x/(x^2 + 4)$ (Figure 5–26).

SOLUTION

(i) *Extent* We note that y is defined for every x, and if we solve the given equation for x in terms of y we obtain successively:

$$x^2 y + 4y = 8x$$
$$x^2 y - 8x + 4y = 0$$
$$x = \frac{8 \pm \sqrt{64 - 16y^2}}{2y} = \frac{4 \pm 2\sqrt{4 - y^2}}{y} \qquad \text{if } y \neq 0$$

But $(0,0)$ is a point on the graph, so x is defined if and only if $4 - y^2 \geq 0$ and hence if and only if $|y| \leq 2$.

(ii) *Symmetry* The graph is not symmetric to the x axis or the y axis, but is symmetric to the origin.

(iii) *Intercepts* The intercepts are $x = 0$ and $y = 0$.

(iv) *Discussion using $f'(x)$* Differentiating in the equation $f(x) = 8x/(x^2 + 4)$, we obtain

$$f'(x) = \frac{(x^2+4)\cdot 8 - 8x \cdot 2x}{(x^2+4)^2} = \frac{32 - 8x^2}{(x^2+4)^2} = \frac{8(2-x)(2+x)}{(x^2+4)^2}.$$

The critical numbers are ± 2. If $x < -2$, $f'(x) < 0$, and hence f is decreasing on the interval $(-\infty, -2]$. If $-2 < x < 2$, $f'(x) > 0$, and hence f is increasing on the interval $[-2, 2]$. If $x > 2$, $f'(x) < 0$, and so f is decreasing on the interval $[2, +\infty)$. By the first derivative test, $f(2) = 2$ is a relative maximum of f, and $f(-2) = -2$ is a relative minimum of f.

(v) *Discussion using $f''(x)$* We next obtain

$$f''(x) = \frac{(x^2 + 4)^2(-16x) - (32 - 8x^2)2(x^2 + 4)2x}{(x^2 + 4)^4}$$

$$= \frac{(x^2 + 4)(-16x) - 4x(32 - 8x^2)}{(x^2 + 4)^3}$$

$$= \frac{16x^3 - 192x}{(x^2 + 4)^3} = \frac{16x(x - 2\sqrt{3})(x + 2\sqrt{3})}{(x^2 + 4)^3}.$$

If $x < -2\sqrt{3}$, or $0 < x < 2\sqrt{3}$, $f''(x) < 0$ and the graph is concave downward. If $-2\sqrt{3} < x < 0$, or $x > 2\sqrt{3}$, $f''(x) > 0$ and the graph is concave upward. The graph has inflection points at $(0, 0)$, $(2\sqrt{3}, \sqrt{3})$, and $(-2\sqrt{3}, -\sqrt{3})$.

(vi) *Vertical asymptotes* There are no vertical asymptotes as y is defined for every x.

(vii) *Horizontal or oblique asymptotes* Since

$$\lim_{x \to +\infty} \frac{8x}{x^2 + 4} = 0 = \lim_{x \to -\infty} \frac{8x}{x^2 + 4},$$

the line $y = 0$ is a horizontal asymptote.

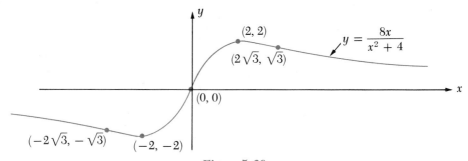

Figure 5-26

Example 3 Sketch the graph of the equation $y^2(x^2 - 1) = (2x^2 - 4)^2$ (Figure 5-27).

SOLUTION

(i) *Extent* Solving for y in terms of x, we obtain

$$y = \pm \frac{2x^2 - 4}{\sqrt{x^2 - 1}}$$

and note that y is defined if $|x| > 1$. Because of the difficulty of solving for x in terms of y, we will bypass this part of the discussion of extent.

(ii) *Symmetry* The graph is symmetric to the x axis, y axis, and origin.

(iii) *Intercepts* $x = \pm\sqrt{2}$. There are no y intercepts.

(iv) *Discussion of $f'(x)$* If we let $f(x) = (2x^2 - 4)/(\sqrt{x^2 - 1})$ we then obtain

$$f'(x) = \frac{\sqrt{x^2 - 1} \cdot 4x - (2x^2 - 4)(\frac{1}{2})(x^2 - 1)^{-1/2} \cdot 2x}{x^2 - 1}$$

$$= \frac{4x(x^2 - 1) - x(2x^2 - 4)}{(x^2 - 1)^{3/2}} = \frac{2x^3}{(x^2 - 1)^{3/2}}.$$

Since $f'(x) > 0$ for $x > 1$, f is increasing on the interval $(1, +\infty)$. Since $f'(x) < 0$ for $x < -1$, f is decreasing on the interval $(-\infty, -1)$. There are no relative extrema.

(v) *Discussion of* $f''(x)$ We next obtain

$$f''(x) = \frac{(x^2 - 1)^{3/2}6x^2 - 2x^3 \cdot \frac{3}{2}(x^2 - 1)^{1/2}2x}{(x^2 - 1)^3}$$

$$= \frac{6x^2(x^2 - 1) - 6x^4}{(x^2 - 1)^{5/2}} = \frac{-6x^2}{(x^2 - 1)^{5/2}}.$$

If $x > 1$ or $x < -1$, $f''(x) < 0$ and the graph is concave downward.

(vi) *Vertical asymptotes*   The vertical asymptotes are the lines $x = \pm 1$ as

$$\lim_{x \to 1^+} \frac{2x^2 - 4}{\sqrt{x^2 - 1}} = -\infty \quad \text{and} \quad \lim_{x \to -1^-} \frac{2x^2 - 4}{\sqrt{x^2 - 1}} = -\infty.$$

(vii) *Horizontal or oblique asymptotes*   Since

$$\lim_{x \to +\infty} \frac{2x^2 - 4}{\sqrt{x^2 - 1}} = +\infty \quad \text{and} \quad \lim_{x \to -\infty} \frac{2x^2 - 4}{\sqrt{x^2 - 1}} = +\infty,$$

there are no horizontal asymptotes. However, we note that when x is large "positively," $\sqrt{x^2 - 1} \approx x$ and hence

$$\frac{2x^2 - 4}{\sqrt{x^2 - 1}} \approx \frac{2x^2 - 4}{x} = 2x - \frac{4}{x}.$$

Thus $y = 2x$ is a suspected oblique asymptote. However, if x is large "negatively," $\sqrt{x^2 - 1} \approx -x$ and hence

$$\frac{2x^2 - 4}{\sqrt{x^2 - 1}} \approx \frac{2x^2 - 4}{-x} = -2x + \frac{4}{x},$$

so $y = -2x$ is also a suspected oblique asymptote. To prove this contention for $y = 2x$, we consider

$$\frac{2x^2 - 4}{\sqrt{x^2 - 1}} - 2x = \frac{2x(x - \sqrt{x^2 - 1})}{\sqrt{x^2 - 1}} - \frac{4}{\sqrt{x^2 - 1}}. \qquad (5.16)$$

If we rationalize the numerator of the first term in the right member of (5.16) by multiplying the numerator and denominator there by $x + \sqrt{x^2 - 1}$, we obtain

$$\frac{2x^2 - 4}{\sqrt{x^2 - 1}} - 2x = \frac{2x[x^2 - (x^2 - 1)]}{\sqrt{x^2 - 1}(x + \sqrt{x^2 - 1})} - \frac{4}{\sqrt{x^2 - 1}}$$

$$= \frac{2x}{x\sqrt{x^2 - 1} + x^2 - 1} - \frac{4}{\sqrt{x^2 - 1}}$$

$$= \frac{2}{\sqrt{x^2 - 1} + x - 1/x} - \frac{4}{\sqrt{x^2 - 1}}. \quad (5.17)$$

We can show from (5.17) that

$$\lim_{x \to +\infty} \left(\frac{2x^2 - 4}{\sqrt{x^2 - 1}} - 2x \right) = 0$$

and so the line $y = 2x$ is an oblique asymptote. Taking into account the symmetry of the graph with respect to the y axis, we find that the line $y = -2x$ is also an oblique asymptote.

Thus we obtain the sketch of the graph of $y = (2x^2 - 4)/\sqrt{x^2 - 1}$.

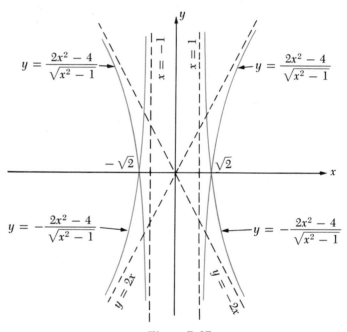

Figure 5-27

This graph consists of two branches whose sketches are labeled in Figure 5-27 by $y = (2x^2 - 4)/\sqrt{x^2 - 1}$. Since the graphs of $y = (2x^2 - 4)/\sqrt{x^2 - 1}$ and $y = -(2x^2 - 4)/\sqrt{x^2 - 1}$ are symmetric to the x axis, a sketch of the graph of the latter equation can be readily made. The graph of the equation $y^2(x^2 - 1) = (2x^2 - 4)^2$ is the union of the graphs of $y = (2x^2 - 4)/\sqrt{x^2 - 1}$ and $y = -(2x^2 - 4)/\sqrt{x^2 - 1}$.

Exercise Set 5.8

Discuss and sketch the graphs of the following equations.

1. $y = 27x - x^3$

2. $y = x^{3/2}(x - 2)^4$

3. $y = \dfrac{x^2 - 9}{x^2 - 4}$

4. $y = \dfrac{3x - 5}{x - 1}$

5. $y = \dfrac{1}{x^2 - 3x}$

6. $x^2y - 4y = 1 - x^2$

7. $y = \dfrac{x^2 - 3x + 6}{x + 2}$

8. $y = \dfrac{x}{x^2 + 4x + 3}$

9. $y = \dfrac{x - 2}{\sqrt{x^2 - 9}}$

10. $y = \dfrac{x^2 - 12}{\sqrt{x^2 + 4}}$

11. $y = x\sqrt{\dfrac{x}{x - 2}}$

12. $y = \pm x\sqrt{x^2 - 3}$

13. $y^2 = \dfrac{x^2(2 - x)}{2 + x}$

14. $y^2(x - 2) = x^3$

15. $y^2(x^2 - 1) = x^4$

6. Further Applications of Derivatives

6.1 Absolute Extrema

In Definition 5.1.1 we defined the maximum and minimum values of a function on a set. We now define the *absolute extrema* of a function.

6.1.1 Definition

Let c be a number in the domain, $\mathcal{D}_f$, of a function f.

(a) $f(c)$ is an *absolute maximum* (*value*) of f if and only if $f(c)$ is the maximum (value) of f on $\mathcal{D}_f$.

(b) $f(c)$ is an *absolute minimum* (*value*) of f if and only if $f(c)$ is the minimum (value) of f on $\mathcal{D}_f$.

An absolute maximum or absolute minimum of a function is called an *absolute extremum* of the function.

From Definition 5.1.1, $f(c)$ is an absolute maximum of f if and only if $f(x) \leq f(c)$ for every x in $\mathcal{D}_f$. Also, $f(c)$ is an absolute minimum of f if and only if $f(x) \geq f(c)$ for every x in $\mathcal{D}_f$ (Figure 6–1).

The absolute maximum of a function, when it exists, is the greatest value that the function can ever assume. Similarly, the absolute minimum of a function is the least value that the function can ever assume.

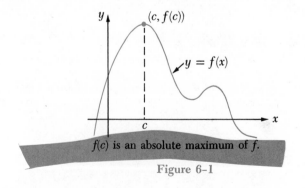

$f(c)$ is an absolute maximum of f.

Figure 6–1

194

A function may or may not have absolute extrema. The function defined by $f(x) = 1/x^2$ has no absolute maximum as $\lim_{x \to 0} 1/x^2 = +\infty$ and therefore $1/x^2$ can be made greater than any positive number simply by choosing x sufficiently close to zero. Also, we can show that this function has no absolute minimum. If M is any positive number, $1/x^2 < M$ if $x > 1/\sqrt{M}$ or $x < -1/\sqrt{M}$, and thus M cannot be an absolute minimum of f. Since $1/x^2$ never assumes nonpositive values, no nonpositive number could be a minimum value of f. Thus the function has no absolute maximum or absolute minimum.

An absolute maximum (or absolute minimum) of a function which is assumed at an interior point of the domain of a function is always a relative maximum (or relative minimum) of the function. However, not every relative extremum is also an absolute extremum. For example, the function f where $f(x) = x^3 - 3x^2 + 4$ has a relative minimum $f(2) = 0$ that is not an absolute minimum (Figure 5–18).

If a function is continuous and its domain is a finite closed interval, by Theorem 5.1.2 the function has both an absolute maximum and an absolute minimum. These absolute extrema can then be located by examining the values of the function at any critical numbers and at the endpoints of the domain of the function, the same procedure that was followed in the solution of Examples 2 and 3 of Section 5.1.

If a function is continuous and its domain is an interval that is not a finite closed interval, the function *may have* an absolute extremum at a critical number or at an endpoint of its domain. When such a function has only one critical number in the interior of its domain, then Theorem 6.1.3, below, is useful because it allows us to use the first and second derivative tests for relative extrema in seeking absolute extrema of the function. In the proof of Theorem 6.1.3, we shall utilize the following preliminary theorem.

6.1.2   Theorem

If f is differentiable on $[a, b]$ and if $f'(a)$ and $f'(b)$ are of opposite signs, then there exists a number $c \in (a, b)$ such that $f'(c) = 0$.

PROOF   We can suppose that $f'(a) < 0$ and $f'(b) > 0$ (Figure 6–2) as a similar proof suffices when the signs of these values of f' are reversed. Since f is differentiable on $[a, b]$, by Theorem 4.3.2 the function is also continuous on the interval. Therefore by Theorem 5.1.2 f has a minimum $f(c)$ at some number $c \in [a, b]$. It will be shown that c cannot be a or b. Since

$$\lim_{h \to 0} f'(a) = \lim_{h \to 0} \frac{f(a + h) - f(a)}{h} < 0,$$

by Theorem 3.3.3 when $h > 0$ is sufficiently close to 0,

$$\frac{f(a + h) - f(a)}{h} < 0. \tag{6.1}$$

Hence from (6.1) using Theorems 1.3.5 and 1.3.6,

$$f(a + h) < f(a).$$

Thus $f(a)$ cannot be the minimum of f on $[a, b]$. Similarly, since $f'(b) > 0$, the reader can prove (Exercise 11, Exercise Set 6.1) when $h < 0$ but sufficiently close to 0 that

$$f(b + h) < f(b).$$

Hence $f(b)$ cannot be the minimum of f on $[a, b]$. Since c cannot be a or b, c is an interior point of $[a, b]$. Therefore by Theorem 5.1.3 $f'(c) = 0$.

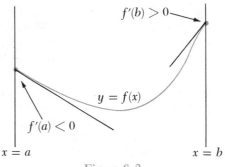

Figure 6-2

6.1.3   Theorem

Suppose f satisfies the following conditions:

  (i) *The domain of f is an interval.*
 (ii) *f is continuous on $\mathfrak{D}_f$ and differentiable in the interior of $\mathfrak{D}_f$.*
(iii) *There is exactly one number c in the interior of $\mathfrak{D}_f$ where f' has the value 0.*

If $f(c)$ is a relative maximum of f (or a relative minimum of f), then $f(c)$ is also an absolute maximum of f (or an absolute minimum of f).

PROOF   We shall prove the theorem when $f(c)$ is a relative maximum and leave the analogous proof when $f(c)$ is a relative minimum as an exercise. First we shall prove that $f'(x)$ is always of the same sign in each of the intervals $I_1 = \{x \in \mathfrak{D}_f : x \leq c\}$ and $I_2 = \{x \in \mathfrak{D}_f : x \geq c\}$. Suppose, for example, there were numbers x_1 and x_2 in, say, I_1 for which $f'(x_1)$ and $f'(x_2)$ were of opposite signs. Then by Theorem 6.1.2 there would be a number c_1 between x_1 and x_2 such that $f'(c_1) = 0$, which is impossible by (iii). Hence, since $f(c)$ is a relative maximum, by Theorem 5.3.4(c) it would be impossible for $f'(x)$ to be of the same sign on both I_1 and I_2. Also it would be impossible for $f'(x)$ to be negative on I_1 and positive on I_2, for otherwise, $f(c)$ would be a relative minimum by Theorem 5.3.4(b). Thus, the only remaining possibility is that $f'(x)$ is positive on I_1 and negative on I_2. Then by Theorem 5.3.2(a) f is increasing on I_1 and by Theorem 5.3.2(b) f is decreasing on I_2. Hence $f(c) \geq f(x)$ for every $x \in I_1 \cup I_2 = \mathfrak{D}_f$, which proves the theorem.

Example 1   Find the absolute extrema of the function f given by $f(x) = 6x - x^2$ if these extrema exist (Figure 6–3).

SOLUTION   We note that $\mathcal{D}_f = (-\infty, +\infty)$. To obtain the critical numbers for f, we solve the equation

$$f'(x) = 6 - 2x = 0$$

and obtain $x = 3$. Since $f'(3) = 0$ and $f''(3) = -2 < 0$, $f(3) = 9$ is a relative maximum by the second derivative test. Hence by Theorem 6.1.3, $f(3) = 9$ is also an absolute maximum of f. Here, f has no absolute minimum because $\lim_{x \to +\infty} (6x - x^2) = -\infty$ (and also $\lim_{x \to -\infty} (6x - x^2) = -\infty$) and therefore $f(x)$ can be made as large "negatively" as desired by choosing x sufficiently large.

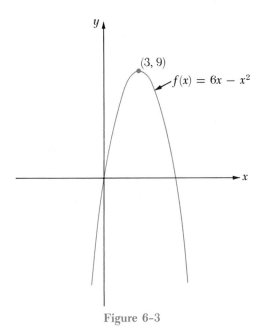

Figure 6–3

Example 2   Find the absolute extrema of the function f given by $f(x) = x\sqrt{2 + x}$ if these extrema exist (Figure 6–4).

SOLUTION   We note that $\mathcal{D}_f = [-2, +\infty)$. To obtain the critical numbers of f, we form

$$f'(x) = \sqrt{2 + x} + \tfrac{1}{2}x(2 + x)^{-1/2} \cdot 1$$

$$= \frac{2(2 + x) + x}{2(2 + x)^{1/2}} = \frac{4 + 3x}{2(2 + x)^{1/2}}.$$

The critical numbers of f are $-\tfrac{4}{3}$ and -2, but only $-\tfrac{4}{3}$ is in the interior of $\mathcal{D}_f$ and $f'(-\tfrac{4}{3}) = 0$. Since

$$f'(x) < 0 \quad \text{if } -2 < x < -\tfrac{4}{3}$$

and

$$f'(x) > 0 \quad \text{if } x > -\tfrac{4}{3},$$

$f(-\tfrac{4}{3}) = -\tfrac{4}{3}\sqrt{\tfrac{2}{3}}$ is a relative minimum of f by the first derivative test. Hence by Theorem 6.1.3 $f(-\tfrac{4}{3})$ is also an absolute minimum of f. There is no absolute maximum for f since $\lim_{x \to +\infty} x\sqrt{2 + x} = +\infty$, and therefore $f(x)$ can be made as large as desired when x is sufficiently large.

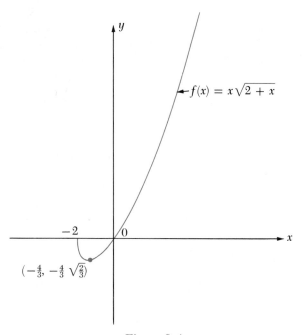

Figure 6–4

Exercise Set 6.1

Obtain the absolute extrema of the functions defined in Exercises 1–10 if they exist.

1. $f(x) = 12 - 4x - x^2$

2. $g(x) = 2x^3 - 3x^2 - 12x + 5 \quad$ if $x \in [-2, 4]$

3. $g(x) = 2x^3 - 3x^2 - 12x + 5 \quad$ if $x \in (0, 4)$

4. $y = x^{2/3} - \tfrac{2}{3}x \quad$ if $x \in [-1, 1]$

5. $y = x^{2/3} - \tfrac{2}{3}x \quad$ if $x \in [-1, 1)$

6. $f(x) = \dfrac{x}{x^2 + 4} \quad$ if $x \in [-1, +\infty)$

7. $G(x) = \dfrac{6 - x}{\sqrt{x}}$

8. $F(x) = \dfrac{\sqrt{x^2 + 36}}{4} + \dfrac{10 - x}{5} \quad$ if $x \in [0, 10]$

9. $\phi(x) = 2x\sqrt{9 - x^2}$

10. $f(x) = \begin{cases} -(x - 2)^3 & \text{if } x < 2 \\ (x - 2)^2 & \text{if } 2 \le x < 4 \end{cases}$

11. In the proof given for Theorem 6.1.2, show that $f(b + h) < f(b)$ when $h < 0$ but sufficiently close to 0.

12. Prove Theorem 6.1.3 when $f(c)$ is a relative minimum of f.

6.2   Applications of Absolute Extrema

There are a wide variety of problems in which the greatest or least value of a quantity is desired. For example, we might wish to find the dimensions of the rectangle with fixed perimeter that has the largest area, the number of units of an item which must be produced by a factory in order to obtain the greatest profit, or the point on the graph of an equation that is closest to some point not in the graph.

In such problems we have a quantity denoted by Z that is to be a maximum or a minimum. Suppose Z can be expressed in terms of one or more variables x_1, x_2, x_3, and so on, which denote quantities that are either explicitly mentioned or implicit in the statement of the problem. Suppose also that each of the variables x_1, x_2, x_3, . . . , can be expressed in terms of a variable x (which may incidentally be one of the variables x_1, x_2, x_3, ... Then Z can be expressed in the form

$$Z = f(x),$$

and the problem of finding the required maximum or minimum value of Z is that of finding the corresponding absolute extremum of the function f.

In the applications considered in this section f has a domain that is an interval, and the required absolute extremum can be found using Theorem 6.1.3.

Example 1   Find two numbers whose sum is 20 and whose product is a maximum.

SOLUTION   If x and y are the required numbers, their product Z is

$$Z = xy. \tag{6.2}$$

(Here x and y correspond to the variables x_1 and x_2 mentioned in the preceding discussion). Also x and y satisfy

$$x + y = 20. \tag{6.3}$$

From (6.3) one obtains $y = 20 - x$ and a substitution of this expression for y in (6.2) yields

$$Z = f(x) = x(20 - x) = 20x - x^2. \tag{6.4}$$

The domain here is the interval $(-\infty, +\infty)$. Differentiating in (6.4) gives

$$Z' = f'(x) = 20 - 2x. \tag{6.5}$$

We note from (6.5) that $f'(x) = 0$ if $x = 10$, and also

$$Z'' = f''(x) = -2 \quad \text{for every } x.$$

Since $f'(10) = 0$ and $f''(10) < 0$, $f(10) = 100$ is a relative maximum of f by the second derivative test. Hence by Theorem 6.1.2 $f(10) = 100$ is also an absolute maximum of f. Thus $x = 10$ and $y = 10$ are the required numbers having a maximum product.

Example 2   At 10:00 A.M. a southbound ship A is 100 miles north of an eastbound ship B. If the speed of A is 20 mph and the speed of B is 10 mph, find the time at which the ships are nearest each other and the smallest distance between the ships (Figure 6–5).

SOLUTION   It will be convenient to let A_0 and B_0 denote the locations of ships A and B at 10:00 A.M. and A_t and B_t their respective positions t hours later. Then

$$|A_0 A_t| = 20t, \qquad |B_0 B_t| = 10t.$$

We want to find t such that the distance Z between the ships is a minimum, or saying the same thing, when Z^2 is a minimum. By the Pythagorean theorem

$$Z^2 = f(t) = (10t)^2 + (100 - 20t)^2 \quad \text{if } t \geq 0. \tag{6.6}$$

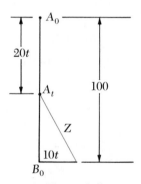

Figure 6–5

Differentiating in (6.6) gives

$$(Z^2)' = f'(t) = 200t + 2(100 - 20t)(-20)$$
$$= 1000t - 4000. \tag{6.7}$$

From (6.7) $f'(t) = 0$ if $t = 4$ and also $(Z^2)'' = 1000$ for every $t \geq 0$. By the second derivative test $f(4)$ is a relative minimum and hence by Theorem 6.1.3 $f(4)$ is an absolute minimum. From (6.6) the smallest distance between the ships attained, when $t = 4$ (at 2 P.M.), is

$$Z = \sqrt{(10 \cdot 4)^2 + (100 - 20 \cdot 4)^2} = 20\sqrt{5} \text{ mi.}$$

Example 3   Find the dimensions of the right circular cylinder of maximum volume which can be inscribed in a given right circular cone (Figure 6–6).

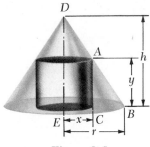

Figure 6-6

SOLUTION   We shall denote the radius of the base and the altitude of the cone by r and h, respectively. The volume V of the cylinder is

$$V = \pi x^2 y \tag{6.8}$$

where x and y are the radius and altitude, respectively, of the cylinder. Since right triangles ABC and DBE (Figure 6–6) are similar,

$$\frac{y}{r - x} = \frac{h}{r}. \tag{6.9}$$

From (6.9) $y = \dfrac{h}{r}(r - x)$, and if we substitute for y in (6.8), we have

$$V = f(x) = \frac{\pi h}{r} x^2 (r - x)$$

$$= \pi h x^2 - \frac{\pi h}{r} x^3 \quad \text{if } 0 \le x \le r. \tag{6.10}$$

Differentiating in (6.10) we obtain

$$V' = f'(x) = 2\pi h x - \frac{3\pi h}{r} x^2, \tag{6.11}$$

and from (6.11) $f'(x) = 0$ if $x = 0$ or $x = \frac{2}{3}r$. We consider the values of f at the endpoints 0 and r and also at the critical number $\frac{2}{3}r$. Since

$$f(0) = 0, \quad f\left(\frac{2}{3}r\right) = \frac{\pi h}{r}\left(\frac{2}{3}r\right)^2\left(\frac{1}{3}r\right) = \frac{4}{27}\pi r^2 h, \quad f(r) = 0,$$

from our discussion in Section 5.1, $f(\frac{2}{3}r)$ is an absolute maximum of f. From (6.9) if $x = \frac{2}{3}r$, then $y = \frac{1}{3}h$. Thus the required cylinder has radius $\frac{2}{3}r$ and altitude $\frac{1}{3}h$.

The preceding problem can also be solved using *implicit differentiation.* From (6.9)

$$y = \frac{h}{r}(r - x), \tag{6.12}$$

and hence y is a differentiable function of x. If we differentiate with respect to x in (6.8), we have

$$V' = \pi(x^2 y' + 2xy). \tag{6.13}$$

Letting $V' = 0$ and solving (6.13) for y', we obtain

$$y' = -\frac{2y}{x}. \tag{6.14}$$

Also from (6.12)

$$y' = -\frac{h}{r}. \tag{6.15}$$

From (6.14) and (6.15)

$$\frac{2y}{x} = \frac{h}{r}. \tag{6.16}$$

Solving simultaneously (6.12) and (6.16), we obtain

$$x = \tfrac{2}{3}r \qquad y = \tfrac{1}{3}h$$

as the required dimensions.

In general, when using this approach, we have a quantity denoted by Z which is to be a maximum or minimum. Z can be expressed in terms of two variables, say x and y, by an equation that we might denote as equation (i). (In the preceding problem (6.8) is equation (i).) Also, the variables x and y are related by another equation, which we denote by (ii). (Equation (6.9) or (6.12) would be equation (ii) in the previous problem.) We assume that *(ii) defines y as a differentiable function of x, that (i) and (ii) together define Z as some differentiable function f of x, and that Z has the required absolute extremum at some critical number in the interior of* $\mathfrak{D}_f$.

Differentiating implicitly in (i) with respect to x, we obtain an equation (iii) which expresses Z' in terms of x, y, and y' (Equation (6.13) above). Since the absolute extremum for Z is assumed to occur in the interior of the domain of f, there we have $Z' = 0$ by Theorem 5.1.3. From solving the equation $Z' = 0$ for y', an equation (iv) is obtained (Equation (6.14) above). Another equation (v) expressing y' in terms of x and y is derived by differentiating, implicitly if necessary, with respect to x in (ii) (Equation (6.15) above). By equating the expressions for y' from (iv) and (v) and solving this equation simultaneously with (ii), we obtain values for x and y that can then be substituted into (i) to give the desired extreme value of Z.

To attempt to prove in general that the implicit differentiation method for finding an absolute extremum actually gives this value would be too advanced for our purposes here. Therefore, we will not be concerned with giving such proofs, preferring instead to concentrate on the use of the implicit differentiation method to find absolute extrema. Even so, it should be noted that if the quantity Z, whose extreme value is found by this method, can be expressed in the form $Z = f(x)$, then the theory of Section 5.1 or 6.1 may provide a rigorous justification of the absolute extremum obtained for Z.

Example 4 Suppose a rectangular box having a square base and no top is to have a specific volume. Find the ratio of the height of the box to the length of the edge of the base if the total surface area of the box is to be a minimum.

SOLUTION The total surface area S is given in terms of x, the length of the edge of the base, and y, the height of the box, by the equation

$$S = x^2 + 4xy. \tag{6.17}$$

Also, the volume V of the box, which is constant, is given by

$$V = x^2 y. \tag{6.18}$$

From (6.18) y is a differentiable function of x, and upon differentiating in (6.17) and (6.18) with respect to x, one obtains

$$S' = 2x + 4y + 4xy' \tag{6.19}$$

and

$$V' = x^2 y' + 2xy. \tag{6.20}$$

Since S is to be a minimum, we set $S' = 0$ in (6.19) and obtain

$$y' = -\frac{x + 2y}{2x}. \tag{6.21}$$

Also, since V is constant, $V' = 0$ and hence from (6.20)

$$y' = -\frac{2y}{x}. \tag{6.22}$$

If the expressions for y' in (6.21) and (6.22) are equated, the following equations are obtained:

$$-\frac{2y}{x} = -\frac{x + 2y}{2x}$$

and hence

$$\frac{y}{x} = \frac{1}{2}. \tag{6.23}$$

Equation (6.23) expresses the ratio of the height of the box to the length of the edge of its base.

Exercise Set 6.2

1. Divide the number 20 into two parts such that the sum of their squares is a minimum.

2. Find the positive number whose positive cube root exceeds three times the given number by the greatest number.

3. Find the dimensions of the rectangle of perimeter 24 that has the greatest area.

4. Find the dimensions of the right circular cone with slant height 12 inches that has the greatest volume.

5. A rectangular box with a square base and an open top is to contain 100 in.^{3}. Find the dimensions of the box if it is to be made from the least amount of material.

6. A cattle rancher wishes to fence a rectangular plot of area 40,000 square rods and divide it into three pens of equal area by cross-fences parallel to one side of the plot. If the total length of fencing used is to be a minimum, find the overall dimensions of the rectangular plot.

7. Find the point on the altitude to the base of an isosceles triangle having the property that the sum of the distances from the point to the vertices of the triangle is a minimum.

8. An irrigation ditch has a cross section that is an isosceles trapezoid. The legs of the trapezoid and the lower base are each 10 ft in length. What should be the length of the upper base if the cross-section area is to be a maximum?

9. Show that the minimum distance from the point (x_1, y_1) to the line $y = mx + b$ is $|mx_1 - y_1 + b| / \sqrt{m^2 + 1}$.

10. Find the greatest distance from a point on the arc of the parabola $y = x^2$ connecting the points $(0, 0)$ and $(2, 4)$ to the chord of the arc. HINT: Use the expression obtained in Exercise 9.

11. Find the nearest point on the graph of the circle $x^2 + y^2 = 25$ to the point $(1, 2)$.

12. Find the dimensions of the largest rectangle that can be inscribed in a right triangle with legs 3 and 4 (a) if two adjacent sides of the rectangle lie along the legs of the triangle; (b) if one side of the rectangle lies along the hypotenuse.

13. Find the dimensions of the right circular cylinder of maximum lateral area that can be inscribed in a right circular cone of radius r and altitude h.

14. Find the dimensions of the right circular cylinder of maximum volume that can be inscribed in a sphere of radius a.

15. Find the dimensions of the right circular cone of maximum volume that can be inscribed in a sphere of radius a.

16. A 20,000-gallon water tank has a circular top, a side that is the lateral surface of a right circular cylinder, and a bottom that is a hemisphere. Find the ratio of the altitude of the tank to its radius if the tank is to be built with the least amount of material.

17. (a) A hiker at point A on a straight road wishes to reach in the shortest time a spring B located 6 miles from the road and 10 miles from A. If his hiking speed on the road is 4 miles per hour but owing to the rougher terrain only 2 miles per hour off the road, how far should he continue on the road before heading in a straight line for the spring?

    (b) Do the same problem as in (a) if the hiker's speed on the road is only $2\frac{1}{4}$ miles per hour.

18. Two ships A and B are anchored at distances 3 and 4 miles, respectively, off a straight beach. C and D are the nearest points on the beach to A and B, respectively, and $|CD| = 5$ miles. A boat will leave A, land a passenger on the beach and then proceed to B. Where should the boat land on the beach if the total distance it travels is to be a minimum?

19. A repair crew is needed to complete a regularly scheduled maintenance job on the machines in a certain plant. Because of space limitations, if the crew is too large some of the men will be idle part of the time. The number of hours needed to do the job is $4 + 10/x$ where x is the number of men in the crew. If each man in the crew is paid \$2.50 per hour, and the crew uses equipment worth \$4.00 per hour regardless of the size of the crew, what would be the most economical size for the crew.

20. Find the smallest area for a triangle formed by a tangent to the graph of the ellipse $x^2/a^2 + y^2/b^2 = 1$ and the coordinate axes.

21. If a container in the form of a right circular cone with a circular top is to have a specific volume, find the ratio of the altitude of the container to its base radius if the total surface area of the container is to be a minimum.

6.3 Applications to Business Problems

The usual objective of a business firm is to maximize its profits. If the firm is engaged in producing and marketing a commodity, its *profit P*, *total revenue R*, and *total cost C* for a period of time are related by the equation

$$P = R - C. \tag{6.24}$$

In this section some examples will be considered in which these variables are functions of a single variable.

Suppose R and C are differentiable functions of x, the number of units of the commodity produced (and marketed) by the firm. Then P also is a differentiable function of x. We will assume that x assumes non-negative values and need not be an integer, although in applications x is often rounded off to the nearest integer. We note that $R = 0$ when $x = 0$.

In a normal business situation if

$$C = \phi(x), \tag{6.25}$$

then ϕ is an increasing function and if x is sufficiently large, $\phi''(x) > 0$. The reason for this assumption is that the total cost of producing an extremely large number of units becomes prohibitive, and hence the graph of ϕ becomes concave upward for such values of x (Figure 6–7). $\phi(0)$ is termed the *overhead expense* and is present regardless of the production level.

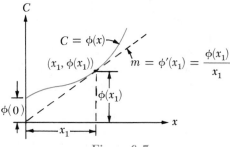

Figure 6–7

From (6.24)

$$P' = R' - C',$$

the primes denoting derivatives with respect to x. To obtain the maximum value of P, we solve the equation

$$P' = R' - C' = 0 \qquad (6.26)$$

for x. Generally in applications we obtain only one solution for x, and it can be verified using theory discussed in Section 5.1 or 6.1 that P is a maximum for this number x.

Example 1 A certain commodity is marketed by a business firm at a price of \$400 per unit. The total cost of marketing x units of the commodity is $C = 0.02x^2 + 160x + 400{,}000$. How many units of the commodity must be sold for maximum profit?

SOLUTION The total revenue is $R = 400x$ and hence from (6.24)

$$P = 400x - (0.02x^2 + 160x + 400{,}000)$$
$$= -0.02x^2 + 240x - 400{,}000.$$

Differentiating with respect to x we obtain

$$P' = -0.04x + 240. \qquad (6.27)$$

Solving the equation $P' = 0$, we obtain

$$x = \frac{240}{0.04} = 6000.$$

From (6.27)

$$P'' = -0.04$$

for every x and by Theorems 6.1.3 and 5.5.1(b), P is a maximum when $x = 6000$ units are sold.

R' and C', the derivatives of R and C with respect to x, are termed the *marginal revenue* and *marginal cost*, respectively. It is left as an exercise to show (Exercise 6, Exercise Set 6.3) that if x is any non-negative integer, $C' = \phi'(x)$ is the approximate change in the total cost resulting from marketing the $(x + 1)$st unit of the commodity.

In Figure 6–8, the profit P for any x is the vertical "signed" distance from the graph of $C = \phi(x)$ to $R = \psi(x)$. From (6.26), if the maximum profit occurs at $x = x_1$, then at x_1 *the marginal revenue and marginal cost are equal*. This is illustrated geometrically by the fact that the tangents to the graphs of $R = \psi(x)$ and $C = \phi(x)$ at $x = x_1$ (which have respective slopes $R' = \psi'(x_1)$ and $C' = \phi'(x_1)$) are parallel.

The number of commodity units that can be sold by a business firm is limited by the consumer demand for the commodity. Normally, the number of units

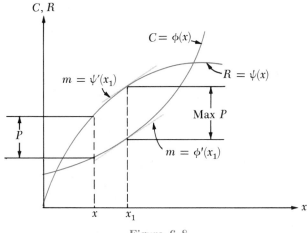

$$C, R$$

Figure 6-8

x demanded by consumers is a decreasing function of p, the price per unit, and we can express the relationship between x and p by writing $x = g(p)$ or

$$p = f(x). \tag{6.28}$$

The function f defined by (6.28) is called a *demand function*.

Example 2 Assume that the same conditions hold as in Example 1 except that the price per unit p is determined from the demand equation $p = 560 - 0.03x$. How many units must be sold for maximum profit? What should be the price per unit for maximum profit? What is the maximum profit?

SOLUTION Here the total revenue is

$$R = xp = 560x - 0.03x^2.$$

The profit is then

$$P = R - C = 560x - 0.03x^2 - (0.02x^2 + 160x + 400{,}000)$$
$$= -0.05x^2 + 400x - 400{,}000.$$

Differentiation with respect to x gives

$$P' = -0.10x + 400. \tag{6.29}$$

Solving the equation $P' = 0$, we obtain

$$x = \frac{400}{0.10} = 4000.$$

From (6.29)

$$P'' = -0.10$$

for every x, and by Theorems 6.1.3 and 5.5.1(b), P is a maximum when $x = 4000$ units are sold. The price per unit for maximum profit is

$$p = 560 - 0.03(4000) = \$440,$$

and the maximum profit is

$$P = -0.05(4000)^2 + 400(4000) - 400,000 = \$400,000.$$

Example 3 Assume that the same conditions hold as in Example 2 except that in addition the government imposes a sales tax of \$20 per unit on the business firm. How many units must be sold for maximum profit? What should be the price per unit for maximum profit? What is the maximum profit?

SOLUTION The total revenue is

$$R = xp = 560x - 0.03x^2.$$

The total cost C is now

$$C = 0.02x^2 + 160x + 400,000 + 20x$$
$$= 0.02x^2 + 180x + 400,000,$$

and hence the profit is

$$P = R - C = -0.05x^2 + 380x - 400,000.$$

Then

$$P' = -0.10x + 380,$$

and we find that P is a maximum if

$$x = \frac{380}{0.10} = 3800 \text{ units.}$$

The price per unit for maximum profit is

$$p = 560 - 0.03(3800) = \$446,$$

and the maximum profit is

$$P = -0.05(3800)^2 + 380(3800) - 400,000 = \$322,000.$$

It is interesting to note here that only \$6 of the \$20 increase in the price per unit should be passed on to the consumer.

The *average cost* of marketing x units of a commodity is defined as C/x where C is the total cost of selling the x units. If C is given by (6.25), then normally $\phi''(x)$ always exists, and it can be shown (Exercise 12 below) that C/x is a minimum at $x = x_1$ if

$$\phi'(x_1) = \frac{\phi(x_1)}{x_1} \qquad \text{and} \qquad \phi''(x_1) > 0. \tag{6.30}$$

From the equation in (6.30) when $x = x_1$, the marginal cost equals the average cost (Figure 6–7).

We also include an example in which P, R, and C are functions of a different variable.

Example 4 A cattle rancher buys a steer at an auction. The steer weighed 600 lb, and the farmer paid 30 cents per lb for the animal. It costs 15 cents per day to feed the steer, and he gains 1 lb per day. However, every day the rancher keeps the steer, the sale price per lb in dollars, S, realized will decline in accordance with the formula $S = 0.45 - 0.00025t$ where t is the number of days that elapse before the animal is sold. In how many days should he sell the steer for maximum profit?

SOLUTION The total cost C is the sum of the initial cost and the cost of feeding. Hence in dollars

$$C = 0.30(600) + 0.15t.$$

The total revenue R is the product of S and the weight of the steer. Hence in dollars

$$R = (0.45 - 0.00025t)(600 + t)$$
$$= -0.00025t^2 + 0.30t + 270,$$

and

$$P = R - C = -0.00025t^2 + 0.15t + 90.$$

Then differentiating with respect to t gives

$$P' = -0.0005t + 0.15,$$

and if we solve the equation $P' = 0$, we obtain

$$t = \frac{0.15}{0.0005} = 300 \text{ days}$$

that must elapse before the steer is sold at the maximum profit.

Exercise Set 6.3

1. The price of a commodity is determined using the demand function defined by $p = 24 - 0.4x$.

   (a) Find the maximum total revenue from the sale of the commodity.

   (b) Using the same coordinate axes, sketch the graphs of the demand function and the total revenue and marginal revenue curves.

2. The total cost from the sale of a commodity is determined from the equation $C = x^2 + 4x + 5$.

   (a) What is the minimum average cost?

   (b) Using the same coordinate axes, sketch the graphs of the total cost, marginal cost, and average cost curves.

3. Find the maximum profit if the demand equation is $p = 36 - 4x$ and the total cost is given by $C = 2x^2 + 5$.

4. Find the maximum profit if $p = (x - 9)^2$ and $C = x^3 - 8x^2 + x + 5$.

5. Solve Example 3 if, instead of a tax of $20 per unit sold, the government gives a *subsidy* of $20 per unit sold.

6. Verify that the value of the marginal cost C' at x is the approximate change in the total cost resulting from marketing the $(x + 1)$st unit. HINT: Apply the mean value theorem to the function C in the interval $[x, x + 1]$.

7. A speculator buys 10,000 bushels of wheat at $2.25 per bushel. The cost of storage of x bushels of wheat in dollars is $10 + 0.0002x$ per day. If the price in dollars per bushel of the wheat after t weeks is $2.25 + 0.1t - 0.01t^2$, when should he sell the wheat for maximum profit?

8. An apartment-house manager can rent each of his 50 apartments if he charges $150 per month rent. For each $5 per month rent he charges above $150 he has one more vacancy. What rental should he charge per month for maximum revenue?

9. A gasoline station owner normally sells 2500 gallons of "regular" grade gasoline daily. This gasoline costs him 30¢ per gallon and he sells it at a normal price 40¢ per gallon. The owner knows from experience that he will sell 200 more (or fewer) gallons daily for each 1¢ reduction (or increase) in the price per gallon from the 40¢ figure. Also, he will sell 300 fewer (or more) gallons for every penny by which his price per gallon exceeds (or is less than) that of a nearby competitor for regular gasoline. At what price per gallon should he sell this gasoline if his competitor charges 35¢ per gallon for regular gasoline?

10. At a certain small factory a worker is paid $2.00 for each acceptable piece he produces and nothing for each unacceptable piece produced. The cost of materials to the factory is 40¢ per piece produced, and there is an overhead cost of $10 per worker per day, regardless of production. Experience has shown that on the average the number of rejected pieces is $0.02x^2 + 0.2x$, where x is the total number produced. The factory can sell each acceptable piece for $7.00. What number of pieces x should be produced by each worker per day if (a) the worker is to earn the maximum wage, (b) the factory is to have the maximum daily profit per worker?

11. If in Example 3 instead of a tax of $20 on each commodity unit that tax t is imposed which gives the government the maximum revenue, find t. HINT: Express the total revenue from taxation as a function of t.

12. If C is given by (6.25), prove that the average cost is a minimum at $x = x_1$ if the conditions (6.30) hold.

6.4   Straight Line Motion

If a particle moves in a straight line in accordance with the equation of motion

$$s = f(t) \qquad (6.31)$$

where s is the coordinate of the location of the particle on the line at time t, its *velocity* v at any time t can be redefined in view of Definitions 4.2.1 and 4.3.1 by writing

$$v = x' = f'(t). \qquad (6.32)$$

From (6.32) v can be called the rate of change of x with respect to t and since v is the limiting value of $(f(t + h) - f(t))/h$, which is measured in units of the form (unit length/unit time), v will also be measured in these units.

We shall consider problems in straight line motion in this section rather than in Section 4.2 because the differentiation formulas we require had not yet been derived in that section.

Example 1   If the equation of motion of a particle moving in a straight line is $s = -5t^3 + 30t^2$ where s is in feet and t is in seconds, find the velocity when $t = 5$.

SOLUTION   Since $s = -5t^3 + 30t^2$, we obtain from (6.32)

$$v = D_t s = -15t^2 + 60t,$$

and when $t = 5$,

$$v = -15(5)^2 + 60 \cdot 5 = -75 \text{ ft/sec.}$$

If $v > 0$ for every t in an interval I, by Theorem 5.3.2(a) the function f in (6.31) is increasing on I, and the *particle is moving in the positive direction* when t is in I.

Similarly, if $v < 0$ for every t in I, by Theorem 5.3.2(b) the function f in (6.31) is decreasing on I, and the *particle is moving in the negative direction* when t is in I.

In contrast to the notion of the velocity of a particle, we are also concerned with the *speed* of a particle at an instant of time. For the particle described above by Equation (6.31), the speed of the particle at time t is defined as $|f'(t)|$, the absolute value of the velocity at time t. For example, in Example 1, the speed of the particle at time $t = 5$ is $|-75| = 75$ ft/sec.

In a study of straight line motion we are also concerned with the *acceleration* of a particle at any time t. However, before we can intelligently discuss acceleration we must define this term. We first note that the *average acceleration* of a particle over a time interval between times t and $t + h$, where $h \neq 0$, could be expressed using (6.32) by writing

$$\text{Average acceleration} = \frac{\text{Change in velocity}}{\text{Elapsed time}} = \frac{f'(t + h) - f'(t)}{h}. \qquad (6.33)$$

Our discussion here is analogous to that leading to the definition of velocity in Section 4.2. Intuition suggests that once we have defined the acceleration at time t, the average acceleration (6.33) can be made to differ from the acceleration at time t by as little as we please if $h \neq 0$ is taken sufficiently small. Thus we do actually define the acceleration at time t as the limit of the average acceleration (6.33) at $h = 0$.

6.4.1   Definition

If a particle moves in a straight line in accordance with the equation of motion (6.31), and $f''(t)$ exists, the *acceleration* a of the particle at time t is given by

$$a = D_t v = f''(t).$$

We note that $f''(t)$ is the rate of change of v with respect to t. Since a is the limiting value of the right member of (6.33), which is measured in units of (unit length/unit time)/(unit time), or saying the same thing, (unit length)/(unit time)^2, a is also measured in these units.

Example 2 Find the acceleration at time $t = \frac{3}{2}$ for the particle mentioned in Example 1.

SOLUTION   Since $s = -5t^3 + 30t^2$, from Definition 6.4.1

$$a = D_t{}^2 s = -30t + 60,$$

and when $t = \frac{3}{2}$,

$$a = -30(\tfrac{3}{2}) + 60 = 15 \text{ ft/sec}^2.$$

Again by Theorem 5.3.2(a), a particle in straight line motion is gaining speed on a time interval I if the rate of change of its speed, $D_t|f'(t)|$, is positive for all t in I. Similarly, by Theorem 5.3.2(b) if $D_t|f'(t)| < 0$ for every t in I, the particle is losing speed on the interval I. Now by Theorem 4.6.4

$$D_t|f'(t)| = \frac{f'(t) \cdot f''(t)}{|f'(t)|}. \tag{6.34}$$

Since $|f'(t)| = f'(t)$ if $f'(t) > 0$, but $|f'(t)| = -f'(t)$ if $f'(t) < 0$, from (6.34)

$$D_t|f'(t)| = \begin{cases} f''(t) & \text{if } f'(t) > 0, \\ -f''(t) & \text{if } f'(t) < 0. \end{cases} \tag{6.35}$$

From (6.35) $D_t|f'(t)| > 0$ if $f'(t) > 0$ and $f''(t) > 0$ or if $f'(t) < 0$ and $f''(t) < 0$. Also, $D_t|f'(t)| < 0$ if $f'(t) > 0$ and $f''(t) < 0$ or if $f'(t) < 0$ and $f''(t) > 0$. Thus in the interval I:

> *The particle is gaining speed if v and a have the same sign.*
> *The particle is losing speed if v and a have opposite signs.*

Example 3 Describe the motion of the particle moving in accordance with the equation $s = -5t^3 + 30t^2$.

SOLUTION   Since $s = -5t^3 + 30t^2$,

$$v = D_t s = -15t^2 + 60t = -15t(t - 4)$$
$$a = D_t{}^2 s = -30t + 60 = -30(t - 2).$$

Noting that $v = 0$ if $t = 0$ or 4 and $a = 0$ if $t = 2$, we shall study the motion of the particle at these numbers where either the velocity or acceleration is zero, and also on the subintervals $(-\infty, 0)$, $(0, 2)$, $(2, 4)$, $(4, +\infty)$ formed from the set R by these numbers. This motion is described in the following table using the results of our comments about the algebraic signs of the velocity and acceleration.

t	s	v	a	*Description of motion*
$t \in (-\infty, 0)$		$-$	$+$	moving in negative direction, losing speed
$t = 0$	0	0	$+$	at rest
$t \in (0, 2)$		$+$	$+$	moving in positive direction, gaining speed
$t = 2$	80	$+$	0	moving in positive direction, coasting
$t \in (2, 4)$		$+$	$-$	moving in positive direction, losing speed
$t = 4$	160	0	$-$	at rest
$t \in (4, +\infty)$		$-$	$-$	moving in negative direction, gaining speed

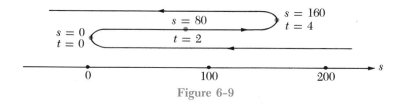

Figure 6-9

A sketch showing the location of the particle at any time t can also be made (Figure 6-9). In this drawing the values of s are indicated for $t = 0, 2$, and 4.

Example 4   Describe the motion of the particle moving in a straight line in accordance with the equation $s = 2t/(t^2 + 1)$.

SOLUTION   Since $s = \dfrac{2t}{t^2 + 1}$,

$$v = D_t s = \frac{(t^2 + 1) \cdot 2 - 2t(2t)}{(t^2 + 1)^2} = \frac{2 - 2t^2}{(t^2 + 1)^2} = \frac{2(1 - t)(1 + t)}{(t^2 + 1)^2}.$$

Hence $v = 0$ when $t = -1$ or 1. Also,

$$a = D_t^2 s = \frac{(t^2 + 1)^2(-4t) - (2 - 2t^2)2(t^2 + 1)2t}{(t^2 + 1)^4}$$

$$= \frac{-4t(t^2 + 1) - 4t(2 - 2t^2)}{(t^2 + 1)^3}$$

$$= \frac{4t^3 - 12t}{(t^2 + 1)^3}$$

$$= \frac{4t(t - \sqrt{3})(t + \sqrt{3})}{(t^2 + 1)^3},$$

and $a = 0$ when $t = 0, -\sqrt{3}$, or $\sqrt{3}$. We shall therefore study the motion of the particle at $t = -\sqrt{3}, -1, 0, 1$, and $\sqrt{3}$ and also on the subintervals $(-\infty, -\sqrt{3})$, $(-\sqrt{3}, -1)$, $(-1, 0)$, $(0, 1)$, $(1, \sqrt{3})$, and $(\sqrt{3}, +\infty)$. This motion is described in the following table.

t	s	v	a	Description of motion
$(-\infty, -\sqrt{3})$		$-$	$-$	moving in negative direction, gaining speed
$-\sqrt{3}$	$-\dfrac{\sqrt{3}}{2}$	$-$	0	moving in negative direction, coasting
$(-\sqrt{3}, -1)$		$-$	$+$	moving in negative direction, losing speed
-1	-1	0	$+$	at rest
$(-1, 0)$		$+$	$+$	moving in positive direction, gaining speed
0	0	$+$	0	moving in positive direction, coasting
$(0, 1)$		$+$	$-$	moving in positive direction, losing speed
1	1	0	$-$	at rest
$(1, \sqrt{3})$		$-$	$-$	moving in negative direction, gaining speed
$\sqrt{3}$	$\dfrac{\sqrt{3}}{2}$	$-$	0	moving in negative direction, coasting
$(\sqrt{3}, +\infty)$		$-$	$+$	moving in negative direction, losing speed

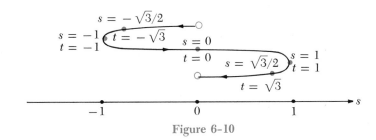

Figure 6-10

We also include a sketch (Figure 6–10) showing the location of the particle at any time t.

Exercise Set 6.4

In Exercises 1–8 discuss the motion of the particle having the given equation of motion. Draw a table and sketch similar to those used in Examples 1 and 2.

1. $s = -16t^2 + 64t$ if $0 \le t \le 4$

2. $s = -16t^2 + 108t$ if $0 \le t \le 6$

3. $s = 2t + \dfrac{54}{t^2}$ if $t \ge 1$

4. $s = t^2 + \dfrac{16}{t} + 4$ if $t \ge \dfrac{1}{2}$

5. $s = t^3 - 3t^2 - 45t - 1$

6. $s = 6t^3 - t^4$

7. $s = \dfrac{1}{t^2 + 4}$

8. $s = \dfrac{t^2 - 3}{t^4}$ if $t \ge 1$

9. The height s in feet above the ground of a toy rocket t seconds after it has been launched straight upward is given by the equation

$$s = -16t^2 + 192t + 6 \quad \text{if } t \geq 0.$$

(a) What is the upward velocity of the rocket at $t = 2$ sec?
(b) How high will the rocket go?
(c) With what initial velocity was the rocket launched?
(d) With what velocity will the rocket strike the ground?

10. A ball rolls on a straight level track in accordance with the equation of motion

$$s = -2t^2 + 40t \quad \text{if } 0 \leq t \leq 10.$$

What distance does the ball roll in slowing from a velocity of 30 ft/sec to a velocity of 10 ft/sec?

11. The legal speed limit for a particular highway is 45 mi/hr (66 ft/sec). A policeman gave a student a citation for speeding because he had been clocked at 4 sec traveling a distance of 300 ft on the highway. The student, who had taken calculus, admitted that his average speed over the 300 foot distance may have exceeded 45 mi/hr, but claimed the policeman had not proved that his speed at any instant of time ever exceeded 45 mi/hr. The policeman, who had also taken calculus, was able to convince the student that at some time during the 4 sec interval his speed had exceeded 45 mi/hr. What argument did the policeman use?

6.5 Related Rates

We consider problems here in which two or more related variables depend upon time t, and we are asked to solve for the rate of change of one or more of the variables with respect to t. The following steps are helpful in the solution of these problems and should be followed in the *order given*.

(i) Determine the important variables mentioned in the problem which vary with time t.
(ii) Write an equation involving those variables mentioned in step (i).
(iii) Differentiate with respect to t (either explicitly or implicitly) in the equation written in step (ii).
(iv) Substitute given and calculated values in the equation obtained in step (iii) and compute the desired quantity.

Example 1 A man sitting on a pier 9 feet above the water pulls in on a rope attached to a boat at the waterline at the rate of 2 feet per second. At what rate is the boat approaching the pier when 15 feet of rope are out (Figure 6–11)?

SOLUTION We follow the steps (i)–(iv) mentioned above.

(i) The important variables mentioned in the problem which depend

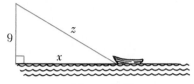

Figure 6-11

upon t are x, the horizontal distance from the boat to the pier, and z, the length of rope that is out.

(ii) x and z are related by the equation $x^2 + 9^2 = z^2$.

(iii) Differentiating implicitly with respect to t, we have

$$2xD_t x = 2zD_t z,$$
$$xD_t x = zD_t z. \tag{6.36}$$

(iv) We let $z = 15$ ft, $D_t z = -2$ ft/sec. (NOTE: $D_t z < 0$ since z is decreasing with t). From the equation in (ii) we calculate $x = 12$. Substituting these values in (6.36) we have

$$12D_t x = 15(-2),$$

and solving for $D_t x$, we have

$$D_t x = \tfrac{-30}{12} = -\tfrac{5}{2} \text{ ft/sec.}$$

Thus the boat is approaching the pier at $\tfrac{5}{2}$ ft/sec at the instant $z = 15$.

The student must not substitute 15 for z in the equation in step (ii). To do so would be making the erroneous assumption that $z = 15$ for *every* t.

Example 2   At noon ship A leaves port O steaming due south at 10 mi/hr. At 2 P.M. ship B leaves O going S $60°$E at 20 mi/hr. Find the rate at which the ships are separating at 5 P.M. (Figure 6–12.)

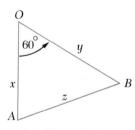

Figure 6-12

SOLUTION

(i) The important variables that depend upon t are $x = |OA|, y = |OB|$, and $z = |AB|$.

(ii) x, y, and z are related by the law of cosines

$$z^2 = x^2 + y^2 - 2xy \cos 60°,$$

and since $\cos 60° = \frac{1}{2}$

$$z^2 = x^2 + y^2 - xy. \tag{6.37}$$

(iii) Differentiating implicitly with respect to t in (6.37), we have

$$2zD_tz = 2xD_tx + 2yD_ty - xD_ty - yD_tx.$$

(iv) At 5 P.M., $x = 50$, $y = 60$ mi, and by a calculation from (6.37), $z = 10\sqrt{31}$ mi. Also $D_tx = 10$ mi/hr and $D_ty = 20$ mi/hr. Substituting these numbers in the equation in (iii), we have

$$2 \cdot 10\sqrt{31}D_tz = 2 \cdot 50 \cdot 10 + 2 \cdot 60 \cdot 20 - 50 \cdot 20 - 60 \cdot 10$$
$$20\sqrt{31}D_tz = 1800$$
$$D_tz = \frac{90}{\sqrt{31}} \text{ mi/hr.}$$

The ships are separating at $90/\sqrt{31}$ mi/hr at 5 P.M.

ALTERNATE SOLUTION   Letting $t = 0$ at noon and expressing x and y in terms of t, we have

$$x = 10t \qquad y = 20(t - 2).$$

From (6.37)

$$z = \sqrt{100t^2 + 400(t - 2)^2 - 200t(t - 2)}$$
$$= 10\sqrt{3t^2 - 12t + 16}.$$

Differentiating then gives

$$D_tz = \frac{5(6t - 12)}{\sqrt{3t^2 - 12t + 16}}.$$

At 5 P.M. $t = 5$ and hence $D_tz = 90/\sqrt{31}$ mi/hr.

Example 3   A trough 10 feet long has a cross section that is an isosceles triangle having a base of 4 feet and altitude 3 feet. If water pours into the trough at 6 ft³/min, how fast is the depth of the water changing when the depth is 1 ft? (Figure 6–13)

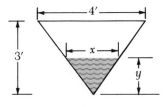

Figure 6–13

SOLUTION

(i) The variables depending upon t are y, the depth of the water, and V, the volume of the water.

(ii) To write an equation involving V and y, we must first involve x, the width of the surface of the water:

$$V = \tfrac{1}{2}10xy = 5xy. \tag{6.38}$$

Since the triangles in Figure 6–13 are similar,

$$\frac{x}{y} = \frac{4}{3}$$

or

$$x = \frac{4y}{3}. \tag{6.39}$$

Substituting from (6.39) into (6.38) gives

$$V = \tfrac{20}{3}y^2.$$

(iii) Differentiating gives

$$D_t V = \tfrac{40}{3}yD_t y.$$

(iv) Letting $D_t V = 6$ ft^3/min and $y = 1$ ft,

$$D_t y = \frac{3D_t V}{40y} = \frac{3 \cdot 6}{40 \cdot 1} = \frac{9}{20} \text{ ft/min.}$$

Exercise Set 6.5

1. If the radius of a sphere is increasing at 2 in./sec, find the rate of change of its volume when the radius is 8 inches.

2. As a balloon rises vertically at the rate of 20 ft/sec it is observed from a point 100 feet from a point on the ground directly beneath the balloon. Find the rate of change of the distance between the balloon and the observer when the balloon is 125 ft above ground.

3. A pile of sand being dumped forms a right circular cone in which the altitude is $\tfrac{2}{3}$ the diameter. If the sand is dumped at 3 ft^3/sec, find the rate of increase of the diameter of the pile when it is 6 feet high.

4. A ship proceeding on a straight course at 12 mi/hr attains its least distance from a lighthouse of 6 mi at noon. Find the rate of change of the distance between the ship and lighthouse at 1:30 P.M.

5. A ladder 17 ft long is leaning against a wall. If the upper end of the ladder is sliding down the wall at 2 ft/sec, find the rate at which the lower end is sliding away from the wall when the upper end is 8 ft from the ground.

6. A pipe 10 ft long with radius 6 inches is covered with a shell of ice. If the ice is melting at 50 in.^{3}/min, at what rate is the thickness of the ice decreasing when it is 4 inches thick?

7. Water is leaking from a conical tank at the rate of 2 ft^3/min. If the altitude and diameter of the tank are respectively 10 ft and 6 ft, at what depth is the

rate of decrease of the depth 3 ft/min? Find the rate of change of the area of the water surface at the same instant.

8. A cylindrical tank with its axis vertical has diameter 4 ft and height 10 ft and is filled with water. Then water suddenly begins pouring out a small leak 3 ft above the base of the tank at the rate of $(\pi/384)\sqrt{h}$ ft³/sec where h (ft) is the height of the water in the tank above the leak. Find the rate of change of the depth of the water in the tank when the depth is 7 ft.

9. A man 6 ft tall walks toward a street light 10 ft above the street at 4 ft/sec. At what rate is the length of his shadow changing when he is 15 ft from a point on the ground beneath the light?

10. A light 20 ft above the ground illuminates a vertical palm tree 15 ft from the light. If a monkey climbs the tree at 2 ft/sec, how fast does his shadow move along the ground when he is 10 ft above the ground?

11. A particle moves in the first quadrant along the parabola $y^2 = 2x$ toward the origin in such a way that $D_t x = -2$ (constant). Find the rate of approach to the origin of the particle when $x = 3$.

12. At 3 P.M. ship A, which is proceeding south at 20 mi/hr, is 100 miles north of ship B, which is proceeding east at 15 mi/hr. Find (a) the rate of change of the distance between the ships at 6 P.M. and (b) the minimum distance between the ships and the time at which this distance is attained.

13. A weight is attached to a rope 60 ft long which passes over a pulley 20 ft above the ground. The other end of the rope is attached to a truck at a point 3 ft above the ground. If the truck begins moving in a straight line away from the weight at 6 ft/sec, at what rate is the weight rising when it is 8 ft above the ground?

6.6 Approximations

Our purpose in this section is to develop a technique for approximating the value of a function f at a number in its domain if the values of the function and its derivative at a nearby number are known. Suppose $f(c)$ and $f'(c)$ exist, and we wish to approximate $f(c + h)$ where $h \neq 0$ is close to zero. From Equation (4.30) with h in place of k and c in place of s

$$f(c + h) - f(c) = f'(c) \cdot h + \phi(h) \cdot h \qquad (6.40)$$

where

$$\lim_{h \to 0} \phi(h) = 0. \qquad (6.41)$$

An illustration of (6.40) which applies when $f(c + h) > f(c) > 0$ if $h > 0$ is given in Figure 6–14.

From (6.40)

$$f(c + h) - [f(c) + f'(c) \cdot h] = h\phi(h). \qquad (6.42)$$

Because of (6.42) and (6.41), $f(c + h) - [f(c) + f'(c) \cdot h]$ not only approaches 0 as h does, but actually tends to 0 much more rapidly than h. Thus, because of (6.42)

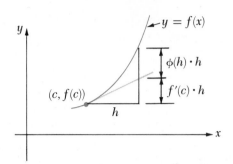

Figure 6–14

and (6.41), one states that $f(c) + f'(c) \cdot h$ is a "good approximation" of $f(c + h)$. This approximation relationship can be expressed by writing

$$f(c + h) \approx f(c) + f'(c) \cdot h \tag{6.43}$$

or

$$f(c + h) - f(c) \approx f'(c) \cdot h, \tag{6.44}$$

where the symbol $\approx$ is read "is approximately equal to."

Example 1 Approximate $(10.27)^3$.

SOLUTION We let $f(x) = x^3$, so $f'(x) = 3x^2$ and hence from (6.43)

$$(c + h)^3 \approx c^3 + 3c^2 \cdot h. \tag{6.45}$$

Since $(10.27)^3$ is near $10^3 = 1000$, we let $c = 10$ and $h = 0.27$ in (6.45). Then

$$(10.27)^3 \approx 10^3 + 3 \cdot 10^2 \cdot (0.27)$$
$$\approx 1000 + 81 = 1081.$$

The actual value of $(10.27)^3$ correct to the nearest tenth is 1083.2.

Example 2 Approximate $\sqrt{35}$.

SOLUTION We let $f(x) = \sqrt{x}$, so $f'(x) = 1/2\sqrt{x}$ and hence from (6.43)

$$\sqrt{c + h} \approx \sqrt{c} + \frac{1}{2\sqrt{c}} \cdot h. \tag{6.46}$$

Since $\sqrt{35}$ is near $\sqrt{36} = 6$, we let $c = 36$ and $h = -1$ in (6.46). Then

$$\sqrt{35} \approx \sqrt{36} + \frac{1}{2\sqrt{36}}(-1)$$
$$\approx 6 - \tfrac{1}{12} = \tfrac{71}{12} \approx 5.9167,$$

which compares with 5.9161, the actual value of $\sqrt{35}$ correct to the fourth decimal place.

Example 3 Approximate the root of the equation $x^3 + x - 3 = 0$ which is near 1.

SOLUTION Since $f(x) = x^3 + x - 3$, $f'(x) = 3x^2 + 1$ and from (6.43)

$$(c + h)^3 + (c + h) - 3 \approx c^3 + c - 3 + (3c^2 + 1) \cdot h. \qquad (6.47)$$

We let $c = 1$ and also let $c + h = 1 + h$ be the desired root of the equation. Then (6.47) can be written

$$0 \approx 1^3 + 1 - 3 + (3 \cdot 1^2 + 1) \cdot h$$
$$0 \approx -1 + 4h$$

and hence $h \approx \frac{1}{4}$ and the desired root is $1 + h \approx \frac{5}{4}$.

Example 4 A cylindrical shell of thickness t has an inside diameter D and altitude h. Find its approximate volume.

SOLUTION Since the volume of the cylindrical shell is the difference of the volumes of an outside cylinder of radius $(D/2) + t$ and an inside cylinder of radius $D/2$, where both cylinders have altitude h, it will be desirable to let $f(x) = \pi x^2 h$, the volume of a cylinder of radius x and altitude h. Then the volume of the cylindrical shell is $f((D/2) + t) - f(D/2)$ which, if t is small, is approximately equal to

$$f'\left(\frac{D}{2}\right) \cdot t = 2\pi\left(\frac{D}{2}\right)ht = \pi Dht$$

because of (6.44).

Whenever any approximation method is used, it is natural to wonder about the accuracy of results obtained by the method. From (6.42) we know that the usefulness of $f(c) + f'(c) \cdot h$ as an approximation for $f(c + h)$ improves as h approaches zero; however, we need some idea of the size of the error caused by using a given number h in this approximation. This error, E, is defined by

$$E = f(c + h) - [f(c) + f'(c) \cdot h], \qquad (6.48)$$

In order to prove a useful theorem about the size of E, it will be convenient to first obtain the following corollary of the mean value theorem.

6.6.1 Theorem (Generalized Mean Value Theorem)

If the following conditions are satisfied:

(i) f and g are continuous on the interval $[a, b]$
(ii) f and g are differentiable on the interval (a, b)
(iii) $g'(x) \neq 0$ if $x \in (a, b)$,

there is a c $\in$ (a, b) such that

$$\frac{f(b) - f(a)}{g(b) - g(a)} = \frac{f'(c)}{g'(c)}.$$

PROOF We first note that g satisfies the hypotheses of the mean value theorem on $[a, b]$. Now $g(b) \neq g(a)$ for otherwise by that theorem there would be some number $c \in (a, b)$ such that

$$g'(c) = \frac{g(b) - g(a)}{b - a} = 0,$$

which is impossible by (iii). The proof can be completed by applying Rolle's theorem to the function ϕ where

$$\phi(x) = f(x) - f(a) - \frac{f(b) - f(a)}{g(b) - g(a)}[g(x) - g(a)].$$

The details are left as an exercise.

6.6.2   Theorem

If the following conditions are satisfied:

> (i) *f and its derivative f' are continuous in the closed interval I with endpoints c and c + h*
> (ii) *$f''(x)$ exists if x is in the interior of I*
> (iii) *E is defined by (6.48)*
> (iv) *There is an M such that $|f''(x)| \leq M$ if x is in the interior of I,*

then

$$|E| \leq \tfrac{1}{2}Mh^2.$$

PROOF We let

$$F(x) = f(x) + f'(x) \cdot (c + h - x), \qquad G(x) = (c + h - x)^2. \qquad (6.49)$$

Then F and G satisfy the hypotheses of Theorem 6.6.1 on I. By Theorem 6.6.1 there is an m in the interior of I (m here plays the role of c in the statement of Theorem 6.6.1) such that

$$\frac{F(c + h) - F(c)}{G(c + h) - G(c)} = \frac{F'(m)}{G'(m)}. \qquad (6.50)$$

From (6.49)

$$F(c + h) - F(c) = [f(c + h) + f'(c + h) \cdot 0] - [f(c) + f'(c) \cdot h]$$
$$= f(c + h) - f(c) - f'(c) \cdot h \qquad (6.51)$$

and

$$G(c + h) - G(c) = -h^2. \qquad (6.52)$$

Differentiating in (6.49), we have

$$F'(x) = f'(x) - f'(x) + f''(x) \cdot (c + h - x)$$
$$= f''(x) \cdot (c + h - x) \tag{6.53}$$

and

$$G'(x) = -2(c + h - x). \tag{6.54}$$

Letting $x = m$ in (6.53) and (6.54) and substituting the resulting expressions along with (6.51) and (6.52) in (6.50), we obtain

$$\frac{f(c + h) - f(c) - f'(c) \cdot h}{-h^2} = \frac{f''(m)}{-2}. \tag{6.55}$$

From (6.48) and (6.55)

$$E = \tfrac{1}{2}h^2 f''(m), \tag{6.56}$$

and finally from (iv) and (6.56)

$$|E| \le \tfrac{1}{2}Mh^2.$$

We shall call $|E|$ the *absolute error* arising from the use of $f(c) + f'(c) \cdot h$ as an approximation for $f(c + h)$. The number $\tfrac{1}{2}Mh^2$ is an *upper bound* on $|E|$ in the sense that $|E|$ cannot exceed $\tfrac{1}{2}Mh^2$.

Theorem 6.6.2 is employed to obtain upper bounds for the absolute errors introduced in the approximations in Examples 1 and 2.

Example 5 Find an upper bound to the absolute error introduced in using 1081 as an approximation for $(10.27)^3$ in Example 1 of this section.

SOLUTION Here $f(x) = x^3$, so $f'(x) = 3x^2$, and $f''(x) = 6x$. If x is in the interval $(10, 10.27)$, $|6x| < 6(10.27) = 61.62 = M$, and by Theorem 6.6.2 since $h = 0.27$

$$|E| \le \tfrac{1}{2}(61.62)(0.27)^2 < 2.247.$$

It should be noted that the number M here is not unique. In fact, any number ≥ 61.62 could have been used for M.

Example 6 Find an upper bound to the absolute error introduced in using the approximation $\tfrac{71}{12}$ for $\sqrt{35}$ in Example 2.

SOLUTION Here $f(x) = x^{1/2}, f'(x) = \tfrac{1}{2}x^{-1/2}$ and hence $f''(x) = -1/4x^{3/2}$. If x is in the interval $(35, 36)$,

$$\left| -\frac{1}{4x^{3/2}} \right| = \frac{1}{4x^{3/2}} < \frac{1}{4 \cdot 35^{3/2}}.$$

As $35^{3/2}$ is cumbersome to express decimally, we shall write

$$\left| -\frac{1}{4x^{3/2}} \right| < \frac{1}{4 \cdot 35^{3/2}} < \frac{1}{4 \cdot 25^{3/2}} = \frac{1}{500} = M,$$

and by Theorem 6.6.2 since $h = -1$,

$$|E| \leq \tfrac{1}{2}(\tfrac{1}{500})(-1)^2 = 0.001.$$

Exercise Set 6.6

1. In each of the following exercises approximate $f(c + h)$ using (6.43).

   (a) $f(x) = x^3 + 2x^2$ $c = 2$ $c + h = 1.98$

   (b) $f(x) = \dfrac{1}{x^2 - 1}$ $c = -3$ $c + h = -2.99$

   (c) $f(x) = \sqrt[3]{x}$ $c = 8$ $c + h = 8.2$

2. Using Theorem 6.6.2, obtain an upper bound for the absolute error, $|E|$, in each part of Exercise 1 arising from the use of the approximation formula (6.43).

3. Approximate the following numbers.

   (a) $63^{4/3}$

   (b) $33^{-2/5}$

   (c) $(623.2)^{1/4}$

4. Obtain an upper bound for the absolute error for each part of Exercise 3 when the approximation (6.43) is used.

5. Approximate the given roots of the following equations.

   (a) $x^3 + 3x^2 - 1 = 0$ root near -3

   (b) $x + \dfrac{5}{x} = 5$ root near 4

   (c) $3x - (x - 1)^{3/2} = 0$ root near 10

   (d) $(x - 1)^3(x - 3) = -1$ root near 3

6. A man flies around the world always traveling at a distance of $\frac{1}{2}$ mile above the earth's surface at the equator. How much farther does he fly than the circumference of the earth at the equator?

7. An automobile engine cylinder has a bore (diameter) of 4.000 in. and a stroke (length) of 3.000 in. If the cylinder is rebored to a diameter of 4.015 in., find the approximate increase in volume of the cylinder.

8. Prove that the volume of a spherical shell of radius r and thickness h is $4\pi r^2 h$.

9. The period T in seconds of a simple pendulum of length L in feet can be given by

$$T = 2\pi \sqrt{\frac{L}{g}}$$

   where $g \approx 32$ ft/(sec)^2. If L is increased by 1% and g remains constant, what is the approximate percentage change in T?

10. The attractive force between unlike electrically charged particles is given by

$F = k/x^2$ where x is the distance between the particles. If x is increased by 1 percent, find the approximate percentage change in F.

11. The volume of a spherical balloon of diameter 8 in. is increased by 5 in.[3]. Find the approximate increase in the surface area of the balloon.

12. Complete the proof of Theorem 6.6.1.

6.7 Differentials

In this section it will be desirable to consider the topic of *differentials* since it introduces a notation that will be utilized in the study of the definite integral and the process of integration (Chapter 7). It will also be noted that the derivative of a function and the familiar differentiation formulas can be expressed using this notation. To begin our discussion, suppose x and y are variables that are related by the equation

$$y = f(x)$$

and $f'(x)$ exists. Then if h is any number, we define dy, the *differential of y*, by writing

$$dy = f'(x) \cdot h. \tag{6.57}$$

The variable dy is a function of the two variables h and x in the sense that there is a unique value for dy corresponding to every pair of values assigned to x and h, respectively.

From (6.44), if h is close to zero, then dy is an approximation of Δy (read "delta y"), which is defined by

$$\Delta y = f(x + h) - f(x). \tag{6.58}$$

Example 1 Let $y = 2x^2 - 3x + 5$. Find dy and Δy when $x = 2$ and $h = -0.03$.

SOLUTION Letting $x = 2$ and $h = -0.03$, from (6.57)

$$dy = f'(x) \cdot h = (4x - 3) \cdot h$$
$$= (4 \cdot 2 - 3)(-0.03) = -0.15,$$

and from (6.58)

$$\Delta y = f(x + h) - f(x)$$
$$= 2(x + h)^2 - 3(x + h) + 5 - (2x^2 - 3x + 5)$$
$$= (4x - 3)h + 2h^2$$
$$= -0.15 + 0.0018 = -0.1482.$$

Example 2 Let $y = \sqrt{4x - 3}$. Find dy and Δy when $x = 3$ and $h = 0.06$.

SOLUTION Letting $x = 3$ and $h = 0.06$, from (6.57)

$$dy = \frac{4h}{2\sqrt{4x - 3}} = \frac{2h}{\sqrt{4x - 3}}$$

$$= \frac{2(0.06)}{\sqrt{4 \cdot 3 - 3}} = 0.04.$$

$$\Delta y = \sqrt{4(x + h) - 3} - \sqrt{4x - 3}.$$

Letting $x = 3$ and $h = 0.06$

$$\Delta y = \sqrt{9.24} - 3 \approx 0.0397.$$

We will also use $df(x)$ in place of dy and $\Delta f(x)$ in place of Δy.

It will also be convenient to define dx, the differential of the independent variable x, and to do so in a manner consistent with the definition of dy.

If $y = x$, then from (6.57) $dy = 1 \cdot h = h$. Replacing y by x gives $dx = h$. Hence we define the *differential of the independent variable x* by

$$dx = h. \tag{6.59}$$

Thus (6.57) can be restated

$$dy = f'(x)\, dx, \tag{6.60}$$

and $f'(x)$ can be represented in differential notation by

$$f'(x) = \frac{dy}{dx} = \frac{df(x)}{dx} \quad \text{if } dx \neq 0. \tag{6.61}$$

Besides using dy/dx in place of $f'(x)$ when $y = f(x)$, we can also write the familiar differentiation formulas using differential notation. If $u = f(x)$ and $v = g(x)$, the sum and product formulas, for example, can be written:

$$\frac{d}{dx}(u + v) = \frac{du}{dx} + \frac{dv}{dx}$$

in place of $D_x(f(x) + g(x)) = D_x f(x) + D_x g(x)$;

$$\frac{d}{dx}(uv) = u\frac{dv}{dx} + v\frac{du}{dx}$$

in place of $D_x(f(x)g(x)) = f(x)D_x g(x) + g(x)D_x f(x)$.

If the equations $y = f(u)$ and $u = g(x)$ define the composite function $f \circ g$, and the chain rule equation

$$D_x(f(g(x))) = f'(g(x)) \cdot g'(x) \tag{6.62}$$

applies, Equation (6.62) can be expressed in differential notation by

$$\frac{dy}{dx} = \frac{dy}{du} \cdot \frac{du}{dx} \quad \text{if } dx \neq 0 \text{ and } du \neq 0. \tag{6.63}$$

Further, if the composite function $f \circ (g \circ \phi)$ is defined by the equations $y = f(t)$, $t = g(u)$, and $u = \phi(x)$, and from the chain rule

$$D_x(f \circ (g \circ \phi))(x) = f'((g \circ \phi)(x)) \cdot D_x(g \circ \phi)(x)$$
$$= f'((g \circ \phi)(x)) \cdot g'(\phi(x)) \cdot \phi'(x), \tag{6.64}$$

then in terms of differentials (6.64) can be written

$$\frac{dy}{dx} = \frac{dy}{dt} \cdot \frac{dt}{du} \cdot \frac{du}{dx} \quad \text{if } dx, \ du, \text{ and } dt \neq 0. \tag{6.65}$$

The formulas (6.63) and (6.65) are useful as easily remembered versions of the formulas (6.62) and (6.64), respectively, since the apparent cancellation of the differentials du in (6.63) and of dt and du in (6.65) would result in the identity $dy/dx = dy/dx$.

Also, if $y = f(x)$, the higher derivatives of f are frequently expressed in differential notation:

$$\left.\begin{aligned}
\frac{dy}{dx} &= f'(x) \\[1em]
\frac{d^2y}{dx^2} &= \frac{d}{dx}\left(\frac{dy}{dx}\right) = f''(x) \\[1em]
\frac{d^3y}{dx^3} &= \frac{d}{dx}\left(\frac{d^2y}{dx^2}\right) = f'''(x) \\[1em]
\frac{d^ny}{dx^n} &= \frac{d}{dx}\left(\frac{d^{n-1}y}{dx^{n-1}}\right) = f^{(n)}(x) \quad n = 2, 3, 4, \ldots
\end{aligned}\right\} \tag{6.66}$$

However, unlike dy and dx, we will not assign separate meanings to d^ny (read "d nth y") and dx^n (read "dx nth") for $n = 2, 3, 4, \ldots$.

As an illustration of (6.66) if $y = x^4 - 3x^3 + 5x - 2$:

$$\frac{dy}{dx} = 4x^3 - 9x^2 + 5$$

$$\frac{d^2y}{dx^2} = 12x^2 - 18x$$

$$\frac{d^3y}{dx^3} = 24x - 18$$

$$\frac{d^4y}{dx^4} = 24$$

Leibniz, the co-inventor of calculus, was the first to use differentials. He also utilized differential notation to express the first derivative and higher derivatives of a function, as is commonly done today, especially in applications of calculus. People in engineering and physical sciences commonly deal with differentials because it is usually easier if, say, $y = f(x)$ to work with dy than its counterpart $f(x + h) - f(x)$. This is particularly true if y is some composite function of the variable x and one wishes to obtain the change in y corresponding to a given change in x.

Exercise Set 6.7

In Exercises 1–4, (a) express dy in terms of x and dx and (b) obtain dy and Δy when $x = 4$ and $dx = -0.02$.

1. $y = 2x^3$

2. $y = (3x - 1)^2$

3. $y = x^{1/2}$

4. $y = \dfrac{2x - 3}{6x + 1}$

5. Find $\dfrac{dy}{dx}$ and $\dfrac{d^2y}{dx^2}$ if:

    (a) $y = x^3 + 2x^2 - 7x - 4$

    (b) $y = (4 - x^2)^{3/2}$

    (c) $y = \dfrac{2x}{x - 3}$

6. Express dy in terms of t and dt if:

    (a) $y = x^2 - 4x + 1$ and $x = t^2 + 2t$

    (b) $y = \sqrt{x^2 + 1}$ and $x = 5 - 3t^2$

7. Antidifferentiation; Definite Integrals

7.1   Antidifferentiation

In previous chapters we have been concerned with obtaining the derivative f' of a given function f. Now, we will consider the inverse problem: given a function f, find a function F such that $F' = f$. We first state the basic definition.

7.1.1   Definition

Let the functions f and F be defined on an interval I. Then F is an *antiderivative* of f on I if and only if

$$F'(x) = f(x) \quad \text{for every } x \in I. \tag{7.1}$$

When (7.1) is satisfied, we shall also call $F(x)$ an *antiderivative of $f(x)$ on I*. The process of obtaining an antiderivative of a function is called *antidifferentiation*.

Consider the following example of Definition 7.1.1. If $f(x) = 5x^4$, the functions F_1, F_2, and F_3 where

$$F_1(x) = x^5 \qquad F_2(x) = x^5 + 2 \qquad F_3(x) = x^5 - 107$$

are antiderivatives of f on the interval $(-\infty, +\infty)$ since $F_1'(x) = F_2'(x) = F_3'(x) = 5x^4$ for every x. In fact, if C is any constant, it would appear that a function F defined by $F(x) = x^5 + C$ is always an antiderivative of f on $(-\infty, +\infty)$. This conjecture is verified by the next theorem.

7.1.2   Theorem

Suppose the following conditions are satisfied:

    (i) *F is any antiderivative of f on an interval I.*
    (ii) *C is any constant.*
    (iii) *$G(x) = F(x) + C$ whenever $x \in I$.*

Then G is an antiderivative of f on I.

The proof of Theorem 7.1.2 is left as an exercise.

By Theorem 7.1.2, if a function has an antiderivative on an interval, it has an infinite number of antiderivatives all of which differ by a constant on the interval. However, one might wonder if the function could have some antiderivatives that did not differ by a constant on the interval. For example, while we know that if $F(x) = x^5 + C$ and $f(x) = 5x^4$, then F is an antiderivative of f on $(-\infty, +\infty)$, could there also be an antiderivative G of this function f on some interval such that $G(x) \neq x^5 + C$ for any C? Such a function G could not exist by the next theorem, which says that any two antiderivatives of a function on an interval must differ by a constant on the interval.

7.1.3   Theorem

If F and G are antiderivatives of f on an interval I, then there is a number C such that

$$G(x) = F(x) + C \quad \text{for all } x \in I.$$

PROOF   Let

$$H(x) = G(x) - F(x) \quad \text{for } x \in I. \tag{7.2}$$

Then

$$H'(x) = G'(x) - F'(x) = f(x) - f(x) = 0.$$

If H is not constant on I, then there exist numbers x_1 and x_2 in I with $x_2 > x_1$ such that

$$H(x_2) \neq H(x_1). \tag{7.3}$$

Also, F and G satisfy the hypotheses of the mean value theorem on the interval $[x_1, x_2]$, and so does H. Hence there is an $m \in (x_1, x_2)$ such that

$$\frac{H(x_2) - H(x_1)}{x_2 - x_1} = H'(m) = 0.$$

Hence $H(x_2) - H(x_1) = 0$ and so

$$H(x_2) = H(x_1),$$

a contradiction of (7.3). Thus our supposition that H is not constant on I is false, and there is some C such that

$$H(x) = C \quad \text{for } x \in I.$$

Then from (7.2), $G(x) - F(x) = C$ for $x \in I$ and hence

$$G(x) = F(x) + C \quad \text{for } x \in I,$$

which proves the theorem.

In the course of proving Theorem 7.1.3, the following corollary of the mean value theorem is also proved.

7.1.4   Theorem

If $H'(x) = 0$ for every x in the interval I, then H is constant on I.

We shall use the symbolism $\int f(x)\, dx$ in denoting an arbitrary antiderivative of f on an interval.

7.1.5   Definition

Let C be any constant and let f be defined on some interval I. Then

$$\int f(x)\, dx = F(x) + C \quad \text{for every } x \in I$$

if and only if

$$F'(x) = f(x) \quad \text{for every } x \in I,$$

that is, if and only if

$$dF(x) = f(x)\, dx \quad \text{for every } x \in I.$$

$\int f(x)\, dx$ is called the *general antiderivative* of $f(x)$ and C is termed the *constant of antidifferentiation*.

The symbol $\int$ will be called an *antidifferentiation sign* pending the discussion of the definite integral in Section 7.8.

If in writing an equation of the type $\int f(x)\, dx = F(x) + C$, no interval I is specified, it is assumed that the function used in place of F is an antiderivative of the function used in place of f on *any interval* on which both of these functions are defined.

The process of antidifferentiation is facilitated by utilizing formulas, which are readily derived using Definition 7.1.5 and appropriate differentiation formulas. The first such formula is concerned with the form for an antiderivative of a constant function.

7.1.6   Theorem

If k is any number, then

$$\int k\, dx = kx + C. \tag{7.4}$$

PROOF   For every $x \in (-\infty, +\infty)$

$$D_x(kx) = k,$$

so the conclusion follows from Definition 7.1.5.

In applying Definition 7.1.5 in this proof, note that kx and k are respectively the $F(x)$ and $f(x)$ mentioned in the definition.

7.1.7   Theorem

If k is any number and r is any rational number $\neq -1$, then

$$\int kx^r \, dx = \frac{kx^{r+1}}{r+1} + C. \tag{7.5}$$

(It is assumed here that x^r exists and $r > 0$ when $x = 0$.)

OUTLINE OF THE PROOF   If x^r is defined for every x in $(-\infty, +0)$, $[0, +\infty)$, or $(0, +\infty)$, the conclusion is easily obtained by differentiating on the right side of (7.5) and invoking Definition 7.1.5.

As illustrations of (7.5):

$$\int 3x^4 \, dx = \frac{3x^5}{5} + C.$$

$$\int 6x^{-3} \, dx = \frac{6x^{-2}}{-2} + C = -\frac{3}{x^2} + C.$$

$$\int \frac{1}{\sqrt{2x}} \, dx = \int \frac{1}{\sqrt{2}} x^{-1/2} \, dx = \frac{1}{\sqrt{2}} \frac{x^{1/2}}{\frac{1}{2}} + C = \sqrt{2x} + C.$$

The proofs of the next two theorems are left as exercises.

7.1.8   Theorem

If k is any number, and F is an antiderivative of f on an interval I, then

$$\int kf(x) \, dx = kF(x) + C \quad \text{for all } x \in I.$$

The conclusion can also be expressed by writing

$$\int kf(x) \, dx = k \int f(x) \, dx, \tag{7.6}$$

which says in words that any antiderivative of $kf(x)$ is k times an antiderivative of $f(x)$. By Theorem 7.1.8 a constant can be moved from one side of the $\int$ sign to the other, but this is never permissible with a variable factor. For example, $\int x^4 \, dx \neq x \int x^3 \, dx$ since $\int x^4 \, dx = x^5/5 + C$ but $x \int x^3 \, dx = x^5/4 + Cx$.

7.1.9   Theorem

If $F_1, F_2, \ldots, F_n$ are antiderivatives of $f_1, f_2, \ldots, f_n$ on an interval I, then

$$\int [f_1(x) + f_2(x) + \cdots + f_n(x)] \, dx = F_1(x) + F_2(x) + \cdots + F_n(x) + C \quad \text{for } x \in I.$$

The conclusion can also be written

$$\int [f_1(x) + f_2(x) + \cdots + f_n(x)]\, dx = \int f_1(x)\, dx + \int f_2(x)\, dx + \cdots + \int f_n(x)\, dx$$
$$(7.7)$$

By Equation (7.7) an antiderivative of the sum of two or more functions is the sum of antiderivatives of the functions.

Example 1 Find $\int (x^3 - 4x^2 + 10)\, dx$.

SOLUTION   Since

$$\int x^3\, dx = \frac{x^4}{4} + C_1 \qquad \int (-4x^2)\, dx = -\frac{4x^3}{3} + C_2 \qquad \int 10\, dx = 10x + C_3,$$

by Theorem 7.1.9 we can antidifferentiate term by term. Hence

$$\int (x^3 - 4x^2 + 10)\, dx = \frac{x^4}{4} + C_1 - \frac{4x^3}{3} + C_2 + 10x + C_3$$

$$= \frac{x^4}{4} - \frac{4x^3}{3} + 10x + C,$$

where $C = C_1 + C_2 + C_3$.

Example 2 Find $\int \dfrac{(x^2 - 4)^2}{x^{1/2}}\, dx$.

SOLUTION

$$\int \frac{(x^2 - 4)^2}{x^{1/2}}\, dx = \int \frac{x^4 - 8x^2 + 16}{x^{1/2}}\, dx = \int (x^{7/2} - 8x^{3/2} + 16x^{-1/2})\, dx,$$

and antidifferentiating term by term

$$\int \frac{(x^2 - 4)^2}{x^{1/2}}\, dx = \frac{2}{9}x^{9/2} - \frac{16}{5}x^{5/2} + 32x^{1/2} + C.$$

7.1.10   Theorem (Chain Rule for Antidifferentiation)

If g is a differentiable function on an interval I, and f has an antiderivative F on an interval J that contains all numbers g(x) when x is in I, then

$$\int f(g(x))g'(x)\, dx = F(g(x)) + C \quad \text{for all } x \in I.$$
$$(7.8)$$

PROOF   By hypothesis $F'(s) = f(s)$ if $s \in J$. Then by the chain rule

$$D_x(F \circ g)(x) = F'(g(x)) \cdot g'(x) = f(g(x)) \cdot g'(x) \quad \text{if } x \in I.$$

The conclusion then follows from Definition 7.1.5.

Theorem 7.1.10 allows us to obtain $\int f(g(x))g'(x)\,dx$ by making the substitutions

$$u = g(x) \tag{7.9}$$

and

$$du = g'(x)\,dx, \tag{7.10}$$

and then writing

$$\int f(g(x))g'(x)\,dx = \int f(u)\,du = F(u) + C. \tag{7.11}$$

Replacing u in the right member of (7.11) by $g(x)$, we obtain the equation in (7.8).

The next theorem is a corollary of Theorem 7.1.10.

7.1.11 Theorem (General Power Formula for Antidifferentiation)

If g is a differentiable function on an interval I, k is any number, and r is any rational number $\neq -1$, then for all $x \in I$

$$\int k(g(x))^r g'(x)\,dx = \frac{k(g(x))^{r+1}}{r + 1} + C. \tag{7.12}$$

From (7.12) we can let $u = g(x)$ and $du = g'(x)\,dx$ and thus write

$$\int k(g(x))^r g'(x)\,dx = \int ku^r\,du = \frac{ku^{r+1}}{r + 1} + C$$

$$= \frac{k(g(x))^{r+1}}{r + 1} + C.$$

The proof of Theorem 7.1.11 is left as an exercise.

In applying (7.12), we look for a factor of the expression to the right of the antidifferentiation sign which is a power of some $g(x)$. Hopefully, the rest of the expression will then be a constant multiple of $g'(x)\,dx$. By substituting from (7.9) and (7.10), the general antiderivative can then be expressed in the form $\int$ (constant) $u^r\,du$, which is readily evaluated.

Example 3 Find $\int \sqrt[3]{x^2 + 1} \cdot 2x\,dx$.

SOLUTION Here we let $x^2 + 1$ play the role of $g(x)$ in (7.12). Letting $u = x^2 + 1$, we fortunately have $du = 2x\,dx$. Then

$$\int \sqrt[3]{x^2 + 1} \cdot 2x\,dx = \int u^{1/3}\,du = \frac{u^{4/3}}{\frac{4}{3}} + C$$

$$= \frac{3(x^2 + 1)^{4/3}}{4} + C.$$

Example 4   Find $\int \dfrac{x^2\,dx}{\sqrt{x^3 + 8}}$.

SOLUTION   Here $x^3 + 8$ plays the role of $g(x)$ in (7.12) so let $u = x^3 + 8$. Then $du = 3x^2\,dx$ and $x^2\,dx = \frac{1}{3}\,du$, and thus

$$\int \frac{x^2\,dx}{\sqrt{x^3 + 8}} = \int \frac{\frac{1}{3}\,du}{u^{1/2}} = \frac{1}{3} u^{-1/2}\,du = \frac{1}{3} \frac{u^{1/2}}{\frac{1}{2}} + C$$

$$= \frac{2\sqrt{x^3 + 8}}{3} + C.$$

Example 5   Find $\int (5x - 4)^{10}\,dx$.

SOLUTION   Since $5x - 4$ is the $g(x)$ in (7.12), let $u = 5x - 4$. Then $du = 5\,dx$ and $dx = \frac{1}{5}\,du$. Hence

$$\int (5x - 4)^{10}\,dx = \int \tfrac{1}{5} u^{10}\,du$$

$$= \frac{1}{5} \cdot \frac{u^{11}}{11} + C = \frac{(5x - 4)^{11}}{55} + C.$$

Example 6   Find $\int \dfrac{(12x - 9)\,dx}{(2x^2 - 3x + 5)^6}$.

SOLUTION   Letting $u = 2x^2 - 3x + 5$, we have $du = (4x - 3)\,dx$ and hence $(12x - 9)\,dx = 3\,du$. Then

$$\int \frac{(12x - 9)\,dx}{(2x^2 - 3x + 5)^6} = \int 3u^{-6}\,du = -\frac{3u^{-5}}{5} + C$$

$$= -\frac{3}{5(2x^2 - 3x + 5)^5} + C.$$

Exercise Set 7.1

Perform the indicated antidifferentiations.

1. $\int 3x\,dx$

2. $\int \dfrac{5\,dx}{x^3}$

3. $\int \sqrt[3]{4x}\,dx$

4. $\int (3x - 1)\,dx$

5. $\int (t^4 - t^2 - 5t + 7)\,dt$

6. $\int y(3 - 2y)\,dy$

7. $\int \dfrac{x^2 - 3x + 1}{x^{3/2}}\,dx$

8. $\int (x^{1/3} + 2)^2\,dx$

9. $\int \left(1 - \dfrac{1}{x^2}\right)\left(2 - \dfrac{3}{x^2}\right) dx$ 10. $\int \dfrac{x^2 - x - 6}{x - 3} dx$

11. $\int x^3 \sqrt{x^4 + 1}\ dx$ 12. $\int \dfrac{x\,dx}{\sqrt{3x^2 - a^2}}$

13. $\int (2s - 1)^8\ ds$ 14. $\int \dfrac{dx}{\sqrt[3]{5 - 4x}}$

15. $\int \dfrac{(2t + 1)\,dt}{\sqrt{3 - t - t^2}}$ 16. $\int \left(y - \dfrac{1}{y}\right)^5\left(1 + \dfrac{1}{y^2}\right) dy$

17. Find $\int (2x - 1)^2\ dx$ by two methods (a) by expanding $(2x - 1)^2$ and antidifferentiating term by term and (b) by using (7.12).

In Exercises 18–22 prove the indicated theorems.

18. Theorem 7.1.2   19. Theorem 7.1.7

20. Theorem 7.1.8   21. Theorem 7.1.9

22. Theorem 7.1.11

23. If $F_1(x) = x^{1/3}$ and $F_2(x) = \begin{cases} x^{1/3} + 1 & \text{if } x > 0 \\ x^{1/3} - 1 & \text{if } x < 0 \end{cases}$ it can be shown that

$$F_1'(x) = F_2'(x) = \dfrac{1}{3x^{2/3}} \quad \text{if } x \neq 0.$$

However, $F_1(x)$ and $F_2(x)$ do not differ by a constant for every $x \neq 0$. Is this result a contradiction of Theorem 7.1.3? Why?

24. If $f(x) = [x]$, find an antiderivative of f on the interval $[0, 3]$. Note from this exercise that a function which is discontinuous on an interval may have an antiderivative on the interval.

7.2   Applications of Antidifferentiation

An equation of the form

$$\dfrac{dy}{dx} = f(x) \tag{7.13}$$

is an example of a *differential equation of first order*. Recalling from Section 6.7 that dy/dx represents the derivative of y with respect to x, a function F is a *solution* of (7.13) on an interval I if and only if $F'(x) = f(x)$ for every x in I.

Suppose the function F is a solution of (7.13) on the interval I. From Theorem 7.1.2 if C is any number, the function defined by

$$y = F(x) + C \tag{7.14}$$

is also a solution of (7.13) on I, and by Theorem 7.1.3 all solutions of (7.13) on I can be expressed in the form (7.14). Thus, in order to find all of the solutions of (7.13) on I, we obtain the general antiderivative of f on I.

The equation (7.14) corresponds to a family of curves (Figure 7–1), each of which corresponds to a different value of C. On each curve where $x = a$, the slope of the curve is $f'(a)$. Since the curves have the same slope at their intersection with a vertical line, we sometimes speak of the curves as "parallel curves." If x_0

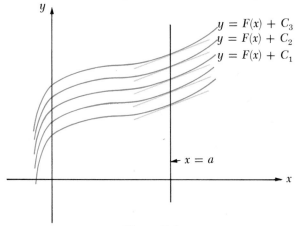

Figure 7-1

is in I, and y_0 is any number, among the family of curves defined by (7.14) there is a unique curve passing through (x_0, y_0), the curve for which C is given by

$$C = y_0 - F(x_0). \tag{7.15}$$

Thus the solution expressed by (7.14) in which C is obtained from (7.15) is the particular solution of (7.13) such that $y = y_0$ when $x = x_0$.

The condition that $y = y_0$ when $x = x_0$ is often spoken of as an *initial condition,* a terminology borrowed from problems involving motion in which the velocity or position of a particle is specified at some initial time $t = t_0$.

Example 1   If the slope of a curve is given by $dy/dx = -3x^{-1/2}$ and the curve passes through the point $(4, -1)$, obtain an equation for the curve.

SOLUTION   Antidifferentiating, we obtain the equation

$$y = -6x^{1/2} + C \tag{7.16}$$

for the family of solutions of $dy/dx = -3x^{-1/2}$. Since (7.16) is satisfied by $(4, -1)$,

$$-1 = -6 \cdot 2 + C \text{ so that } C = 11.$$

Substituting 11 for C in (7.16), we obtain the desired equation

$$y = -6x^{1/2} + 11.$$

We shall also solve *second order* differential equations of the form

$$\frac{d^2y}{dx^2} = f(x)$$

by antidifferentiating twice. With each antidifferentiation a new arbitrary constant is introduced.

Example 2 If at every point on a graph, $d^2y/dx^2 = -4$, and the graph passes through the points $(1, 2)$ and $(-1, 4)$, find an equation of the graph.

SOLUTION From $d^2y/dx^2 = -4$, we first obtain by antidifferentiation

$$\frac{dy}{dx} = -4x + C_1.$$

Antidifferentiating again gives

$$y = -2x^2 + C_1 x + C_2. \tag{7.17}$$

Since the graph of (7.17) passes through the given points, the substitution of the coordinates of these points for x and y, respectively, gives:

$$2 = -2 \cdot 1^2 + C_1 \cdot 1 + C_2$$
$$4 = -2(-1)^2 + C_1(-1) + C_2$$

If we solve these equations for C_1 and C_2 we obtain

$$C_1 = -1 \quad \text{and} \quad C_2 = 5,$$

and upon substituting these numbers in (7.17), we have the equation

$$y = -2x^2 - x + 5.$$

If a particle moves vertically near the surface of earth, it is subjected to a downward acceleration because of the influence of gravity of g where $g \approx 32$ ft/sec^2. If gravity is the only force exerted on the particle, the acceleration of the particle is given by

$$a = -g.$$

The minus sign is chosen here since we have arbitrarily selected the positive direction to be upward and the negative direction to be downward.

Example 3 A ball is thrown vertically upward with an initial velocity of 48 ft/sec from a height of 64 ft above the ground. (i) How high does the ball rise? (ii) At what velocity does the ball strike the ground after reaching its maximum height?

SOLUTION Here we shall obtain a motion equation $s = f(t)$ where s is the vertical directed distance in feet of the ball from a selected origin at the instant the time is t sec. It will be convenient to let $s = 0$ at the ground level, and, as mentioned, have the positive direction upward (Figure 7–2). For convenience, we let $t = 0$ at the instant the ball is thrown.

Starting with

$$a = \frac{dv}{dt} = -32,$$

we antidifferentiate and obtain for the velocity,

$$v = -32t + C_1. \tag{7.18}$$

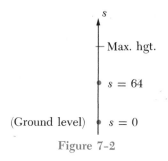

Figure 7-2

C_1 can be evaluated using the fact that $v = 48$ when $t = 0$. Thus

$$48 = -32 \cdot 0 + C_1 \text{ and here } C_1 = 48.$$

Substituting 48 for C_1 in (7.18) gives

$$v = \frac{ds}{dt} = -32t + 48. \tag{7.19}$$

Antidifferentiating again, we obtain

$$s = -16t^2 + 48t + C_2. \tag{7.20}$$

C_2 can be evaluated since $s = 64$ when $t = 0$. Thus $C_2 = 64$, and we can rewrite (7.20) to obtain

$$s = -16t^2 + 48t + 64. \tag{7.21}$$

The maximum height is reached when $v = 0$. Thus from (7.19)

$$0 = -32t + 48,$$

and hence

$$t = \tfrac{3}{2} \text{ sec}$$

when the maximum height is attained. The maximum height, obtained by letting $t = \tfrac{3}{2}$ in (7.21), is

$$s = -16(\tfrac{3}{2})^2 + 48 \cdot \tfrac{3}{2} + 64 = 100 \text{ ft.}$$

To find when the ball reaches the ground, let $s = 0$ in (7.21) and obtain

$$0 = -16t^2 + 48t + 64$$
$$t^2 - 3t - 4 = 0$$
$$(t - 4)(t + 1) = 0$$
$$t = 4, -1$$

The time $t = -1$ is not relevant since it occurs before the ball is thrown. The ball therefore reaches the ground at time $t = 4$ sec. The velocity obtained from (7.19) at $t = 4$ is

$$v = -32 \cdot 4 + 48 = -80 \text{ ft/sec.}$$

We have noted that the acceleration of a particle moving in a straight line in accordance with the equation of motion $s = f(t)$ can be expressed by

$$a = \frac{dv}{dt} = \frac{d^2s}{dt^2}.$$

If $v = F(s)$ where F is differentiable at $s = f(t)$, and f is differentiable at t, we can also express a using the chain rule by

$$a = \frac{dv}{dt} = \frac{dv}{ds}\frac{ds}{dt} = v\frac{dv}{ds}$$

$$= \frac{d(\frac{1}{2}v^2)}{ds}. \tag{7.22}$$

Example 4 Solve Example 3, starting with a differential equation in which a is given by (7.22).

SOLUTION We let $a = d(\frac{1}{2}v^2)/ds = -32$. Then antidifferentiation yields

$$\frac{1}{2}v^2 = -32s + \frac{C}{2}.$$

The antidifferentiation constant has been chosen to be $C/2$ here, rather than C, so that multiplication by 2 will give a simpler form in (7.23). Then

$$v^2 = -64s + C. \tag{7.23}$$

Since $v = 48$ when $s = 64$,

$$48^2 = -64 \cdot 64 + C \qquad \text{and hence} \qquad C = 6400.$$

Substituting for C in (7.23) gives

$$v^2 = -64s + 6400. \tag{7.24}$$

If $v = 0$, from (7.24), the maximum height is

$$s = \frac{6400}{64} = 100 \text{ ft.}$$

When the ball reaches the ground, $s = 0$ and $v < 0$. Hence from (7.24)

$$v = -80 \text{ ft/sec.}$$

Exercise Set 7.2

In Exercises 1–4 obtain the complete solution of the given differential equations.

1. $\dfrac{dy}{dx} = 3x^2 - 4x - 2$

2. $\dfrac{dy}{dx} = \dfrac{1}{\sqrt{2x + 3}}$

3. $\dfrac{d^2s}{dt^2} = -\dfrac{1}{t^3}$

4. $\dfrac{d^2y}{dx^2} = (x + 1)^5$

In Exercises 5–11 obtain the particular solution of the given differential equations satisfying the given conditions.

5. $\dfrac{dy}{dx} = \dfrac{x^2}{\sqrt{x^3 + 1}}$, $y = \tfrac{1}{2}$ when $x = 2$

6. $\dfrac{dx}{dt} = \dfrac{t}{\sqrt[3]{2t^2 + 1}}$, $x = -1$ when $t = 0$

7. $\dfrac{dy}{dx} = (2x + 3)^4$, $y = 5$ when $x = -1$

8. $\dfrac{dy}{dx} = (x^2 - 2x + 5)^{3/2}(1 - x)$, $y = 0$ when $x = 1$

9. $\dfrac{d^2y}{dt^2} = -t^2$, $y = -1$ and $\dfrac{dy}{dt} = 2$ when $t = 0$

10. $\dfrac{d^2y}{dx^2} = x^{1/3}(x + 1)$, $y = 1$ and $\dfrac{dy}{dx} = -\tfrac{2}{3}$ when $x = -1$

11. $\dfrac{d^2y}{dx^2} = x^{-3/2}$, $y = 2$ when $x = 1$ and $y = -4$ when $x = 9$

12. Find an equation for the curve passing through the point $(-1, 4)$ if the equation of the tangent at the point is $3x + y - 1 = 0$ and $D_x^2 y = 1 - x$ at any point (x, y) on the curve.

13. Find an equation for the curve $y = f(x)$ if $f(5) = -3$ is a relative minimum of f and $f''(x) = 4$ for every x.

14. If a curve with equation $y = f(x)$ passes through the points $(2, 2)$ and $(1, -4)$ and $f''(x) = x^2$ for every x, find the slope of the curve at the point $(1, -4)$.

15. A ball is thrown upward with an initial velocity of 48 ft/sec from a height of 24 ft above the ground. How high will the ball rise? In how many seconds will the ball reach the ground?

16. A ball is thrown downward from a height of 500 ft above the ground with an initial velocity of 10 ft/sec. In how many seconds will the ball strike the ground?

17. With what initial velocity must an object be projected upward from the ground to reach a height of 480 ft?

18. A car is moving with a velocity of 66 ft/sec when it begins a constant deceleration of 20 ft/sec². How far does it move during the first four seconds of deceleration?

19. An airplane takes off from a landing field with a horizontal velocity of 200 ft/sec after traveling 2400 feet along the runway at a constant acceleration. Find the magnitude of this acceleration.

20. In a car traveling at 60 mi/hr, the brakes are applied and the car is brought to rest in 200 ft. How far has the car moved by the time its velocity is reduced to 30 mi/hr, assuming a constant deceleration?

21. A sprinter who runs the 100 meter dash in 10.2 sec accelerates at a constant rate for the first 25 meters and continues at a constant speed for the rest of the race. Find his acceleration.

22. A particle moves in a straight line with constant acceleration a and its position and velocity at time $t = 0$ are given by $s = s_0$ and $v = v_0$, respectively. Derive the following equations describing the motion of the particle:

   (a) $v = v_0 + at$

   (b) $s = s_0 + v_0 t + \frac{1}{2}at^2$

   (c) $v^2 = v_0^2 + 2as$

23. According to Newton's gravitation law and his second law of motion, the force of attraction exerted on an object of mass m by the earth, neglecting air resistance and other gravitational influence is

$$f(x) = m\frac{d^2x}{dt^2} = -\frac{GmM}{x^2}, \tag{7.25}$$

   where G is the universal gravitational constant, M is the mass of the earth and x is the distance between the object and the center of the earth. Here $f(x)$ and d^2x/dt^2 are negative since the force and the acceleration are directed toward the center of the earth. We mentioned earlier in this section that at the earth's surface $d^2x/dt^2 = -g$. Thus if $x = R$, the radius of the earth,

$$f(R) = -mg = -\frac{GmM}{R^2}. \tag{7.26}$$

   Substitution from (7.26) into (7.25) gives

$$m\frac{d^2x}{dt^2} = -\frac{mgR^2}{x^2}. \tag{7.27}$$

   Using (7.27) and the values $R = 4000$ miles and $g = \frac{1}{165}$ mi/sec^2, find the minimum initial velocity at which the object must be projected upward: (a) if it is to rise to a height of 1000 miles above the surface of the earth, (b) if the object is to continue forever from the earth. The answer obtained in part (b) is called the *escape velocity* from the earth's gravitation.

7.3 Summation Notation

The process of antidifferentiation, which was introduced in Section 7.2, will be used in Section 7.7 to evaluate the *definite integral* of a continuous function on a closed interval. To facilitate the statement of the definition of the definite integral, it is convenient to introduce *summation notation*. The use of this notation permits one to express sums of many terms in a concise form.

Let f be a function whose domain is a subset of the set of all integers. Then if $p > 0$ and n are integers, the symbol $\sum_{i=n}^{n+p} f(i)$, which is read "sum of f at i as i goes from n to $n + p$," is defined by the equation

$$\sum_{i=n}^{n+p} f(i) = f(n) + f(n + 1) + f(n + 2) + \cdots + f(n + p). \tag{7.28}$$

The symbol $\sum_{i=n}^{n+p} f(i)$ is called *summation notation* and also *sigma notation* since it contains the capital Greek letter sigma, "Σ," which is suggestive of the first letter "S" in the word "Sum." n is called the *lower limit* of the sum, $n + p$ the *upper limit* of the sum, and i is called the *index* of the sum. The letter i is used here since it is short for the word "index"; however, note that this use of i has no connection whatever with the imaginery number $\sqrt{-1}$. Other letters, for example j or k, could also have been used in place of i. Thus

$$\sum_{i=n}^{n+p} f(i) = \sum_{j=n}^{n+p} f(j) = \sum_{k=n}^{n+p} f(k).$$

It should be noted that the terms to be added in the expansion of $\sum_{i=n}^{n+p} f(i)$ can be formed successively by replacing i first by the lower limit n, then by $n + 1$, then by $n + 2$, and so on, until finally i is replaced by $n + p$. The number of terms in the expansion of (7.28) is obtained by calculating

$$\text{(upper limit)} - \text{(lower limit)} + 1 = p + 1.$$

As an illustration of (7.28)

$$\sum_{i=2}^{5} \frac{2^i}{3i + 1} = \frac{2^2}{3 \cdot 2 + 1} + \frac{2^3}{3 \cdot 3 + 1} + \frac{2^4}{3 \cdot 4 + 1} + \frac{2^5}{3 \cdot 5 + 1}.$$

$$= \frac{4}{7} + \frac{8}{10} + \frac{16}{13} + \frac{32}{16}.$$

Using j as the index of summation, we could write

$$\sum_{j=-2}^{2} (j + 1)^2 = (-1)^2 + 0^2 + 1^2 + 2^2 + 3^2 = 15.$$

The following sum formulas are readily verified for every positive integer n by mathematical induction. Equation (7.30) below was verified in Example 1 of Section 1.2.

$$\sum_{i=1}^{n} i = 1 + 2 + 3 + \cdots + n = \frac{n(n + 1)}{2}. \tag{7.29}$$

$$\sum_{i=1}^{n} i^2 = 1^2 + 2^2 + 3^2 + \cdots + n^2 = \frac{n(n + 1)(2n + 1)}{6}. \tag{7.30}$$

$$\sum_{i=1}^{n} i^3 = 1^3 + 2^3 + 3^3 + \cdots + n^3 = \frac{n^2(n + 1)^2}{4}. \tag{7.31}$$

$$\sum_{i=1}^{n} i^4 = 1^4 + 2^4 + 3^4 + \cdots + n^4 = \frac{n(n + 1)(6n^3 + 9n^2 + n - 1)}{30}. \tag{7.32}$$

Some properties of summation notation that can be readily proved follow:

$$\sum_{i=n}^{n+p} k = k(p+1) \quad k \text{ any number} \tag{7.33}$$

$$\sum_{i=n}^{n+p} kf(i) = k \sum_{i=n}^{n+p} f(i) \quad k \text{ any number} \tag{7.34}$$

$$\sum_{i=n}^{n+p} [f(i) + g(i)] = \sum_{i=n}^{n+p} f(i) + \sum_{i=n}^{n+p} g(i) \tag{7.35}$$

$$\sum_{i=n}^{n+p} [f(i) - f(i-1)] = f(n+p) - f(n-1). \tag{7.36}$$

PROOF OF (7.34)

$$\sum_{i=n}^{n+p} kf(i) = kf(n) + kf(n+1) + kf(n+2) + \cdots + kf(n+p)$$

$$= k[f(n) + f(n+1) + f(n+2) + \cdots + f(n+p)]$$

$$= k \sum_{i=n}^{n+p} f(i).$$

Example 1   Evaluate $\sum_{i=1}^{50} (2i^2 - 2i + 1)$.

SOLUTION   From (7.34) and (7.35)

$$\sum_{i=1}^{50} (2i^2 - 2i + 1) = 2 \sum_{i=1}^{50} i^2 - 2 \sum_{i=1}^{50} i + \sum_{i=1}^{50} 1.$$

Hence from (7.30), (7.29), and (7.33)

$$\sum_{i=1}^{50} (2i^2 - 2i + 1) = \frac{2 \cdot 50(50+1)(2 \cdot 50 + 1)}{6} - \frac{2 \cdot 50(50+1)}{2} + 50$$

$$= 83{,}350.$$

Example 2   Evaluate $\sum_{k=1}^{48} (\sqrt{k+1} - \sqrt{k})$.

SOLUTION   We note that if we let $f(k) = \sqrt{k+1}$, then $f(k-1) = \sqrt{k}$ and from (7.36),

$$\sum_{k=1}^{48} (\sqrt{k+1} - \sqrt{k}) = \sqrt{49} - \sqrt{1} = 6.$$

Exercise Set 7.3

In Exercises 1–10 calculate the indicated sum.

1. $\displaystyle\sum_{i=1}^{5} (3i + 2)$

2. $\displaystyle\sum_{j=2}^{6} \frac{1}{2j - 1}$

3. $\displaystyle\sum_{i=-1}^{4} (-1)^{i+1} \cdot 3^{i-1}$

4. $\displaystyle\sum_{i=0}^{6} \frac{6!}{i!(6 - i)!}$

5. $\displaystyle\sum_{k=1}^{30} (2k + 1)$

6. $\displaystyle\sum_{i=1}^{40} i(3i - 4)$

7. $\displaystyle\sum_{i=1}^{20} (i + 1)(i - 3)$

8. $\displaystyle\sum_{k=2}^{99} \left(\frac{1}{k + 1} - \frac{1}{k} \right)$

9. $\displaystyle\sum_{j=1}^{n} (2^j - 2^{j-1})$

10. $\displaystyle\sum_{i=1}^{13} (\sqrt[3]{2i + 1} - \sqrt[3]{2i - 1})$

11. Prove property (7.33).

12. Prove property (7.35).

13. Prove property (7.36).

14. If

$$\bar{x} = \frac{\sum_{i=1}^{n} x_i}{n},$$

prove that $\sum_{i=1}^{n} (x_i - \bar{x})^2 = \sum_{i=1}^{n} x_i^2 - n\bar{x}^2$.

7.4   Introduction to Area

As a motivation for the concept of the definite integral, we concern ourselves with the problem of defining and obtaining the *area* of the region T in the xy plane bounded by the lines $x = a$ and $x = b$, the y axis, and the graph of $y = f(x)$ where f is continuous and assumes non-negative values on the interval $[a, b]$ (Figure 7–3). We shall call T the *region under the graph* of f from a to b.

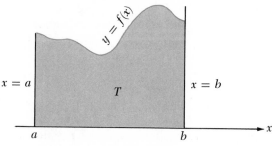

Figure 7–3

Since the graph of $y = f(x)$ for $x \in [a, b]$ is not, in general, a line segment and does not consist of connected line segments, the methods of elementary geometry will not suffice in finding the area of T.

To obtain a satisfactory definition for the area of T, denoted by A_T, we will require that the following criteria be satisfied:

(i) If A_T exists, it is a non-negative number.
(ii) If T is a rectangle with dimensions h and k, then $A_T = hk$.
(iii) If $E \subset F$ and the areas of E and F exist, then $A_E \leq A_F$.
(iv) If a region E can be subdivided into subregions E_1, E_2, $\ldots$, E_n, no two of which have points in common except possibly on their boundaries, and the area of each of the subregions exists, then the area A_E exists, and

$$A_E = \sum_{i=1}^{n} A_{E_i}.$$

Since f is continuous on $[a, b]$, by Theorem 5.1.2 there are numbers m and M in $[a, b]$ for which $f(m)$ and $f(M)$ are respectively the minimum and maximum values of f on $[a, b]$. In Figure 7–4, $f(m)$ is the altitude of rectangle $ABCD$, which is contained in T, and $f(M)$ is the altitude of rectangle $ABEF$, which contains T. Since the areas of $ABCD$ and $ABEF$ are respectively $f(m)(b - a)$ and $f(M)(b - a)$, then from (iii) if A_T exists,

$$f(m)(b - a) \leq A_T \leq f(M)(b - a). \tag{7.37}$$

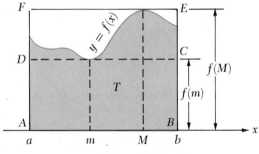

Figure 7–4

Intuition suggests that if A_T exists, it will be better approximated than in (7.37) if we use the following scheme. First choose numbers $c_0, c_1, c_2, \ldots, c_n$ in $[a, b]$ where $a = c_0 < c_1 < c_2 < \cdots < c_{n-1} < c_n = b$. These numbers divide the interval into n subintervals, and therefore the set

$$P = \{a, c_1, c_2, \ldots, c_{n-1}, b\} \tag{7.38}$$

is called a *partition* of $[a, b]$. Then using these subintervals as bases, we construct *inscribed* rectangles as in Figure 7–5(a) and *circumscribed* rectangles as in Figure 7–5(b). In each subinterval the minimum value of f is the altitude of the corresponding inscribed rectangle, and the maximum value of f is the altitude of the corresponding circumscribed rectangle.

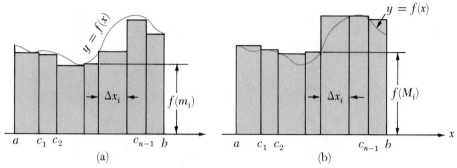

Figure 7-5

In the ith subinterval $[c_{i-1}, c_i]$, where $i = 1, 2, \ldots, n$, of the partition P, given by (7.38)

$$
\left.\begin{aligned}
\Delta x_i &= c_i - c_{i-1} = \text{length of the subinterval } [c_{i-1}, c_i]; \\
f(m_i) &= \text{minimum of } f \text{ on } [c_{i-1}, c_i]; \\
f(M_i) &= \text{maximum of } f \text{ on } [c_{i-1}, c_i].
\end{aligned}\right\}
\tag{7.39}
$$

The sum of the areas of the inscribed rectangles, $\underline{S}_P$, and the sum of the areas of the circumscribed rectangles, $\overline{S}_P$, are given by

$$
\underline{S}_P = \sum_{i=1}^{n} f(m_i)\,\Delta x_i \qquad \text{and} \qquad \overline{S}_P = \sum_{i=1}^{n} f(M_i)\,\Delta x_i.
\tag{7.40}
$$

Since T is contained in the union of the circumscribed rectangles and contains the union of the inscribed rectangles, from (iii)

$$
\underline{S}_p \le A_T \le \overline{S}_P, \quad \text{if } A_T \text{ exists.}
\tag{7.41}
$$

The numbers $\underline{S}_P$ and $\overline{S}_P$ are called the *lower sum* and *upper sum* of f, respectively, for the partition P.

Example 1 If T is the region bounded by the graphs of the equations $y = f(x) = (x - 2)^2$, $x = 1$, $x = 5$, and $y = 0$, and using $P = \{1, 2, \frac{10}{3}, \frac{9}{2}, 5\}$ as a partition of the interval $[1, 5]$, calculate $\underline{S}_P$ and $\overline{S}_P$ for f (Figure 7–6(a) and (b)). Thus find bounds for A_T if it exists.

SOLUTION The calculation can be arranged in a tabular form. From (7.39) letting $f(x) = (x - 2)^2$, we have the table on page 248. Then from (7.40)

$$
\underline{S}_P = \sum_{i=1}^{4} f(m_i)\,\Delta x_i = 0 \cdot 1 + 0 \cdot \tfrac{4}{3} + \tfrac{16}{9} \cdot \tfrac{7}{6} + \tfrac{25}{4} \cdot \tfrac{1}{2} = \tfrac{1123}{216},
$$

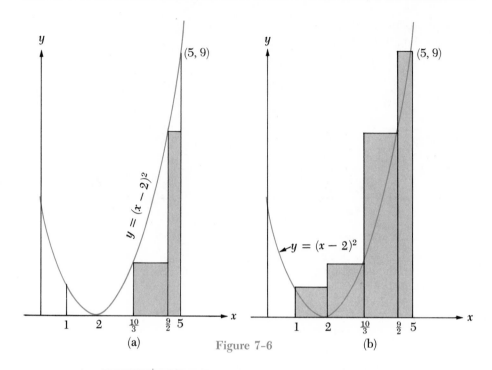

Figure 7-6

	c_i	Δx_i	m_i	M_i	$f(m_i)$	$f(M_i)$
$i = 0$	1					
1	2	1	2	1	0	1
2	$\frac{10}{3}$	$\frac{4}{3}$	2	$\frac{10}{3}$	0	$\frac{16}{9}$
3	$\frac{9}{2}$	$\frac{7}{6}$	$\frac{10}{3}$	$\frac{9}{2}$	$\frac{16}{9}$	$\frac{25}{4}$
4	5	$\frac{1}{2}$	$\frac{9}{2}$	5	$\frac{25}{4}$	9

$$\overline{S}_P = \sum_{i=1}^{4} f(M_i)\,\Delta x_i = 1\cdot 1 + \tfrac{16}{9}\cdot\tfrac{4}{3} + \tfrac{25}{4}\cdot\tfrac{7}{6} + 9\cdot\tfrac{1}{2} = \tfrac{3275}{216},$$

and if A_T exists, by (7.41),

$$\tfrac{1123}{216} \le A_T \le \tfrac{3275}{216}.$$

From (7.39), we define Δ, the *norm* of the partition P given by (7.38), by

$$\Delta_P = \text{greatest of the numbers } \Delta x_1, \Delta x_2, \ldots, \Delta x_n.$$

A partition Q of $[a, b]$ is *finer* than a partition P of $[a, b]$ if and only if $\Delta_Q < \Delta_P$. We next rework Example 1 using a finer partition of $[1, 5]$.

Example 2 For the partition $Q = \{1, \frac{3}{2}, 2, \frac{5}{2}, 3, \frac{7}{2}, 4, \frac{9}{2}, 5\}$ of $[1, 5]$ (Figure 7-7), calculate $\underline{S}_Q$ and $\overline{S}_Q$ for the function f given in Example 1. Again find bounds for A_T if it exists.

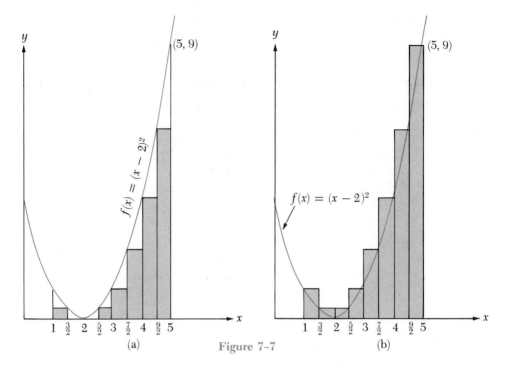

Figure 7-7

SOLUTION We again present the calculation in tabular form.

	c_i	Δx_i	m_i	M_i	$f(m_i)$	$f(M_i)$
$= 0$	1					
1	$\frac{3}{2}$	$\frac{1}{2}$	$\frac{3}{2}$	1	$\frac{1}{4}$	1
2	2	$\frac{1}{2}$	2	$\frac{3}{2}$	0	$\frac{1}{4}$
3	$\frac{5}{2}$	$\frac{1}{2}$	2	$\frac{5}{2}$	0	$\frac{1}{4}$
4	3	$\frac{1}{2}$	$\frac{5}{2}$	3	$\frac{1}{4}$	1
5	$\frac{7}{2}$	$\frac{1}{2}$	3	$\frac{7}{2}$	1	$\frac{9}{4}$
6	4	$\frac{1}{2}$	$\frac{7}{2}$	4	$\frac{9}{4}$	4
7	$\frac{9}{2}$	$\frac{1}{2}$	4	$\frac{9}{2}$	4	$\frac{25}{4}$
8	5	$\frac{1}{2}$	$\frac{9}{2}$	5	$\frac{25}{4}$	9

From (7.40)

$$\underline{S}_Q = \sum_{i=1}^{8} f(m_i)\,\Delta x_i = \tfrac{1}{4}\cdot\tfrac{1}{2} + 0\cdot\tfrac{1}{2} + 0\cdot\tfrac{1}{2} + \tfrac{1}{4}\cdot\tfrac{1}{2} + 1\cdot\tfrac{1}{2} + \tfrac{9}{4}\cdot\tfrac{1}{2} + 4\cdot\tfrac{1}{2} + \tfrac{25}{4}\cdot\tfrac{1}{2} = 7,$$

$$\overline{S}_Q = \sum_{i=1}^{8} f(M_i)\,\Delta x_i = 1\cdot\tfrac{1}{2} + \tfrac{1}{4}\cdot\tfrac{1}{2} + \tfrac{1}{4}\cdot\tfrac{1}{2} + 1\cdot\tfrac{1}{2} + \tfrac{9}{4}\cdot\tfrac{1}{2} + 4\cdot\tfrac{1}{2} + \tfrac{25}{4}\cdot\tfrac{1}{2} + 9\cdot\tfrac{1}{2} = 12,$$

and if A_T exists, then by (7.41),

$$7 \le A_T \le 12.$$

It will be noted that $\underline{S}_Q$ and $\bar{S}_Q$ are closer together than $\underline{S}_P$ and $\bar{S}_P$ in Example 1.

From Examples 1 and 2 intuition suggests that if the area A of a region under a graph exists, the associated sums $\underline{S}_P$ and $\bar{S}_P$ can be made as close together as is desired simply by taking the norm Δ_P sufficiently small. Since we have

$$\underline{S}_P \leq A \leq \bar{S}_P \quad \text{for every partition } P,$$

intuition would also suggest that A is the only number which could be the middle member of this inequality for every partition P.

We are now ready to state the definition of area promised at the beginning of this section.

7.4.1 Definition

Let f be continuous and assume non-negative values on the interval $[a, b]$, and let T be the region given by

$$T = \{(x, y) : x \in [a, b] \text{ and } y \in [0, f(x)]\}. \tag{7.42}$$

Suppose there is a unique number A such that for every partition P of $[a, b]$ the sums $\underline{S}_P$ and $\bar{S}_P$ satisfy

$$\underline{S}_P \leq A \leq \bar{S}_P. \tag{7.43}$$

Then the area A_T of the region T is given by

$$A = A_T.$$

With this definition the question immediately arises, "How can we be certain that this unique number A actually exists?" The answer is provided by the next two theorems. The proof of the first theorem belongs more properly in an advanced calculus course and is omitted here. In the hypothesis of this theorem we remove the requirement that the function f assume non-negative values on $[a, b]$ and merely require that f be continuous on the interval.

7.4.2 Theorem

If f is continuous on $[a, b]$, then

    (a) *there is an A such that for every partition P of $[a, b]$, the sums $\underline{S}_P$ and $\bar{S}_P$ satisfy (7.43),*

and

    (b) *for every $\epsilon > 0$ there is a $\delta > 0$ such that*

$$\bar{S}_P - \underline{S}_P < \epsilon \quad \text{if } \Delta_P < \delta.$$

7.4.3   Theorem

If f is continuous on [a, b], then the number A mentioned in Theorem 7.4.2 is unique.

PROOF   Suppose there exist numbers A and A' for which

$$\underline{S}_P \leq A \leq \overline{S}_P \quad \text{and} \quad \underline{S}_P \leq A' \leq \overline{S}_P \tag{7.44}$$

for every partition P of $[a, b]$. Then from (7.44)

$$-(\overline{S}_P - \underline{S}_P) \leq A - A' \leq \overline{S}_P - \underline{S}_P \quad \text{for every } P. \tag{7.45}$$

We next define ϵ' by letting

$$\epsilon' = |A - A'|. \tag{7.46}$$

Suppose $\epsilon' > 0$. Then by Theorem 7.4.2 there is a $\delta > 0$ such that

$$\overline{S}_P - \underline{S}_P < \epsilon' \quad \text{if } \Delta_P < \delta.$$

Hence from (7.45) if $\Delta_P < \delta$,

$$-\epsilon' < -(\overline{S}_P - \underline{S}_P) \leq A - A' \leq \overline{S}_P - \underline{S}_P < \epsilon',$$

and by Theorem 1.5.3(a)

$$|A - A'| < \epsilon',$$

a contradiction of (7.46). Thus our supposition that $\epsilon' > 0$ is invalid, and from (7.46) $\epsilon' = 0$. Therefore $A = A'$; that is, the number A mentioned in Theorem 7.4.2 is unique.

Example 3   Find the area of the region bounded by the x axis and the graphs of the equations $y = x^2$, $x = a$, and $x = b$ where $b > a \geq 0$ (Figure 7–8).

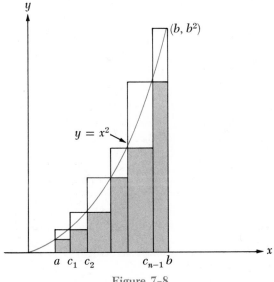

Figure 7-8

SOLUTION We consider an arbitrary partition $P = \{a, c_1, c_2, \ldots, c_{n-1}, b\}$ of the interval $[a, b]$. The lower and upper sums associated with $f(x) = x^2$ on $[a, b]$ are

$$\underline{S}_P = a^2(c_1 - a) + c_1^2(c_2 - c_1) + c_2^2(c_3 - c_2) + \cdots + c_{n-1}^2(b - c_{n-1})$$

$$= \sum_{i=1}^{n} c_{i-1}^2(c_i - c_{i-1}) \tag{7.47}$$

$$\overline{S}_P = c_1^2(c_1 - a) + c_2^2(c_2 - c_1) + \cdots + b^2(b - c_{n-1})$$

$$= \sum_{i=1}^{n} c_i^2(c_1 - c_{i-1}) \tag{7.48}$$

Since the function f here is increasing on $[a, b]$, for every $i, 0 \le c_{i-1} < c_i$, and hence we can prove (Exercise 3, Exercise Set 7.4)

$$c_{i-1}^2 < \frac{c_{i-1}^2 + c_{i-1}c_i + c_i^2}{3} < c_i^2. \tag{7.49}$$

Multiplying each member of (7.49) by $c_i - c_{i-1}$, we have

$$c_{i-1}^2(c_i - c_{i-1}) < \frac{c_i^3 - c_{i-1}^3}{3} < c_i^2(c_i - c_{i-1}). \tag{7.50}$$

Summing (7.50) over $i = 1, 2, \ldots, n$ we obtain, using (7.47) and (7.48)

$$\underline{S}_P < \sum_{i=1}^{n} \frac{c_i^3 - c_{i-1}^3}{3} < \overline{S}_P. \tag{7.51}$$

An expansion of the middle member of (7.51) gives

$$\frac{c_1^3 - a^3}{3} + \frac{(c_2^3 - c_1^3)}{3} + \frac{(c_3^3 - c_2^3)}{3} + \cdots + \frac{(b^3 - c_{n-1}^3)}{3} = \frac{b^3 - a^3}{3},$$

so (7.51) can be rewritten

$$\underline{S}_P < \frac{b^3 - a^3}{3} < \overline{S}_P. \tag{7.52}$$

Since the function f here is continuous on $[a, b]$, the number $(b^3 - a^3)/3$ is the number A mentioned in Theorem 7.4.2(a). Therefore by Theorem 7.4.3 and Definition 7.4.1, $(b^3 - a^3)/3$ is the area of the given region.

Exercise Set 7.4

1. If $f(x) = mx$, where $m > 0$, on the interval $[0, b]$, find $\underline{S}_P$ and $\overline{S}_P$ for f on $[0, b]$ if the partition P is given by:

(a) $P = \left\{0, \dfrac{b}{2}, b\right\}$

(b) $P = \left\{ 0, \dfrac{b}{8}, \dfrac{b}{4}, \dfrac{b}{2}, b \right\}$

(c) $P = \left\{ 0, \dfrac{b}{n}, \dfrac{2b}{n}, \ldots, \dfrac{(n-1)b}{n}, b \right\}$

2. If $f(x) = 4 + 2x - x^2$ on the interval $[-1, 2]$, find $\underline{S}_P$ and $\overline{S}_P$ for f on $[-1, 2]$ if P is given by:

(a) $P = \{ -1, \tfrac{1}{2}, 2 \}$

(b) $P = \{ -1, 0, 1, 2 \}$

(c) $P = \{ -1, -\tfrac{1}{2}, 0, 1, \tfrac{4}{3}, 2 \}$

(d) P, a partition of $[-1, 2]$ forming n subintervals of equal length

3. Prove (7.49) where $c_i > c_{i-1} \geq 0$.

4. Prove that for any positive integer n

$$c_{i-1}^n < \frac{\sum_{j=0}^{n} c_{i-1}^{n-j} c_i^{j}}{n+1} < c_i^{n} \quad \text{if } c_i > c_{i-1} \geq 0. \tag{7.53}$$

5. Using statement (7.53) if required, find the area of the region bounded by the graphs of the equations $y = f(x)$, $x = a$, $x = b$ where $b > a \geq 0$, and $y = 0$ if:

(a) $f(x) = k, \quad k > 0$

(b) $f(x) = mx, \quad m > 0$

(c) $f(x) = x^3$

7.5   The Definite Integral

In Theorem 7.4.3 we saw that if a function f is continuous on $[a, b]$, there is a unique number A that separates the upper sums, $\overline{S}_P$, and the lower sums, $\underline{S}_P$, of f on $[a, b]$. Moreover by Theorem 7.4.2, as the norm of the partition P, Δ_P, approaches zero, $\overline{S}_P$ approaches A from above and $\underline{S}_P$ approaches A from below.

Suppose that instead of forming the sum $\overline{S}_P = \sum_{i=1}^{n} f(M_i)\, \Delta x_i$ or $\underline{S}_P = \sum_{i=1}^{n} f(m_i)\, \Delta x_i$ in which we select the maximum or minimum value of f on each subinterval $[c_{i-1}, c_i]$, we consider a sum $\sum_{i=1}^{n} f(x_i)\Delta x_i$ where x_i is *any* number in the subinterval $[c_{i-1}, c_i]$. If f is non-negative on $[a, b]$, this sum may be regarded as the sum of the areas of rectangles where the ith rectangle in the sum has dimensions Δx_i and $f(x_i)$ (Figure 7–9). Here i runs from 1 to n. The sum $\sum_{i=1}^{n} f(x_i)\, \Delta x_i$ depends upon the partition P of the interval $[a, b]$ and the numbers $x_1, x_2, \ldots, x_n$ which are selected in the respective subintervals $[c_{i-1}, c_i]$.

If $f(m_i)$ and $f(M_i)$ are defined by (7.39), then for $i = 1, 2, \ldots, n$

$$f(m_i) \leq f(x_i) \leq f(M_i)$$
$$f(m_i)\, \Delta x_i \leq f(x_i)\, \Delta x_i \leq f(M_i)\, \Delta x_i.$$

Hence, if we sum from 1 to n,

$$\underline{S}_P \leq \sum_{i=1}^{n} f(x_i)\, \Delta x_i \leq \overline{S}_P. \tag{7.54}$$

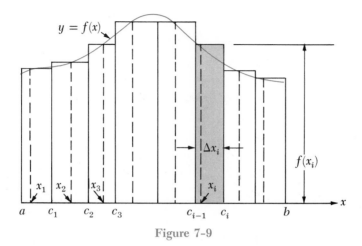

Figure 7-9

From (7.54) intuition suggests that since $\underline{S}_P$ and $\overline{S}_P$ approach A as Δ_P approaches zero, the sum $\sum_{i=1}^{n} f(x_i)\,\Delta x_i$ also approaches A as Δ_P approaches zero.

Before continuing further with this discussion we shall define a Riemann sum and give an example. It will be noted in the definition that we do not require the function f to be continuous on $[a, b]$.

7.5.1   Definition

Let f be *defined* on $[a, b]$ and let $P = \{a = c_0, c_1, c_2, \ldots, c_{n-1}, c_n = b\}$ be any partition of $[a, b]$. Also let $c_{i-1} \leq x_i \leq c_i$ for $i = 1, 2, \ldots, n$. The sum

$$\sum_{i=1}^{n} f(x_i)\,\Delta x_i = f(x_1)\,\Delta x_1 + f(x_2)\,\Delta x_2 + \cdots + f(x_n)\,\Delta x_n$$

is called a *Riemann sum* of f associated with P.

Example 1 If $f(x) = 4 - x^2$ find the Riemann sum of f on $[a, b]$ where $P = \{-1, 0, \frac{1}{2}, \frac{5}{4}, 2\}$ and $x_1 = -\frac{3}{4}$, $x_2 = \frac{1}{4}$, $x_3 = \frac{5}{4}$, and $x_4 = \frac{5}{4}$ (Figure 7–10).

SOLUTION   The calculations are shown in the table on page 255. Then

$$\sum_{i=1}^{4} f(x_i)\,\Delta x_i = f(x_1)\,\Delta x_1 + f(x_2)\,\Delta x_2 + f(x_3)\,\Delta x_3 + f(x_4)\,\Delta x_4$$

$$= \tfrac{55}{16}\cdot 1 + \tfrac{63}{16}\cdot\tfrac{1}{2} + \tfrac{39}{16}\cdot\tfrac{3}{4} + \tfrac{39}{16}\cdot\tfrac{3}{4}$$

$$= \tfrac{145}{16}$$

is the required Riemann sum.

We next prove the theorem about Riemann sums for continuous functions which was suggested by our intuitive discussion preceding Definition 7.5.1.

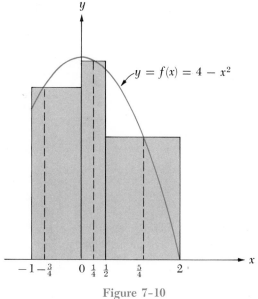

Figure 7-10

Figure 7-10

	c_i	x_i	$f(x_i)$	Δx_i
$i = 0$	-1			
1	0	$-\frac{3}{4}$	$\frac{55}{16}$	1
2	$\frac{1}{2}$	$\frac{1}{4}$	$\frac{63}{16}$	$\frac{1}{2}$
3	$\frac{5}{4}$	$\frac{5}{4}$	$\frac{39}{16}$	$\frac{3}{4}$
4	2	$\frac{5}{4}$	$\frac{39}{16}$	$\frac{3}{4}$

7.5.2   Theorem

Let f be continuous on [a, b], and let A be the unique number obtained in Theorem 7.4.3. For every $\epsilon > 0$, there is a $\delta > 0$ such that for any partition P of [a, b] with $\Delta_P < \delta$, $\Sigma_{i=1}^{n} f(x_i)\, \Delta x_i$, an arbitrary Riemann sum of f associated with P, satisfies the inequality

$$\left| \sum_{i=1}^{n} f(x_i)\, \Delta x_i - A \right| < \epsilon.$$

PROOF   From Theorem 7.4.2(a), for every P

$$\underline{S}_P \leq A \leq \overline{S}_P, \tag{7.55}$$

and from (7.54) and (7.55)

$$-(\overline{S}_P - \underline{S}_P) \leq \sum_{i=1}^{n} f(x_i)\, \Delta x_i - A \leq \overline{S}_P - \underline{S}_P. \tag{7.56}$$

Let $\epsilon > 0$ be given. By Theorem 7.4.2(b) there is a $\delta > 0$ such that

$$\overline{S}_P - \underline{S}_P < \epsilon \quad \text{if } \Delta_P < \delta. \tag{7.57}$$

From (7.56) and (7.57)

$$-\epsilon < \sum_{i=1}^{n} f(x_i)\,\Delta x_i - A < \epsilon \quad \text{if } \Delta_P < \delta,$$

and hence

$$\left| \sum_{i=1}^{n} f(x_i)\,\Delta x_i - A \right| < \epsilon \quad \text{if } \Delta_P < \delta.$$

We next define the definite integral of a function on a finite closed interval in terms of Riemann sums. The *definite integral of f on* $[a, b]$ will be denoted for the present by the symbol

$$\mathrm{I}_{x=a}^{b} f(x),$$

although other variables, such as y, t, s, and so on, may be used here in place of x.

7.5.3   Definition

Let f be *defined* on $[a, b]$. Then $\mathrm{I}_{x=a}^{b} f(x) = A$ if and only if for every $\epsilon > 0$ there is a $\delta > 0$ such that for any partition P satisfying $\Delta_P < \delta$, we have

$$\left| \sum_{i=1}^{n} f(x_i)\,\Delta x_i - A \right| < \epsilon \tag{7.58}$$

for any Riemann sum $\sum_{i=1}^{n} f(x_i)\,\Delta x_i$ of f associated with P.

When $\mathrm{I}_{x=a}^{b} f(x)$ exists, we say f is *integrable* on $[a, b]$. We also call $f(x)$ the *integrand*, a the *lower limit*, and b the *upper limit* of the definite integral.

We leave as an exercise the proof that if the definite integral $\mathrm{I}_{x=a}^{b} f(x)$ exists, it is unique.

The next theorem unites the concepts of the definite integral and the area of a region under a graph.

7.5.4   Theorem

Let f be continuous and assume non-negative values on the interval $[a, b]$. *The area* A_T *of the region T where*

$$T = \{(x, y) : x \in [a, b] \text{ and } y \in [0, f(x)]\}$$

is given by

$$A_T = \int_{x=a}^{b} f(x). \tag{7.59}$$

PROOF By Theorems 7.4.2 and 7.4.3 there is a unique number A such that for every partition P of $[a, b]$, the sums $\underline{S}_P$ and $\overline{S}_P$ associated with P and f satisfy

$$\underline{S}_P \leq A \leq \overline{S}_P,$$

and by Definition 7.4.1

$$A = A_T. \tag{7.60}$$

By Theorem 7.5.2 for any $\epsilon > 0$ there is a $\delta > 0$ such that if P is any partition of $[a, b]$ for which $\Delta_P < \delta$, and S is any Riemann sum of f associated with P, we have

$$|S - A| < \epsilon.$$

Thus by Definition 7.5.3,

$$A_T = A = \int_{x=a}^{b} f(x). \tag{7.61}$$

In Example 3 of Section 7.4 we showed that the area of the region bounded by the graphs of the equations $y = x^2$, $y = 0$, $x = a$, and $x = b$, where $b > a \geq 0$ is $(b^3 - a^3)/3$. Thus from (7.59) if $f(x) = x^2$, we have

$$\int_{x=a}^{b} x^2 = \frac{b^3 - a^3}{3} \qquad \text{if } b > a \geq 0. \tag{7.62}$$

In the next example we obtain a definite integral of a constant function.

Example 2 Prove: If $f(x) = k$ for every $x \in [a, b]$, then

$$\int_{x=a}^{b} k = k(b - a).$$

SOLUTION For any partition P of $[a, b]$, a Riemann sum of f is of the form

$$\sum_{i=1}^{n} f(x_i) \, \Delta x_i = \sum_{i=1}^{n} k \, \Delta x_i = k \sum_{i=1}^{n} \Delta x_i = k(b - a).$$

Then for every $\epsilon > 0$,

$$\left| \sum_{i=1}^{n} f(x_i) \, \Delta x_i - k(b - a) \right| = 0 < \epsilon$$

regardless of how Δ_P is selected. The conclusion then follows from Definition 7.5.3.

Note from this example that if $k > 0$, then $I^b_{x=a} k$ is the area of the rectangular region formed by the lines $y = k$, $y = 0$, $x = a$, and $x = b$. However if $k < 0$, the definite integral is the negative of the area of this region.

In general, from Theorem 7.5.2 and Definition 7.5.3, we have the following theorem.

7.5.5 Theorem

If f is continuous on $[a, b]$, then $I^b_{x=a} f(x)$ exists.

As an illustration of Theorem 7.5.5, $I^4_{x=1} \sqrt{x}$ exists since the function f where $f(x) = \sqrt{x}$ is continuous on the interval $[1, 4]$. Even though the theorem does not enable us to evaluate the definite integral directly, it will be useful in deriving the formula for the evaluation of definite integrals of continuous functions.

The converse of Theorem 7.5.5 is not true. In Exercise 9 of Exercise Set 7.5 we have an example of a function that is not continuous on a closed interval and yet is integrable on the interval.

It can easily be proved that if $I^b_{x=a} f(x)$ exists, then for every $\epsilon > 0$ there is a partition P of $[a, b]$ such that for any two Riemann sums S_1 and S_2 of f associated with P,

$$|S_1 - S_2| < \epsilon$$

(Exercise 13, Exercise Set 7.5). Thus, if there exists some $\epsilon > 0$ such that for every partition P of $[a, b]$ there are Riemann sums S_1 and S_2 of f associated with P such that

$$|S_1 - S_2| \geq \epsilon,$$

then $I^b_{x=a} f(x)$ fails to exist.

Example 3 Show that if $f(x) = \begin{cases} 3 & \text{if } x \text{ is a rational number} \\ -1 & \text{if } x \text{ is irrational} \end{cases}$

then $I^2_{x=0} f(x)$ does not exist.

SOLUTION Let P be any partition of $[0, 2]$. If in each subinterval of $[0, 2]$ formed by P we choose the numbers x_i to be rational, we have the Riemann sum

$$S_1 = \sum_{i=1}^{n} f(x_i)\, \Delta x_i = \sum_{i=1}^{n} 3\, \Delta x_i = 3 \sum_{i=1}^{n} \Delta x_i = 6.$$

However, if the numbers x_i are chosen to be irrational,

$$S_2 = \sum_{i=1}^{n} f(x_i)\, \Delta x_i = \sum_{i=1}^{n} (-1)\, \Delta x_i = -\sum_{i=1}^{n} \Delta x_i = -2.$$

Hence

$$|S_1 - S_2| = 8,$$

and from our discussion above $I^2_{x=0} f(x)$ does not exist.

7.5.6   Theorem

If f is integrable on $[a, b]$, then there is an $M > 0$ such that $|f(x)| \leq M$—that is, f is bounded on $[a, b]$.

PROOF   Let P be any partition of $[a, b]$. If f were *unbounded* on $[a, b]$, f would be unbounded on some subinterval $[c_{k-1}, c_k]$ on $[a, b]$, formed by P. Thus, in the Riemann sum $\Sigma_{i=1}^{n} f(x_i) \Delta x_i$, $|f(x_k)|$ could be chosen as large as desired by a proper choice of x_k. In particular, if $I_{x=a}^{b} f(x) = A$, we could choose x_k so that either

$$f(x_k) > \frac{A + 1 - \Sigma_{i=1}^{k-1} f(x_i) \Delta x_i - \Sigma_{i=k+1}^{n} f(x_i) \Delta x_i}{\Delta x_k},$$

or

$$f(x_k) < \frac{A - 1 - \Sigma_{i=1}^{k-1} f(x_i) \Delta x_i - \Sigma_{i=k+1}^{n} f(x_i) \Delta x_i}{\Delta x_k}.$$

Hence either

$$f(x_k) \Delta x_k > A + 1 - \sum_{i=1}^{k-1} f(x_i) \Delta x_i - \sum_{i=k+1}^{n} f(x_i) \Delta x_i, \tag{7.63}$$

or

$$f(x_k) \Delta x_k < A - 1 - \sum_{i=1}^{k-1} f(x_i) \Delta x_i - \sum_{i=k+1}^{n} f(x_i) \Delta x_i, \tag{7.64}$$

and so if we transpose terms in (7.63), we have

$$\sum_{i=1}^{n} f(x_i) \Delta x_i - A > 1, \tag{7.65}$$

or in (7.64) we have

$$\sum_{i=1}^{n} f(x_i) \Delta x_i - A < -1. \tag{7.66}$$

From (7.65) and (7.66)

$$\left| \sum_{i=1}^{n} f(x_i) \Delta x_i - A \right| > 1.$$

Thus if $\epsilon = 1$, for every partition P there is an associated Riemann sum $\Sigma_{i=1}^{n} f(x_i) \Delta x_i$ for which (7.58) does not hold. Hence $I_{x=a}^{b} f(x)$ would not exist. However, as this conclusion would contradict our hypothesis, f must be bounded on $[a, b]$.

By a restatement of Theorem 7.5.6, *if f is unbounded on $[a, b]$, then f is not integrable on the interval.*

Example 4   Show that $I_{x=0}^{3} 1/x$ fails to exist.

SOLUTION We note that for every $B > 0$ when $0 < x < 1/B$, then $1/x > B$ (Figure 7–11) and hence the function f where $f(x) = 1/x$ is unbounded on $[0, 3]$. Then by Theorem 7.5.6 $I^3_{x=0} 1/x$ fails to exist.

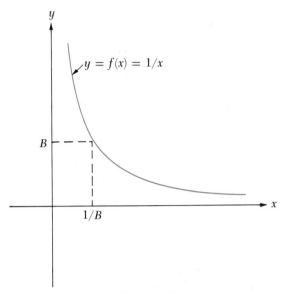

Figure 7-11

Exercise Set 7.5

In Exercises 1–4 find the required Riemann sums of f on the indicated closed intervals.

1. $f(x) = x^2 + 3x$, $x \in [-2, 3]$ $P = \{-2, -1, \frac{1}{2}, \frac{3}{2}, 3\}$

   $x_1 = -1$, $x_2 = -\frac{1}{2}$, $x_3 = 1$, $x_4 = 2$

2. $f(x) = x - 1$, $x \in [-1, 4]$ $P = \{-1, 0, \frac{3}{4}, \frac{5}{4}, 2, \frac{13}{4}, 4\}$

   $x_1 = -1$, $x_2 = \frac{1}{2}$, $x_3 = 1$, $x_4 = \frac{3}{2}$, $x_5 = 3$, $x_6 = \frac{15}{4}$

3. $f(x) = x^2 + 3x$, $x \in [-2, 3]$ $P = \{-2, -1, \frac{1}{2}, \frac{3}{2}, 3\}$

   Each x_i is the midpoint of the subinterval in which it is chosen.

4. $f(x) = x - 1$, $x \in [-1, 4]$ $P = \{-1, 0, \frac{3}{4}, \frac{5}{4}, 2, \frac{13}{4}, 4\}$

   Each x_i is the midpoint of the subinterval in which it is chosen.

5. Does $\displaystyle I^4_{x=1} \frac{3x + 2}{2x + 1}$ exist? Why?

6. Does $\displaystyle I^2_{x=-2} |x|$ exist? Why?

7. Does $\displaystyle I^2_{x=0} \frac{1}{\sqrt{2 - x}}$ exist? Why?

8. Does $\displaystyle\overset{5}{\underset{x=1}{\text{I}}} f(x)$ exist if $f(x) = \begin{cases} x+1 & \text{if } x \geq 3 \\ 2x-2 & \text{if } x < 3 \end{cases}$? Why?

9. If $f(x) = \begin{cases} 0 & \text{if } x = 0 \\ a & \text{if } x \neq 0 \end{cases}$ find $\displaystyle\overset{1}{\underset{x=0}{\text{I}}} f(x)$.

10. Find $\text{I}^2_{x=0}\, x$.

11. Find $\text{I}^1_{x=0}\, x^2$.

12. If $f(x) = \begin{cases} x & \text{if } x = \dfrac{1}{n} \quad \text{where } n = 1, 2, 3, \ldots \\ 0 & \text{if } x \neq \dfrac{1}{n} \quad \text{where } n = 1, 2, 3, \ldots \end{cases}$ find $\displaystyle\overset{1}{\underset{x=0}{\text{I}}} f(x)$.

13. Prove the statement made preceding Example 3: If $\text{I}^b_{x=a}\, f(x)$ exists, then for every $\epsilon > 0$ there is a partition P of $[a, b]$ such that for any two Riemann sums S_1 and S_2 of f associated with P, $|S_1 - S_2| < \epsilon$.

14. Prove that if $\text{I}^b_{x=a}\, f(x)$ exists, it is unique.

7.6   Properties of the Definite Integral

We began in Section 7.4 by seeking a definition for the area of a region under the graph of a continuous function f between two lines $x = a$ and $x = b > a$ where f assumes non-negative values on $[a, b]$. From Theorems 7.4.2 and 7.4.3 we know this area exists for such a region. Then by Theorem 7.5.4 this area turned out to be the definite integral of f on $[a, b]$. However, we saw that a function may assume negative values on $[a, b]$ and still have a definite integral on $[a, b]$.

In this section we will develop some additional properties of the definite integral which lead to a formula in Section 7.8 for readily evaluating definite integrals.

The first theorem was proved in Example 2 of Section 7.5.

7.6.1   Theorem

If $f(x) = k$, a constant, for every $x \in [a, b]$, then

$$\overset{b}{\underset{x=a}{\text{I}}}\, k = k(b - a).$$

7.6.2   Theorem

If $f(x) \geq 0$ for every $x \in [a, b]$ and $\text{I}^b_{x=a}\, f(x)$ exists, then $\text{I}^b_{x=a}\, f(x) \geq 0$.

PROOF   Since $f(x) \geq 0$ for every $x \in [a, b]$, for any Riemann sum of f on $[a, b]$, $\sum_{i=1}^n f(x_i)\, \Delta x_i$, we have

$$\sum_{i=1}^n f(x_i)\, \Delta x_i \geq 0. \tag{7.67}$$

Suppose $I^b_{x=a} f(x) < 0$. From Definition 7.5.3 there would be a $\delta > 0$ such that

$$\left| \sum_{i=1}^{n} f(x_i) \Delta x_i - I^b_{x=a} f(x) \right| < - I^b_{x=a} f(x)$$

when $\Delta_P < \delta$. Hence, if $\Delta_P < \delta$,

$$\sum_{i=1}^{n} f(x_i) \Delta x_i - I^b_{x=a} f(x) < - I^b_{x=a} f(x),$$

and we would have

$$\sum_{i=1}^{n} f(x_i) \Delta x_i < 0 \quad \text{if } \Delta_P < \delta. \tag{7.68}$$

However, (7.68) is a contradiction of (7.67) so our supposition that $I^b_{x=a} f(x) < 0$ is invalid, and the conclusion is obtained.

7.6.3   Theorem

If k is constant and $I^b_{x=a} f(x)$ exists, then

$$I^b_{x=a} kf(x) = k \, I^b_{x=a} f(x).$$

PROOF   If $k = 0$,

$$I^b_{x=a} kf(x) = I^b_{x=a} 0 = 0 = k \, I^b_{x=a} f(x)$$

and the theorem is therefore true. Suppose $k \neq 0$. Since $I^b_{x=a} f(x)$ exists, by Definition 7.5.3 for every $\epsilon > 0$ there is a $\delta > 0$ such that for any Riemann sum $\sum_{i=1}^{n} f(x_i) \Delta x_i$ on $[a, b]$,

$$\left| \sum_{i=1}^{n} f(x_i) \Delta x_i - I^b_{x=a} f(x) \right| < \frac{\epsilon}{|k|}$$

if $\Delta_P < \delta$. Then if $\Delta_P < \delta$,

$$\left| \sum_{i=1}^{n} kf(x_i) \Delta x_i - k \, I^b_{x=a} f(x) \right| = \left| k \left[\sum_{i=1}^{n} f(x_i) \Delta x_i - I^b_{x=a} f(x) \right] \right|$$

$$= |k| \left| \sum_{i=1}^{n} f(x_i) \Delta x_i - I^b_{x=a} f(x) \right|$$

$$< |k| \frac{\epsilon}{|k|} = \epsilon,$$

and the conclusion follows from Definition 7.5.3.

As an illustration of Theorem 7.6.3, from (7.62)

$$\mathbf{I}_{x=1}^{3}\ 4x^2 = 4\ \mathbf{I}_{x=1}^{3}\ x^2 = 4\frac{3^3 - 1^3}{3} = \frac{104}{3}.$$

7.6.4   Theorem

If $\mathbf{I}_{x=a}^{b}\ f(x)$ and $\mathbf{I}_{x=a}^{b}\ g(x)$ exist, then

$$\mathbf{I}_{x=a}^{b}\ [f(x) \pm g(x)] = \mathbf{I}_{x=a}^{b}\ f(x) \pm \mathbf{I}_{x=a}^{b}\ g(x).$$

PROOF   We shall give the proof for $\mathbf{I}_{x=a}^{b}\ [f(x) + g(x)]$. Suppose $\epsilon > 0$ is given. Since $\mathbf{I}_{x=a}^{b}\ f(x)$ exists, there is a $\delta_1 > 0$ such that

$$\left| \sum_{i=1}^{n} f(x_i)\,\Delta x_i - \mathbf{I}_{x=a}^{b}\ f(x) \right| < \frac{\epsilon}{2} \qquad (7.69)$$

if $\Delta_P < \delta_1$. Also since $\mathbf{I}_{x=a}^{b}\ g(x)$ exists, there is a $\delta_2 > 0$ such that

$$\left| \sum_{i=1}^{n} g(x_i)\,\Delta x_i - \mathbf{I}_{x=a}^{b}\ g(x) \right| < \frac{\epsilon}{2} \qquad (7.70)$$

if $\Delta_P < \delta_2$. Then

$$\left| \sum_{i=1}^{n} [f(x_i) + g(x_i)]\,\Delta x_i - \left[\mathbf{I}_{x=a}^{b}\ f(x) + \mathbf{I}_{x=a}^{b}\ g(x) \right] \right|$$

$$= \left| \left[\sum_{i=1}^{n} f(x_i)\,\Delta x_i - \mathbf{I}_{x=a}^{b}\ f(x) \right] + \left[\sum_{i=1}^{n} g(x_i)\,\Delta x_i - \mathbf{I}_{x=a}^{b}\ g(x) \right] \right|$$

$$\leq \left| \sum_{i=1}^{n} f(x_i)\,\Delta x_i - \mathbf{I}_{x=a}^{b}\ f(x) \right| + \left| \sum_{i=1}^{n} g(x_i)\,\Delta x_i - \mathbf{I}_{x=a}^{b}\ g(x) \right|. \qquad (7.71)$$

Now if $\Delta_P < \delta = $ smaller of δ_1 and δ_2, from (7.69), (7.70), and (7.71)

$$\left| \sum_{i=1}^{n} [f(x_i) + g(x_i)]\,\Delta x_i - \left[\mathbf{I}_{x=a}^{b}\ f(x) + \mathbf{I}_{x=a}^{b}\ g(x) \right] \right| < \frac{\epsilon}{2} + \frac{\epsilon}{2} = \epsilon.$$

The conclusion then follows from Definition 7.5.3.

The proof that

$$\mathbf{I}_{x=a}^{b}\ [f(x) - g(x)] = \mathbf{I}_{x=a}^{b}\ f(x) - \mathbf{I}_{x=a}^{b}\ g(x)$$

is left as an exercise.

As an illustration of Theorem 7.6.4,

$$\overset{3}{\underset{x=1}{\text{I}}} (4x^2 + 5) = \overset{3}{\underset{x=1}{\text{I}}} 4x^2 + \overset{3}{\underset{x=1}{\text{I}}} 5,$$

and from Theorem 7.6.1 and the example following Theorem 7.6.3

$$\overset{3}{\underset{x=1}{\text{I}}} (4x^2 + 5) = \tfrac{104}{3} + 10 = \tfrac{134}{3}.$$

Theorem 7.6.4 can be generalized and therefore

$$\overset{b}{\underset{x=a}{\text{I}}} [f_1(x) + f_2(x) + \cdots + f_n(x)] =$$

$$\overset{b}{\underset{x=a}{\text{I}}} f_1(x) + \overset{b}{\underset{x=a}{\text{I}}} f_2(x) + \cdots + \overset{b}{\underset{x=a}{\text{I}}} f_n(x) \quad (7.72)$$

provided the integrals on the right exist.

7.6.5   Theorem

If $f(x) \geq g(x)$ for every $x \in [a, b]$, and $\text{I}_{x=a}^{b} f(x)$ and $\text{I}_{x=a}^{b} g(x)$ exist, then

$$\overset{b}{\underset{x=a}{\text{I}}} f(x) \geq \overset{b}{\underset{x=a}{\text{I}}} g(x).$$

PROOF   Since $f(x) \geq g(x)$ if $x \in [a, b]$, $f(x) - g(x) \geq 0$ if $x \in [a, b]$. By Theorems 7.6.2 and 7.6.4

$$\overset{b}{\underset{x=a}{\text{I}}} [f(x) - g(x)] = \overset{b}{\underset{x=a}{\text{I}}} f(x) - \overset{b}{\underset{x=a}{\text{I}}} g(x) \geq 0,$$

and hence

$$\overset{b}{\underset{x=a}{\text{I}}} f(x) \geq \overset{b}{\underset{x=a}{\text{I}}} g(x).$$

As an illustration of Theorem 7.6.5 since

$$\sqrt{x} \geq 2 \quad \text{when } x \geq 4,$$

we have

$$\overset{7}{\underset{x=4}{\text{I}}} \sqrt{x} \geq \overset{7}{\underset{x=4}{\text{I}}} 2 = 2(7 - 4) = 6.$$

The next theorem is a corollary of Theorem 7.6.5.

7.6.6   Theorem

If f is continuous on $[a, b]$, then

$$\left| \overset{b}{\underset{x=a}{\text{I}}} f(x) \right| \leq \overset{b}{\underset{x=a}{\text{I}}} |f(x)|. \quad (7.73)$$

PROOF If f is continuous on $[a, b]$, then by Theorem 3.8.3 the function g where $g(x) = |f(x)|$ is also continuous on $[a, b]$. Then by Theorem 7.5.5 $I^b_{x=a} f(x)$ and $I^b_{x=a} |f(x)|$ exist. From Theorem 1.5.6

$$-|f(x)| \le f(x) \le |f(x)| \quad \text{if } x \in [a, b], \tag{7.74}$$

and hence from Theorem 7.6.5 and Theorem 7.6.3 (with $k = -1$)

$$- \mathop{I}_{x=a}^{b} |f(x)| = \mathop{I}_{x=a}^{b} - |f(x)| \le \mathop{I}_{x=a}^{b} f(x) \le \mathop{I}_{x=a}^{b} |f(x)|.$$

Then by Theorem 1.5.3(a) and Theorem 1.5.3(b)

$$\left| \mathop{I}_{x=a}^{b} f(x) \right| \le \mathop{I}_{x=a}^{b} |f(x)|.$$

7.6.7 Theorem

If f is defined on $[a, b]$, $a < c < b$, and $I^c_{x=a} f(x)$ and $I^b_{x=c} f(x)$ exist, then $I^b_{x=a} f(x)$ exists, and

$$\mathop{I}_{x=a}^{c} f(x) + \mathop{I}_{x=c}^{b} f(x) = \mathop{I}_{x=a}^{b} f(x). \tag{7.75}$$

PROOF Let $P = \{a = c_0, c_1, c_2, \ldots, c_{n-1}, c_n = b\}$ be any partition of $[a, b]$. We note that c may or may not be an element of P. In any event, there is some positive integer k such that $c \in (c_{k-1}, c_k]$. An arbitrary Riemann sum of f on $[a, b]$ associated with P is of the form

$$S = \sum_{i=1}^{n} f(x_i)\, \Delta x_i.$$

Now, the sum

$$S_1 = \sum_{i=1}^{k-1} f(x_i)\, \Delta x_i + f(c)(c - c_{k-1})$$

is a Riemann sum of f on $[a, c]$. Also,

$$S_2 = f(c)(c_k - c) + \sum_{i=k+1}^{n} f(x_i)\, \Delta x_i$$

is a Riemann sum of f on $[c, b]$. Note that

$$\left| S - \left[\mathop{I}_{x=a}^{c} f(x) + \mathop{I}_{x=c}^{b} f(x) \right] \right|$$

$$= \left| \left[S_1 - \mathop{I}_{x=a}^{c} f(x) \right] + \left[S_2 - \mathop{I}_{x=c}^{b} f(x) \right] - f(c)(c - c_{k-1}) \right.$$

$$\left. - f(c)(c_k - c) + f(x_k)\, \Delta x_k \right|$$

$$\le \left| S_1 - \mathop{I}_{x=a}^{c} f(x) \right| + \left| S_2 - \mathop{I}_{x=c}^{b} f(x) \right| + 2|f(c)|\, \Delta x_k + |f(x_k)|\, \Delta x_k. \tag{7.76}$$

If we can show that the left member of (7.76) can be made as small as desired when Δ_P is sufficiently small, then $I_{x=a}^{b} f(x)$ exists and (7.75) is true.

Let P_1 and P_2 be the partitions associated with S_1 and S_2, respectively, and let $\epsilon > 0$ be given. By hypothesis, there are numbers δ_1 and δ_2 both > 0 such that

$$\left| S_1 - I_{x=a}^{c} f(x) \right| < \frac{\epsilon}{5} \tag{7.77}$$

if $\Delta_{P_1} < \delta_1$ and

$$\left| S_2 - I_{x=c}^{b} f(x) \right| < \frac{\epsilon}{5} \tag{7.78}$$

if $\Delta_{P_2} < \delta_2$.

Since $I_{x=a}^{c} f(x)$ and $I_{x=c}^{b} f(x)$ exist, by Theorem 7.5.6 there is an $M > 0$ such that

$$|f(x)| \leq M \quad \text{for } x \in [a, b] \tag{7.79}$$

and since $\Delta x_k \leq \Delta_P$, from (7.79),

$$2|f(c)| \, \Delta x_k + |f(x_k)| \, \Delta x_k \leq 3M \, \Delta_P. \tag{7.80}$$

If we let $\delta =$ smallest of the numbers δ_1, δ_2, and $\epsilon/5M$, and choose $\Delta_P < \delta$, then from (7.76), (7.77), (7.78), and (7.80)

$$\left| S - \left[I_{x=a}^{c} f(x) + I_{x=c}^{b} f(x) \right] \right| < \frac{\epsilon}{5} + \frac{\epsilon}{5} + 3M \cdot \frac{\epsilon}{5M} = \epsilon.$$

Hence from Definition 7.5.3 $I_{x=a}^{b} f(x)$ exists and (7.75) is obtained.

If f is continuous on $[a, b]$ and assumes non-negative values on the interval (Figure 7–12), by Theorem 7.6.7 the area of the region under the graph of f from a to b is the sum of the areas of the regions under the graph from a to c and from c to b.

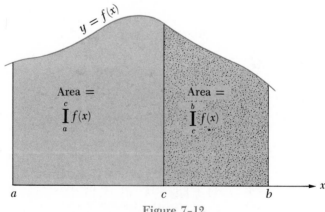

Figure 7–12

It will be convenient to define the definite integral when the upper limit is not greater than the lower limit.

7.6.8   Definition

(a) If f is defined at a,

$$\mathop{\mathrm{I}}_{x=a}^{a} f(x) = 0.$$

(b) If $a > b$, and f is integrable on $[b, a]$,

$$\mathop{\mathrm{I}}_{x=a}^{b} f(x) = -\mathop{\mathrm{I}}_{x=b}^{a} f(x).$$

Using Definition 7.6.8 it can be shown that (7.75) is valid regardless of the ordering of the numbers a, b, and c. If for example,

$$c < b < a,$$

by Theorem 7.6.7

$$\mathop{\mathrm{I}}_{x=c}^{a} f(x) = \mathop{\mathrm{I}}_{x=c}^{b} f(x) + \mathop{\mathrm{I}}_{x=b}^{a} f(x),$$

and if we transpose terms

$$-\mathop{\mathrm{I}}_{x=b}^{a} f(x) = -\mathop{\mathrm{I}}_{x=c}^{a} f(x) + \mathop{\mathrm{I}}_{x=c}^{b} f(x).$$

Then if we apply Definition 7.6.8(b) to the terms in this equation which are preceded by a minus sign,

$$\mathop{\mathrm{I}}_{x=a}^{b} f(x) = \mathop{\mathrm{I}}_{x=a}^{c} f(x) + \mathop{\mathrm{I}}_{x=c}^{b} f(x).$$

Exercise Set 7.6

1. What modification can be made in the proof of Theorem 7.6.4 to prove that $\mathrm{I}_{x=a}^{b} [f(x) - g(x)] = \mathrm{I}_{x=a}^{b} f(x) - \mathrm{I}_{x=a}^{b} g(x)$?

2. Derive Equation (7.75) if we replace the hypothesis $a < c < b$ by the statement given below:
   (a) $b < a < c$
   (b) $b = c < a$
   (c) $c < a < b$

3. Using Theorems 7.6.1 and 7.6.5, prove that if $m \le f(x) \le M$ for every $x \in [a, b]$, and $\mathrm{I}_{x=a}^{b} f(x)$ exists, then

$$m(b - a) \le \mathop{\mathrm{I}}_{x=a}^{b} f(x) \le M(b - a).$$

7.7   First Fundamental Theorem of Calculus

We first prove a theorem that unites the concepts of the derivative and the definite integral, which heretofore have been treated as entirely separate ideas.

7.7.1   Theorem (First Fundamental Theorem of Calculus)

If f is continuous on an interval I containing a, and

$$G(x) = \overset{x}{\underset{t=a}{\text{I}}}\ f(t)\quad \text{if } x \in I,\tag{7.81}$$

then

$$G'(x) = f(x)\quad \text{if } x \in I.$$

PROOF   We first note that since f is continuous on I, by Theorem 7.5.5 and Definition 7.6.8, $G(x)$ exists regardless of the ordering of a and x. (If $x > a$ and f assumes non-negative values on the interval $[a, x]$, $G(x)$ is the area of the region under the graph of f from a to x (Figure 7–13).)

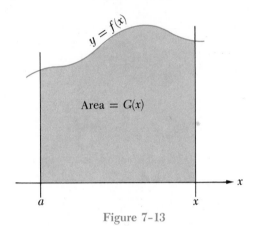

Figure 7–13

To prove the theorem, we shall utilize the definition of the derivative and show that

$$\lim_{h \to 0} \frac{G(x + h) - G(x)}{h} = f(x).$$

If $h \neq 0$ and x and $x + h$ are in I, the quotient in the left member of this equation can be rewritten using (7.81). Thus

$$\frac{G(x + h) - G(x)}{h} = \frac{\text{I}_{t=a}^{x+h} f(t) - \text{I}_{t=a}^{x} f(t)}{h}.\tag{7.82}$$

Since by Theorem 7.6.7

$$\overset{x}{\underset{t=a}{\text{I}}}\ f(t) + \overset{x+h}{\underset{t=x}{\text{I}}}\ f(t) = \overset{x+h}{\underset{t=a}{\text{I}}}\ f(t),\tag{7.83}$$

a substitution from (7.83) for $\text{I}_{t=a}^{x+h} f(t)$ in (7.82) gives

$$\frac{G(x + h) - G(x)}{h} = \frac{\text{I}_{t=x}^{x+h} f(t)}{h}.\tag{7.84}$$

By Theorem 7.6.1 $f(x) = (1/h)\,\text{I}_{t=x}^{x+h} f(x)$ since $f(x)$ is a constant with respect to the variable of integration t. Hence from (7.84)

$$\frac{G(x + h) - G(x)}{h} - f(x) = \frac{I_{t=x}^{x+h} f(t) - I_{t=x}^{x+h} f(x)}{h}$$

$$\left| \frac{G(x + h) - G(x)}{h} - f(x) \right| = \left| \frac{I_{t=x}^{x+h} [f(t) - f(x)]}{h} \right|. \tag{7.85}$$

Since f is continuous on I, given $\epsilon > 0$ there is a $\delta > 0$ such that

$$|f(t) - f(x)| < \epsilon \tag{7.86}$$

if $t \in I$ and $|t - x| < \delta$. Moreover, (7.86) is true if $|t - x| < |h|$ and $0 < |h| < \delta$. Then if $0 < h < \delta$, by Theorem 7.6.6 and Theorem 7.6.5, using (7.85) and (7.86),

$$\left| \frac{G(x + h) - G(x)}{h} - f(x) \right| \le \frac{I_{t=x}^{x+h} |f(t) - f(x)|}{h} < \frac{I_{t=x}^{x+h} \epsilon}{h}.$$

Since $I_{t=x}^{x+h} \epsilon = \epsilon h$ by Theorem 7.6.1,

$$\left| \frac{G(x + h) - G(x)}{h} - f(x) \right| < \epsilon \tag{7.87}$$

when $0 < h < \delta$. It can also be proved that (7.87) holds when $-\delta < h < 0$ (Exercise 13, Exercise Set 7.7). Thus from the definition of the limit of a function

$$G'(x) = \lim_{h \to 0} \frac{G(x + h) - G(x)}{h} = f(x).$$

It will be observed that $G'(x)$ here is obtained by substituting the upper limit x for t in the integrand. Also, we note that $G'(x)$ is independent of the lower limit a.

Example 1 Find $G'(x)$ if $G(x) = I_{t=0}^{x} \dfrac{1}{1 + t^4}$.

SOLUTION Since the function f where $f(t) = 1/(1 + t^4)$ is continuous on the interval $(-\infty, +\infty)$, by Theorem 7.5.5

$$G(x) = I_{t=0}^{x} \frac{1}{1 + t^4}$$

exists for every x, and by Theorem 7.7.1

$$G'(x) = \frac{1}{1 + x^4} \quad \text{for every } x.$$

Example 2 Find $G'(x)$ if $G(x) = I_{t=-1}^{x^3} (t^2 + 4)$.

SOLUTION We note that $G(x) = F(x^3)$ where $F(u) = I_{t=-1}^{u} (t^2 + 4)$. By Theorem 7.7.1

$$F'(u) = u^2 + 4, \tag{7.88}$$

and from the chain rule and (7.88)

$$G'(x) = D_x F(x^3) = F'(x^3) \cdot D_x x^3$$
$$= [(x^3)^2 + 4] \cdot 3x^2 = 3x^2(x^6 + 4).$$

Example 3   Find $D_x \left[\displaystyle\mathop{\text{I}}_{t=\sqrt{x}}^{1} (3t - 1)^{10} \right].$

SOLUTION   We let

$$G(x) = \mathop{\text{I}}_{t=\sqrt{x}}^{1} (3t - 1)^{10} = - \mathop{\text{I}}_{t=1}^{\sqrt{x}} (3t - 1)^{10} = -F(\sqrt{x}),$$

where $F(u) = \text{I}_{t=1}^{u} (3t - 1)^{10}$. Hence by Theorem 7.7.1

$$F'(u) = (3u - 1)^{10}. \tag{7.89}$$

From the chain rule and (7.89)

$$G'(x) = D_x(-F(\sqrt{x})) = -D_x(F(\sqrt{x}))$$
$$= -F'(\sqrt{x}) \cdot D_x \sqrt{x}$$
$$= -(3\sqrt{x} - 1)^{10} \cdot \frac{1}{2\sqrt{x}} = -\frac{(3\sqrt{x} - 1)^{10}}{2\sqrt{x}}.$$

     Theorem 7.7.1 will be used in the next section to derive the second fundamental theorem of calculus, which provides a formula for evaluating definite integrals of continuous functions. In the meantime we will use Theorem 7.7.1 to prove two useful corollaries.

     The proof of the following corollary of Theorem 7.7.1 is left as an exercise. (See Exercise 12 of Exercise Set 7.7.)

7.7.2   Theorem (Mean Value Theorem of Integral Calculus)

If f is continuous on the interval [a, b], then there is a c ∈ [a, b] such that

$$\mathop{\text{I}}_{x=a}^{b} f(x) = (b - a)f(c). \tag{7.90}$$

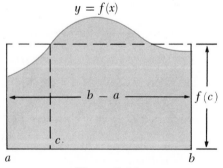

Figure 7-14

If f assumes non-negative values on $[a, b]$, Theorem 7.7.2 says that the area of the region under the graph of f from a to b is equal to the area of a rectangle with base of length $b - a$ and height $f(c)$ (Figure 7–14).

The number $f(c)$ in (7.90) is called the *mean value* of f on the interval $[a, b]$.

A second corollary of Theorem 7.7.1 can also be proved.

7.7.3   Theorem (Intermediate Value Theorem)

If f is continuous on $[a, b]$, $f(a) \neq f(b)$, and k is any number between $f(a)$ and $f(b)$, then there is a number c in the interval (a, b) such that $f(c) = k$ (Figure 7–15).

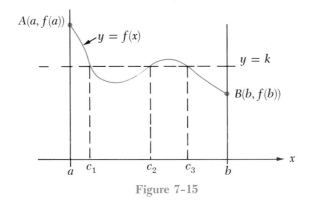

Figure 7–15

PROOF   Let $G(x) = \mathrm{I}^x_{t=a} f(t)$. From Theorem 7.7.1, $G'(x) = f(x)$ if $x \in [a, b]$, and hence G is continuous on the interval. Also, the function F, where

$$F(x) = G(x) - kx, \tag{7.91}$$

is continuous on $[a, b]$ and hence has a maximum and a minimum on $[a, b]$. We shall suppose that

$$f(a) < k < f(b)$$

as the proof is analogous when $f(a) > k > f(b)$. From (7.91)

$$F'(x) = G'(x) - k = f(x) - k \quad \text{for } x \in [a, b] \tag{7.92}$$

and by hypothesis and (7.92):

$$F'(a) = f(a) - k \quad \text{and} \quad F'(b) = f(b) - k$$
$$< 0 \qquad\qquad\qquad > 0 \tag{7.93}$$

Hence, from Theorem 6.1.2 and (7.92) there is a $c \in (a, b)$ such that

$$0 = F'(c) = f(c) - k.$$

Thus $f(c) = k$ and the theorem is proved.

Geometrically, Theorem 7.7.3 states that if the portion of the graph of f connecting the points $A(a, f(a))$ and $B(b, f(b))$ is unbroken, and A is above the line $y = k$ and B is below the line $y = k$, the graph must cross the line $y = k$ somewhere between A and B.

It is entirely possible that the graph of f might intersect the line $y = k$ in more than one point. The graph of f in Figure 7–15, for example, intersects the line $y = k$ at the points (c_1, k), (c_2, k), and (c_3, k), and either c_1, c_2, or c_3 could suffice as the number c in the intermediate value theorem.

Example 4 Let $f(x) = 4 - x^2$. We note that $f(-3) = -5$ and $f(1) = 3$. Using Theorem 7.7.3 can we say there is a number c in the interval $(-3, 1)$ such that $f(c) = 2$? If so, find this number c (Figure 7–16).

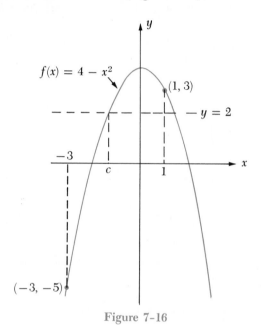

Figure 7-16

SOLUTION Since f is continuous on $[-3, 1]$ and since 2 is between $f(-3)$ and $f(1)$, by Theorem 7.7.3 there is some number $c \in (-3, 1)$ such that $f(c) = 2$. To find c, we solve the equation

$$f(c) = 4 - c^2 = 2$$

and obtain

$$c = \pm\sqrt{2}.$$

Now $\sqrt{2} \notin (-3, 1)$, but $-\sqrt{2} \in (-3, 1)$. Therefore $c = -\sqrt{2}$ is the required number.

Example 5 Show that the equation $x^3 + x - 4 = 0$ has a root between 1 and 2.

SOLUTION We consider the function f where $f(x) = x^3 + x - 4$, noting that f is continuous on the interval $[1, 2]$. Since $f(1) = -2 < 0$ and $f(2) = 6 > 0$, by Theorem 7.7.3 there is a c in $(1, 2)$ for which $f(c) = c^3 + c - 4 = 0$. This number c is the required root of the equation $x^3 + x - 4 = 0$.

Exercise Set 7.7

Find $F'(x)$ in Exercises 1–6.

1. $F(x) = \displaystyle\int_{t=-3}^{x} (t^2 - 9)$

2. $F(x) = \displaystyle\int_{u=2}^{x} \dfrac{1}{u^{3/2} + 1}$

3. $F(x) = \displaystyle\int_{t=x}^{5} |t^{1/2} - 1|$

4. $F(x) = \displaystyle\int_{u=0}^{x^2} (u^2 + 9)$

5. $F(x) = \displaystyle\int_{s=|x|}^{-1} \dfrac{1 - s}{1 + s}$

6. $F(x) = \displaystyle\int_{t=x}^{x^2} \dfrac{t}{t^4 + 4}$.   HINT: $\displaystyle\int_{t=0}^{x^2} \dfrac{t}{t^4 + 4} = \displaystyle\int_{t=0}^{x} \dfrac{t}{t^4 + 4} + \displaystyle\int_{t=x}^{x^2} \dfrac{t}{t^4 + 4}$

7. Using formula (7.90) find the mean value of f where $f(x) = x^2$ when $x \in [2, 5]$.

8. If $f(x) = \sqrt{2x + 3}$ is there a number $c \in (3, 8)$ such that $f(c) = 4$? Why? If so, find this number.

9. If $f(x) = x^2 - 3x - 4$ is there a number $c \in (-1, 5)$ such that $f(c) = 2$? Why? If so, find this number.

10. If $f(x) = 2/(x + 1)$, $f(-2) = -2$ and $f(4) = \frac{2}{5}$. We note that $f(-2) < -1 < f(4)$. Does Theorem 7.7.3 guarantee that there is a $c \in (-2, 4)$ such that $f(c) = -1$? Why?

11. Locate between consecutive integers the real roots of the equation $x^3 - 2x^2 - 5x + 1 = 0$.

12. Prove Theorem 7.7.2 by applying Theorem 5.2.2 to the function G defined in (7.81).

13. Obtain (7.87) in the proof of Theorem 7.7.1 when $-\delta < h < 0$.

7.8   Evaluation of Definite Integrals

We now derive the formula for evaluating definite integrals of continuous functions.

7.8.1   Theorem (Second Fundamental Theorem of Calculus)

If f is continuous on an interval I containing the numbers a and b, and F is any antiderivative of f on I, then

$$\int_{x=a}^{b} f(x) = F(b) - F(a).$$

PROOF Since f is continuous on I and $a \in I$, from Theorems 7.7.1 and 7.1.3 any antiderivative F of f on I is of the form

$$F(x) = \mathop{\mathrm{I}}_{t=a}^{x} f(t) + C \quad \text{if } x \in I. \tag{7.94}$$

From (7.94)

$$F(b) = \mathop{\mathrm{I}}_{t=a}^{b} f(t) + C \tag{7.95}$$

and

$$F(a) = \mathop{\mathrm{I}}_{t=a}^{a} f(t) + C = 0 + C = C. \tag{7.96}$$

Hence from (7.95) and (7.96)

$$F(b) - F(a) = \mathop{\mathrm{I}}_{t=a}^{b} f(t). \tag{7.97}$$

Since the definite integral is independent of the variable used to denote the integrand, we could just as well have written

$$F(b) - F(a) = \mathop{\mathrm{I}}_{x=a}^{b} f(x),$$

thereby obtaining the conclusion.

In applications of Theorem 7.8.1 we often let $F(b) - F(a)$ be denoted by $F(x)|_b^a$. Thus by Theorem 7.8.1, since $\int f(x)\, dx$ is an antiderivative of $f(x)$ when $x \in I$,

$$\mathop{\mathrm{I}}_{x=a}^{b} f(x) = \int f(x)\, dx \Big|_a^b \tag{7.98}$$

or

$$\mathop{\mathrm{I}}_{x=a}^{b} f(x) = \int_a^b f(x)\, dx.$$

The expression $\int_a^b f(x)\, dx$ is suggested by the right member of (7.98), and we will use it henceforth in denoting the definite integral of f from a to b. Since the operation of antidifferentiation is utilized here to evaluate a definite integral, we shall henceforth call this operation *integration* and refer to the symbol $\int$ as an *integral sign*. Also, an antiderivative of a function that is continuous on an interval will now be called an *integral* of a function on the interval. Further, we will henceforth term $\int f(x)\, dx$ the *indefinite integral* of the *integrand* $f(x)$ on the appropriate interval.

Example 1 Find $\displaystyle\int_1^5 2x^3\, dx.$

SOLUTION An integral of f where $f(x) = 2x^3$ on the interval $[1, 5]$ is given by $F(x) = x^4/2$. Hence by Theorem 7.8.1,

$$\int_1^5 2x^3\, dx = \frac{x^4}{2}\Big|_1^5 = \frac{5^4}{2} - \frac{1^4}{2} = 312.$$

It should be noted that if we had used $F(x) = x^4/2 + C$, C being any number, in integrating $2x^3$, we would have obtained the same result, as

$$\int_1^5 2x^3 \, dx = \left(\frac{x^4}{2} + C\right)\Big|_1^5 = \left(\frac{5^4}{2} + C\right) - \left(\frac{1^4}{2} + C\right) = 312$$

since the C's drop out after the parentheses are removed.

Example 2 Find $\int_{-1}^2 (x^2 - 3x + 5) \, dx$.

SOLUTION Since $\int (x^2 - 3x + 5) \, dx = \frac{x^3}{3} - \frac{3x^2}{2} + 5x + C$, by Theorem 7.8.1

$$\int_{-1}^2 (x^2 - 3x + 5) \, dx = \left(\frac{x^3}{3} - \frac{3x^2}{2} + 5x\right)\Big|_{-1}^2$$

$$= \left(\frac{2^3}{3} - \frac{3 \cdot 2^2}{2} + 5 \cdot 2\right) - \left(\frac{(-1)^3}{3} - \frac{3(-1)^2}{2} + 5(-1)\right)$$

$$= \tfrac{8}{3} - 6 + 10 + \tfrac{1}{3} + \tfrac{3}{2} + 5 = \tfrac{27}{2}.$$

Example 3 Find $\int_0^2 \frac{dx}{(3x + 2)^2}$.

SOLUTION If we let $u = 3x + 2$, then $du = 3 \, dx$, and $dx = \tfrac{1}{3} \, du$. Hence

$$\int \frac{dx}{(3x + 2)^2} = \int \frac{1}{3} u^{-2} \, du = -\frac{1}{3u} + C = -\frac{1}{3(3x + 2)} + C.$$

Then by Theorem 7.8.1

$$\int_0^2 \frac{dx}{(3x + 2)^2} = -\frac{1}{3(3x + 2)}\Big|_0^2 = -\tfrac{1}{24} + \tfrac{1}{6} = \tfrac{1}{8}.$$

This integral could also be evaluated using the following change of variable formula.

In applying this formula we make the substitutions

$$u = g(x) \qquad du = g'(x) \, dx$$

in the integral to be evaluated, and replace the original limits of integration, $x = a$ and $x = b$, by $u = g(a)$ and $u = g(b)$, respectively.

7.8.2 Theorem

If f and g are functions such that g' is continuous on an interval I containing a and b, and f is continuous on an interval J which contains every number $g(x)$ such that $x \in I$, then

$$\int_a^b f(g(x))g'(x)\,dx = \int_{g(a)}^{g(b)} f(u)\,du.$$

PROOF Since f is continuous on J, f has an integral F on J by Theorem 7.7.1. Since $g(a)$ and $g(b)$ are numbers in J, by Theorem 7.8.1

$$\int_{g(a)}^{g(b)} f(u)\,du = F(g(b)) - F(g(a)). \tag{7.99}$$

Since $F'(u) = f(u)$ if $u \in J$, by the chain rule

$$D_x F(g(x)) = F'(g(x))g'(x) = f(g(x))g'(x) \quad \text{if } x \in I.$$

Thus $F \circ g$ is an integral of $(f \circ g)g'$ on I. Since g' is continuous on I, so is g; and by Theorem 3.8.4 $f \circ g$ is also continuous on I. Hence by Theorem 3.7.3 $(f \circ g)g'$ is continuous on I. Again by Theorem 7.8.1

$$\int_a^b f(g(x))g'(x)\,dx = F(g(b)) - F(g(a)), \tag{7.100}$$

and from (7.99) and (7.100) the conclusion is obtained.

   In particular from Theorem 7.8.2, when $f(u) = u^r$ where $r \neq -1$ is a rational number,

$$\int_a^b (g(x))^r g'(x)\,dx = \int_{g(a)}^{g(b)} u^r\,du. \tag{7.101}$$

Example 4 Evaluate $\int_0^2 dx/(3x+2)^2$ using (7.101). (Compare Example 3.)

SOLUTION We let $u = 3x + 2$ and hence $du = 3\,dx$ and $\frac{1}{3}\,du = dx$. We also note that if $x = 0$, then $u = 2$ and if $x = 2$, then $u = 8$. From (7.101)

$$\int_0^2 \frac{du}{(3x+2)^2} = \int_2^8 \frac{1}{3u^2}\,du = -\frac{1}{3u}\Big|_2^8$$

$$= -\tfrac{1}{24} + \tfrac{1}{6} = \tfrac{1}{8}.$$

Example 5 Evaluate $\int_1^4 \sqrt{2x^2 + 8x}(x+2)\,dx$.

SOLUTION We let $u = 2x^2 + 8x$ and hence $du = (4x + 8)\,dx = 4(x + 2)\,dx$. Therefore $(x + 2)\,dx = \frac{1}{4}\,du$. We also note that $u = 10$ when $x = 1$ and $u = 64$ when $x = 4$. Hence by (7.101)

$$\int_1^4 \sqrt{2x^2 + 8x}(x+2)\,dx = \int_{10}^{64} \tfrac{1}{4}u^{1/2}\,du = \tfrac{1}{4}\cdot\tfrac{2}{3}u^{3/2}\Big|_{10}^{64}$$

$$= \tfrac{1}{6}(512 - 10\sqrt{10}) = \tfrac{1}{3}(256 - 5\sqrt{10}).$$

Thus far, we have been concerned almost entirely with definite integrals in which the function was continuous on the appropriate closed interval. The next theorem enables us to evaluate definite integrals of functions that are continuous on a closed interval except possibly at the endpoints where the function is defined but has a "finite jump."

7.8.3   Theorem

If the following conditions are satisfied:

(i) $\int_a^b g(x)\, dx$ *exists,*

(ii) f *is defined on* $[a, b]$,

(iii) $f(x) = g(x)$ *for all* $x \in (a, b)$,

then $\int_a^b f(x)\, dx$ *exists and*

$$\int_a^b f(x)\, dx = \int_a^b g(x)\, dx.$$

PROOF   From hypothesis (i) g is defined on $[a, b]$ and hence we define the function ϕ by letting

$$\phi(x) = f(x) - g(x) \quad \text{if } x \in [a, b]. \tag{7.102}$$

From (iii)

$$\phi(x) = 0 \quad \text{if } x \in (a, b). \tag{7.103}$$

If P is any partition of $[a, b]$, let $\sum_{i=1}^n \phi(x_i)\, \Delta x_i$ be a Riemann sum of ϕ associated with P. From (7.103) all terms in this sum except possibly the first and last terms are 0, and so

$$\sum_{i=1}^n \phi(x_i)\, \Delta x_i = \phi(x_1)\, \Delta x_1 + \phi(x_n)\, \Delta x_n. \tag{7.104}$$

We shall prove that

$$\int_a^b \phi(x)\, dx = 0. \tag{7.105}$$

If $\epsilon > 0$ is given and $\Delta_P < \epsilon/2M$, where M is some positive number that is greater than the larger of $|\phi(x_1)|$ and $|\phi(x_n)|$, then from (7.104)

$$\left| \sum_{i=1}^n \phi(x_i)\, \Delta x_i \right| \leq |\phi(x_1)|\, \Delta x_1 + |\phi(x_n)|\, \Delta x_n$$

$$< M \cdot \frac{\epsilon}{2M} + M \cdot \frac{\epsilon}{2M}$$

$$< \epsilon.$$

Equation (7.105) then follows from Definition 7.5.3. From (7.102)

$$f(x) = g(x) + \phi(x)$$

and hence from (7.105) using Theorem 7.6.4,

$$\int_a^b f(x)\,dx = \int_a^b [g(x) + \phi(x)]\,dx = \int_a^b g(x)\,dx + \int_a^b \phi(x)\,dx$$
$$= \int_a^b g(x)\,dx + 0 = \int_a^b g(x)\,dx.$$

Example 6   Evaluate $\int_1^3 f(x)\,dx$ where $f(x) = \begin{cases} 0 & \text{if } x = 1 \\ 2x & \text{if } x \in (1,3) \\ 2 & \text{if } x = 3. \end{cases}$

SOLUTION   We note that the function f is defined on the interval $[1, 3]$, $f(x) = 2x$ for every $x \in (1, 3)$, and $\int_1^3 2x\,dx = x^2\big|_1^3 = 8$. Hence, by Theorem 7.8.3

$$\int_1^3 f(x)\,dx = \int_1^3 2x\,dx = 8.$$

Example 7   Evaluate $\int_0^5 h(x)\,dx$ where $h(x) = \begin{cases} 3x + 2 & \text{if } x \le 1 \\ x - 4 & \text{if } x > 1. \end{cases}$

SOLUTION   By Theorem 7.8.3

$$\int_0^1 h(x)\,dx = \int_0^1 (3x + 2)\,dx = \left(\frac{3x^2}{2} + 2x\right)\bigg|_0^1 = \frac{7}{2}$$

and

$$\int_1^5 h(x)\,dx = \int_1^5 (x - 4)\,dx = \frac{(x-4)^2}{2}\bigg|_1^5 = -4.$$

Then by Theorem 7.6.7

$$\int_0^5 h(x)\,dx = \int_0^1 h(x)\,dx + \int_1^5 h(x)\,dx = \tfrac{7}{2} - 4 = -\tfrac{1}{2}.$$

Exercise Set 7.8

In Exercises 1–26 evaluate the given definite integrals.

1. $\displaystyle\int_2^4 x^2\,dx$

2. $\displaystyle\int_1^4 \frac{2}{x\sqrt{x}}\,dx$

3. $\displaystyle\int_0^3 \sqrt{3t}\,dt$

4. $\displaystyle\int_{-2}^1 (x^2 + 3x - 4)\,dx$

5. $\displaystyle\int_1^9 y(2\sqrt{y} + 3)\,dy$

6. $\displaystyle\int_1^8 \frac{\sqrt[3]{x} - 2}{x^2}\,dx$

7. $\displaystyle\int_2^4 \left(1 - \frac{1}{x^2}\right)\left(1 + \frac{2}{x^2}\right) dx$

8. $\displaystyle\int_{-1}^1 (3x^{1/3} + 1)^2 \, dx$

9. $\displaystyle\int_{-1}^1 (x^{1/3} - x^{-1/3})^2 \, dx$

10. $\displaystyle\int_0^2 \frac{t^4 + t^2 + 1}{t^2 + t + 1} \, dt$

11. $\displaystyle\int_{-4}^1 |2x + 5| \, dx$

12. $\displaystyle\int_{-2}^2 (|x| + |x + 1|) \, dx$

13. $\displaystyle\int_0^1 [3x - 2] \, dx$

14. $\displaystyle\int_0^3 x[x + 1] \, dx$

15. $\displaystyle\int_1^4 [2x + 1] \, dx$

16. $\displaystyle\int_{-1}^5 f(x) \, dx$ if $f(x) = \begin{cases} x^2 + 1 & \text{if } x \le 1 \\ 4x - 2 & \text{if } x > 1 \end{cases}$

17. $\displaystyle\int_0^7 \frac{dx}{\sqrt[3]{x + 1}}$

18. $\displaystyle\int_0^{3/2} (x - 1)^7 \, dx$

19. $\displaystyle\int_1^2 \frac{du}{(2u - 5)^3}$

20. $\displaystyle\int_1^2 (4 - 3t)^5 \, dt$

21. $\displaystyle\int_0^1 3t \sqrt[3]{(t^2 + 1)^4} \, dt$

22. $\displaystyle\int_0^1 \frac{5y^2 \, dy}{(y^3 + 1)^3}$

23. $\displaystyle\int_1^2 (x + 1)\sqrt{x^2 + 2x - 1} \, dx$

24. $\displaystyle\int_0^a \frac{x \, dx}{(a^2 + x^2)^4}$

25. $\displaystyle\int_1^8 \frac{dx}{\sqrt[3]{x^2}\sqrt{1 + \sqrt[3]{x}}}$

26. $\displaystyle\int_4^9 \frac{\sqrt{\sqrt{x} - 2}}{\sqrt{x}} \, dx$

8. Applications of the Definite Integral

8.1 Area of a Plane Region (continued)

We recall from Theorem 7.5.4 that if the function f is continuous on $[a, b]$ and assumes non-negative values on the interval, then the area A_T of the region T under the graph of f from a to b is given by

$$A_T = \int_a^b f(x)\, dx. \tag{8.1}$$

An example illustrating the use of formula (8.1) can be given now that the evaluation of definite integrals has been discussed.

Example 1 Find the area A of the region bounded by the graph of $y = 4 + 3x - x^2$ and the x axis (Figure 8–1).

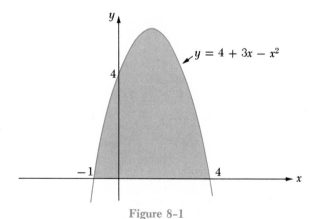

$y = 4 + 3x - x^2$

Figure 8–1

SOLUTION Note that $y = 4 + 3x - x^2 = (4 - x)(1 + x)$. Thus if $x \in [-1, 4]$, $4 - x$ and $1 + x$ are of the same sign and hence $y \geq 0$. Then by (8.1)

$$A = \int_{-1}^{4} (4 + 3x - x^2)\, dx$$

$$= \left(4x + \frac{3x^2}{2} - \frac{x^3}{3} \right)\Bigg|_{-1}^{4}$$

$$= (16 + 24 - \tfrac{64}{3}) - (-4 + \tfrac{3}{2} + \tfrac{1}{3})$$

$$= \tfrac{125}{6}.$$

Next consider the problem of defining the area of the *region between two graphs*. We shall require that this area, when defined, satisfy not only the criteria (i)–(iv) for area given in Section 7.4, but also the following requirement:

(v) If each point in a plane region T (Figure 8–2) having area A_T is moved the same distance in the same direction to form a new plane region T', then $A_{T'}$, the area of T', exists and

$$A_{T'} = A_T.$$

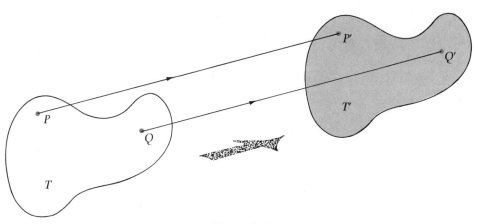

Figure 8–2

In Figure 8–2, if P and Q are arbitrary points in T, and P' and Q' are the respective corresponding points in T', then $|PP'| = |QQ'|$ and $PP' \parallel QQ'$.

If the functions f and g are continuous on $[a, b]$ and $f(x) \geq g(x)$ for every x in the interval, we denote by T (Figure 8–3(a)), the region given by

$$T = \{(x, y): x \in [a, b] \text{ and } y \in [g(x), f(x)]\}. \tag{8.2}$$

We note here that $y = f(x)$ is an equation of the *upper boundary* of T and $y = g(x)$ is associated with the *lower boundary* of T.

Since g is continuous on $[a, b]$, there is a number

$$k = \text{minimum value of } g \text{ on } [a, b]. \tag{8.3}$$

(In Figure 8–3(a) k is negative and hence the distance between the line $y = k$ and the x axis is $-k$, which is positive.) If $k \neq 0$, the graphs of $y = f(x) - k$ and $y = g(x) - k$ are obtained by moving the graphs of $y = f(x)$ and $y = g(x)$, respectively, *downward* a distance k if $k > 0$ or *upward* a distance $-k$ if $k < 0$. Between

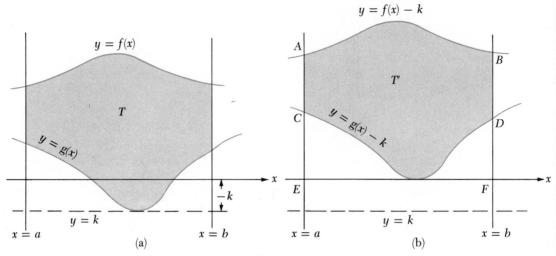

Figure 8-3

the lines $x = a$ and $x = b$, we denote the region under the graph of $y = f(x) - k$ by $ABFE$ and the region under the graph of $y = g(x) - k$ by $CDFE$ (Figure 8–3(b)).
From (8.1),

$$\text{Area of } ABFE = \int_a^b [f(x) - k]\, dx, \tag{8.4}$$

$$\text{Area of } CDFE = \int_a^b [g(x) - k]\, dx. \tag{8.5}$$

If T' is the region given by

$$T' = \{(x, y)\colon x \in [a, b] \text{ and } y \in [g(x) - k, f(x) - k]\},$$

and if $A_{T'}$, the area of T', were defined, from criterion (iv) in Section 7.4,

$$\text{Area of } CDFE + A_{T'} = \text{area of } ABFE,$$

and so

$$A_{T'} = \text{area of } ABFE - \text{area of } CDFE. \tag{8.6}$$

From (8.4), (8.5), and (8.6),

$$A_{T'} = \int_a^b [f(x) - k]\, dx - \int_a^b [g(x) - k]\, dx$$

$$= \int_a^b \{[f(x) - k] - [g(x) - k]\}\, dx$$

$$= \int_a^b [f(x) - g(x)]\, dx.$$

Then if A_T, the area of T, were defined, by criterion (v) of this section, we would have

$$A_T = A_{T'} = \int_a^b [f(x) - g(x)]\, dx.$$

Thus we have the following statement:

8.1.1   Definition

Let f and g be continuous on $[a, b]$ and let $f(x) \geq g(x)$ for every x in the interval. The area A_T of the region T (Figure 8–3(a)) that is defined by (8.2) is

$$A_T = \int_a^b [f(x) - g(x)]\, dx. \tag{8.7}$$

We note that if $g(x) = 0$ for every x in $[a, b]$—this is, the lower boundary of the region T in (8.2) is the x axis—the formula (8.1) for the area of the region under the graph of $y = f(x)$ is obtained as a special case of (8.7).

Example 2   Find the area A of the region bounded by the graphs of the equations $y = x + 6$, $y = \frac{1}{2}x^2$, $x = 1$, and $x = 4$ (Figure 8–4).

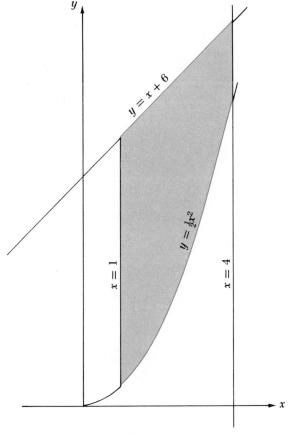

Figure 8-4

SOLUTION   Note that the region bounded by the graphs extends from $x = 1$ to $x = 4$. Since $x + 6 \geq \frac{1}{2}x^2$ for every x in $[1, 4]$, in the terminology of Definition 8.1.1, $f(x) = x + 6$, $g(x) = \frac{1}{2}x^2$, $a = 1$, and $b = 4$. Thus from (8.7) the area of the region is

$$A = \int_1^4 [(x + 6) - \tfrac{1}{2}x^2]\, dx$$

$$= \left(\frac{x^2}{2} + 6x - \tfrac{1}{6}x^3\right)\Big|_1^4$$

$$= (8 + 24 - \tfrac{64}{6}) - (\tfrac{1}{2} + 6 - \tfrac{1}{6})$$

$$= 15.$$

Example 3 Find the area A of the region bounded by the graphs of $y = 2 + x$ and $y = 4 - x^2$ (Figure 8-5).

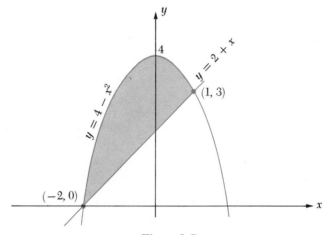

Figure 8-5

SOLUTION One can obtain the intersection points of the graphs after eliminating y between the equations of the graphs and solving the resulting equation. Letting

$$2 + x = 4 - x^2$$

$$x^2 + x - 2 = 0$$

$$(x + 2)(x - 1) = 0$$

we obtain

$$x = -2 \qquad \text{or} \qquad x = 1.$$

The intersection points are $(-2, 0)$ and $(1, 3)$, and therefore the region extends from $x = -2$ to $x = 1$. Since $4 - x^2 \geq 2 + x$ for every x in $[-2, 1]$, in the terminology of (8.7), $f(x) = 4 - x^2$, $g(x) = 2 + x$, $a = -2$, and $b = 1$. Hence from (8.7)

$$A = \int_{-2}^1 (2 - x - x^2)\, dx$$

$$= \left(2x - \frac{x^2}{2} - \frac{x^3}{3}\right)\Big|_{-2}^1$$

$$= (2 - \tfrac{1}{2} - \tfrac{1}{3}) - (-4 - 2 + \tfrac{8}{3})$$

$$= \tfrac{9}{2}.$$

Example 4   Find the total area A of the region bounded by the graph of $y = x^3 - x^2 - 6x$ and the x axis (Figure 8–6).

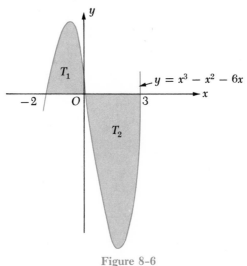

Figure 8–6

SOLUTION   We note that since

$$y = x^3 - x^2 - 6x = x(x - 3)(x + 2),$$

$y \geq 0$ if $x \in [-2, 0]$ or $x \geq 3$, and $y \leq 0$ if $x \leq -2$ or $x \in [0, 3]$.
We let A_1 be the area of the subregion T_1 which extends from $x = -2$ to $x = 0$ and A_2 be the area of the subregion T_2 which extends from $x = 0$ to $x = 3$. Then

$$A = A_1 + A_2. \tag{8.8}$$

From (8.1)

$$A_1 = \int_{-2}^{0} (x^3 - x^2 - 6x)\, dx$$

$$= \left(\frac{x^4}{4} - \frac{x^3}{3} - 3x^2 \right)\Big|_{-2}^{0} = -\left(4 + \frac{8}{3} - 12 \right)$$

$$= \tfrac{16}{3}. \tag{8.9}$$

To find A_2, we first note that the upper boundary of T_2, the x axis, has the equation $y = 0$ and the lower boundary of T_2 is the graph of the equation $y = x^3 - x^2 - 6x$. Hence from (8.7), upon letting $f(x) = 0$ and $g(x) = x^3 - x^2 - 6x$,

$$A_2 = \int_{0}^{3} [0 - (x^3 - x^2 - 6x)]\, dx$$

$$= \int_{0}^{3} (-x^3 + x^2 + 6x)\, dx$$

$$= \left(-\frac{x^4}{4} + \frac{x^3}{3} + 3x^2 \right)\Big|_{0}^{3} = -\frac{81}{4} + 9 + 27$$

$$= \tfrac{63}{4}. \tag{8.10}$$

From (8.8), (8.9), and (8.10) the total area of the region is

$$A = \tfrac{16}{3} + \tfrac{63}{4} = \tfrac{253}{12}.$$

If g is continuous on $[a, b]$ and assumes *negative values* on the interval, the area A of the region (Figure 8–7)

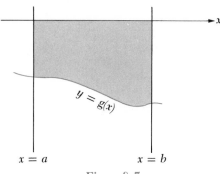

Figure 8-7

$$\{(x, y): x \in [a, b] \quad \text{and} \quad y \in [g(x), 0]\}$$

can be obtained from (8.7) by letting $f(x) = 0$ on $[a, b]$. Thus

$$A = -\int_a^b g(x) \, dx. \qquad (8.11)$$

Note that the area A_2 in Example 4 could have been obtained from (8.11) by letting $g(x) = x^3 - x^2 - 6x$, $a = 0$, and $b = 3$.

Next, suppose we wanted to calculate the area of the region T bounded by the graphs of the equations $y^2 = 6 + 3x$ and $y = 3x$ (shaded in Figure 8–8). Our first reaction might be to utilize the integral formula (8.7). To use this formula, we will need an equation of the form $y = f(x)$ that is satisfied by all points in the upper boundary of T, the arc of the graph of $y^2 = 6 + 3x$ which joins the points $(-2, 0)$ and $(1, 3)$. This equation, which is obtained by solving $y^2 = 6 + 3x$ for y when $y \geq 0$, is found to be $y = \sqrt{6 + 3x}$. Now, the lower boundary of T is the union of the arc of the graph of $y^2 = 6 + 3x$ joining $(-2, 0)$ and $(-\tfrac{2}{3}, -2)$ and the line segment connecting $(-\tfrac{2}{3}, -2)$ and $(1, 3)$. The points in the arc satisfy the equation $y = -\sqrt{6 + 3x}$ and those in the line segment satisfy $y = 3x$. Thus, along the lower boundary of T:

$$y = g(x) = \begin{cases} -\sqrt{6 + 3x} & \text{if } -2 \leq x \leq -\tfrac{2}{3} \\ 3x & \text{if } -\tfrac{2}{3} \leq x \leq 1 \end{cases}$$

The area A of the region T can then be obtained from (8.7).

$$A = \int_{-2}^{1} [f(x) - g(x)] \, dx$$

$$= \int_{-2}^{-2/3} [\sqrt{6 + 3x} - (-\sqrt{6 + 3x})] \, dx + \int_{-2/3}^{1} (\sqrt{6 + 3x} - 3x) \, dx. \qquad (8.12)$$

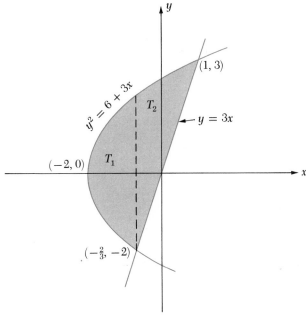

Figure 8-8

The first integral in the right member of (8.12) gives the area of the subregion T_1 (Figure 8–8) of T and the second integral gives the area of T_2. The first integral will simplify to $2\int_{-2}^{-2/3} \sqrt{6 + 3x}\, dx$. This integral can also be obtained by noting that since the upper and lower boundaries of T_1 are symmetric to the x axis, it suffices to obtain twice the area of the portion of T_1 lying above the x axis.

Rather than obtain A by (8.12), which would be rather laborious, we will approach the problem of finding this area from another direction. We note that along the right-hand boundary of T, the line segment connecting $(-\frac{2}{3}, -2)$ and $(1, 3)$, the equation

$$x = \frac{y}{3}$$

is satisfied. Along the left-hand boundary of T, the graph of $y^2 = 6 + 3x$, the equation

$$x = \frac{y^2 - 6}{3}$$

is obtained. It will be seen that T is an example of a region which is described in the next paragraph.

By analogy with Definition 8.1.1, if f and g are continuous on the interval $[a, b]$, the area A of the region bounded by the graphs of $x = f(y)$, $x = g(y)$, $y = a$, and $y = b$, where $f(y) \geq g(y)$ for every $y \in [a, b]$ (Figure 8–9), is defined by

$$A = \int_a^b [f(y) - g(y)]\, dy. \tag{8.13}$$

A comparison of (8.13) and (8.7) shows that the only change is the replacement of x as the variable of integration by y. It will be noted in Figure 8–9 that the equation $x = f(y)$ is satisfied along the right-hand boundary and $x = g(y)$ is satisfied on the left-hand boundary of the region.

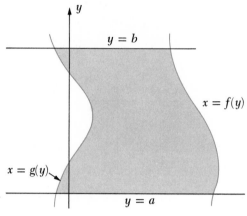

Figure 8–9

The area problem leading to (8.12) will now be solved using (8.13). It will be observed that the integral giving this area is a much simpler expression than that in (8.12).

Example 5 Find the area A of the region T bounded by the graphs of the equations $y^2 = 6 + 3x$ and $y = 3x$ (Figure 8–8).

SOLUTION We note that the region extends from $y = -2$ to $y = 3$. It was mentioned that $x = y/3$ along the right-hand boundary of T and that $x = (y^2 - 6)/3$ along the left-hand boundary of T. Hence from (8.13)

$$A = \int_{-2}^{3} \left(\frac{y}{3} - \frac{y^2 - 6}{3} \right) dy = \int_{-2}^{3} \left(-\frac{1}{3}y^2 + \frac{1}{3}y + 2 \right) dy$$

$$= \left(-\frac{y^3}{9} + \frac{y^2}{6} + 2y \right) \Big|_{-2}^{3} = \left(-3 + \frac{3}{2} + 6 \right) - \left(\frac{8}{9} + \frac{2}{3} - 4 \right)$$

$$= \tfrac{125}{18}.$$

We might inquire at this point whether the area formulas we have discussed could be used to find the area of a circular region. Suppose we wished to find the area of the region enclosed by the circle $x^2 + y^2 = r^2$ (Figure 8–10). Note that the upper semicircle has the equation $y = \sqrt{r^2 - x^2}$ and the lower semicircle has the equation $y = -\sqrt{r^2 - x^2}$.
From (8.7) the area A of the region bounded by the circle is

$$A = 2 \int_{-r}^{r} \sqrt{r^2 - x^2} \, dx. \qquad (8.14)$$

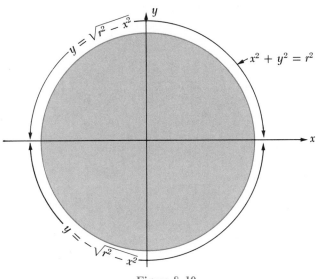

Figure 8–10

If we consider the symmetry of the region with respect to the y axis

$$A = 4 \int_0^r \sqrt{r^2 - x^2} \, dx. \tag{8.15}$$

The integral in (8.14) cannot be evaluated using techniques we have already discussed. However, if the substitution $u = x/r$ is made in (8.15), A can be written

$$A = 4 \int_0^1 \sqrt{r^2 - r^2 u^2} \, r \, du = 4r^2 \int_0^1 \sqrt{1 - u^2} \, du. \tag{8.16}$$

We would like to have $A = \pi r^2$, the familiar result from elementary geometry for the area of a circular region, so we will give the following definition of π.

8.1.2   Definition

$$\pi = 4 \int_0^1 \sqrt{1 - u^2} \, du. \tag{8.17}$$

Now, from (8.16) and (8.17), the area of the circular region is

$$A = \pi r^2.$$

Hence from (8.15) and (8.14), respectively

$$\int_0^r \sqrt{r^2 - x^2} \, dx = \frac{\pi r^2}{4} \tag{8.18}$$

and

$$\int_{-r}^{r} \sqrt{r^2 - x^2} \, dx = \frac{\pi r^2}{2}. \tag{8.19}$$

From the definition (8.17) for π it will be shown (see Section 15.4) that the circumference C of a circle of radius r is given by

$$C = 2\pi r$$

and that

$$\pi \approx 3.1416$$

(Exercise 18, Exercise Set 14.8).

Exercise Set 8.1

In each of the following exercises obtain the area of the region bounded by the graphs of the given equations. For each solution sketch the graph of the region.

1. $y = 2x + 3$, $y = 0$, $x = 2$, and $x = 6$
2. $y = 9x - x^2$ and $y = 0$
3. $x = y^2 - 2y - 3$ and $x = 0$
4. $x^2 = y^3$, $x = 0$, and $y = 4$ (in first quadrant)
5. $x^{1/2} + y^{1/2} = a^{1/2}$, $x = 0$, and $y = 0$
6. $y = x^2 - 3x - 4$ and $y = 0$
7. $y = x^2 - 2x + 4$, $y = 2 - \frac{1}{2}x$, $x = 0$, and $x = 4$
8. $x = 4y^2 - y^3$, $x = -1 - y$, $y = 0$, and $y = 3$
9. $2y = x^2$ and $y = 2 - x^2$
10. $y^2 = 4x$ and $x + y = 3$
11. $x = (y - 2)^2$ and $x = y$
12. $y = x^3 - 4x$ and $y = 2x + x^2$
13. $x = 5y$ and $x = y^3 - 2y^2 - 3y$
14. $y = x^3 - x^2$ and $y = 2x^2 - 4$
15. Find the area of one loop of the graph of $y^2 = x^2(4 - x^2)$.

8.2   Volumes by Slicing

In this section we shall discuss the calculation of the volume of a certain type of three-dimensional region which uses the definite integral of a function of one variable. However, before giving the definition of this volume, some preliminary remarks should be made. We first define a *cylindrical solid*.

8.2.1   Definition

Let $h > 0$ and T be a plane region with area A, and let Q be any point in T (Figure 8–11). Also let L_Q be the line segment which consists of all points P on the same side of T such that PQ is perpendicular to T and $|PQ| \leq h$. The set

$$S = \{L_Q : Q \in T\}$$

is called a *cylindrical solid with base T and height h*. The *volume*, V, of S is defined by writing

$$V = hA. \tag{8.20}$$

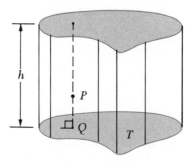

Figure 8–11

If the base T is a circular disc, S is called a *right circular cylinder of height h*, and if T is a rectangle, S is termed a *rectangular parallelepiped*.

We also mention that a *plane section* of a solid region is the intersection of the region and a plane. The plane sections of a cylindrical solid, formed by planes that are parallel to the plane of the base of the solid, have a constant area. The areas of all such plane sections are the same as the area of the base.

In this section we shall be concerned with volumes of solid regions that are bounded by two planes perpendicular to a coordinate axis and such that any *cross section* of the solid—that is, any plane section of the solid taken perpendicular to the axis—has a known area. For example, suppose the solid region S is bounded by planes perpendicular to the x axis at $x = a$ and $x = b$ where $b > a$. Suppose also that the cross section of S that is perpendicular to the x axis at an arbitrary point $(x, 0)$ has an area that we denote by $A(x)$. We assume here that the *area function A* is continuous and assumes non-negative values on $[a, b]$. In Figure 8–12 the cross sections of S that are perpendicular to the x axis at the points with coordinates a, x, and b are shaded. The areas of these cross sections are, respectively, $A(a)$, $A(x)$, and $A(b)$.

Since S is a three-dimensional region, we have included in Figure 8–12 the z *axis*, which is perpendicular to the x and y axes at their intersection point. The arrow on the z axis indicates that the positive direction on this number line is upward. A complete discussion of three-dimensional rectangular coordinates will be given in Chapter 17.

In order to obtain the volume V of the solid region S, we will require

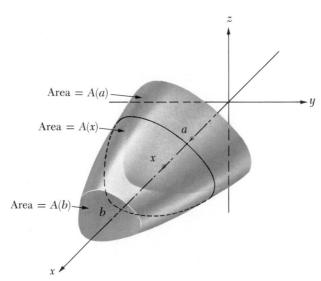

Figure 8-12

that the following criteria be satisfied. It will be noted that these requirements are analogous to those given for area in Section 7.4:

(i) If V exists, it is a non-negative number.
(ii) If S is a cylindrical solid with base area A and height h, then $V = hA$.
(iii) If E and F are solid regions whose volumes exist and $E \subset F$, then

volume of $E \leq$ volume of F.

(iv) If a solid region E can be subdivided into subregions $E_1, E_2, \ldots, E_n$, no two of which have points in common except possibly in their boundaries, and the volumes of the subregions exist, then the volume of E exists and

$$\text{volume of } E = \sum_{i=1}^{n} \text{volume of } E_i.$$

To define the volume V of S, we proceed as in Section 7.4 by first forming a partition $P = \{a = c_0, c_1, c_2, \ldots, c_{n-1}, b\}$ of $[a, b]$. As before, we let

$$\Delta x_i = c_i - c_{i-1} \quad \text{for } i = 1, 2, \ldots, n$$
$$A(m_i) = \text{minimum of } A \text{ on } [c_{i-1}, c_i]$$
$$A(M_i) = \text{maximum of } A \text{ on } [c_{i-1}, c_i].$$

Suppose ΔS_i is the subregion of S that extends from $x = c_{i-1}$ to $x = c_i$ (Figure 8-13). ΔS_i is contained in a solid cylinder with altitude Δx_i and cross section area $A(M_i)$ and contains a solid cylinder with altitude Δx_i and cross section area $A(m_i)$. From (8.20) the volumes of these solid cylinders are respectively $A(M_i) \Delta x_i$ and $A(m_i) \Delta x_i$.

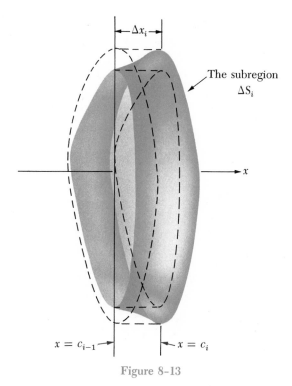

The subregion
ΔS_i

$x = c_{i-1}$ $x = c_i$

Figure 8–13

Hence if the volume ΔV_i of ΔS_i *were defined*, then

$$A(m_i)\,\Delta x_i \le \Delta V_i \le A(M_i)\,\Delta x_i.$$

If V were defined, we would have $V = \sum_{i=1}^{n} \Delta V_i$, and therefore

$$\sum_{i=1}^{n} A(m_i)\,\Delta x_i \le V \le \sum_{i=1}^{n} A(M_i)\,\Delta x_i. \tag{8.21}$$

The left and right members of (8.21) are Riemann sums of the function A associated with the partition P. From Definition 7.5.3 these Riemann sums can be made to differ from the definite integral $\int_a^b A(x)\,dx$ by as little as desired when Δ_P, the norm of P, is sufficiently small. Thus we actually define V by the following statement:

8.2.2   Definition

Let S be a solid region bounded by the planes perpendicular to the x axis at $x = a$ and $x = b$ and suppose the area of a typical plane section of S perpendicular to the x axis at $(x, 0)$ is given by $A(x)$. Suppose also that the function A is continuous and assumes non-negative values on $[a, b]$. The volume V of S is

$$V = \int_a^b A(x)\,dx. \tag{8.22}$$

Equation (8.22) is called the *volume-by-slicing* formula. From this formula the volume is obtained by integrating the cross section area of the solid.

Example 1 Let S be the solid having for its base the region in the first quadrant bounded by the circle $x^2 + y^2 = 25$. If every plane section of the solid taken perpendicular to the x axis is a square, find V, the volume of S (Figure 8–14).

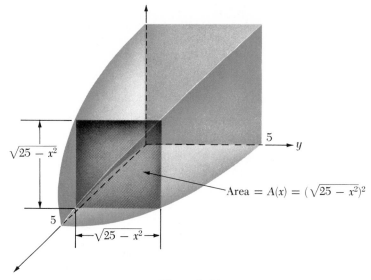

Figure 8-14

SOLUTION We note that the solid extends from $x = 0$ to $x = 5$. A cross section of the solid which is perpendicular to the x axis at any number x is a square with edge $\sqrt{25 - x^2}$. The area of this cross section is

$$A(x) = (\sqrt{25 - x^2})^2 = 25 - x^2.$$

Hence by (8.22)

$$V = \int_0^5 (25 - x^2)\, dx$$

$$= \left(25x - \frac{x^3}{3}\right)\Big|_0^5$$

$$= 125 - \tfrac{125}{3} = \tfrac{250}{3}.$$

Example 2 Find the volume of a right circular cone having radius r and altitude h (Figure 8–15).

SOLUTION For convenience the cone is oriented so that its vertex is at the origin, and its axis lies in the positive x axis. The cone therefore extends from $x = 0$ to $x = h$. We note that a cross section of the cone at any number x is a

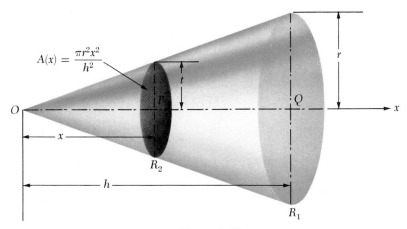

Figure 8-15

circular region. To obtain $A(x)$, the area of such a cross section, we first must find its radius, which has been denoted by t. Since triangles POR_2 and QOR_1 in Figure 8–15 are similar,

$$\frac{t}{x} = \frac{r}{h}. \tag{8.23}$$

Then from (8.23)

$$A(x) = \pi t^2 = \frac{\pi r^2 x^2}{h^2}$$

and hence from (8.22) the volume of the cone is

$$V = \int_0^h \frac{\pi r^2 x^2}{h^2}\, dx = \frac{\pi r^2}{h^2}\frac{x^3}{3}\Big|_0^h$$

$$= \frac{\pi r^2 h}{3}.$$

A right circular cone is an example of a *solid of revolution*. A solid of revolution is a solid region that is generated by revolving a region in a plane about some line, called its *axis of revolution*, that does not intersect the plane region except possibly on its boundary. The right circular cone in Figure 8–15 was generated by revolving right triangle OR_1Q about the x axis.

A cross section of a solid of revolution is either a circular region or an *annulus*, a plane region bounded by two concentric circles. If the inner and outer radii of an annulus are respectively r_1 and r_2 (Figure 8–16), its area A is

$$A = \pi(r_2{}^2 - r_1{}^2). \tag{8.24}$$

In the next example we calculate the volume of a solid of revolution where a typical cross section is an annulus.

Example 3   Find the volume of the solid formed by revolving the region bounded by the graphs of $y = x^2$ and $x = y^2$ about the line $y = 1$ (Figures 8–17(a), 8–17(b)).

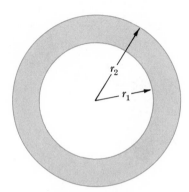

Figure 8–16

SOLUTION The graphs of $y = x^2$ and $x = y^2$ intersect at $(0, 0)$ and $(1, 1)$, and the region bounded by these graphs lies in the first quadrant of the xy plane. We also note that at any point (x, y) on the first quadrant portion of the graph of $x = y^2$, we have $y = \sqrt{x}$.

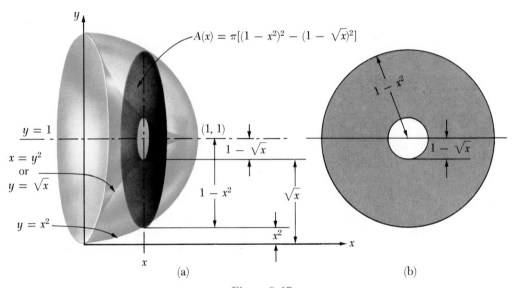

Figure 8–17

A cross section of the solid taken perpendicular to the line $y = 1$ at any number x is an annulus with inside radius $1 - \sqrt{x}$ and outside radius $1 - x^2$ (Figure 8–17(b)). The area $A(x)$ of the annulus from (8.24) is

$$A(x) = \pi[(1 - x^2)^2 - (1 - \sqrt{x})^2]$$
$$= \pi(x^4 - 2x^2 - x + 2\sqrt{x}).$$

Hence from (8.22)

$$V = \int_0^1 \pi(x^4 - 2x^2 - x + 2\sqrt{x})\, dx$$

$$= \pi \left(\frac{x^5}{5} - \frac{2x^3}{3} - \frac{x^2}{2} + \frac{4x^{3/2}}{3} \right) \Big|_0^1$$

$$= \tfrac{11}{30}\pi.$$

Suppose S is a solid region bounded by planes perpendicular to the y axis at $y = a$ and $y = b$ where $b > a$, and the area of a typical cross section taken perpendicular to the y axis is $A(y)$ where the function A is continuous and non-negative on $[a, b]$. By analogy with (8.22) in Definition 8.2.2 the volume V of S is defined to be

$$V = \int_a^b A(y)\, dy. \qquad (8.25)$$

We shall use (8.25) in the next example.

Example 4   Find the volume of the solid generated when the region bounded by the x axis and the graphs of $y = x^2$ and $x = 2$ is revolved about the line $x = 2$ (Figure 8–18).

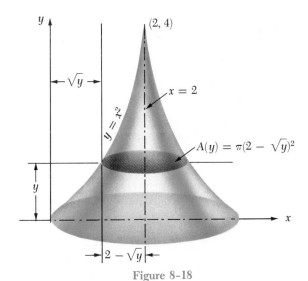

Figure 8-18

SOLUTION   By solving the equation $y = x^2$ for x, we find that on the boundary $y = x^2$, $x = \sqrt{y}$. Thus a cross section of the solid located a distance y above the x axis is a circular region with radius $2 - \sqrt{y}$. Hence the area of the cross section is

$$A(y) = \pi(2 - \sqrt{y})^2.$$

Since the solid is bounded by planes perpendicular to the y axis at $y = 0$ and $y = 4$, from (8.25)

$$V = \int_0^4 \pi(2 - \sqrt{y})^2\, dy$$

$$= \int_0^4 \pi(4 - 4\sqrt{y} + y)\, dy$$

$$= \pi(4y - \tfrac{8}{3}y^{3/2} + \tfrac{1}{2}y^2)\Big|_0^4$$

$$= \tfrac{8}{3}\pi.$$

Exercise Set 8.2

1. A solid has for its base the plane region bounded by the circle $x^2 + y^2 = 25$. Find the volume of the solid if every plane section taken perpendicular to the x axis is

   (a) a semicircle;

   (b) an isosceles right triangle with its hypotenuse in the base;

   (c) an isosceles right triangle with its leg in the base;

   (d) an equilateral triangle.

2. Find the volume of a sphere of radius r. SUGGESTION: Locate the sphere so that its center is at the origin. Note that a typical plane section of the sphere which is perpendicular to a diameter of the sphere is a circle.

3. A hemispherical bowl of radius r is filled to a depth h with water. Find the volume of water in the bowl.

4. A tetrahedron has a base which is a right triangle with legs of length 3 in. and 4 in., respectively. The remaining vertex is 6 in. directly above the vertex of the right angle in the base. Find the volume of the tetrahedron.

5. Find the volume of a pyramid that has an altitude of length h and a base that is a square having an edge of length a.

In Exercises 6–17 find the volume of the solid of revolution generated when the plane region bounded by the graphs of the given equation is revolved about the given line. In each exercise sketch the solid generated.

6. $y = x^2$, $y = 0$, $x = a > 0$; x axis

7. $y = 4x - x^2$, $y = 0$; x axis

8. $x^2 - y^2 = 9$ and $x = 5$;   (a) x axis   (b) y axis

9. $y = 4x^2$, $x = 2$, $y = 0$;   (a) x axis   (b) y axis   (c) $x = 2$ (d) $x = -1$

10. $x = y^{1/2}/2$, $y = 4$, $x = 0$;   (a) x axis   (b) y axis   (c) $y = 4$

11. $x^2 = y^3$, $y = 4$;   (a) x axis   (b) y axis   (c) $y = 4$ (d) $x = -2$

12. $y = x^2$, $y = 3x$;   (a) x axis   (b) y axis   (c) $x = 3$ (d) $y = -2$

13. $2(y - 2)^2 = x$, $(y - 2)^2 = x - 1$; y axis

14. $y = 9 - x^2$, $y = 2x^2$; x axis

15. $x^2 = y^2 - y^4$ (one loop); y axis

16. $x^2 + y^2 = 10$, $x + y = 4$; x axis

17. $x^2 + y^2 = r^2$; $y = k$ where $k > r > 0$. HINT: Use either (8.18) or (8.19).

18. If a hole of radius r is drilled through the center of a sphere of radius a, find the volume of the material remaining in the solid.

19. A wedge is cut from a right circular cylinder of radius r. If the wedge is bounded by a horizontal plane P_1 and another plane P_2 that intersects P_1 along a diameter of the cylinder, find the volume of the wedge if P_1 and P_2 intersect at an angle of 45°.

8.3 Volumes of Solids of Revolution (Cylindrical Shell Method)

Suppose we wanted to calculate the volume of the solid of revolution generated when the region bounded by the graphs of $y = 9 - \frac{1}{2}x^2$ and $y = (x - 3)^2$ is revolved about the y axis (Figure 8–19). We would find the method of slicing exceedingly difficult here as one cannot readily express $A(y)$, the area of a typical plane section of the solid taken perpendicular to the y axis.

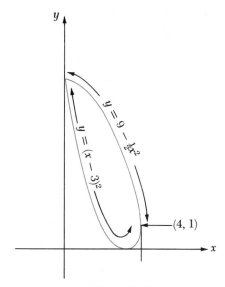

Figure 8–19

To solve a problem such as this, we introduce another scheme for defining the volume of a solid of revolution, using *cylindrical shells*. Suppose a rectangle $ABCD$ is revolved about a line L which is parallel to AB and does not intersect the rectangle. Then a solid region called a cylindrical shell is generated by the rectangle (Figure 8–20).
Letting

$$r = \text{distance between } AB \text{ and } L$$
$$h = |BC| = |AD|$$
$$k = |AB| = |CD|,$$

the inner radius of the cylindrical shell is r, the outer radius is $r + h$, and hence the area of the base of the cylindrical shell is

$$\pi[(r + h)^2 - r^2].$$

Since the height of the cylindrical shell is k, its volume is

$$V = (\text{base area})(\text{height})$$
$$= \pi k[(r + h)^2 - r^2] = \pi k(2rh + h^2)$$
$$= 2\pi kh\left(r + \frac{h}{2}\right).$$

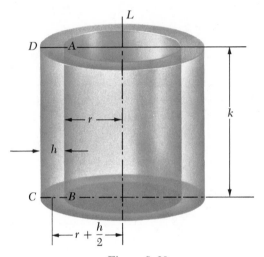

Figure 8-20

If we call $r + (h/2)$ the average radius of the cylindrical shell, then

$$V = 2\pi(\text{area of rect. } ABCD)(\text{avg. radius of cyl. shell}). \qquad (8.26)$$

Suppose f and g are continuous on $[a, b]$. Then formula (8.26) will be utilized to define the volume V of a solid of revolution generated by revolving the region

$$T = \{(x, y): x \in [a, b] \quad \text{and} \quad y \in [g(x), f(x)]\} \qquad (8.27)$$

about a line $x = c$ where c is not between a and b (Figure 8-21).

We again utilize the now-familiar procedure of forming an arbitrary partition $P = \{a = c_0, c_1, c_2, \ldots, c_{n-1}, c_n = b\}$ of $[a, b]$ and let

$$x_i = \frac{c_{i-1} + c_i}{2}$$
$$\Delta x_i = c_i - c_{i-1} \qquad \text{where } i = 1, 2, 3, \ldots, n.$$

If the rectangle with height $f(x_i) - g(x_i)$ and thickness Δx_i is revolved about the line $x = c$, from (8.26) a cylindrical shell is generated with volume

$$\Delta V_i = 2\pi[f(x_i) - g(x_i)]\,\Delta x_i \cdot |x_i - c|. \qquad (8.28)$$

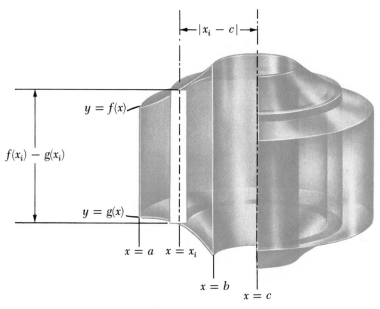

Intuition suggests that if the norm Δ_P of the partition P of $[a, b]$ is sufficiently small, the sum

$$\sum_{i=1}^{n} \Delta V_i = \sum_{i=1}^{n} 2\pi |x_i - c|[f(x_i) - g(x_i)]\, \Delta x_i \qquad (8.29)$$

would be a good approximation of V if V were defined. Furthermore, the smaller Δ_P is chosen, the better (8.29) would seem as an approximation of V. Note that (8.29) is a Riemann sum of the function given by $F(x) = 2\pi |x - c|[f(x) - g(x)]$, which is associated with the partition P of $[a, b]$. Since f and g are continuous on $[a, b]$, so is F, and hence the integral $\int_a^b 2\pi |x - c|[f(x) - g(x)]\, dx$ exists. Therefore by Definition 7.5.3, the sum (8.29) can be made to differ from this integral by as little as we please when Δ_P is sufficiently small, and we are led to the following definition.

8.3.1 Definition

Let f and g be continuous on $[a, b]$ and let $f(x) \geq g(x)$ for every x in the interval. The volume V of the solid of revolution generated by revolving the region (8.27) about the line $x = c$ is given by

$$V = \int_a^b 2\pi |x - c|[f(x) - g(x)]\, dx. \qquad (8.30)$$

Formula (8.30) is called the *cylindrical shell volume formula*. The formula may be remembered by thinking of the definite integral as being approximated by a Riemann sum of the form (8.29) where ΔV_i as given by (8.28) is the volume of a cylindrical shell with average radius $|x_i - c|$, height $f(x_i) - g(x_i)$, and thickness Δx. The integrand in (8.30) may then be readily obtained from ΔV_i by replacing x_i by x and Δx_i by dx.

When the volume of a solid of revolution can be obtained by both the slicing technique and the cylindrical shell formula, it can be shown that the results will agree, although we will not concern ourselves here with the proof of this assertion. The consistency of the two definitions of volume is illustrated in the following problem, which was earlier solved by the slicing method (Example 4, Section 8.2) and is now solved using cylindrical shells.

. **Example 1**   Find the volume of the solid generated when the region bounded by the x axis and the graphs of $y = x^2$ and $x = 2$ is revolved about the line $x = 2$ (Figure 8–22).

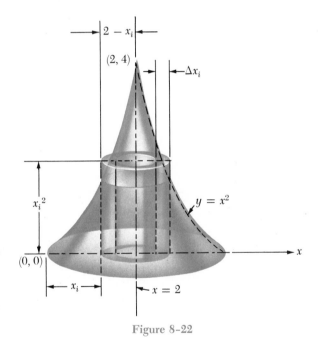

Figure 8-22

SOLUTION   We note that the plane region that is revolved extends from $x = 0$ to $x = 2$. A typical cylindrical shell resulting from a partition of $[0, 2]$ has an average radius $2 - x_i$ and is generated by revolving a rectangle of height x_i^2 and thickness Δx_i about the line $x = 2$. Thus ΔV_i, the volume of the cylindrical shell is given by

$$\Delta V_i = 2\pi(2 - x_i)x_i^2 \, \Delta x_i. \tag{8.31}$$

Using (8.31) and replacing x_i and Δx_i by x and dx respectively, we obtain for the volume V of the solid

$$V = \int_0^2 2\pi(2 - x)x^2 \, dx$$

$$= \int_0^2 2\pi(2x^2 - x^3) \, dx$$

$$= 2\pi \left(\frac{2x^3}{3} - \frac{x^4}{4} \right) \Big|_0^2$$

$$= \frac{8\pi}{3}.$$

We next use the cylindrical shell method to solve the volume problem that was introduced at the beginning of this section.

Example 2   Find the volume of the solid generated when the region bounded by the graphs of $y = 9 - \frac{1}{2}x^2$ and $y = (x - 3)^2$ is revolved about the y axis (Figure 8–23).

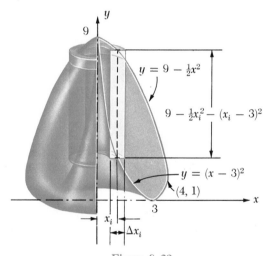

Figure 8–23

SOLUTION   We note that the plane region revolved extends from $x = 0$ to $x = 4$. A typical cylindrical shell resulting from a partition of $[0, 4]$ has an average radius x_i and is generated by revolving a rectangle of height $9 - \frac{1}{2}x_i^2 - (x_i - 3)^2$ and thickness Δx_i about the y axis. The volume ΔV_i of this cylindrical shell is then

$$\Delta V_i = 2\pi x_i [9 - \tfrac{1}{2}x_i^2 - (x_i - 3)^2] \, \Delta x_i$$
$$= \pi(12x_i^2 - 3x_i^3) \, \Delta x_i$$

and the volume V of the solid is

$$V = \int_0^4 \pi(12x^2 - 3x^3) \, dx$$

$$= \pi(4x^3 - \tfrac{3}{4}x^4) \Big|_0^4 = 64\pi.$$

By analogy with Definition 8.3.1 the volume V obtained by revolving the region T bounded by the graphs of $x = f(y)$, $x = g(y)$, $y = a$, and $y = b$ about the line $y = c$ where $c \notin (a, b)$ is defined as

$$V = \int_a^b 2\pi |y - c| [f(y) - g(y)]\, dy. \tag{8.32}$$

Example 3 Find the volume of the solid generated when the region bounded by the graphs of $y = x^2$ and $x = y^2$ is revolved about the line $y = 1$ (Figure 8–24).

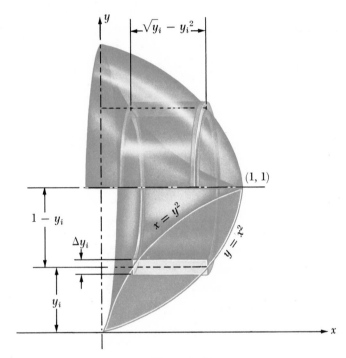

Figure 8-24

SOLUTION The plane region revolved extends from $y = 0$ to $y = 1$. A typical cylindrical shell resulting from a partition of $[0, 1]$ has an average radius $1 - y_i$ and is generated by revolving a rectangle of length $\sqrt{y_i} - y_i^2$ and width Δy_i about the line $y = 1$. The volume ΔV_i of the cylindrical shell is therefore

$$\Delta V_i = 2\pi(1 - y_i)(\sqrt{y_i} - y_i^2)\, \Delta y_i,$$

and the volume of the solid is

$$V = \int_0^1 2\pi(1 - y)(\sqrt{y} - y^2)\, dy$$

$$= \int_0^1 2\pi(y^{1/2} - y^{3/2} - y^2 + y^3)\, dy$$

$$= 2\pi(\tfrac{2}{3}y^{3/2} - \tfrac{2}{5}y^{5/2} - \tfrac{1}{3}y^3 + \tfrac{1}{4}y^4)\Big|_0^1$$

$$= 2\pi\left(\frac{11}{60}\right) = \frac{11\pi}{30}.$$

The student may compare this solution with that in Example 3 of Section 8.2.

Exercise Set 8.3

In each of the following exercises obtain the volume of the solid generated by revolving the region bounded by the graphs of the given equations about the given line following the semicolon. Also, sketch the appropriate solid and give the dimensions of a typical cylindrical shell having volume ΔV_i.

1. $y = 2x + 3$, $y = 0$, $x = 2$, and $x = 6$; (a) y axis (b) $x = 2$

2. $y = 4x - x^2$; (a) y axis (b) $x = 4$

3. Exercise 9 of Exercise Set 8.2

4. Exercise 10 of Exercise Set 8.2

5. Exercise 11 of Exercise Set 8.2

6. $2(y - 2)^2 = x$, $(y - 2)^2 = x - 1$; x axis

7. $y = 9 - x^2$, $y = 2x^2$; y axis

8. Exercise 16 of Exercise set 8.2

9. Exercise 17 of Exercise Set 8.2

10. Exercise 18 of Exercise Set 8.2

8.4 Work

In physics, if F is a constant force in a certain direction and acts upon a body so as to move it a distance d in that direction, the work W done by the force is defined as

$$W = Fd. \tag{8.33}$$

If, for example, a body weighing 40 lb is raised vertically a distance of 6 ft, the force F required would be 40 lb, $d = 6$ ft, and hence from (8.33)

$$W = 40 \cdot 6 = 240 \text{ ft-lb.}$$

The formula (8.33) cannot be used directly to calculate work if the force F is variable because of the uncertainty about the number to be used in place of F.

In this section, therefore, we seek to define the work done by a variable force in moving a particle a given distance in a straight line. Suppose the particle moves from $x = a$ to $x = b$, where $b > a$, under the influence of a force whose magnitude is denoted by $f(x)$, to indicate that the magnitude depends upon the x coordinate of the particle. Suppose also that the function f is continuous on $[a, b]$.

To define the work W done by $f(x)$ in moving the particle from $x = a$ to $x = b$, we first form a partition $P = \{a = c_0, c_1, c_2, \ldots, c_{n-1}, c_n = b\}$ of $[a, b]$. As before, let the length of an arbitrary subinterval $[c_{i-1}, c_i]$ be Δx_i. We shall make the assumption that W, if defined, is the sum of the work done by $f(x)$ moving the particle from $x = a$ to $x = c_1$, moving the particle from $x = c_1$ to $x = c_2$, and so on, and the work done moving the particle from $x = c_{n-1}$ to $x = b$.

If the work done by $f(x)$ in moving the particle from $x = c_{i-1}$ to $x = c_i$ were defined, from (8.33) this work could be approximated by

$$\Delta W_i = f(x_i)\, \Delta x_i \tag{8.34}$$

where x_i is arbitrarily chosen in $[c_{i-1}, c_i]$ (Figure 8–25). Naturally, the smaller we

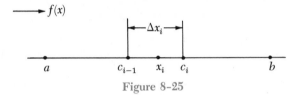

Figure 8–25

choose the norm of P the more accurate the approximation (8.34) would seem to be. Furthermore, if W were defined, from (8.34) W could be approximated by

$$\sum_{i=1}^{n} \Delta W_i = \sum_{i=1}^{n} f(x_i)\, \Delta x_i. \tag{8.35}$$

Intuition suggests that the smaller we choose the norm of P, the better the approximation (8.35) is of W. Thus we define W by saying

$$W = \int_a^b f(x)\, dx. \tag{8.36}$$

Example 1 The force necessary to stretch or compress a spring a distance x from its natural length is given by

$$f(x) = kx \tag{8.37}$$

where k is a constant called the *spring constant*. Equation (8.37) expresses *Hooke's law*. If the natural length of a spring is 12 inches, and a force of 20 lb is required to stretch the spring to a length of 15 inches, find the work done in stretching the spring from its natural length to a length of 18 inches.

SOLUTION We first evaluate k in (8.37). When the spring is stretched to a length of 15 inches, $x = 3$. Thus from (8.37)

$$20 = k \cdot 3$$

and

$$k = \tfrac{20}{3} \text{ lb/in.}$$

Thus, in this example (8.37) has the form

$$f(x) = \tfrac{20}{3}x.$$

Since $x = 6$ when the spring has a length of 18 inches, from (8.36) the required work is

$$W = \int_0^6 \frac{20}{3} x \, dx = \frac{10}{3} x^2 \Big|_0^6$$

$$= 120 \text{ in.-lb.}$$

Example 2   A cable is wound around a windlass so that 60 ft of cable hangs down from the windlass. If the cable weighs 4 lb/ft, find the work done in winding up the cable (Figure 8–26).

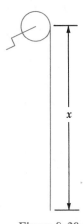

Figure 8–26

SOLUTION   It will be convenient to have the positive x axis extending downward from the origin, which will be at the point where the cable joins the windlass. We note that in the beginning the cable extends from $x = 0$ to $x = 60$.

When x ft of cable are hanging down, the force used in winding up the cable,

$$f(x) = \text{weight of cable} = (\text{density})(\text{length})$$

$$= 4x. \tag{8.38}$$

Then from (8.38) and (8.36), the work W done in winding up the cable is

$$W = \int_0^{60} 4x \, dx$$

$$= 2x^2 \Big|_0^{60} = 7200 \text{ ft-lb.}$$

We now consider another problem in calculating the work done in moving a mass when not all points of the mass are moved the same distance. This situation is met in the next example in which all of the water in a tank is to be pumped upward to the same level. We assume here that if a mass S can be subdivided into discrete submasses $S_1, S_2, \ldots, S_n$ and W_1 is the work done in moving

S_1, W_2 is the work done in moving S_2, and so on, then the work W done in moving S is given by

$$W = \sum_{i=1}^{n} W_i. \tag{8.39}$$

Example 3 A tank 30 ft in length has a cross section that is a semicircle of radius 12 ft and is filled to a depth of 9 ft with water of density δ lb/$\overline{\text{ft}}^3$. Find the work done in pumping the water to the top of the tank. (See Figure 8–27.)

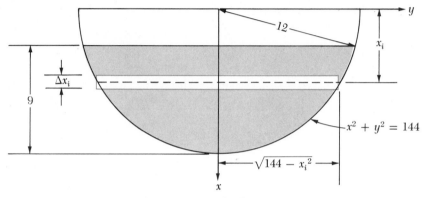

Figure 8–27

SOLUTION It will be convenient to locate the x and y axes at one end of the tank as shown in Figure 8–27 with the origin at the center of the semicircle, the y axis at the top of the semicircle, and the x axis extending downward. Since the radius of the semicircle is 12 ft, an equation for the semicircle is $x^2 + y^2 = 144$.

We note that the water in the tank extends from $x = 3$ to $x = 12$, and so we consider a partition $P = \{3 = c_0, c_1, c_2, \ldots, c_{n-1}, c_n = 12\}$ of the interval $[3, 12]$.

As before, we denote an arbitrary subinterval formed from $[3, 12]$ using P by $[c_{i-1}, c_i]$ and let $\Delta x_i = c_i - c_{i-1}$. The weight of the "slab" of water between the planes $x = c_{i-1}$ and $x = c_i$ is

$$\Delta w_i = (\text{density})(\text{volume})$$
$$= \delta \cdot 30 \cdot 2\sqrt{144 - x_i^2}\, \Delta x_i \text{ lb} \tag{8.40}$$

where x_i is some number between c_{i-1} and c_i. Since the force necessary to pump the slab to the top of the tank is its weight Δw_i, from (8.33) and (8.40) the work done in pumping this slab to the top, if defined, would be approximately

$$\Delta W_i = x_i \Delta w_i$$
$$= 60\, \delta x_i \sqrt{144 - x_i^2}\, \Delta x_i \text{ ft-lb.} \tag{8.41}$$

Hence from (8.39) and (8.41) the work W done in pumping all of the water in the tank to the top, if defined, would be approximately

$$\sum_{i=1}^{n} \Delta W_i = \sum_{i=1}^{n} 60\, \delta x_i \sqrt{144 - x_i^2}\, \Delta x_i. \tag{8.42}$$

Intuition suggests that the smaller we choose the norm of P, the more accurate is the approximation (8.42) of the work W. Thus, we define W as

$$W = \int_3^{12} 60\, \delta x \sqrt{144 - x^2}\; dx$$

$$= -60\, \delta \cdot \tfrac{1}{2} \cdot \tfrac{2}{3}(144 - x^2)^{3/2}\Big|_3^{12}$$

$$= 20\, \delta(135)^{3/2}$$

$$= 8100\, \delta \sqrt{15} \text{ ft-lb.}$$

This approach of finding the work done in moving a mass by subdividing the mass into discrete submasses and summing the work done in moving each submass can also be used to solve Example 2.

Exercise Set 8.4

1. A spring has a natural length of 20 in., and a force of 18 lb will stretch the spring to a length of 25 in. Find: (a) the work required to stretch the spring from a length of 20 in. to 25 in., (b) the work required to stretch the spring from a length of 25 in. to 30 in., (c) the force stretching the spring when 540 in.-lb have been expended in stretching the spring beyond its natural length.

2. A spring has a natural length of 30 in., and a force of 50 lb will compress it to a length of 20 in. Find: (a) the work required to compress the spring from its natural length to a length of 20 in., (b) the work required to compress the spring from a length of 20 in. to a length of 10 in., (c) the length c of the spring in inches, where $10 < c < 30$, such that the work required to compress the spring from a length of 30 in. to c in. is the same as the work required to compress the spring from a length of c in. to 10 in.

3. If 1.5 times as much work is done stretching a spring from a length of 12 in. to 14 in. as in stretching a spring from a length of 10 in. to 12 in., find the natural length of the spring.

4. The pressure p exerted by a gas in a cylinder on a piston is given by

$$p = \frac{k}{v^\alpha}$$

where $\alpha > 1$ and k are constants, and v is the volume of the gas in the cylinder. Show that the work done by the gas when its volume increases from v_1 to v_2 is

$$\frac{k}{1 - \alpha}(v_2^{1-\alpha} - v_1^{1-\alpha}).$$

HINT: Assume that the cross section area of the piston is A, which is constant, and that x is the distance from the closed end of the cylinder to the piston.

5. The force of attraction between two particles of mass m_1 and m_2 is given by $f(x) = Gm_1m_2/x^2$ where G is constant and x is the distance between the particles. If the particles are originally a distance a apart, find the work done against

the force of attraction if one particle is fixed and (a) the other is moved so that the distance between the particles is b, where $b > a$ and (b) the other particle is moved indefinitely far away.

6. What would be the work done in Example 2 if the cable is attached to a 200 lb weight and is wound up completely?

7. What would be the work done if the cable in Example 2 is wound up until 20 ft are hanging down?

8. A rope 50 ft long and weighing 0.15 lb/ft is attached to a bucket containing 75 lb of water. If the rope is raised vertically at the rate of 3 ft/sec, and the water leaks at the rate of 1 lb/sec, find the work done in raising the bucket 50 ft.

9. A tank has the shape of a right circular cylinder with radius 20 ft and height 30 ft. If it is filled to a height of 25 ft with gasoline of density 42 lb/$\overline{ft^3}$, find the work done in pumping the gasoline to the top of the tank.

10. A hemispherical tank of radius 10 ft is filled with a liquid of density δ lb/$\overline{ft^3}$. Find the work required to pump the liquid to a height of 10 ft above the top of the tank.

11. A tank in the shape of a right circular cone with vertex downward has a radius of 6 ft and an altitude of 10 ft. If it is filled with water of density 62.4 lb/$\overline{ft^3}$ to a depth of 8 ft, find the work done in pumping the water to a height of 5 ft above the top of the tank.

12. Find the work done by gravity in emptying the tank in Exercise 11 if the water runs out the bottom.

13. Find the work done by gravity in emptying the tank in Example 3 when full if the water runs out the bottom.

14. Find the work done by the earth's gravity in moving a 1 lb object from a height of 1000 miles above the earth to the earth's surface. HINT: Use Equations (7.25) and (7.26) in Exercises 23, Exercise Set 7.2.

8.5 Fluid Force

Suppose that a thin horizontal plate is submerged at a certain depth in a liquid and we wish to calculate the total force exerted by the liquid on one side of the plate. This force is actually the weight of the column of the liquid that rests on the plate. Thus, if a plate has area A $\overline{ft^2}$ and is submerged at a depth of x ft in a liquid of density δ lb/$\overline{ft^3}$, the total force on the plate is F lb where

$$F = \delta x A. \tag{8.43}$$

We could, of course, measure F, δ, x, and A in other compatible units involving length and force.

As an illustration of (8.43) we can calculate the total force exerted by the water in a tank on its base if the base is horizontal. Suppose the tank is a rectangular parallelopiped that has a base 10 ft × 12 ft and is filled to a depth of 4 ft. Here

$$\delta = 62.4 \text{ lb}/\overline{\text{ft}^3} \text{ for water}$$

$$x = 4 \text{ ft}$$

$$A = 10(12) = 120 \ \overline{\text{ft}^2}$$

and from (8.43) the total force on the base is

$$F = (62.4)(4)(120) = 29,952 \text{ lb}.$$

The *pressure p* of a liquid at a depth x is defined to be

$$p = \delta x \tag{8.44}$$

where δ is the density of the liquid. If x is in ft, and δ is in $\text{lb}/\overline{\text{ft}^3}$, the p will be in $\text{lb}/\overline{\text{ft}^2}$. Thus from (8.43) and (8.44) the total force on the horizontal plate of area A is

$$F = pA. \tag{8.45}$$

From (8.45) $p = F/A$, and hence the pressure at any depth is the force exerted per unit area at that depth by the liquid.

From the study of physics we know that *the pressure at any depth is the same in every direction.* Thus at any depth the pressure on a submerged *vertical plane surface* can be calculated from (8.44). We shall consider the problem of defining and calculating the total force exerted by the liquid on one side of such a surface. This force could not be calculated using (8.43) or (8.45) because the pressure on the surface is not constant, but instead varies as the depth.

Suppose first the coordinate axes are chosen so that the x axis extends vertically downward and the surface of the liquid is in the line $x = k$, k being some constant. We wish to develop a general formula for the total force on one side of a vertical surface that is the plane region

$$T = \{(x, y) : x \in [a, b] \text{ and } y \in [g(x), f(x)]\} \tag{8.46}$$

where f and g are continuous on $[a, b]$ and $f(x) \geq g(x)$ for every x in the interval (Figure 8–28).

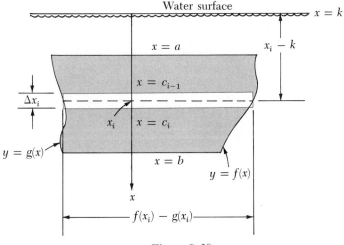

Figure 8–28

As before, we form a partition $P = \{a = c_0, c_1, c_2, \ldots, c_{n-1}, c_n = b\}$ of $[a, b]$ and construct lines $x = c_1$, $x = c_2$, $\ldots$, $x = c_{n-1}$, which divide T into n strips where $\Delta x_i = c_i - c_{i-1}$ is the width of the ith strip.

In each strip a rectangle is constructed with area

$$\Delta A_i = [f(x_i) - g(x_i)] \, \Delta x_i, \tag{8.47}$$

x_i being a suitably chosen number in the interval $[c_{i-1}, c_i]$. Suppose δ is the density of the liquid. If the total force on one side of the ith rectangle were defined from (8.44) and (8.45), this force would be approximately

$$\Delta F_i = (\text{pressure})(\text{area of rect.}) = (\text{density})(\text{depth})(\text{area of rect.})$$
$$= \delta(x_i - k)[f(x_i) - g(x_i)] \, \Delta x_i. \tag{8.48}$$

Hence the total force on one side of the vertical surface would be approximately

$$\sum_{i=1}^{n} \Delta F_i = \sum_{i=1}^{n} \delta(x_i - k)[f(x_i) - g(x_i)] \, \Delta x_i. \tag{8.49}$$

Intuitively, we feel that the sum in (8.49) ought to differ from F, the total force on one side of T, by as little as we please when the norm of P is sufficiently small. Also, if the norm of P is small enough, the right member of (8.49) differs from the integral $\int_a^b \delta(x - k)[f(x) - g(x)] \, dx$ by as little as we please.

Thus we define the total force F on one side of the surface T given by (8.46) to be

$$F = \int_a^b \delta(x - k)[f(x) - g(x)] \, dx. \tag{8.50}$$

In applications of (8.50) it is helpful to first express ΔF_i in terms of x_i and Δx_i, using (8.48).

Example 1 A trough 15 ft in length has a cross section that is an isosceles trapezoid with upper base 5 ft, lower base 3 ft, and depth 2 ft (Figure 8–29). Find the total force on one end of the trough if it is filled with water.

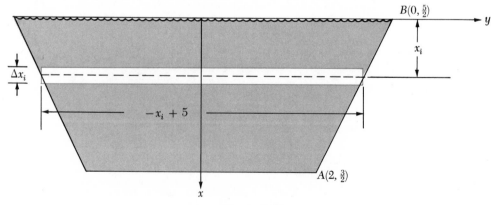

Figure 8–29

SOLUTION   We select the y axis to be at the water surface and for convenience choose the x axis to be the axis of symmetry of the end of the trough. The water extends from $x = 0$ to $x = 2$ with the points A and B being $(2, \frac{3}{2})$ and $(0, \frac{5}{2})$, respectively, and the side AB of the trapezoid has the equation

$$y = -\tfrac{1}{2}x + \tfrac{5}{2}.$$

If we construct the rectangle shown in Figure 8–29 with thickness Δx_i and length $2y_i = 2(-\frac{1}{2}x_i + \frac{5}{2}) = -x_i + 5$, the force on the rectangle, using (8.48), is

$$
\begin{aligned}
\Delta F_i &= (\text{density})(\text{depth})(\text{area}) \\
&= 62.4 x_i(-x_i + 5)\,\Delta x_i \\
&= 62.4(-x_i^2 + 5x_i)\,\Delta x_i \qquad\qquad (8.51)
\end{aligned}
$$

where $\delta = 62.4\ \text{lb}/\overline{\text{ft}^3}$ is the density of water. From (8.51) the total force F on one end of the trough is

$$
\begin{aligned}
F &= \int_0^2 62.4(-x^2 + 5x)\,dx \\[2mm]
&= 62.4\left(-\tfrac{1}{3}x^3 + \tfrac{5}{2}x^2\right)\Big|_0^2 = 62.4\left(\tfrac{22}{3}\right) \\[2mm]
&= 457.6\ \text{lb}.
\end{aligned}
$$

Example 2   Find the total force on one side of a vertical circular plate of radius 3 ft which is submerged in water so that the center of the plate is 10 ft below the surface of the water (Figure 8–30).

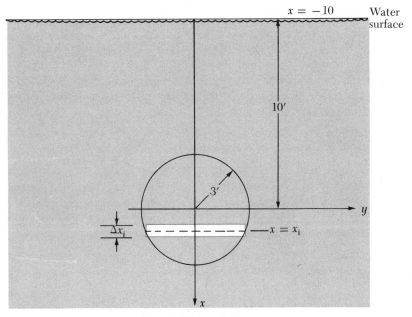

Figure 8–30

SOLUTION   It will be convenient here to have the y axis passing through the center of the circle. The equation of the circle is then $x^2 + y^2 = 9$ and the plate extends from $x = -3$ to $x = 3$. We construct the rectangle shown in Figure 8–30 with thickness Δx_i and length $2\sqrt{9 - x_i^2}$. From (8.48) the force on one side of the rectangle is

$$\Delta F_i = (\text{density})(\text{depth})(\text{area})$$
$$= 62.4(x_i + 10)2\sqrt{9 - x_i^2}\,\Delta x_i \tag{8.52}$$

From (8.52) the total force on one side of the circular plate is

$$F = \int_{-3}^{3} 62.4(x + 10)2\sqrt{9 - x^2}\,dx$$

$$= 62.4\left(\int_{-3}^{3} 2x\sqrt{9 - x^2}\,dx + \int_{-3}^{3} 20\sqrt{9 - x^2}\,dx\right). \tag{8.53}$$

The first integral on the right in (8.53) can be evaluated as follows:

$$\int_{-3}^{3} 2x\sqrt{9 - x^2}\,dx = -\tfrac{2}{3}(9 - x^2)^{3/2}\Big|_{-3}^{3} = 0.$$

From (8.18) the second integral on the right in (8.53) is

$$\int_{-3}^{3} 20\sqrt{9 - x^2}\,dx = 40\int_{0}^{3} \sqrt{9 - x^2}\,dx = 40\frac{9\pi}{4} = 90\pi.$$

Then from (8.53)

$$F = (62.4)(90\pi) \approx 17{,}600 \text{ lb.}$$

Exercise Set 8.5

1. A dam has a vertical rectangular gate that is 20 ft wide and 15 ft deep. Find the total force of the water on the gate if (a) the water level is at the top of the gate, (b) the water level is 5 ft below the top of the gate, (c) the water level is 10 ft above the top of the gate.

2. What would be the maximum allowable depth of the water behind the gate in Exercise 1 if it is built to withstand a total force of 100,000 lb?

3. A vertical semicircular plate of radius 5 ft is submerged in water so that its straight side is parallel to the surface of the water and its curved side extends downward. Find the total force of the water on one side of the plate if (a) its straight side is in the water surface, (b) the straight side is 2 ft below the water surface.

4. A vertical dam in the shape of a parabola is 80 ft across the top and is 20 ft high. Find the force on the dam if the surface of the water is at the top of the dam.

5. A tank that is a right circular cylinder of radius 6 ft with its axis horizontal contains gasoline of density 42 lb/ft³. Find the total force exerted by the gasoline

on the end of the tank if (a) the tank is completely full, (b) the tank is filled to a depth of 6 ft.

6. Rework Example 1 if the y axis is (a) moved vertically upward 1 ft, (b) moved vertically downward 1 ft.

7. A swimming pool is 90 ft long and 50 ft wide. Find the total force on the bottom of the pool if in a lengthwise cross section of the pool the depth increases linearly from 3 ft to 12 ft.

8. A plate in the shape of an isosceles triangle with its vertex downward has a base of 10 ft and an altitude of 12 ft, and the axis of symmetry of the plate forms an angle of 30° with a vertical line. Find the total force on one side of the plate if the base is in the surface of the water.

9. Exponential and Logarithmic Functions

9.1 Natural Logarithm Function

The student may recall from high school algebra that if $x > 0$ and $b > 0$ but $\neq 1$ that the logarithm of x to the base b, written $\log_b x$, was defined by saying

$$\log_b x = y \quad \text{if and only if} \quad b^y = x. \tag{9.1}$$

In problems of interest, which were usually concerned with computations involving multiplication, division, and obtaining powers and roots of numbers, logarithms to the base 10, the *common logarithms*, were usually employed.

In high school algebra one assumes without proof that given b, then for every $x > 0$ there is a unique y such that $b^y = x$, or saying the same thing, there is a unique function L such that $y = L(x) = \log_b x$. Such a function L would have the interval $(0, +\infty)$ as its domain, and would satisfy the following properties of logarithms:

$$L(x_1 x_2) = L(x_1) + L(x_2) \tag{9.2}$$

$$L\left(\frac{x_1}{x_2}\right) = L(x_1) - L(x_2) \tag{9.3}$$

$$L(x^r) = rL(x) \quad \text{if } r \text{ is a rational number} \tag{9.4}$$

$$L(1) = 0 \tag{9.5}$$

$$L(x_2) > L(x_1) \quad \text{if } x_2 > x_1 \tag{9.6}$$

It will be our purpose in this chapter to show that such a function L exists for each $b > 0$ but $\neq 1$. In order to confirm the existence of this function, we shall proceed by "working backwards" from the properties (9.2)–(9.5), hoping to discover a suitable function. If L exists, then from (9.3) and (9.5) when $h \neq 0$ but sufficiently close to 0,

$$\frac{L(x + h) - L(x)}{h} = \frac{L((x + h)/x)}{h} = \frac{L(1 + (h/x)) - L(1)}{h}$$

$$= \frac{1}{x} \frac{L(1 + (h/x)) - L(1)}{h/x} \quad \text{if } x > 0. \tag{9.7}$$

If $L'(1)$ exists, intuition suggests that if h approaches zero when x is fixed, so does h/x, and hence

$$\lim_{h \to 0} \frac{L(1 + (h/x)) - L(1)}{h/x} = L'(1).$$

(This limit can be verified using Theorem 3.8.1.) Thus if $L'(1)$ exists, from taking limits at $h = 0$ on each side of (9.7),

$$L'(x) = \frac{1}{x} L'(1) \quad \text{for } x > 0.$$

Then by Theorem 3.4.6 L' would be continuous on $(0, +\infty)$, and from Theorems 7.7.1 and 7.1.3 for $x > 0$, we would have

$$L(x) = \int_1^x \frac{1}{t} L'(1) \, dt + C. \tag{9.8}$$

In order to satisfy (9.5), we must then have

$$L(1) = L'(1) \int_1^1 \frac{1}{t} \, dt + C = 0.$$

Since $\int_1^1 (1/t) \, dt = 0$, then $C = 0$, and from (9.8)

$$L(x) = \int_1^x \frac{1}{t} L'(1) \, dt.$$

If $L'(1) = 1$, then

$$L(x) = \int_1^x \frac{1}{t} \, dt \quad \text{if } x > 0. \tag{9.9}$$

In obtaining (9.9) we have shown that if there is a function L that satisfies the logarithm properties (9.3) and (9.5) and, also, if $L'(1) = 1$, then L must be given by (9.9). We will consider the converse question: Does the function L in (9.9) satisfy the properties (9.3) and (9.5), as well as (9.2), (9.4), and (9.6)? Furthermore, is there some base b such that $L(x) = y$ if and only if $b^y = x$?

First we state a definition, which is suggested by (9.9).

9.1.1   Definition

The *natural logarithm function*, ln, is defined by writing

$$\ln x = \int_1^x \frac{1}{t} \, dt \quad \text{if } x > 0.$$

The number $\ln x$ is called the natural logarithm of x.

Since $1/t > 0$ for $t > 0$, when $x > 1$, $\ln x$ is the area of the region in the *ty* plane bounded by the graphs of the equations $y = 1/t$, $y = 0$, $t = 1$, and $t = x$. This region is shaded in Figure 9–1.

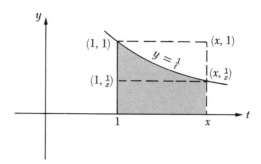

Figure 9-1

If $t \in [1, x]$, then $1/x \leq 1/t \leq 1$, and from Theorem 7.6.5

$$\int_1^x \frac{1}{x}\, dt \leq \int_1^x \frac{1}{t}\, dt \leq \int_1^x 1\, dt.$$

Integrating the left and right members of this equation, we obtain

$$\frac{x-1}{x} \leq \ln x \leq x - 1 \quad \text{if } x \geq 1. \tag{9.10}$$

Since $(x - 1)/x > 0$ when $x > 1$, from (9.10)

$$\ln x > 0 \quad \text{if } x > 1, \tag{9.11}$$

and in particular if $x = 2$ in (9.10)

$$\tfrac{1}{2} \leq \ln 2 \leq 1.$$

Using methods that will be introduced in Section 14.8, it can be shown that correct to 5 decimal places

$$\ln 2 \approx 0.69315.$$

We leave as an exercise the proof that

$$\frac{x-1}{x} \leq \ln x \leq x - 1 \quad \text{if } 0 < x < 1. \tag{9.12}$$

(See Exercises 22 and 24, Exercise Set 9.1). From (9.12) since $x - 1 < 0$ when $x < 1$

$$\ln x < 0 \quad \text{if } 0 < x < 1. \tag{9.13}$$

Also, from Definition 9.1.1 we obtain

$$\ln 1 = \int_1^1 \frac{1}{t}\, dt = 0, \tag{9.14}$$

so (9.5) is satisfied by the function ln.

Before showing that ln also satisfies properties (9.2), (9.3), (9.4), and (9.6), we shall derive the differentiation formula for ln.

9.1.2 Theorem

$$D_x \ln x = \frac{1}{x} \quad if\ x > 0.$$

PROOF   From Theorem 7.7.1,

$$D_x \ln x = D_x \int_1^x \frac{1}{t}\, dt = \frac{1}{x}$$

for every x in the interval $(0, +\infty)$.

The following corollary of Theorem 9.1.2 contains some important properties of ln.

9.1.3 Theorem

(a) *The function* ln *is continuous on its domain.*
(b) ln *is increasing on its domain.*
(c) *The graph of* ln *is concave downward at every point of the graph.*

PROOF OF (a)   By Theorem 9.1.2, ln is differentiable on its domain and by Theorem 4.3.2 ln is continuous on its domain.

PROOF OF (b)   Since $D_x \ln x = 1/x > 0$ if $x > 0$, by Theorem 5.3.2(a) ln is increasing on its domain.

PROOF OF (c)   Since $D_x \ln x = 1/x$, $D_x^2 \ln x = -1/x^2 < 0$, and hence by Theorem 5.4.2(b) the graph of ln is concave downward at every point on the graph.

As a consequence of Theorem 9.1.3(b)

$$\ln x_2 > \ln x_1 \quad if\ x_2 > x_1, \tag{9.15}$$

so (9.6) is satisfied by ln.
Also from Theorem 9.1.2, if $g'(x)$ exists and $g(x) > 0$, then by the chain rule

$$D_x \ln (g(x)) = \frac{1}{g(x)} \cdot g'(x). \tag{9.16}$$

Example 1   Find $D_x \ln (x^2 + 3x - 5)$.

SOLUTION   From (9.16)

$$D_x \ln (x^2 + 3x - 5) = \frac{1}{x^2 + 3x - 5} D_x (x^2 + 3x - 5)$$

$$= \frac{2x + 3}{x^2 + 3x - 5}$$

Example 2 Find y' if $y = \ln \sqrt[3]{\dfrac{4x-1}{x+1}}$.

SOLUTION From (9.16)

$$y' = \frac{1}{\sqrt[3]{\dfrac{4x-1}{x+1}}} D_x \sqrt[3]{\frac{4x-1}{x+1}}$$

$$= \sqrt[3]{\frac{x+1}{4x-1}} \cdot \frac{1}{3}\left(\frac{4x-1}{x+1}\right)^{-2/3} D_x\left(\frac{4x-1}{x+1}\right).$$

After performing the indicated differentiation in the right member and simplifying,

$$y' = \frac{x+1}{3(4x-1)} \cdot \frac{(x+1)\cdot 4 - (4x-1)\cdot 1}{(x+1)^2}$$

$$= \frac{5}{3(x+1)(4x-1)}.$$

Example 2 will be reworked after we show that the function ln satisfies properties (9.3) and (9.4). To prove that ln satisfies these properties as well as (9.2), formula (9.16) will be invoked. If $x_1 > 0$, then from (9.16)

$$D_x \ln (x_1 x) = \frac{1}{x_1 x} \cdot D_x(x_1 x).$$

$$= \frac{1}{x_1 x} \cdot x_1 = \frac{1}{x} = D_x \ln x.$$

Hence from Theorem 7.1.3 there is a C such that

$$\ln (x_1 x) = \ln x + C \quad \text{if } x \in (0, +\infty). \tag{9.17}$$

In particular (9.17) is true if $x = 1$. Hence

$$\ln x_1 = \ln 1 + C,$$

and from (9.14)

$$\ln x_1 = C,$$

so (9.17) can be rewritten

$$\ln (x_1 x) = \ln x_1 + \ln x.$$

In particular, this equation is true if $x = x_2 > 0$. Hence

$$\ln (x_1 x_2) = \ln x_1 + \ln x_2 \quad \text{if } x_1 > 0 \text{ and } x_2 > 0, \tag{9.18}$$

so ln satisfies (9.2).

If r is a rational number, from (9.16)

$$D_x \ln (x^r) = \frac{1}{x^r} \cdot D_x x^r$$

$$= \frac{1}{x^r} \cdot rx^{r-1}$$

$$= \frac{r}{x}.$$

But also since

$$D_x \left(r \ln x \right) = \frac{r}{x},$$

there is a C such that

$$\ln \left(x^r \right) = r \ln x + C \quad \text{if } x \in (0, +\infty). \tag{9.19}$$

In particular (9.19) is true if $x = 1$; then $\ln 1 = r \cdot \ln 1 + C$ and $C = 0$. Hence if $x > 0$

$$\ln \left(x^r \right) = r \ln x \quad \text{where } r \text{ is any rational number}, \tag{9.20}$$

so ln satisfies (9.4). It is left for the student to show from (9.18) and (9.20) that

$$\ln \left(\frac{x_1}{x_2} \right) = \ln x_1 - \ln x_2 \quad \text{if } x_1 > 0 \text{ and } x_2 > 0. \tag{9.21}$$

Thus ln satisfies (9.3).

In the next example another solution is given for the differentiation problem in Example 2. In this solution properties (9.20) and (9.21) of ln are used to first express y in a more manageable form.

Example 3   Find y' if $y = \ln \sqrt[3]{\dfrac{4x - 1}{x + 1}}$.

SOLUTION   From (9.20) and (9.21)

$$y = \ln \sqrt[3]{\frac{4x - 1}{x + 1}} = \frac{1}{3} \ln \frac{4x - 1}{x + 1}$$
$$= \tfrac{1}{3} [\ln \left(4x - 1 \right) - \ln \left(x + 1 \right)].$$

From (9.16)

$$y' = \frac{1}{3} \left[\frac{4}{4x - 1} - \frac{1}{x + 1} \right]$$

$$= \frac{5}{3(x + 1)(4x - 1)}.$$

We continue now by giving some limit properties of ln. Since ln is an increasing function on its domain, it is reasonable to inquire whether the function is also bounded on its domain. Suppose there were some $B > 0$ such that

$$\ln x \le B$$

for every $x \in (0, +\infty)$. If $a > 1$, then for every positive integer n it would be true that

$$\ln a^n = n \ln a \leq B. \tag{9.22}$$

Since $a > 1$, $\ln a > 0$ by (9.15), and therefore from (9.22)

$$n \leq \frac{B}{\ln a}$$

for every positive integer n, which is obviously impossible. Thus ln is unbounded on its domain, and hence for every $B > 0$ there is some number $M > 1$ such that

$$\ln M > B.$$

If $x > M$, then since ln is an increasing function

$$\ln x > \ln M > B.$$

Thus from Definition 5.7.1 we have proved the first part of the following theorem.

9.1.4   Theorem

(a) $\lim_{x \to +\infty} \ln x = +\infty$.
(b) $\lim_{x \to 0+} \ln x = -\infty$.
(c) *The range of* ln *is the set* R.

PROOF OF (b)   Let $\epsilon > 0$ be given. If $0 < x < 2^{-n}$ where n is an integer such that $n > 1/\epsilon \ln 2$, we have

$$\ln x < \ln 2^{-n} = -n \ln 2 < -\frac{1}{\epsilon} < 0.$$

Hence

$$0 > \frac{1}{\ln x} > -\epsilon,$$

and by Definition 3.6.1 $\lim_{x \to 0+} 1/\ln x = 0$. The conclusion is then obtained using Definition 3.9.1(b).

PROOF OF (c)   Suppose k is any number in R. Then by Theorem 9.1.4(a) and Definition 5.7.1 there is certainly some $b > 1$ such that

$$\ln b > |k|.$$

Then by Theorem 1.3.7

$$\ln \frac{1}{b} = -\ln b < -|k|.$$

From these inequalities and Theorem 1.5.6

$$\ln \frac{1}{b} < -|k| \leq k \leq |k| < \ln b.$$

By Theorem 9.1.3(a) ln is continuous on $[1/b, b]$, and hence by the intermediate value theorem there is a $c \in (1/b, b)$ such that $\ln c = k$. Thus any number k is in the range of $\ln$, and the theorem is proved.

From Theorem 9.1.4(b) the y axis is a vertical asymptote of the graph of $y = \ln x$.

From Theorem 9.1.4(c) there is some positive number e such that

$$\ln e = 1.$$

Since $\ln$ is an increasing function, there could not be an $x \neq e$ such that $\ln x = 1$. Thus we have the following definition.

9.1.5   Definition

e is the *unique* positive number such that

$$\ln e = 1.$$

We leave as an exercise the proof that $2 < e < 3$. Using a numerical method, it will be shown in Section 13.2 that

$$e \approx 2.71828.$$

A sketch of the graph of the equation $y = \ln x$ is shown in Figure 9–2.

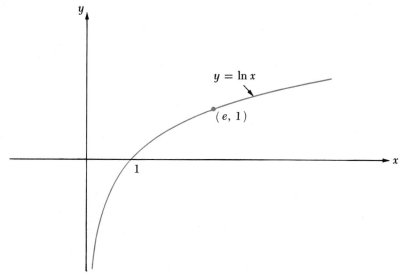

Figure 9–2

Exercise Set 9.1

1. If $\ln 2 \approx 0.69315$ and $\ln 3 \approx 1.09861$, approximate the following numbers:
   (a) $\ln 6$ (b) $\ln 8$ (c) $\ln \frac{1}{2}$ (d) $\ln \frac{1}{4}$ (e) $\ln 18$

In Exercises 2–9 differentiate the functions defined by the given equations.

2. $f(x) = \ln(x^2 + 4)$ 3. $y = x^2 \ln 3x$

4. $\phi(t) = \dfrac{\ln(t + 1)}{2}$ 5. $F(x) = (\ln(x + 1)^2)^3$

6. $y = \ln\sqrt{x^2 + 2x}$ 7. $f(x) = \ln\sqrt[4]{\dfrac{x^2 + 1}{x^2 - 1}}$

8. $f(x) = \ln(\ln x)$ 9. $F(x) = \sqrt{x^2 + 1} - \ln(x + \sqrt{x^2 + 1})$

In Exercises 10 and 11 find dy/dx using implicit differentiation.

10. $\ln(x + y) - \ln(x - y) = 1$ 11. $y = \ln(x + y + 2)$

In Exercises 12–16 sketch the graph of the given equation making use of information gained from the study of analytic geometry and calculus.

12. $y = \ln(x + 1)$ 13. $y = x - \ln x$

14. $y = \ln|x|$ 15. $y = \ln 1/x$

16. $x = \ln y$

17. Derive (9.21) from (9.18) and (9.20).

18. Prove that $\ln 2 < 1$ (we have already proved that $\ln 2 \leq 1$).

19. Prove: For every positive integer n, $D_x^n(\ln x) = \dfrac{(-1)^{n-1}(n-1)!}{x^n}$.

20. Prove that $\ln 3 > 1$. HINT: Consider the areas of trapezoids formed under the curve by tangents to the curve.

21. Prove that $2 < e < 3$. HINT: Use the results of Exercises 18 and 20.

22. Prove that $x - 1 \geq \ln x$ for every $x > 0$. HINT: Consider the function f where $f(x) = x - 1 - \ln x$. Does it have any extrema?

23. Prove: $\lim_{x \to 0+} x \ln x = 0$. HINT: If $0 < x \leq 1$, then $\displaystyle\int_x^1 t^{-1}\, dt \leq \int_x^1 t^{-3/2}\, dt$.

24. Prove that $(x - 1)/x \leq \ln x$ for every $x > 0$.

9.2   Logarithmic Differentiation; Integrals

In Section 9.1 we derived the formula

$$D_x \ln g(x) = \frac{1}{g(x)} D_x g(x) \quad \text{if } g(x) > 0 \text{ and } g'(x) \text{ exists.} \tag{9.23}$$

This formula is useful in facilitating the differentiation of complicated expressions involving products, quotients, or powers. To differentiate such an expression, one first expresses the natural logarithms of the given expression as a sum of terms involving the ln of simpler expressions. The resulting process for differentiating the given expression, which is termed *logarithmic differentiation*, is illustrated in the following example.

Example 1 Find $g'(x)$ using logarithmic differentiation if

$$g(x) = \frac{x^4 \sqrt{3x+1}}{\sqrt[3]{x-2}}.$$

SOLUTION Using properties (9.18), (9.20), and (9.21), when $g(x) > 0$, $\ln g(x)$ can be expressed in the form

$$\ln g(x) = 4 \ln x + \tfrac{1}{2} \ln (3x+1) - \tfrac{1}{3} \ln (x-2).$$

Then if $g'(x)$ exists, from (9.23)

$$\frac{1}{g(x)} g'(x) = \frac{4}{x} + \frac{3}{2(3x+1)} - \frac{1}{3(x-2)}$$

$$g'(x) = g(x) \left[\frac{24(x-2)(3x+1) + 9x(x-2) - 2x(3x+1)}{6x(x-2)(3x+1)} \right]$$

$$= \frac{x^4 \sqrt{3x+1}}{\sqrt[3]{x-2}} \left[\frac{75x^2 - 140x - 48}{6x(x-2)(3x+1)} \right]$$

$$= \frac{x^3 (75x^2 - 140x - 48)}{6(x-2)^{4/3}(3x+1)^{1/2}}. \tag{9.24}$$

The student may have observed that the theoretical justification for logarithmic differentiation is a little shaky. First of all, in Example 1 $g'(x)$ was obtained after assuming without proof that it already existed. Also, $g'(x)$ was obtained here only for those numbers x for which $g(x) > 0$—that is, $x > 2$. However, one can directly obtain the expression for $g'(x)$ in (9.24) for every $x > -\tfrac{1}{3}$ except $x = 2$ and without the initial assumption that $g'(x)$ exists. Such a derivation, though, would admittedly be more laborious than the solution in Example 1. Similar complicated derivations can be used to justify the method of logarithmic differentiation on the other examples we consider. Thus we will think of the method as a convenient shortcut to the solution which bypasses the usual theory required for its justification. From (9.23) the following theorem can be proved.

9.2.1 Theorem

If $g(x) \neq 0$ *and* $g'(x)$ *exists, then*

$$D_x \ln |g(x)| = \frac{1}{g(x)} D_x g(x). \tag{9.25}$$

PROOF From (9.23) and Theorem 4.6.4

$$D_x \ln |g(x)| = \frac{1}{|g(x)|} D_x |g(x)| = \frac{g(x) g'(x)}{|g(x)|^2}. \tag{9.26}$$

Since $|g(x)|^2 = (g(x))^2$ in the right member of (9.26),

$$D_x \ln |g(x)| = \frac{g'(x)}{g(x)},$$

which proves the theorem.

The student should note the similarity of the formulas (9.23) and (9.25).

Example 2   Find $D_x \ln |x^2 - 1|$.

SOLUTION   From (9.25)

$$D_x \ln |x^2 - 1| = \frac{1}{x^2 - 1} D_x(x^2 - 1) = \frac{2x}{x^2 - 1}.$$

Also from (9.25), we have the corresponding integration formula

$$\int \frac{g'(x)\, dx}{g(x)} = \ln |g(x)| + C, \tag{9.27}$$

and as a special case of (9.27)

$$\int \frac{dx}{x} = \ln |x| + C. \tag{9.28}$$

In view of (9.27) we can assert that if r is any rational number then

$$\int (g(x))^r g'(x)\, dx = \begin{cases} \dfrac{(g(x))^{r+1}}{r + 1} + C & \text{if } r \neq -1 \\[2mm] \ln |g(x)| + C & \text{if } r = -1. \end{cases} \tag{9.29}$$

In performing an integration based upon (9.27), it will be useful to make the customary substitutions $u = g(x)$ and $du = g'(x)\, dx$.

Example 3   Find $\int \dfrac{x^2\, dx}{x^3 + 1}$.

SOLUTION   Letting $u = x^3 + 1$,   then $du = 3x^2\, dx$,   and   hence $x^2\, dx = \frac{1}{3}\, du$. Then from (9.27)

$$\int \frac{x^2\, dx}{x^3 + 1} = \int \frac{1}{3} \frac{du}{u} = \frac{1}{3} \ln |u| + C$$

$$= \tfrac{1}{3} \ln |x^3 + 1| + C.$$

Example 4   Find $\int \dfrac{x^2 - x + 1}{x + 2}\, dx$.

SOLUTION   To integrate a rational function where the degree of the numerator is not less than the degree of the denominator, it is worthwhile to first divide the numerator by the denominator. Then

$$\int \frac{x^2 - x + 1}{x + 2} \, dx = \int \left(x - 3 + \frac{7}{x + 2} \right) dx = \int (x - 3) \, dx + \int \frac{7 \, dx}{x + 2} \quad (9.30)$$

To evaluate $\int \dfrac{7 \, dx}{x + 2}$, let $u = x + 2$ and obtain $du = dx$. Then

$$\int \frac{7 \, dx}{x + 2} = \int \frac{7 \, du}{u} = 7 \ln |u| + C'$$

$$= 7 \ln |x + 2| + C', \quad (9.31)$$

where C' is the constant of integration. From (9.30) and (9.31)

$$\int \frac{x^2 - x + 1}{x + 2} \, dx = \frac{x^2}{2} - 3x + 7 \ln |x + 2| + C,$$

where the constant C is obtained by combining C' and the constant of integration from evaluating $\int (x - 3) \, dx$.

Example 5   Find $\displaystyle\int_1^3 \frac{dx}{2x - 7}$.

   SOLUTION   Letting $u = 2x - 7$, we have $du = 2 \, dx$, and hence $dx = \frac{1}{2} \, du$. Since $u = -5$ when $x = 1$ and $u = -1$ when $x = 3$,

$$\int_1^3 \frac{dx}{2x - 7} = \int_{-5}^{-1} \frac{1}{2} \frac{du}{u} = \frac{1}{2} \ln |u| \,\Big|_{-5}^{-1}$$

$$= \tfrac{1}{2} \ln |-1| - \tfrac{1}{2} \ln |-5|$$

$$= -\tfrac{1}{2} \ln 5.$$

Exercise Set 9.2

In Exercises 1–6 differentiate using logarithmic differentiation.

1. $y = (3x + 1)(2x - 5)(x + 2)^2$

2. $f(x) = \sqrt[4]{\dfrac{2x - 1}{2x + 1}}$

3. $f(x) = \dfrac{(x^2 + 4)^{1/2}(3x - 4)}{\sqrt[3]{(2x + 3)}}$

4. $x = \dfrac{t^3 \sqrt{t^2 + 2}}{\sqrt[4]{3t + 1}}$

5. $y = \dfrac{\ln (x - 1)}{\ln (x + 1)}$

6. $f(t) = \sqrt{t^2 - 1} \ln (t^2 + 1)$

Perform the indicated integration in Exercises 7–12.

7. $\displaystyle\int \frac{dx}{4x + 3}$

8. $\displaystyle\int \frac{3x \, dx}{x^2 + 2}$

9. $\int \dfrac{2x + 3}{3x + 4}\,dx$

10. $\int \dfrac{x^3 + 1}{x - 1}\,dx$

11. $\int \dfrac{\ln x\,dx}{2x}$

12. $\int \dfrac{dx}{x \ln x}$

Evaluate the definite integrals in Exercises 13–16.

13. $\displaystyle\int_0^1 \dfrac{dx}{4 - 3x}$

14. $\displaystyle\int_2^5 \dfrac{x\,dx}{1 - x^2}$

15. $\displaystyle\int_1^2 \dfrac{3x + 1}{x + 2}\,dx$

16. $\displaystyle\int_2^3 \dfrac{x^2 + x + 2}{x + 3}\,dx$

9.3 Inverse Functions

It will be desirable at this point to introduce the notion of *inverse functions* in order to be able to discuss the exponential function, which is the inverse function of the natural logarithm function.

9.3.1 Definition

The functions f and g are inverse functions, and either function is called the *inverse function* of the other if and only if the following condition is true:

$$(x, y) \in f \quad \text{if and only if} \quad (y, x) \in g.$$

Suppose f and g are inverse functions. From Definition 9.3.1 the domain of either of the functions is the range of the other. Also, since the line $y = x$ is the perpendicular bisector of any line segment with endpoints (x, y) and (y, x) (Figure 9–3), the graphs of f and g are symmetric to the line $y = x$.

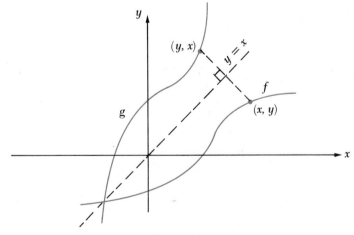

Figure 9–3

The following theorem gives a basic and useful characterization of inverse functions.

9.3.2   Theorem

f and g are inverse functions if and only if $f(g(x)) = x$ for every $x \in \mathcal{D}_g$ and $g(f(x)) = x$ for every $x \in \mathcal{D}_f$.

PROOF   We first prove the "only if" part. Suppose f and g are inverse functions. If $x \in \mathcal{D}_f$, then the ordered pair $(x, f(x))$ is an element of f, and therefore by Definition 9.3.1, $(f(x), x)$ is an element of g. From the latter statement $g(f(x)) = x$ for any $x \in \mathcal{D}_f$. It is left for the student to furnish the details of the analogous proof that $f(g(x)) = x$ for any $x \in \mathcal{D}_g$ thereby completing the proof of the "only if" part.

To prove the "if" part, we suppose that $f(g(x)) = x$ for every $x \in \mathcal{D}_g$ and that $g(f(x)) = x$ for every $x \in \mathcal{D}_f$. If $(x, y) \in f$, then since $y = f(x)$, $g(y) = g(f(x)) = x$; that is, $(y, x) \in g$. If $(y, x) \in g$, then since $x = g(y)$, we have $f(x) = f(g(y)) = y$; that is, $(x, y) \in f$.

Example 1   If $f(x) = x^3$ and $g(x) = x^{1/3}$, are f and g inverse functions?

SOLUTION   For every x, $f(g(x)) = f(x^{1/3}) = (x^{1/3})^3 = x$. Also for every x, $g(f(x)) = g(x^3) = (x^3)^{1/3} = x$. Hence by Theorem 9.3.2, f and g are inverse functions.

Example 2   If $f(x) = x^2$ and $g(x) = x^{1/2}$, are f and g inverse functions?

SOLUTION   If $x \geq 0$, $f(g(x)) = (x^{1/2})^2 = x$. However, if $x < 0$, $g(f(x)) = (x^2)^{1/2} = |x| = -x$. Hence by Theorem 9.3.2, f and g are not inverse functions.

When f and g are inverse functions, we usually write

$$f^{-1} = g \quad\text{and}\quad g^{-1} = f.$$

f^{-1} is read "f inverse." The numeral -1 in the exponent position following a letter representing a function is used to denote the inverse function of the function indicated by the letter. The reader may also have seen this notation used to denote the inverse element of a set of numbers with respect to an operation (for example, addition or multiplication) on the set or the inverse of a matrix. It should be mentioned that $f^{-1}(x)$ does *not* have the same meaning as $(f(x))^{-1} = 1/f(x)$.

If f^{-1} exists, $x \in \mathcal{D}_f$, and $y \in \mathcal{D}_{f^{-1}}$, then from Theorem 9.3.2

$$f(f^{-1}(y)) = y \qquad f^{-1}(f(x)) = x. \tag{9.32}$$

Thus from (9.32)

$$y = f(x) \quad\text{if and only if}\quad x = f^{-1}(y). \tag{9.33}$$

Hence, if the inverse function of a function f exists, the function f^{-1} can be defined if the equation $y = f(x)$ can be solved for x as a function of y.

At this point the question might be asked: "What kind of function has an inverse function?" Certainly, the function f having the graph shown in Figure 9–4, where $x_1 \neq x_2$, *would not* have an inverse function. For if f^{-1} did exist, the ordered pairs (k, x_1) and (k, x_2) would both be elements of f^{-1}, which would be a

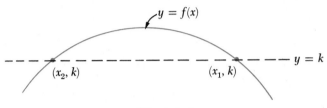

Figure 9–4

violation of the definition of a function. According to the next theorem, a function has an inverse function if and only if every line perpendicular to the y axis intersects the graph of the given function in at most one point.

9.3.3 Theorem

The function f has an inverse function f^{-1} if and only if $x_1 = x_2$ whenever $f(x_1) = f(x_2)$.

PROOF To prove the "if" part we assume that $x_1 = x_2$ whenever $f(x_1) = f(x_2)$. Consider the set $g = \{(x, y): x = f(y)\}$. If g were not a function, it would contain two ordered pairs (x, a_1) and (x, a_2), where $a_1 \neq a_2$. But then $f(a_1) = f(a_2) = x$, so by hypothesis $a_1 = a_2$. Thus g is a function. From the definition of g, $x = f(g(x))$ for every $x \in \mathcal{D}_g$. Also for every $y \in \mathcal{D}_f$, $y = g(x) = g(f(y))$. Hence by Theorem 9.3.2, f and g are inverse functions. The proof of the "only if" part is left as an exercise.

Theorem 9.3.3 can be utilized to determine if a given function f has an inverse function. When f^{-1} exists, a characterization of the function can often be obtained using (9.33). The next example is illustrative of our remarks here.

Example 3 Does the function f where $f(x) = (2x - 4)/(x + 3)$ (Figure 9–5) have an inverse function? If so, define this function.

SOLUTION If $f(x_1) = f(x_2)$ $\left(\text{that is, if } \dfrac{2x_1 - 4}{x_1 + 3} = \dfrac{2x_2 - 4}{x_2 + 3}\right)$, then

$$(2x_1 - 4)(x_2 + 3) = (x_1 + 3)(2x_2 - 4)$$
$$2x_1x_2 - 4x_2 + 6x_1 - 12 = 2x_1x_2 + 6x_2 - 4x_1 - 12$$
$$x_1 = x_2.$$

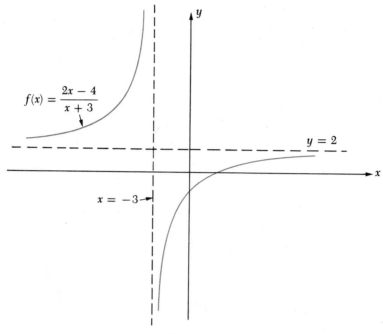

$$f(x) = \frac{2x - 4}{x + 3}$$

$$y = 2$$

$$x = -3$$

Figure 9-5

Therefore by Theorem 9.3.3 the function f^{-1} exists. To find f^{-1}, we solve the equation $y = (2x - 4)/(x + 3)$ for x in terms of y. Thus

$$x = f^{-1}(y) = \frac{-3y - 4}{y - 2},$$

or saying the same thing

$$y = f^{-1}(x) = \frac{-3x - 4}{x - 2}.$$

Example 4 Does the function f where $f(x) = x^2 - 4x + 1$ have an inverse function? If so, define this function.

SOLUTION Since $f(0) = f(4) = 1$, f^{-1} does not exist by Theorem 9.3.3.

Exercise Set 9.3

In Exercises 1–15 determine if the function defined in the exercise has an inverse function. Define this inverse function when it exists.

1. $\{(-2, 1), (3, 0), (4, 2), (5, 1)\}$

2. $\{(4, 2), (1, -3), (5, 0), (0, 5)\}$

3. $y = 7 - 2x$

4. $y = 2x + 4$

5. $y = \dfrac{2x}{x + 1}$

6. $y = \dfrac{3x + 5}{x - 3}$

7. $y = \sqrt{\dfrac{x-1}{x+1}}$

8. $y = \dfrac{2-x^2}{x}$

9. $y = x^2 - 6x - 7 \quad$ if $x \geq 3$

10. $y = x^2 - 6x - 7 \quad$ if $x < 0$

11. $y = x^2 - 6x - 7 \quad$ if $x > 0$

12. $y = \dfrac{1}{(x-1)^2} \quad$ if $0 \leq x < 1$

13. $y = \begin{cases} x^2 & \text{if } x \leq 0 \\ x^2 - 1 & \text{if } 0 < x \leq 1 \end{cases}$

14. $y = \begin{cases} \dfrac{1}{x+1} & \text{if } x < -1 \\ -x - 1 & \text{if } x \geq -1 \end{cases}$

15. $y = x + \dfrac{1}{x}$

16. Let a, b, c, and d be any numbers such that c and d are not both zero and

$$f(x) = \frac{ax+b}{cx+d}.$$

(a) Under what condition will the function f have an inverse function?

(b) Under what conditions will f be its own inverse function?

17. Complete the proof of the "only if" part of Theorem 9.3.2.

18. Prove the "only if" part of Theorem 9.3.3.

9.4 Calculus of Inverse Functions

In this section we continue our discussion of inverse functions by considering two theorems that deal with the continuity and differentiability, respectively, of these functions. The first theorem utilizes Theorem 9.3.3, which can be restated in the following equivalent form:

The function f has an inverse function f^{-1} if and only if for any two distinct numbers x_1 and x_2 in $\mathcal{D}_f$ it is true that $f(x_1) \neq f(x_2)$.

In this rendition of Theorem 9.3.3 the clause "$x_1 = x_2$ whenever $f(x_1) = f(x_2)$" has been replaced by its *contrapositive*: "for any two distinct numbers x_1 and x_2 in $\mathcal{D}_f$ it is true that $f(x_1) \neq f(x_2)$."

9.4.1 Theorem

Suppose the domain of the function f is an interval. If f is continuous and increasing (or decreasing) on $\mathcal{D}_f$, then its inverse function f^{-1} is continuous and increasing (or decreasing) on the set $\mathcal{R}_f$, which is also an interval.

PROOF It will suffice in this proof to assume that f is increasing on $\mathcal{D}_f$ as the proof is analogous when f is decreasing. Then for any pair of distinct numbers x_1 and x_2 in $\mathcal{D}_f$ either $f(x_1) > f(x_2)$ or $f(x_1) < f(x_2)$—that is, $f(x_1) \neq f(x_2)$. Hence by Theorem 9.3.3 f has an inverse function f^{-1} on $\mathcal{R}_f$.

To prove that f^{-1} is increasing on $\mathcal{R}_f$ when f is increasing on $\mathcal{D}_f$, consider any two numbers y_1 and y_2 in $\mathcal{R}_f$ where $y_2 > y_1$. The corresponding numbers $f^{-1}(y_1)$ and $f^{-1}(y_2)$ are, of course, in $\mathcal{D}_f$ and we shall give an indirect proof that $f^{-1}(y_2) > f^{-1}(y_1)$. Suppose $f^{-1}(y_2) \leq f^{-1}(y_1)$. Then since f is increasing on $\mathcal{D}_f$, we have $f(f^{-1}(y_2)) \leq f(f^{-1}(y_1))$, that is, $y_2 \leq y_1$, which contradicts our assumption that $y_2 > y_1$. Thus $f^{-1}(y_2) > f^{-1}(y_1)$, and hence f^{-1} is increasing on $\mathcal{R}_f$.

We next outline the proof that $\mathcal{R}_f$ is an interval. Suppose $\mathcal{D}_f$ is the finite closed interval $[a, b]$. Since f is increasing on $[a, b]$, the numbers in $\mathcal{R}_f$ may range from $f(a)$ to $f(b)$; that is, $\mathcal{R}_f \subset [f(a), f(b)]$. By the intermediate value theorem, if v is any number between $f(a)$ and $f(b)$, there is some number $u \in (a, b)$ such that $f(u) = v$. Thus $\mathcal{R}_f = [f(a), f(b)]$. By a similar proof it can be shown that if $\mathcal{D}_f$ is some other kind of interval (see Section 1.4), then $\mathcal{R}_f$ is the same kind of interval.

To prove that f^{-1} is continuous on $\mathcal{R}_f$, it must be shown that if $c \in \mathcal{R}_f$ and $\epsilon > 0$ are arbitrarily chosen, there is a $\delta > 0$ such that

$$|f^{-1}(y) - f^{-1}(c)| < \epsilon \quad \text{if} \quad y \in \mathcal{R}_f \text{ and } |y - c| < \delta,$$

or equivalently,

$$f^{-1}(c) - \epsilon < f^{-1}(y) < f^{-1}(c) + \epsilon \quad \text{if } y \in \mathcal{R}_f \text{ and}$$
$$c - \delta < y < c + \delta. \qquad (9.34)$$

Let $\epsilon > 0$ be given. If $f^{-1}(c)$ is an *interior point* of $\mathcal{D}_f$, there is an ϵ_0, where $\epsilon \geq \epsilon_0 > 0$ such that the interval $(f^{-1}(c) - \epsilon_0, f^{-1}(c) + \epsilon_0) \in \mathcal{D}_f$. Since f is increasing on $\mathcal{D}_f$

$$f(f^{-1}(c) - \epsilon_0) < f(f^{-1}(c)) = c < f(f^{-1}(c) + \epsilon_0).$$

Let

$$\delta = \text{smaller of } f(f^{-1}(c) + \epsilon_0) - c \text{ and } c - f(f^{-1}(c) - \epsilon_0)).$$

(In Figure 9-6 δ would equal $c - f(f^{-1}(c) - \epsilon_0)$.)

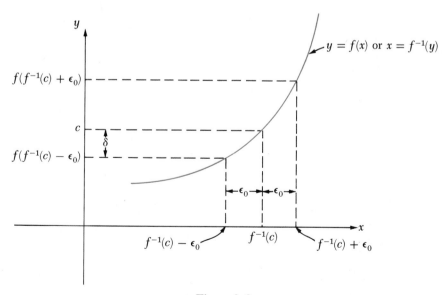

Figure 9-6

Suppose y is chosen so that

$$f(f^{-1}(c) - \epsilon_0) \le c - \delta < y < c + \delta \le f(f^{-1}(c) + \epsilon_0). \qquad (9.35)$$

Since f^{-1} is increasing on $\Re_f$, from (9.35)

$$f^{-1}(f(f^{-1}(c) - \epsilon_0)) \le f^{-1}(y) \le f^{-1}(f(f^{-1}(c) + \epsilon_0)),$$

and hence by (9.32)

$$f^{-1}(c) - \epsilon_0 < f^{-1}(y) < f^{-1}(c) + \epsilon_0. \qquad (9.36)$$

Since $\epsilon \ge \epsilon_0 > 0$, from (9.36)

$$f^{-1}(c) - \epsilon \le f^{-1}(c) - \epsilon_0 < f^{-1}(y) < f^{-1}(c) + \epsilon_0 \le f^{-1}(c) + \epsilon$$

which was to be proved (compare (9.34)). Thus f^{-1} is continuous at any interior point of $\Re_f$.

A modification of the foregoing proof can be used to show that f^{-1} is continuous at an endpoint of $\Re_f$ if the endpoint is an element of $\Re_f$. If c is a left endpoint of $\Re_f$, then $f^{-1}(c)$ is a left endpoint of $\mathcal{D}_f$, and there is certainly an ϵ_0 where $\epsilon \ge \epsilon_0 > 0$ such that the interval $(f^{-1}(c), f^{-1}(c) + \epsilon_0) \subset \mathcal{D}_f$. Suppose

$$\delta = f(f^{-1}(c) + \epsilon_0) - c.$$

If y satisfies

$$c - \delta < c \le y < c + \delta = f(f^{-1}(c) + \epsilon_0),$$

since f^{-1} is increasing on $\Re_f$

$$f^{-1}(c) \le f^{-1}(y) < f^{-1}(c) + \epsilon_0. \qquad (9.37)$$

Since $\epsilon \ge \epsilon_0 > 0$, from (9.37)

$$f^{-1}(c) - \epsilon < f^{-1}(c) \le f^{-1}(y) < f^{-1}(c) + \epsilon_0 < f^{-1}(c) + \epsilon.$$

Thus f^{-1} is continuous at a left endpoint of $\Re_f$ if this endpoint is also an element of $\Re_f$. By a similar proof, which is left as an exercise, f^{-1} can be shown to be continuous at a right endpoint of $\Re_f$ if this endpoint is an element of $\Re_f$.

By Theorem 9.1.3 the function ln is continuous and increasing on its domain, the interval $(0, +\infty)$, and by Theorem 9.1.4(c) the range of ln is the set R. Hence by Theorem 9.4.1, $\ln^{-1}$ exists and is continuous and increasing on R.

We next derive a differentiation formula for the derivative of the inverse of a function.

9.4.2   Theorem

Suppose the following conditions hold:

> *(i) f has an inverse function f^{-1}*
> *(ii) f^{-1} is continuous at c*
> *(iii) f has a nonzero derivative at $f^{-1}(c)$*

Then

$$(f^{-1})'(c) = \frac{1}{f'(f^{-1}(c))}. \tag{9.38}$$

PROOF   If $(f^{-1})'(c)$ exists, then

$$(f^{-1})'(c) = \lim_{k \to 0} \frac{f^{-1}(c + k) - f^{-1}(c)}{k}. \tag{9.39}$$

To show that this limit exists we shall rewrite the quotient in the right member of (9.39). Suppose we let

$$\phi(k) = f^{-1}(c + k) - f^{-1}(c) \tag{9.40}$$

where c and $c + k \in \Re_f$. If $k \neq 0$, $f^{-1}(c + k) \neq f^{-1}(c)$, for otherwise from (9.32) we would have $f(f^{-1}(c + k)) = f(f^{-1}(c))$ or $c + k = c$, and hence $k = 0$.
Since k is given by the equation

$$k = f(f^{-1}(c + k)) - f(f^{-1}(c)), \tag{9.41}$$

when $k \neq 0$

$$\frac{f^{-1}(c + k) - f^{-1}(c)}{k} = \frac{1}{\dfrac{f(f^{-1}(c + k)) - f(f^{-1}(c))}{f^{-1}(c + k) - f^{-1}(c)}},$$

and from (9.40)

$$\frac{f^{-1}(c + k) - f^{-1}(c)}{k} = \frac{1}{\dfrac{f(f^{-1}(c) + \phi(k)) - f(f^{-1}(c))}{\phi(k)}}. \tag{9.42}$$

If we define the function F by

$$F(k) = \begin{cases} \dfrac{f(f^{-1}(c) + k) - f(f^{-1}(c))}{k} & \text{if } k \neq 0 \\[3mm] f'(f^{-1}(c)) & \text{if } k = 0, \end{cases} \tag{9.43}$$

from (9.42)

$$\frac{f^{-1}(c + k) - f^{-1}(c)}{k} = \frac{1}{F(\phi(k))} \quad \text{if } k \neq 0. \tag{9.44}$$

Since f is differentiable at $f^{-1}(c)$, from (9.43)

$$\lim_{k \to 0} F(k) = \lim_{k \to 0} \frac{f(f^{-1}(c) + k) - f(f^{-1}(c))}{k} = f'(f^{-1}(c))$$

$$= F(0) \tag{9.45}$$

and therefore F is continuous at 0. Since f^{-1} is continuous on $\Re_f$, $\lim_{k \to 0} f^{-1}(c + k) = f^{-1}(c)$ and from (9.40)

$$\lim_{k \to 0} \phi(k) = 0.$$

Then from (9.45) and (9.43) using the composite function limit theorem,

$$\lim_{k \to 0} F(\phi(k)) = F\left(\lim_{k \to 0} \phi(k)\right) = F(0)$$
$$= f'(f^{-1}(c)).$$

Finally from (9.39), (9.44), and the preceding equation

$$(f^{-1})'(c) = \lim_{k \to 0} \frac{1}{F(\phi(k))} = \frac{1}{f'(f^{-1}(c))}.$$

Example 1   If $f(x) = x + 2\sqrt{x}$, define the derivative of f^{-1} and find $(f^{-1})'(8)$.

SOLUTION   Since $f'(x) = 1 + x^{-1/2}$ for every $x > 0$, f is continuous and increasing on the interval $[0, +\infty)$. By Theorem 9.4.1, f^{-1} exists on $\mathcal{R}_{f^{-1}} = \mathcal{D}_f = [0, +\infty)$. Then by Theorem 9.4.2 if $y \in [0, +\infty)$,

$$(f^{-1})'(y) = \frac{1}{f'(f^{-1}(y))} = \frac{1}{1 + (f^{-1}(y))^{-1/2}}. \qquad (9.46)$$

In order to find $(f^{-1})'(8)$, we must first obtain $f^{-1}(8)$. From the equation defining f, $x = f^{-1}(8)$ if

$$x + 2\sqrt{x} = 8. \qquad (9.47)$$

To solve this equation, one obtains

$$x + 2\sqrt{x} - 8 = 0$$
$$(\sqrt{x} + 4)(\sqrt{x} - 2) = 0,$$

and hence $\sqrt{x} = 2$ or $\sqrt{x} = -4$. But $\sqrt{x} \neq -4$ since $\sqrt{x}$ cannot be negative. Then from the equation $\sqrt{x} = 2$, we obtain

$$x = 4,$$

which is the only solution of (9.47).

Therefore, $f^{-1}(8) = 4$ and hence from (9.46)

$$(f^{-1})'(8) = \frac{1}{1 + (f^{-1}(8))^{-1/2}} = \frac{1}{1 + 4^{-1/2}} = \frac{2}{3}.$$

In the discussion following Theorem 9.4.1 it was shown that the function $\ln^{-1}$ is continuous and increasing on its domain, the set R. Since $D_x \ln x = 1/x$ for every $x \in \mathcal{D}_{\ln} = (0, +\infty)$, by Theorem 9.4.2

$$(\ln^{-1})'(y) = \frac{1}{\ln'(\ln^{-1}(y))} = \frac{1}{1/\ln^{-1} y} = \ln^{-1} y \qquad (9.48)$$

for any number y. We therefore have the unexpected result that $\ln^{-1}$ is its own derivative.

Exercise Set 9.4

In Exercises 1–10 use theory that has been discussed to prove that the function f defined by the given equation has an inverse function. Give the indicated values of the derivative of f^{-1}.

1. $f(x) = 3x + 4$, $(f^{-1})'(-2)$

2. $f(x) = 1 - 5x$, $(f^{-1})'(3)$

3. $f(x) = (x - 1)^3$, $(f^{-1})'(1)$

4. $f(x) = \dfrac{1}{x^3 + 1}$, $(f^{-1})'(1)$

5. $f(x) = x^3 + 4x + 5$, $(f^{-1})'(5)$

6. $f(x) = x^5 + 4x^3 + 4x + 2$, $(f^{-1})'(2)$

7. $f(x) = x^2 - 2x + 2$ if $x \geq 1$, $(f^{-1})'(5)$

8. $f(x) = x^2 + 2x + 4$ if $x < -1$, $(f^{-1})'(4)$

9. $f(x) = |x| + 2x$, $(f^{-1})'(-1)$

10. $f(x) = x + [x]$, $(f^{-1})'(\frac{5}{2})$

In Exercises 11 and 12 prove that f has an inverse function f^{-1} and find the derivative of f^{-1}.

11. $f(x) = \displaystyle\int_1^x \sqrt{1 + t^2}\, dt$

12. $f(x) = \displaystyle\int_x^{-1} \frac{t^2 + 2}{t^4 + 1}\, dt$

13. Complete the proof of Theorem 9.4.1 by showing that if f is continuous and increasing on an interval with c as its right endpoint, then f^{-1} is continuous at $f(c)$.

9.5 The Exponential Function

From the discussion following Theorem 9.4.1 the following theorem can be stated.

9.5.1 Theorem

$\ln^{-1}$ *exists and is continuous and increasing on its domain, the set R. The range of* $\ln^{-1}$ *is the interval* $(0, +\infty)$.

We shall discuss some other properties of $\ln^{-1}$. From (9.32)

$$\ln (\ln^{-1} x) = x \quad \text{for every } x. \tag{9.49}$$

$$\ln^{-1} (\ln x) = x \quad \text{if } x \in (0, +\infty). \tag{9.50}$$

Since $\ln 1 = 0$, from (9.50)

$$\ln^{-1} 0 = \ln^{-1} (\ln 1) = 1. \tag{9.51}$$

Since $\ln e = 1$, also from (9.50)

$$\ln^{-1} 1 = \ln^{-1} (\ln e) = e. \tag{9.52}$$

From (9.33) if x_1 and x_2 are any numbers and

$$y_1 = \ln^{-1} x_1 \qquad y_2 = \ln^{-1} x_2$$

then

$$x_1 = \ln y_1 \qquad x_2 = \ln y_2.$$

Using these relationships and (9.50)

$$(\ln^{-1} x_1)(\ln^{-1} x_2) = y_1 y_2 = \ln^{-1} (\ln y_1 y_2)$$
$$= \ln^{-1} (\ln y_1 + \ln y_2)$$
$$= \ln^{-1} (x_1 + x_2). \tag{9.53}$$

By a similar development one can prove that

$$\frac{\ln^{-1} x_1}{\ln^{-1} x_2} = \ln^{-1} (x_1 - x_2), \tag{9.54}$$

and as a particular case of (9.54) where $x_1 = 0$ and $x_2 = x$,

$$\frac{1}{\ln^{-1} x} = \ln^{-1} (-x). \tag{9.55}$$

If r is any rational number

$$(\ln^{-1} x_1)^r = y_1{}^r = \ln^{-1} (\ln y_1{}^r)$$
$$= \ln^{-1} (r \ln y_1)$$
$$= \ln^{-1} (rx_1). \tag{9.56}$$

In particular if $x_1 = 1$ in (9.56), then from (9.52)

$$e^r = \ln^{-1} r. \tag{9.57}$$

Now since the value of $\ln^{-1}$ at any rational number r is e^r, one might inquire if (9.56) is also true when r is an *irrational* number. However, e^r and, for that matter, the rth power of any positive number have not hitherto been defined when r is irrational.

Since $\ln^{-1} x$ exists for every x, whether rational or irrational, we shall extend (9.57) with the following definition of e^x.

9.5.2   Definition

If x is any real number,

$$e^x = \ln^{-1} x.$$

Because of Definition 9.5.2 we call $\ln^{-1}$ *the exponential function*, and from the definition, properties (9.49) through (9.56) of $\ln^{-1}$ can be restated as follows:

$$\ln e^x = x \quad \text{for every } x \tag{9.58}$$
$$e^{\ln x} = x \quad \text{for every } x > 0 \tag{9.59}$$
$$e^0 = 1 \tag{9.60}$$
$$e^1 = e \tag{9.61}$$

$$e^{x_1}e^{x_2} = e^{x_1+x_2} \quad \text{for every } x_1 \text{ and } x_2 \tag{9.62}$$

$$\frac{e^{x_1}}{e^{x_2}} = e^{x_1-x_2} \quad \text{for every } x_1 \text{ and } x_2 \tag{9.63}$$

$$\frac{1}{e^x} = e^{-x} \quad \text{for every } x \tag{9.64}$$

$$(e^{x_1})^r = e^{rx_1} \quad \text{for every } x_1 \text{ and } r. \tag{9.65}$$

We next prove an important limit theorem.

9.5.3   Theorem

(a) $\lim\limits_{x \to +\infty} e^x = +\infty.$

(b) $\lim\limits_{x \to -\infty} e^x = 0.$

PROOF OF (a)   By Theorem 9.5.1 $\ln^{-1}$ is increasing on its domain, the set R, and has $(0, +\infty)$ as its range. Hence by (9.58), for any $B > 0$

$$e^x > B = e^{\ln B} \tag{9.66}$$

whenever $x > \ln B$. Moreover, (9.66) is true if M is any positive number and $x > M > \ln B$. The conclusion then follows by Definition 5.7.1.

The proof of (b) is left as an exercise.
From Theorem 9.5.3(b) the x axis is a horizontal asymptote of the graph of the exponential function. A sketch of the graph of this function is shown in Figure 9–7.

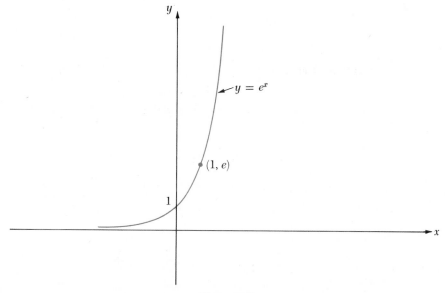

Figure 9-7

The differentiation formula for $\ln^{-1}$, (9.48), can be restated using Definition 9.5.2.

9.5.4 Theorem

$$D_x e^x = e^x.$$

Also the following generalization of Theorem 9.5.4 can be proved using the chain rule.

9.5.5 Theorem

If $g'(x)$ exists,

$$D_x e^{g(x)} = e^{g(x)} g'(x). \tag{9.67}$$

Example 1 Find $f'(x)$ if $f(x) = e^{\sqrt{x^2+1}}$.

SOLUTION From (9.67)

$$f'(x) = e^{\sqrt{x^2+1}} D_x \sqrt{x^2 + 1}$$

$$= \frac{xe^{\sqrt{x^2+1}}}{\sqrt{x^2 + 1}}.$$

From Theorem 9.5.5 we obtain the integration formula

$$\int e^{g(x)} g'(x) \, dx = e^{g(x)} + C. \tag{9.68}$$

Example 2 Evaluate $\int xe^{1-x^2} \, dx$.

SOLUTION We let $u = g(x) = 1 - x^2$ and then $du = g'(x) \, dx = -2x \, dx$. Hence $x \, dx = -\frac{1}{2} du$ and by (9.68)

$$\int xe^{1-x^2} \, dx = \int -\tfrac{1}{2} e^u \, du = -\tfrac{1}{2} e^u + C$$

$$= -\tfrac{1}{2} e^{1-x^2} + C.$$

Exercise Set 9.5

1. Simplify: (a) $\ln e^{1/x}$ (b) $e^{\ln(1/x)}$ (c) $\ln(1/e^x)$ (d) $e^{3\ln x}$ (e) $\ln((e^x)^3)$
   (f) $e^{2\ln x - 3\ln y}$ (g) $e^{(1+\ln x^2)/2}$

In Exercises 2–11 find $f'(x)$.

2. $f(x) = e^{x^3}$

3. $f(x) = e^{-(x-1)^2/2}$

4. $f(x) = e^{\sqrt{x}}$

5. $f(x) = \dfrac{e^{ax} + e^{-ax}}{2}$

6. $f(x) = x^2 e^{-x}$

7. $f(x) = e^{1/x}$

8. $f(x) = e^{(x-1)/(x+2)}$

9. $f(x) = \dfrac{e^{2x}}{x^2 + 1}$

10. $f(x) = \ln(e^x - e^{-x})$

11. $f(x) = \ln\dfrac{e^x + e^{-x}}{e^x - e^{-x}}$

In Exercises 12 and 13 find $D_x y$ by implicit differentiation.

12. $x^2 e^{3y} + y^2 e^{3x} = 1$

13. $e^x - e^y = e^{x-y}$

In Exercises 14–25 obtain the required integral.

14. $\displaystyle\int e^{3x}\, dx$

15. $\displaystyle\int \frac{1}{e^{2x}}\, dx$

16. $\displaystyle\int e^{3x} e^{4x}\, dx$

17. $\displaystyle\int x^2 e^{x^3}\, dx$

18. $\displaystyle\int e^{(5-x)/2}\, dx$

19. $\displaystyle\int \frac{e^x\, dx}{e^x + 1}$

20. $\displaystyle\int \frac{e^{2x} - 1}{e^x}\, dx$

21. $\displaystyle\int (e^x - e^{-x})^2\, dx$

22. $\displaystyle\int \frac{e^x - e^{-x}}{e^x + e^{-x}}\, dx$

23. $\displaystyle\int \frac{e^x\, dx}{(1 - 3e^x)^2}$

24. $\displaystyle\int \frac{e^{2x}\, dx}{(1 + 2e^{2x})^3}$

25. $\displaystyle\int \frac{e^{2x}}{1 + e^x}\, dx$

26. Prove that the graph of the exponential function is concave upward at every point on the graph.

27. Derive Equation (9.54).

28. Find the inflection points in the graph of the function given by $f(x) = e^{-x^2/2}$.

29. Sketch the graph of the equation $y = x^2 e^{-x}$, using calculus.

30. Prove: $\displaystyle\lim_{x\to 0} \frac{e^x - 1}{x} = 1$. HINT: Use the definition of the derivative.

31. Prove: If $x \neq 0$, then $e^x > 1 + x$. HINT: Apply the mean value theorem to the closed interval with endpoints 0 and x or use the results of Exercise 22, Exercise Set 9.1.

32. Prove Theorem 9.5.3(b). HINT: Use Definition 5.6.2 and Theorem 9.5.1.

9.6   Other Exponential and Logarithmic Functions

If $a > 0$ and r is a rational number, by (9.59)

$$a^r = e^{\ln a^r}$$

and hence from (9.20)

$$a^r = e^{r \ln a} \quad r, \text{ any rational number.}$$

An extension of this equation defines a^x where x is *any* number.

9.6.1   Definition

Let x be any number and a be any positive number. Then:

$$a^x = e^{x \ln a} \tag{9.69}$$

The function f defined by

$$f(x) = a^x \quad \text{where } a > 0 \text{ but } a \neq 1 \tag{9.70}$$

is called the *exponential function to the base a.* The usual properties that one would expect of this exponential function are listed below and can be derived using Definition 9.6.1 and the corresponding properties of the exponential function in Section 9.5.

$$a^0 = 1 \tag{9.71}$$

$$a^1 = a \tag{9.72}$$

$$a^{x_1} a^{x_2} = a^{x_1 + x_2} \quad \text{for every } x_1 \text{ and } x_2 \tag{9.73}$$

$$\frac{a^{x_1}}{a^{x_2}} = a^{x_1 - x_2} \quad \text{for every } x_1 \text{ and } x_2 \tag{9.74}$$

$$\frac{1}{a^x} = a^{-x} \quad \text{for every } x \tag{9.75}$$

$$(a^x)^r = a^{rx} \quad \text{for every } x \text{ and } r \tag{9.76}$$

$$(ab)^x = a^x b^x \quad \text{for every } x \text{ if } b > 0 \tag{9.77}$$

To prove (9.73), for example, we use (9.69) and (9.62). Then

$$a^{x_1} a^{x_2} = e^{x_1 \ln a} e^{x_2 \ln a} = e^{x_1 \ln a + x_2 \ln a}$$
$$= e^{(x_1 + x_2) \ln a} = a^{x_1 + x_2}$$

We now give some limit properties of the exponential function to the base a. The first two properties are analogous to those in Theorem 9.5.3.

9.6.2   Theorem

(a) If $a > 1$, then $\lim\limits_{x \to +\infty} a^x = +\infty$.

(b) If $a > 1$, then $\lim\limits_{x \to -\infty} a^x = 0$.

(c) *If* $0 < a < 1$, *then* $\lim\limits_{x \to +\infty} a^x = 0$.

(d) *If* $0 < a < 1$, *then* $\lim\limits_{x \to -\infty} a^x = +\infty$.

PROOF OF (a)   First note that since $a > 1$,

$$\ln a > \ln 1 = 0.$$

Next suppose $B > 0$. Since $\lim_{x \to +\infty} e^x = +\infty$ (Theorem 9.5.3(a)), by Definition 5.7.1 there is an $M' > 0$ such that

$$e^{x \ln a} = a^x > B$$

if $x \ln a > M'$, or, equivalently if

$$x > \frac{M'}{\ln a} = M > 0.$$

The conclusion then follows by Definition 5.7.1.
  The proofs of (b), (c), and (d) are left as exercises.

  From Definition 9.6.1 the following differentiation formula is easily derived.

9.6.3   Theorem

If $a > 0$,

$$D_x a^x = a^x \ln a.$$

PROOF $D_x a^x = D_x e^{x \ln a} = e^{x \ln a} D_x (x \ln a)$
$$= a^x \ln a.$$

This formula can be generalized by the chain rule.

9.6.4   Theorem

If $a > 0$ *and* $g'(x)$ *exists*

$$D_x a^{g(x)} = a^{g(x)}(\ln a)g'(x). \tag{9.78}$$

Example 1   Find $f'(x)$ if $f(x) = 10^{-5x}$.

SOLUTION   From Theorem 9.6.4

$$f'(x) = D_x 10^{-5x} = 10^{-5x}(\ln 10)(-5) = -5(\ln 10)10^{-5x}.$$

Example 2   Find $f'(x)$ if $f(x) = x^x$.

SOLUTION   From Definition 9.6.1 when $x > 0$

$$f(x) = x^x = e^{x \ln x},$$

and from Theorem 9.5.5

$$f'(x) = D_x e^{x \ln x} = e^{x \ln x}(x \cdot 1/x + 1 \cdot \ln x)$$
$$= x^x(1 + \ln x).$$

Also from Theorem 9.6.4 we have the integration formula

$$\int a^{g(x)} g'(x) \, dx = \frac{a^{g(x)}}{\ln a} + C. \tag{9.79}$$

Example 3   Obtain $\int 3^{x^2} x \, dx$.

SOLUTION   Let $u = g(x) = x^2$. Since $du = 2x \, dx$, $x \, dx = \frac{1}{2} du$, and from (9.79)

$$\int 3^{x^2} x \, dx = \int \frac{1}{2} \cdot 3^u \, du = \frac{3^u}{2 \ln 3} + C$$
$$= \frac{3^{x^2}}{2 \ln 3} + C.$$

The exponential function to the base a has domain R since a^x is defined for every x. It can also be proved (Exercise 5, Exercise Set 9.6) that the range of this function is the interval $(0, +\infty)$. From Theorem 9.6.3 the derivative of the function exists on R where it is positive-valued if $a > 1$ or negative-valued if $0 < a < 1$. Thus the exponential function to the base a is continuous and monotonic on R and by Theorem 9.4.1 has an inverse function that is continuous and monotonic on $(0, +\infty)$. This inverse function, which is called the *logarithmic function to the base a* and will be denoted by $\log_a$, is defined in the next statement.

9.6.5   Definition

Let $a \neq 1$ and x be positive numbers. Then

$$y = \log_a x \quad \text{if and only if} \quad a^y = x. \tag{9.80}$$

The following properties of this logarithmic function can be derived from (9.80) and the proofs are left as exercises.

$$a^{\log_a x} = x \quad \text{if } x > 0 \tag{9.81}$$
$$\log_a 1 = 0 \tag{9.82}$$
$$\log_a a = 1 \tag{9.83}$$

$$\log_a x_1 x_2 = \log_a x_1 + \log_a x_2 \quad \text{if } x_1, x_2 > 0 \tag{9.84}$$

$$\log_a \frac{x_1}{x_2} = \log_a x_1 - \log_a x_2 \quad \text{if } x_1, x_2 > 0 \tag{9.85}$$

$$\log_a x^r = r \log_a x \quad \text{if } x > 0 \tag{9.86}$$

We can also express $\log_a x$ in terms of natural logarithms. If $y = \log_a x$, then $a^y = x$ and

$$\ln a^y = \ln x$$
$$y \ln a = \ln x$$
$$y = \frac{\ln x}{\ln a}$$

or saying the same thing,

$$\log_a x = \frac{\ln x}{\ln a}. \tag{9.87}$$

Since $\ln e = 1$, from (9.87)

$$\log_a e = \frac{1}{\ln a}, \tag{9.88}$$

and therefore by (9.87) and (9.88)

$$\log_a x = (\log_a e)(\ln x). \tag{9.89}$$

Differentiating in (9.89),

$$D_x \log_a x = \frac{\log_a e}{x}. \tag{9.90}$$

If $g'(x)$ exists, then from (9.90) and the chain rule

$$D_x \log_a g(x) = \frac{\log_a e}{g(x)} \cdot g'(x). \tag{9.91}$$

Example 4   If $y = \log_{10} \dfrac{x^2 - 1}{x}$, find $D_x y$.

SOLUTION   From (9.85)

$$y = \log_{10} (x^2 - 1) - \log_{10} x,$$

and hence from (9.89)

$$D_x y = \frac{\log_{10} e}{x^2 - 1} \cdot 2x - \frac{\log_{10} e}{x}$$

$$= \log_{10} e \left(\frac{2x}{x^2 - 1} - \frac{1}{x} \right)$$

$$= \frac{\log_{10} e(x^2 + 1)}{x(x^2 - 1)}.$$

The next theorem gives an interesting limit which is sometimes used to define the number e.

9.6.6 Theorem

$$\lim_{x \to 0} (1 + x)^{1/x} = e. \tag{9.92}$$

PROOF   Since ln is differentiable on its domain, and $\ln 1 = 0$,

$$\ln'(1) = \lim_{x \to 0} \frac{\ln(1 + x) - \ln 1}{x}$$

$$= \lim_{x \to 0} \ln(1 + x)^{1/x}. \tag{9.93}$$

Also, by Theorem 9.1.2

$$\ln'(1) = 1. \tag{9.94}$$

Equating the expressions for $\ln'(1)$ in (9.93) and (9.94) gives

$$\lim_{x \to 0} \ln(1 + x)^{1/x} = 1. \tag{9.95}$$

Since $(1 + x)^{1/x} = e^{\ln(1+x)^{1/x}}$ when $x > -1$, from (9.95) using the composite function limit theorem

$$\lim_{x \to 0} (1 + x)^{1/x} = \lim_{x \to 0} e^{\ln(1+x)^{1/x}}$$

$$= e^{\lim_{x \to 0} \ln(1+x)^{1/x}} = e^1 = e.$$

From Theorem 9.6.6 another limit can be derived, which has a useful application.

9.6.7 Theorem

Suppose r is any number. Then

$$\lim_{x \to +\infty} \left(1 + \frac{r}{x}\right)^{xt} = e^{rt}. \tag{9.96}$$

PROOF   If $r = 0$, the limit (9.96) is immediately obtained since, from (9.69), $1^{xt} = 1$ for every x. If $r \neq 0$, we note that:

$$\left(1 + \frac{r}{x}\right)^{xt} = e^{\ln(1+(r/x))^{xt}} = e^{\ln[(1+(r/x))^{x/r}]^{rt}}$$

$$= e^{rt \ln(1+(r/x))^{x/r}} \tag{9.97}$$

Thus in order to obtain $\lim_{x \to +\infty} (1 + (r/x))^{xt}$ from (9.97), the limit $\lim_{x \to +\infty} \ln(1 + (r/x))^{x/r}$ is required. To obtain the latter limit, we consider the function f given by

$$f(x) = \begin{cases} \ln(1 + x)^{1/x} & \text{if } x \neq 0 \\ 1 & \text{if } x = 0. \end{cases}$$

By (9.95) and Theorem 3.5.1

$$\lim_{x \to 0} f(x) = \lim_{x \to 0} \ln (1 + x)^{1/x} = 1$$

$$= f(0).$$

Hence f is continuous at 0. Then by Theorem 5.6.9

$$\lim_{x \to +\infty} f\left(\frac{r}{x}\right) = f\left(\lim_{x \to +\infty} \frac{r}{x}\right) = f(0) = 1,$$

and hence

$$\lim_{x \to +\infty} \ln \left(1 + \frac{r}{x}\right)^{x/r} = \lim_{x \to +\infty} f\left(\frac{r}{x}\right) = 1. \tag{9.98}$$

Then from (9.97) using Theorem 5.6.9,

$$\lim_{x \to +\infty} \left(1 + \frac{r}{x}\right)^{xt} = e^{\lim_{x \to +\infty} rt \ln (1+(r/x))^{x/r}}$$

and from (9.98)

$$\lim_{x \to +\infty} \left(1 + \frac{r}{x}\right)^{xt} = e^{rt}.$$

If $r > 0$, the limit (9.96) has an interesting application, the accumulation of money when interest is *continuously compounded*. If P dollars is invested at $100r\%$ interest per year *compounded yearly* (and the interest earned is not withdrawn), this amount of money will accumulate as noted below after the indicated numbers of years. In each case the accumulated amount after a particular year is obtained by summing the accumulation at the beginning of the year and the interest earned during the year.

After 1 year: $P + Pr = P(1 + r)$
After 2 years: $P(1 + r) + rP(1 + r) = P(1 + r)^2$
After 3 years: $P(1 + r)^2 + Pr(1 + r)^2 = P(1 + r)^3$

Then after n years the accumulation A_n can be shown by mathematical induction to be

$$A_n = P(1 + r)^n \text{ dollars.} \tag{9.99}$$

Next suppose the interest rate is $100r\%$ per year compounded every $1/m$ year where m is a positive integer; that is, the rate is $(100r/m)\%$ per $1/m$ year compounded each $1/m$ year. Then from (9.99) the accumulation after $t = n/m$ years, where n is a positive integer, is

$$A_n = A_{mt} = P\left(1 + \frac{r}{m}\right)^{mt} \text{ dollars.} \tag{9.100}$$

For example, if the interest rate is 6% per year compounded quarterly, which is equivalent to an interest rate of 1.5% per quarter compounded quarterly, the accumulation from \$100 after $2\frac{1}{2}$ years (10 quarters) is

$$A_{10} = 100(1.015)^{10} \approx \$116.05.$$

If in (9.100) we allow m to increase indefinitely, then $1/m$ approaches 0, and A_{mt} approaches what would seem to be the amount of money accumulated after $t = n/m$ years from an investment of P dollars at $100r\%$ per year compounded *at every instant of time!* Now when m is sufficiently large, A_{mt} can be made to differ from

$$\lim_{x \to +\infty} P\left(1 + \frac{r}{x}\right)^{xt} = Pe^{rt}$$

(see Theorem 9.6.7) by as little as we please. Therefore the following definition can be stated.

9.6.8   Definition

Let t and r be any positive numbers. The accumulation after t years from an initial investment of P dollars at $100r\%$ per year compounded continuously is Pe^{rt} dollars.

From Definition 9.6.8 an investment of \$100 for $2\frac{1}{2}$ years at 6% interest compounded continuously will amount to

$$100e^{.06(2.5)} = 100(1.1618) = \$116.18.$$

It may be surprising to note that the accumulation here is only 13 cents more than in the example following (9.100).

Exercise Set 9.6

1. Approximate the following numbers using a table of natural logarithms:
   (a) $\log_{10} 5$ (b) $\log_{4.7} 3$ (c) $\log_2 \sqrt[5]{3}$

2. Sketch the graph of the equation $y = \log_4 x$.

3. Derive the following properties of the exponential function to the base a:
   (a) property (9.71)   (b) property (9.72)   (c) property (9.74)   (d) property (9.75)   (e) property (9.76)   (f) property (9.77)

4. Derive the following properties of the logarithmic function to the base a:
   (a) property (9.81)   (b) property (9.82)   (c) property (9.83)   (d) property (9.84)   (e) property (9.85)   (f) property (9.86)

5. Prove that the range of the exponential function to the base a is $(0, +\infty)$.

In Exercises 6–23 find $f'(x)$.

6. $f(x) = 5^x$

7. $f(x) = 10^{2x+3}$

8. $f(x) = 3^{\sqrt{x}}$

9. $f(x) = 3^{x^2} 2^{4x}$

10. $f(x) = x^3 2^{1/x}$

11. $f(x) = \dfrac{2^{3x}}{x^2 + 1}$

12. $f(x) = x^{\sqrt{x}}$

13. $f(x) = x^{1/x}$

14. $f(x) = x^{x^2}$

15. $f(x) = xe^{-x}$

16. $f(x) = (x^2 + 1)^x$

17. $f(x) = (x + 2)^{\ln x}$

18. $f(x) = \log_2 (4x - 5)$

19. $f(x) = \log_{10} (x^2 - 3x - 5)$

20. $f(x) = \log_a \sqrt{1 - x}$

21. $f(x) = \log_a \dfrac{x - 1}{x}$

22. $f(x) = \sqrt[3]{\log_2 \dfrac{x}{x + 1}}$

23. $f(x) = \log_{10} (\log_{10} x)$

Obtain the integrals in Exercises 24–31.

24. $\displaystyle \int 2^{5x} \, dx$

25. $\displaystyle \int \left(\frac{1}{2^x}\right)^3 dx$

26. $\displaystyle \int 3^{x^2} x \, dx$

27. $\displaystyle \int \frac{(2x + 3) \, dx}{10^{x^2 + 3x - 1}}$

28. $\displaystyle \int \frac{2^{\ln x}}{x} \, dx$

29. $\displaystyle \int (3^x + 3^{-x})^2 \, dx$

30. $\displaystyle \int 3^{2x} 5^{2x} \, dx$

31. $\displaystyle \int a^{x - a^x} \, dx$

32. Find $\displaystyle \lim_{x \to 0} (1 + 2x)^{1/x}$.

33. Prove Theorem 9.6.2(b). HINT: Use Theorem 9.5.3(b).

34. Prove Theorem 9.6.2(c). HINT: Use Theorem 9.6.2(b).

35. Prove Theorem 9.6.2(d).

36. In how many years will an investment double itself if money is worth $5\frac{1}{2}\%$ compounded continuously?

37. In 1624 the Dutch bought Manhattan Island from the Indians for $24. If this money had been invested at 5% compounded continuously, to what amount would it accumulate in 1974?

9.7 Exponential Growth and Decay

There are numerous examples in science where at any time the rate of change of a certain substance with respect to time is proportional to the amount of the substance at that time. For example, the rate of decay of a radioactive element at any time is proportional to the mass of the element at that time. Also, under favorable conditions the rate of growth of a human population at any time is proportional to the number of people at the time.

In such situations, if $y > 0$ is the amount of the substance at time t, then there is some k such that

$$\frac{dy}{dt} = ky. \tag{9.101}$$

If the function f is a solution of the differential equation (9.101) on an interval I (that is, $f'(t) = kf(t)$ when $t \in I$, then

$$\frac{f'(t)}{f(t)} = k,$$

which may be rewritten

$$D_t \ln f(t) = k. \tag{9.102}$$

To solve (9.102) and hence (9.101), we integrate in (9.102) and obtain

$$\ln f(t) = kt + C_1 \tag{9.103}$$

where C_1 is the constant of integration. Since there is some $C > 0$ such that $C_1 = \ln C$, (9.103) may be rewritten successively as

$$\ln f(t) = kt + \ln C$$
$$= \ln e^{kt} + \ln C$$
$$= \ln Ce^{kt},$$

from which we obtain

$$f(t) = Ce^{kt} \quad \text{if } t \in I. \tag{9.104}$$

Thus if f is a solution of (9.101) on I, f is given by (9.104). In other words, there can be no solution of (9.101) on I other than a solution given by (9.104). Conversely, it can be easily shown by direct substitution that a function f given by (9.104) is a solution of (9.101) on I.

If $t = t_0 \in I$, then from (9.104) $f(t_0) = Ce^{kt_0}$, and

$$C = e^{-kt_0}f(t_0).$$

Accordingly, (9.104) can be rewritten:

$$y = f(t) = f(t_0)e^{k(t-t_0)} \tag{9.105}$$

The equations (9.104) and (9.105) are said to express the *law of exponential growth* if $k > 0$ and the *law of exponential decay* if $k < 0$.

We next consider some applications leading to the solution of differential equations of the form (9.101).

Example 1 One of the waste products of a nuclear explosion is the radioactive isotope strontium-90. This substance, which behaves chemically like calcium, has a half-life of approximately 25 years; that is, one half of the substance will remain after 25 years. If the rate of decay of strontium-90 is given by (9.101), and if 20 mg of the isotope are present now, find: (a) how much of the isotope will remain in 15 years, (b) in how many years only 5 mg will remain.

SOLUTION If y is the mass of strontium-90 present t years from now, from (9.101)

$$\frac{dy}{dt} = ky.$$

This differential equation has the solution on the interval $[0, +\infty)$ given by

$$y = Ce^{kt}. \qquad (9.106)$$

Since $y = 20$ when $t = 0$, from (9.106) $20 = Ce^{k \cdot 0}$, and

$$C = 20. \qquad (9.107)$$

Substituting from (9.107) into (9.106) gives

$$y = 20e^{kt}. \qquad (9.108)$$

Since the half-life is 25 years, $y = 10$ when $t = 25$. Hence from (9.108)

$$10 = 20e^{25k}$$
$$e^{25k} = \tfrac{1}{2}$$
$$25k = \ln \tfrac{1}{2} = -\ln 2$$
$$k = -.04 \ln 2. \qquad (9.109)$$

A substitution for k in (9.108) from (9.109) gives:

$$y = 20e^{(-.04 \ln 2)t} \qquad (9.110)$$

If $t = 15$

$$y = 20e^{(-.04 \ln 2)(15)} \approx 20e^{-.4159}$$
$$\approx 13.2 \text{ mg,}$$

which is the required solution in part (a).

To obtain the solution in part (b), let $y = 5$ in (9.110) and solve for t. Then:

$$5 = 20e^{(-.04 \ln 2)t} \qquad (9.111)$$

From (9.111)

$$e^{(-.04 \ln 2)t} = \tfrac{1}{4}$$

and hence

$$t(-.04 \ln 2) = \ln \tfrac{1}{4} = -\ln 4$$
$$t = \frac{-\ln 4}{-.04 \ln 2} = 50 \text{ years.}$$

Thus 5 mg of the substance will remain after 50 years.

Example 2 Suppose the electric generating capacity of a certain utility company is increasing according to the exponential law of growth. If the generating capacity increased by 50% from 1965 to 1971, how will it increase from 1965 to 1977?

SOLUTION By hypothesis, if y is the generating capacity (in kilowatts) at time t years after 1965, then

$$y = Ce^{kt}. \qquad (9.112)$$

If $t = 0$ (in 1965), then from (9.112) $y = C$. Thus $y = 1.5C$ when $t = 6$ (in 1971) and hence from (9.112)

$$1.5C = Ce^{6k}$$
$$1.5 = e^{6k}$$
$$6k = \ln 1.5$$
$$k = \tfrac{1}{6} \ln 1.5. \tag{9.113}$$

Substituting from (9.113) into (9.112) gives:

$$y = Ce^{(1/6 \ln 1.5)t}$$

Then if $t = 12$ (in 1977)

$$y = Ce^{(1/6 \ln 1.5)(12)} = Ce^{2 \ln 1.5}$$
$$= Ce^{\ln (1.5)^2} = (1.5)^2 C$$
$$= 2.25C.$$

Thus the generating capacity of the company will increase from C to $2.25C$ from 1965 to 1977, a gain of 125%.

Exercise Set 9.7

1. When a steel bar is heated, the rate of change of its length L with respect to its centigrade temperature T is given by $dL/dT = (1.2)(10^{-5})L$. Suppose the length of the bar is 39 ft when $T = 0°$ C. Express the length of the bar as a function of its centigrade temperature.

2. The radioactive element polonium has a half-life of 140 days. (a) How long would be required for 10 mg of this substance to decay to 2 mg? (b) How much of this substance would remain after 100 days?

3. At any altitude above the earth the rate of change of the atmospheric pressure with respect to the altitude is proportional to the atmospheric pressure. If the atmospheric pressure is 30 in. at sea level and 15 in. at an altitude of 3 miles, find: (a) the atmospheric pressure at an altitude of 4 miles, (b) the altitude when the atmospheric pressure is 7.5 in.

4. Suppose an object is in a surrounding medium having a constant temperature. According to Newton's law of cooling from physics, if x is the difference between the temperature of the object and the temperature of the medium, the rate of change of x with respect to time t at any time is proportional to x at that time. A bowl of soup at temperature 180° F is placed in a room at temperature 70° F. If the temperature of the soup is 150° F after two minutes, when will the soup temperature be 100° F?

5. In a certain experiment a culture of bacteria increases according to the law of exponential growth. If there were 180 bacteria after the second day of the experiment and 300 bacteria after the fourth day of the experiment, how many bacteria were in the culture originally?

6. At any time, the rate of depreciation of an automobile is proportional to the value of the automobile at that time. If a car is worth $4000 new and loses 20% of its value during the fourth year of its life, what was its value after the third year of its life?

7. If only 10% of a radioactive element remains after one year, what is the half-life of the element?

8. Suppose the rate of change of y with respect to x is proportional to y. If the values x_1, x_2, and x_3 of x are consecutive terms in an arithmetic progression, prove that their corresponding y values, y_1, y_2, and y_3, are consecutive terms in a geometric progression.

9. In 1970 the world population was estimated to be 3.6 billion. Find the population in the years 2020 and 2070 assuming a continuous growth rate of (a) 1%, (b) 2%.

10. A tank is filled with 2000 gallons of a brine solution that contains 100 lb of dissolved salt. If fresh water is allowed to enter the tank at the rate of 50 gal/min at time $t = 0$ and is thoroughly mixed with the contents of the tank and the resulting mixture is withdrawn at 50 gal/min: (a) Find the weight of the dissolved salt in the tank when $t = 12$ min. (b) At what time will only 25 lb of salt be in solution in the tank? HINT: Let $x =$ lb of dissolved salt in the tank at time t min. Then

$$\frac{dx}{dt} = \left(\begin{array}{c}\text{lb/min of salt}\\\text{entering the tank}\end{array}\right) - \left(\begin{array}{c}\text{lb/min of salt}\\\text{leaving the tank}\end{array}\right).$$

The first term in the right member of this equation is zero, as fresh water contains no salt. Hence

$$\frac{dx}{dt} = -\left(\begin{array}{c}\text{lb/min of salt}\\\text{leaving the tank}\end{array}\right).$$

10. Trigonometric and Inverse Trigonometric Functions; Hyperbolic Functions

10.1 Review of Trigonometric Functions

It will be presumed here that the student has a certain familiarity with the nature and use of trigonometric functions. Nevertheless, at the beginning of this chapter we will review some of this material, including the fundamental identities, which are important in a treatment of the calculus of the trigonometric functions.

We shall begin by considering the unit circle

$$x^2 + y^2 = 1 \tag{10.1}$$

and the point $Q(1, 0)$ on this circle. It is then assumed that for any number θ there is a unique point $P(x, y)$ on the circle such that the arc length $|\overset{\frown}{QP}|$ on the circle is $|\theta|$ where P is located by the following procedure: If $\theta > 0$, as in Figure 10–1, then $\overset{\frown}{QP}$ is traversed counterclockwise around the circle from Q to P. However, if $\theta < 0$, $\overset{\frown}{QP}$ is traversed in the clockwise direction on the circle. It is also possible to have $|\theta| > 2\pi$, the circumference of the circle (10.1), in which case $\overset{\frown}{QP}$ is traversed after completing at least one full revolution of the circle. The functions cosine and sine, abbreviated cos and sin, respectively, are then defined by stipulating that

$$\cos\theta \text{ is the } x \text{ coordinate of point } P,$$
$$\sin\theta \text{ is the } y \text{ coordinate of point } P. \tag{10.2}$$

It will be noted that the foregoing discussion is essentially geometric in nature. The notions of arc length on a circle and counterclockwise and clockwise directions, which are intuitively evident, have not been described analytically, nor has the well-known formula for the circumference of a circle been verified. Although it is possible to define sine and cosine analytically (see Section 15.4), the analytic definitions of these functions are not intuitively obvious, and therefore for pedagogical reasons we have given the more familiar statements in (10.2).

Several properties of sine and cosine are immediately derived from (10.2). First, note that the point $Q(1, 0)$ is associated with $\theta = 0$, and therefore

$$\cos 0 = 1 \qquad \sin 0 = 0. \tag{10.3}$$

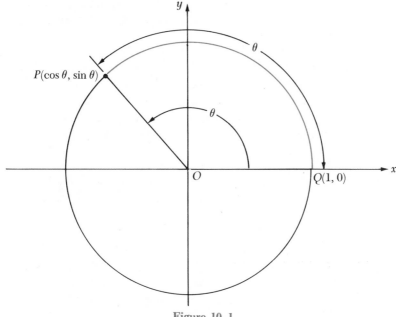

Figure 10-1

Also, since the arc length from $Q(1, 0)$ to either of the points $(0, 1)$ or $(0, -1)$ is $\frac{\pi}{2}$ ($\frac{1}{4}$ of the circumference of the circle (10.1)), we have:

$$\left.\begin{array}{cc} \cos \dfrac{\pi}{2} = 0 & \sin \dfrac{\pi}{2} = 1 \\[2mm] \cos\left(-\dfrac{\pi}{2}\right) = 0 & \sin\left(-\dfrac{\pi}{2}\right) = -1 \end{array}\right\} \tag{10.4}$$

Since the point P in (10.2) is on the circle $x^2 + y^2 = 1$,

$$\cos^2 \theta + \sin^2 \theta = 1. \tag{10.5}$$

Also note from (10.2) that

$\left.\begin{array}{l} \sin \theta \text{ increases from } 0 \text{ to } 1 \\ \cos \theta \text{ decreases from } 1 \text{ to } 0 \end{array}\right\}$ as θ increases from 0 to $\dfrac{\pi}{2}$;

$\left.\begin{array}{l} \sin \theta \text{ decreases from } 1 \text{ to } 0 \\ \cos \theta \text{ decreases from } 0 \text{ to } -1 \end{array}\right\}$ as θ increases from $\dfrac{\pi}{2}$ to π;

$\left.\begin{array}{l} \sin \theta \text{ decreases from } 0 \text{ to } -1 \\ \cos \theta \text{ increases from } -1 \text{ to } 0 \end{array}\right\}$ as θ increases from π to $\dfrac{3\pi}{2}$;

$\left.\begin{array}{l} \sin \theta \text{ increases from } -1 \text{ to } 0 \\ \cos \theta \text{ increases from } 0 \text{ to } 1 \end{array}\right\}$ as θ increases from $\dfrac{3\pi}{2}$ to 2π.

Since the point $P(\cos \theta, \sin \theta)$ can also be located by traversing a "signed" distance of $\theta + 2m\pi$, m being any integer, along the circle from $Q(1, 0)$,

$$\cos (\theta + 2m\pi) = \cos \theta \qquad \sin (\theta + 2m\pi) = \sin \theta. \tag{10.6}$$

The equations (10.6) illustrate the periodic nature of the functions cosine and sine. A function f is said to be *periodic* with *period* p if and only if there is some $p \neq 0$ such that

$$f(x + p) = f(x) \quad \text{for every } x \in \mathcal{D}_f.$$

Thus for any integer $m \neq 0$, $2\pi m$ is a period of cosine and sine because of (10.6).

The graphs of sine and cosine are sketched in Figure 10–2 and Figure 10–3, respectively. These drawings suggest that sine and cosine are everywhere continuous—properties which will be proved in Section 10.2.

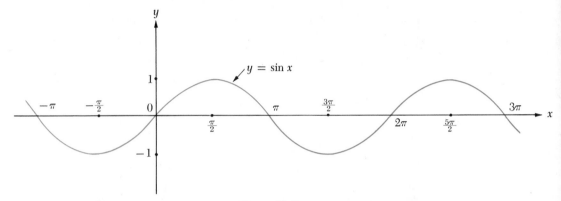

Figure 10-2

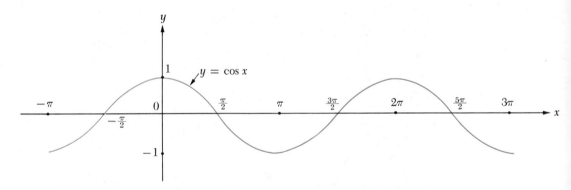

Figure 10-3

We will derive other important properties of sine and cosine as soon as the difference formula

$$\cos (\theta_1 - \theta_2) = \cos \theta_1 \cos \theta_2 + \sin \theta_1 \sin \theta_2 \tag{10.7}$$

has been proved.

PROOF OF (10.7) If $\theta_1 = \theta_2$, then from (10.5) and (10.4)

$$\cos \theta_1 \cos \theta_2 + \sin \theta_1 \sin \theta_2 = \cos^2 \theta_1 + \sin^2 \theta_1$$
$$= 1 = \cos 0$$
$$= \cos (\theta_1 - \theta_2).$$

Next we consider the case where $\theta_1 > \theta_2 > 0$ (Figure 10–4). If the points P_1, P_2, P, and Q on the circle (10.1) have coordinates as given in Figure 10–4, then $|\overgroup{P_2P_1}| = |\overgroup{QP}| = \theta_1 - \theta_2$. Hence by a theorem in elementary geometry, the arcs $\overgroup{QP}$ and $\overgroup{P_1P_2}$ have chords of equal length. Thus

$$|QP| = |P_1P_2|,$$

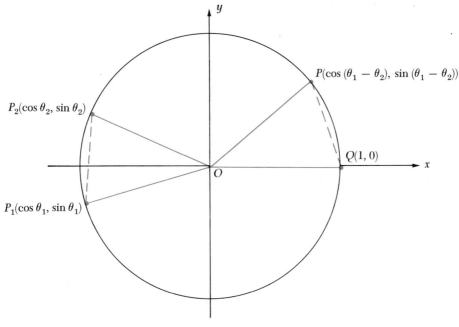

Figure 10–4

and by the two-point distance formula

$$\sqrt{[\cos(\theta_1 - \theta_2) - 1]^2 + \sin^2(\theta_1 - \theta_2)} = \sqrt{(\cos\theta_1 - \cos\theta_2)^2 + (\sin\theta_1 - \sin\theta_2)^2}$$

After squaring each member of this equation and expanding the squares of binomials obtained,

$$\cos^2(\theta_1 - \theta_2) - 2\cos(\theta_1 - \theta_2) + 1 + \sin^2(\theta_1 - \theta_2)$$
$$= \cos^2\theta_1 - 2\cos\theta_1\cos\theta_2 + \cos^2\theta_2 + \sin^2\theta_1$$
$$- 2\sin\theta_1\sin\theta_2 + \sin^2\theta_2.$$

Then by (10.5) each member of this equation can be simplified and

$$2 - 2\cos(\theta_1 - \theta_2) = 2 - 2\cos\theta_1\cos\theta_2 - 2\sin\theta_1\sin\theta_2,$$

from which is obtained

$$\cos(\theta_1 - \theta_2) = \cos\theta_1\cos\theta_2 + \sin\theta_1\sin\theta_2.$$

Similar proofs can be given for the other cases involving θ_1 and θ_2 although the details will be omitted.

Using properties (10.2) through (10.7), analytic proofs can then be given for the following properties of cosine and sine. Some of these proofs of properties in this list will utilize earlier identities that are also in the list.

$$\cos\left(\frac{\pi}{2} - \theta\right) = \sin\theta \tag{10.8}$$

$$\sin\left(\frac{\pi}{2} - \theta\right) = \cos\theta \tag{10.9}$$

$$\cos(-\theta) = \cos\theta \tag{10.10}$$

$$\sin(-\theta) = -\sin\theta \tag{10.11}$$

$$\cos(\theta_1 + \theta_2) = \cos\theta_1 \cos\theta_2 - \sin\theta_1 \sin\theta_2 \tag{10.12}$$

$$\sin(\theta_1 - \theta_2) = \sin\theta_1 \cos\theta_2 - \cos\theta_1 \sin\theta_2 \tag{10.13}$$

$$\sin(\theta_1 + \theta_2) = \sin\theta_1 \cos\theta_2 + \cos\theta_1 \sin\theta_2 \tag{10.14}$$

$$\sin 2\theta = 2\sin\theta \cos\theta \tag{10.15}$$

$$\cos 2\theta = \cos^2\theta - \sin^2\theta \tag{10.16}$$

$$\cos 2\theta = 2\cos^2\theta - 1 \tag{10.17}$$

$$\cos 2\theta = 1 - 2\sin^2\theta \tag{10.18}$$

$$\sin^2\frac{1}{2}\theta = \frac{1 - \cos\theta}{2} \tag{10.19}$$

$$\cos^2\frac{1}{2}\theta = \frac{1 + \cos\theta}{2} \tag{10.20}$$

$$\sin\theta_1 \cos\theta_2 = \tfrac{1}{2}[\sin(\theta_1 - \theta_2) + \sin(\theta_1 + \theta_2)] \tag{10.21}$$

$$\sin\theta_1 \sin\theta_2 = \tfrac{1}{2}[\cos(\theta_1 - \theta_2) - \cos(\theta_1 + \theta_2)] \tag{10.22}$$

$$\cos\theta_1 \cos\theta_2 = \tfrac{1}{2}[\cos(\theta_1 - \theta_2) + \cos(\theta_1 + \theta_2)] \tag{10.23}$$

We shall verify properties (10.10) and (10.13) here and leave the proofs of the others as exercises.

PROOF OF (10.10)   In (10.7) let $\theta_1 = 0$ and $\theta_2 = \theta$. Then

$$\cos(-\theta) = \cos(0 - \theta) = \cos 0 \cos\theta + \sin 0 \sin\theta$$
$$= \cos\theta$$

because of (10.3).

PROOF OF (10.13)   From (10.8)

$$\sin(\theta_1 - \theta_2) = \cos\left[\frac{\pi}{2} - (\theta_1 - \theta_2)\right] = \cos\left[\left(\frac{\pi}{2} - \theta_1\right) + \theta_2\right]$$

and by (10.12)

$$\sin(\theta_1 - \theta_2) = \cos\left(\frac{\pi}{2} - \theta_1\right)\cos\theta_2 - \sin\left(\frac{\pi}{2} - \theta_1\right)\sin\theta_2.$$

Then by (10.8) and (10.9)
$$\sin(\theta_1 - \theta_2) = \sin\theta_1 \cos\theta_2 - \cos\theta_1 \sin\theta_2.$$

The remaining trigonometric functions—tangent, cotangent, secant, cosecant—can now be defined in terms of cosine and sine:

$$\tan\theta = \frac{\sin\theta}{\cos\theta} \qquad \cot\theta = \frac{\cos\theta}{\sin\theta} \tag{10.24}$$

$$\sec\theta = \frac{1}{\cos\theta} \qquad \csc\theta = \frac{1}{\sin\theta} \tag{10.25}$$

From these definitions and some of the properties from (10.5) through (10.20) the following identities can also be obtained.

$$1 + \tan^2\theta = \sec^2\theta \tag{10.26}$$

$$1 + \cot^2\theta = \csc^2\theta \tag{10.27}$$

$$\tan(-\theta) = -\tan\theta \tag{10.28}$$

$$\tan(\theta_1 + \theta_2) = \frac{\tan\theta_1 + \tan\theta_2}{1 - \tan\theta_1 \tan\theta_2} \tag{10.29}$$

$$\tan(\theta_1 - \theta_2) = \frac{\tan\theta_1 - \tan\theta_2}{1 + \tan\theta_1 \tan\theta_2} \tag{10.30}$$

$$\tan 2\theta = \frac{2\tan\theta}{1 - \tan^2\theta} \tag{10.31}$$

$$\tan\frac{1}{2}\theta = \frac{1 - \cos\theta}{\sin\theta} \tag{10.32}$$

$$\tan\frac{1}{2}\theta = \frac{\sin\theta}{1 + \cos\theta} \tag{10.33}$$

PROOF OF (10.26)   From (10.24)

$$1 + \tan^2\theta = 1 + \frac{\sin^2\theta}{\cos^2\theta}$$

and using (10.5) and (10.25)

$$1 + \tan^2\theta = \frac{\cos^2\theta + \sin^2\theta}{\cos^2\theta} = \frac{1}{\cos^2\theta}$$

$$= \sec^2\theta.$$

We shall also comment on the notion of a "trigonometric function of an angle," which is encountered in applications. First, consider the points $P(\cos\theta, \sin\theta)$ and $Q(1, 0)$ on the circle $x^2 + y^2 = 1$ (Figure 10–1). If O is the center of this circle, then the number θ is called a *radian measure* of the central angle QOP. In fact, since P can have an arbitrarily chosen coordinate representation of the form $(\cos(\theta + 2m\pi), \sin(\theta + 2m\pi))$ where m is any integer, any number of the form $\theta + 2m\pi$ is a radian measure of QOP. The relationship between radian measure and the more familiar degree measure for an angle is given by the formula

$$2\pi \text{ rad} = 360°$$

or

$$1 \text{ rad} = \left(\frac{180}{\pi}\right)^{\circ}.$$

An angle that is equal to QOP has the same radian measure as QOP. Thus when one speaks, for example, of "the sine of angle A" where A has radian measure θ, he means simply $\sin\theta$. An analogous meaning applies when referring to any trigonometric function of angle A.

The values of the trigonometric functions at any multiple of $\frac{\pi}{6}$ or $\frac{\pi}{4}$ that is not an integral multiple of $\frac{\pi}{2}$ can be readily obtained by invoking geometric properties associated with an angle of $\frac{\pi}{6}$ radian or $\frac{\pi}{4}$ radian. For example, to obtain the exact values of the trigonometric functions at $\frac{2\pi}{3}$, we can construct the central angle $POQ = \frac{2\pi}{3}$ rad of circle (10.1) as shown in Figure 10–5. In a right triangle with an acute angle of $30° = \frac{\pi}{6}$ rad the length of the leg opposite the angle is one-half the length of the hypotenuse. Thus in Figure 10–5 $|OS| = \frac{1}{2}$ and hence $|PS| = \sqrt{3}/2$. Then P will have coordinates $(-\frac{1}{2}, \sqrt{3}/2)$ and therefore

$$\cos\frac{2\pi}{3} = -\frac{1}{2} \qquad \sin\frac{2\pi}{3} = \frac{\sqrt{3}}{2}$$

Also

$$\tan\frac{2\pi}{3} = -\sqrt{3} \qquad \cot\frac{2\pi}{3} = -\frac{1}{\sqrt{3}}$$

$$\sec\frac{2\pi}{3} = -2 \qquad \csc\frac{2\pi}{3} = \frac{2}{\sqrt{3}}$$

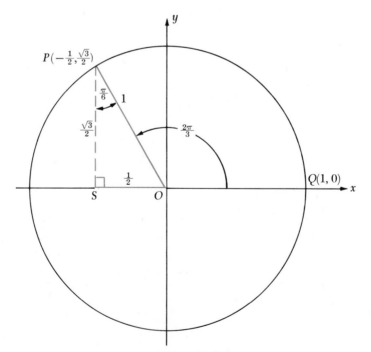

Figure 10–5

Some values of the trigonometric functions at certain multiples of $\frac{\pi}{6}$ and $\frac{\pi}{4}$ are given in the following table:

	$\theta = 0$	$\frac{\pi}{6}$	$\frac{\pi}{4}$	$\frac{\pi}{3}$	$\frac{\pi}{2}$	$\frac{2\pi}{3}$	$\frac{3\pi}{4}$	$\frac{5\pi}{6}$	π	$\frac{3\pi}{2}$	2π
$\sin\theta$	0	$\frac{1}{2}$	$\frac{1}{\sqrt{2}}$	$\frac{\sqrt{3}}{2}$	1	$\frac{\sqrt{3}}{2}$	$\frac{1}{\sqrt{2}}$	$\frac{1}{2}$	0	-1	0
$\cos\theta$	1	$\frac{\sqrt{3}}{2}$	$\frac{1}{\sqrt{2}}$	$\frac{1}{2}$	0	$-\frac{1}{2}$	$-\frac{1}{\sqrt{2}}$	$-\frac{\sqrt{3}}{2}$	-1	0	1
$\tan\theta$	0	$\frac{1}{\sqrt{3}}$	1	$\sqrt{3}$	$\circ$	$-\sqrt{3}$	-1	$-\frac{1}{\sqrt{3}}$	0	$\circ$	0
$\csc\theta$	$\circ$	2	$\sqrt{2}$	$\frac{2}{\sqrt{3}}$	1	$\frac{2}{\sqrt{3}}$	$\sqrt{2}$	2	$\circ$	-1	$\circ$
$\sec\theta$	1	$\frac{2}{\sqrt{3}}$	$\sqrt{2}$	2	$\circ$	-2	$-\sqrt{2}$	$-\frac{2}{\sqrt{3}}$	-1	$\circ$	1
$\cot\theta$	$\circ$	$\sqrt{3}$	1	$\frac{1}{\sqrt{3}}$	0	$-\frac{1}{\sqrt{3}}$	-1	$-\sqrt{3}$	$\circ$	0	$\circ$

Function values indicated by $\circ$ in this table are undefined.

Exercise Set 10.1

1. Obtain the exact values of the trigonometric functions at the following numbers.

(a) $\frac{4\pi}{3}$ (b) $\frac{7\pi}{4}$ (c) $\frac{11\pi}{6}$ (d) $\frac{8\pi}{3}$ (e) $\frac{13\pi}{6}$

In Exercises 2–20 derive the given formula.

2. (10.11) HINT: Express $\cos\left(\frac{\pi}{2}+\theta\right)$ as $\cos\left(\frac{\pi}{2}-(-\theta)\right)$ and $\cos\left(\theta-\left(-\frac{\pi}{2}\right)\right)$.

3. (10.12)   4. (10.14)

5. (10.15)   6. (10.16)

7. (10.17)   8. (10.18)

9. (10.19)   10. (10.20)

11. (10.27)   12. (10.28)

13. (10.29)   14. (10.30)

15. (10.31)   16. (10.32)

17. (10.33)   18. (10.21)

19. (10.22)   20. (10.23)

10.2 Derivatives of Sine and Cosine

In this section the differentiation formulas for sine and cosine will be derived and the graphs of these functions will be analyzed using the techniques that were discussed in Chapter 5. In obtaining the differentiation formula for sine, Theorem 10.2.1(c) below will be utilized. Note that this theorem is not immediately evident since the limit in the denominator is 0. Theorem 10.2.1(a) and (b) is used in the derivation of Theorem 10.2.1(c).

10.2.1   Theorem

(a) $\lim_{\theta \to 0} \sin \theta = 0.$

(b) $\lim_{\theta \to 0} \cos \theta = 1.$

(c) $\lim_{\theta \to 0} \dfrac{\sin \theta}{\theta} = 1.$

PROOF   If $0 < \theta < \frac{\pi}{2}$, the corresponding point $P(\cos \theta, \sin \theta)$ is on the first quadrant portion of the graph of the circle $x^2 + y^2 = 1$ (Figure 10–6). If Q is the point $(1, 0)$ and $T(1, \tan \theta)$ is the intersection point of a perpendicular to the x axis at Q and the extension of OP, then

$$\text{triangular region } OQP \subset \text{sector } OQP \subset \text{triangular region } OQT \quad (10.34)$$

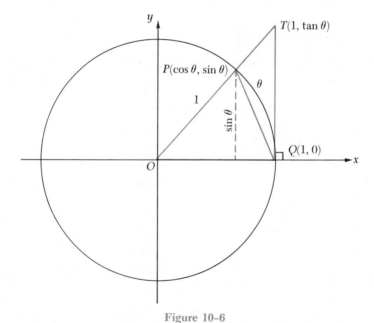

Figure 10–6

The area of triangular regions OQP and OQT are, respectively, $\frac{1}{2} \sin \theta$ and $\frac{1}{2} \tan \theta$. Since the area of a sector of a circle having radius r and central angle θ in radians is $\frac{1}{2}r^2\theta$, as will be proved in Exercise 16, Section 15.4, the area of sector OPQ is $\frac{1}{2}\theta$. Hence from (10.34)

$$\tfrac{1}{2} \sin \theta < \tfrac{1}{2}\theta < \tfrac{1}{2} \tan \theta. \qquad (10.35)$$

Multiplying each member of (10.35) by $2/\sin\theta > 0$, and remembering that $\tan\theta = \sin\theta/\cos\theta$, we have

$$1 < \frac{\theta}{\sin\theta} < \frac{1}{\cos\theta},$$

and hence by Theorem 1.3.10

$$1 > \frac{\sin\theta}{\theta} > \cos\theta > 0 \quad \text{if } 0 < \theta < \frac{\pi}{2} \tag{10.36}$$

It will now be proved that (10.36) is also true if $0 > \theta > -\frac{\pi}{2}$. For such numbers θ, $0 < -\theta < \frac{\pi}{2}$, so by (10.36)

$$1 > \frac{\sin(-\theta)}{-\theta} > \cos(-\theta). \tag{10.37}$$

By (10.10) and (10.11) we have

$$\frac{\sin(-\theta)}{-\theta} = \frac{-\sin\theta}{-\theta} = \frac{\sin\theta}{\theta},$$

and $\cos(-\theta) = \cos\theta$. Thus (10.37) can be written as

$$1 > \frac{\sin\theta}{\theta} > \cos\theta > 0 \quad \text{if } 0 > \theta > -\frac{\pi}{2}.$$

If this inequality is combined with (10.36), one can assert that

$$1 > \frac{\sin\theta}{\theta} > \cos\theta > 0 \quad \text{if } 0 < |\theta| < \frac{\pi}{2}. \tag{10.38}$$

From (10.38)

$$\left|\frac{\sin\theta}{\theta}\right| < 1$$

and

$$|\sin\theta| < |\theta| \quad \text{if } 0 < |\theta| < \frac{\pi}{2}.$$

From this statement, if $\epsilon > 0$ is arbitrarily chosen and $0 < |\theta| < \delta = $ smaller of ϵ and $\frac{\pi}{2}$, then $|\sin\theta| < \epsilon$. Hence by Definition 3.3.1 $\lim_{\theta\to 0}\sin\theta = 0$, and Theorem 10.2.1(a) is proved.

To prove Theorem 10.2.1(b) note from (10.38) that $\cos\theta > 0$ if $0 < |\theta| < \frac{\pi}{2}$, and hence from (10.5)

$$\cos\theta = \sqrt{1 - \sin^2\theta} \quad \text{if } 0 < |\theta| < \frac{\pi}{2}.$$

Then by Theorem 3.8.2 and Theorem 10.2.1(a),

$$\lim_{\theta\to 0}\cos\theta = \lim_{\theta\to 0}\sqrt{1 - \sin^2\theta} = \sqrt{\lim_{\theta\to 0}(1 - \sin^2\theta)}$$

$$= 1. \tag{10.39}$$

Since $\lim_{\theta \to 0} 1 = 1$ and (10.39) is true, from (10.38) we obtain $\lim_{\theta \to 0} \sin \theta / \theta = 1$ by Theorem 3.5.3. Thus Theorem 10.2.1(c) has been proved.

Example 1   Find $\lim_{\theta \to 0} \tan \theta / \theta$.

SOLUTION   Since

$$\frac{\tan \theta}{\theta} = \frac{\sin \theta}{\theta} \frac{1}{\cos \theta},$$

using Theorem 10.2.1(b) and (c)

$$\lim_{\theta \to 0} \frac{\tan \theta}{\theta} = \lim_{\theta \to 0} \frac{\sin \theta}{\theta} \frac{1}{\cos \theta}$$

$$= \left(\lim_{\theta \to 0} \frac{\sin \theta}{\theta} \right) \left(\lim_{\theta \to 0} \frac{1}{\cos \theta} \right)$$

$$= 1 \cdot 1 = 1.$$

The following limit is also used in the derivation of the differentiation formula for sine.

10.2.2   Theorem

$$\lim_{\theta \to 0} \frac{1 - \cos \theta}{\theta} = 0.$$

PROOF   We first write

$$\frac{1 - \cos \theta}{\theta} = \frac{(1 - \cos \theta)(1 + \cos \theta)}{\theta(1 + \cos \theta)} = \frac{1 - \cos^2 \theta}{\theta(1 + \cos \theta)}$$

$$= \frac{\sin^2 \theta}{\theta(1 + \cos \theta)}$$

$$= \frac{\sin \theta}{\theta} \cdot \frac{\sin \theta}{1 + \cos \theta}.$$

Then

$$\lim_{\theta \to 0} \frac{1 - \cos \theta}{\theta} = \lim_{\theta \to 0} \frac{\sin \theta}{\theta} \lim_{\theta \to 0} \frac{\sin \theta}{1 + \cos \theta}$$

$$= 1 \cdot \frac{\lim_{\theta \to 0} \sin \theta}{\lim_{\theta \to 0} (1 + \cos \theta)}$$

$$= 1 \cdot 0 = 0.$$

We are now ready to obtain the differentiation formula for sine.

10.2.3   Theorem

$$D_x \sin x = \cos x.$$

PROOF   Let $f(x) = \sin x$. Then

$$\frac{f(x + h) - f(x)}{h} = \frac{\sin (x + h) - \sin x}{h}$$

and by (10.14)

$$\frac{f(x + h) - f(x)}{h} = \frac{\sin x \cos h + \cos x \sin h - \sin x}{h}$$

$$= \cos x \frac{\sin h}{h} - \sin x \frac{1 - \cos h}{h}.$$

Then by Theorem 10.2.1(c) and Theorem 10.2.2,

$$f'(x) = \lim_{h \to 0} \frac{f(x + h) - f(x)}{h}$$

$$= \lim_{h \to 0} \left(\cos x \frac{\sin h}{h} \right) - \lim_{h \to 0} \left(\sin x \frac{1 - \cos h}{h} \right)$$

$$= \cos x \lim_{h \to 0} \frac{\sin h}{h} - \sin x \lim_{h \to 0} \frac{1 - \cos h}{h}$$

$$= (\cos x) \cdot 1 - (\sin x) \cdot 0$$

$$= \cos x.$$

Theorem 10.2.3 is generalized using the chain rule in the next theorem.

10.2.4   Theorem

If $g'(x)$ exists, then

$$D_x \sin g(x) = \cos g(x) D_x g(x). \tag{10.40}$$

Example 2   Find $f'(x)$ if $f(x) = \sin (x^2 + 1)$.

SOLUTION   From (10.40)

$$D_x \sin (x^2 + 1) = \cos (x^2 + 1) D_x (x^2 + 1)$$
$$= 2x \cos (x^2 + 1).$$

The differentiation formula for cosine can be derived from (10.40).

10.2.5   Theorem

$$D_x \cos x = -\sin x.$$

PROOF   From (10.9)

$$\cos x = \sin \left(\frac{\pi}{2} - x \right),$$

and hence by (10.40)

$$D_x \cos x = D_x \sin\left(\frac{\pi}{2} - x\right)$$

$$= \cos\left(\frac{\pi}{2} - x\right) D_x\left(\frac{\pi}{2} - x\right)$$

$$= -\cos\left(\frac{\pi}{2} - x\right).$$

Then from (10.8)

$$D_x \cos x = -\sin x.$$

The next formula is derived from Theorem 10.2.5 using the chain rule.

10.2.6   Theorem

If g'(x) exists, then

$$D_x \cos g(x) = -\sin g(x) \, D_x \, g(x). \tag{10.41}$$

Example 3   Find $D_x \cos^3 2x$.

SOLUTION   By the general power formula (4.40)

$$D_x \cos^3 2x = 3 \cos^2 2x \, D_x \cos 2x,$$

and by (10.41)

$$D_x \cos^3 2x = 3 \cos^2 2x(-\sin 2x \, D_x \, 2x)$$
$$= -6 \cos^2 2x \sin 2x.$$

Because of Theorem 10.2.3 and Theorem 10.2.5 sine and cosine are everywhere differentiable and hence everywhere continuous. Thus we can solve the familiar kinds of problems involving limits and derivatives for functions that are formed from sine and cosine.

Example 4   Find the relative extrema of the function f where $f(x) = 2 \sin x + \sin 2x$ (Figure 10–7).

SOLUTION   Since $f'(x)$ exists for every x, any relative extrema of f will occur at numbers x where

$$f'(x) = 2 \cos x + 2 \cos 2x = 0.$$

Solving this equation we obtain successively:

$$2 \cos x + 2(2 \cos^2 x - 1) = 0$$
$$2 \cos^2 x + \cos x - 1 = 0$$

$$(2 \cos x - 1)(\cos x + 1) = 0$$
$$\cos x = \tfrac{1}{2} \qquad \cos x = -1$$

The solutions of these equations, which are the critical numbers of f, are $x = \frac{\pi}{3} + 2m\pi$, $x = -\frac{\pi}{3} + 2m\pi$, and $x = \pi + 2m\pi$, where m is any integer. The functional values at these critical numbers may be relative extrema. We shall investigate these critical numbers using the second derivative test, and therefore we compute

$$f''(x) = -2 \sin x - 4 \sin 2x.$$

Then

$$f''\left(\frac{\pi}{3} + 2m\pi\right) = -2\left(\frac{\sqrt{3}}{2}\right) - 4\left(\frac{\sqrt{3}}{2}\right) = -3\sqrt{3}$$

$$f''\left(-\frac{\pi}{3} + 2m\pi\right) = -2\left(-\frac{\sqrt{3}}{2}\right) - 4\left(-\frac{\sqrt{3}}{2}\right) = 3\sqrt{3}$$

$$f''(\pi + 2m\pi) = 0.$$

Hence by the second derivative test

$$f\left(\frac{\pi}{3} + 2m\pi\right) = \frac{3\sqrt{3}}{2} \text{ is a relative maximum;}$$

$$f\left(-\frac{\pi}{3} + 2m\pi\right) = -\frac{3\sqrt{3}}{2} \text{ is a relative minimum.}$$

However, the second derivative test fails to disclose whether $f(\pi + 2m\pi) = 0$ is a relative extremum. But $f'(x)$ can be expressed in the form

$$f'(x) = 2 \cos x + 2(2 \cos^2 x - 1)$$
$$= 2(2 \cos^2 x + \cos x - 1)$$
$$= 2(2 \cos x - 1)(\cos x + 1),$$

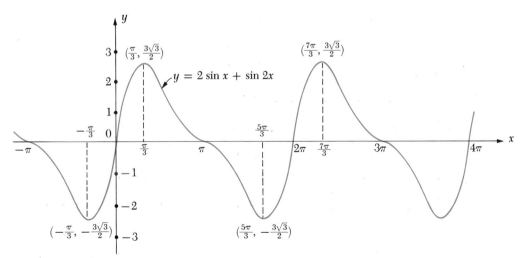

Figure 10-7

and for $x \in \left(\frac{\pi}{2} + 2m\pi, \pi + 2m\pi \right) \cup \left(\pi + 2m\pi, \frac{3\pi}{2} + 2m\pi \right)$

$$2 \cos x - 1 < 0 \qquad \text{and} \qquad \cos x + 1 > 0.$$

Thus, there is no sign change in $f'(x)$ at $x = \pi + 2m\pi$, and by the first derivative test $f(\pi + 2m\pi) = 0$ is not a relative extremum.

Exercise Set 10.2

In Exercises 1–12 obtain the derivative of the given function.

1. $f(x) = \cos (x^2 + 3x)$

2. $g(x) = \sin \sqrt{x^3 + 1}$

3. $f(x) = \sin^2 2x$

4. $F(x) = \dfrac{1 - \cos x}{\sin x}$

5. $g(y) = 2ye^{-\cos y}$

6. $\psi(y) = \sin^3 e^{2y}$

7. $f(x) = x^3 \sin \dfrac{1}{x}$

8. $f(x) = (1 + \sin x)^3$

9. $\phi(x) = (1 + \cos^2 2x)^{3/2}$

10. $f(x) = \ln (\cos x - \sin x)$

11. $G(x) = \sin^3 x \cos^4 x$

12. $F(x) = x^{\sin x}$

In Exercises 13 and 14 find dy/dx by implicit differentiation.

13. $\sin (x - y) = y \cos x$

14. $\cos xy = x \cos (x - y)$

In Exercises 15–22 obtain the given limit, or show that it fails to exist.

15. $\displaystyle\lim_{x \to 0} \dfrac{\cos^2 x - 1}{\sec x - 1}$

16. $\displaystyle\lim_{x \to \pi/2} \dfrac{\cos x}{1 - \sin x}$

17. $\displaystyle\lim_{x \to 0} \dfrac{1 - \cos x}{x^2}$

18. $\displaystyle\lim_{x \to 0} \dfrac{\sin^2 (x + \frac{\pi}{6}) - \sin^2 \frac{\pi}{6}}{x}$

19. $\displaystyle\lim_{x \to 0} \dfrac{\sin 2x}{3x}$

20. $\displaystyle\lim_{x \to 0} \dfrac{\sin 3x}{4x}$

21. $\displaystyle\lim_{x \to 0} \dfrac{2x}{\tan 5x}$

22. $\displaystyle\lim_{x \to 1} \dfrac{\sin (x - 1)}{x^2 + x - 2}$

23. Obtain $f'(x)$ if $f(x) = \cos x$, using Definition 4.3.1.

In Exercises 24–27 obtain the relative extrema of the given functions.

24. $f(x) = \cos^2 x + \sin x$

25. $f(x) = \sin^3 x \cos x$

26. $f(x) = \cos 2x + 2 \sin x$

27. $f(x) = 4 \cos x - \cos 2x$

28. If $f(x) = \begin{cases} x \sin (1/x) & \text{if } x \neq 0 \\ 0 & \text{if } x = 0 \end{cases}$ is f continuous at 0? Is f differentiable at 0?

29. If $f(x) = \begin{cases} x^2 \sin (1/x) & \text{if } x \neq 0 \\ 0 & \text{if } x = 0 \end{cases}$ is f continuous at 0? Is f differentiable at 0?

10.3   Derivatives of the Other Trigonometric Functions

We recall that the functions tangent, cotangent, secant, and cosecant were defined in Section 10.1 in terms of sine and cosine. Since

$$\tan x = \frac{\sin x}{\cos x} \qquad \sec x = \frac{1}{\cos x},$$

the functions tangent and secant are continuous on their domains, the set $\left\{x : x \neq \dfrac{2n + 1}{2} \pi \text{ where } n \text{ is any integer}\right\}$. Also, since

$$\cot x = \frac{\cos x}{\sin x} \qquad \csc x = \frac{1}{\sin x},$$

the functions cotangent and cosecant are continuous on their domains, the set $\{x : x \neq n\pi \text{ where } n \text{ is any integer}\}$.

The next theorem gives the differentiation formulas for tangent, cotangent, secant, and cosecant.

10.3.1   Theorem

(a) $D_x \tan x = \sec^2 x$ (10.42)

(b) $D_x \cot x = -\csc^2 x$ (10.43)

(c) $D_x \sec x = \sec x \tan x$ (10.44)

(d) $D_x \csc x = -\csc x \cot x$ (10.45)

PROOF   We will obtain the formulas (10.42) and (10.45) and leave the derivation of (10.43) and (10.44) as exercises. First,

$$D_x \tan x = D_x \frac{\sin x}{\cos x} = \frac{\cos x \, D_x \sin x - \sin x \, D_x \cos x}{\cos^2 x}$$

$$= \frac{\cos x \cos x - \sin x(-\sin x)}{\cos^2 x}$$

$$= \frac{\cos^2 x + \sin^2 x}{\cos^2 x} = \frac{1}{\cos^2 x}$$

$$= \sec^2 x.$$

Also

$$D_x \csc x = D_x (\sin x)^{-1} = -1(\sin x)^{-2} D_x \sin x = -\frac{\cos x}{\sin^2 x}$$

$$= -\frac{1}{\sin x} \frac{\cos x}{\sin x}$$

$$= -\csc x \cot x.$$

From Theorem 10.3.1 if $D_x g(x)$ exists

$$D_x \tan g(x) = \sec^2 g(x) \, D_x g(x)$$ (10.46)

$$D_x \cot g(x) = -\csc^2 g(x) \, D_x g(x)$$ (10.47)

$$D_x \sec g(x) = \sec g(x) \tan g(x) D_x g(x) \tag{10.48}$$

$$D_x \csc g(x) = -\csc g(x) \cot g(x) D_x g(x). \tag{10.49}$$

Example 1   Find $D_x \tan^3 2x$.

SOLUTION   From (10.46) and the general power formula for derivatives

$$D_x \tan^3 2x = 3 \tan^2 2x \, D_x \tan 2x$$
$$= 3 \tan^2 2x \sec^2 2x \, D_x 2x$$
$$= 6 \sec^2 2x \tan^2 2x.$$

Example 2   Find $f'(x)$ if $f(x) = \ln (\csc 3x - \cot 3x)$.

SOLUTION   From (10.47) and (10.49)

$$f'(x) = \frac{D_x (\csc 3x - \cot 3x)}{\csc 3x - \cot 3x} = \frac{D_x \csc 3x - D_x \cot 3x}{\csc 3x - \cot 3x}$$

$$= \frac{-3 \csc 3x \cot 3x + 3 \csc^2 3x}{\csc 3x - \cot 3x}$$

$$= \frac{3 \csc 3x(\csc 3x - \cot 3x)}{\csc 3x - \cot 3x}$$

$$= 3 \csc 3x.$$

We will discuss the graphs of these trigonometric functions, beginning with the tangent function. Since $D_x \tan x = \sec^2 x$, tan is continuous and increasing on $(-\frac{\pi}{2}, \frac{\pi}{2})$. Also since

$$D_x^2 \tan x = D_x \sec^2 x = 2 \sec x \, D_x \sec x = 2 \sec^2 x \tan x,$$

$D_x^2 \tan x > 0$ if $x \in (0, \frac{\pi}{2})$ and $D_x^2 \tan x < 0$ if $x \in (-\frac{\pi}{2}, 0)$. Thus when $x \in (0, \frac{\pi}{2})$ the graph of $y = \tan x$ is concave upward, and when $x \in (-\frac{\pi}{2}, 0)$ the graph is concave downward. The point $(0, 0)$ is an inflection point of the graph. Since

$$\lim_{x \to -\pi/2^+} \frac{1}{\tan x} = \lim_{x \to -\pi/2} \frac{\cos x}{\sin x} = 0,$$

and $\tan x < 0$ when $x \in (-\frac{\pi}{2}, 0)$, we have $\lim_{x \to -\pi/2^+} \tan x = -\infty$. Thus the line $x = -\frac{\pi}{2}$ is a vertical asymptote of the graph of tan. Also,

$$\lim_{x \to -\pi/2^-} \frac{1}{\tan x} = 0$$

and since $\tan x > 0$ when $x \in (0, \frac{\pi}{2})$, $\lim_{x \to \pi/2^-} \tan x = +\infty$. Thus the line $x = \frac{\pi}{2}$ is also a vertical asymptote of the graph. Since $\tan \pi = 0$, by (10.29)

$$\tan (x + \pi) = \tan x,$$

and hence π is a period of the tangent function. Thus the behavior of tan in $(-\frac{\pi}{2}, \frac{\pi}{2})$ is reproduced in $(\frac{\pi}{2}, \frac{3\pi}{2})$ and also in any interval $\left(\dfrac{2m-1}{2}\pi, \dfrac{2m+1}{2}\pi\right)$, where m is an integer (Figure 10–8).

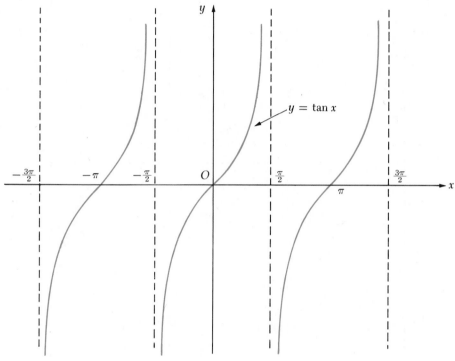

Figure 10–8

The graph of the cotangent function can be sketched using the identity

$$\cot x = -\tan\left(x + \frac{\pi}{2}\right),\tag{10.50}$$

which can be derived from (10.24) (Exercise 16, Exercise Set 10.3). The sketch of the graph is shown in Figure 10–9.

Since $\sec x = 1/\cos x$, the secant function is defined at every number except $\pm\frac{\pi}{2}$, $\pm\frac{3\pi}{2}$, $\pm\frac{5\pi}{2}$, and so on. Also, since

$$\sec(x + 2\pi) = \frac{1}{\cos(x + 2\pi)} = \frac{1}{\cos x} = \sec x,$$

2π is a period of sec. Thus it suffices to study the function in the interval $(-\frac{\pi}{2}, \frac{3\pi}{2})$, since its behavior in any interval $\left(\dfrac{4m-1}{2}\pi, \dfrac{4m+3}{2}\pi\right)$, where m is any integer, would be the same. If $x \in (-\frac{\pi}{2}, \frac{\pi}{2})$ then $\sec x \geq 1$ and if $x \in (\frac{\pi}{2}, \frac{3\pi}{2})$, $\sec x \leq -1$.

Since $D_x \sec x = \sec x \tan x$, sec is differentiable and hence continuous at

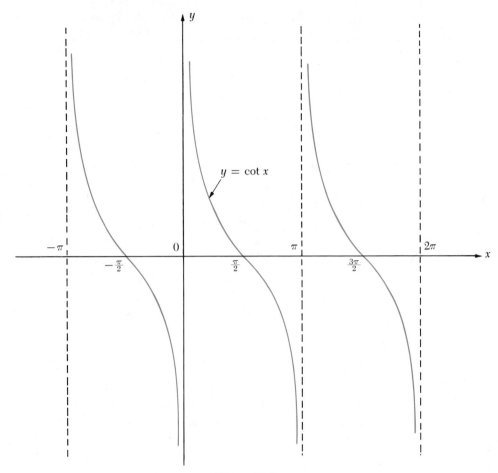

$y = \cot x$

Figure 10-9

every number in its domain. The function has critical numbers at 0 and π since $\sec'(0) = \sec'(\pi) = 0$. Differentiating again,

$$\sec''(x) = \sec x \, D_x \tan x + \tan x \, D_x \sec x$$
$$= \sec^3 x + \sec x \tan^2 x$$
$$= \sec x(\sec^2 x + \tan^2 x), \qquad (10.51)$$

and by the second derivative test, $\sec 0 = 1$ is a relative minimum and $\sec \pi = -1$ is a relative maximum. From (10.51) the graph of sec is concave upward at points $(x, \sec x)$ if $x \in (-\frac{\pi}{2}, \frac{\pi}{2})$ and concave downward at $(x, \sec x)$ if $x \in (\frac{\pi}{2}, \frac{3\pi}{2})$.

We also note that $\lim_{x \to -\pi/2+} \sec x = +\infty$, $\lim_{x \to \pi/2-} \sec x = +\infty$, $\lim_{x \to \pi/2+} \sec x = -\infty$, and $\lim_{x \to 3\pi/2-} \sec x = -\infty$. Hence the lines $x = -\frac{\pi}{2}$, $x = \frac{\pi}{2}$, and $x = \frac{3\pi}{2}$, and so on, are vertical asymptotes.

A sketch of the graph of sec is given in Figure 10–10.

It can be proved (Exercise 17, Exercise Set 10.3) that

$$\csc x = \sec\left(x - \frac{\pi}{2}\right), \qquad (10.52)$$

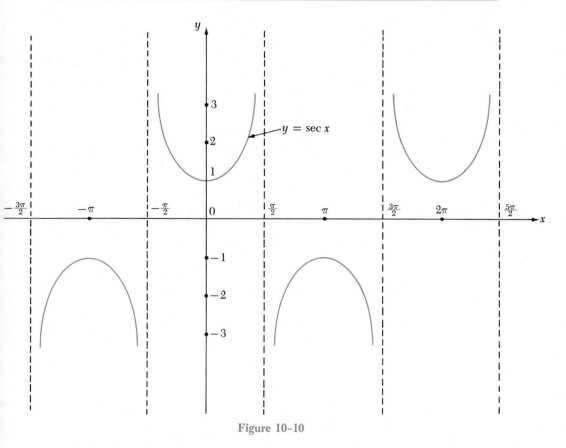

Figure 10–10

and hence the graph of the cosecant function can be obtained by sliding the graph of the secant function $\frac{\pi}{2}$ units to the right. A sketch of the graph of csc is shown in Figure 10–11.

Exercise Set 10.3

In Exercises 1–13 obtain the derivative of the given function.

1. $f(x) = \tan x^2$

2. $g(x) = \sec \sqrt{x}$

3. $F(x) = \ln \cot x$

4. $\phi(x) = e^{-\csc 2x}$

5. $f(x) = \sqrt{\sec 3x}$

6. $G(\theta) = \cot^3(\sin \theta)$

7. $\psi(t) = \csc \dfrac{t}{t+1}$

8. $F(t) = \dfrac{\sec t - \tan t}{\sec t + \tan t}$

9. $f(x) = \csc x \cot x$

10. $f(x) = (2 + \csc^2 x)^{1/2}$

11. $G(x) = \sec^2 x \tan^2 x$

12. $g(x) = (\tan x)^{2x}$

13. $h(x) = \ln |\csc x - \cot x|$

In Exercises 14 and 15 find dy/dx by implicit differentiation.

14. $y = \cot xy$

15. $\sec(x + y) - \tan(x - y) = 1$

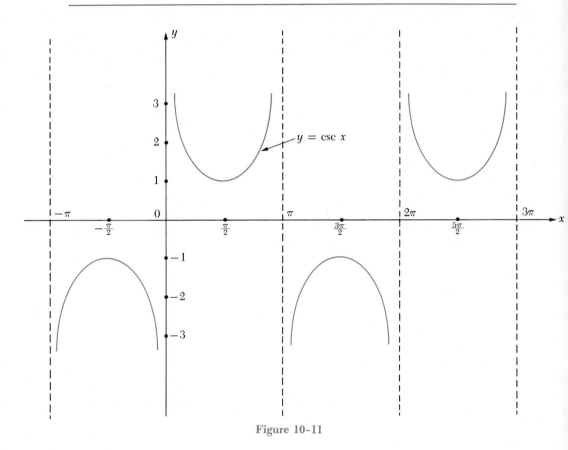

Figure 10-11

16. Prove (10.50).

17. Prove (10.52).

18. Derive differentiation formula (10.43).

19. Derive differentiation formula (10.44).

20. Obtain $D_x \tan x$ using the definition of the derivative (Definition 4.3.1).

21. Find $\lim_{x \to \pi/2} (\sec x - \tan x)$.

22. Find the minimum value for $f(x) = \sec x + \csc x$ on $(0, \frac{\pi}{2})$.

23. Find the relative extrema for $f(x) = 3 \tan x - 4x$ on $[0, \pi]$.

10.4 Applications of Derivatives of Trigonometric Functions

We will consider some applications of derivatives of trigonometric functions. The first problem involves related rates and is concerned with finding the rate of change of the measure of an angle.

Example 1 A man 6 ft tall walks on level ground toward a flagpole of height 78 ft at the rate of 5 ft/sec. Find the rate of change of the angle of elevation of the top of the flagpole from the top of the man's head when he is 48 ft from the pole.

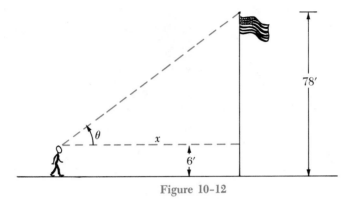

Figure 10-12

SOLUTION   The important quantities that change with respect to time are the distance x between the man and the pole and the angle of elevation θ (Figure 10–12). These variables are related by the equation

$$\cot \theta = \frac{x}{72}. \tag{10.53}$$

Considering that θ and x are differentiable functions of time t, we differentiate in (10.53), and get

$$-\csc^2 \theta \, \frac{d\theta}{dt} = \frac{1}{72} \frac{dx}{dt}. \tag{10.54}$$

When $x = 48$, then $\cot \theta = \frac{48}{72} = \frac{2}{3}$, and hence $\csc^2 \theta = 1 + \cot^2 \theta = \frac{13}{9}$. Also $dx/dt = -5$, since the man is walking toward the pole at 5 ft/sec. The substitution of these values in (10.54) gives

$$-\frac{13}{9} \frac{d\theta}{dt} = \frac{1}{72}(-5),$$

from which we obtain

$$\frac{d\theta}{dt} = \frac{5}{104} \text{ rad/sec.}$$

Thus the angle is increasing at $\frac{5}{104}$ rad/sec when he is 48 ft from the flagpole.

In a maximum-minimum problem, which is geometric in nature, it may be useful to represent the quantity that is to have an extreme value as a function of the measure of some angle.

Example 2 Find the dimensions of the right circular cone of maximum volume having a given slant height L (Figure 10–13).

SOLUTION   If we let the vertex angle of the cone be 2θ, then the altitude and radius of the cone are, respectively,

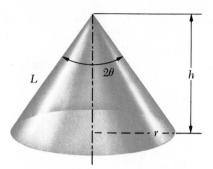

Figure 10-13

$$h = L \cos \theta \qquad r = L \sin \theta$$

where $0° < \theta < 90°$. The volume of the cone is therefore

$$V = \frac{\pi}{3} r^2 h = \frac{\pi}{3} L^3 \sin^2 \theta \cos \theta.$$

Differentiation with respect to θ then gives

$$\frac{dV}{d\theta} = \frac{\pi}{3} L^3 [-\sin^3 \theta + 2 \sin \theta \cos^2 \theta].$$

Letting $dV/d\theta = 0$ and dividing out the factor of $\frac{\pi}{3} L^3$, we have

$$-\sin^3 \theta + 2 \sin \theta \cos^2 \theta = 0. \tag{10.55}$$

If the substitution $\cos^2 \theta = 1 - \sin^2 \theta$ is made in (10.55), one obtains

$$-\sin^3 \theta + 2 \sin \theta (1 - \sin^2 \theta) = 0$$
$$-3 \sin^3 \theta + 2 \sin \theta = 0$$
$$-\sin \theta (3 \sin^2 \theta - 2) = 0. \tag{10.56}$$

For Equation (10.56) we obtain solutions given by $\sin \theta = 0$, $\sqrt{6}/3$, and $-\sqrt{6}/3$. However, since θ is restricted to $(0, \frac{\pi}{2})$, only those solutions of (10.56) for which $\sin \theta > 0$ can be considered. Thus, the only solution of interest, $\sin \theta = \sqrt{6}/3$, is associated with the maximum value of V. Then $\cos \theta = \sqrt{3}/3$ and the required dimensions are

$$r = \frac{\sqrt{6}}{3} L \qquad h = \frac{\sqrt{3}}{3} L.$$

We next consider a physical problem that leads to a differential equation having solutions involving sines and cosines. In seeking to solve this problem, we shall utilize Newton's second law from physics. According to this law, if ΣF_i is the resultant of all the forces acting on a body of mass m in a certain direction and a is the acceleration of the body in that direction, then

$$\sum F_i = ma. \tag{10.57}$$

If the body has no acceleration in that direction, then

$$\sum F_i = 0.$$

Suppose that a weightless spring is fixed at one end and is allowed to hang down (Figure 10–14(a)). If a weight of mass m is attached to its lower end, the spring is stretched an additional distance y and will be at rest (Figure 10–14(b)).

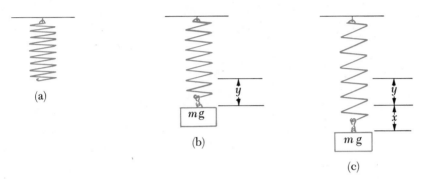

Figure 10–14

We shall take the downward direction as positive, and hence $y > 0$. With the addition of the weight, the forces acting on the spring are mg, which is due to gravity, and $-ky$, where $k > 0$, which, according to Hooke's law, is the force tending to restore the spring to its natural length. This restoring force is considered negative since it acts in the upward direction. Since the spring is at rest, the resultant force in the vertical direction is

$$mg - ky = 0. \tag{10.58}$$

If the weight is pulled down an additional distance and then released, we let x denote the distance of the weight below the equilibrium position illustrated in Figure 10–14(b) at time t after the weight is released. Thus at time t the spring is extended a distance $y + x$ beyond its natural length (Figure 10–14(a)), and therefore by Hooke's law the restoring force on the spring is $-k(y + x)$. At time t the only forces on the spring are $-k(y + x)$ and mg. Therefore from (10.57), the resultant force in the vertical direction is

$$m\frac{d^2x}{dt^2} = mg - k(y + x)$$

$$= mg - ky - kx.$$

From (10.58), this differential equation simplifies to

$$m\frac{d^2x}{dt^2} = -kx \tag{10.59}$$

or

$$\frac{d^2x}{dt^2} + \frac{k}{m}x = 0. \tag{10.60}$$

Then if $\omega = \sqrt{k/m}$, (10.60) can be rewritten

$$\frac{d^2x}{dt^2} + \omega^2 x = 0. \tag{10.61}$$

The differential equation (10.61) is satisfied by $x = \sin \omega t$ and $x = \cos \omega t$. In fact, it will be shown (see Exercise 13, Exercise Set 10.4 and Exercise 35, Exercise Set 10.5) that every solution of (10.61) is of the form

$$x = A \cos \omega t + B \sin \omega t. \tag{10.62}$$

The motion of the weight described by (10.62) or (10.61) is periodic with period T, given by

$$T = \frac{2\pi}{\omega}$$

and is termed *simple harmonic motion*. From (10.62) the velocity of the weight at any time t is

$$\frac{dx}{dt} = -A\omega \sin \omega t + B\omega \cos \omega t.$$

We leave as an exercise the proof that $\sqrt{A^2 + B^2}$ is the maximum displacement of the weight, the *amplitude* of the motion.

Example 3   Suppose that the spring in the preceding discussion is stretched a distance of 6 in. when a 5 lb weight is attached to the spring. If the weight is pulled down an additional 2 in. and then given a downward velocity of 3 ft/sec, find the equation of motion of the weight.

SOLUTION   From (10.58) if $y = 6$ in. $= \frac{1}{2}$ ft,

$$mg = 5 = k \cdot \tfrac{1}{2}$$

and hence

$$k = 10 \text{ lb/ft.}$$

Since the 5 lb weight has mass $m = \dfrac{5}{32} \dfrac{\text{lb}}{\text{ft/sec}^2}$,

$$\omega = \sqrt{\frac{k}{m}} = \sqrt{\frac{10}{\frac{5}{32}}} = 8 \text{ sec}^{-1},$$

and from (10.62) the equation of motion is of the form

$$x = A \cos 8t + B \sin 8t. \tag{10.63}$$

Since $x = 2$ in. $= \frac{1}{6}$ ft when $t = 0$, from (10.63) $A = \frac{1}{6}$. Differentiating in (10.63) gives

$$\frac{dx}{dt} = -8A \sin 8t + 8B \cos 8t. \tag{10.64}$$

Since $dx/dt = 3$ ft/sec when $t = 0$, from (10.64) $8B = 3$ and $B = \frac{3}{8}$. The substitution in (10.63) of the values obtained for A and B gives the required equation of motion

$$x = \tfrac{1}{6} \cos 8t + \tfrac{3}{8} \sin 8t.$$

Exercise Set 10.4

1. A man sitting on a pier 9 ft above the water pulls in on a rope attached to a boat at the waterline at the rate of 2 ft/sec. At what rate is the angle between the rope and the water changing when 15 ft of rope are out?

2. A balloon ascends vertically at the rate of 25 ft/sec from a point on level ground 200 ft from an observer who is also on the ground. Find the rate of change of the angle of elevation of the balloon from the observer when the balloon has risen to a height of 800 ft.

3. An airplane flies at a height of $\frac{1}{2}$ mile above the ground at a speed of 500 mi/hr toward an observer on the ground. Find the rate of change of the angle of elevation of the plane from the observer when the angle of elevation is 45°.

4. A triangle has two sides with lengths 8 in. and 10 in. If the included angle between these sides is increasing at 3 deg/sec, find the rate of increase of (a) the length of the third side and (b) the area of the triangle when the included angle is 60°.

5. A crank OP of length 6 in. revolves counterclockwise about point O at the rate of 20 rev/sec (Figure 10–15). PQ, a connecting rod of length 15 in., is attached to a piston at Q, which is constrained to move back and forth. Find the velocity of Q when $\theta = 120°$.

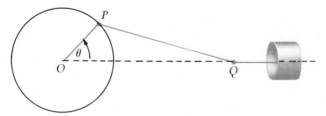

Figure 10–15

In Exercises 6–10 use the measure of an angle as the independent variable.

6. Find the dimensions of the rectangle of maximum area having a diagonal of given length L.

7. A corridor of width 5 ft intersects a corridor of width 10 ft at right angles. Find the length of the longest board that can be carried horizontally from the smaller corridor to the larger corridor. HINT: See Figure 10–16.

8. A ladder of length 27 ft has its foot on level ground and rests against the top of a fence 8 ft high. What should be the angle between the ladder and ground if the horizontal projection of the ladder on the other side of the fence is to be a maximum?

9. Rework Exercise 18, Exercise Set 6.2. HINT: Express the distance in terms of two angles. Use the implicit function method for finding absolute extrema.

10. Find the dimensions of the right circular cone of minimum volume in which a sphere of radius a can be inscribed.

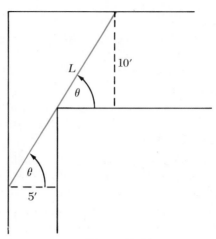

Figure 10–16

11. Suppose that a weightless spring is suspended from a fixed point as in Example 3. Suppose also that the spring constant is $k = 1.5$ gm/sec^2 and the mass of a weight attached to the spring is $m = 6$ gm. If the spring is compressed 2 cm and then released with no initial velocity, find the period and amplitude of the motion and also the velocity $\frac{5\pi}{6}$ sec after the weight was released.

12. Show that the maximum displacement of the weight in the system having the equation of motion (10.62) is $\sqrt{A^2 + B^2}$.

13. We observed that $\sin \omega t$ is a solution of the differential equation (10.61). Suppose $f(t)$ is any other solution of (10.61). There is certainly some function v such that $f(t) = v(t) \sin \omega t$ for all t for which $\sin \omega t \neq 0$. (In fact $v(t) = f(t)/\sin \omega t$). (a) Show that $2\omega v'(t) \cos \omega t + v''(t) \sin \omega t = 0$. (b) Show that $D_t[v'(t) \sin^2 \omega t] = 0$. HINT: Multiply each member of the equation in (a) by $\sin \omega t$.

10.5 Integrals of Trigonometric Functions

From the differentiation formulas

$$D_x \cos g(x) = -\sin g(x) D_x g(x)$$
$$D_x \sin g(x) = \cos g(x) D_x g(x)$$
$$D_x \tan g(x) = \sec^2 g(x) D_x g(x)$$
$$D_x \cot g(x) = -\csc^2 g(x) D_x g(x)$$
$$D_x \sec g(x) = \sec g(x) \tan g(x) D_x g(x)$$
$$D_x \csc g(x) = -\csc g(x) \cot g(x) D_x g(x)$$

respectively, the following integration formulas are immediately derived.

$$\int \sin g(x)g'(x)\,dx = -\cos g(x) + C \tag{10.65}$$

$$\int \cos g(x)g'(x)\,dx = \sin g(x) + C \tag{10.66}$$

$$\int \sec^2 g(x)g'(x)\,dx = \tan g(x) + C \tag{10.67}$$

$$\int \csc^2 g(x)g'(x)\,dx = -\cot g(x) + C \tag{10.68}$$

$$\int \sec g(x)\tan g(x)g'(x)\,dx = \sec g(x) + C \tag{10.69}$$

$$\int \csc g(x)\cot g(x)g'(x)\,dx = -\csc g(x) + C \tag{10.70}$$

Example 1 Obtain $\int x \sin x^2\,dx$.

SOLUTION Using (10.65), we let $u = x^2$. Then $du = 2x\,dx$ and $x\,dx = \frac{1}{2}\,du$, so

$$\int x \sin x^2\,dx = \int \tfrac{1}{2}\sin u\,du = -\tfrac{1}{2}\cos u + C = -\tfrac{1}{2}\cos x^2 + C.$$

Example 2 Obtain $\int \sec^2 3x\,dx$.

SOLUTION From (10.67), letting $u = 3x$, we have $du = 3\,dx$, and

$$\int \sec^2 3x\,dx = \int \tfrac{1}{3}\sec^2 u\,du = \tfrac{1}{3}\tan u + C$$
$$= \tfrac{1}{3}\tan 3x + C.$$

Example 3 Obtain $\int \frac{1}{\sqrt{x}}\csc \sqrt{x}\cot \sqrt{x}\,dx$.

SOLUTION Letting $u = \sqrt{x}$, $du = \tfrac{1}{2}\sqrt{x}\,dx$, and hence from (10.70)

$$\int \frac{1}{\sqrt{x}}\csc \sqrt{x}\cot \sqrt{x}\,dx = \int 2\csc u\cot u\,du = -2\csc u + C$$
$$= -2\csc \sqrt{x} + C.$$

The integrands of the indefinite integrals in Examples 1–3 are of the form $\int D_x(f(g(x)))\,dx$, where f is some trigonometric function. We next consider some integrals of the form

$$\int (f(g(x)))^r\, D_x f(g(x))\,dx \quad r \text{ any rational number}.$$

These integrals will be evaluated using the general power formula (Theorem 7.1.11).

Example 4   Obtain $\int \sin^5 \frac{1}{2}x \cos \frac{1}{2}x \, dx$.

SOLUTION   If $u = \sin \frac{1}{2}x$, then $du = \frac{1}{2} \cos \frac{1}{2}x \, dx$, and $\cos \frac{1}{2}x \, dx = 2 \, du$.
Then

$$\int \sin^5 \frac{1}{2}x \cos \frac{1}{2}x \, dx = \int 2u^5 \, du = \frac{u^6}{3} + C$$

$$= \frac{\sin^6 \frac{1}{2}x}{3} + C.$$

Example 5   Obtain $\int \dfrac{\sec^3 (\ln x) \tan (\ln x) \, dx}{x}$.

SOLUTION   If $u = \sec (\ln x)$, then $du = \dfrac{\sec (\ln x) \tan (\ln x)}{x} \, dx$, and
hence

$$\int \frac{\sec^3 (\ln x) \tan (\ln x) \, dx}{x} = \int u^2 \, du = \frac{u^3}{3} + C$$

$$= \frac{\sec^3 (\ln x)}{3} + C.$$

Example 6   Obtain $\int \cot^3 2x \csc^2 2x \, dx$.

SOLUTION   If $u = \cot 2x$, then $du = -2 \csc^2 2x \, dx$ and hence
$\csc^2 2x \, dx = -\frac{1}{2} \, du$. Then

$$\int \cot^3 2x \csc^2 2x \, dx = \int -\tfrac{1}{2}u^3 \, du = -\tfrac{1}{8}u^4 + C$$

$$= -\frac{\cot^4 2x}{8} + C.$$

Besides (10.65) and (10.66) we also have formulas for evaluating
$\int \tan g(x)g'(x) \, dx$, $\int \cot g(x)g'(x) \, dx$, $\int \sec g(x)g'(x) \, dx$, and $\int \csc g(x)g'(x) \, dx$. The first
of these formulas follows.

$$\int \tan g(x)g'(x) \, dx = -\ln |\cos g(x)| + C. \tag{10.71}$$

PROOF $\int \tan g(x)g'(x) \, dx = \int \dfrac{\sin g(x)g'(x) \, dx}{\cos g(x)}$,

and if we let $u = \cos g(x)$, then $du = -\sin g(x)g'(x) \, dx$. Hence

$$\int \tan g(x)g'(x) \, dx = \int -\frac{du}{u} = -\ln |u| + C$$

$$= -\ln |\cos g(x)| + C.$$

Since $|\sec g(x)| = \dfrac{1}{|\cos g(x)|}$,

$$\ln |\sec g(x)| = \ln 1 - \ln |\cos g(x)| = -\ln |\cos g(x)|,$$

and (10.71) can be rewritten

$$\int \tan g(x)g'(x)\, dx = \ln |\sec g(x)| + C.$$

Example 7   Obtain $\int e^{3x} \tan e^{3x}\, dx.$

SOLUTION   From (10.71), if we let $u = e^{3x}$, then $du = 3e^{3x}\, dx$ and

$$\int e^{3x} \tan e^{3x}\, dx = \int \tfrac{1}{3} \tan u \, du = -\tfrac{1}{3} \ln |\cos u| + C$$

$$= -\tfrac{1}{3} \ln |\cos e^{3x}| + C.$$

The derivation of the next integration formula is similar to the proof for (10.71) and is left as an exercise.

$$\int \cot g(x)g'(x)\, dx = \ln |\sin g(x)| + C \qquad (10.72)$$

We next derive the integration formula for $\int \sec g(x)g'(x)\, dx.$

$$\int \sec g(x)g'(x)\, dx = \ln |\sec g(x) + \tan g(x)| + C \qquad (10.73)$$

PROOF   If the numerator and denominator of the integrand are multiplied by $\sec g(x) + \tan g(x)$, then

$$\int \sec g(x)g'(x)\, dx = \int \frac{[\sec^2 g(x) + \sec g(x) \tan g(x)]g'(x)\, dx}{\sec g(x) + \tan g(x)}.$$

Letting $u = \sec g(x) + \tan g(x)$, we have $du = [\sec^2 g(x) + \sec g(x) \tan g(x)]g'(x)\, dx$ and

$$\int \sec g(x)g'(x)\, dx = \int \frac{du}{u} = \ln |u| + C$$

$$= \ln |\sec g(x) + \tan g(x)| + C.$$

A similar derivation, which is left as an exercise, gives the formula

$$\int \csc g(x)g'(x)\, dx = -\ln |\csc g(x) + \cot g(x)| + C. \qquad (10.74)$$

Example 8   Obtain $\int \dfrac{dx}{\sin 4x}.$

SOLUTION   Since $\displaystyle\int \frac{dx}{\sin 4x} = \int \csc 4x\, dx$, from (10.74) if $u = 4x$ and $du = 4\,dx$,

$$\int \frac{dx}{\sin 4x} = \int \csc 4x\, dx = \int \frac{1}{4} \csc u\, du = -\frac{1}{4}\ln|\csc u + \cot u| + C$$

$$= -\tfrac{1}{4}\ln|\csc 4x + \cot 4x| + C.$$

Exercise Set 10.5

In Exercises 1–18 obtain the indefinite integral.

1. $\displaystyle\int \cos 2x\, dx$

2. $\displaystyle\int \frac{\sin \sqrt{x}\, dx}{\sqrt{x}}$

3. $\displaystyle\int \sec 3x \tan 3x\, dx$

4. $\displaystyle\int \frac{dy}{\sin^2 (3y + 1)}$

5. $\displaystyle\int \frac{x\, dx}{\cot x^2}$

6. $\displaystyle\int \sec (\sin x) \cos x\, dx$

7. $\displaystyle\int \cos^3 2y \sin 2y\, dy$

8. $\displaystyle\int \sin^2 \frac{3x + 1}{2} \cos \frac{3x + 1}{2}\, dx$

9. $\displaystyle\int \tan^2 4x \sec^2 4x\, dx$

10. $\displaystyle\int x \sec^3 x^2 \tan x^2\, dx$

11. $\displaystyle\int \csc^5 \frac{1}{2}x \cot \frac{1}{2}x\, dx$

12. $\displaystyle\int \frac{\cot^4 (\ln x) \csc^2 (\ln x)\, dx}{x}$

13. $\displaystyle\int \sqrt[3]{2 + \sin 3x} \cos 3x\, dx$

14. $\displaystyle\int \frac{\sin x\, dx}{1 + \cos x}$

15. $\displaystyle\int \frac{\csc^2 t\, dt}{1 + 2 \cot t}$

16. $\displaystyle\int \sqrt{1 + \tan 2t} \sec^2 2t\, dt$

17. $\displaystyle\int \frac{(1 - \sin t)\, dt}{\sin t}$

18. $\displaystyle\int \frac{(2 + \tan \theta)\, d\theta}{\cos \theta}$

In Exercises 19–26 evaluate the definite integral.

19. $\displaystyle\int_0^{\pi/2} \sin 2x\, dx$

20. $\displaystyle\int_0^{\pi/12} \tan 3x\, dx$

21. $\displaystyle\int_{\pi/8}^{\pi/4} \cot 2x\, dx$

22. $\displaystyle\int_{-\pi/6}^{\pi/6} \sec 2x\, dx$

23. $\displaystyle\int_0^{\pi/2} \sin^2 x \cos x\, dx$

24. $\displaystyle\int_0^{\pi/4} \sec^2 x \tan^2 x\, dx$

25. $\displaystyle\int_{\pi/3}^{\pi/2} \frac{\cot x\, dx}{1 + \cot^2 x}$

26. $\displaystyle\int_{\pi/4}^{\pi/2} \frac{\cos^2 \theta\, d\theta}{\sin \theta}$

27. Verify formula (10.72).

28. Verify formula (10.74).

29. Find the area of the region bounded by one arch of the graph of sine and the x axis.

30. Find the area of the region bounded by the graphs of $y = \cos x$ and $y = \sin x$ for $x \in [0, \pi]$.

31. Prove that $\int \sec g(x)g'(x) = -\ln |\sec g(x) - \tan g(x)| + C$.

32. How is the indefinite integral obtained in Exercise 31 reconciled with the result from formula (10.73)?

33. Using two different approaches, obtain $\int \sec^2 x \tan x \, dx = (\sec^2 x/2) + C$ and $\int \sec^2 x \tan x \, dx = (\tan^2 x/2) + C$. How do we reconcile these two different-appearing evaluations?

34. Prove that $\int \csc g(x)g'(x) \, dx = \ln |\csc g(x) - \cot g(x)| + C$.

35. Using the results of Exercise 13, Exercise Set 10.4, show that $f(t)$ is of the form $A \cos \omega t + B \sin \omega t$ if $\sin \omega t \neq 0$. (If $f(t)$ is a solution of (10.61) on some interval, then f is certainly continuous on the interval. Hence, for some A and B, $f(t) = A \cos \omega t + B \sin \omega t$ for every t in the interval, even where $\sin \omega t = 0$.)

10.6 Inverse Sine and Inverse Cosine

Despite the title of this section, the functions sine and cosine do not have inverse functions. This assertion follows from Theorem 9.3.3 since, for example, $\sin 0 = \sin \pi = 0$ and $\cos 0 = \cos 2\pi = 1$. An inspection of the graph of the sine function (Figure 10–2) suggests, however, that it is possible to choose a subset of this function which *does* possess an inverse function. Using Theorem 9.3.3, all one must do is to choose this subset so that it is intersected in at most one point by any line of the form $y = k$. In particular, therefore, it would seem that the function

$$\left\{(x, \sin x): x \in \left[-\frac{\pi}{2}, \frac{\pi}{2}\right]\right\} \tag{10.75}$$

has an inverse function (Figure 10–17). The function (10.75), which is continuous and increasing on $[-\frac{\pi}{2}, \frac{\pi}{2}]$ and has range $[-1, 1]$ does, indeed, have an inverse function (by Theorem 9.4.1) which is continuous and increasing on $[-1, 1]$ and has range $[-\frac{\pi}{2}, \frac{\pi}{2}]$. This inverse function is called *inverse sine* or *arcsine*, abbreviated $\sin^{-1}$ or arcsin, and its value at any number x in its domain is denoted by

$$\sin^{-1} x \qquad \text{or} \qquad \arcsin x.$$

The use of the terminology "inverse sine" for an inverse function of a subset of sine rather than sine itself may seem inappropriate, but this usage is common in mathematical literature and will be used here as well.

10.6.1 Definition

$$\sin^{-1} = \left\{(x, \sin x): x \in \left[-\frac{\pi}{2}, \frac{\pi}{2}\right]\right\}^{-1}.$$

Definition 10.6.1 can be rephrased as follows:

$$y = \sin^{-1} x \quad \text{if and only if} \quad x = \sin y \text{ and } y \in \left[-\frac{\pi}{2}, \frac{\pi}{2}\right]. \qquad (10.76)$$

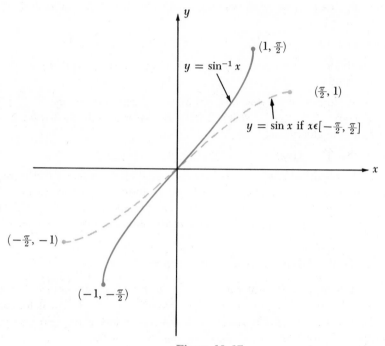

Figure 10–17

In Figure 10–17 the graph of $\sin^{-1}$ is represented by a continuous curve and the graph of its inverse function (10.75) by a dashed curve. Since $\frac{\pi}{6} \in [-\frac{\pi}{2}, \frac{\pi}{2}]$, by (10.76)

$$\sin^{-1}\frac{1}{2} = \frac{\pi}{6} \quad \text{since} \quad \frac{1}{2} = \sin\frac{\pi}{6}.$$

Also,

$$\sin^{-1}(-1) = -\frac{\pi}{2} \qquad \sin^{-1} 0 = 0 \qquad \sin^{-1} 1 = \frac{\pi}{2}$$

and from (10.76)

$$\sin(\sin^{-1} x) = x \quad \text{for } x \in [-1, 1]. \qquad (10.77)$$

In applications of $\sin^{-1}$ involving calculus a differentiation formula is needed.

10.6.2   Theorem

$$D_x \sin^{-1} x = \frac{1}{\sqrt{1 - x^2}} \quad \text{for } x \in (-1, 1). \qquad (10.78)$$

PROOF   By (9.38) for every x for which $\sin'(\sin^{-1} x) = \cos(\sin^{-1} x) \neq 0$

$$D_x \sin^{-1} x = \frac{1}{\cos(\sin^{-1} x)}. \qquad (10.79)$$

Since $\sin^{-1} x \in [-\frac{\pi}{2}, \frac{\pi}{2}]$, $\cos(\sin^{-1} x) \geq 0$ and hence from (10.5) and (10.77)

$$\cos(\sin^{-1} x) = \sqrt{1 - \sin^2(\sin^{-1} x)}$$
$$= \sqrt{1 - x^2} \quad \text{for } x \in [-1, 1]. \qquad (10.80)$$

Since $\cos(\sin^{-1} x) \neq 0$, if and only if $x \in (-1, 1)$, a substitution from (10.80) into (10.79) gives the conclusion (10.78).

Hence if g is a function such that $g(x) \in (-1, 1)$ and $g'(x)$ exists, from (10.78)

$$D_x \sin^{-1} g(x) = \frac{D_x g(x)}{\sqrt{1 - (g(x))^2}}. \qquad (10.81)$$

Example 1   Find $f'(x)$ if $f(x) = \sin^{-1}(1/x)$.

SOLUTION   From (10.81)

$$D_x \sin^{-1} \frac{1}{x} = \frac{D_x \dfrac{1}{x}}{\sqrt{1 - \dfrac{1}{x^2}}} = \frac{-\dfrac{1}{x^2}}{\dfrac{\sqrt{x^2 - 1}}{|x|}}.$$

Then, since $x^2 = |x|^2$

$$D_x \sin^{-1} \frac{1}{x} = -\frac{1}{|x|\sqrt{x^2 - 1}}.$$

In considering a possible definition for inverse cosine, it will be recalled that the function cos is continuous and decreasing on $[0, \pi]$. Of course, cos is also monotonic on other intervals of length π as well, so the interval $[0, \pi]$ is selected quite arbitrarily here. By Theorem 9.4.1 the function $\{(x, \cos x): x \in [0, \pi]\}$, which has the interval $[-1, 1]$ as its range, has an inverse function that is continuous and decreasing on $[-1, 1]$ and has $[0, \pi]$ as its range. This inverse function is called *inverse cosine* or *arccosine*, abbreviated $\cos^{-1}$ or arccos, and is defined as follows:

10.6.3   Definition

$$\cos^{-1} = \{(x, \cos x): x \in [0, \pi]\}^{-1}.$$

Thus

$$y = \cos^{-1} x \quad \text{if and only if} \quad x = \cos y \text{ and } y \in [0, \pi]. \qquad (10.82)$$

In particular, $\cos^{-1}\left(-\frac{1}{2}\right) = \frac{2\pi}{3}$, and

$$\cos^{-1} 1 = 0 \qquad \cos^{-1}(-1) = \pi,$$

and from (10.82)

$$\cos\left(\cos^{-1} x\right) = x \qquad \text{if } x \in [-1, 1]. \tag{10.83}$$

A sketch of the graph of $\cos^{-1}$ is shown in Figure 10–18.

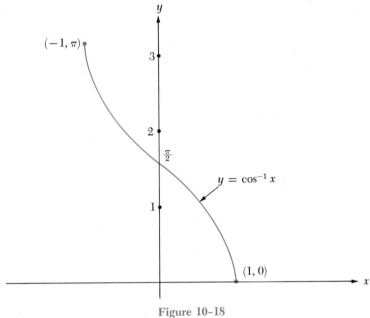

Figure 10-18

A useful relation between $\cos^{-1}$ and $\sin^{-1}$ can be derived without using calculus.

10.6.4   Theorem

$$\sin^{-1} x + \cos^{-1} x = \frac{\pi}{2} \quad \text{if } x \in [-1, 1]. \tag{10.84}$$

PROOF   Let

$$y = \cos^{-1} x. \tag{10.85}$$

Then $0 \le y \le \pi$, and hence $-\pi \le -y \le 0$ and $-\frac{\pi}{2} \le \frac{\pi}{2} - y \le \frac{\pi}{2}$. Therefore from (10.9) and (10.83)

$$\sin\left(\frac{\pi}{2} - y\right) = \cos y = \cos\left(\cos^{-1} x\right) = x,$$

and from (10.76)

$$\frac{\pi}{2} - y = \sin^{-1} x \tag{10.86}$$

since $\frac{\pi}{2} - y \in [-\frac{\pi}{2}, \frac{\pi}{2}]$. Equation (10.84) then follows from (10.85) and (10.86).

We will now use (10.84) to obtain the derivative of $\cos^{-1}$.

10.6.5 Theorem

$$D_x \cos^{-1} x = -\frac{1}{\sqrt{1 - x^2}} \quad \text{if } x \in (-1, 1). \tag{10.87}$$

PROOF From (10.84) $\cos^{-1} x = \frac{\pi}{2} - \sin^{-1} x$, and hence if $x \in (-1, 1)$

$$D_x \cos^{-1} x = D_x \left(\frac{\pi}{2} - \sin^{-1} x\right) = -D_x \sin^{-1} x$$

$$= -\frac{1}{\sqrt{1 - x^2}}.$$

If g is a function such that $g(x) \in (-1, 1)$ and $g'(x)$ exists, then

$$D_x \cos^{-1} g(x) = -\frac{D_x g(x)}{\sqrt{1 - (g(x))^2}}. \tag{10.88}$$

Example 2 Find $f'(x)$ if $f(x) = \cos^{-1} x^3$.

SOLUTION From (10.88)

$$D_x \cos^{-1} x^3 = -\frac{D_x x^3}{\sqrt{1 - (x^3)^2}} = -\frac{3x^2}{\sqrt{1 - x^6}}.$$

Example 3 A ladder of length 13 ft is leaning against a wall. If the lower end of the ladder slides away from the wall at the rate of 3 ft/sec, at what rate is the inclination of the ladder with respect to the ground changing when the lower end is 12 ft from the wall (Figure 10–19)?

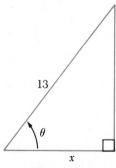

Figure 10-19

SOLUTION The important quantities that are changing with respect to time are x, the distance from the foot of the ladder to the wall and θ, the inclination of the ladder to the ground. Since

$$\theta = \cos^{-1} \frac{x}{13},$$

differentiation with respect to time t gives

$$\frac{d\theta}{dt} = -\frac{D_t\left(\frac{x}{13}\right)}{\sqrt{1 - \left(\frac{x}{13}\right)^2}} = -\frac{\frac{1}{13}\frac{dx}{dt}}{\sqrt{1 - \frac{x^2}{169}}}$$

$$= -\frac{\frac{dx}{dt}}{\sqrt{169 - x^2}}. \tag{10.89}$$

Now $dx/dt = 3$ ft/sec and, since $x = 12$, from (10.89)

$$\frac{d\theta}{dt} = -\frac{3}{5} \text{ rad/sec.}$$

It is probably astounding to note that if $x = \sqrt{168.9991}$, which is slightly less than 13, then from (10.89)

$$\frac{d\theta}{dt} = \frac{-3}{\sqrt{.0009}} = -100 \text{ rad/sec!!}$$

Exercise Set 10.6

1. Evaluate (a) $\sin^{-1}(\sqrt{3}/2)$ (b) $\cos^{-1}(-\sqrt{2}/2)$ (c) $\sin^{-1}(-\frac{1}{2})$.

2. Evaluate (a) $\cos \cos^{-1}\frac{2}{3}$ (b) $\cot \sin^{-1}\frac{4}{5}$ (c) $\sin \cos^{-1}(-\frac{1}{4})$.

3. Evaluate (a) $\sin^{-1} \sin \frac{3\pi}{4}$ (b) $\cos (2 \cos^{-1}\frac{1}{3})$ (c) $\cos [\sin^{-1}\frac{3}{5} - \sin^{-1}(-\frac{5}{13})]$
   (d) $\sin [\cos^{-1}(-\frac{1}{2}) + \sin^{-1}(-\frac{4}{5})]$.

4. Sketch the graph of $y = \sin^{-1}\frac{1}{2}x$.

5. Sketch the graph of $y = \cos^{-1}(x - 2)$.

In Exercises 6–11 obtain the derivatives of the given functions.

6. $f(x) = \sin^{-1} 3x$

7. $f(x) = \cos^{-1}(2x - 1)$

8. $F(y) = \cos^{-1}\dfrac{1}{\sqrt{1 + y^2}}$

9. $g(x) = \sin^{-1}\dfrac{x}{2} - \dfrac{1}{\sqrt{4 - x^2}}$

10. $g(t) = t \cos^{-1}\dfrac{t}{2}$

11. $f(x) = x^3 \sin^{-1}\dfrac{3}{x}$

In Exercises 12 and 13 obtain $D_x y$ by implicit differentiation.

12. $\sin^{-1}(x + y) = 2y$

13. $\cos^{-1} xy = x - y$

14. Find $\lim\limits_{x \to 0} \dfrac{\sin^{-1} x}{x}$.

15. Rework Exercise 1, Exercise Set 10.4, using the derivative of an inverse trigonometric function.

16. Derive formula (10.87) from Definition 10.6.3 using Theorem 9.4.2.

10.7 Inverse Tangent and Inverse Secant

The third inverse trigonometric function that will be considered is $\tan^{-1}$. It will be recalled that the function $\tan$ is continuous and increasing on $\left(-\frac{\pi}{2}, \frac{\pi}{2}\right)$ and has the set R as its range. Hence by Theorem 9.4.1 the function $\{(x, \tan x) : x \in \left(-\frac{\pi}{2}, \frac{\pi}{2}\right)\}$ has an inverse function that is continuous and increasing on the set R and has $\left(-\frac{\pi}{2}, \frac{\pi}{2}\right)$ as its range. This function, $\tan^{-1}$, is therefore defined as follows:

10.7.1 Definition

$$\tan^{-1} = \left\{(x, \tan x) : x \in \left(-\frac{\pi}{2}, \frac{\pi}{2}\right)\right\}^{-1}.$$

In other words

$$y = \tan^{-1} x \quad \text{if and only if} \quad x = \tan y \text{ and } y \in \left(-\frac{\pi}{2}, \frac{\pi}{2}\right). \qquad (10.90)$$

For example,

$$\tan^{-1} 1 = \frac{\pi}{4} \qquad \tan^{-1} 0 = 0 \qquad \tan^{-1}(-1) = -\frac{\pi}{4}$$

and also

$$\tan(\tan^{-1} x) = x \quad \text{for every } x. \qquad (10.91)$$

A sketch of the graph of $\tan^{-1}$ is given in Figure 10–20. We next derive the differentiation formula for $\tan^{-1}$.

10.7.2 Theorem

$$D_x \tan^{-1} x = \frac{1}{1 + x^2}.$$

PROOF Since $D_t \tan t = \sec^2 t > 0$ when $t \in \left(-\frac{\pi}{2}, \frac{\pi}{2}\right)$, by Theorem 9.4.2

$$D_x \tan^{-1} x = \frac{1}{\tan'(\tan^{-1} x)} = \frac{1}{\sec^2(\tan^{-1} x)}$$

$$= \frac{1}{1 + \tan^2(\tan^{-1} x)},$$

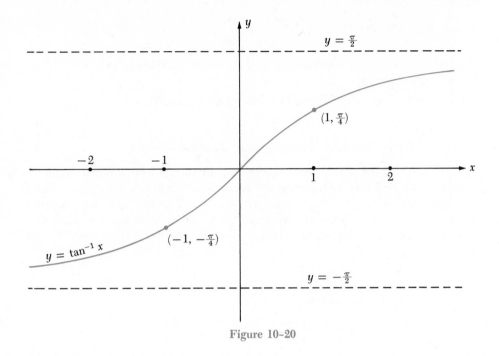

Figure 10-20

and from (10.91)

$$D_x \tan^{-1} x = \frac{1}{1 + x^2}.$$

If g is a function such that $g'(x)$ exists, then

$$D_x \tan^{-1} g(x) = \frac{D_x \, g(x)}{1 + (g(x))^2}. \qquad (10.92)$$

Example 1 Find $f'(x)$ if $f(x) = \tan^{-1}(x/3)$.

SOLUTION From (10.92)

$$D_x \tan^{-1} \frac{x}{3} = \frac{D_x \left(\dfrac{x}{3}\right)}{1 + \left(\dfrac{x}{3}\right)^2} = \frac{\dfrac{1}{3}}{1 + \dfrac{x^2}{9}}$$

$$= \frac{3}{9 + x^2}.$$

Example 2 A country road intersects a highway at right angles at C. Along one side of the road is a row of closely planted trees AB $\frac{3}{4}$ mile in length with the nearest tree B in the row being $\frac{1}{4}$ mile from C. How far from the intersection should an observer be on the highway to obtain the best view of the trees? HINT: The best view is obtained when the angle at the observer subtended by the row of trees is a maximum (Figure 10–21).

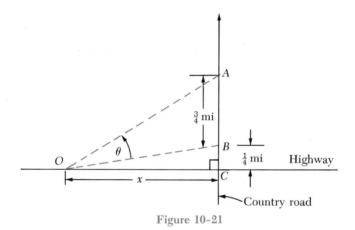

Figure 10–21

SOLUTION We let θ be the angle subtended by the row of trees and $x > 0$ be the distance of the observer from the country road. Then

$$\theta = \text{angle } AOC - \text{angle } BOC$$

$$= \tan^{-1}\frac{1}{x} - \tan^{-1}\frac{\frac{1}{4}}{x}.$$

Differentiation with respect to x gives

$$\frac{d\theta}{dx} = D_x \tan^{-1}\frac{1}{x} - D_x \tan^{-1}\frac{1}{4x}$$

$$= \frac{-\dfrac{1}{x^2}}{1 + \dfrac{1}{x^2}} - \frac{-\dfrac{1}{4x^2}}{1 + \dfrac{1}{16x^2}} = -\frac{1}{x^2 + 1} + \frac{4}{16x^2 + 1}$$

$$= \frac{-12x^2 + 3}{(x^2 + 1)(16x^2 + 1)}.$$

Hence $d\theta/dx = 0$ if $-12x^2 + 3 = 0$—that is, if

$$x = \tfrac{1}{2}.$$

By the first derivative test, θ is a maximum if $x = \tfrac{1}{2}$, and therefore the observer should be $\tfrac{1}{2}$ mile from the intersection.

Of the remaining three inverse trigonometric functions, only the inverse secant function, $\sec^{-1}$, will be discussed here.

It will be recalled that the function sec is continuous and increasing on $[0, \frac{\pi}{2})$ and every number ≥ 1 is the value sec x for some x in the interval. Also sec is continuous and decreasing on $[\pi, \frac{3\pi}{2})$ and every number ≤ -1 is the value sec x for some $x \in [\pi, \frac{3\pi}{2})$. Hence by Theorem 9.3.3 there is a function $\sec^{-1}$ as defined in the next statement.

10.7.3   Definition

$$\sec^{-1} = \left\{ (x, \sec x) : x \in \left[0, \frac{\pi}{2}\right) \cup \left[\pi, \frac{3\pi}{2}\right) \right\}^{-1}.$$

Thus

$$y = \sec^{-1} x \quad \text{if and only if} \quad x = \sec y \text{ and } y \in \left[0, \frac{\pi}{2}\right) \cup \left[\pi, \frac{3\pi}{2}\right). \quad (10.93)$$

A sketch of the graph of $\sec^{-1}$ is shown in Figure 10–22.

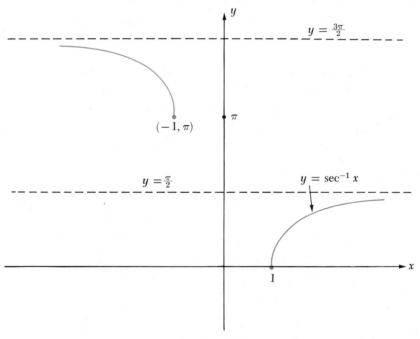

Figure 10-22

In obtaining the differentiation formula for $\sec^{-1}$ the following formula will be used.

10.7.4   Theorem

$$\sec^{-1} x = \begin{cases} \cos^{-1} \dfrac{1}{x} & \text{if } x \geq 1 \\[2mm] \pi + \cos^{-1}\left(-\dfrac{1}{x}\right) & \text{if } x \leq -1. \end{cases}$$

PROOF   Let $y = \sec^{-1} x$; then $x = \sec y$, and $\cos y = 1/x$. If $x \geq 1$, then from the definition of y, $y \in [0, \frac{\pi}{2})$. Hence $y = \cos^{-1}(1/x) = \sec^{-1} x$. If $x \leq -1$, then $y \in [\pi, \frac{3\pi}{2})$, and $y - \pi \in [0, \frac{\pi}{2})$. Then $\cos(y - \pi) = -\cos y = -1/x$, and hence $\cos^{-1}(-1/x) = y - \pi = \sec^{-1} x - \pi$; that is, $\pi + \cos^{-1}(-1/x) = \sec^{-1} x$.

We next state and derive the differentiation formula for $\sec^{-1}$.

10.7.5   Theorem

$$D_x \sec^{-1} x = \frac{1}{x\sqrt{x^2 - 1}} \quad \text{if } |x| > 1. \tag{10.94}$$

PROOF   If $x > 1$, from Theorems 10.7.4 and 4.3.3

$$D_x \sec^{-1} x = D_x \cos^{-1} \frac{1}{x} = -\frac{D_x\left(\dfrac{1}{x}\right)}{\sqrt{1 - \dfrac{1}{x^2}}}$$

$$= \frac{\dfrac{1}{x^2}}{\dfrac{\sqrt{x^2 - 1}}{x}}$$

$$= \frac{1}{x\sqrt{x^2 - 1}}.$$

The verification of (10.94) when $x < -1$ is left as an exercise.

Instead of using Definition 10.7.3, some texts define $\sec^{-1} x$ by saying that $\sec^{-1} x = \cos^{-1}(1/x)$ when $|x| \geq 1$, thereby giving $\sec^{-1}$ the range $[0, \frac{\pi}{2}) \cup (\frac{\pi}{2}, \pi]$. Such a definition facilitates the derivation of a differentiation formula for $\sec^{-1}$, but results in the formula $D_x(\sec^{-1} x) = 1/|x|\sqrt{x^2 - 1}$ for $|x| > 1$. Because of the factor $|x|$ in the denominator of the right member, this formula is a bit more complicated than (10.94). Another advantage in using (10.93) is that if $y = \sec^{-1} x$, then $\sqrt{x^2 - 1} = \sqrt{\sec^2 y - 1} = \tan y$ rather than $|\tan y|$, since $\tan y \geq 0$ when $y \in [0, \frac{\pi}{2}) \cup [\pi, \frac{3\pi}{2})$. This fact will prove useful in some of the work in integration in Section 11.4.

If $g(x) > 1$ or $g(x) < -1$ and $g'(x)$ exists, then from (10.94)

$$D_x \sec^{-1} g(x) = \frac{D_x\, g(x)}{g(x)\sqrt{(g(x))^2 - 1}}. \tag{10.95}$$

Example 3   Find $f'(x)$ if $f(x) = \sec^{-1} x^2$.

SOLUTION   From (10.95) if $x^2 > 1$,

$$f'(x) = \frac{D_x x^2}{x^2\sqrt{(x^2)^2 - 1}} = \frac{2x}{x^2\sqrt{x^4 - 1}}$$

$$= \frac{2}{x\sqrt{x^4 - 1}}.$$

Exercise Set 10.7

1. Evaluate (a) $\tan^{-1}(\sqrt{3}/3)$ (b) $\tan^{-1}(-1)$ (c) $\sec^{-1} 2$ (d) $\sec^{-1}(-2)$.

2. Evaluate (a) $\cot\tan^{-1}\frac{2}{3}$ (b) $\sin\sec^{-1}\frac{5}{3}$ (c) $\tan\sec^{-1}(-3)$ (d) $\cos\tan^{-1}(-\frac{7}{4})$.

3. Evaluate (a) $\tan(\tan^{-1}(-\frac{1}{4}) + \tan^{-1}\frac{1}{3})$ (b) $\sin(2\sec^{-1}\frac{5}{2})$.

4. Sketch the graph of $y = 2\tan^{-1} x$.

5. Sketch the graph of $y = \sec^{-1}(x + 1)$.

In Exercises 6–13 find the derivative of the given function.

6. $f(x) = \tan^{-1} x^2$

7. $g(x) = \sec^{-1}\dfrac{1}{x}$

8. $f(x) = \sec^{-1}\dfrac{3x - 4}{4}$

9. $g(t) = \tan^{-1}\dfrac{t - 1}{t + 1}$

10. $F(y) = \tan^{-1}(\cot 2y)$

11. $f(x) = \sec^{-1}\dfrac{\sqrt{1 + x^2}}{x}$

12. $\phi(x) = \dfrac{\sqrt{x^2 - 9}}{x^2} + \dfrac{1}{2}\sec^{-1}\dfrac{x}{3}$

13. $h(x) = \dfrac{\tan^{-1} 2x}{1 + 4x^2}$

14. Find $\displaystyle\lim_{x\to 0}\dfrac{\tan^{-1} 2x}{x}$.

15. A billboard is 15 ft high and its base is 20 ft above the ground. How far away should an observer stand in order to have the best view of the billboard?

16. Derive formula (10.94) from Definition 10.7.3 and Theorem 9.4.2.

17. An object falls from a height of 150 ft and is viewed by an observer on the ground 100 ft from the point on the ground directly beneath the object. Assuming that the object falls in accordance with the formula $s = 16t^2$, where s ft is the distance the object falls in t sec, what is the rate of change of the angle of elevation of the object from the observer when $t = 2$ sec?

18. Let A and B be the points $(6, 1)$ and $(3, 4)$, respectively. Using calculus, find the point P on the x axis such that the angle between PA and PB is a maximum.

10.8   Integrals Yielding Inverse Trigonometric Functions

From the differentiation formulas

$$D_x \sin^{-1} g(x) = \frac{D_x g(x)}{\sqrt{1 - (g(x))^2}}$$

$$D_x \tan^{-1} g(x) = \frac{D_x \, g(x)}{1 + (g(x))^2}$$

$$D_x \sec^{-1} g(x) = \frac{D_x \, g(x)}{g(x) \sqrt{(g(x))^2 - 1}}$$

the following integration formulas can be derived:

$$\int \frac{g'(x) \, dx}{\sqrt{1 - (g(x))^2}} = \sin^{-1} g(x) + C \qquad (10.96)$$

$$\int \frac{g'(x) \, dx}{1 + (g(x))^2} = \tan^{-1} g(x) + C \qquad (10.97)$$

$$\int \frac{g'(x) \, dx}{g(x) \sqrt{(g(x))^2 - 1}} = \sec^{-1} g(x) + C \qquad (10.98)$$

It would also be possible to rewrite (10.96) as

$$\int \frac{g'(x) \, dx}{\sqrt{1 - (g(x))^2}} = -\cos^{-1} g(x) + C$$

since by (10.88) $D_x (-\cos^{-1} g(x)) = \dfrac{D_x \, g(x)}{\sqrt{1 - (g(x))^2}}$.

From (10.96), (10.97), and (10.98), respectively,

$$\int \frac{g'(x) \, dx}{\sqrt{a^2 - (g(x))^2}} = \sin^{-1} \frac{g(x)}{a} + C \quad \text{if } a > 0 \qquad (10.99)$$

$$\int \frac{g'(x) \, dx}{a^2 + (g(x))^2} = \frac{1}{a} \tan^{-1} \frac{g(x)}{a} + C \qquad (10.100)$$

$$\int \frac{g'(x) \, dx}{g(x) \sqrt{(g(x))^2 - a^2}} = \frac{1}{a} \sec^{-1} \frac{g(x)}{a} + C \quad \text{if } a > 0. \qquad (10.101)$$

PROOF OF (10.99)   If $a > 0$

$$\int \frac{g'(x) \, dx}{\sqrt{a^2 - (g(x))^2}} = \int \frac{\dfrac{g'(x)}{a} \, dx}{\sqrt{1 - \left(\dfrac{g(x)}{a}\right)^2}}.$$

Hence, if $u = \dfrac{g(x)}{a}$ and $du = \dfrac{g'(x)}{a} \, dx$, then from (10.96)

$$\int \frac{g'(x) \, dx}{\sqrt{a^2 - (g(x))^2}} = \int \frac{du}{\sqrt{1 - u^2}} = \sin^{-1} u + C$$

$$= \sin^{-1} \frac{g(x)}{a} + C.$$

The derivations of (10.100) and (10.101) are left as exercises.

Example 1   Obtain $\int \dfrac{dx}{\sqrt{7 - 4x^2}}$.

SOLUTION   We let $u = 2x$ and hence $du = 2\,dx$. Then $dx = \frac{1}{2}\,du$, and from (10.99)

$$\int \frac{dx}{\sqrt{7 - 4x^2}} = \int \frac{\frac{1}{2}\,du}{\sqrt{7 - u^2}} = \frac{1}{2} \sin^{-1} \frac{u}{\sqrt{7}} + C$$

$$= \frac{1}{2} \sin^{-1} \frac{2x}{\sqrt{7}} + C.$$

Example 2   Obtain $\int \dfrac{dx}{2x^2 + 3x + 10}$.

SOLUTION   We first factor out 2 in the denominator. Then

$$\int \frac{dx}{2x^2 + 3x + 10} = \frac{1}{2} \int \frac{dx}{x^2 + \frac{3}{2}x + 5}.$$

Completing the square in $x^2 + \frac{3}{2}x$ in the denominator gives

$$(x^2 + \tfrac{3}{2}x) + 5 = (x^2 + \tfrac{3}{2}x + \tfrac{9}{16}) + 5 - \tfrac{9}{16} = (x + \tfrac{3}{4})^2 + \tfrac{71}{16},$$

and the original integral can therefore be rewritten

$$\int \frac{dx}{2x^2 + 3x + 10} = \frac{1}{2} \int \frac{dx}{(x + \frac{3}{4})^2 + \frac{71}{16}}. \qquad (10.102)$$

Letting $u = x + \frac{3}{4}$, and $du = dx$ in (10.102), from (10.100)

$$\int \frac{dx}{2x^2 + 3x + 10} = \frac{1}{2} \int \frac{du}{u^2 + \frac{71}{16}} = \frac{1}{2} \cdot \frac{4}{\sqrt{71}} \tan^{-1} \frac{u}{\sqrt{71}/4} + C$$

$$= \frac{2}{\sqrt{71}} \tan^{-1} \frac{4x + 3}{\sqrt{71}} + C.$$

Example 3   Obtain $\int \dfrac{(5x^2 - 4)\,dx}{3x\sqrt{4x^2 - 9}}$.

SOLUTION   We first express the integral as a difference of integrals.

$$\int \frac{(5x^2 - 4)\,dx}{3x\sqrt{4x^2 - 9}} = \int \frac{\frac{5}{3}x\,dx}{\sqrt{4x^2 - 9}} - \int \frac{4\,dx}{3x\sqrt{4x^2 - 9}}.$$

In the first integral, let $u = 4x^2 - 9$. Then $du = 8x\,dx$, and $\frac{5}{3}x\,dx = \frac{5}{24}\,du$. Hence

$$\int \frac{\frac{5}{3}x\,dx}{\sqrt{4x^2 - 9}} = \int \frac{\frac{5}{24}\,du}{u^{1/2}} = \frac{5}{24} \cdot \frac{u^{1/2}}{\frac{1}{2}} + C_1$$

$$= \tfrac{5}{12}\sqrt{4x^2 - 9} + C_1. \qquad (10.103)$$

In the second integral, let $u = 2x$, $du = 2\,dx$. Then by (10.101)

$$\int \frac{4\,dx}{3x\sqrt{4x^2-9}} = \int \frac{2\,du}{\dfrac{3u}{2}\sqrt{u^2-9}} = \frac{4}{3}\int \frac{du}{u\sqrt{u^2-9}}$$

$$= \frac{4}{3}\cdot\frac{1}{3}\sec^{-1}\frac{u}{3} + C_2$$

$$= \frac{4}{9}\sec^{-1}\frac{2x}{3} + C_2. \tag{10.104}$$

Hence from (10.103) and (10.104)

$$\int \frac{(5x^2-4)\,dx}{3x\sqrt{4x^2-9}} = \frac{5}{12}\sqrt{4x^2-9} - \frac{4}{9}\sec^{-1}\frac{2x}{3} + C.$$

Example 4   Obtain $\displaystyle\int \frac{(x+1)\,dx}{\sqrt{4-6x-x^2}}$.

SOLUTION   Completing the square in the radicand, we write

$$4 - 6x - x^2 = 4 - (x^2+6x) = 13 - (x^2+6x+9) = 13 - (x+3)^2.$$

Then

$$\int \frac{(x+1)\,dx}{\sqrt{4-6x-x^2}} = \int \frac{(x+3)-2}{\sqrt{13-(x+3)^2}}\,dx$$

$$= \int \frac{(x+3)\,dx}{\sqrt{13-(x+3)^2}} - \int \frac{2\,dx}{\sqrt{13-(x+3)^2}}. \tag{10.105}$$

In the first integral let $u = 13 - (x+3)^2$. Then $du = -2(x+3)\,dx$, and hence

$$\int \frac{(x+3)\,dx}{\sqrt{13-(x+3)^2}} = \int -\frac{1}{2}u^{-1/2}\,du = -u^{1/2} + C_1$$

$$= -\sqrt{13-(x+3)^2} + C_1$$

$$= -\sqrt{4-6x-x^2} + C_1. \tag{10.106}$$

In the second integral, let $u = x+3$. Then $du = dx$, and by (10.99)

$$\int \frac{2\,dx}{\sqrt{13-(x+3)^2}} = \int \frac{2\,du}{\sqrt{13-u^2}} = 2\sin^{-1}\frac{u}{\sqrt{13}} + C_2$$

$$= 2\sin^{-1}\frac{x+3}{\sqrt{13}} + C_2. \tag{10.107}$$

Hence from (10.106) and (10.107), (10.105) can be expressed as

$$\int \frac{(x+1)\,dx}{\sqrt{4-6x-x^2}} = -\sqrt{4-6x-x^2} - 2\sin^{-1}\frac{x+3}{\sqrt{13}} + C.$$

Exercise Set 10.8

1. Derive formula (10.100).

2. Derive formula (10.101).

In Exercises 3–18 perform the indicated integration.

3. $\displaystyle\int \frac{dx}{\sqrt{25 - 9x^2}}$

4. $\displaystyle\int \frac{dx}{4x^2 + 9}$

5. $\displaystyle\int \frac{dx}{x\sqrt{4x^2 - 1}}$

6. $\displaystyle\int \frac{x^2\,dx}{\sqrt{5 - x^6}}$

7. $\displaystyle\int \frac{dy}{\tan^{-1} y(y^2 + 1)}$

8. $\displaystyle\int \frac{\sin^{-1} x\,dx}{\sqrt{1 - x^2}}$

9. $\displaystyle\int \frac{2\sin x\,dx}{\cos^2 x + 1}$

10. $\displaystyle\int \frac{dx}{3x\sqrt{x^4 - 3}}$

11. $\displaystyle\int \frac{dx}{e^x + 4e^{-x}}$

12. $\displaystyle\int \frac{dx}{x\sqrt{4 - (\ln x)^2}}$

13. $\displaystyle\int \frac{dt}{(t + 2)\sqrt{t^2 + 4t - 7}}$

14. $\displaystyle\int \frac{x\,dx}{x^2 - x + 2}$

15. $\displaystyle\int \frac{dt}{\sqrt{8 - 2x - x^2}}$

16. $\displaystyle\int \frac{dt}{3t^2 + t + 5}$

17. $\displaystyle\int \frac{(x + 1)\,dx}{2x^2 - x + 3}$

18. $\displaystyle\int \frac{(x - 2)\,dx}{\sqrt{10 - 4x - 2x^2}}$

In Exercises 19–22 evaluate the definite integrals.

19. $\displaystyle\int_{-1/4}^{1/4} \frac{dx}{\sqrt{1 - 4x^2}}$

20. $\displaystyle\int_{2}^{2\sqrt{2}} \frac{dx}{x\sqrt{x^2 - 2}}$

21. $\displaystyle\int_{0}^{2} \frac{dx}{x^2 + 2x + 4}$

22. $\displaystyle\int_{1}^{3} \frac{dy}{\sqrt{4y - y^2}}$

10.9   Hyperbolic Functions

In this section we consider the *hyperbolic functions*, whose behavior is remarkably analogous to that of the trigonometric functions. Because of the similarities between the two types of function, the names hyperbolic sine, hyperbolic cosine, hyperbolic tangent, and so on, are given to the hyperbolic functions. The notations

sinh for hyperbolic sine,
cosh for hyperbolic cosine,
tanh for hyperbolic tangent,
and so on,

are used to denote these functions. We will see that the name hyperbolic arises because of the association of the functions sinh and cosh with the hyperbola $x^2 - y^2 = 1$.

10.9.1 Definition

For every x

$$\sinh x = \frac{e^x - e^{-x}}{2} \qquad \cosh x = \frac{e^x + e^{-x}}{2} \tag{10.108}$$

The sketches of the graphs of sinh and cosh, which are given in Figure 10–23(a) and (b), illustrate the definitions in (10.108).

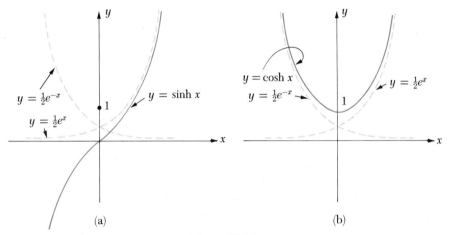

Figure 10-23

From these definitions

$$\sinh 0 = 0 \qquad \cosh 0 = 1$$

and, for example,

$$\sinh 1 = \frac{e - e^{-1}}{2} \approx 1.1752 \qquad \cosh \frac{1}{2} = \frac{e^{1/2} + e^{-1/2}}{2} \approx 1.1276.$$

Also the following inequalities can be obtained.

$$\cosh x \geq 1 \quad \text{for every } x \tag{10.109}$$
$$\sinh x > 0 \text{ if } x > 0 \quad \text{and} \quad \sinh x < 0 \text{ if } x < 0 \tag{10.110}$$

PROOF OF (10.109)

$$\cosh x - 1 = \frac{e^x + e^{-x}}{2} - 1 = \frac{e^x - 2 + e^{-x}}{2} = \frac{(e^{x/2} - e^{-x/2})^2}{2} \geq 0.$$

Hence (10.109) follows.

The inequalities in (10.110) follow from the fact that the exponential function is positive and increasing and $e^0 = 1$.

The following identities, which are the hyperbolic analogues of the trigonometric identities (10.11), (10.10), and (10.5), respectively, can be derived from the definitions of sinh and cosh.

$$\sinh(-x) = -\sinh x \tag{10.111}$$

$$\cosh(-x) = \cosh x \tag{10.112}$$

$$\cosh^2 x - \sinh^2 x = 1 \tag{10.113}$$

Because of (10.109) and (10.113) an arbitrary point of the form $(\cosh t, \sinh t)$ is on the right branch of the hyperbola $x^2 - y^2 = 1$ (Figure 10–24):

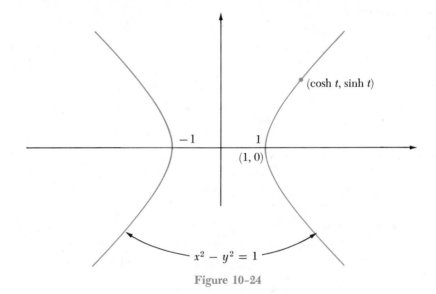

Figure 10-24

We also note that the differentiation formulas stated below for sinh and cosh are analogous to those given for sine and cosine.

10.9.2   Theorem

For every x,

$$D_x \sinh x = \cosh x. \tag{10.114}$$

PROOF $D_x \sinh x = D_x\left(\dfrac{e^x - e^{-x}}{2}\right) = \dfrac{1}{2}D_x(e^x - e^{-x}) = \dfrac{1}{2}(e^x + e^{-x})$

$$= \cosh x$$

10.9.3   Theorem

For every x,

$$D_x \cosh x = \sinh x. \tag{10.115}$$

The proof of (10.115) is left as an exercise.

Since sinh and cosh are everywhere differentiable, they are also everywhere continuous. From (10.114) and (10.109) sinh is everywhere increasing. On the other hand, from (10.115) and (10.110) cosh is decreasing on $(-\infty, 0]$ but is increasing on $[0, +\infty)$.

From (10.114) and (10.115) if $g'(x)$ exists, then

$$D_x \sinh g(x) = \cosh g(x) \, D_x \, g(x) \tag{10.116}$$

and

$$D_x \cosh g(x) = \sinh g(x) \, D_x \, g(x). \tag{10.117}$$

Example 1   Find $D_x \sinh 3x$.

SOLUTION   From (10.116)

$$D_x \sinh 3x = \cosh 3x \cdot D_x \, 3x = 3 \cosh 3x.$$

Example 2   Find $D_x \cosh \sqrt{x^2 + 1}$.

SOLUTION   From (10.117)

$$D_x \cosh \sqrt{x^2 + 1} = \sinh \sqrt{x^2 + 1} \cdot D_x \sqrt{x^2 + 1}$$

$$= \frac{x}{\sqrt{x^2 + 1}} \sinh \sqrt{x^2 + 1}.$$

The following integration formulas can be obtained from (10.116) and (10.117), respectively.

$$\int \cosh g(x) g'(x) \, dx = \sinh g(x) + C \tag{10.118}$$

$$\int \sinh g(x) g'(x) \, dx = \cosh g(x) + C \tag{10.119}$$

The student should note that formulas (10.116) and (10.118) are exactly parallel to the corresponding trigonometric formulas, while in (10.117) and (10.119) there is a difference in sign from the corresponding trigonometric formulas.

Example 3   Obtain $\int \sinh^4 x \cosh^3 x \, dx$.

SOLUTION   From (10.113) $\cosh^2 x = 1 + \sinh^2 x$ and

$$\int \sinh^4 x \cosh^3 x \, dx = \int \sinh^4 x (1 + \sinh^2 x) \cosh x \, dx$$

$$= \int \sinh^4 x \cosh x \, dx + \int \sinh^6 x \cosh x \, dx.$$

Both integrals on the right can be evaluated by letting $u = \sinh x$ and $du = \cosh x \, dx$. Then

$$\int \sinh^4 x \cosh^3 x \, dx = \frac{\sinh^5 x}{5} + \frac{\sinh^7 x}{7} + C.$$

The remaining hyperbolic functions are defined in terms of sinh and cosh.

10.9.4 Definition

(a) $\tanh x = \dfrac{\sinh x}{\cosh x}$

(b) $\coth x = \dfrac{\cosh x}{\sinh x}$

(c) $\operatorname{sech} x = \dfrac{1}{\cosh x}$

(d) $\operatorname{csch} x = \dfrac{1}{\sinh x}$

The differentiation formulas for these hyperbolic functions are

$$D_x \tanh x = \operatorname{sech}^2 x \qquad\qquad (10.120)$$
$$D_x \coth x = -\operatorname{csch}^2 x \qquad\qquad (10.121)$$
$$D_x \operatorname{sech} x = -\operatorname{sech} x \tanh x \qquad\qquad (10.122)$$
$$D_x \operatorname{csch} x = -\operatorname{csch} x \coth x \qquad\qquad (10.123)$$

PROOF OF (10.120)

$$D_x \tanh x = D_x \frac{\sinh x}{\cosh x} = \frac{\cosh x \, D_x \sinh x - \sinh x \, D_x \cosh x}{\cosh^2 x}$$

$$= \frac{\cosh^2 x - \sinh^2 x}{\cosh^2 x} = \frac{1}{\cosh^2 x} = \operatorname{sech}^2 x.$$

The derivations of (10.121), (10.122), and (10.123) are left as exercises. Since $\operatorname{sech}^2 x > 0$ for every x, tanh is continuous and increasing on its domain, the set R. Since $\tanh 0 = 0$, $\tanh x > 0$ when $x > 0$ and $\tanh x < 0$ when $x < 0$. As

$$\lim_{x \to +\infty} \tanh x = \lim_{x \to +\infty} \frac{e^x - e^{-x}}{e^x + e^{-x}} = \lim_{x \to +\infty} \frac{e^{2x} - 1}{e^{2x} + 1}$$

$$= \lim_{x \to +\infty} \frac{1 - \dfrac{1}{e^{2x}}}{1 + \dfrac{1}{e^{2x}}} = 1,$$

the line $y = 1$ is a horizontal asymptote of the graph of tanh. Also, $\lim_{x \to -\infty}$

$(\tanh x) = -1$, as the student will be asked to prove, and therefore the line $y = -1$ is a horizontal asymptote of the graph of tanh (Figure 10–25). The range of tanh is the interval $(-1, 1)$.

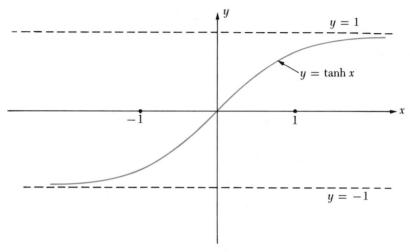

Figure 10–25

Exercise Set 10.9

In Exercises 1–7 prove the given identity.

1. (10.111)   2. (10.112)   3. (10.113)

4. $\sinh 2x = 2 \sinh x \cosh x$

5. $\cosh 2x = \cosh^2 x + \sinh^2 x$

6. $1 - \tanh^2 x = \operatorname{sech}^2 x$

7. $\coth^2 x - 1 = \operatorname{csch}^2 x$

In Exercises 8–11 derive the given differentiation formula.

8. (10.115)   9. (10.121)   10. (10.123)   11. (10.122)

12. Prove that for every positive integer n
$$(\cosh x + \sinh x)^n = \cosh nx + \sinh nx.$$

In Exercises 13–18 obtain the derivative of the given function.

13. $f(x) = \cosh \sqrt{x}$

14. $f(t) = \tanh \dfrac{1}{t}$

15. $F(y) = \ln (\sinh 2x)$

16. $f(x) = \tan^{-1} (\sinh x)$

17. $h(x) = \sin^{-1} (\tanh x)$

18. $\psi(x) = \dfrac{\sinh 2x}{1 - \cosh 2x}$

In Exercises 19–26 obtain the indefinite integral.

19. $\int \sinh 3x \, dx$ 20. $\int \cosh \frac{1}{2}x \, dx$

21. $\int \sinh 2x \cosh 2x \, dx$ 22. $\int \tanh \frac{1}{3}y \, dy$

23. $\int \sinh^3 4s \, ds$ 24. $\int \cosh^2 3x \sinh^3 3x \, dx$

25. $\int \tanh^2 x \, dx$ 26. $\int \sqrt{1 + \cosh x} \, dx$

27. Show that the differential equation $d^2x/dt^2 = k^2x$ is satisfied by $x = A \cosh kt + B \sinh kt$.

28. Show that the differential equation $(1/g)(d^2x/dt^2) = 1 - (1/a^2)(dx/dt)^2$ is satisfied by $x = (a^2/g) \ln \cosh gt/a$.

29. Prove: If $x = \sinh y$, then $y = \ln (x + \sqrt{x^2 + 1})$.
    (Thus $\sinh^{-1} x = \ln (x + \sqrt{x^2 + 1})$.) HINT: Obtain e^y in terms of x.

30. Prove: If $x = \tanh y$, then $y = \frac{1}{2} \ln (1 + x)/(1 - x)$.
    (Thus $\tanh^{-1} x = \frac{1}{2} \ln (1 + x)/(1 - x)$.)

31. Prove: $\lim\limits_{x \to -\infty} \tanh x = -1$.

11. Methods of Integration

11.1 Integration by Parts

Thus far we have discussed the derivatives and integrals of *elementary functions*. An elementary function is a function formed from algebraic, exponential, or trigonometric functions by a finite number of operations of addition, subtraction, multiplication, division, composition of functions, or taking the inverse of a function. We have seen that the derivative of an elementary function is an elementary function. However, the indefinite integral of an elementary function may not be an elementary function. For example, the indefinite integrals

$$\int e^{-x^2}\, dx \qquad \int \frac{dx}{\sqrt{1-x^4}}$$

exist since the functions given by the integrands are certainly continuous on some interval (by Theorem 7.7.1 and Definition 7.1.5). However, it can be shown with some effort that these integrals cannot be expressed in terms of elementary functions.

The indefinite integrals of certain elementary functions that have been already derived in the text are given in formulas (1)–(21) listed in the Table of Integrals. However, there are many integrals that cannot be evaluated using methods previously discussed in this text. One such integral, which we will consider in this section is

$$\int x^3 \sqrt{x^2 + 1}\, dx. \tag{11.1}$$

The purpose of this chapter, therefore, is to expand the student's repertoire of methods of integration.

The first method to be considered is *integration by parts*. If f and g are differentiable on some interval I, then by the product rule

$$D_x\left[f(x)g(x)\right] = f(x)g'(x) + f'(x)g(x) \quad \text{for all } x \in I.$$

Therefore, if C' is the constant of integration, then for all $x \in I$

$$\int [f(x)g'(x) + f'(x)g(x)] \, dx = f(x)g(x) + C'$$

$$\int f(x)g'(x) \, dx + \int f'(x)g(x) \, dx = f(x)g(x) + C'$$

$$\int f(x)g'(x) \, dx = f(x)g(x) - \int f'(x)g(x) \, dx + C' \qquad (11.2)$$

In practice we let $u = f(x)$ and $v = g(x)$ and hence $du = f'(x) \, dx$ and $dv = g'(x) \, dx$. Then from (11.2)

$$\int u \, dv = uv - \int v \, du + C'. \qquad (11.3)$$

We have denoted the arbitrary constant in (11.3) by C' rather than C, preferring to let C be the constant eventually obtained by combining with C' the constant of integration from the evaluation of $\int v \, du$. Equation (11.3) is called the *integration-by-parts formula*.

In applying (11.3) no general rule for choosing u and dv can be given. However, it is usually a good plan to first select dv in the expression to be integrated and to have dv contain as much of the integrand as possible and still be easy to integrate. Once dv has been chosen, the factor that remains in the integrand is u. Generally, dv will *not* be of the form $x^n \, dx$. For if dv were of this form, then v would contain $x^{n+1}/(x + 1)$ and the resulting integral $\int v \, du$ would likely be more complicated than the original integral $\int u \, dv$. (For an example, see the discussion following Example 1 below.) Also, if the integrand contains factors that arise from logarithmic or inverse trigonometric functions, then u usually contains such factors. Hopefully, once dv and u have been designated, the integral $\int v \, du$, which is obtained, either can be readily evaluated or is at least in a simpler form than $\int u \, dv$. Formula (11.3) is illustrated in the following example in which the integral (11.1) is evaluated.

Example 1   Evaluate $\int x^3 \sqrt{x^2 + 1} \, dx$.

SOLUTION   We shall let

$$dv = x\sqrt{x^2 + 1} \, dx \qquad \text{and} \qquad u = x^2.$$

Then

$$v = \int x\sqrt{x^2 + 1} \, dx = \tfrac{1}{2} \cdot \tfrac{2}{3}(x^2 + 1)^{3/2} + \overline{C} = \tfrac{1}{3}(x^2 + 1)^{3/2} + \overline{C}$$

and

$$du = 2x \, dx.$$

For v we shall choose the integral of $x\sqrt{x^2 + 1}$ for which $\overline{C} = 0$. (Following this solution we will show that it is immaterial what value is assigned to $\overline{C}$; however, as will be noted, our work is simplified with the choice $\overline{C} = 0$.) Thus we shall let $v = \tfrac{1}{3}(x^2 + 1)^{3/2}$. Then from (11.3)

$$\int x^3 \sqrt{x^2 + 1} \, dx = \frac{1}{3}x^2(x^2 + 1)^{3/2} - \int \frac{2x}{3}(x^2 + 1)^{3/2} \, dx + C'. \qquad (11.4)$$

The integral on the right in (11.4) can be easily evaluated. Thus

$$\int x^3 \sqrt{x^2 + 1} \; dx = \tfrac{1}{3}x^2(x^2 + 1)^{3/2} - \tfrac{2}{3} \cdot \tfrac{1}{2} \cdot \tfrac{2}{5}(x^2 + 1)^{5/2} + C$$

$$= \tfrac{1}{3}x^2(x^2 + 1)^{3/2} - \tfrac{2}{15}(x^2 + 1)^{5/2} + C.$$

Here C is the sum of the constants of integration on the right.

If in the foregoing solution we had used

$$v = \tfrac{1}{3}(x^2 + 1)^{3/2} + \bar{C}$$

instead of $v = \tfrac{1}{3}(x^2 + 1)^{3/2}$, then from (11.3) we would have

$$\int x^3 \sqrt{x^2 + 1} \; dx = x^2[\tfrac{1}{3}(x^2 + 1)^{3/2} + \bar{C}]$$

$$- \int 2x[\tfrac{1}{3}(x^2 + 1)^{3/2} + \bar{C}] \; dx + C'$$

$$= \tfrac{1}{3}x^2(x^2 + 1)^{3/2} + \bar{C}x^2$$

$$- \int \frac{2x}{3}(x^2 + 1)^{3/2} \; dx - \int 2\bar{C}x \; dx + C'$$

$$= \tfrac{1}{3}x^2(x^2 + 1)^{3/2} + \bar{C}x^2$$

$$- \int \frac{2x}{3}(x^2 + 1)^{3/2} \; dx - \bar{C}x^2 + C_1$$

where the constant C_1 is obtained by combining C' and the constant of integration from the evaluation of $-\int 2\bar{C}x \; dx$. Since the terms involving $\bar{C}$ in the preceding equation cancel each other, one obtains

$$\int x^3 \sqrt{x^2 + 1} \; dx = \frac{1}{3}x^2(x^2 + 1)^{3/2} - \int \frac{2x}{3}(x^2 + 1)^{3/2} \; dx + C_1,$$

which is of the form (11.4).

In the solution of Example 1 other choices might have been considered for dv and u. For example, if we had let

$$dv = x^3 \; dx \qquad \text{and} \qquad u = \sqrt{x^2 + 1}, \qquad\qquad (11.5)$$

then

$$v = \frac{x^4}{4} \qquad \text{and} \qquad du = \frac{x \; dx}{\sqrt{x^2 + 1}},$$

and by (11.3)

$$\int x^3 \sqrt{x + 1} \; dx = \frac{x^4 \sqrt{x^2 + 1}}{4} - \int \frac{x^5 \; dx}{4\sqrt{x^2 + 1}} + C'. \qquad\qquad (11.6)$$

However, the integral on the right in (11.6) cannot be readily evaluated and is actually in a more complicated form than the integral on the left. Thus the choice of dv and u in (11.5) would not have been suitable in the solution of Example 1.

Example 2   Evaluate $\int x^2 e^{3x}\, dx$.

SOLUTION   In accordance with the suggestions given for forming u and dv we let $dv = e^{3x}\, dx$ and $u = x^2$. Then $v = \frac{1}{3}e^{3x}$ and $du = 2x\, dx$. Then from (11.3)

$$\int x^2 e^{3x}\, dx = \tfrac{1}{3}x^2 e^{3x} - \int \tfrac{2}{3}x e^{3x}\, dx + C'. \tag{11.7}$$

The integral on the right in (11.7) is an improvement on the given integral since the factor of x^2 in the integrand has been replaced by x. This suggests that if (11.3) were again applied, the integral $\int e^{3x}\, dx$ would be obtained on the right. Thus we let $dv = e^{3x}\, dx$ and $u = \frac{2}{3}x$. Then $v = \frac{1}{3}e^{3x}$ and $du = \frac{2}{3}\, dx$, and from (11.3)

$$\int x^2 e^{3x}\, dx = \tfrac{1}{3}x^2 e^{3x} - \left(\tfrac{2}{9}x e^{3x} - \int \tfrac{2}{9}e^{3x}\, dx\right) + C''$$

$$= \tfrac{1}{3}x^2 e^{3x} - \tfrac{2}{9}x e^{3x} + \tfrac{2}{27}e^{3x} + C.$$

The next example is a little more complicated.

Example 3   Evaluate $\int \sec^3 x\, dx$.

SOLUTION   In accordance with our original suggestion that dv contain as much as possible of the integrand and still be easy to integrate, we let $dv = \sec^2 x\, dx$ and hence $u = \sec x$. Then $v = \tan x$ and $du = \sec x \tan x\, dx$, and from (11.3)

$$\int \sec^3 x\, dx = \sec x \tan x - \int \sec x \tan^2 x\, dx + C'. \tag{11.8}$$

Since $\tan^2 x = \sec^2 x - 1$, (11.8) can be written as

$$\int \sec^3 x\, dx = \sec x \tan x - \int \sec^3 x\, dx + \int \sec x\, dx + C'. \tag{11.9}$$

Note that the integral $\int \sec^3 x\, dx$ occurs in both members of (11.9). If in (11.9) we transpose $-\int \sec^3 x\, dx$ to the left, we have

$$2\int \sec^3 x\, dx = \sec x \tan x + \ln|\sec x + \tan x| + C''$$

and

$$\int \sec^3 x\, dx = \tfrac{1}{2}\sec x \tan x + \tfrac{1}{2}\ln|\sec x + \tan x| + C. \tag{11.10}$$

In Exercise 16, Exercise Set 11.1 the student will derive the formula

$$\int \csc^3 x\, dx = -\tfrac{1}{2}\csc x \cot x + \tfrac{1}{2}\ln|\csc x - \cot x| + C. \tag{11.11}$$

Exercise Set 11.1

Obtain the following indefinite integrals using (11.3). Check your solution by differentiating the expression obtained.

1. $\int x \cos 2x \, dx$

2. $\int x \sin 3x \, dx$

3. $\int y^3 e^{-y^2} \, dy$

4. $\int \sin^{-1} 2x \, dx$

5. $\int \tan^{-1} x \, dx$

6. $\int x \tan^{-1} x \, dx$

7. $\int x \sqrt{3x + 5} \, dx$

8. $\int \frac{t \, dt}{(2t - 1)^9}$

9. $\int \ln x \, dx$

10. $\int (\ln x)^2 \, dx$

11. $\int \frac{x^2 \, dx}{\sqrt{2x + 1}}$

12. $\int y^2 \cosh 3y \, dy$

13. $\int r^2 \sin 2r \, dr$

14. $\int x \sin^2 2x \, dx$

15. $\int \sin (\ln t) \, dt$

16. $\int \csc^3 x \, dx$

17. $\int e^x \sin 3x \, dx$

18. $\int e^{2x} \cos x \, dx$

19. Prove that if $a > 0$ and $n > 1$, then

$$\int \frac{1}{x^n} \sin^{-1} \frac{x}{a} \, dx = -\frac{1}{(n-1)x^{n-1}} \sin^{-1} \frac{x}{a} + \frac{1}{n-1} \int \frac{dx}{x^{n-1}\sqrt{a^2 - x^2}}.$$

20. Prove that $\int x^m (\ln x)^n \, dx = \frac{x^{m+1}(\ln x)^n}{m+1} - \frac{n}{m+1} \int x^m (\ln x)^{n-1} \, dx$ if $m \neq -1$.

21. Use the formula in Exercise 20 to evaluate $\int x^4 \ln x \, dx$.

22. Find the area of the region bounded by the graphs of $y = \ln x$, $x = e^{3/2}$ and the x axis.

23. Find the volume obtained when the region in Exercise 22 is revolved about the x axis.

24. Find the volume obtained when the region bounded by graphs of $y = \cos^{-1} x$, $x = -1$, and $y = 0$ is revolved about the x axis.

25. Prove that if $n \geq 2$, then

$$\int \sin^n x \, dx = -\frac{\sin^{n-1} x \cos x}{n} + \frac{n-1}{n} \int \sin^{n-2} x \, dx.$$

26. Prove that if $n \geq 2$, then

$$\int \cos^n x \, dx = \frac{\cos^{n-1} x \sin x}{n} + \frac{n-1}{n} \int \cos^{n-2} x \, dx.$$

27. Prove that if n is any positive even integer, then

$$\int_0^{\pi/2} \sin^n x \, dx = \int_0^{\pi/2} \cos^n x \, dx = \frac{1 \cdot 3 \cdot 5 \cdots (n-1)}{2 \cdot 4 \cdot 6 \cdots n} \cdot \frac{\pi}{2}.$$

HINT: Let $n = 2m$ and use induction on m.

28. Prove that if $n = 3, 5, 7, \ldots$, then

$$\int_0^{\pi/2} \sin^n x \, dx = \int_0^{\pi/2} \cos^n x \, dx = \frac{2 \cdot 4 \cdot 6 \cdots (n-1)}{3 \cdot 5 \cdot 7 \cdots n}.$$

11.2   Some Integrals Involving Powers of Sine and Cosine

In this section we first consider integrals of the form

$$\int \sin^m g(x) \cos^n g(x) g'(x) \, dx \tag{11.12}$$

where m and n are non-negative and one of the following situations exists:

(i) m or n is an odd integer
(ii) m and n are even integers

These integrals that are of the type (i) can be evaluated using the identity

$$\cos^2 g(x) + \sin^2 g(x) = 1. \tag{11.13}$$

First suppose that m is odd. Then $m - 1$ is even and $(m - 1)/2$ is a non-negative integer. From (11.13)

$$\sin^2 g(x) = 1 - \cos^2 g(x).$$

Hence

$$\int \sin^m g(x) \cos^n g(x) g'(x) \, dx$$

$$= \int (\sin^2 g(x))^{(m-1)/2} \cos^n g(x) \sin g(x) g'(x) \, dx$$

$$= \int (1 - \cos^2 g(x))^{(m-1)/2} \cos^n g(x) \sin g(x) g'(x) \, dx$$

$$= \int (\text{polynomial in } \cos g(x)) \sin g(x) g'(x) \, dx.$$

This integral can now be evaluated by letting $u = \cos g(x)$ and $du = -\sin g(x) g'(x) \, dx$.

If n is odd, using the substitution

$$\cos^2 g(x) = 1 - \sin^2 g(x),$$

the integral (11.12) is written in the form $\int (\text{polynomial in } \sin g(x)) \cos g(x) g'(x) \, dx$, which can then be evaluated by letting $u = \sin g(x)$ and $du = \cos g(x) g'(x) \, dx$.

Example 1   Obtain $\int \sin^5 x \cos^4 x \, dx$.

SOLUTION   Since the power of $\sin x$ is odd, we rewrite the integral letting $\sin^2 x = 1 - \cos^2 x$. Then

$$\int \sin^5 x \cos^4 x \, dx = \int (1 - \cos^2 x)^2 \cos^4 x \sin x \, dx$$

$$= \int (\cos^4 x - 2 \cos^6 x + \cos^8 x) \sin x \, dx$$

$$= \int \cos^4 x \sin x \, dx - \int 2 \cos^6 x \sin x \, dx + \int \cos^8 x \sin x \, dx.$$

Each of the integrals on the right is evaluated by letting $u = \cos x$ and $du = - \sin x \, dx$. Then

$$\int \sin^5 x \cos^4 x \, dx = - \frac{\cos^5 x}{5} + \frac{2 \cos^7 x}{7} - \frac{\cos^9 x}{9} + C.$$

Example 2   Obtain $\int \sin^{1/2} 2x \cos^3 2x \, dx$.

SOLUTION   Since the power of $\cos$ is odd, the integrand is rewritten by letting $\cos^2 2x = 1 - \sin^2 2x$. Hence

$$\int \sin^{1/2} 2x \cos^3 2x \, dx = \int \sin^{1/2} 2x (1 - \sin^2 2x) \cos 2x \, dx$$

$$= \int \sin^{1/2} 2x \cos 2x \, dx - \int \sin^{5/2} 2x \cos 2x \, dx.$$

Each of the integrals can be evaluated by letting $u = \sin 2x$ and $du = 2 \cos 2x \, dx$. Then $\cos 2x \, dx = \frac{1}{2} du$ and

$$\int \sin^{1/2} 2x \cos^3 2x \, dx = \int \tfrac{1}{2} u^{1/2} \, du - \int \tfrac{1}{2} u^{5/2} \, du$$

$$= \tfrac{1}{2} \cdot \tfrac{2}{3} u^{3/2} - \tfrac{1}{2} \cdot \tfrac{2}{7} u^{7/2} + C$$

$$= \tfrac{1}{3} \sin^{3/2} 2x - \tfrac{1}{7} \sin^{7/2} 2x + C.$$

If m and n are even in (11.12), substitutions for $\cos^2 g(x)$ and $\sin^2 g(x)$ from (10.19) and (10.20) can be made.

Example 3   Obtain $\int \cos^2 3x \, dx$.

SOLUTION   Since

$$\cos^2 3x = \frac{1 + \cos 6x}{2},$$

we have

$$\int \cos^2 3x \, dx = \int (\tfrac{1}{2} + \tfrac{1}{2} \cos 6x) \, dx$$

$$= \int \tfrac{1}{2} \, dx + \int \tfrac{1}{2} \cos 6x \, dx$$

$$= \tfrac{1}{2}x + \tfrac{1}{12} \sin 6x + C.$$

Example 4   Obtain $\int \sin^4 x \cos^2 x \, dx$.

SOLUTION   Since $\sin^2 x = \dfrac{1 - \cos 2x}{2}$ and $\cos^2 x = \dfrac{1 + \cos 2x}{2}$,

$$\int \sin^4 x \cos^2 x \, dx = \int \left(\frac{1 - \cos 2x}{2} \right)^2 \left(\frac{1 + \cos 2x}{2} \right) dx$$

$$= \int \left(\frac{1 - \cos 2x - \cos^2 2x + \cos^3 2x}{8} \right) dx$$

$$= \int \tfrac{1}{8} \, dx - \int \tfrac{1}{8} \cos 2x \, dx - \int \tfrac{1}{8} \cos^2 2x \, dx$$

$$+ \int \tfrac{1}{8} \cos^3 2x \, dx$$

$$= \tfrac{1}{8}x - \tfrac{1}{16} \sin 2x - \tfrac{1}{8} \int \cos^2 2x \, dx$$

$$+ \tfrac{1}{8} \int \cos^3 2x \, dx. \tag{11.14}$$

In (11.14) let $\cos^2 2x = \dfrac{1 + \cos 4x}{2}$ and $\cos^3 2x = (1 - \sin^2 2x) \cos 2x$. Then

$$\int \sin^4 x \cos^2 x \, dx = \tfrac{1}{8}x - \tfrac{1}{16} \sin 2x - \tfrac{1}{8}(\tfrac{1}{2}x + \tfrac{1}{8} \sin 4x)$$

$$+ \tfrac{1}{8} \int \cos 2x \, dx - \tfrac{1}{8} \int \sin^2 2x \cos 2x \, dx$$

$$= \tfrac{1}{16}x - \tfrac{1}{64} \sin 4x - \tfrac{1}{48} \sin^3 2x + C.$$

If the integral is of the form $\int \sin mx \cos nx \, dx$, $\int \sin mx \sin nx \, dx$, or $\int \cos mx \cos nx \, dx$, the product formulas (10.21), (10.22), and (10.23) are useful.

Example 5   Obtain $\int \sin 4x \cos 3x \, dx$.

SOLUTION   From (10.21) $\sin 4x \cos 3x = \tfrac{1}{2}(\sin x + \sin 7x)$ so

$$\int \sin 4x \cos 3x \, dx = \int \tfrac{1}{2} \sin x \, dx + \int \tfrac{1}{2} \sin 7x \, dx$$

$$= -\tfrac{1}{2} \cos x - \tfrac{1}{14} \cos 7x + C.$$

Exercise Set 11.2

In Exercises 1–18 obtain the indefinite integral.

1. $\displaystyle\int \cos^5 \tfrac{1}{3}x \sin \tfrac{1}{3}x \, dx$

2. $\displaystyle\int e^{-x} \sin^3 e^{-x} \cos e^{-x} \, dx$

3. $\displaystyle\int (\cos 3t \cos t + \sin 3t \sin t) \, dt$

4. $\displaystyle\int (\sin 3t - \cos 3t)^2 \, dt$

5. $\displaystyle\int \sin^3 2x \, dx$

6. $\displaystyle\int \cos^5 x \, dx$

7. $\displaystyle\int \sin^3 y \cos^3 y \, dy$

8. $\displaystyle\int \cos^2 y \sin^5 y \, dy$

9. $\displaystyle\int \sin^3 x \sqrt{\cos x} \, dx$

10. $\displaystyle\int \frac{\cos^5 x \, dx}{\sqrt{\sin x}}$

11. $\displaystyle\int \sin^2 \tfrac{1}{3}x \, dx$

12. $\displaystyle\int \cos^2 2x \sin^2 2x \, dx$

13. $\displaystyle\int \sin^2 x \cos^4 x \, dx$

14. $\displaystyle\int \sin 6x \sin 4x \, dx$

15. $\displaystyle\int \cos 5x \cos 3x \, dx$

16. $\displaystyle\int (\sin 3x + \cos 2x)^2 \, dx$

17. $\displaystyle\int \sin 2x \cos 4x \, dx$

18. $\displaystyle\int \cos 3x \cos x \cos 5x \, dx$

11.3   Some Integrals Involving Powers of Secant and Tangent

Indefinite integrals of the type

$$\int \tan^n g(x) g'(x) \, dx \tag{11.15}$$

where n is an integer ≥ 2 can be evaluated by rewriting the integrand using the identity

$$1 + \tan^2 g(x) = \sec^2 g(x). \tag{11.16}$$

The following example illustrates the method involved.

Example 1   Obtain $\displaystyle\int \tan^5 x \, dx$.

SOLUTION   Using (11.16) twice,

$$\int \tan^5 x \, dx = \int \tan^3 x \tan^2 x \, dx = \int \tan^3 x \, (\sec^2 x - 1) \, dx$$

$$= \int \tan^3 x \sec^2 x \, dx - \int \tan^3 x \, dx$$

$$= \int \tan^3 x \sec^2 x \, dx - \int \tan x \, (\sec^2 x - 1) \, dx$$

$$= \int \tan^3 x \sec^2 x \, dx - \int \tan x \sec^2 x \, dx + \int \tan x \, dx.$$

To evaluate the first two integrals on the right the general power formula can be used with $u = \tan x$ and $du = \sec^2 x \, dx$. Then

$$\int \tan^5 x \, dx = \frac{\tan^4 x}{4} - \frac{\tan^2 x}{2} - \ln|\cos x| + C.$$

In general, the application of (11.16) to an integral of type (11.15) yields an integral of the form $\int \tan^{n-2} g(x)g'(x) \, dx$. Applying (11.16) to this integral will give another integral of the form $\int \tan^{n-4} g(x)g'(x) \, dx$, and so on until finally the integral $\int \tan g(x)g'(x) \, dx$ is obtained if n is odd, or $\int \tan^2 g(x)g'(x) \, dx$ if n is even. These integrals can be easily evaluated either by using (10.71) or by rewriting $\int \tan^2 g(x)g'(x) \, dx$ as

$$\int \tan^2 g(x)g'(x) \, dx = \int (\sec^2 g(x) - 1)g'(x) \, dx$$

$$= \int \sec^2 g(x)g'(x) \, dx - \int g'(x) \, dx.$$

Thus by (10.67)

$$\int \tan^2 g(x)g'(x) \, dx = \tan g(x) - g(x) + C.$$

A similar approach utilizing the identity

$$1 + \cot^2 g(x) = \csc^2 g(x) \tag{11.17}$$

can be employed to evaluate an integral of the form $\int \cot^n g(x)g'(x) \, dx$ where n is an integer ≥ 2.

Example 2   Obtain $\int \cot^4 x \, dx.$

SOLUTION   From (11.17)

$$\int \cot^4 x \, dx = \int \cot^2 x \cot^2 x \, dx = \int \cot^2 x(\csc^2 x - 1) \, dx$$

$$= \int \cot^2 x \csc^2 x \, dx - \int \cot^2 x \, dx$$

$$= \int \cot^2 x \csc^2 x \, dx - \int (\csc^2 x - 1) \, dx$$

$$= \int \cot^2 x \csc^2 x \, dx - \int \csc^2 x \, dx + \int dx.$$

The first integral on the right can be evaluated by letting $u = \cot x$ and $du = -\csc^2 x \, dx$. Then

$$\int \cot^4 x \, dx = -\frac{\cot^3 x}{3} + \cot x + x + C.$$

We next consider integrals of the form

$$\int \sec^m g(x) \tan^n g(x) g'(x) \, dx \tag{11.18}$$

where m and n are non-negative integers and either of the following conditions is true:

(i) $m \geq 2$ is even
(ii) $n \geq 1$ is odd

If $m \geq 2$ is even, then $m - 2$ is even, and the integrand can be rewritten using (11.16):

$$\int \sec^m g(x) \tan^n g(x) g'(x) \, dx = \int (\sec^2 g(x))^{(m-2)/2} \tan^n g(x) \sec^2 g(x) g'(x) \, dx$$
$$= \int (1 + \tan^2 g(x))^{(m-2)/2} \tan^n g(x) \sec^2 g(x) g'(x) \, dx$$
$$= \int (\text{polynomial in } \tan g(x)) \sec^2 g(x) g'(x) \, dx$$

The integral can be evaluated after making the substitutions

$$u = \tan g(x) \qquad du = \sec^2 g(x) g'(x) \, dx.$$

Example 3   Obtain $\displaystyle\int \sec^6 2x \tan^2 2x \, dx.$

SOLUTION   Using (11.16)

$$\int \sec^6 2x \tan^2 2x \, dx = \int \sec^4 2x \tan^2 2x \sec^2 2x \, dx$$

$$= \int (1 + \tan^2 2x)^2 \tan^2 2x \sec^2 2x \, dx$$

$$= \int (\tan^2 2x + 2 \tan^4 2x + \tan^6 2x) \sec^2 2x \, dx.$$

Letting $u = \tan 2x$, $du = 2 \sec^2 2x \, dx$ and $\sec^2 2x \, dx = \frac{1}{2} \, du$. Then

$$\int \sec^6 2x \tan^2 2x \, dx = \int \frac{1}{2}(u^2 + 2u^4 + u^6) \, du$$

$$= \frac{u^3}{6} + \frac{u^5}{5} + \frac{u^7}{14} + C$$

$$= \frac{\tan^3 2x}{6} + \frac{\tan^5 2x}{5} + \frac{\tan^7 2x}{14} + C.$$

If $n \geq 1$ is odd, then $n - 1$ is even and from (11.16)

$$\int \sec^m g(x) \tan^n g(x)g'(x) \, dx$$

$$= \int \sec^{m-1} g(x)(\tan^2 g(x))^{(n-1)/2} \sec g(x) \tan g(x)g'(x) \, dx$$

$$= \int \sec^{m-1} g(x)(\sec^2 g(x) - 1)^{(n-1)/2} \sec g(x) \tan g(x)g'(x) \, dx$$

$$= \int (\text{polynomial in } \sec g(x)) \sec g(x) \tan g(x)g'(x) \, dx.$$

The integral can then be evaluated after making the substitutions

$$u = \sec g(x) \qquad du = \sec g(x) \tan g(x)g'(x) \, dx.$$

Example 4   Obtain $\int \sec^4 3x \tan^5 3x \, dx$.

SOLUTION   Using (11.16)

$$\int \sec^4 3x \tan^5 3x \, dx = \int \sec^3 3x(\tan^2 3x)^2 \sec 3x \tan 3x \, dx$$

$$= \int \sec^3 3x(\sec^2 3x - 1)^2 \sec 3x \tan 3x \, dx$$

$$= \int (\sec^7 3x - 2 \sec^5 3x + \sec^3 3x) \sec 3x \tan 3x \, dx.$$

Letting $u = \sec 3x$, $du = 3 \sec 3x \tan 3x \, dx$, and $\sec 3x \tan 3x \, dx = \frac{1}{3} du$,

$$\int \sec^4 3x \tan^5 3x \, dx = \int \frac{1}{3}(u^7 - 2u^5 + u^3) \, du$$

$$= \frac{u^8}{24} - \frac{u^6}{9} + \frac{u^4}{12} + C$$

$$= \frac{\sec^8 3x}{24} - \frac{\sec^6 3x}{9} + \frac{\sec^4 3x}{12} + C.$$

Integrals of the form

$$\int \csc^m g(x) \cot^n g(x)g'(x) \, dx$$

where m and n are non-negative integers and either of the following conditions is true:

(i) $m \geq 2$ is even
(ii) $n \geq 1$ is odd

can be evaluated by a method that utilizes the identity (11.17) and is analogous to a method used to evaluate integrals of the type (11.18).

Example 5   Obtain $\int \csc^3 x \cot^3 x \, dx$.

SOLUTION   Since this integral is of the type (ii), we write

$$\int \csc^3 x \cot^3 x \, dx = \int \csc^2 x \cot^2 x \csc x \cot x \, dx$$

$$= \int \csc^2 x (\csc^2 x - 1) \csc x \cot x \, dx$$

$$= \int (\csc^4 x - \csc^2 x) \csc x \cot x \, dx.$$

Then letting $u = \csc x$ and $du = -\csc x \cot x \, dx$,

$$\int \csc^3 x \cot^3 x \, dx = \int (u^4 - u^2)(-du) = -\frac{u^5}{5} + \frac{u^3}{3} + C$$

$$= -\frac{\csc^5 x}{5} + \frac{\csc^3 x}{3} + C.$$

Exercise Set 11.3

In Exercises 1–16 evaluate the indefinite integral.

1. $\int \cot^3 x \, dx$

2. $\int \tan^4 x \, dx$

3. $\int \csc^4 3x \, dx$

4. $\int \csc^3 x \cot^3 x \, dx$

5. $\int \sec^3 x \tan^3 x \, dx$

6. $\int \sec^7 2x \tan^3 2x \, dx$

7. $\int \sec^4 2x \tan^2 2x \, dx$

8. $\int \csc^4 x \cot^4 x \, dx$

9. $\int \csc^5 \frac{1}{2}x \cot^3 \frac{1}{2}x \, dx$

10. $\int \sec^6 2x \, dx$

11. $\int (\tan 2y - \cot 2y)^2 \, dy$

12. $\int (\sec 3\theta + \tan 3\theta)^2 \, d\theta$

13. $\int \frac{\sin^3 t}{\cos^7 t} \, dt$

14. $\int \tan^{1/2} x \sec^4 x \, dx$

15. $\int \frac{\tan^3 x \, dx}{\sec x - \tan x}$

16. $\int \sec^4 y \cot^3 y \, dy$

17. Evaluate $\int_0^{\pi/4} \sec^2 x \tan^2 x \, dx.$ 18. Evaluate $\int_{\pi/6}^{\pi/3} \tan^3 x \, dx.$

19. Evaluate $\int_{\pi/6}^{\pi/2} \csc^{3/2} x \cot^3 x \, dx.$

20. If $n \geq 2$, prove that $\int \tan^n x \, dx = \dfrac{\tan^{n-1} x}{n - 1} - \int \tan^{n-2} x \, dx.$

21. If $n \geq 2$, derive a formula similar to that in Exercise 20 for $\int \cot^n x \, dx$.

11.4   Integration by Trigonometric Substitution

We frequently encounter integrals in which the integrand involves $\sqrt{a^2 - (g(x))^2}$, $\sqrt{a^2 + (g(x))^2}$, or $\sqrt{(g(x))^2 - a^2}$ where $a > 0$. Such integrals can often be evaluated following a substitution involving trigonometric functions. The following directions can be given:

If the integrand contains $\sqrt{a^2 - (g(x))^2}$, we let $u = \sin^{-1}\dfrac{g(x)}{a}$.

If the integrand contains $\sqrt{(g(x))^2 + a^2}$, we let $u = \tan^{-1}\dfrac{g(x)}{a}$.

If the integrand contains $\sqrt{(g(x))^2 - a^2}$, we let $u = \sec^{-1}\dfrac{g(x)}{a}$.

The foregoing substitutions are based on the trigonometric identities

$$\cos^2\theta + \sin^2\theta = 1 \qquad \text{and} \qquad 1 + \tan^2\theta = \sec^2\theta.$$

If, for example, the integrand contains $\sqrt{a^2 - (g(x))^2}$, then from

$$u = \sin^{-1}\frac{g(x)}{a} \tag{11.20}$$

we have

$$g(x) = a\sin u$$
$$a^2 - (g(x))^2 = a^2 - a^2\sin^2 u = a^2(1 - \sin^2 u)$$
$$= a^2\cos^2 u. \tag{11.21}$$

From (11.20), $u \in [-\frac{\pi}{2}, \frac{\pi}{2}]$ and therefore $\cos u > 0$. Hence from (11.21)

$$\sqrt{a^2 - (g(x))^2} = \sqrt{a^2\cos^2 u} = a\cos u.$$

A substitution of the form (11.20) is utilized in the following example.

Example 1   Obtain $\displaystyle\int \frac{\sqrt{9 - x^2}\ dx}{x^2}$.

SOLUTION   Since the factor $\sqrt{9 - x^2}$ is of the form $\sqrt{a^2 - (g(x))^2}$ we make the substitution

$$u = \sin^{-1}\frac{x}{3}. \tag{11.22}$$

Then from (11.22)

$$\sin u = \frac{x}{3} \qquad \text{and} \qquad u \in \left[-\frac{\pi}{2}, \frac{\pi}{2}\right], \tag{11.23}$$

and hence

$$x = 3\sin u \qquad dx = 3\cos u\ du. \tag{11.24}$$

The other trigonometric functions of u can be readily obtained from a right triangle sketch based upon (11.23) (Figure 11–1). In this sketch u denotes an acute angle of the triangle, x is the length of the leg opposite u, and 3 is the length of the hypotenuse. Then by the Pythagorean theorem, the other leg is of length $\sqrt{9 - x^2}$. From Figure 11–1

$$\sqrt{9 - x^2} = 3 \cos u. \tag{11.25}$$

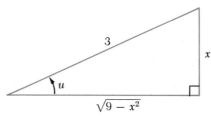

Figure 11–1

Even though in Figure 11–1 u is shown to be an acute angle, the relations (11.24) and (11.25) are still valid even if $-\frac{\pi}{2} < u \leq 0$.

After a substitution using (11.24) and (11.25)

$$\int \frac{\sqrt{9 - x^2}\, dx}{x^2} = \int \frac{9 \cos^2 u\, du}{9 \sin^2 u} = \int \cot^2 u\, du$$

$$= \int (\csc^2 u - 1)\, du = -\cot u - u + C.$$

From Figure 11–1 $\cot u = \sqrt{9 - x^2}/x$ and thus

$$\int \frac{\sqrt{9 - x^2}\, dx}{x^2} = -\frac{\sqrt{9 - x^2}}{x} - \sin^{-1}\frac{x}{3} + C.$$

As we noted at the beginning of this section, when the integrand involves $a^2 + (g(x))^2$ the substitution

$$u = \tan^{-1}\frac{g(x)}{a} \tag{11.26}$$

is used. From (11.26)

$$g(x) = a \tan u$$

$$a^2 + (g(x))^2 = a^2 + a^2 \tan^2 u = a^2(1 + \tan^2 u)$$

$$= a^2 \sec^2 u. \tag{11.27}$$

Also from (11.26), $u \in (-\frac{\pi}{2}, \frac{\pi}{2})$ and therefore $\sec u > 0$. Hence from (11.27)

$$\sqrt{a^2 + (g(x))^2} = \sqrt{a^2 \sec^2 u} = a \sec u.$$

A substitution of the form (11.26) is used in Example 2.

Example 2 Obtain $\displaystyle\int \frac{dx}{\sqrt{4x^2 + 5}}$.

SOLUTION The denominator is of the form $\sqrt{(g(x))^2 + a^2}$, and therefore we make the substitution

$$u = \tan^{-1} \frac{2x}{\sqrt{5}}. \tag{11.28}$$

Then from (11.28)

$$\tan u = \frac{2x}{\sqrt{5}} \quad \text{and} \quad u \in \left(-\frac{\pi}{2}, \frac{\pi}{2}\right)$$

and hence

$$x = \frac{\sqrt{5}}{2} \tan u \qquad dx = \frac{\sqrt{5}}{2} \sec^2 u \, du. \tag{11.29}$$

The other trigonometric functions of u besides $\tan u$ can be obtained from the sketch of the right triangle in Figure 11–2. We see that

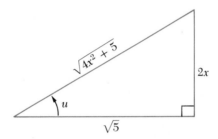

Figure 11–2

$$\sqrt{4x^2 + 5} = \sqrt{5} \sec u. \tag{11.30}$$

After substituting from (11.29) and (11.30)

$$\int \frac{dx}{\sqrt{4x^2 + 5}} = \int \frac{(\sqrt{5}/2) \sec^2 u \, du}{\sqrt{5} \sec u} = \int \frac{1}{2} \sec u \, du$$

$$= \tfrac{1}{2} \ln |\sec u + \tan u| + C'$$

$$= \frac{1}{2} \ln \left| \frac{\sqrt{4x^2 + 5}}{\sqrt{5}} + \frac{2x}{\sqrt{5}} \right| + C'$$

$$= \frac{1}{2} \ln \left| \frac{\sqrt{4x^2 + 5} + 2x}{\sqrt{5}} \right| + C'. \tag{11.31}$$

The expression obtained in (11.31) can be simplified. Since $\sqrt{4x^2 + 5} + 2x$ $\geq |2x| + 2x \geq 0$, we may drop the absolute value bars enclosing $(\sqrt{4x^2 + 5} + 2x)/\sqrt{5}$ since it is non-negative. Also, since

$$\frac{1}{2} \ln \frac{\sqrt{4x^2 + 5} + 2x}{\sqrt{5}} = \frac{1}{2} \ln (\sqrt{4x^2 + 5} + 2x) - \frac{1}{2} \ln \sqrt{5},$$

C' can be combined with $-\tfrac{1}{2} \ln \sqrt{5}$ to give a constant C. Hence

$$\int \frac{dx}{\sqrt{4x^2 + 5}} = \frac{1}{2} \ln \left(\sqrt{4x^2 + 5} + 2x\right) + C.$$

From our remarks at the beginning of this section when the integrand involves $\sqrt{(g(x))^2 - a^2}$, the substitution

$$u = \sec^{-1} \frac{g(x)}{a} \tag{11.32}$$

is used. Then

$$g(x) = a \sec u$$
$$(g(x))^2 - a^2 = a^2 \sec^2 u - a^2 = a^2(\sec^2 u - 1)$$
$$= a^2 \tan^2 u. \tag{11.33}$$

Also from (11.32) $u \in [0, \frac{\pi}{2}) \cup [\pi, \frac{3\pi}{2})$ and therefore $\tan u \geq 0$. Hence from (11.33)

$$\sqrt{(g(x))^2 - a^2} = \sqrt{a^2 \tan^2 u} = a \tan u.$$

The substitution (11.32) will be used in the next example.

Example 3   Evaluate $\displaystyle\int \frac{\sqrt{3x^2 - 4}}{x} \, dx.$

SOLUTION   Since the factor $\sqrt{3x^2 - 4}$ is of the form $\sqrt{(g(x))^2 - a^2}$, we let

$$u = \sec^{-1} \frac{x\sqrt{3}}{2}. \tag{11.34}$$

From (11.34)

$$\sec u = \frac{x\sqrt{3}}{2} \qquad \text{and} \qquad u \in \left[0, \frac{\pi}{2}\right) \cup \left[\pi, \frac{3\pi}{2}\right)$$

and hence

$$x = \frac{2}{\sqrt{3}} \sec u \qquad dx = \frac{2}{\sqrt{3}} \sec u \tan u \, du. \tag{11.35}$$

With the aid of the right triangle sketch in Figure 11–3 we obtain

$$\sqrt{3x^2 - 4} = 2 \tan u. \tag{11.36}$$

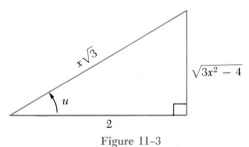

Figure 11–3

Substituting from (11.35) and (11.36),

$$\int \frac{\sqrt{3x^2 - 4}}{x}\, dx = \int \frac{2 \tan u \cdot \dfrac{2}{\sqrt{3}} \sec u \tan u \, du}{\dfrac{2}{\sqrt{3}} \sec u}$$

$$= \int 2 \tan^2 u \, du$$

$$= \int 2(\sec^2 u - 1) \, du$$

$$= 2 \tan u - 2u + C$$

$$= \sqrt{3x^2 - 4} - 2 \sec^{-1} \frac{x\sqrt{3}}{2} + C.$$

Whenever the integrand contains an expression of the form

$$\sqrt{Ax^2 + Bx + C}, \tag{11.37}$$

the procedure of completing the square can be utilized to rewrite the expression in one of the forms $\sqrt{a^2 - (g(x))^2}$, $\sqrt{a^2 + (g(x))^2}$, or $\sqrt{(g(x))^2 - a^2}$. To verify this assertion, we obtain

$$Ax^2 + Bx + C = A\left(x^2 + \frac{B}{A}x\right) + C$$

$$= A\left(x^2 + \frac{B}{A}x + \frac{B^2}{4A^2}\right) + C - \frac{B^2}{4A}$$

$$= A\left(x + \frac{B}{2A}\right)^2 + \left(C - \frac{B^2}{4A}\right).$$

Thus whenever an integrand contains an expression of the form (11.37), the use of a trigonometric substitution should be considered as a possible technique for evaluating the integral.

Example 4 Evaluate $\displaystyle\int \frac{dx}{(7 + 4x - 2x^2)^{3/2}}$.

SOLUTION Completing the square for $7 + 4x - 2x^2$ gives

$$7 + 4x - 2x^2 = 9 - 2(x - 1)^2.$$

Since $9 - 2(x - 1)^2$ is of the form $a^2 - (g(x))^2$, we use the substitution

$$u = \sin^{-1} \frac{(x - 1)\sqrt{2}}{3}. \tag{11.38}$$

From (11.38)

$$\sin u = \frac{(x - 1)\sqrt{2}}{3}, \tag{11.39}$$

and hence

$$x = \frac{3}{\sqrt{2}} \sin u + 1 \qquad dx = \frac{3}{\sqrt{2}} \cos u \, du. \qquad (11.40)$$

From (11.38) since $u \in [-\frac{\pi}{2}, \frac{\pi}{2}]$

$$\sec u = \frac{3}{\sqrt{7 + 4x - 2x^2}} \qquad (11.41)$$

and

$$\frac{\sec^3 u}{27} = \frac{1}{(7 + 4x - 2x^2)^{3/2}}. \qquad (11.42)$$

Hence from (11.40) and (11.42)

$$\int \frac{dx}{(7 + 4x - 2x^2)^{3/2}} = \int \frac{\sec^3 u}{27} \cdot \frac{3}{\sqrt{2}} \cos u \, du$$

$$= \int \frac{1}{9\sqrt{2}} \sec^2 u \, du$$

$$= \frac{1}{9\sqrt{2}} \tan u + C. \qquad (11.43)$$

We can obtain

$$\tan u = \frac{(x - 1)\sqrt{2}}{\sqrt{7 + 4x - 2x^2}} \qquad (11.44)$$

either by using a right triangle diagram or by obtaining $\tan u = \sin u \sec u$ from (11.39) and (11.41). Hence by substituting from (11.44) into (11.43),

$$\int \frac{dx}{(7 + 4x - 2x^2)^{3/2}} = \frac{x - 1}{9\sqrt{7 + 4x - 2x^2}} + C.$$

The derivations of the following useful integrals are left as exercises. Here $a > 0$.

$$\int \sqrt{a^2 - x^2} \, dx = \frac{1}{2}x\sqrt{a^2 - x^2} + \frac{1}{2}a^2 \sin^{-1}\frac{x}{a} + C. \qquad (11.45)$$

$$\int (a^2 - x^2)^{3/2} \, dx = \frac{1}{4}x(a^2 - x^2)^{3/2} + \frac{3}{8}a^2x\sqrt{a^2 - x^2}$$

$$+ \frac{3}{8}a^4 \sin^{-1}\frac{x}{a} + C. \qquad (11.46)$$

$$\int x^2\sqrt{a^2 - x^2} \, dx = -\frac{1}{4}x(a^2 - x^2)^{3/2} + \frac{1}{8}a^2x\sqrt{a^2 - x^2}$$

$$+ \frac{1}{8}a^4 \sin^{-1}\frac{x}{a} + C. \qquad (11.47)$$

It should be realized that Theorem 7.1.10 (chain rule for antidifferentiation) does not provide the justification for the trigonometric substitutions discussed in this section. Here an indefinite integral of the form

$$\int f(x)\, dx \tag{11.48}$$

is evaluated by making a substitution of the form

$$u = \psi(x). \tag{11.49}$$

If (11.49) can be solved, giving

$$x = \psi^{-1}(u) = \phi(u)$$

and hence

$$dx = \phi'(u)\, du,$$

then in place of (11.48) we shall consider $\int f(\phi(u)\phi'(u)\, du$. If

$$\int f(\phi(u))\phi'(u)\, du = H(u) + C,$$

it will be proved in the following theorem that

$$\int f(x)\, dx = H(\phi^{-1}(x)) + C.$$

11.4.1   Theorem

Suppose the following conditions are satisfied.

> *(i) f is continuous on an interval I.*
> *(ii) ϕ has a continuous derivative on an interval J and $\phi'(u) \neq 0$ for every $u \in J$.*
> *(iii) $\{\phi(u): u \in J\} \subset I$*

Then the function $(f \circ \phi)\phi'$ has an integral H on J and

$$\int f(x)\, dx = H(\phi^{-1}(x)) + C$$

on the set $\{\phi(u): u \in J\}$, which is an interval.

PROOF   From (i) f is continuous on I, and from (ii) and (iii) ϕ is continuous on J and has its range in I. Thus by Theorem 3.8.4 $f \circ \phi$ is continuous on J. Since ϕ' is continuous on J, $(f \circ \phi)\phi'$ is also continuous on J. Then by the first fundamental theorem there is a function H such that

$$H'(u) = f(\phi(u))\phi'(u) \quad \text{for every } u \in J. \tag{11.50}$$

Also from (ii), $\phi'(u)$ is always of the same sign for every $u \in J$, for otherwise by Theorem 6.1.2 there would be a $c \in J$ for which $\phi'(c) = 0$. Thus by Theorem 5.3.2 ϕ is monotonic on J, and by Theorem 9.4.1 there is an inverse function ϕ^{-1} that is continuous on the set $\{\phi(u): u \in J\}$ and moreover this set is an interval. By Theorem 9.4.2

$$(\phi^{-1})'(x) = \frac{1}{\phi'(\phi^{-1}(x))} \quad \text{for } x \in \{\phi(u): u \in J\}. \tag{11.51}$$

Then from (11.50) and (11.51) if u is replaced by $\phi^{-1}(x)$,

$$H'(\phi^{-1}(x)) = f(x)\phi'(\phi^{-1}(x))$$

$$= \frac{f(x)}{(\phi^{-1})'(x)}. \tag{11.52}$$

By the chain rule

$$D_x H(\phi^{-1}(x)) = H'(\phi^{-1}(x))(\phi^{-1})'(x)$$

and hence from (11.52)

$$D_x H(\phi^{-1}(x)) = f(x) \quad \text{for } x \in \{\phi(u): u \in J\}.$$

Thus the conclusion is obtained.

From Theorem 11.4.1 we shall derive a second change of variable formula for evaluating a definite integral (compare Theorem 7.8.2). This formula is useful when the integrand can be rewritten by a trigonometric substitution.

11.4.2 Theorem

Suppose f and ϕ satisfy the hypotheses of Theorem 11.4.1. Then if a and b are any numbers in the set $\{\phi(u): u \in J\}$,

$$\int_a^b f(x)\, dx = \int_{\phi^{-1}(a)}^{\phi^{-1}(b)} f(\phi(u))\phi'(u)\, du. \tag{11.53}$$

PROOF From the proof of Theorem 11.4.1 the function ϕ^{-1} is continuous on the set $\{\phi(u): u \in J\}$, which is an interval. Then from (11.50) by the second fundamental theorem of calculus (Theorem 7.8.1),

$$\int_{\phi^{-1}(a)}^{\phi^{-1}(b)} f(\phi(u))\phi'(u)\, du = H(\phi^{-1}(b)) - H(\phi^{-1}(a)). \tag{11.54}$$

Also by Theorem 11.4.1 and Theorem 7.8.1

$$\int_a^b f(x)\, dx = H(\phi^{-1}(b)) - H(\phi^{-1}(a)). \tag{11.55}$$

The conclusion is then obtained by equating the left members of (11.54) and (11.55).

Example 5 Evaluate $\displaystyle \int_0^2 \frac{dx}{(4x^2 + 9)^{3/2}}$.

SOLUTION Let $u = \tan^{-1} \dfrac{2x}{3}$. Then $\tan u = \dfrac{2x}{3}$ and

$$x = \phi(u) = \tfrac{3}{2} \tan u,$$

$$dx = \tfrac{3}{2} \sec^2 u \, du,$$
$$4x^2 + 9 = 9(\tan^2 u + 1) = 9 \sec^2 u.$$

Also

$$u = 0 \quad \text{when } x = 0$$
$$u = \tan^{-1}\tfrac{4}{3} \quad \text{when } x = 2.$$

Hence if (11.53) is applied to the given definite integral,

$$\int_0^2 \frac{dx}{(4x^2 + 9)^{3/2}} = \int_0^{\tan^{-1}(4/3)} \frac{\tfrac{3}{2} \sec^2 u \, du}{(9 \sec^2 u)^{3/2}}$$

$$= \int_0^{\tan^{-1}(4/3)} \tfrac{1}{18} \cos u \, du$$

$$= \tfrac{1}{18} \sin u \Big|_0^{\tan^{-1}(4/3)} = \tfrac{1}{18} \cdot \tfrac{4}{5}$$

$$= \tfrac{2}{45}.$$

Exercise Set 11.4

In Exercises 1–24 obtain the indefinite integrals using a trigonometric substitution.

1. $\displaystyle\int \frac{dx}{(4 - x^2)^{3/2}}$

2. $\displaystyle\int \frac{x^3 \, dx}{\sqrt{5 - x^2}}$

3. $\displaystyle\int \frac{dy}{\sqrt{y^2 + 3}}$

4. $\displaystyle\int \frac{dy}{(9y^2 - 2)^{3/2}}$

5. $\displaystyle\int \sqrt{4x^2 - 25} \, dx$

6. $\displaystyle\int \frac{x^2 \, dx}{\sqrt{3x^2 - 16}}$

7. $\displaystyle\int \frac{x^2 \, dx}{\sqrt{9x^2 + 16}}$

8. $\displaystyle\int \frac{dx}{(x^2 + 1)^2}$

9. $\displaystyle\int \frac{dx}{x^2 \sqrt{x^2 - 4}}$

10. $\displaystyle\int \frac{\sqrt{v^2 + 9}}{v} \, dv$

11. $\displaystyle\int \frac{dt}{(t^2 + 4t + 11)^2}$

12. $\displaystyle\int \frac{dx}{\sqrt{5 - 2x - x^2}}$

13. $\displaystyle\int \frac{(x + 1) \, dx}{\sqrt{4 - 3x - x^2}}$

14. $\displaystyle\int \frac{y \, dy}{\sqrt{y^2 - y + 1}}$

15. $\displaystyle\int \frac{(2x - 3) \, dx}{\sqrt{4x^2 - 8x - 20}}$

16. $\displaystyle\int \frac{(2x - 1) \, dx}{(3x^2 - 9x - 27)^{3/2}}$

17. $\displaystyle\int \frac{e^{3x} \, dx}{\sqrt{4 - e^{2x}}}$

18. $\displaystyle\int \frac{(2 \sinh t - 5) \cosh t \, dt}{9 \sinh^2 t + 1}$

19. $\displaystyle\int \frac{dt}{t[(\ln t)^2 - 1]^{3/2}}$

20. $\displaystyle\int \frac{dx}{\sqrt{4e^{-2x} - 9}}$

21. $\displaystyle \int \frac{dx}{(x+2)\sqrt{x^2-4}}$

22. $\displaystyle \int \frac{(x-2)^2\,dx}{\sqrt{x^2-4x+5}}$

23. $\displaystyle \int \frac{\sqrt{9-x^2}}{x^3}\,dx$

24. $\displaystyle \int \sec^{-1} x\,dx$

In Exercises 25–30 evaluate the given definite integrals.

25. $\displaystyle \int_0^1 \frac{x^2\,dx}{(25-9x^2)^{3/2}}$

26. $\displaystyle \int_0^{\sqrt{2}} \frac{dx}{(4x^2+1)^{5/2}}$

27. $\displaystyle \int_0^{(1/2)\ln 2} \frac{dx}{e^x\sqrt{2e^{2x}-1}}$

28. $\displaystyle \int_0^1 \frac{x\,dx}{\sqrt{6-x-x^2}}$

29. $\displaystyle \int_0^6 \frac{dx}{(2x^2+4x+4)^{3/2}}$

30. $\displaystyle \int_1^5 \frac{dx}{\sqrt{3x^2+6x+5}}$

31. Derive formula (11.45).

32. Derive formula (11.46).

33. Derive formula (11.47).

34. Find the area of the region bounded by the hyperbola $4x^2 - y^2 = 4$ and the line $x = \sqrt{5}$.

35. Find the area of the region enclosed by the ellipse

$$\frac{x^2}{a^2} + \frac{y^2}{b^2} = 1.$$

11.5 Integration by Partial Fractions

Suppose we wanted to evaluate the indefinite integral

$$\int \frac{(x^2 - 7x + 9)\,dx}{x^3 - 3x + 2} \tag{11.56}$$

where it will be noted that the denominator of the integrand,

$$x^3 - 3x + 2 = (x-1)^2(x+2). \tag{11.57}$$

It can be shown that

$$\frac{x^2 - 7x + 9}{x^3 - 3x + 2} = -\frac{2}{x-1} + \frac{1}{(x-1)^2} + \frac{3}{x+2}. \tag{11.58}$$

Hence

$$\int \frac{x^2 - 7x + 9}{x^3 - 3x + 2}\,dx = \int -\frac{2\,dx}{x-1} + \int \frac{dx}{(x-1)^2} + \int \frac{3\,dx}{x+2}$$

$$= -2\ln|x-1| - \frac{1}{x-1} + 3\ln|x+2| + C$$

$$= -\frac{1}{x-1} + \ln\frac{|x+2|^3}{|x-1|^2} + C,$$

and since $|x - 1|^2 = (x - 1)^2$,

$$\int \frac{x^2 - 7x + 9}{x^3 - 3x + 2} \, dx = -\frac{1}{x - 1} + \ln \frac{|x + 2|^3}{(x - 1)^2} + C.$$

The usefulness of this approach for integrating rational functions depends upon a method for formulating the equation (11.58), in which the integrand is expressed as the sum of fractions each of which can be integrated. From the following theorem, which is proved in algebra, the equation (11.58) can be written.

11.5.1   Theorem

Suppose the following conditions are satisfied:

     (i) *$P(x)$ and $Q(x)$ are polynomials with real coefficients and the degree of $P(x)$ is less than the degree of $Q(x)$.*
    (ii) *The leading coefficient of $Q(x)$ is 1.*
   (iii) *a is any number and n is a positive integer.*
   (iv) *There is a polynomial $g(x)$ with real coefficients such that $Q(x) = (x - a)^n g(x)$, but $x - a$ is not a factor of $g(x)$.*

Then there are numbers $A_1, A_2, \ldots, A_n$ and a polynomial $f(x)$ with real coefficients having a degree that is less than the degree of $g(x)$ such that

$$\frac{P(x)}{Q(x)} = \frac{A_1}{x - a} + \frac{A_2}{(x - a)^2} + \cdots + \frac{A_n}{(x - a)^n} + \frac{f(x)}{g(x)}.$$

The fractions $\dfrac{A_1}{x - a}, \dfrac{A_2}{(x - a)^2}, \ldots, \dfrac{A_n}{(x - a)^n}$ are called *partial fractions* of $P(x)/Q(x)$. If $Q(x)$ factors completely into the product $(x - a_1)^{n_1}(x - a_2)^{n_2} \cdots (x - a_m)^{n_m}$, then $P(x)/Q(x)$ can be completely expanded as a sum of partial fractions. For each factor $(x - a_i)^{n_i}$ of $Q(x)$ the contribution to this sum will be n_i partial fractions, with respective denominators $x - a_i, (x - a_i)^2, \ldots, (x - a_i)^{n_i}$.

    Thus from (11.57) and Theorem 11.5.1, the integrand of (11.56) has the partial fraction representation

$$\frac{x^2 - 7x + 9}{x^3 - 3x + 2} = \frac{A_1}{x - 1} + \frac{A_2}{(x - 1)^2} + \frac{B}{x + 2}. \tag{11.59}$$

If $x \neq 1$ and -2, then (11.59) is equivalent to

$$x^2 - 7x + 9 = A_1(x - 1)(x + 2) + A_2(x + 2) + B(x - 1)^2$$
$$= A_1(x^2 + x - 2) + A_2(x + 2) + B(x^2 - 2x + 1).$$

If the terms on the right are regrouped,

$$x^2 - 7x + 9 = (A_1 + B)x^2 + (A_1 + A_2 - 2B)x + (-2A_1 + 2A_2 + B). \tag{11.60}$$

By a theorem in algebra, if two polynomials of degree n in x are equal for more than n values of x, then their corresponding coefficients are equal. Thus, if the corresponding coefficients of the polynomials on each side of (11.60) are equated,

$$\text{for the } x^2 \text{ coefficients:} \quad 1 = A_1 + B$$
$$\text{for the } x \text{ coefficients:} \quad -7 = A_1 + A_2 - 2B$$
$$\text{for the constant terms:} \quad 9 = -2A_1 + 2A_2 + B$$

If these equations are solved simultaneously, the solution

$$A_1 = -2 \qquad A_2 = 1 \qquad B = 3 \tag{11.61}$$

is obtained. If the numbers obtained in (11.61) are substituted in (11.59), the equation (11.58) is obtained.

Another illustration of integration by partial fractions using Theorem 11.5.1 follows.

Example 1 Obtain $\displaystyle\int \frac{x^5 - x^4 - 3x^3 - 2x^2 + 2x + 5}{x^4 + x^3} \, dx.$

SOLUTION The integrand is the quotient of two polynomials with real coefficients, but the degree of the numerator is not less than the degree of the denominator, so Theorem 11.5.1 cannot be applied immediately. However, by long division

$$\frac{x^5 - x^4 - 3x^3 - 2x^2 + 2x + 5}{x^4 + x^3} = x - 2 + \frac{-x^3 - 2x^2 + 2x + 5}{x^4 + x^3}$$

and therefore

$$\int \frac{x^5 - x^4 - 3x^3 - 2x^2 + 2x + 5}{x^4 + x^3} \, dx$$

$$= \int x \, dx - \int 2 \, dx + \int \frac{-x^3 - 2x^2 + 2x + 5}{x^4 + x^3} \, dx. \tag{11.62}$$

Since the degree of the numerator of the integrand in the third integral on the right is less than the degree of the denominator, and since $x^4 + x^3 = x^3(x + 1)$, by Theorem 11.5.1 the integrand has a partial fraction representation of the form

$$\frac{-x^3 - 2x^2 + 2x + 5}{x^4 + x^3} = \frac{A}{x} + \frac{B}{x^2} + \frac{C}{x^3} + \frac{D}{x + 1}. \tag{11.63}$$

If $x \neq 0$ and -1, then (11.63) is equivalent to the following equations:

$$-x^3 - 2x^2 + 2x + 5 = Ax^2(x + 1) + Bx(x + 1) + C(x + 1) + Dx^3$$
$$-x^3 - 2x^2 + 2x + 5 = (A + D)x^3 + (A + B)x^2 + (B + C)x + C \tag{11.64}$$

From equating the corresponding coefficients of the polynomials in each member of (11.64), we obtain the equations:

$$A + D = -1$$
$$A + B = -2$$

$$B + C = 2$$
$$C = 5$$

If these equations are solved simultaneously, the solution

$$A = 1 \qquad B = -3 \qquad C = 5 \qquad D = -2 \tag{11.65}$$

is obtained. If the numbers from (11.65) are substituted into (11.63), then from (11.63) and (11.62)

$$\int \frac{(x^5 - x^4 - 3x^3 - 2x^2 + 2x + 5)\, dx}{x^4 + x^3}$$

$$= \frac{x^2}{2} - 2x + \ln|x| + \frac{3}{x} - \frac{5}{2x^2} - 2\ln|x+1| + C$$

$$= \frac{x^2}{2} - 2x + \frac{3}{x} - \frac{5}{2x^2} + \ln\frac{|x|}{(x+1)^2} + C.$$

It is proved in algebra that every nonconstant polynomial with real coefficients can be represented as a product in which each factor is a polynomial that is linear or quadratic and has real coefficients. The following theorem from algebra is useful in the integration of the quotient of two polynomials when the denominator has at least one irreducible quadratic factor.

11.5.2   Theorem

Suppose the following conditions are satisfied:

    (i) $P(x)$ and $Q(x)$ satisfy conditions (i) and (ii) of Theorem 11.5.1.
    (ii) n is a positive integer and b and c are any numbers such that $b^2 - 4c < 0$.
    (iii) There is a polynomial $g(x)$ with real coefficients such that $Q(x) = (x^2 + bx + c)^n g(x)$, but $x^2 + bx + c$ is not a factor of $g(x)$.

Then there are numbers $A_1, A_2, \ldots, A_n$, and $B_1, B_2, \ldots, B_n$ and a polynomial $f(x)$ with real coefficients having a degree that is less than the degree of $g(x)$ such that

$$\frac{P(x)}{Q(x)} = \frac{A_1 x + B_1}{x^2 + bx + c} + \frac{A_2 x + B_2}{(x^2 + bx + c)^2} + \cdots + \frac{A_n x + B_n}{(x^2 + bx + c)^n} + \frac{f(x)}{g(x)}.$$

The condition $b^2 - 4c < 0$ in (ii) guarantees that $x^2 + bx + c$ is irreducible.

Suppose $Q(x)$ factors completely into a product in which each factor is of the form $x - a$ or $x^2 + bx + c$ where a, b, and c are real numbers and $x^2 + bx + c$ is irreducible. Then $P(x)/Q(x)$ can be expressed as the sum of partial fractions each of which is of one of the types mentioned in Theorem 11.5.1 or Theorem 11.5.2.

Example 2   Obtain $\int \dfrac{(2x^4 + x^3 + 7x^2 + 2)\, dx}{x^5 + 2x^3 + x}$.

SOLUTION   Since $x^5 + 2x^3 + x = x(x^2 + 1)^2$, the integrand has a partial fraction representation

$$\frac{2x^4 + x^3 + 7x^2 + 2}{x^5 + 2x^3 + x} = \frac{A}{x} + \frac{Bx + C}{x^2 + 1} + \frac{Dx + E}{(x^2 + 1)^2}, \qquad (11.66)$$

which if $x \neq 0$ is equivalent to

$$2x^4 + x^3 + 7x^2 + 2 = A(x^2 + 1)^2 + (Bx + C)x(x^2 + 1) + x(Dx + E)$$
$$= (A + B)x^4 + Cx^3 + (2A + B + D)x^2 + (C + E)x + A.$$

Equating corresponding coefficients of the polynomials on each side of this equation, we obtain the equations:

$$A + B = 2$$
$$C = 1$$
$$2A + B + D = 7$$
$$C + E = 0$$
$$A = 2$$

If these equations are solved simultaneously, the solution

$$A = 2 \qquad B = 0 \qquad C = 1 \qquad D = 3 \qquad E = -1 \qquad (11.67)$$

is obtained. Hence from (11.66) and (11.67)

$$\int \frac{(2x^4 + x^3 + 7x^2 + 2)\, dx}{x^5 + 2x^3 + x} = \int \frac{2\, dx}{x} + \int \frac{dx}{x^2 + 1} + \int \frac{(3x - 1)\, dx}{(x^2 + 1)^2} \qquad (11.68)$$

The third integral on the right in (11.68) can be expressed as follows

$$\int \frac{(3x - 1)\, dx}{(x^2 + 1)^2} = \int \frac{3x\, dx}{(x^2 + 1)^2} - \int \frac{dx}{(x^2 + 1)^2}. \qquad (11.69)$$

To evaluate $\int \dfrac{3x\, dx}{(x^2 + 1)^2}$, let $u = x^2 + 1$ and hence $du = 2x\, dx$. Then

$$\int \frac{3x\, dx}{(x^2 + 1)^2} = \int \frac{\frac{3}{2}\, du}{u^2} = -\frac{3}{2u} + C_1$$

$$= -\frac{3}{2(x^2 + 1)} + C_1. \qquad (11.70)$$

To evaluate $\int \dfrac{dx}{(x^2 + 1)^2}$, let $u = \tan^{-1} x$, and then $x = \tan u$ and $dx = \sec^2 u\, du$. Thus

$$\int \frac{dx}{(x^2 + 1)^2} = \int \frac{\sec^2 u\, du}{\sec^4 u} = \int \cos^2 u\, du = \int \left(\frac{1}{2} + \frac{1}{2} \cos 2u \right) du$$

$$= \tfrac{1}{2}u + \tfrac{1}{4}\sin 2u + C_2$$
$$= \tfrac{1}{2}\tan^{-1}x + \tfrac{1}{4}\sin(2\tan^{-1}x) + C_2. \tag{11.71}$$

Since $\tan u = x$, $\sin u = x/\sqrt{x^2+1}$, and $\cos u = 1/\sqrt{x^2+1}$,

$$\sin(2\tan^{-1}x) = \sin 2u = 2\sin u \cos u = \frac{2x}{x^2+1},$$

and hence (11.71) can be rewritten

$$\int \frac{dx}{(x^2+1)^2} = \frac{1}{2}\tan^{-1}x + \frac{x}{2(x^2+1)} + C_2. \tag{11.72}$$

Using the results of (11.70) and (11.72), we obtain from (11.68)

$$\int \frac{(2x^4 + x^3 + 7x^2 + 2)\,dx}{x^5 + 2x^3 + x}$$

$$= 2\ln|x| + \tan^{-1}x - \frac{3}{2(x^2+1)} - \frac{1}{2}\tan^{-1}x - \frac{x}{2(x^2+1)} + C$$

$$= 2\ln|x| + \frac{1}{2}\tan^{-1}x - \frac{x+3}{2(x^2+1)} + C.$$

The following formulas which will be used in our later work can be derived using partial fractions:

$$\int \frac{g'(x)\,dx}{(g(x))^2 - a^2} = \frac{1}{2a}\ln\left|\frac{g(x)-a}{g(x)+a}\right| + C \tag{11.73}$$

$$\int \frac{g'(x)\,dx}{a^2 - (g(x))^2} = \frac{1}{2a}\ln\left|\frac{g(x)+a}{g(x)-a}\right| + C \tag{11.74}$$

Exercise Set 11.5

In Exercises 1–20 obtain the given indefinite integral.

1. $\displaystyle\int \frac{dx}{x^2 - 9}$

2. $\displaystyle\int \frac{dx}{x^2 - 3x}$

3. $\displaystyle\int \frac{(10 - x)\,dx}{x^2 + x - 2}$

4. $\displaystyle\int \frac{(5x - 19)\,dx}{x^2 - 7x + 10}$

5. $\displaystyle\int \frac{(-5x - 5)\,dx}{2x^2 - 3x - 2}$

6. $\displaystyle\int \frac{(-x + 33)\,dx}{2x^2 - 18}$

7. $\displaystyle\int \frac{(x^2 + 11x - 6)\,dx}{x^3 - 3x^2}$

8. $\displaystyle\int \frac{(-2x^2 - 8x + 36)\,dx}{x^3 - x^2 - 12x}$

9. $\displaystyle\int \frac{(x^2 - 3)\,dx}{(x - 2)^3}$

10. $\displaystyle\int \frac{(x^2 + 4)\,dx}{(x + 2)^3}$

11. $\displaystyle\int \frac{(3x^3 + 15x^2 + 3x)\,dx}{x^2 + 2x + 1}$

12. $\displaystyle\int \frac{(x^2 - x - 15)\,dx}{x^2 - 5x + 6}$

13. $\displaystyle\int \frac{(x^3 - 8)\,dx}{x^3 + 8}$

14. $\displaystyle\int \frac{(-x^3 - 7x^2 + 40x - 16)\,dx}{x^4 - 8x^3 + 16x^2}$

15. $\displaystyle\int \frac{x^2\,dx}{x^4 - 81}$

16. $\displaystyle\int \frac{(-13x + 22)\,dx}{x^3 - 2x^2 + 4x - 8}$

17. $\displaystyle\int \frac{(2x^2 + 7x + 7)\,dx}{x^3 + x - 2}$

18. $\displaystyle\int \frac{(5x^2 + 7x + 2)\,dx}{x^3 + 2x^2 - 2x + 3}$

19. $\displaystyle\int \frac{(-5x^4 + 3x^3 - 27x^2 + 16x - 12)\,dx}{(x^2 + 4)^2(x - 1)}$

20. $\displaystyle\int \frac{(x^4 + 7x^3 - 5x^2 + 13x - 10)\,dx}{(x^2 + 2x + 3)(x - 1)^2}$

11.6   Miscellaneous Substitutions

Sometimes an integral can be evaluated only after a substitution that eliminates a radical or fractional power of an expression in the integrand. Thus, if the integrand contains a term of the form $(g(x))^{m/n}$ where $n > 1$ and m are positive integers, the rationalizing substitution

$$u = (g(x))^{1/n}$$

is often useful.

Example 1   Obtain $\displaystyle\int (x + 3)\sqrt{2x - 1}\,dx.$

SOLUTION   To eliminate the radical from the integrand, let $u = \sqrt{2x - 1}$. Then $u^2 = 2x - 1$, $x = (u^2 + 1)/2$, and $dx = u\,du$. Hence

$$\int (x + 3)\sqrt{2x - 1}\,dx = \int \left(\frac{u^2 + 1}{2} + 3\right)u \cdot u\,du = \int \left(\frac{1}{2}u^4 + \frac{7}{2}u^2\right)du$$

$$= \frac{u^5}{10} + \frac{7u^3}{6} + C$$

$$= \frac{(2x - 1)^{5/2}}{10} + \frac{7(2x - 1)^{3/2}}{6} + C.$$

Example 2   Obtain $\displaystyle\int \frac{dx}{4 - (x + 1)^{2/3}}.$

SOLUTION   To eliminate the fractional power from the integrand, let $u = (x + 1)^{1/3}$. Then $u^3 = x + 1$, $x = u^3 - 1$, and $dx = 3u^2\,du$. Hence

$$\int \frac{dx}{4 - (x + 1)^{2/3}} = \int \frac{3u^2\,du}{4 - u^2}$$

and after long division in the integrand of the right member

$$\int \frac{dx}{4 - (x + 1)^{2/3}} = \int \left[-3 + \frac{12}{4 - u^2} \right] du = \int -3\, du + \int \frac{12\, du}{4 - u^2}.$$

Applying (11.74) to the second integral on the right,

$$\int \frac{dx}{4 - (x + 1)^{2/3}} = -3u + 3 \ln \left| \frac{u + 2}{u - 2} \right| + C$$

$$= -3 \sqrt[3]{x + 1} + 3 \ln \left| \frac{\sqrt[3]{x + 1} + 2}{\sqrt[3]{x + 1} - 2} \right| + C.$$

If the integrand is a quotient of polynomials involving the values of trigonometric functions at x, the substitution

$$u = \tan \frac{x}{2} \tag{11.75}$$

is frequently employed. The function tan has a continuous derivative on any interval $((2m - 1)\pi/2, (2m + 1)\pi/2)$, m being some integer, but is undefined at any number that is an odd multiple of $\frac{\pi}{2}$. Since $(2m - 1)\pi/2 < x/2 < (2m + 1)\pi/2$, if and only if $(2m - 1)\pi < x < (2m + 1)\pi$, it will be tacitly assumed when using the substitution (11.75) that the integrand is defined on some subset of an interval $((2m - 1)\pi, (2m + 1)\pi)$.

From (11.75)

$$\sec^2 \frac{x}{2} = 1 + \tan^2 \frac{x}{2} = 1 + u^2,$$

and also

$$du = \frac{1}{2} \sec^2 \frac{x}{2}\, dx = \frac{1}{2}(1 + u^2)\, dx.$$

Hence

$$dx = \frac{2\, du}{1 + u^2}. \tag{11.76}$$

The values of the other trigonometric functions at x can also be obtained in terms of u.

$$\sin x = 2 \sin \frac{x}{2} \cos \frac{x}{2} = 2 \tan \frac{x}{2} \cos^2 \frac{x}{2} = \frac{2 \tan \frac{x}{2}}{\sec^2 \frac{x}{2}}$$

$$\sin x = \frac{2u}{1 + u^2} \tag{11.77}$$

$$\cos x = 2 \cos^2 \frac{x}{2} - 1 = \frac{2}{\sec^2 \frac{x}{2}} - 1 = \frac{2}{1 + u^2} - 1$$

$$\cos x = \frac{1 - u^2}{1 + u^2} \tag{11.78}$$

$$\tan x = \frac{\sin x}{\cos x}$$

$$\tan x = \frac{2u}{1 - u^2}. \tag{11.79}$$

The expressions for $\csc x$, $\sec x$, and $\cot x$ in terms of u are, respectively, the reciprocals of the expressions in (11.77), (11.78), and (11.79).

A useful device for recalling the substitutions (11.77), (11.78), and (11.79) is the right triangle in Figure 11–4.

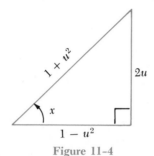

Figure 11–4

Example 3 Obtain $\displaystyle\int \frac{dx}{\sin x - \cos x}$.

SOLUTION From (11.76), (11.77), and (11.78)

$$\int \frac{dx}{\sin x - \cos x} = \int \frac{\dfrac{2\,du}{1 + u^2}}{\dfrac{2u}{1 + u^2} - \dfrac{1 - u^2}{1 + u^2}} = \int \frac{2\,du}{u^2 + 2u - 1}$$

$$= \int \frac{2\,du}{(u + 1)^2 - 2}.$$

Then, letting $z = u + 1$, $dz = du$ and

$$\int \frac{dx}{\sin x - \cos x} = \int \frac{2\,dz}{z^2 - 2}.$$

From (11.73)

$$\int \frac{dx}{\sin x - \cos x} = \frac{1}{\sqrt{2}} \ln \left| \frac{z - \sqrt{2}}{z + \sqrt{2}} \right| + C = \frac{1}{\sqrt{2}} \ln \left| \frac{u + 1 - \sqrt{2}}{u + 1 + \sqrt{2}} \right| + C$$

$$= \frac{1}{\sqrt{2}} \ln \left| \frac{\tan \dfrac{x}{2} + 1 - \sqrt{2}}{\tan \dfrac{x}{2} + 1 + \sqrt{2}} \right| + C.$$

Example 4 Obtain $\displaystyle\int \frac{dx}{2 + \cos 2x}$.

SOLUTION Since the only trigonometric function present in the integrand is evaluated at $2x$, instead of the substutution (11.75), we will find it easier to use

$$u = \tan x. \tag{11.80}$$

Then we obtain $dx = \dfrac{du}{1 + u^2}$, and as in (11.78), $\cos 2x = \dfrac{1 - u^2}{1 + u^2}$. Thus

$$\int \frac{dx}{2 + \cos 2x} = \int \frac{\dfrac{du}{1 + u^2}}{2 + \dfrac{1 - u^2}{1 + u^2}} = \int \frac{du}{3 + u^2}.$$

Recalling (11.80), we have

$$\int \frac{dx}{2 + \cos 2x} = \frac{1}{\sqrt{3}} \tan^{-1} \left(\frac{\tan x}{\sqrt{3}} \right) + C$$

The student should realize that many other substitutions are possible depending upon the nature of the integrand. The ability to recognize an appropriate substitution in an integration problem is often a matter of luck and experience and therefore no general guidelines can be given to cover all situations.

Exercise Set 11.6

In Exercises 1–26 obtain the given indefinite integral.

1. $\displaystyle\int \frac{dx}{1 + \sqrt[4]{x}}$

2. $\displaystyle\int \frac{\sqrt{x}\, dx}{1 - \sqrt{x}}$

3. $\displaystyle\int x^2 \sqrt{x + 2}\, dx$

4. $\displaystyle\int \frac{\sqrt{3x + 2}}{x - 1}\, dx$

5. $\displaystyle\int \frac{dx}{\sqrt{e^x - 1}}$

6. $\displaystyle\int \sqrt{e^\theta + 2}\, d\theta$

7. $\displaystyle\int \frac{dx}{x^{1/2} + x^{1/4}}$

8. $\displaystyle\int \frac{dy}{y^{4/3} - y^{5/6}}$

9. $\displaystyle\int \sqrt{2 - \sqrt{x + 1}}\, dx$

10. $\displaystyle\int \frac{\sqrt{x}}{\sqrt{x} - 4}\, dx$

11. $\displaystyle\int \frac{dt}{\sqrt{2t + 4} - \sqrt{t - 1}}$

12. $\displaystyle\int \frac{dx}{\sqrt{x + 1} + \sqrt{2x}}$

13. $\displaystyle\int \frac{dx}{x\sqrt{x^2 + 2x - 1}}$. Let $u = \dfrac{1}{x}$

14. $\displaystyle\int \frac{dx}{x^2 \sqrt{x^2 + 2x - 1}}$

15. $\displaystyle\int \frac{dx}{2 - \sin x}$

16. $\displaystyle\int \frac{dx}{5 - 4\cos x}$

17. $\displaystyle\int \frac{\sec x\, dx}{\tan x - 2}$

18. $\displaystyle\int \frac{\sin x\, dx}{\sin x + 3}$

19. $\displaystyle\int \frac{dx}{1 - \sin x + \cos x}$

20. $\displaystyle\int \frac{\csc x\, dx}{1 - \csc^2 x}$

21. $\int \dfrac{dx}{1 - \cos 3x}$

22. $\int \dfrac{dx}{\sin 2x + \tan 2x}$

23. $\int \dfrac{dx}{3 \cos x + 4 \sin x}$

24. $\int \dfrac{dx}{\cos x - \sqrt{3} \sin x}$

25. $\int \dfrac{dx}{5 \csc x + 4}$

26. $\int \dfrac{dx}{3 \cos^2 x - \sin^2 x}$

27. Using the method introduced in this section, show that

$$\int \sec x \, dx = \ln \sqrt{\dfrac{1 + \sin x}{1 - \sin x}} + C.$$

Reconcile this expression for $\int \sec x \, dx$ with that obtained from formula (13) in the Table of Integrals.

28. Using the method introduced in this section, evaluate $\int \csc x \, dx$. Reconcile this expression with that obtained from formula (14) in the Table of Integrals.

12. Conics

12.1 The Parabola

In this chapter we will study some important curves that are called *conics* or *conic sections* since they occur as the intersection of a plane and a right circular cone (of two nappes). However, the major emphasis in our study of these curves will be on their analytic properties, and the discussion of their geometric significance will be left to Section 12.3.

The first conic to be considered is the *parabola*.

12.1.1 Definition

A *parabola* is the set of all points in the plane equidistant from a given line (called the *directrix*) and a given point (called the *focus*) that is not on the line.

Suppose F is the focus and L the directrix of a parabola, and d is the distance from an arbitrary point P to L. By the distance from a point to a line we mean the length of a perpendicular line segment from the point to the line. From Definition 12.1.1 P is on the parabola (Figure 12–1) if and only if

$$|FP| = d. \tag{12.1}$$

In particular, the midpoint of the perpendicular from the focus to the directrix (the point labeled V in Figure 12–1) is on the parabola. This point is called the *vertex* of the parabola.

We will derive equations for parabolas whose directrices are perpendicular to the coordinate axes.

12.1.2 Theorem

Let $p \neq 0$, h, and k be any numbers. An equation of the parabola with focus $F(h, k + p)$ and directrix $y = k - p$ is

$$(x - h)^2 = 4p(y - k). \tag{12.2}$$

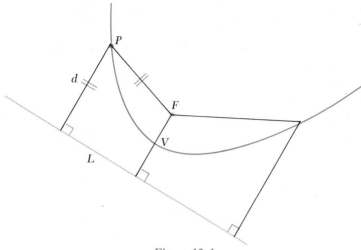

Figure 12-1

PROOF   A point $P(x, y)$ is in the parabola if and only if $|FP| = d$ where d is the distance between P and the line $y = k - p$. (Figure 12–2(a) applies if $p > 0$ and Figure 12–2(b) applies when $p < 0$).

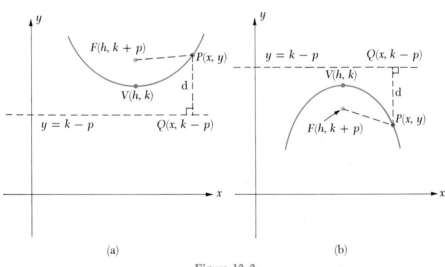

(a)   (b)

Figure 12-2

Thus from (12.1) $P(x, y)$ is in the parabola if and only if the following equivalent equations hold.

$$\sqrt{(x - h)^2 + (y - k - p)^2} = \sqrt{(y - k + p)^2}$$
$$(x - h)^2 + (y - k - p)^2 = (y - k + p)^2$$
$$(x - h)^2 = 4py - 4pk$$
$$(x - h)^2 = 4p(y - k),$$

which proves the theorem.

The parabola in Theorem 12.1.2 has the vertex (h, k). If the vertex is the origin, then (12.2) simplifies to

$$x^2 = 4py.$$

The focus of the parabola (12.2) is above the vertex if $p > 0$ and is below the vertex if $p < 0$. In either case the distance between the focus and vertex is $|p|$ and between the vertex and directrix is also $|p|$. The farther the focus is from the vertex, the larger is $|p|$ and the fatter is the parabola.

Example 1 Write an equation for the parabola having the focus $(4, -5)$ and directrix $y = 1$ (Figure 12–3).

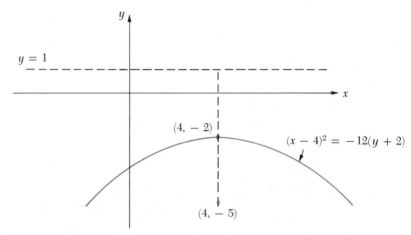

Figure 12-3

SOLUTION The distance from the focus to the directrix is $2|p| = 6$ and hence $|p| = 3$. Since the focus is below the directrix, $p = -3$. The vertex, which is midway between the point $(4, -5)$ and the line $y = 1$, is $(h, k) = (4, -2)$. From (12.2) an equation of the parabola is

$$(x - 4)^2 = -12(y + 2).$$

Before considering some other examples, we will discuss in detail the parabola (12.2). See Figure 12–4. We first note that if $(h + c, y)$ is in the graph, then from (12.2) $c^2 = 4p(y - k)$. Hence $(h - c, y)$ is also in the graph, and the parabola is symmetric to the line $x = h$, which is called the *axis* of the parabola. The chord L_1L_2 of the parabola which is perpendicular to the axis at the focus F is called the *latus rectum*. We leave as an exercise the proof that the length of the latus rectum is $4|p|$ (Exercise 20, Exercise Set 12.1).

From (12.2) since $p \neq 0$

$$y = \frac{(x - h)^2 + 4pk}{4p}, \tag{12.3}$$

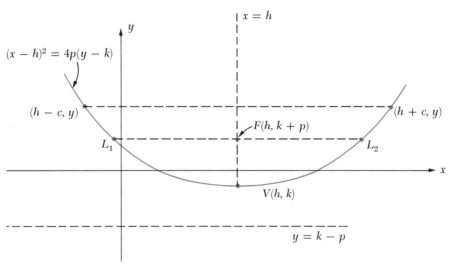

$$(x - h)^2 = 4p(y - k)$$

Figure 12-4

and hence

$$D_x y = \frac{x - h}{2p}. \qquad (12.4)$$

If $p > 0$ (as in Figure 12–4), then from (12.4), $D_x y > 0$ when $x > h$ and $D_x y < 0$ when $x < h$. The function defined by (12.3) is therefore decreasing on $(-\infty, h]$ and increasing on $[h, +\infty)$. Similarly if $p < 0$, the function defined by (12.3) is increasing on $(-\infty, h]$ and decreasing on $[h, +\infty)$.

From (12.4)

$$D_x^2 y = \frac{1}{2p}.$$

Thus if $p > 0$, the parabola is concave upward at each of its points; and if $p < 0$, the graph is concave downward at each of its points.

Example 2   Find the vertex, focus, directrix, and the endpoints of the latus rectum of the parabola $(x + 2)^2 = 10(y - 3)$. Sketch the graph of the equation.

SOLUTION   Since the equation of the parabola is of the form (12.2), the vertex is $(h, k) = (-2, 3)$. Also, $4p = 10$ and $p = \frac{5}{2}$. Since $p > 0$, the focus is a distance $\frac{5}{2}$ above the vertex, and is therefore $(-2, \frac{11}{2})$. The directrix is $\frac{5}{2}$ units below the vertex and thus has the equation $y = \frac{1}{2}$. The length of the latus rectum is $4|p| = 10$, and hence each endpoint of the latus rectum is 5 units from the focus. The endpoints of the latus rectum are $(-7, \frac{11}{2})$ and $(3, \frac{11}{2})$. The sketch of the graph is shown in Figure 12–5.

If in the statement of Theorem 12.1.2 the roles of x and y are interchanged, then the same reasoning used in the proof of Theorem 12.1.2 can be used to prove the following theorem.

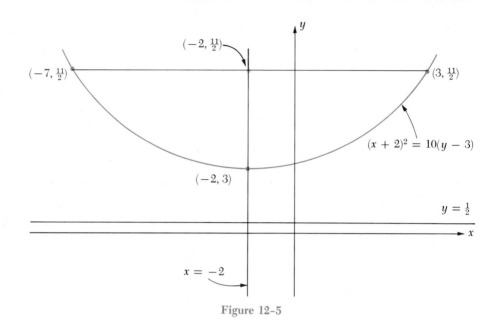

12.1.3   Theorem

Let $p \neq 0$, h, and k be any numbers. An equation of the parabola with focus $F(h + p, k)$ and directrix $x = h - p$ is

$$(y - k)^2 = 4p(x - h). \tag{12.5}$$

The vertex of the parabola (12.5) is also (h, k). If the origin is the vertex, then (12.5) reduces to

$$y^2 = 4px.$$

The axis of the parabola (12.5) is the line $y = k$ and, again, the length of the latus rectum is $4|p|$. If $p > 0$, the graph opens to the right with the focus a distance $|p|$ to the right of the vertex (Figure 12–6(a)). If $p < 0$, the graph opens to the left and the focus is a distance $|p|$ to the left of the vertex (Figure 12–6(b)).

Example 3   Find the vertex, focus, directrix, and endpoints of the latus rectum of the parabola $y^2 + 14x - 6y - 47 = 0$. Sketch the graph of the equation.

SOLUTION   The given equation can be rewritten in the standard form (12.5). First we write

$$y^2 - 6y = -14x + 47.$$

Then, completing the square in y gives

$$y^2 - 6y + 9 = -14x + 56$$
$$(y - 3)^2 = -14(x - 4). \tag{12.6}$$

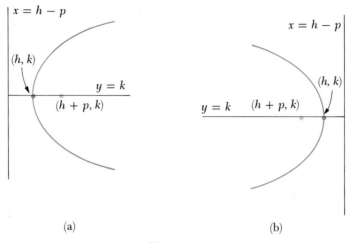

(a)   (b)

Figure 12-6

Since equation (12.6) is of the form (12.5), the vertex is $(4, 3)$. Also, $4p = -14$, and hence $p = -\frac{7}{2}$. The focus is $\frac{7}{2}$ units to the left of the vertex, and is therefore $(\frac{1}{2}, 3)$. The directrix is $\frac{7}{2}$ units to the right of the vertex, and is therefore the line $x = \frac{15}{2}$. The length of the latus rectum is $4|p| = 14$ and hence the endpoints of the latus rectum are $(\frac{1}{2}, 10)$ and $(\frac{1}{2}, -4)$. A sketch of the graph is shown in Figure 12-7.

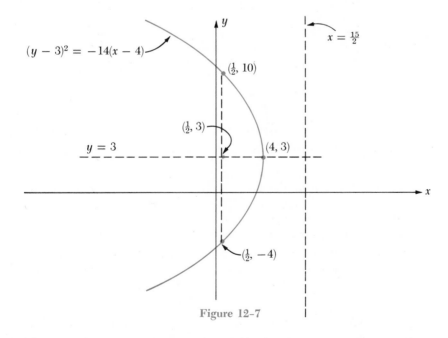

Figure 12-7

From equation (12.2) or (12.5) it is possible to obtain an equation of the form

$$Ax^2 + Cy^2 + Dx + Ey + F = 0 \qquad (12.7)$$

where $A = 0$ or $C = 0$, but A and C are not both zero, for any parabola whose axis is perpendicular to a coordinate axis. Conversely, the following theorem can be proved.

12.1.4 Theorem

If in (12.7) $AC = 0$, but A and C are not both zero, then (12.7) is an equation of a parabola, two parallel lines, or a single line, or the equation has no graph.

        OUTLINE OF THE PROOF If $A \neq 0, C = 0$, and $E \neq 0$ or if $A = 0, C \neq 0$, and $D \neq 0$, the student can easily prove after completing the square in x or y that an equation of the form (12.2) or (12.5) for a parabola is obtained. If $A \neq 0, C = 0$, and $E = 0$, then from (12.7)

$$Ax^2 + Dx + F = 0$$

$$x^2 + \frac{D}{A}x = -\frac{F}{A},$$

and after completing the square in the left member

$$x^2 + \frac{D}{A}x + \frac{D^2}{4A^2} = \frac{D^2}{4A^2} - \frac{F}{A}$$

$$\left(x + \frac{D}{2A}\right)^2 = \frac{D^2}{4A^2} - \frac{F}{A}. \tag{12.8}$$

If the right member of (12.8) is zero, we have another equation for the line $x = -D/2A$. If the right member is positive, we have an equation of the parallel lines

$$x = -\frac{D}{2A} \pm \sqrt{\frac{D^2}{4A^2} - \frac{F}{A}}.$$

If the right member is negative, the equation (12.8) has no graph.

        If $A = 0, C \neq 0$, and $D = 0$, it may similarly be shown that the graph consists of a single line, parallel lines, or is empty.

        Before introducing the next example, we mention a fact from physics. If a ray of light L_1 meets a reflecting surface C at a point P, the ray is reflected along a path L_2 such that the *angle of incidence* (the angle α_1 between the incident ray L_1 and the tangent to C at P) is equal to the *angle of reflection* (the angle α_2 between the tangent and the reflected ray L_2) (Figure 12–8). The next example illustrates the reflection property of the parabola and is utilized in parabolic reflectors.

        Example 4 Suppose P is any point in a parabola with focus F (Figure 12–9), and T is the tangent to the parabola at P. If PQ is parallel to the axis of the parabola, prove that the angle α between T and PQ is equal to the angle β between FP and T.

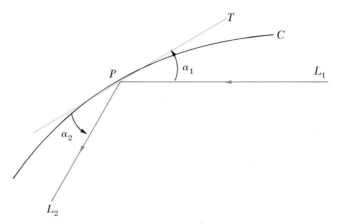

Figure 12-8

SOLUTION   For convenience we shall select the coordinate axes so that the origin is the vertex of the parabola, the y axis is its axis, and the parabola is concave upward. Thus the curve has an equation $x^2 = 4py$ where $p > 0$ and F is the point $(0, p)$. We shall denote P by (x_1, y_1).

From differentiation in the equation $x^2 = 4py$, we obtain $D_x y = x/2p$, and hence the inclination θ of the tangent T satisfies

$$\tan \theta = (D_x y)_{x=x_1} = \frac{x_1}{2p}.$$

If the point P is in the first quadrant, as in Figure 12–9, then α and θ are complementary and hence

$$\tan \alpha = \tan (90° - \theta)$$

$$= \cot \theta = \frac{2p}{x_1}. \tag{12.9}$$

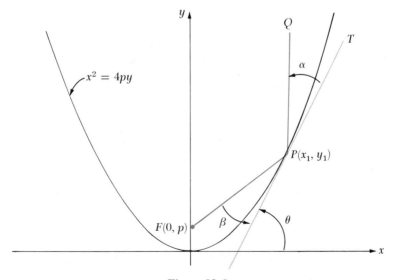

Figure 12-9

From (4.55), if we let m_1 and m_2 be the slopes of FP and T, respectively,

$$\tan \beta = \frac{m_2 - m_1}{1 + m_1 m_2} = \frac{\dfrac{x_1}{2p} - \dfrac{y_1 - p}{x_1}}{1 + \dfrac{y_1 - p}{x_1} \dfrac{x_1}{2p}}$$

$$= \frac{x_1{}^2 - 2py_1 + 2p^2}{x_1 y_1 + px_1}.$$

Since $P(x_1, y_1)$ is on the parabola $x^2 = 4py$, $x_1{}^2 = 4py_1$, and

$$\tan \beta = \frac{4py_1 - 2py_1 + 2p^2}{x_1 y_1 + px_1} = \frac{2p(y_1 + p)}{x_1(y_1 + p)}$$

$$= \frac{2p}{x_1}. \tag{12.10}$$

From (12.9) and (12.10), $\tan \alpha = \tan \beta$ and therefore $\alpha = \beta$.

In the case where P is in the second quadrant we similarly obtain $\alpha = \beta$. If P is the vertex, then the x axis is the tangent T to the parabola at P and $\alpha = \beta = 90°$.

The reflection property discussed in Example 4 is utilized in a parabolic mirror. Such a mirror has the shape of a *paraboloid of revolution,* a surface that is formed by revolving a parabola about its axis. The parabolic mirror in an automobile headlight reflects light rays coming from a light bulb at the focus of the parabola to form a beam of light that is parallel to the axis of the parabola. On the other hand, a parabolic mirror for a reflecting telescope, when properly aligned, receives parallel light rays from a distant point source such as a star, and reflects them to form a sharply defined image at the focus. A radar antenna, which is usually in the shape of a paraboloid of revolution or a parabolic cylinder, serves both to transmit a microwave beam and to reflect the returning echo waves from an object in the path of the transmitted beam. The transmitted waves originate at a focal point of the antenna and the reflected waves are concentrated at the same point, from which they are sent to a receiver.

Exercise Set 12.1

In Exercises 1–9 obtain an equation of the parabola satisfying the given conditions.

1. Focus $(-1, 4)$, directrix $x = 5$
2. Focus $(6, 5)$, directrix $x = 1$
3. Vertex $(2, -3)$, focus $(2, 1)$
4. Vertex $(-4, 2)$, directrix $x = -1$
5. Directrix $y = 3$, axis $x = 4$, $p = -2$
6. Endpoints of the latus rectum are $(-3, -4)$ and $(4, -4)$; $p > 0$
7. Axis perpendicular to the y axis, vertex $(1, 2)$, contains the point $(-4, 7)$

8. Axis perpendicular to the x axis; contains the points $(9, -1)$, $(3, -4)$, and $(-9, 8)$

9. Contains the points $(4, 9)$, $(-2, 3)$, and $(-5, \frac{9}{2})$ (two solutions)

In Exercises 10–19 find the vertex, focus, directrix, and endpoints of the latus rectum of the given parabola. Sketch the graph of the equation.

10. $y^2 = 6x$

11. $x^2 = -10y$

12. $x^2 - 4y + 8 = 0$

13. $y^2 + 6x - 12 = 0$

14. $y^2 - 8y + 12x - 8 = 0$

15. $x^2 + 6x - 8y + 33 = 0$

16. $x^2 + 4x + 5y - 7 = 0$

17. $y^2 - 6x + 10y + 10 = 0$

18. $3x^2 - 18x + 8y + 23 = 0$

19. $5y^2 + 12x - 10y - 19 = 0$

20. Prove that the length of the latus rectum of the parabola (12.2) is $4|p|$.

21. Complete the proof of Theorem 12.1.4 by proving that if $A \neq 0$, $C = 0$, and $E \neq 0$ in (12.7), then the graph of (12.7) is a parabola.

22. Show that the area of the parabolic segment shown in Figure 12–10 is $\frac{2}{3}bh$.

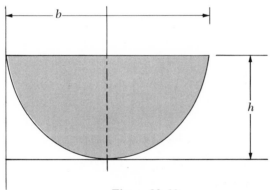

Figure 12-10

23. Write an equation in simplest form for the parabola with focus $(3, 1)$ and directrix the line $x - y = -2$. HINT: Use the formula in Exercise 9, Exercise Set 6.2.

24. Prove that the parabolas $y^2 = 4\alpha(x + \alpha)$ and $y^2 = 4\beta(x + \beta)$ intersect orthogonally if α and β are of opposite signs, but do not intersect if $\alpha \neq \beta$ and α and β have the same sign.

25. If the width of a parabolic arch is 120 ft and the height of the arch is 36 ft, at what distance from one end of the arch is the height 20 ft?

26. Prove that the tangent to the parabola $y^2 = 4px$ at a point (x_1, y_1) intersects the x axis at the point $(-x_1, 0)$.

27. Prove that there is one and only one line of slope $m \neq 0$ tangent to the parabola $y^2 = 4px$ and find an equation of the line.

28. Prove that tangents to a parabola at the endpoints of a chord passing through a focus intersect at right angles on the directrix of the parabola.

12.2   The Ellipse

The next conic that we will consider is the *ellipse*.

12.2.1   Definition

An *ellipse* is the set of all points P in a plane such that the sum of the distances from P to two given points (called *foci*) in the plane is constant.

Because of Definition 12.2.1 an ellipse can be constructed by fastening the ends of a length of string at its foci F_1 and F_2, and then tracing the complete curve with a pencil P (Figure 12–11) always keeping the string taut with the pencil. We obtain the sketch of an ellipse from this construction since $|F_1P| + |F_2P|$, the length of the string, is constant.

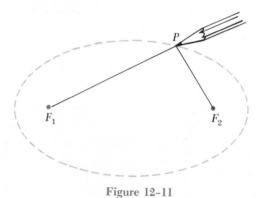

Figure 12-11

The point P and the foci F_1 and F_2 must satisfy the inequality $|F_1P| + |F_2P| \geq |F_1F_2|$, regardless of whether the three points are collinear or are vertices of a triangle. If P is on F_1F_2, then $|F_1P| + |F_2P| = |F_1F_2|$, which is constant, and the ellipse is just the line segment F_1F_2.

A fairly simple equation is obtained from Definition 12.2.1 when F_1 and F_2 are in a line perpendicular to a coordinate axis and any point P on the ellipse satisfies

$$|F_1P| + |F_2P| > |F_1F_2|. \qquad (12.11)$$

To be specific, let the foci be $F_1(h - c, k)$ and $F_2(h + c, k)$ where $c \geq 0$, and h and k are any numbers. Then suppose the ellipse is the set of all points P such that

$$|F_1P| + |F_2P| = 2a. \qquad (12.12)$$

Since $|F_1F_2| = 2c$, from (12.11) and (12.12)

$$a > c \geq 0. \qquad (12.13)$$

Using the two-point distance formula, (12.12) becomes

$$\sqrt{[x - (h - c)]^2 + (y - k)^2} + \sqrt{[x - (h + c)]^2 + (y - k)^2} = 2a,$$

from which one obtains

$$\sqrt{[(x-h)+c]^2 + (y-k)^2} = 2a - \sqrt{[(x-h)-c]^2 + (y-k)^2}. \quad (12.14)$$

If $x - h$ and $y - k$ are kept intact, then after squaring each member of (12.14), and transposing terms and squaring again, the student can obtain

$$(a^2 - c^2)(x-h)^2 + a^2(y-k)^2 = a^2(a^2 - c^2). \quad (12.15)$$

Since $a > c$, we shall define b by

$$b = \sqrt{a^2 - c^2}. \quad (12.16)$$

A substitution from (12.16) in (12.15) then gives

$$b^2(x-h)^2 + a^2(y-k)^2 = a^2 b^2. \quad (12.17)$$

Thus from (12.17)

$$\frac{(x-h)^2}{a^2} + \frac{(y-k)^2}{b^2} = 1. \quad (12.18)$$

We have shown that if $P(x, y)$ is any point on the ellipse defined by (12.12), then (x, y) satisfies (12.18). Conversely, after a proof that is left as an exercise (Exercise 32, Exercise Set 12.2), if (x, y) satisfies (12.18), then $P(x, y)$ is on this ellipse. Thus (12.18) is an equation of the ellipse.

If $c = 0$ (that is, F_1 and F_2 coincide at (h, k)), then from (12.16) $a = b$, and the ellipse (12.18) is a circle with center (h, k) and radius a. If $c > 0$, then of course, $a > b$.

The point (h, k), which is the midpoint of the line segment joining the foci, is called the *center* of the ellipse. If the center is the origin, then $h = k = 0$ and (12.18) becomes

$$\frac{x^2}{a^2} + \frac{y^2}{b^2} = 1.$$

From (12.18) one obtains the solution for y in terms of x:

$$y = k \pm \frac{b}{a}\sqrt{a^2 - (x-h)^2}. \quad (12.19)$$

Hence y is defined if and only if $(x - h)^2 \le a^2$—that is, if and only if $x \in [h - a, h + a]$. We note from (12.18) or (12.19) that the points $V_1(h - a, k)$ and $V_2(h + a, k)$ are on the ellipse. Similarly, one can find that x is defined if and only if $y \in [k - b, k + b]$. Also, we note that the points $M_1(h, k - b)$ and $M_2(h, k + b)$ are on the ellipse. A sketch of the graph of the ellipse (12.18) is shown in Figure 12–12.

Since the endpoints of $V_1 V_2$ are each a units from the center, and the endpoints of $M_1 M_2$ are each b units from the center, where $b \le a$, $V_1 V_2$ is called the *major axis* and $M_1 M_2$ the *minor axis* of the ellipse. We note that the foci are always on the major axis. The endpoints V_1 and V_2 of the major axis are called the *vertices* of the ellipse.

Thus (12.18) is *an equation of the ellipse with center (h, k), major axis*

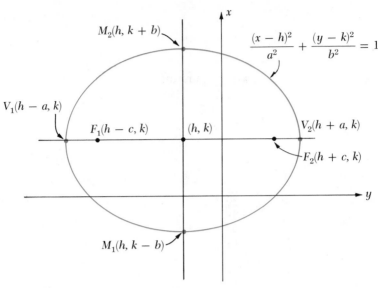

Figure 12-12

on the line $y = k$, and with major axis and minor axis of lengths $2a$ and $2b$, respectively.

In graphing an ellipse, the endpoints of the major and minor axes should be plotted and also the endpoints of each *latus rectum*. A latus rectum (plural *latera recta*) of an ellipse, which is defined the same way as for a parabola, is a chord of the ellipse which is perpendicular to the major axis at a focus. The student can easily prove that the length of a latus rectum of the ellipse (12.18) is $2b^2/a$ (Exercise 26, Exercise Set 12.2).

Example 1 Find an equation of the ellipse with foci $(-2, -3)$ and $(4, -3)$ and vertices $(-4, -3)$ and $(6, -3)$. Graph the ellipse using the endpoints of the major and minor axes and the latera recta.

SOLUTION Since the foci are on a line perpendicular to the y axis, the standard equation (12.18) applies. The center is the midpoint of the line segment joining $(-2, -3)$ and $(4, -3)$. Hence $(h, k) = (1, -3)$. The distance from the center to either focus is $c = 3$. The distance from the center to either vertex is $a = 5$. From (12.16)

$$b^2 = 25 - 9 = 16.$$

Hence from (12.18) an equation of the ellipse is

$$\frac{(x - 1)^2}{25} + \frac{(y + 3)^2}{16} = 1.$$

The endpoints of the minor axis are $(1, 1)$ and $(1, -7)$. Since each latus rectum endpoint is a distance $b^2/a = \frac{16}{5}$ from the corresponding focus, the endpoints of the latera recta are $(-2, \frac{1}{5})$, $(-2, -\frac{31}{5})$, $(4, \frac{1}{5})$, and $(4, -\frac{31}{5})$. A sketch of the ellipse is shown in Figure 12-13.

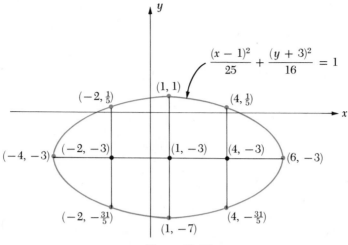

Figure 12–13

　　　If the foci of the ellipse are $F_1(h, k - c)$ and $F_2(h, k + c)$ where $c \geq 0$ and (12.12) holds, an equation for the ellipse can be obtained by interchanging $x - h$ and $y - k$ in (12.18) while retaining the definition (12.16) for b. The equation is therefore

$$\frac{(y - k)^2}{a^2} + \frac{(x - h)^2}{b^2} = 1. \tag{12.20}$$

The ellipse (12.20) can be described as *the ellipse with center (h, k), major axis on the $x = h$, and with major axis and minor axis of lengths $2a$ and $2b$, respectively.* A sketch of the graph of (12.20) is shown in Figure 12–14.

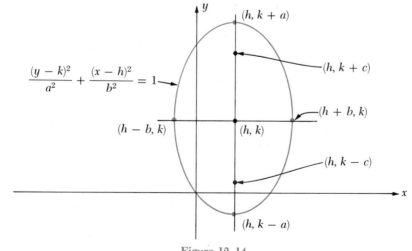

Figure 12–14

　　　Example 2   Discuss the graph of the equation $9x^2 + 5y^2 + 36x - 30y + 36 = 0$ (Figure 12–15).

SOLUTION   If the equation is rewritten, and the square is completed in x and y, we have successively

$$9(x^2 + 4x) + 5(y^2 - 6y) = -36$$
$$9(x^2 + 4x + 4) + 5(y^2 - 6y + 9) = -36 + 36 + 45$$
$$9(x + 2)^2 + 5(y - 3)^2 = 45$$
$$\frac{(x + 2)^2}{5} + \frac{(y - 3)^2}{9} = 1, \tag{12.21}$$

which is an equation in the standard form (12.20). Thus the graph is an ellipse with center $(-2, 3)$. Since 9 is the larger denominator and 5 the smaller denominator in the left member of (12.21), we let $a^2 = 9$ and $b^2 = 5$, and hence $a = 3$ and $b = \sqrt{5}$. Since the fraction involving $(y - 3)^2$ has the larger denominator, by comparison with the ellipse (12.20) the major axis is in the line $x = -2$. The vertices, which are each a distance $a = 3$ from the center, are therefore $(-2, 0)$ and $(-2, 6)$. The endpoints of the minor axis, which are on the line $y = 3$ a distance $b = \sqrt{5}$ from the center are $(-2 - \sqrt{5}, 3)$ and $(-2 + \sqrt{5}, 3)$.

From (12.16) $c = \sqrt{a^2 - b^2} = \sqrt{9 - 5} = 2$. Hence the foci, which are on the major axis a distance $c = 2$ from the center, are $(-2, 5)$ and $(-2, 1)$. Since each latus rectum endpoint is a distance $b^2/a = \frac{5}{3}$ from the corresponding focus, the endpoints of the latera recta are $(-\frac{11}{3}, 1)$, $(-\frac{1}{3}, 1)$, $(-\frac{11}{3}, 5)$, and $(-\frac{1}{3}, 5)$.

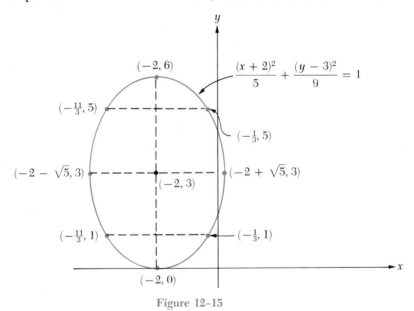

Figure 12-15

From equation (12.18) or (12.20) it is possible to obtain an equation of the form

$$Ax^2 + Cy^2 + Dx + Ey + F = 0 \tag{12.22}$$

where $AC > 0$ for any ellipse whose major axis is on a line which is perpendicular to a coordinate axis. Conversely, we have the following theorem.

12.2.2 Theorem

If $AC > 0$, then (12.22) is an equation of an ellipse (which may be a circle or a set consisting of one point), or the equation has no graph.

      The theorem can be proved by completing the square in x and y and then considering the sign of the right member of the equation obtained. The reader may wish to carry out the details for himself.

      An important illustration of the ellipse occurs in planetary motion. Johannes Kepler (1571–1630), a German astronomer, after 20 years of labor, discovered that each planet moves in an elliptical orbit in which the sun is one of the foci. This discovery upset the Copernican view that the planets traveled in circular paths, and provided the stimulus for the discovery of the universal gravitational law by Newton. Other applications of the ellipse occur in the design of "whispering galleries" (Exercise 33, Exercise Set 12.2) and elliptical gears.

Exercise Set 12.2

In Exercises 1–10 obtain an equation of the ellipse satisfying the given conditions.

1. Vertices $(\pm 6, 0)$, foci $(\pm 4, 0)$

2. Vertex $(0, -5)$, focus $(0, -3)$, center $(0, 1)$

3. Center $(4, -1)$, focus $(4, 4)$, minor axis of length 8

4. Center $(-2, 1)$, vertex $(-2, -7)$, minor axis of length 6

5. Sum of the distances from $(-3, 2)$ and $(5, 2)$ to any point on the ellipse is 12.

6. Foci $(2, 0)$ and $(2, 6)$, passes through $(3, 7)$

7. Vertices $(-3 - 2\sqrt{3}, -1)$ and $(-3 + 2\sqrt{3}, -1)$, passes through $(-1, 1)$

8. Center $(2, 5)$, focus $(2, 9)$, length of the latus rectum is $\dfrac{4\sqrt{5}}{5}$.

9. Center at the origin, major axis which is on the y axis has a length twice that of the minor axis, passes through $(1, 4)$.

10. Vertex $(2, 1)$, nearer focus $(0, 1)$, minor axis of length $4\sqrt{3}$

In Exercises 11–22 find the foci, vertices, endpoints of minor axis, and endpoints of the latera recta of the ellipse having the given equation. Sketch the graph of the equation.

11. $\dfrac{x^2}{25} + \dfrac{y^2}{16} = 1$

12. $\dfrac{x^2}{7} + \dfrac{y^2}{16} = 1$

13. $\dfrac{(x - 5)^2}{\frac{16}{9}} + \dfrac{(y + 2)^2}{4} = 1$

14. $\dfrac{(x + 4)^2}{4} + \dfrac{(y - 3)^2}{3} = 1$

15. $400(x - 3)^2 + 225(y - 2)^2 = 144$

16. $4(x - 1)^2 + 6(y - 4)^2 = 12$

17. $3x^2 + y^2 - 4y - 23 = 0$ 18. $3x^2 + 2y^2 + 6x - 33 = 0$

19. $4x^2 + 9y^2 - 8x - 54y - 59 = 0$ 20. $4x^2 + 3y^2 + 16x + 6y - 89 = 0$

21. $9x^2 + 16y^2 - 72x - 32y + 16 = 0$ 22. $9x^2 + 5y^2 + 54x - 40y - 19 = 0$

23. Find an equation of the tangent to the ellipse $4x^2 + 3y^2 = 16$ at $(1, -2)$.

24. Find an equation of the normal to the ellipse $2x^2 + 9y^2 = 27$ at $(-3, 1)$.

25. Show that the tangent at any point (x_1, y_1) in the ellipse $(x^2/a^2) + (y^2/b^2) = 1$ has the equation $(xx_1/a^2) + (yy_1/b^2) = 1$.

26. Show that the length of the latus rectum of the ellipse (12.18) is $2b^2/a$.

27. Prove: the longest chord of an ellipse through the center is the major axis.

28. Obtain Equation (12.15) from Equation (12.14).

29. Prove Theorem 12.2.2.

30. Find equations for the tangents to the ellipse $4x^2 + 9y^2 = 36$ with slope $-\frac{2}{3}$.

31. Find equations for the tangents to the ellipse $x^2 + 25y^2 = 100$ with slope $\frac{1}{2}$.

32. Prove that if (x, y) satisfies (12.18), then the sum of the distances from the point $P(x, y)$ to the foci $F_1(h - c, k)$ and $F_2(h + c, k)$ is $2a$.

33. Prove the reflection property of ellipses: If $P(x_1, y_1)$ is any point on the ellipse $(x^2/a^2) + (y^2/b^2) = 1$ with foci $F_1(-c, 0)$ and $F_2(c, 0)$, the angle α between F_2P and the tangent T at P equals the angle β between T and F_1P (Figure 12–16). (This principle is utilized in the construction of ellipsoidal "whispering galleries" in which a person speaking softly at one focus can be clearly heard by another person, at the other focus a considerable distance away.)

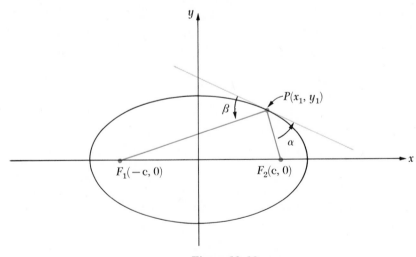

Figure 12–16

34. Find the maximum area for a rectangle that can be inscribed in the ellipse $(x^2/a^2) + (y^2/b^2) = 1$ with its sides parallel to the coordinate axes.

35. Prove that the nearest point in an ellipse to a focus is the nearer vertex and that the farthest point in the ellipse from the focus is the farther vertex.

36. The moon's orbit about the earth is an ellipse with the earth at one focus. If the major and minor axes of the ellipse have lengths 474,000 miles and 473,000 miles, respectively, what are the greatest and least distances from the earth to the moon?

12.3   The Hyperbola

The last conic which we will study is the *hyperbola*.

12.3.1   Definition

A *hyperbola* is the set of all points P in a plane such that the absolute value of the difference of the distances from P to two distinct points (called *foci*) in the plane is constant.

Using Definition 12.3.1 we will derive an equation for a hyperbola in which the foci are in a line that is perpendicular to a coordinate axis. It will be noted that the steps in the derivation of the equation are similar to those used earlier to obtain equation (12.18) for the ellipse. Suppose the foci of the hyperbola are $F_1(h - c, k)$ and $F_2(h + c, k)$ where $c > 0$ and h and k are any numbers. Then if $a > 0$, suppose that the hyperbola (Figure 12–17) is the set of all points P such that

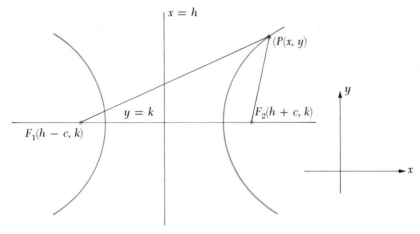

Figure 12-17

$$||F_1P| - |F_2P|| = 2a. \tag{12.23}$$

From (12.23) $P(x, y)$ is in the hyperbola if and only if

$$|\sqrt{(x - h - c)^2 + (y - k)^2} - \sqrt{(x - h + c)^2 + (y - k)^2}| = 2a, \tag{12.24}$$

and from (12.24)

$$\sqrt{(x - h - c)^2 + (y - k)^2} - \sqrt{(x - h + c)^2 + (y - k)^2} = \pm 2a$$
$$\sqrt{[(x - h) - c]^2 + (y - k)^2} = \pm 2a + \sqrt{[(x - h) + c]^2 + (y - k)^2}. \tag{12.25}$$

If $x - h$ and $y - k$ are kept intact, the student can prove from (12.25) that

$$(a^2 - c^2)(x - h)^2 + a^2(y - k)^2 = a^2(a^2 - c^2). \tag{12.26}$$

Before obtaining an equation in simplified form, we digress to prove that $c \geq a$. If we supposed that $c < a$, then $2c < 2a$, and hence $|F_1F_2| < ||F_1P| - |F_2P||$. Then by Theorem 1.5.3(c), either $|F_1F_2| < |F_1P| - |F_2P|$ or $|F_1P| - |F_2P| < -|F_1F_2|$. Hence, either $|F_2P| + |F_1F_2| < |F_1P|$ or $|F_1P| + |F_1F_2| < |F_2P|$. Each of these inequalities is impossible because it contradicts the triangle inequality, regardless of whether F_1, F_2, and P are collinear or are vertices of a triangle.

If $c = a$, then $2c = 2a$ and from (12.23) either $|F_1P| - |F_2P| = |F_1F_2|$ or $|F_2P| - |F_1P| = |F_1F_2|$. In this case the hyperbola consists of those points $P(x, y)$ in the line $y = k$ where $x \notin (h - c, h + c)$.

In deriving an equation for the hyperbola (12.23), we shall confine ourselves to the case where

$$c > a > 0.$$

Note that the order relation between a and c for a hyperbola is different from that for an ellipse, where $a > c$. If b is defined by

$$b = \sqrt{c^2 - a^2}, \tag{12.27}$$

then from (12.27) $b^2 = c^2 - a^2$ and $-b^2 = a^2 - c^2 < 0$. Hence from (12.26)

$$\frac{(x - h)^2}{a^2} - \frac{(y - k)^2}{b^2} = 1. \tag{12.28}$$

We have proved that if $P(x, y)$ is any point on the hyperbola (12.23), then (x, y) satisfies (12.28). Conversely, after a proof that is left as an exercise, if (x, y) satisfies (12.28), then $P(x, y)$ is on the hyperbola. Thus (12.28) is an equation of hyperbola (12.23).

The point (h, k), which is the midpoint of the line segment joining the foci, is called the *center* of the hyperbola. If the center is the origin, (12.28) becomes

$$\frac{x^2}{a^2} - \frac{y^2}{b^2} = 1.$$

From (12.28) one obtains

$$y = k \pm \frac{b}{a} \sqrt{(x - h)^2 - a^2}. \tag{12.29}$$

Hence y is defined if and only if $(x - h)^2 \geq a^2$—that is, if and only if $x \geq h + a$ or $x \leq h - a$. The hyperbola (12.28) therefore consists of two branches, one extending to the right of $x = h + a$ and the other extending to the left from $x = h - a$.

From (12.28) or (12.29) we find that the hyperbola intersects the line $y = k$ in the points $V_1(h - a, k)$ and $V_2(h + a, k)$, which are called the *vertices* of the hyperbola (Figure 12–18). However, the hyperbola does not intersect the line $x = h$, as y is undefined when $x \in (h - a, h + a)$. Unlike the situation with the ellipse, there is no order relation between a and b and hence the terms major axis and minor axis are not used in connection with hyperbolas. The line segment V_1V_2, however, is called the *transverse axis*. It should be noted that the foci are on the line containing the transverse axis. The line segment of length $2b$ which joins the points $(h, k - b)$ and $(h, k + b)$ is called the *conjugate axis*.

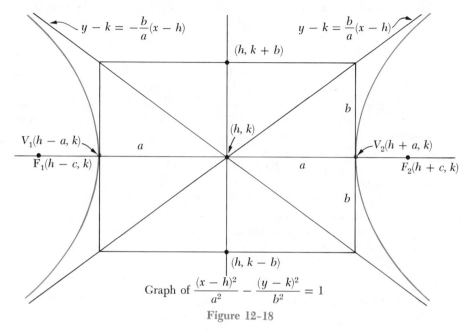

$$\text{Graph of } \frac{(x - h)^2}{a^2} - \frac{(y - k)^2}{b^2} = 1$$

Figure 12-18

Thus (12.28) *is an equation of the hyperbola with center* (h, k), *transverse axis on the line* $y = k$, *and with transverse axis and conjugate axis of lengths* $2a$ *and* $2b$, *respectively.*

It is intuitively apparent that the lines $y - k = \pm\dfrac{b}{a}(x - h)$ are oblique asymptotes of the hyperbola (12.28). To show that the line $y - k = (b/a)(x - h)$ is such an asymptote consider the difference

$$\left[k + \frac{b}{a}\sqrt{(x - h)^2 - a^2}\right] - \left[k + \frac{b}{a}(x - h)\right] = \frac{b}{a}\left[\sqrt{(x - h)^2 - a^2} - (x - h)\right]$$

$$= \frac{b}{a}\frac{[(x - h)^2 - a^2] - (x - h)^2}{\sqrt{(x - h)^2 - a^2} + (x - h)} = \frac{-ab}{\sqrt{(x - h)^2 - a^2} + (x - h)}$$

$$= \frac{\dfrac{-ab}{x - h}}{\sqrt{1 - \dfrac{a^2}{(x - h)^2}} + 1} \qquad \text{when } x > h.$$

Since

$$\lim_{x \to +\infty}\left\{\left[k + \frac{b}{a}\sqrt{(x - h)^2 - a^2}\right] - \left[k + \frac{b}{a}(x - h)\right]\right\}$$

$$= \lim_{x \to +\infty}\frac{\dfrac{-ab}{x - h}}{\sqrt{1 - \dfrac{a^2}{(x - h)^2}} + 1} = 0,$$

by Definition 5.8.2 the line $y - k = (b/a)(x - h)$ is an oblique asymptote of the graph of $y = k + (b/a)\sqrt{(x - h)^2 - a^2}$. Similarly, it can be shown that the line is an asymptote of the graph of $y = k - (b/a)\sqrt{(x - h)^2 - a^2}$ by showing that

$$\lim_{x \to -\infty} \left\{ \left[k - \frac{b}{a}\sqrt{(x - h)^2 - a^2} \right] - \left[k + \frac{b}{a}(x - h) \right] \right\} = 0.$$

The proof that the line $y - k = -(b/a)(x - h)$ is an asymptote of the hyperbola (12.28) follows the same scheme and is therefore omitted.

In graphing a hyperbola, the vertices should be plotted and the asymptotes sketched since the branches of the hyperbola approach the asymptotes as one continues out along the branches. To locate the asymptotes in the sketch, it is desirable to sketch the *auxiliary rectangle*, a rectangle having the point (h, k) as a center and having dimensions $2a$ and $2b$ (Figure 12–18). Each asymptote will then pass through (h, k) and the opposite vertices of the rectangle.

Further, the endpoints of each latus rectum can be plotted. A latus rectum of a hyperbola, as one might expect from our previous study of the parabola and ellipse, is a chord that is perpendicular to the transverse axis at a focus. The student can verify that the length of the latus rectum of the hyperbola (12.28) is $2b^2/a$ (Exercise 27, Exercise Set 12.3).

Example 1 Find an equation of the hyperbola with foci $(-2, 2)$ and $(8, 2)$ and vertices $(0, 2)$ and $(6, 2)$. Graph the hyperbola.

SOLUTION Since the foci are in a line perpendicular to the y axis, the standard equation (12.28) applies. The center, which is the midpoint of the line segment joining the foci, is $(3, 2)$. The distance from the center to either focus is $c = 5$. The distance from the center to either vertex is $a = 3$. From (12.27)

$$b^2 = c^2 - a^2 = 25 - 9 = 16.$$

Hence from (12.28) an equation of the hyperbola is

$$\frac{(x - 3)^2}{9} - \frac{(y - 2)^2}{16} = 1.$$

The asymptotes are the lines $y - 2 = \pm\frac{4}{3}(x - 3)$. Since $b^2/a = \frac{16}{3}$ is the distance between an endpoint of a latus rectum and the corresponding focus, the endpoints of the latera recta are $(-2, \frac{22}{3})$, $(-2, -\frac{10}{3})$, $(8, \frac{22}{3})$, and $(8, -\frac{10}{3})$. A sketch of the hyperbola is shown in Figure 12–19.

If the foci are $F_1(h, k - c)$ and $F_2(h, k + c)$ where $c > 0$, and (12.23) holds, an equation for the hyperbola can be derived by interchanging $x - h$ and $y - k$ in (12.28) while retaining the definition (12.27) for b. This equation is

$$\frac{(y - k)^2}{a^2} - \frac{(x - h)^2}{b^2} = 1, \tag{12.30}$$

which can be described as an equation of the hyperbola with center (h, k), transverse axis on the line $x = h$, and with transverse axis and conjugate axis of lengths $2a$ and $2b$, respectively.

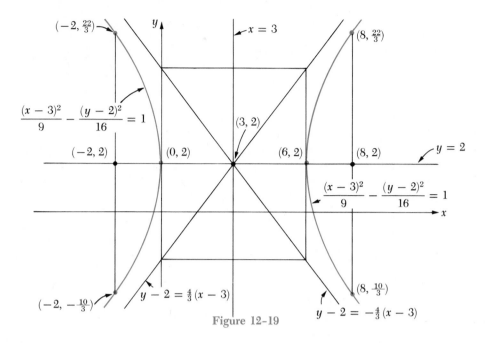

Figure 12-19

A sketch of the graph of (12.30) is shown in Figure 12–20. We note that the asymptotes of the hyperbola (12.30) are the lines $y - k = \pm(a/b)(x - h)$.

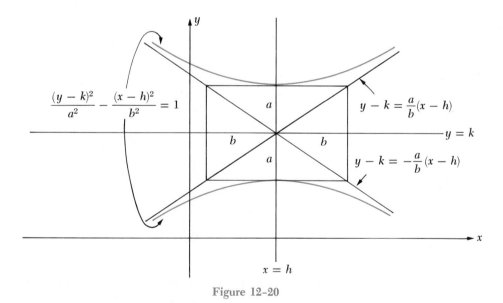

Figure 12-20

Example 2   Discuss the graph of the equation $4y^2 - 8y - 9x^2 - 36x - 68 = 0$.

SOLUTION   If the equation is rewritten, and we complete the square in x and y we have successively

$$4(y^2 - 2y) - 9(x^2 + 4x) = 68$$
$$4(y^2 - 2y + 1) - 9(x^2 + 4x + 4) = 68 + 4 - 36$$
$$4(y - 1)^2 - 9(x + 2)^2 = 36$$
$$\frac{(y - 1)^2}{9} - \frac{(x + 2)^2}{4} = 1,$$

which is of the standard form (12.30) for a hyperbola. The center of the hyperbola is $(-2, 1)$. The denominator of the fraction involving $(y - 1)^2$ is $a^2 = 9$ and hence $a = 3$. The vertices, which are therefore 3 units from the center, are $(-2, 4)$ and $(-2, -2)$. The denominator of the fraction involving $(x + 2)^2$ is $b^2 = 4$. Hence $b = 2$. The asymptotes are the lines

$$y - 1 = \pm\tfrac{3}{2}(x + 2).$$

From (12.27), $c = \sqrt{a^2 + b^2} = \sqrt{4 + 9} = \sqrt{13}$. The foci, which are in the line containing the vertices, are a distance $\sqrt{13}$ from the center. Hence the foci are $(-2, 1 + \sqrt{13})$ and $(-2, 1 - \sqrt{13})$. Since each endpoint of a latus rectum is a distance $b^2/a = \tfrac{4}{3}$ from its corresponding focus, the endpoints of the latera recta are $(-\tfrac{2}{3}, 1 + \sqrt{13})$, $(-\tfrac{10}{3}, 1 + \sqrt{13})$, $(-\tfrac{2}{3}, 1 - \sqrt{13})$, and $(-\tfrac{10}{3}, 1 - \sqrt{13})$. A sketch of the graph of the given equation is shown in Figure 12-21.

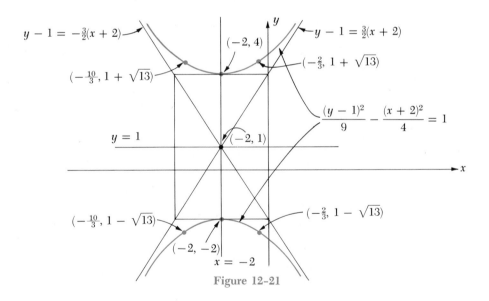

Figure 12-21

From Equation (12.28) or (12.30) it is possible to obtain an equation of the form

$$Ax^2 + Cy^2 + Dx + Ey + F = 0 \qquad (12.31)$$

for any hyperbola whose transverse axis is in a line that is perpendicular to a coordinate axis. The proof of the following theorem is left as an exercise.

12.3.2   Theorem

If $AC < 0$ *in* (12.31), *then* (12.31) *is an equation of a hyperbola or a pair of intersecting lines.*

An interesting application of the hyperbola was developed during World War I to locate enemy artillery. Suppose three listening stations that are in contact with each other and are using synchronized clocks note the exact time that a report from an enemy gun is heard. From noting the difference in time for the sound to reach each pair of stations and knowing the speed of sound, it is possible to calculate the difference in the distances from the gun to each pair of stations. Thus, with each pair of stations as foci there will be a branch of a hyperbola which is the locus of possible positions of the gun. The location of the gun can then be determined graphically from the intersection of two of these branches.

Essentially the same principle is used in the LORAN system of navigation which was developed during World War II. In this system shore stations working in pairs transmit pulse signals that can be received by a ship (or aircraft). There is a time interval between these signals, which can be determined by electronic equipment on the ship, and also the velocity of the signals is known. Therefore it is possible to determine the difference of the distances between the ship and the stations. Thus with each pair of stations as foci there will be a branch of a hyperbola

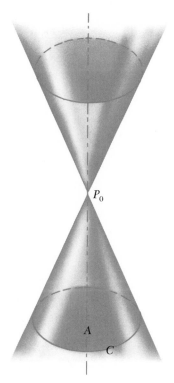

Figure 12-22

which is the locus of possible positions of the ship. The location of the ship can then be determined graphically from the intersection of two of these branches.

We mentioned at the beginning of this chapter that the parabola, ellipse, and hyperbola are called conic sections since they can be obtained as the intersection of a plane and a *right circular cone*. A right circular cone is a surface in three-dimensional space which may be defined as follows: Suppose C is a circle with center A and P_0 is a point on a line that is perpendicular to the plane of the circle at A. The set of points on all lines that pass through P_0 and a point of C is called a right circular cone (Figure 12–22). These lines are called the *generators* of the cone. The point P_0 is termed the *vertex* of the cone and the line through the points P_0 and A is the *axis* of the cone. The part of the cone on one side of the vertex is called a *nappe* of the cone. If a plane is parallel to only one generator of the cone, its intersection with the cone is a parabola (Figure 12–23(a)). Note that such a plane will intersect only one nappe of the cone. If the plane intersects only one nappe of the cone, but is not parallel to a generator, the intersection is an ellipse (Figure 12–23(b)). (If this plane is perpendicular to the axis, the intersection with the cone is a circle, which is a special case of an ellipse.) If the plane intersects both nappes, the intersection is a hyperbola (Figure 12–23(c)).

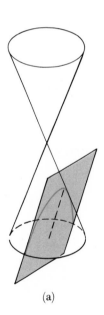
(a)

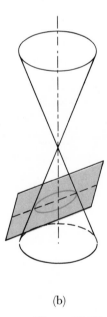
(b)

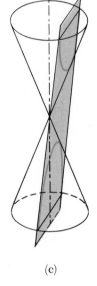
(c)

Figure 12–23

Exercise Set 12.3

In Exercises 1–10 obtain an equation of the hyperbola satisfying the given conditions.

1. Foci $(\pm 6, 0)$, vertices $(\pm 4, 0)$

2. Vertex $(0, -2)$, focus $(0, -5)$, center $(0, 2)$

3. Center $(3, -4)$, focus $(3, 4)$, transverse axis of length 10

4. Absolute value of the difference of the distances from $(-2, 5)$ and $(4, 5)$ to any point on the hyperbola is 4.

5. Foci $(1, 2)$ and $(1, 12)$, conjugate axis of length $2\sqrt{10}$

6. Vertices $(2, -1)$ and $(6, -1)$, passes through $(9, \frac{1}{2})$

7. Foci $(-1 \pm 2\sqrt{3}, 2)$, passes through $(3, 4)$

8. Center $(1, 2)$, focus $(1, 7)$, slopes of the asymptotes are $\pm\frac{1}{2}$.

9. Center $(-5, 2)$, focus $(-5, 5)$, length of the latus rectum is 5.

10. Passing through $(2, 3)$, $(4, 1)$, $(1, 6)$, $(5, -2)$

In Exercises 11–22 find the foci, vertices, endpoints of the latera recta and asymptotes of the hyperbola having the given equation. Sketch the graph of the equation.

11. $\dfrac{x^2}{9} - \dfrac{y^2}{27} = 1$ 12. $\dfrac{y^2}{25} - \dfrac{x^2}{25} = 1$

13. $\dfrac{(y-4)^2}{16} - \dfrac{(x-1)^2}{36} = 1$ 14. $\dfrac{(x+3)^2}{4} - \dfrac{(y-2)^2}{3} = 1$

15. $36(x-5)^2 - 49(y-4)^2 = 25$ 16. $144(x+1)^2 - 25(y+2)^2 = 81$

17. $x^2 - 4y^2 + 8y - 20 = 0$ 18. $-x^2 + 2y^2 - 24y + 40 = 0$

19. $25x^2 - 4y^2 + 150x + 16y + 109 = 0$

20. $4x^2 - 9y^2 - 24x - 90y - 225 = 0$

21. $-4x^2 + 5y^2 + 16x - 30y + 9 = 0$

22. $3x^2 - 5y^2 - 12x + 10y - 23 = 0$

23. Find an equation of the tangent to the hyperbola $2x^2 - 3y^2 = 42$ at $(6, -\sqrt{10})$.

24. Find an equation of the normal to the hyperbola $5x^2 - 4y^2 = 16$ at $(2, 1)$.

25. Derive equation (12.26) from equation (12.25).

26. Show that the tangent at any point (x_1, y_1) in the hyperbola $\dfrac{x^2}{a^2} - \dfrac{y^2}{b^2} = 1$

has an equation $\dfrac{xx_1}{a^2} - \dfrac{yy_1}{b^2} = 1$.

27. Prove that the length of a latus rectum of the hyperbola (12.28) is $2b^2/a$.

28. Prove Theorem 12.3.2.

29. A tangent is drawn to a point $P_1(x_1, y_1)$ in the right-hand branch of the hyperbola $(x^2/a^2) - (y^2/b^2) = 1$. Prove that a line from the focus $(c, 0)$ which is perpendicular to the tangent intersects the line through the origin and P_1 on the line $x = a^2/c$.

30. Prove that every line parallel to an asymptote of the hyperbola $(x^2/a^2) - (y^2/b^2) = 1$ intersects the hyperbola in exactly one point.

31. One pair of LORAN shore stations are located at the points $A_1(0, 3)$ and

$A_2(0, -3)$ and another pair of stations are at $B_1(2, 0)$ and $B_2(10, 0)$. Here 1 unit $= 50$ miles. A ship determines that it is 100 miles closer to A_1 than to A_2 and is also $100\sqrt{6}$ miles closer to B_2 than to B_1. Find the coordinates of the ship.

32. Prove a property for hyperbolas like the reflection property of ellipses (Exercise 33, Exercise Set 12.2).

12.4   Translation and Rotation of Axes

Sometimes a simpler equation of a graph can be found by choosing a different coordinate system from the usual xy system. One type of transformation of coordinates which is used to simplify an equation is called *translation of axes*. In this transformation a line $y = k$ is selected as an x' axis and a line $x = h$ as a y' axis, and their intersection point O' becomes the origin of the new $x'y'$ system (Figure 12–24). We require that the x and x' axes have the same positive direction and the same for the y and y' axes.

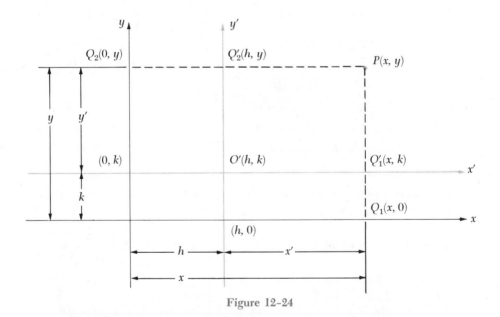

Figure 12–24

In Figure 12–24 and this paragraph, the coordinates given for each point are in the xy system. Suppose $P(x, y)$ is any point in the plane. Then perpendiculars drawn from P will intersect the x axis and y axis in the points $Q_1(x, 0)$ and $Q_2(0, y)$, respectively. These perpendiculars, extended if necessary, intersect the x' axis and the y' axis in the points $Q_1'(x, k)$ and $Q_2'(h, y)$, respectively.

As is suggested by Figure 12–24, the x' coordinate of Q_1 is $x - h$ and the y' coordinate of Q_2 is $y - k$. Thus the coordinates (x', y') for P are given by

$$x' = x - h \qquad y' = y - k. \tag{12.32}$$

In particular from (12.32), the $x'y'$ coordinates for O' are $(0, 0)$. Note that the same construction was used here to obtain the $x'y'$ coordinates of P as was employed in Section 2.1 to obtain the Cartesian coordinates of a point. Equations (12.32) and the equivalent equations

$$x = x' + h \qquad y = y' + k \qquad\qquad (12.33)$$

are called the *translation equations* for translating axes to the point $(x, y) = (h, k)$.

Example 1 If the x and y axes are translated to a new origin at $(4, -2)$ in the xy system, find the x', y' coordinates of the point $P(-1, 5)$ in the xy system (Figure 12–25).

SOLUTION Since $h = 4$ and $k = -2$, from the translation equations (12.33)

$$x' = x - 4 = -1 - 4 = -5 \qquad y' = y + 2 = 5 + 2 = 7.$$

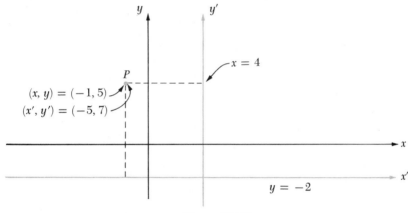

Figure 12-25

An equation obtained from a given equation by translation of axes or any other transformation of coordinates is called a *transform* of the given equation.

Example 2 Using translation of axes, obtain a transform of the equation $xy - 2x + 3y - 18 = 0$ in which there are no x' or y' terms.

SOLUTION From (12.33) the given equation can be rewritten.

$$(x' + h)(y' + k) - 2(x' + h) + 3(y' + k) - 18 = 0$$
$$x'y' + (k - 2)x' + (h + 3)y' - 18 + hk - 2h + 3k = 0. \qquad (12.34)$$

From (12.34) the x' and y' terms will be eliminated if h and k are chosen so that

$$k - 2 = 0 \qquad \text{and} \qquad h + 3 = 0,$$

that is, if

$$k = 2 \qquad \text{and} \qquad h = -3. \qquad\qquad (12.35)$$

If the values obtained from (12.35) are substituted into (12.34) the equation

$$x'y' - 12 = 0 \qquad (12.36)$$

is obtained.

Note that a transformation of coordinates, for example a translation of axes, changes the equation of a graph, but leaves the graph itself unaffected. Thus, if a transformation of coordinates results in a simpler equation for a graph, as in this example, the process of graphing the given equation is facilitated. To graph the given equation in Example 2, we first plot the x and y axes. Then the x' and y' axes are plotted with the origin at $(-3, 2)$ in the xy system. A sketch of the graph of the given equation is then drawn by graphing equation (12.36) with respect to the x' and y' axes (Figure 12–26).

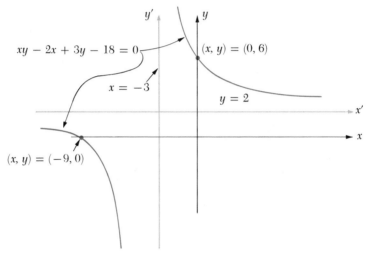

Figure 12–26

The equations in standard form which were obtained in Sections 12.1 to 12.3 using the technique of completing the square in x or y could also have been obtained by translation of axes. The student can show (Exercise 2, Exercise Set 12.4) that the equation

$$9x^2 + 5y^2 + 36x - 30y + 36 = 0, \qquad (12.37)$$

which was discussed in Example 2, Section 12.2, has the transform

$$\frac{x'^2}{5} + \frac{y'^2}{9} = 1 \qquad (12.38)$$

if h and k in the translation equations are chosen so as to eliminate the x' and y' terms. The standard form equation (12.21) for the ellipse (12.37) can then be derived from (12.38).

A second transformation of coordinates which is used to obtain a simpler equation for a graph is called *rotation of axes*. In this transformation a line L_1 containing the origin O of the xy system is chosen as the x' axis. Suppose θ is the angle between the x axis and x' axis (in the sense of Definition 4.10.1(a)). Then the y' axis is that line containing O such that the angle between the y axis and the line is also θ (Figure 12–27). Thus the x' *and* y' axes are perpendicular to each other at O.

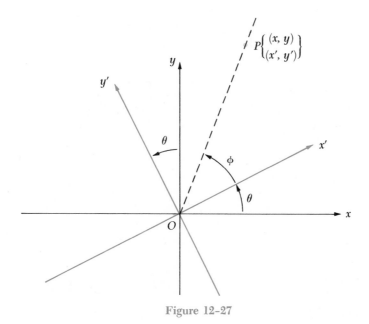

Figure 12–27

If P is any point in the plane, and ϕ is the angle between the x' axis and OP, then the coordinates (x, y) of P are given by

$$x = |OP| \cos (\theta + \phi) \qquad y = |OP| \sin (\theta + \phi). \tag{12.39}$$

Similarly, the coordinates (x', y') of P satisfy

$$x' = |OP| \cos \phi \qquad y' = |OP| \sin \phi. \tag{12.40}$$

Hence from (12.39) and (12.40)

$$x = |OP|(\cos \theta \cos \phi - \sin \theta \sin \phi) \qquad y = |OP|(\sin \theta \cos \phi + \cos \theta \sin \phi)$$
$$x = x' \cos \theta - y' \sin \theta \qquad y = x' \sin \theta + y' \cos \theta. \tag{12.41}$$

The equations (12.41) are called the *rotation equations* for rotating axes through an angle θ. If these equations are solved for x' and y' in terms of x and y, (Exercise 8, Exercise Set 12.4), we have

$$x' = x \cos \theta + y \sin \theta \qquad y' = -x \sin \theta + y \cos \theta. \tag{12.42}$$

Example 3 If the point P has coordinates $(4, -\sqrt{3})$ in the xy system, find its coordinates in the $x'y'$ system if axes are rotated through an angle of $60°$.

SOLUTION   From (12.42)

$$x' = 4 \cos 60° - \sqrt{3} \sin 60° \qquad y' = -4 \sin 60° - \sqrt{3} \cos 60°$$

$$= 4 \cdot \frac{1}{2} - \sqrt{3}\left(\frac{\sqrt{3}}{2}\right) \qquad\qquad = -4\left(\frac{\sqrt{3}}{2}\right) - \sqrt{3}\left(\frac{1}{2}\right)$$

$$= \frac{1}{2}. \qquad\qquad\qquad\qquad = -\frac{5\sqrt{3}}{2}.$$

Therefore P has coordinates $\left(\dfrac{1}{2}, -\dfrac{5\sqrt{3}}{2}\right)$ in the $x'y'$ system.

Example 4   Obtain a transform of the equation $3x^2 - 10xy + 3y^2 - 32 = 0$ if the coordinate axes are rotated through an angle $\theta = 45°$. Graph the given equation and show both sets of coordinates axes in the sketch.

SOLUTION   Since $\theta = 45°$, $\cos \theta = \sin \theta = 1/\sqrt{2}$, and therefore from (12.41) the rotation equations are

$$x = \frac{x' - y'}{\sqrt{2}} \qquad y = \frac{x' + y'}{\sqrt{2}}. \tag{12.43}$$

After substituting from (12.43) into the given equation, we have

$$3\left(\frac{x' - y'}{\sqrt{2}}\right)^2 - 10\left(\frac{x' - y'}{\sqrt{2}}\right)\left(\frac{x' + y'}{\sqrt{2}}\right) + 3\left(\frac{x' + y'}{\sqrt{2}}\right)^2 - 32 = 0$$

which simplifies to

$$\frac{-4x'^2 + 16y'^2}{2} - 32 = 0$$

and hence to

$$4y'^2 - x'^2 = 16.$$

The equation of the transform in the standard form for a hyperbola is

$$\frac{y'^2}{4} - \frac{x'^2}{16} = 1. \tag{12.44}$$

A sketch of the graph of (12.44) and hence of the given equation is shown in Figure 12–28.

Exercise Set 12.4

1. If the x and y axes are translated to x' and y' axes having an origin at $(-5, 3)$ in the xy system, find the x', y' coordinates of the following points, which are expressed in the xy system:
   (a) $(-2, 7)$ (b) $(5, 2)$ (c) $(-8, 5)$ (d) $(-3, -6)$.

In Exercises 2–7 translate axes so as to eliminate the indicated terms.

2. $9x^2 + 5y^2 + 36x - 30y + 36 = 0$;   first-degree terms

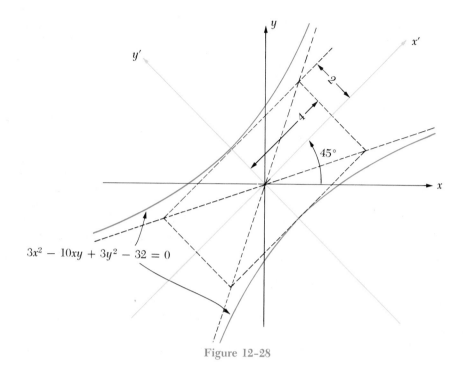

$3x^2 - 10xy + 3y^2 - 32 = 0$

Figure 12–28

3. $4x^2 - y^2 + 24x + 8y + 30 = 0$;   first-degree terms

4. $x^2 - 3xy + 4y^2 - 2x - 4y + 1 = 0$;   first-degree terms

5. $2x^2 + xy - y^2 - 4x + 26y - 5 = 0$;   first-degree terms

6. $y = x^3 - 9x + 6$; x^2 term, constant

7. $y = x^3 + 3x^2 - 4x + 1$; x^2 term, constant

8. Derive equations (12.42) by solving equations (12.41) for x' and y'.

9. If the x and y axes are rotated through an angle of $150°$, find the x', y' coordinates of the following points expressed in the xy system:
   (a) $(2, -3)$ (b) $(1, -\sqrt{3})$ (c) $(2\sqrt{3}, 2)$.

In Exercises 10–15 obtain the transform of the given equation after rotating axes through the given angle.

10. $2x - 3y = 4$; $\theta = \tan^{-1} 2$

11. $2x - 3y = 4$; $\theta = \tan^{-1} \frac{3}{2}$

12. $xy = 12$; $\theta = 45°$

13. $x^2 + 2xy + y^2 - 5x + y = 17$; $\theta = 45°$

14. $5x^2 - 3xy + y^2 = 11$; $\theta = \cos^{-1}\left(-\dfrac{3}{\sqrt{10}}\right)$

15. $5x^2 + 3xy + y^2 = 11$; $\theta = \cos^{-1}\dfrac{3}{\sqrt{10}}$

16. Find the transform of the equation $x^2 + y^2 = r^2$ if the axes are rotated through any angle θ.

17. Graph the equation in Exercise 15.

18. Graph the equation in Exercise 13.

12.5   Graphing by Rotation of Axes

Suppose we wanted to sketch the graph of a *general equation of second degree* in x and y—that is, an equation of the form

$$Ax^2 + Bxy + Cy^2 + Dx + Ey + F = 0 \qquad (12.45)$$

where A, B, and C are not all zero. The problem of graphing this equation, as written, can be complicated if $B \neq 0$. However if a transform of (12.45) of the form

$$A'x'^2 + C'y'^2 + D'x' + E'y' + F' = 0 \qquad (12.46)$$

can be obtained through a transformation of coordinates, the equation (12.45) can be graphed after rewriting (12.46) using one of the standard forms for a conic section. It turns out that if the rotation equations

$$x = x' \cos \theta - y' \sin \theta \qquad y = x' \sin \theta + y' \cos \theta \qquad (12.47)$$

are used on (12.45), then a transform of the form (12.46) can be obtained. To show this, we substitute from (12.47) into (12.45), and obtain

$$
\begin{aligned}
A(x' \cos \theta - y' \sin \theta)^2 &+ B(x' \cos \theta - y' \sin \theta)(x' \sin \theta + y' \cos \theta) \\
&+ C(x' \sin \theta + y' \cos \theta)^2 + D(x' \cos \theta - y' \sin \theta) \quad (12.48) \\
&+ E(x' \sin \theta + y' \cos \theta) + F = 0.
\end{aligned}
$$

This equation can be rewritten in the form (12.46) where

$$
\begin{aligned}
A' &= A \cos^2 \theta + B \sin \theta \cos \theta + C \sin^2 \theta \\
B' &= 2(C - A) \sin \theta \cos \theta + B(\cos^2 \theta - \sin^2 \theta) \\
C' &= A \sin^2 \theta - B \sin \theta \cos \theta + C \cos^2 \theta \\
D' &= D \cos \theta + E \sin \theta \qquad\qquad\qquad\qquad\qquad (12.49) \\
E' &= E \cos \theta - D \sin \theta \\
F' &= F
\end{aligned}
$$

Since $\sin 2\theta = 2 \sin \theta \cos \theta$ and $\cos 2\theta = \cos^2 \theta - \sin^2 \theta$,

$$B' = B \cos 2\theta - (A - C) \sin 2\theta. \qquad (12.50)$$

We wish to find the angle θ through which axes must be rotated so that $B' = 0$. When $B \neq 0$, we see from (12.50) that $B' = 0$ if $B \cos 2\theta = (A - C) \sin 2\theta$—that is, if $\cot 2\theta = (A - C)/B$. We have therefore proved the following theorem which assures us that if $B \neq 0$, then a transform (12.46) is obtainable by rotation of axes.

12.5.1   Theorem

If B ≠ 0, the equation (12.45) has a transform

$$A'x'^2 + C'y'^2 + D'x' + E'y' + F' = 0 \tag{12.46}$$

when axes are rotated through any angle θ for which

$$\cot 2\theta = \frac{A - C}{B}. \tag{12.51}$$

To sketch the graph of (12.45), we first draw the x and y axes. Then the x' and y' axes are drawn so that the angle θ satisfying (12.51) is between the x and x' axes and between the y and y' axes. The graph of (12.46), and hence of (12.45), is then plotted with respect to the x' and y' axes.

The angle θ through which axes must be rotated to give the transform (12.46) can be chosen so that $0° < 2\theta < 180°$. Then $0° < \theta < 90°$ and $\sin \theta > 0$ and $\cos \theta > 0$. Hence from (10.19) and (10.20)

$$\cos \theta = \sqrt{\frac{1 + \cos 2\theta}{2}} \qquad \sin \theta = \sqrt{\frac{1 - \cos 2\theta}{2}}. \tag{12.52}$$

The value of $\cos 2\theta$ that is substituted in (12.52) can be determined by trigonometry from the value of $\cot 2\theta$ obtained using (12.51).

Example 1   Graph the equation $8x^2 + 4xy + 5y^2 - 80x - 20y + 164 = 0$.

SOLUTION   To obtain a transform of the given equation of the form (12.46), the axes are rotated through an angle θ given by

$$\cot 2\theta = \frac{8 - 5}{4} = \frac{3}{4}.$$

If 2θ is chosen so that $0° < 2\theta < 180°$, then from trigonometry $\cos 2\theta = \frac{3}{5}$. Also, $0° < \theta < 90°$ and hence:

$$\sin \theta = \sqrt{\frac{1 - \cos 2\theta}{2}} \qquad \cos \theta = \sqrt{\frac{1 + \cos 2\theta}{2}}$$

$$= \sqrt{\frac{1 - \frac{3}{5}}{2}} \qquad = \sqrt{\frac{1 + \frac{3}{5}}{2}}$$

$$= \frac{1}{\sqrt{5}} \qquad = \frac{2}{\sqrt{5}}$$

The desired rotation equations are therefore

$$x = \frac{2x' - y'}{\sqrt{5}} \qquad y = \frac{x' + 2y'}{\sqrt{5}}. \tag{12.53}$$

Substituting from (12.53) into the given equation, we have

$$8\left(\frac{2x' - y'}{\sqrt{5}}\right)^2 + 4\left(\frac{2x' - y'}{\sqrt{5}}\right)\left(\frac{x' + 2y'}{\sqrt{5}}\right) + 5\left(\frac{x' + 2y'}{\sqrt{5}}\right)^2$$

$$- 80\left(\frac{2x' - y'}{\sqrt{5}}\right) - 20\left(\frac{x' + 2y'}{\sqrt{5}}\right) + 164 = 0$$

which simplifies to

$$\frac{45x'^2 + 20y'^2}{5} - \frac{180}{\sqrt{5}}x' + \frac{40}{\sqrt{5}}y' + 164 = 0$$

$$9x'^2 + 4y'^2 - 36\sqrt{5}x' + 8\sqrt{5}y' + 164 = 0. \qquad (12.54)$$

Completing the square in x' and y' in (12.54) gives

$$9(x'^2 - 4\sqrt{5}x') + 4(y'^2 + 2\sqrt{5}y') = -164$$
$$9(x'^2 - 4\sqrt{5}x' + 20) + 4(y'^2 + 2\sqrt{5}y' + 5) = 36$$
$$9(x' - 2\sqrt{5})^2 + 4(y' + \sqrt{5})^2 = 36.$$

The equation of the transform in the standard form for an ellipse is

$$\frac{(x' - 2\sqrt{5})^2}{4} + \frac{(y' + \sqrt{5})^2}{9} = 1.$$

A sketch of the graph of the transform (and hence of the given equation) is shown in Figure 12–29.

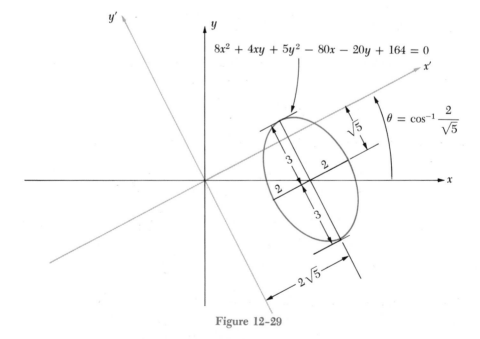

$$8x^2 + 4xy + 5y^2 - 80x - 20y + 164 = 0$$

$$\theta = \cos^{-1}\frac{2}{\sqrt{5}}$$

Figure 12-29

If $A = C$ in (12.45), then from (12.51) $\cot 2\theta = 0$ and hence $\cos 2\theta = 0$. Thus we could choose $2\theta = 90°$ and therefore obtain $\theta = 45°$. Recall that the transform obtained in Example 4, Section 12.4, had no $x'y'$ term and was obtained by rotating axes through an angle of $45°$.

To facilitate a general discussion of equations of the form (12.45), we will require the following theorem which says that the expressions $A + C$ and $B^2 - 4AC$ are *invariant under rotation of axes*.

12.5.2 Theorem

If the transform $A'x'^2 + B'x'y' + C'y'^2 + D'x' + E'y' + F' = 0$ is obtained from (12.45) by a rotation of axes through any angle θ, then

(a) $A' + C' = A + C$.
(b) $B'^2 - 4A'C' = B^2 - 4AC$.

The proofs of both parts of the theorem follow from (12.49) and are left as exercises. The proof of (a) is quite simple, but that of (b) is rather laborious. Theorem 12.5.2 is used to prove the next theorem.

12.5.3 Theorem

The graph of the second-degree equation (12.45) is determined from the following conditions:

(i) *If $B^2 - 4AC > 0$, the graph is a hyperbola or a pair of intersecting lines.*
(ii) *If $B^2 - 4AC = 0$, the graph is a parabola, two parallel lines, a single line, or (12.45) has no graph.*
(iii) *If $B^2 - 4AC < 0$, the graph is an ellipse or (12.45) has no graph.*

PROOF OF (ii) Suppose $B^2 - 4AC = 0$. If the axes are rotated through an angle θ such that $B' = 0$, then from Theorem 12.5.2(b), $-4A'C' = B'^2 - 4A'C' = B^2 - 4AC = 0$. Hence $A'C' = 0$.

We can show that A' and C' are not both zero. If $A' = 0 = C'$, then $A + C = A' + C' = 0$, and $A = -C$. Therefore $B^2 - 4AC = B^2 + 4C^2 = 0$ and hence $B = C = A = 0$, which is impossible since A, B, and C are not simultaneously zero.

Thus $A'C' = 0$, but A' and C' are not both zero. Therefore by Theorem 12.1.3 the graph of (12.46) and hence of (12.45) fulfills the conclusion in (ii). The proofs of (i) and (iii) are left as exercises.

As an illustration of Theorem 12.5.3, the graph of the equation $x^2 - 2xy + y^2 - 3x - 7 = 0$ is a parabola since $B^2 - 4AC = (-2)^2 - 4 \cdot 1 \cdot 1 = 0$. Also, the graph of $x^2 + 3xy + 2y^2 + x - 5y + 6 = 0$ is a hyperbola since $B^2 - 4AC = 3^2 - 4 \cdot 1 \cdot 2 = 1 > 0$.

Exercise Set 12.5

In Exercises 1–10 obtain the graph of the given equation by rotation of axes. In the sketch be sure to include both sets of axes.

1. $2x^2 + 3xy + 2y^2 = 6$

2. $x^2 + 10\sqrt{3}xy + 11y^2 = 16$

3. $11x^2 + 6xy + 19y^2 - 40 = 0$

4. $5x^2 + 6xy + 5y^2 + 16x + 16y - 2 = 0$

5. $7x^2 + 6xy - y^2 + 100x + 20y + 268 = 0$

6. $16x^2 + 24xy + 9y^2 + 30x + 210y - 75 = 0$

7. $x^2 - 4xy + 4y^2 + 40x - 80y + 375 = 0$

8. $3x^2 - 4xy + 30x - 20y + 55 = 0$

9. $x^2 + 2\sqrt{3}xy + 3y^2 - 64y + 32 = 0$

10. $11x^2 - 24xy + 4y^2 + 82x - 24y + 35 = 0$

11. Prove Theorem 12.5.2(a). 　　　　　　12. Prove (i) in Theorem 12.5.3.

13. Prove (iii) in Theorem 12.5.3. 　　　　14. Prove Theorem 12.5.2(b).

13. Numerical Methods; Indeterminate Forms; Improper Integrals

13.1   Newton's Method

This section deals with a numerical method for approximating a real root r of an equation

$$f(x) = 0 \tag{13.1}$$

when the function f is differentiable on an interval containing r. Such a method is important because usually an equation of the form (13.1) cannot be solved by a general formula like the quadratic formula, for example. Generally in problems of interest, the root r is known to exist by the intermediate value theorem (Theorem 7.7.3) because f is continuous on a closed interval and has values at the endpoints of the interval which are opposite in sign.

We begin by making a judicious guess x_1 of the real root r of equation (13.1) which is to be approximated. A perpendicular to the x axis at the point $(x_1, 0)$ intersects the graph of f at $(x_1, f(x_1))$ (Figure 13–1). Then the tangent to the graph

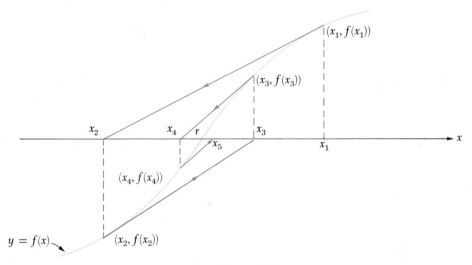

Figure 13-1

of f at $(x_1, f(x_1))$ intersects the x axis at some point $(x_2, 0)$. The procedure starting with the point $(x_1, 0)$ is then repeated beginning with $(x_2, 0)$. A perpendicular to the x axis at $(x_2, 0)$ intersects the graph of f at $(x_2, f(x_2))$. The tangent to the graph of f at $(x_2, f(x_2))$ intersects the x axis at a point $(x_3, 0)$. A continuation of this process gives a set of numbers $x_1, x_2, x_3, \ldots$ which will tend to r provided x_1 is chosen sufficiently close to r and also the function f satisfies certain properties near r. We therefore have a geometric description of *Newton's method* for approximating r.

Since an equation of the tangent at $(x_1, f(x_1))$ is

$$y = f(x_1) + f'(x_1)(x - x_1), \tag{13.2}$$

x_2 can be obtained by letting $y = 0$ in (13.2) and then solving for x. Thus if $f'(x_1) \neq 0$,

$$x_2 = x_1 - \frac{f(x_1)}{f'(x_1)}.$$

(If $f'(x_1) = 0$, either $x_1 = r$, or x_1 is a poor guess for r.) Similarly, x_3 can be found in terms of x_2 from the equation

$$x_3 = x_2 - \frac{f(x_2)}{f'(x_2)},$$

and in general

$$x_{n+1} = x_n - \frac{f(x_n)}{f'(x_n)} \qquad n = 1, 2, 3, \ldots. \tag{13.3}$$

The procedure of obtaining successive approximations for the root r using (13.3) is called *Newton's method*.

The following example illustrates the use of Newton's method.

Example 1 Approximate $\sqrt{34}$ correct to the third decimal place.

SOLUTION   Since $\sqrt{34}$ is the positive root of the equation $x^2 - 34 = 0$, we let

$$f(x) = x^2 - 34 \tag{13.4}$$

and hence $f'(x) = 2x$. Then from (13.3) and (13.4)

$$x_{n+1} = x_n - \frac{x_n{}^2 - 34}{2x_n} \qquad n = 1, 2, 3, \ldots. \tag{13.5}$$

From (13.4) $f(5) < 0$ and $f(6) > 0$, and hence by the intermediate value theorem

$$5 < \sqrt{34} < 6.$$

As a first guess for $\sqrt{34}$, we shall choose $x_1 = 6$. Then if $n = 1$ in (13.5),

$$x_2 = 6 - \tfrac{2}{12} \approx 5.8333.$$

If $n = 2$ and we substitute the value obtained for x_2 in (13.5),

$$x_3 \approx 5.8333 - \tfrac{0.027389}{11.667} \approx 5.8310.$$

Letting $n = 3$ and substituting the value obtained for x_3 in (13.5), we obtain

$$x_4 \approx 5.8310 - \frac{0.00056100}{11.662} \approx 5.8310,$$

our fourth approximation of $\sqrt{34}$. Since the last two approximations agree to the third decimal place, the required accuracy in the problem, we declare that $\sqrt{34} \approx 5.831$.

It was assumed in this solution that the successive approximations x_1, $x_2, x_3, \ldots$ tend to $\sqrt{34}$, a fact that is proved below using Theorem 13.1.1. Generally, however, in problems of this type we will not concern ourselves with a proof that the Newton's method approximations do actually approach the desired root, preferring instead to focus on the technique followed using the method.

It will also be instructive to consider an example in which the Newton's method approximations $x_1, x_2, x_3, \ldots$ do not approach the root r. We note that 0 is a root of the equation

$$x^{1/3} = 0.$$

Suppose $x_1 \neq 0$ is some approximation of 0. From (13.3), using $f(x) = x^{1/3}$ and hence $f'(x) = \frac{1}{3}x^{-2/3}$,

$$x_2 = x_1 - \frac{x_1^{1/3}}{\frac{1}{3}x_1^{-2/3}} = -2x_1.$$

Also from (13.3)

$$x_3 = -2x_2 = 4x_1$$

$$\cdots$$

$$x_{n+1} = -2x_n = (-1)^n 2^n x_1.$$

Since the successive approximations are alternately positive and negative and increasing in absolute value, it is intuitively apparent that they cannot approach any number. The approximations obtained here are illustrated by Figure 13–2.

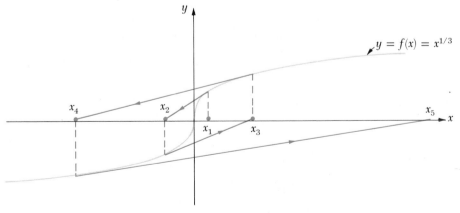

Figure 13–2

The following theorem is useful for determining if the Newton's method approximations tend toward the root r.

13.1.1   Theorem

If the following conditions are satisfied:

(i) *f has first and second derivatives on an open interval I containing a number r where $f(r) = 0$,*
(ii) *there is an $m > 0$ such that $|f'(x)| \geq m$ for every $x \in I$,*
(iii) *there is an $M > 0$ such that $|f''(x)| \leq M$ for every $x \in I$,*
(iv) *$x_n \in I$ and x_{n+1} are successive Newton's method approximations for r,*

then

$$|x_{n+1} - r| \leq \frac{M}{2m}(x_n - r)^2 \quad n = 1, 2, 3, \ldots . \tag{13.6}$$

PROOF   If E is the error which results from using $f(x_n)$ as an approximation for $f(r) = 0$, then from (6.48)

$$E = f(r) - [f(x_n) + f'(x_n)(r - x_n)]$$
$$= -[f(x_n) + f'(x_n)(r - x_n)]. \tag{13.7}$$

Since f satisfies the hypotheses of Theorem 6.6.2 on the closed interval with endpoints x_n and r, by this theorem

$$|E| \leq \tfrac{1}{2}M(x_n - r)^2. \tag{13.8}$$

Since

$$x_{n+1} = x_n - \frac{f(x_n)}{f'(x_n)} \quad \text{for } n = 1, 2, 3, \ldots, \tag{13.3}$$

from (13.7)

$$x_{n+1} - r = x_n - \frac{f(x_n)}{f'(x_n)} - r$$
$$= \frac{-[f(x_n) + f'(x_n)(r - x_n)]}{f'(x_n)}$$
$$= \frac{E}{f'(x_n)}.$$

Then from (13.8)

$$|x_{n+1} - r| \leq \frac{M(x_n - r)^2}{2|f'(x_n)|},$$

and hence by hypothesis (ii)

$$|x_{n+1} - r| \leq \frac{M}{2m}(x_n - r)^2,$$

which was to be proved.

If (13.6) holds, then

$$|x_2 - r| \leq \frac{M}{2m}(x_1 - r)^2$$

$$|x_3 - r| \leq \frac{M}{2m}(x_2 - r)^2 \leq \frac{M^3}{8m^3}(x_1 - r)^4$$

$$\cdots$$

$$|x_{n+1} - r| \leq \left(\frac{M}{2m}\right)^{2^n - 1}(x_1 - r)^{2^n} \qquad n = 1, 2, 3, \ldots. \qquad (13.9)$$

Thus if $M/2m$ and $|x_1 - r|$ are each less than 1, it is clear from (13.9) that the right member of (13.9) tends to 0 as n increases indefinitely and thus x_{n+1} tends to r.

Because of (13.9) we can show that the approximations given by (13.5) do tend to $\sqrt{34}$, as was stated following Example 1. Here, $f(x) = x^2 - 34$ (compare Example 1), and I can be chosen as the interval $(5, 7)$. Then $|f'(x)| \geq 10$ and $|f''(x)| = 2$ whenever $x \in I$. Hence from (13.9) since $M/2m = 2/2 \cdot 10 = \frac{1}{10}$ and $|x_1 - \sqrt{34}| < 1$,

$$|x_{n+1} - \sqrt{34}| \leq (\tfrac{1}{10})^{2^n - 1} \qquad n = 1, 2, 3, \ldots$$

Thus the approximations $x_1, x_2, x_3, \ldots$ in Example 1 do tend to $\sqrt{34}$.

On the other hand, Theorem 13.1.1 cannot be applied to the function given by $f(x) = x^{1/3}$, which was discussed just before the theorem. We note here that hypothesis (iii) is not satisfied on any open interval containing 0 since $f''(x) = -\frac{2}{9}x^{-5/3}$ is not defined when $x = 0$.

Suppose the function f satisfies the hypotheses of Theorem 13.1.1 on some open interval containing r and also $M/m \leq 4$. If x_n is an approximation of r correct to the kth decimal place (that is, if

$$|x_n - r| < \tfrac{1}{2} \cdot 10^{-k}),$$

then from (13.6)

$$|x_{n+1} - r| < 2(\tfrac{1}{2} \cdot 10^{-k})^2 = \tfrac{1}{2} \cdot 10^{-2k}$$

that is, x_{n+1} approximates r correct to the $2k$th decimal place, which is *twice as many* places as when x_n is the approximation of r.

This behavior is illustrated by the approximations x_2 and x_3 that were obtained for $\sqrt{34}$ in Example 1, using $f(x) = x^2 - 34$. Note from Example 1 that x_2 and x_3 approximate $\sqrt{34}$ to two and four decimal places, respectively. Also, x_4 would have been correct to eight decimal places had it been expressed with that accuracy. Such a result is not unexpected since, here, $M/m = \frac{1}{5} \leq 4$ whenever $x \in (5, 7)$.

Exercise Set 13.1

1. In each part find the square root correct to the third decimal place.
   (a) $\sqrt{47}$ (b) $\sqrt{66}$ (c) $\sqrt{103}$

2. Derive the Newton's method formula $x_{n+1} = \frac{2}{3}x_n + M/3x_n^2$, $n = 1, 2, 3, \ldots$, used to approximate $\sqrt[3]{M}$.

3. In each part find the cube root correct to the third decimal place.
   (a) $\sqrt[3]{9}$ (b) $\sqrt[3]{29}$ (c) $\sqrt[3]{61}$

4. Derive a Newton's method formula for approximating a fourth root which is analogous to the equation in Exercise 2.

In Exercises 5–9 obtain the required roots correct to the third decimal place.

5. Positive root of $x^3 + 2x - 5 = 0$.

6. Negative root of $x^3 + 2x^2 + 4 = 0$.

7. The root of $\cos x = x$.

8. The positive root of $\sin x = x^2$.

9. The root of $e^{-x} = \ln x$.

13.2   Taylor's† Formula

From (6.43) a function f that is differentiable at a number a can be approximated at numbers x near a by the approximation formula

$$f(x) \approx f(a) + f'(a)(x - a). \tag{13.10}$$

This formula is said to give a *linear approximation* of $f(x)$. From (6.56) the error E resulting from using $f(a) + f'(a)(x - a)$ to approximate $f(x)$ is given by

$$E = \tfrac{1}{2}f''(c)(x - a)^2 \tag{13.11}$$

where c is between a and x.

Polynomial functions, including linear functions, are useful as approximating functions since they are so easy to work with. Therefore it is pertinent to inquire whether a better approximation of a given $f(x)$ might be obtained using some polynomial in $x - a$ of degree > 1 than the linear polynomial in (13.10). If the polynomial

$$P_n(x) = \sum_{i=0}^{n} c_i(x - a)^i \tag{13.12}$$

were used to approximate $f(x)$, then it would be reasonable to require that

$$P_n(a) = c_0 = f(a), \tag{13.13}$$

and if the first n derivatives of f at a exist, to insist also that

$$P_n^{(k)}(a) = f^{(k)}(a) \quad \text{for } k = 1, 2, 3, \ldots, n. \tag{13.14}$$

---

† Named for Brook Taylor (1685–1731) an English mathematician who authored a book containing the *Taylor series* in 1715. Taylor was a member of the Royal Society and at one time was a member of a committee of the Society to investigate the claims of Newton and Leibniz for discoveries in calculus.

Thus we are requiring that $P_n(x)$ agree in value with $f(x)$ and its first n derivatives at $x = a$. From (13.12) it can be proved by mathematical induction (Exercise 25, Exercise Set 13.2) that

$$P_n^{(k)}(x) = \sum_{i=0}^{n-k} (i + 1)(i + 2) \cdots (i + k)c_{i+k}(x - a)^i. \tag{13.15}$$

Hence $P_n^{(k)}(a) = k!c_k$ and

$$c_k = \frac{P_n^{(k)}(a)}{k!} \quad \text{if} \quad k = 1, 2, 3, \ldots, n. \tag{13.16}$$

From (13.13), (13.14), and (13.16) the approximating polynomial $P_n(x)$ would have the form

$$P_n(x) = \sum_{i=0}^{n} \frac{f^{(i)}(a)}{i!}(x - a)^i \tag{13.17}$$

where $f^{(0)}(a) = f(a)$.

13.2.1   Definition

Suppose f has an nth derivative at some number a. The polynomial (13.17) is called *the nth degree Taylor polynomial of f at a*.

In particular, the linear approximation $f(a) + f'(a)(x - a)$ of $f(x)$ in (13.10) is the first degree Taylor polynomial of f at a.

Example 1   Find the 4th degree Taylor polynomial of cos at $\frac{\pi}{3}$.

SOLUTION   We let $f(x) = \cos x$ and $a = \frac{\pi}{3}$ in (13.17). Since

$$f(x) = \cos x \qquad f\left(\frac{\pi}{3}\right) = \frac{1}{2}$$

$$f'(x) = -\sin x \qquad f'\left(\frac{\pi}{3}\right) = -\frac{\sqrt{3}}{2}$$

$$f''(x) = -\cos x \qquad f''\left(\frac{\pi}{3}\right) = -\frac{1}{2}$$

$$f'''(x) = \sin x \qquad f'''\left(\frac{\pi}{3}\right) = \frac{\sqrt{3}}{2}$$

$$f^{(4)}(x) = \cos x \qquad f^{(4)}\left(\frac{\pi}{3}\right) = \frac{1}{2}$$

then from (13.17)

$$P_4(x) = f\left(\frac{\pi}{3}\right) + \frac{f'(\frac{\pi}{3})(x - \frac{\pi}{3})}{1!} + \frac{f''(\frac{\pi}{3})(x - \frac{\pi}{3})^2}{2!} + \frac{f'''(\frac{\pi}{3})(x - \frac{\pi}{3})^3}{3!}$$

$$+ \frac{f^{(4)}(\frac{\pi}{3})(x - \frac{\pi}{3})^4}{4!}$$

$$= \frac{1}{2} - \frac{\sqrt{3}}{2}\left(x - \frac{\pi}{3}\right) - \frac{1}{4}\left(x - \frac{\pi}{3}\right)^2 + \frac{\sqrt{3}}{12}\left(x - \frac{\pi}{3}\right)^3 + \frac{1}{48}\left(x - \frac{\pi}{3}\right)^4$$

Example 2 Find the 5th degree Taylor polynomial for exp at 0.

SOLUTION We let $f(x) = e^x$ and $a = 0$ in (13.17). Since

$$f(x) = e^x, \qquad\qquad f(0) = 1$$
$$f^{(n)}(x) = e^x, \quad \text{for any } n, \qquad f^{(n)}(0) = 1, \quad \text{for any } n;$$

and from (13.17)

$$P_5(x) = f(0) + \frac{f'(0)x}{1!} + \frac{f''(0)x^2}{2!} + \frac{f'''(0)x^3}{3!} + \frac{f^{(4)}(0)x^4}{4!} + \frac{f^{(5)}(0)x^5}{5!}$$

$$= 1 + \frac{x}{1!} + \frac{x^2}{2!} + \frac{x^3}{3!} + \frac{x^4}{4!} + \frac{x^5}{5!}.$$

The usefulness of $P_n(x)$ as an approximation of $f(x)$ depends upon the size of the *remainder term* $R_n(x)$.

13.2.2 Definition

Suppose f has an nth derivative at some number a, $P_n(x)$ is given by (13.17), and $x \in \mathcal{D}_f$. Then

$$R_n(x) = f(x) - P_n(x). \tag{13.18}$$

The next theorem is useful in obtaining bounds for $R_n(x)$.

13.2.3 Theorem

Suppose a is any number and $x \neq a$. If f has a continuous nth derivative on the closed interval I with endpoints a and x, and $f^{(n+1)}$ is defined in the interior of I, then there is a number c between a and x such that

$$R_n(x) = \frac{f^{(n+1)}(c)}{(n + 1)!}(x - a)^{n+1}. \tag{13.19}$$

PROOF If $t \in I$, let

$$F(t) = f(x) - f(t) - f'(t)(x - t) - \frac{f''(t)(x - t)^2}{2!} - \cdots - \frac{f^{(n)}(t)(x - t)^n}{n!} - R_n(x)$$

$$G(t) = (x - t)^{n+1}.$$

Then

$$F(x) = -R_n(x) \qquad F(a) = f(x) - P_n(x) - R_n(x) = 0 \atop G(x) = 0 \qquad\qquad G(a) = (x - a)^{n+1}. \quad\Bigg\}$$

$$(13.20)$$

If $F(t)$ is differentiated with respect to t, the terms involving $f'(t), f''(t), \ldots, f^{(n)}(t)$ will vanish in pairs and thus

$$F'(t) = -\frac{f^{(n+1)}(t)}{n!}(x - t)^n. \qquad\qquad (13.21)$$

Also

$$G'(t) = -(n + 1)(x - t)^n. \qquad\qquad (13.22)$$

The functions F and G here satisfy the hypotheses of the generalized mean value theorem (Theorem 6.6.1) on the interval I. Hence by this theorem there is a number c between a and x such that

$$\frac{F(x) - F(a)}{G(x) - G(a)} = \frac{F'(c)}{G'(c)}. \qquad\qquad (13.23)$$

Substituting from (13.20), (13.21), and (13.22) into (13.23) gives

$$\frac{-R_n(x)}{-(x - a)^{n+1}} = \frac{-\dfrac{f^{(n+1)}(c)}{n!}(x - c)^n}{-(n + 1)(x - c)^n}$$

from which we obtain

$$R_n(x) = \frac{f^{(n+1)}(c)}{(n + 1)!}(x - a)^{n+1}$$

and the theorem is proved.

The form given in (13.19) is called the *Lagrange*† form for the remainder. It will be noted that $R_1(x)$ is the error E given in (13.11), and that, in general, $R_n(x)$ *is the error when* $P_n(x)$ *is used as an approximation for* $f(x)$. For the functions considered in this section it can be shown, using theory that will be introduced in Section 14.7, that $R_n(x)$ tends to 0 as n increases without bound.

Example 3 Find $R_4(x)$ for $\cos x$ at $\frac{\pi}{3}$. If $x \in [\frac{\pi}{3}, \frac{\pi}{3} + 0.1]$, find an upper bound to the absolute error that arises from the use of $P_4(x)$ in Example 1 as an approximation for $\cos x$.

SOLUTION From (13.19) if $n = 4$,

$$R_4(x) = \frac{f^{(5)}(c)}{5!}(x - a)^5 \quad c \text{ between } a \text{ and } x.$$

---

† Named for Joseph Louis Lagrange (1736–1813), a leading French mathematician who was born in Turin, Italy. Lagrange introduced the notion of a "derived function" (derivative) and was the first to use the notation f' for the derivative of f. Such was the fame of Lagrange that in 1766 Frederick the Great wrote to him that the "greatest king in Europe" wanted at his court in Berlin the "greatest mathematician in Europe." Lagrange accepted the invitation and held this position for 20 years.

In this example $a = \frac{\pi}{3}$ and $f(x) = \cos x$. From Example 1 $f^{(5)}(x) = -\sin x$, and therefore

$$R_4(x) = -\frac{\sin c}{5!}\left(x - \frac{\pi}{3}\right)^5 \quad c \text{ between } \frac{\pi}{3} \text{ and } x.$$

Since $|\sin c| \leq 1$ and $|x - \frac{\pi}{3}| \leq 0.1$,

$$|R_4(x)| \leq \frac{1}{5!}(0.1)^5 < 10^{-7}.$$

To approximate $f(x)$ with a given accuracy, we use that Taylor polynomial $P_n(x)$ associated with the least n for which $|R_n(x)|$ is sufficiently small. The next two examples are illustrative of the method followed.

Example 4 Approximate the number e correct to the fifth decimal place.

SOLUTION   As an extension of our work in Example 2 the Taylor polynomial of degree n for e^x at 0 is

$$P_n(x) = 1 + \frac{x}{1!} + \frac{x^2}{2!} + \cdots + \frac{x^n}{n!}. \tag{13.24}$$

Then if $x = 1$ in (13.24)

$$e \approx P_n(1) = 1 + \frac{1}{1!} + \frac{1}{2!} + \cdots + \frac{1}{n!}.$$

Now, $P_n(1)$ will approximate e correct to the fifth decimal place if n is chosen so that

$$|R_n(1)| < 0.000005.$$

We therefore seek the least n for which this inequality is true. From (13.19) $R_n(1) = e^c/(n + 1)!$ and since $2 < e < 3$ (this inequality was mentioned following Definition 9.1.5), we have

$$|R_n(1)| < \frac{3}{(n + 1)!} \quad \text{for } n = 0, 1, 2, \ldots.$$

After calculating $|R_n(1)|$ for $n = 0, 1, 2, \ldots$, we find that

$$|R_9(1)| < \frac{3}{10!} = \frac{1}{1,209,600} < 0.000005,$$

and therefore e should be approximated by $P_9(1)$. The calculations are as follows:

$$1 = 1.0000000 \qquad \frac{1}{1!} = 1.0000000 \qquad \frac{1}{2!} = 0.5000000$$

$$\frac{1}{3!} \approx 0.1666667 \qquad \frac{1}{4!} \approx 0.0416667 \qquad \frac{1}{5!} \approx 0.0083333$$

$$\frac{1}{6!} \approx 0.0013889 \qquad \frac{1}{7!} \approx 0.0001984 \qquad \frac{1}{8!} \approx 0.0000248$$

$$\frac{1}{9!} \approx 0.0000028$$

$$\sum_{n=1}^{9} \frac{1}{n!} \approx 2.7182816.$$

If this sum is expressed correct to the fifth decimal place, then

$$e \approx 2.71828.$$

Example 5 Approximate $\sqrt[3]{9}$ correct to the third decimal place using a Taylor polynomial of $f(x) = x^{1/3}$ at $a = 8$.

SOLUTION We shall approximate $\sqrt[3]{9}$ by a Taylor polynomial $P_n(9)$ where n is the least non-negative integer such that

$$|R_n(9)| < 0.0005. \tag{13.25}$$

Therefore we first calculate $|R_n(9)|$ for $n = 0, 1, 2, \ldots$ until we find the least n for which (13.25) holds. At the same time the derivatives of f at 8 that are used in forming $P_n(9)$ can be evaluated.

We therefore form

$$f(x) = x^{1/3} \qquad f(8) = 8^{1/3} = 2$$
$$f'(x) = \tfrac{1}{3}x^{-2/3} \qquad f'(8) = \tfrac{1}{3} \cdot \tfrac{1}{4} = \tfrac{1}{12}.$$

Then from (13.19) $R_0(9) = \dfrac{f'(c)}{1!} \cdot (1)$ where $8 < c < 9$. Since f' is decreasing on the interval $[8, 9]$,

$$|R_0(9)| < \tfrac{1}{12}$$

which is too large in view of (13.25). Continuing,

$$f''(x) = -\tfrac{2}{9}x^{-5/3} \qquad f''(8) = -\tfrac{2}{9} \cdot \tfrac{1}{32} = -\tfrac{1}{144}.$$

Then from (13.19) $R_1(9) = \dfrac{-\tfrac{2}{9}c^{-5/3}}{2!} \cdot (1)^2$ and hence

$$|R_1(9)| < \frac{\tfrac{1}{144}}{2!} = \frac{1}{288} \quad \text{(too large)}.$$

Continuing,

$$f'''(x) = \tfrac{10}{27}x^{-8/3} \qquad f'''(8) = \tfrac{10}{27} \cdot \tfrac{1}{256} = \tfrac{5}{3456}$$

$$|R_2(9)| = \left| \frac{\tfrac{10}{27}c^{-8/3}}{3!} \right| < \frac{\tfrac{5}{3456}}{3!} = \frac{5}{20{,}736}.$$

Since $|R_2(9)| < 0.0005$, the condition (13.25) is satisfied when $n = 2$. We therefore approximate $\sqrt[3]{9}$ with

$$P_2(9) = f(8) + \frac{f'(8)}{1!} + \frac{f''(8)}{2!}$$

$$= 2 + \tfrac{1}{12} - \tfrac{1}{288}$$

$$\approx 2.080.$$

If f satisfies the hypotheses of Theorem 13.2.3, then from (13.17), (13.18), and (13.19)

$$f(x) = \sum_{i=0}^{n} \frac{f^{(i)}(a)}{i!}(x-a)^i + \frac{f^{(n+1)}(c)}{(n+1)!}(x-a)^{n+1}. \tag{13.26}$$

The equation (13.26) is called *Taylor's formula* for $f(x)$. When $a = 0$, Taylor's formula becomes

$$f(x) = \sum_{i=0}^{n} \frac{f^{(i)}(0)}{i!}x^i + \frac{f^{(n+1)}(c)}{(n-1)!}x^{n+1}$$

where c is between 0 and x. This equation is sometimes called *Maclaurin's for-mula†* for $f(x)$. From Examples 1 and 3 Taylor's formula for $\cos x$ when $a = \frac{\pi}{3}$ and $n = 4$ is

$$\cos x = \frac{1}{2} - \frac{\sqrt{3}}{2}\left(x - \frac{\pi}{3}\right) - \frac{1}{4}\left(x - \frac{\pi}{3}\right)^2 + \frac{\sqrt{3}}{12}\left(x - \frac{\pi}{3}\right)^3$$

$$+ \frac{1}{48}\left(x - \frac{\pi}{3}\right)^4 - \frac{\sin c}{5!}\left(x - \frac{\pi}{3}\right)^5.$$

Also, from Example 2 and (13.19) Taylor's formula for e^x when $a = 0$ and $n = 5$ is

$$e^x = 1 + \frac{x}{1!} + \frac{x^2}{2!} + \frac{x^3}{3!} + \frac{x^4}{4!} + \frac{x^5}{5!} + \frac{e^c x^6}{6!}.$$

Other forms of the remainder $R_n(x)$ can also be derived besides the Lagrange form. If $f^{(n+1)}$ is continuous on the closed interval I mentioned in Theorem 13.2.3, then

$$R_n(x) = \frac{1}{n!}\int_a^x (x-t)^n f^{(n+1)}(t)\, dt. \tag{13.27}$$

The right member of (13.27) is the *integral form* of the remainder. The derivation of (13.27), following some hints, is left as an exercise.

Exercise Set 13.2

In Exercises 1–12 obtain Taylor's formula using the Lagrange form for the remainder for the given $f(x)$, n, and a.

---

† Named for Colin Maclaurin (1698–1746), a professor of mathematics at the University of Edinburgh.

1. $f(x) = \sin x$, $n = 4$, $a = 0$.

2. $f(x) = \sin x$, $n = 4$, $a = \frac{\pi}{2}$.

3. $f(x) = x^4$, $n = 5$, $a = 2$.

4. $f(x) = x^4$, $n = 5$, $a = 0$.

5. $f(x) = \ln x$, $n = 5$, $a = 1$.

6. $f(x) = \sec x$, $n = 3$, $a = \frac{\pi}{4}$.

7. $f(x) = \sqrt{1 - x}$, $n = 4$, $a = 0$.

8. $f(x) = e^x$, $n = 5$, $a = 1$.

9. $f(x) = \tan^{-1} x$, $n = 3$, $a = 0$.

10. $f(x) = \ln \cos x$, $n = 4$, $a = 0$.

11. $f(x) = \cosh x$, $n = 4$, $a = 0$.

12. $f(x) = (1 + x)^{4/3}$, $n = 4$, $a = 0$.

In Exercises 13–21 approximate the given number correct to fourth decimal place by a Taylor polynomial $P_n(x)$, in which n is chosen so that $|R_n(x)|$ is sufficiently small. Use a convenient number for a in each exercise.

13. $\cos 59°$ NOTE: $1° \approx 0.017453$ radian

14. $\tan 48°$ 15. $\sqrt{38}$ 16. $\sqrt[3]{62}$ 17. $e^{0.5}$

18. $e^{-1.2}$ 19. $\ln 1.2$ 20. $(.98)^{10}$ 21. $\displaystyle\int_0^{0.5} \sin x^2 \, dx$

22. Estimate the error when $\sqrt{1 + x}$ is approximated by $1 + \frac{1}{2}x$ if $0 \le x \le 0.5$.

23. Estimate the error when the following Taylor polynomials are used to approximate $\sin x$ if $|x| < 0.2$.

(a) x (b) $x - \dfrac{x^3}{3!}$

24. Estimate the error when $1 - (x^2/2)$ approximates $\cos x$ if $|x| < 0.2$.

25. Obtain (13.15) by mathematical induction.

26. Derive (13.27) assuming the hypothesis given for $f^{(n+1)}$. HINT: Let F be given by $F(t) = f(x) - f(t) - (x - t)f'(t) - \cdots - \dfrac{(x - t)^n f^{(n)}(t)}{n!} - R_n(x).$

Evaluate $\displaystyle\int_a^x F'(t) \, dt.$

13.3   Trapezoidal Rule

In the usual method of evaluating the definite integral

$$\int_a^b f(x) \, dx, \tag{13.28}$$

we first find an antiderivative F of the function f on the closed interval with endpoints a and b. If the function F cannot be readily evaluated at a and b, we ordinarily approximate the definite integral by a numerical method. Generally in such methods it is assumed that the values of the integrand can be approximated by a polynomial over a small enough subinterval of the interval of integration. The methods are suggested by the fact that the definite integral (13.28) is the area of the region under the graph of f between $x = a$ and $x = b$ when f assumes non-negative values on the interval of integration. However, the methods are still applicable even when f assumes negative values on the interval.

An obvious approximation of (13.28) would be a Riemann sum of f on the interval of integration if the norm of the associated partition is sufficiently small. We have seen that any positive term of a Riemann sum is the area of a rectangle lying above the x axis and any negative term of the sum is the negative of the area of a rectangle lying below the x axis.

Generally a Riemann sum gives a poorer approximation of a definite integral than is obtained by the *trapezoidal rule*. To develop the formula for this method, we suppose that f is continuous on $[a, b]$, and that the interval is divided into n subintervals of *equal* length by the partition $\{a = x_0, x_1, x_2, \ldots, x_{n-1}, x_n = b\}$ (Figure 13–3). The length of each subinterval formed by the partition is

$$h = \frac{b - a}{n}. \tag{13.29}$$

In deriving the trapezoidal rule, we assume that for $i = 1, 2, \ldots, n$

$$\int_{x_{i-1}}^{x_i} f(x)\, dx \approx \frac{h}{2}[f(x_{i-1}) + f(x_i)]. \tag{13.30}$$

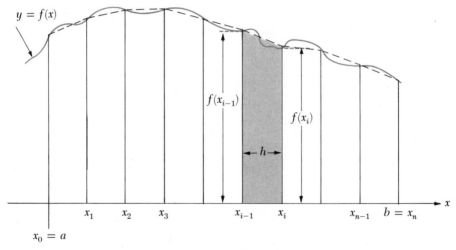

Figure 13–3

If $f(x) \geq 0$ when $x \in [x_{i-1}, x_i]$, then (13.30) states that the area of the region bounded by the x axis and the graphs of the equations $y = f(x)$, $x = x_{i-1}$, and $x = x_i$ is approximately equal to the area of the trapezoid with altitude of length h and bases with lengths $f(x_{i-1})$ and $f(x_i)$. This trapezoid is shaded in Figure 13–3. If (13.30) is summed from $i = 1$ to n, then

$$\int_a^b f(x)\, dx \approx h[\tfrac{1}{2}f(x_0) + f(x_1) + f(x_2) + \cdots + f(x_{n-1}) + \tfrac{1}{2}f(x_n)]. \quad (13.31)$$

The approximation formula (13.31) is called the *trapezoidal rule*.

Example 1   Approximate $\displaystyle\int_{1.5}^3 \frac{1}{x}\, dx$ using (13.31) with $n = 6$ subintervals.

SOLUTION   Since $n = 6$ each interval is of length $h = 0.25$. For $f(x) = 1/x$ the following table gives the calculations leading to the trapezoidal rule approximation of the integral.

$$
\begin{array}{ll}
x_0 = 1.5 & \tfrac{1}{2}f(x_0) = 0.33333 \\
x_1 = 1.75 & f(x_1) = 0.57143 \\
x_2 = 2.0 & f(x_2) = 0.50000 \\
x_3 = 2.25 & f(x_3) = 0.44444 \\
x_4 = 2.5 & f(x_4) = 0.40000 \\
x_5 = 2.75 & f(x_5) = 0.36364 \\
x_6 = 3.0 & \tfrac{1}{2}f(x_6) = 0.16667 \\
 & \sum = 2.77951
\end{array}
$$

$$\int_{1.5}^3 \frac{1}{x}\, dx \approx h \sum = 0.25(2.77951)$$

$$\approx 0.6949.$$

The approximation here has been rounded off to the fourth decimal place to allow for round-off errors in the terms of the sum Σ. The approximation can be compared with the actual value

$$\int_{1.5}^3 \frac{1}{x}\, dx = \ln 3 - \ln 1.5 = \ln \frac{3}{1.5}$$

$$= \ln 2,$$

which correct to the fifth decimal place is 0.69315.

The error E resulting from using the trapezoidal rule approximation for $\int_a^b f(x)\, dx$ is given by

$$E = \int_a^b f(x)\, dx - h[\tfrac{1}{2}f(x_0) + f(x_1) + f(x_2) + \cdots + f(x_{n-1}) + \tfrac{1}{2}f(x_n)]$$

$$= \sum_{i=1}^{n} \int_{x_{i-1}}^{x_i} f(x)\, dx - \sum_{i=1}^{n} \frac{h}{2}[f(x_{i-1}) + f(x_1)]$$

$$= \sum_{i=1}^{n} \left\{ \int_{x_{i-1}}^{x_i} f(x)\, dx - \frac{h}{2}[f(x_{i-1}) + f(x_i)] \right\}. \tag{13.32}$$

An upper bound for $|E|$ is given in the next theorem. It will be noted that the larger we choose n, the smaller will be this upper bound.

13.3.1   Theorem

If f'' is continuous on $[a, b]$ and there is an $M > 0$ such that $|f''(x)| \leq M$ for every $x \in [a, b]$, then the error E, given by (13.32), satisfies

$$|E| \leq \frac{(b-a)^3}{12n^2} M. \tag{13.33}$$

PROOF   Because of the form of the expression in braces in the right member of (13.32), we shall consider the function g where

$$g(x) = \int_{c}^{c+x} f(t)\, dt - \frac{x}{2}[f(c) + f(c + x)] \quad \text{if } x \in [0, h]. \tag{13.34}$$

Here it is assumed that the interval $[c, c + h] \subset [a, b]$. By the chain rule (Theorem 4.5.1) and the first fundamental theorem (Theorem 7.7.1)

$$g'(x) = D_x \int_{c}^{c+x} f(t)\, dt - \frac{x}{2} D_x f(c + x) - \frac{1}{2}[f(c) + f(c + x)]$$

$$= f(c + x)\, D_x\, (c + x) - \frac{x}{2} f'(c + x)\, D_x\, (c + x) - \frac{1}{2}[f(c) + f(c + x)]$$

$$= \tfrac{1}{2}[f(c + x) - xf'(c + x) - f(c)] \quad \text{if } x \in [0, h].$$

Also by the chain rule,

$$g''(x) = \tfrac{1}{2}[D_x f(c + x) - x\, D_x f'(c + x) - f'(c + x)]$$
$$= \tfrac{1}{2}[f'(c + x) \cdot 1 - xf''(c + x) \cdot 1 - f'(c + x)]$$
$$= -\frac{x}{2} f''(c + x) \quad \text{if } x \in [0, h].$$

Then from our hypothesis about the boundedness of f'' on $[a, b]$

$$|g''(x)| = \left| -\frac{x}{2} f''(c + x) \right| \leq \frac{Mx}{2} \quad \text{if } x \in [0, h]. \tag{13.35}$$

Also by hypothesis f'' is continuous on $[a, b]$. Thus g'' and, also, g' are continuous on $[0, h]$. Since $g(0) = g'(0) = 0$, for $x \in [0, h]$

$$\int_{0}^{x} g''(t)\, dt = g'(x) - g'(0) = g'(x) \tag{13.36}$$

and

$$\int_0^x g'(t)\, dt = g(x) - g(0) = g(x). \tag{13.37}$$

Then from (13.35) and (13.36)

$$|g'(x)| = \left| \int_0^x g''(t)\, dt \right| \le \int_0^x |g''(t)|\, dt \le \int_0^x \frac{Mt}{2}\, dt$$

$$\le \frac{Mx^2}{4}. \tag{13.38}$$

Also from (13.37) and (13.38)

$$|g(x)| = \left| \int_0^x g'(t)\, dt \right| \le \int_0^x |g'(t)|\, dt \le \int_0^x \frac{Mt^2}{4}\, dt$$

$$\le \frac{Mx^3}{12}.$$

Since the number c in (13.34) could have been any x_{i-1} where $i = 1, 2, \ldots, n$ and since $x_i = x_{i-1} + h$, we have from (13.34), letting $x = h$,

$$|g(h)| = \left| \int_{x_{i-1}}^{x_i} f(t)\, dt - \frac{h}{2}[f(x_{i-1}) + f(x_i)] \right| \le \frac{Mh^3}{12}.$$

Then from (13.32)

$$|E| = \left| \sum_{i=1}^{n} \left[\int_{x_{i-1}}^{x_i} f(t)\, dt - \frac{h}{2}(f(x_{i-1}) + f(x_i)) \right] \right|$$

$$\le \sum_{i=1}^{n} \left| \int_{x_{i-1}}^{x_i} f(t)\, dt - \frac{h}{2}[f(x_{i-1}) + f(x_i)] \right|$$

$$\le \sum_{i=1}^{n} \frac{Mh^3}{12} = \frac{Mnh^3}{12}.$$

Since $h = \dfrac{b-a}{n}$,

$$|E| \le \frac{M(b-a)^3}{12n^2},$$

which proves the theorem.

Example 2   Find an upper bound to the absolute error introduced by using the trapezoidal rule approximation in Example 1 for $\int_{1.5}^{3} \frac{1}{x}\, dx$.

SOLUTION   Letting $f(x) = 1/x$, we have $f'(x) = -1/x^2$, and

$$f''(x) = \frac{2}{x^3}.$$

Hence for $x \in [1.5, 3]$

$$|f''(x)| \leq |f''(1.5)| = \frac{2}{(1.5)^3}.$$

Letting $a = 1.5$, $b = 3$, and $n = 6$, we have from (13.33)

$$|E| \leq \frac{(1.5)^3}{12 \cdot 6^2} \frac{2}{(1.5)^3} = \frac{1}{216}$$

$$< 0.0047.$$

Exercise Set 13.3

In Exercises 1–8 approximate the given definite integral by the trapezoidal rule using the number of subintervals indicated by n. Round off your results to the fourth decimal place. In Exercises 1–4 check your results against the exact value of the definite integral.

1. $\displaystyle\int_0^{2.5} x^2 \, dx, \quad n = 5$ 2. $\displaystyle\int_0^{\pi} \sin^2 x \, dx, \quad n = 6$

3. $\displaystyle\int_0^1 \frac{dx}{1 + x^2}, \quad n = 10$ (Give an approximation of π from your answer.)

4. $\displaystyle\int_0^2 \sqrt{x^2 + 1} \, dx, \quad n = 5$ 5. $\displaystyle\int_{\pi/2}^{\pi} \sqrt{1 + \cos^2 x} \, dx, \quad n = 6$

6. $\displaystyle\int_0^1 \frac{dx}{1 + x^4}, \quad n = 4$ 7. $\displaystyle\int_0^{0.8} e^{-x^2} \, dx, \quad n = 8$

8. $\displaystyle\int_0^{\pi/2} f(x) \, dx$ where $f(x) = \begin{cases} \dfrac{\sin x}{x} & \text{if } x \neq 0 \\ 1 & \text{if } x = 0 \end{cases}, \quad n = 6$

9–16. Find an upper bound to the absolute error introduced by the use of the trapezoidal rule approximations in Exercises 1–8, respectively.

17. How small should h be chosen so that the trapezoidal rule approximation for $\int_0^{0.8} e^{-x^2} \, dx$ is correct to six decimal places?

18. How small should h be chosen so that the trapezoidal rule approximation for $\displaystyle\int_0^1 \frac{dx}{1 + x^2}$ can be used to give an estimate of π which is correct to six decimal places?

13.4   Simpson's Rule

In the next method for the numerical integration of

$$\int_a^b f(x) \, dx,$$

parabolic arcs instead of line segments are used to approximate the consecutive arcs of the graph of f on the interval $[a, b]$. In this method, which is called *Simpson's rule* or the *parabolic rule*, it is assumed that f is continuous on $[a, b]$ and that the interval is divided into $2n$ subintervals of equal length by the partition $\{a = x_0, x_1, x_2, \ldots, x_{2n} = b\}$. The length of each subinterval is therefore

$$h = \frac{b - a}{2n}.$$

For $k = 0, 1, 2, \ldots, 2n$, we let P_k denote the point $(x_k, f(x_k))$ on the graph of f. We assume that the arc of the curve $y = f(x)$ connecting P_0 and P_2 can be approximated by a parabolic arc through P_0, P_1, and P_2. Similarly the arc of the curve connecting P_2 and P_4 can be approximated by a parabolic arc through P_2, P_3, and P_4, until finally the arc of the curve connecting P_{2n-2} and P_{2n} is approximated by a parabolic arc through P_{2n-2}, P_{2n-1}, and P_{2n} (Figure 13-4).

If $i = 1, 2, \ldots, n$, for convenience in handling the notation, we let $c_i = x_{2i-1}$. Then $x_{2i-2} = c_i - h$ and $x_{2i} = c_i + h$. An equation for the parabolic arc through the points P_{2i-2}, P_{2i-1}, and P_{2i} has the form

$$y = A(x - c_i)^2 + B(x - c_i) + C. \qquad (13.39)$$

Thus we assume that

$$\int_{c_i-h}^{c_i+h} f(x)\, dx \approx \int_{c_i-h}^{c_i+h} [A(x - c_i)^2 + B(x - c_i) + C]\, dx. \qquad (13.40)$$

If y in (13.39) and $f(x)$ are non-negative for every $x \in [c_i - h, c_i + h]$, (13.40) states that the area of the region under the graph of f between the lines $x = c_i - h$ and $x = c_i + h$ can be approximated by the area of the region under the parabolic arc (13.39) and between the same lines.

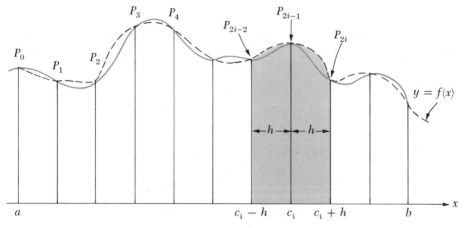

Figure 13-4

If the right member of (13.40) is evaluated,

$$\int_{c_i-h}^{c_i+h} f(x)\, dx \approx \left[\frac{A(x - c_i)^3}{3} + \frac{B(x - c_i)^2}{2} + Cx \right]\Bigg|_{c_i-h}^{c_i+h}$$

$$\approx \tfrac{2}{3}Ah^3 + 2Ch$$
$$\approx \tfrac{1}{3}h(2Ah^2 + 6C). \tag{13.41}$$

Now since the arc given by (13.39) contains the points P_{2i-2}, P_{2i-1}, and P_{2i},

$$\left. \begin{aligned} f(c_i + h) &= Ah^2 + Bh + C \\ f(c_i) &= C \\ f(c_i - h) &= Ah^2 - Bh + C \end{aligned} \right\} \tag{13.42}$$

From (13.42) $f(c_i - h) + 4f(c_i) + f(c_i + h) = 2Ah^2 + 6C$. Hence from (13.41)

$$\int_{c_i-h}^{c_i+h} f(x)\,dx \approx \tfrac{1}{3}h[f(c_i - h) + 4f(c_i) + f(c_i + h)]. \tag{13.43}$$

Summing in (13.43) from $i = 1$ to n, we obtain

$$\int_a^b f(x)\,dx \approx \tfrac{1}{3}h \sum_{i=1}^n [f(c_i - h) + 4f(c_i) + f(c_i + h)]. \tag{13.44}$$

If we let $c_i = x_{2i-1}$, then $c_i + h = x_{2i}$, and $c_i - h = x_{2i-2}$. Hence an expansion of the right member of (13.44) gives

$$\int_a^b f(x)\,dx \approx \tfrac{1}{3}h[f(x_0) + 4f(x_1) + 2f(x_2) + 4f(x_3)$$
$$+ \cdots + 2f(x_{2n-2}) + 4f(x_{2n-1}) + f(x_{2n})]. \tag{13.45}$$

The approximation formula (13.45) is called *Simpson's rule*.

Example 1 Approximate $\displaystyle\int_{1.5}^3 \frac{1}{x}\,dx$ using (13.45) with $2n = 6$ sub-intervals.

SOLUTION   Since $2n = 6$, each subinterval of $[1.5, 3]$ is of length $h = 0.25$. The following table gives the calculations leading to the Simpson's rule approximation of $\displaystyle\int_{1.5}^3 \frac{1}{x}\,dx$.

$$
\begin{aligned}
x_0 &= 1.5 & f(x_0) &= 0.66667 \\
x_1 &= 1.75 & 4f(x_1) &= 2.28571 \\
x_2 &= 2.0 & 2f(x_2) &= 1.00000 \\
x_3 &= 2.25 & 4f(x_3) &= 1.77778 \\
x_4 &= 2.5 & 2f(x_4) &= 0.80000 \\
x_5 &= 2.75 & 4f(x_5) &= 1.45455 \\
x_6 &= 3.0 & f(x_6) &= 0.33333 \\
& & \sum &= 8.31804
\end{aligned}
$$

$$\int_{1.5}^3 \frac{1}{x}\,dx \approx \frac{1}{3}h \sum = \frac{1}{3}(0.25)(8.31804)$$
$$\approx 0.6932.$$

The approximation here can be compared with that obtained using the trapezoidal rule in Example 1, Section 13.3.

The error E resulting from using the Simpson's rule approximation for $\int_a^b f(x)\,dx$ is given by

$$E = \int_a^b f(x)\,dx - \tfrac{1}{3}h[f(x_0) + 4f(x_1) + 2f(x_2) + 4f(x_3)$$

$$+ \cdots + 2f(x_{2n-2}) + 4f(x_{2n-1}) + f(x_{2n})]. \quad (13.46)$$

An upper bound for $|E|$ is given in the next theorem.

13.4.1 Theorem

If $f^{(4)}$ is continuous on $[a, b]$ and $|f^{(4)}(x)| \le M$ for every $x \in [a, b]$, then

$$|E| \le \frac{M(b-a)^5}{180(2n)^4}. \quad (13.47)$$

The proof of this theorem, which is omitted because of its length, is essentially given in Haaser, LaSalle, and Sullivan: *Introduction to Analysis (Vol. 1)*, p. 678 (Ginn, 1959).

Example 2   Find an upper bound for the absolute error introduced by using the Simpson's rule approximation in Example 1 for $\int_{1.5}^3 \dfrac{1}{x}\,dx$.

SOLUTION   Letting $f(x) = 1/x$, we obtain $f'(x) = -1/x^2$, $f''(x) = 2/x^3$, $f'''(x) = -6/x^4$, and

$$f^{(4)}(x) = \frac{24}{x^5}.$$

Then for $x \in [1.5, 3]$

$$|f^{(4)}(x)| \le f^{(4)}(1.5) = \frac{24}{(1.5)^5}.$$

If we let $a = 1.5$, $b = 3$, and $2n = 6$, then from (13.47)

$$|E| \le \frac{\dfrac{24}{(1.5)^5} \cdot (1.5)^5}{180 \cdot 6^4} = \frac{1}{9720}$$

$$< 0.0002.$$

By comparison with the result obtained in Example 2, Section 13.3, we note that Simpson's rule gives the more accurate approximation for $\int_{1.5}^3 \dfrac{1}{x}\,dx$ when the same number of subintervals are used.

Exercise Set 13.4

In Exercises 1–8 approximate the given definite integral by Simpson's rule using the number of subintervals indicated by $2n$. Round off your results to the fourth decimal place.

1. $\displaystyle\int_0^1 \frac{dx}{1+x^2}, \quad 2n = 10$

2. $\displaystyle\int_0^\pi \sin^2 x \, dx, \quad 2n = 6$

3. $\displaystyle\int_{\pi/2}^\pi \sqrt{1 + \cos^2 x} \, dx, \quad 2n = 6$

4. $\displaystyle\int_0^1 \frac{dx}{1+x^4}, \quad 2n = 4$

5. $\displaystyle\int_0^{0.8} e^{-x^2} \, dx, \quad 2n = 8$

6. $\displaystyle\int_0^{\pi/2} f(x) \, dx \quad$ where $f(x) = \begin{cases} \dfrac{\sin x}{x} & \text{if } x \neq 0 \\ 1 & \text{if } x = 0 \end{cases}, \quad 2n = 6$

7. $\displaystyle\int_0^\pi \frac{dx}{2 + \sin x}, \quad 2n = 6$

8. $\displaystyle\int_1^4 \ln x \, dx, \quad 2n = 6$

The definite integrals in Exercises 1–6 were approximated using the trapezoidal rule in Exercise Set 13.3.

In Exercises 9–12 find an upper bound to the absolute error introduced in the given exercises.

9. Exercise 1

10. Exercise 2

11. Exercise 5

12. Exercise 8

13. How small should h be chosen so that the Simpson's rule approximation for $\int_0^{0.8} e^{-x^2} \, dx$ is correct to six decimal places?

14. How small should h be chosen so that the Simpson's rule approximation for

$\displaystyle\int_0^1 \frac{dx}{1+x^2}$ can be used to give an estimate of π which is correct to six decimal places?

15. Using Theorem 13.4.1, prove that the Simpson's rule approximation of $\int_a^b f(x) \, dx$ gives the exact value of the definite integral if $f(x)$ is a polynomial of degree ≤ 3.

16. The following speeds were recorded by an automobile race driver during a 6-minute run.

At $t = 0$ (min)	At $v = 97$ (mi/hr)
1	99
2	102
3	107
4	116
5	112
6	109

Using Simpson's rule obtain an approximation of the total distance traveled by the driver during the run.

13.5 Indeterminate Form 0/0

We now depart from our discussion of numerical methods to consider some methods for evaluating limits that are not amenable to the use of the limit theorems that have already been discussed. It will be recalled from Section 3.5 that if f and g are functions for which

$$\lim_{x \to c} f(x) = 0 = \lim_{x \to c} g(x), \qquad (13.48)$$

then $\lim_{x \to c} f(x)/g(x)$ may or may not exist. For example, from Example 1, Section 3.5

$$\lim_{x \to 1} \frac{x^2 - 1}{x - 1} = \lim_{x \to 1} (x + 1) = 2$$

and by Theorem 10.2.1(c)

$$\lim_{x \to 0} \frac{\sin x}{x} = 1.$$

However, by Theorem 3.5.2

$$\lim_{x \to 2} \frac{x^2 - 4}{(x - 2)^2} \text{ fails to exist.}$$

The expression $f(x)/g(x)$ is sometimes said to be of "the form 0/0" at c when (13.48) holds. The form 0/0 is called an *indeterminate form* since different expressions of this form have different limits.

The following theorem, which is the first of L'Hôpital's† rules, is useful for evaluating limits of expressions of this form when they exist. According to this rule, the limit of the quotient of two functions of this form is the limit of the quotient of their derivatives if this limit exists.

13.5.1 Theorem (L'Hôpital's First Rule)

If the functions f and g satisfy the following conditions:

(i) $\lim_{x \to c} f(x) = 0 = \lim_{x \to c} g(x),$
(ii) $f'(x)$ *and* $g'(x)$ *exist for every* $x \neq c$ *in some open interval I containing c,*
(iii) $g'(x) \neq 0$ *for every* $x \neq c$ *in I,*

then

---

† This theorem is named after Marquis de L'Hôpital (1661–1704), a French nobleman who included it in a text published in 1696. Actually, the rule was discovered by Johann Bernoulli (1667–1748), one of a family of distinguished Swiss mathematicians. Bernoulli, who was a tutor of L'Hôpital, had sent him a proof of the rule in a letter in 1694.

$$\lim_{x \to c} \frac{f(x)}{g(x)} = \lim_{x \to c} \frac{f'(x)}{g'(x)}$$

provided the limit on the right exists.

PROOF   We first define the functions F and G for every $x \in I$ by requiring that

$$F(x) = \begin{cases} f(x) & \text{if } x \neq c \\ 0 & \text{if } x = c \end{cases} \qquad G(x) = \begin{cases} g(x) & \text{if } x \neq c \\ 0 & \text{if } x = c. \end{cases}$$

Since $F'(x) = f'(x)$ and $G'(x) = g'(x)$ for every $x \neq c$ in I, the functions F and G are certainly continuous at every number in I except c. Moreover, since

$$\lim_{x \to c} F(x) = \lim_{x \to c} f(x) = 0 = F(c) \qquad \text{and} \qquad \lim_{x \to c} G(x) = \lim_{x \to c} g(x) = 0 = G(c),$$

the functions F and G are continuous at c as well.

Suppose $x \neq c$ is chosen in I. Since F and G satisfy the hypotheses of the generalized mean value theorem on the closed interval with endpoints c and x, by this theorem there is a number m between c and x such that

$$\frac{F(x) - F(c)}{G(x) - G(c)} = \frac{F'(m)}{G'(m)}$$

or saying the same thing,

$$\frac{f(x)}{g(x)} = \frac{f'(m)}{g'(m)}. \tag{13.49}$$

If $\lim_{x \to c} (f'(x)/g'(x)) = L$, then for every $\epsilon > 0$ there is a $\delta > 0$ such that

$$\left| \frac{f'(x)}{g'(x)} - L \right| < \epsilon \quad \text{if } 0 < |x - c| < \delta. \tag{13.50}$$

If $0 < |x - c| < \delta$, then since m is between c and x, $0 < |m - c| < \delta$, and hence from (13.49) and (13.50)

$$\left| \frac{f(x)}{g(x)} - L \right| = \left| \frac{f'(m)}{g'(m)} - L \right| < \epsilon.$$

Thus $\lim_{x \to c} (f(x)/g(x)) = L$ and the theorem is proved.

Example 1   Prove that $\lim_{x \to 0} \dfrac{e^x - 1}{x} = 1$ using L'Hôpital's first rule.

SOLUTION   It will be recalled that this limit was obtained by a different method in Exercise 30, Exercise Set 9.5. Here we let $f(x) = e^x - 1$ and $g(x) = x$, noting that f and g satisfy the hypotheses of Theorem 13.5.1 if $c = 0$. Hence

$$\lim_{x\to 0} \frac{e^x - 1}{x} = \lim_{x\to 0} \frac{D_x\,(e^x - 1)}{D_x\,x} = \lim_{x\to 0} \frac{e^x}{1} = e^x.$$

Sometimes, more than one application of the rule is necessary to obtain the limit of a quotient.

Example 2   Find $\displaystyle\lim_{x\to 0} \frac{\sin^{-1} x - x}{x^3}$.

SOLUTION

$$\lim_{x\to 0} \frac{\sin^{-1} x - x}{x^3} = \lim_{x\to 0} \frac{\dfrac{1}{\sqrt{1-x^2}} - 1}{3x^2}. \tag{13.51}$$

However, the fraction on the right is still of the form 0/0 at 0. Applying the rule to the limit on the right gives

$$\lim_{x\to 0} \frac{\dfrac{1}{\sqrt{1-x^2}} - 1}{3x^2} = \lim_{x\to 0} \frac{\dfrac{x}{(1-x^2)^{3/2}}}{6x}. \tag{13.52}$$

Rather than apply the rule a third time to evaluate the limit on the right, we simply divide out the factors of x. Thus

$$\lim_{x\to 0} \frac{\dfrac{x}{(1-x^2)^{3/2}}}{6x} = \lim_{x\to 0} \frac{1}{6(1-x^2)^{3/2}} = \frac{1}{6}. \tag{13.53}$$

Hence from (13.51), (13.52), and (13.53) $\displaystyle\lim_{x\to 0} \frac{\sin^{-1} x - x}{x^3} = \frac{1}{6}$.

Of course, Theorem 13.5.1 cannot be used to evaluate the limit of a quotient if the functions involved fail to satisfy the hypotheses of the theorem. For example, $\lim_{x\to 2} (3x - 1)/2x^2 = \frac{5}{8}$, but if one differentiates the numerator and denominator of the fraction, then $\lim_{x\to 2} 3/4x = \frac{3}{8}$ is obtained. Theorem 13.5.1 does not apply here since the functions f and g given by $f(x) = 3x - 1$ and $g(x) = 2x^2$ do not have the limit zero at 2.

If the statement of Theorem 13.5.1 is modified so that in (i) $x \to c$ is replaced by $x \to c^+$ and the interval I in (ii) has c as its left endpoint, the equation

$$\lim_{x\to c^+} \frac{f(x)}{g(x)} = \lim_{x\to c^+} \frac{f'(x)}{g'(x)}$$

can be derived. A similar restatement of the theorem gives an analogous equation for left-hand limits at c.

Another of L'Hôpital's rules is used to evaluate limits of the form 0/0 at $+\infty$.

13.5.2   Theorem (L'Hôpital's Second Rule)

If f and g satisfy the following conditions:

(i) $\lim_{x \to +\infty} f(x) = 0 = \lim_{x \to +\infty} g(x)$,
(ii) $f'(x)$ *and* $g'(x)$ *exist for every* x *in some interval* $(c, +\infty)$,
(iii) $g'(x) \neq 0$ *if* $x \in (c, +\infty)$,

then

$$\lim_{x \to +\infty} \frac{f(x)}{g(x)} = \lim_{x \to +\infty} \frac{f'(x)}{g'(x)}$$

provided the limit on the right exists.

The proof of Theorem 13.5.2 is left as an exercise (Exercise 21, Exercise Set 13.5).

If in the statement of Theorem 13.5.2 "$x \to +\infty$" is replaced by "$x \to -\infty$" and "$(c, +\infty)$" by "$(-\infty, c)$," then the equation

$$\lim_{x \to -\infty} \frac{f(x)}{g(x)} = \lim_{x \to -\infty} \frac{f'(x)}{g'(x)} \tag{13.54}$$

can be derived.

One should never attempt to use L'Hôpital's rules blindly. It is entirely possible that at some stage in the solution the limit can be easily obtained without a further application of one of the rules, as happened in Example 2. Also, we sometimes find that repeated applications of the rules get us nowhere. For example, after one use of Theorem 13.5.1

$$\lim_{x \to 0} \frac{e^{-1/x^2}}{x^2} = \lim_{x \to 0} \frac{\frac{2}{x^3} e^{-1/x^2}}{2x} = \lim_{x \to 0} \frac{e^{-1/x^2}}{x^4}$$

if the limit on the right exists. Not only is the fraction on the right here more unsuitable than the given fraction because of the greater power of x in the denominator, but also it is clear that after additional applications of Theorem 13.5.1 e^{-1/x^2} will remain in the numerator, and the power of x in the denominator will only increase. This limit will, however, be evaluated in an exercise in the next section after the indeterminate form ∞/∞ has been introduced.

Exercise Set 13.5

In Exercises 1–20 evaluate the given limit if it exists.

1. $\lim\limits_{x \to 2} \dfrac{2x^3 - 7x^2 + 5x + 2}{x^3 + 3x^2 - 6x - 8}$

2. $\lim\limits_{x \to -1} \dfrac{x^4 + 5x^2 - 6}{x^4 - 3x - 4}$

3. $\lim\limits_{x \to 0} \dfrac{1 - \cos x}{\sin x}$

4. $\lim\limits_{t \to 0} \dfrac{\tan 2t}{\sin 3t}$

5. $\lim\limits_{x \to 1} \dfrac{\sin \pi x}{\ln x}$

6. $\lim\limits_{x \to 0} \dfrac{e^x - e^{-x}}{3x}$

7. $\lim\limits_{x \to 0} \dfrac{\cos x - \left(1 - \dfrac{x^2}{2}\right)}{x^3}$

8. $\lim\limits_{x \to 0} \dfrac{5^x - 3^x}{2x}$

9. $\lim\limits_{x \to 0} \dfrac{\sinh x}{\sin^{-1} x}$

10. $\lim\limits_{x \to \pi/2} \dfrac{1 - \sin x}{1 + \cos 2x}$

11. $\lim\limits_{x \to -\infty} \dfrac{e^x}{3^x}$

12. $\lim\limits_{x \to 0} \dfrac{\sin x^2}{\sin^2 x}$

13. $\lim\limits_{t \to 0} \dfrac{\sin^3 t}{\sinh t - \sin t}$

14. $\lim\limits_{u \to 0} \dfrac{\cosh u - \cos u}{\cos^3 u}$

15. $\lim\limits_{y \to 0} \dfrac{y - \sin y}{y - \tan y}$

16. $\lim\limits_{x \to 2\pi} \dfrac{(x - 2\pi)^2}{\ln \cos x}$

17. $\lim\limits_{x \to 0} \dfrac{x - \tan^{-1} x}{x^3}$

18. $\lim\limits_{x \to 0} \dfrac{(\sin^{-1} x)^2}{\ln (1 - x^2)}$

19. $\lim\limits_{x \to 0} \dfrac{\displaystyle\int_0^x \sin t^2 \, dt}{\displaystyle\int_0^x (1 - e^{-t^2}) \, dt}$

20. $\lim\limits_{x \to 0} \dfrac{1}{x^4} \displaystyle\int_0^x \tan t^3 \, dt$

21. Prove Theorem 13.5.2. Hint: Recall from Exercise 21, Exercise Set 5.6,

$$\lim_{x \to +\infty} F(x) = L \quad \text{if and only if} \quad \lim_{t \to 0^+} F\left(\frac{1}{t}\right) = L.$$

Then use Theorem 13.5.1, as modified for right-hand limits at c.

13.6   Other Indeterminate Forms

Besides the indeterminate form $0/0$ we shall consider other such forms, the first of which is the form ∞/∞. The expression $f(x)/g(x)$ is sometimes said to be of the "form ∞/∞" at c when $+\infty$ or $-\infty$ is a one-sided or two-sided limit of f and g at c.

We mention two theorems for evaluating limits associated with this indeterminate form.

13.6.1   Theorem (L'Hôpital's Third Rule)

If f and g are functions for which the following conditions hold:

    (i) $\lim_{x \to c} f(x)$ and $\lim_{x \to c} g(x)$ are $+\infty$ or $-\infty$,
    (ii) $f'(x)$ and $g'(x)$ exist for every $x \neq c$ in some open interval I containing c,
    (iii) $g'(x) \neq 0$ for every $x \neq c$ in I,

then

$$\lim_{x \to c} \frac{f(x)}{g(x)} = \lim_{x \to c} \frac{f'(x)}{g'(x)}$$

provided the limit on the right exists.

The proof of this theorem is more properly discussed in advanced calculus (see R. C. Buck, *Advanced Calculus* (McGraw Hill, 2nd ed., 1964)), and is therefore omitted.

If the statement of Theorem 13.6.1 is modified so that in (i) "$x \to c$" is replaced by "$x \to c^+$" and the interval I in (ii) has c as its left endpoint, the equation

$$\lim_{x \to c^+} \frac{f(x)}{g(x)} = \lim_{x \to c^+} \frac{f'(x)}{g'(x)}$$

holds. A similar restatement of the theorem gives an analogous equation for left-hand limits at c.

Example 1   Find $\displaystyle\lim_{x \to \pi/2^-} \frac{\ln \cos x}{\tan x}$.

SOLUTION   We first note that if we let $f(x) = \ln \cos x$ and $g(x) = \tan x$ then $\lim_{x \to \pi/2^-} \ln \cos x = -\infty$ and $\lim_{x \to \pi/2^-} \tan x = +\infty$. Thus the hypotheses of Theorem 13.6.1, as modified for left-hand limits, are satisfied by f and g at $\pi/2$. Hence, by the modification of this theorem,

$$\lim_{x \to \pi/2^-} \frac{\ln \cos x}{\tan x} = \lim_{x \to \pi/2^-} \frac{-\dfrac{\sin x}{\cos x}}{\sec^2 x}$$

provided the limit on the right exists. The fraction in the right member is still of the form ∞/∞ since $\lim_{x \to \pi/2^-} (-\sin x/\cos x) = -\infty$ and $\lim_{x \to \pi/2^-} \sec^2 x = +\infty$. However, this fraction can be rewritten in terms of sine and cosine and one then obtains

$$\lim_{x \to \pi/2^-} \frac{\ln \cos x}{\tan x} = \lim_{x \to \pi/2^-} (-\sin x \cos x) = 0.$$

In the next theorem the form of L'Hôpital's rule for evaluating limits of the form ∞/∞ at $+\infty$ is given.

13.6.2   Theorem (L'Hôpital's Fourth Rule)

If f and g satisfy the following conditions:

(i) $\lim_{x \to +\infty} f(x)$ *and* $\lim_{x \to +\infty} g(x)$ *are* $+\infty$ *or* $-\infty$,

(ii) $f'(x)$ and $g'(x)$ exist for every x in some interval $(c, +\infty)$,

(iii) $g'(x) \neq 0$ if $x \in (c, +\infty)$,

then

$$\lim_{x \to +\infty} \frac{f(x)}{g(x)} = \lim_{x \to +\infty} \frac{f'(x)}{g'(x)}$$

if the limit on the right exists.

The proof is analogous to that of Theorem 13.5.2 and is omitted.

If in the statement of Theorem 13.6.2 "$x \to +\infty$" is replaced by "$x \to -\infty$" and "$(c, +\infty)$" by "$(-\infty, c)$," the equation

$$\lim_{x \to -\infty} \frac{f(x)}{g(x)} = \lim_{x \to -\infty} \frac{f'(x)}{g'(x)}$$

holds.

Example 2   Find $\lim_{x \to +\infty} \dfrac{x^2}{e^x}$.

SOLUTION   Note that if $f(x) = x^2$ and $g(x) = e^x$, then f and g satisfy the hypotheses of Theorem 13.6.2. Hence by Theorem 13.6.2

$$\lim_{x \to +\infty} \frac{x^2}{e^x} = \lim_{x \to +\infty} \frac{2x}{e^x}$$

if the limit on the right exists. The fraction in the right member is still of the form ∞/∞ at $+\infty$. Thus if Theorem 13.6.2 is again applied, we obtain

$$\lim_{x \to +\infty} \frac{x^2}{e^x} = \lim_{x \to +\infty} \frac{2x}{e^x} = \lim_{x \to +\infty} \frac{2}{e^x}$$

$$= 0.$$

As an extension of this example, the reader is asked to prove (Exercise 25, Exercise Set 13.6) that

$$\lim_{x \to +\infty} \frac{x^n}{e^x} = 0 \quad \text{for any positive integer } n.$$

Thus e^x increases much more rapidly than any positive integral power of x as x increases indefinitely.

Limits for indeterminate forms of the type "$\infty - \infty$" and "$\infty \cdot 0$" can often be evaluated after first obtaining the expression in the form $0/0$ or ∞/∞. The expression in the next example is of the form $\infty - \infty$.

Example 3   Find $\lim_{x \to 0} \left(\dfrac{1}{x \sin x} - \dfrac{1}{x^2} \right)$.

SOLUTION   We first note that $\lim_{x \to 0} 1/(x \sin x) = +\infty$ and also $\lim_{x \to 0} 1/x^2 = +\infty$. After combining fractions, it will be observed that L'Hôpital's first rule can be applied. Thus

$$\lim_{x \to 0} \left(\frac{1}{x \sin x} - \frac{1}{x^2} \right) = \lim_{x \to 0} \frac{x - \sin x}{x^2 \sin x} = \lim_{x \to 0} \frac{1 - \cos x}{x^2 \cos x + 2x \sin x}$$

if the limit on the right exists. After two more applications of L'Hôpital's first rule

$$\lim_{x \to 0} \left(\frac{1}{x \sin x} - \frac{1}{x^2} \right) = \lim_{x \to 0} \frac{\sin x}{-x^2 \sin x + 4x \cos x + 2 \sin x}$$

$$= \lim_{x \to 0} \frac{\cos x}{-x^2 \cos x - 6x \sin x + 6 \cos x}$$

$$= \tfrac{1}{6}.$$

Example 4   Find $\lim_{x \to 0^+} x \ln x$.

SOLUTION   Since $\lim_{x \to 0^+} x = 0$ and $\lim_{x \to 0^+} \ln x = -\infty$, the expression $x \ln x$ is of the form $0 \cdot \infty$. Then

$$\lim_{x \to 0^+} x \ln x = \lim_{x \to 0^+} \frac{\ln x}{1/x}$$

and by the modification of L'Hôpital's third rule for right-hand limits

$$\lim_{x \to 0^+} x \ln x = \lim_{x \to 0^+} \frac{1/x}{-1/x^2} = \lim_{x \to 0^+} (-x)$$

$$= 0.$$

In the final example we find the limit for an expression of the form 0^0.

Example 5   Find $\lim_{x \to 0} x^{\sin x}$.

SOLUTION   For $x > 0$, $x^{\sin x} = e^{\ln x^{\sin x}}$. Hence by the composite function limit theorem

$$\lim_{x \to 0} x^{\sin x} = \lim_{x \to 0} e^{\sin x \ln x} = e^{\lim_{x \to 0} \sin x \ln x} \tag{13.55}$$

provided $\lim_{x \to 0} \sin x \ln x$ exists. By L'Hôpital's third rule

$$\lim_{x \to 0} \sin x \ln x = \lim_{x \to 0} \frac{\ln x}{\csc x} = \lim_{x \to 0} \frac{1/x}{-\csc x \cot x}$$

$$= \lim_{x \to 0} \left(-\frac{\sin x \tan x}{x} \right) = \lim_{x \to 0} \left(-\frac{\sin x}{x} \right) \lim_{x \to 0} \tan x$$

$$= (-1)0 = 0.$$

Hence from (13.55), $\lim_{x \to 0} x^{\sin x} = e^0 = 1$.

Exercise Set 13.6

In Exercises 1–24 evaluate the given limit if it exists.

1. $\lim\limits_{x \to +\infty} \dfrac{(\ln x)^2}{x^{1/2}}$

2. $\lim\limits_{x \to +\infty} \dfrac{x^3}{3^x}$

3. $\lim\limits_{t \to \pi/2^-} \dfrac{\ln(1 + \sec t)}{\tan t}$

4. $\lim\limits_{y \to 0^+} \dfrac{2 \cot y - 1}{\csc y}$

5. $\lim\limits_{x \to 0^+} \dfrac{\ln \csc x}{\ln \cot x}$

6. $\lim\limits_{x \to \pi/2^+} \dfrac{\ln \sec x}{\ln \tan x}$

7. $\lim\limits_{x \to +\infty} \dfrac{e^x + 100x}{e^x + \ln x}$

8. $\lim\limits_{x \to +\infty} \dfrac{\ln(e^x + x)}{x + \ln x}$

9. $\lim\limits_{x \to +\infty} x \sin \dfrac{1}{x}$

10. $\lim\limits_{x \to 0^+} \tan^{-1} x \cot x$

11. $\lim\limits_{x \to 0^+} x^2 \cot x$

12. $\lim\limits_{x \to +\infty} x \ln \dfrac{x^2}{x^2 + 1}$

13. $\lim\limits_{x \to 0^+} (\csc x - \cot x)$

14. $\lim\limits_{x \to 0} \left(\dfrac{1}{x} - \dfrac{1}{\ln(x + 1)} \right)$

15. $\lim\limits_{x \to 0} \left(\csc^2 x - \dfrac{1}{x^2} \right)$

16. $\lim\limits_{x \to 0} \left(\dfrac{1}{e^x - 1} - \dfrac{1}{\tan^{-1} x} \right)$

17. $\lim\limits_{y \to +\infty} (1 + 4x)^{2/x}$

18. $\lim\limits_{x \to \pi/2^-} (\sec x)^{\cos x}$

19. $\lim\limits_{x \to 0^+} (\sin x)^{\csc x}$

20. $\lim\limits_{x \to +\infty} (e^x + x^2)^{1/x}$

21. $\lim\limits_{x \to +\infty} (\ln x)^{1/x}$

22. $\lim\limits_{x \to 0^+} x^x$

23. $\lim\limits_{y \to +\infty} \left(\cosh \dfrac{1}{y} \right)^y$

24. $\lim\limits_{x \to +\infty} \left(\cos \dfrac{1}{x} \right)^{x^2}$

25. Prove that $\lim_{x \to +\infty} x^n/e^x = 0$ for every positive integer n.

26. Prove that $\lim_{x \to 0^+} e^{-1/x}/x^n = 0$ for every positive integer n. HINT: Use the limit derived in Exercise 25.

27. Prove that $\lim_{x \to 0} e^{-1/x^2}/x^n = 0$ for every positive integer n.

13.7 Improper Integrals

By Definition 7.5.3 the definite integral of a function f on an interval I can exist only if (i) I is a bounded closed interval and (ii) f is defined on I. Sometimes, however, it is useful to consider integrals of functions for which either of the requirements (i) and (ii) is not fulfilled. Such integrals are called *improper integrals*.

To introduce the first type of improper integral, we recall from Exercise 23, Exercise Set 7.2 that the force of attraction exerted by the earth on a particle of mass m is given by

$$-\frac{mgR^2}{x^2}$$

where x is the distance from the center of the earth to the particle, m is the mass of the particle, R is the radius of the earth, and g is the gravitational constant. In order to project the particle out from the earth's surface, the particle must be subjected to a force greater than

$$f(x) = \frac{mgR^2}{x^2}.$$

The work done by the force $f(x)$ in moving the particle from the earth's surface to a distance t from the center of the earth is

$$\int_R^t f(x)\, dx = \int_R^t \frac{mgR^2}{x^2}\, dx = mgR^2\left(\frac{1}{R} - \frac{1}{t}\right).$$

If the particle is moved indefinitely far from the center of the earth, the work done by $f(x)$ would therefore appear to be

$$\lim_{t \to +\infty} \int_R^t \frac{mgR^2}{x^2}\, dx = mgR$$

although we have not formally defined the work done by a force in moving a particle an infinite distance. The integral $\int_R^{+\infty} mgR^2/x^2\, dx$ is defined by the statement

$$\int_R^{+\infty} \frac{mgR^2}{x^2}\, dx = \lim_{t \to +\infty} \int_R^t \frac{mgR^2}{x^2}\, dx = mgR$$

according to the first of the following two definitions, which are given for improper integrals over infinite intervals.

13.7.1   Definition

Suppose that the definite integral $\int_a^t f(x)\, dx$ exists for every $t > a$. The improper integral $\int_a^{+\infty} f(x)\, dx$ is given by

$$\int_a^{+\infty} f(x)\, dx = \lim_{t \to +\infty} \int_a^t f(x)\, dx \tag{13.56}$$

if this limit exists.

The integral $\int_{-\infty}^b f(x)\, dx$ is analogously defined.

13.7.2   Definition

Suppose $\int_t^b f(x)\, dx$ exists for every $t < b$. The improper integral $\int_{-\infty}^b f(x)\, dx$ is given by

$$\int_{-\infty}^{b} f(x)\, dx = \lim_{t \to -\infty} \int_{t}^{b} f(x)\, dx \qquad (13.57)$$

if this limit exists.

Example 1 Evaluate the improper integral $\int_{0}^{+\infty} e^{-x}\, dx$ if it exists (Figure 13–5).

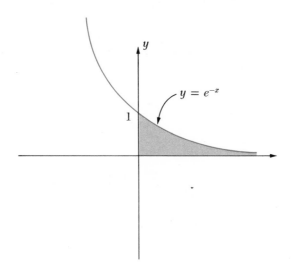

$y = e^{-x}$

1

Figure 13–5

SOLUTION From Definition 13.7.1

$$\int_{0}^{+\infty} e^{-x}\, dx = \lim_{t \to +\infty} \int_{0}^{t} e^{-x}\, dx = \lim_{t \to +\infty} \left. (-e^{-x}) \right|_{0}^{t}$$
$$= \lim_{t \to +\infty} (1 - e^{-t}) = 1.$$

Thus $\int_{0}^{+\infty} e^{-x}\, dx = 1$ could be considered the area of the region $\{(x, y): x \geq 0 \text{ and } y \in [0, e^{-x}]\}$, even though the region extends indefinitely far to the right from the y axis (Figure 13–5).

Example 2 Evaluate the improper integral $\int_{-\infty}^{0} e^{-x}\, dx$ if it exists.

SOLUTION Since

$$\lim_{t \to -\infty} \int_{t}^{0} e^{-x}\, dx = \lim_{t \to -\infty} (e^{-t} - 1) = +\infty$$

by Definition 13.7.2 $\int_{-\infty}^{0} e^{-x}\, dx$ does not exist.

We next define an improper integral over the set R.

13.7.3 Definition

Suppose the function f is integrable on any finite closed interval. The improper integral $\int_{-\infty}^{+\infty} f(x)\,dx$ is given by

$$\int_{-\infty}^{+\infty} f(x)\,dx = \int_{-\infty}^{a} f(x)\,dx + \int_{a}^{+\infty} f(x)\,dx \qquad (13.58)$$

if for some number a the improper integrals on the right in (13.58) exist. If either of these integrals fails to exist, then $\int_{-\infty}^{+\infty} f(x)\,dx$ does not exist.

We leave as an exercise the proof that $\int_{-\infty}^{+\infty} f(x)\,dx$, when it exists, is independent of the choice of the number a (Exercise 28, Exercise Set 13.7).

Example 3 Obtain $\int_{-\infty}^{+\infty} \dfrac{dx}{x^2 + 1}$ if it exists.

SOLUTION From (13.56)

$$\int_{0}^{+\infty} \frac{dx}{x^2 + 1} = \lim_{t \to +\infty} \int_{0}^{t} \frac{dx}{x^2 + 1} = \lim_{t \to +\infty} \tan^{-1} x \Big|_{0}^{t}$$

$$= \lim_{t \to +\infty} \tan^{-1} t = \frac{\pi}{2}$$

and from (13.57)

$$\int_{-\infty}^{0} \frac{dx}{x^2 + 1} = \lim_{t \to -\infty} \tan^{-1} x \Big|_{t}^{0} = \lim_{t \to -\infty} (-\tan^{-1} t) = \frac{\pi}{2}.$$

Hence by Definition 13.7.3

$$\int_{-\infty}^{+\infty} \frac{dx}{x^2 + 1} = \int_{-\infty}^{0} \frac{dx}{x^2 + 1} + \int_{0}^{+\infty} \frac{dx}{x^2 + 1} = \frac{\pi}{2} + \frac{\pi}{2} = \pi.$$

Before introducing another type of improper integral, we recall that

$$\int_{0}^{t} \frac{dy}{\sqrt{1 - y^2}} = \sin^{-1} t$$

and therefore

$$\lim_{t \to 1^-} \int_{0}^{t} \frac{dy}{\sqrt{1 - y^2}} = \lim_{t \to 1^-} \sin^{-1} t = \frac{\pi}{2}. \qquad (13.59)$$

However

$$\int_{0}^{1} \frac{dy}{\sqrt{1 - y^2}} \qquad (13.60)$$

is not a definite integral since the integrand is undefined when $y = 1$.

The integral (13.60) is defined in the second definition that follows.

13.7.4   Definition

Suppose that the definite integral $\int_t^b f(x)\,dx$ exists for every $t \in (a, b)$, but $f(a)$ is undefined. The improper integral $\int_a^b f(x)\,dx$ is given by

$$\int_a^b f(x)\,dx = \lim_{t\to a^+} \int_t^b f(x)\,dx \qquad (13.61)$$

if this limit exists.

13.7.5   Definition

Suppose that the definite integral $\int_a^t f(x)\,dx$ exists for every $t \in (a, b)$, but $f(b)$ is undefined. The improper integral $\int_a^b f(x)\,dx$ is given by

$$\int_a^b f(x)\,dx = \lim_{t\to b^-} \int_a^t f(x)\,dx \qquad (13.62)$$

if this limit exists.

Because of Definition 13.7.5 the integral (13.60) can be evaluated:

$$\int_0^1 \frac{dy}{\sqrt{1-y^2}} = \lim_{t\to 1^-} \int_0^t \frac{dy}{\sqrt{1-y^2}} = \frac{\pi}{2}.$$

Example 4   Evaluate $\displaystyle\int_0^2 \frac{dx}{\sqrt{x}}$ if it exists.

SOLUTION   The integrand $1/\sqrt{x}$ is undefined at 0. From Definition 13.7.4

$$\int_0^2 \frac{dx}{\sqrt{x}} = \lim_{t\to 0^+} \int_t^2 \frac{dx}{\sqrt{x}} = \lim_{t\to 0^+} 2\sqrt{x}\,\Big|_t^2$$
$$= \lim_{t\to 0^+} (2\sqrt{2} - 2\sqrt{t}) = 2\sqrt{2}.$$

The integrals defined by Definitions 13.7.4 and 13.7.5 are improper because the integrand is not defined at an endpoint of the interval of integration. We also define an improper integral where the integrand does not exist at an interior point of the interval of integration.

13.7.6   Definition

Suppose $a < c < b$ and f is defined on $[a, b]$ except at c. The improper integral $\int_a^b f(x)\,dx$ is given by

$$\int_a^b f(x)\,dx = \int_a^c f(x)\,dx + \int_c^b f(x)\,dx \qquad (13.63)$$

if both of the improper integrals on the right in (13.63) exist. If either of these integrals fails to exist, then $\int_a^b f(x)\,dx$ does not exist.

Example 5   Obtain the improper integral $\int_0^9 \dfrac{dx}{(x-1)^{1/3}}$ if it exists.

SOLUTION   We note that the integrand is undefined when $x = 1$. Hence by Definition 13.7.6 if $\int_0^1 \dfrac{dx}{(x-1)^{1/3}}$ and $\int_1^9 \dfrac{dx}{(x-1)^{1/3}}$ exist, then $\int_0^9 \dfrac{dx}{(x-1)^{1/3}}$ exists and

$$\int_0^9 \frac{dx}{(x-1)^{1/3}} = \int_0^1 \frac{dx}{(x-1)^{1/3}} + \int_1^9 \frac{dx}{(x-1)^{1/3}}. \qquad (13.64)$$

By Definition 13.7.5

$$\int_0^1 \frac{dx}{(x-1)^{1/3}} = \lim_{t \to 1^-} \int_0^t \frac{dx}{(x-1)^{1/3}} = \lim_{t \to 1^-} \frac{3}{2}(x-1)^{2/3}\Big|_0^t$$

$$= \lim_{t \to 1^-} [\tfrac{3}{2}(t-1)^{2/3} - \tfrac{3}{2}]$$

$$= -\tfrac{3}{2} \qquad (13.65)$$

and by Definition 13.7.4

$$\int_1^9 \frac{dx}{(x-1)^{1/3}} = \lim_{t \to 1^+} \int_t^9 \frac{dx}{(x-1)^{1/3}} = \lim_{t \to 1^+} \frac{3}{2}(x-1)^{2/3}\Big|_t^9$$

$$= \lim_{t \to 1^+} [\tfrac{3}{2}(8)^{2/3} - \tfrac{3}{2}(t-1)^{2/3}] = 6 - 0$$

$$= 6. \qquad (13.66)$$

Substituting in (13.64) from (13.65) and (13.66), we obtain

$$\int_0^9 \frac{dx}{(x-1)^{1/3}} = -\frac{3}{2} + 6 = \frac{9}{2}$$

On the other hand it can be shown that the improper integral $\int_0^5 \dfrac{dx}{(x-2)^2}$ does not exist. Since its integrand is undefined when $x = 2$, by Definition 13.7.6 this integral can exist only if $\int_0^2 \dfrac{dx}{(x-2)^2}$ and $\int_2^5 \dfrac{dx}{(x-2)^2}$ both exist. However, since

$$\lim_{t \to 2^-} \int_0^t \frac{dx}{(x-2)^2} = \lim_{t \to 2^-} \left(-\frac{1}{x-2}\right)\Big|_0^t$$

$$= +\infty,$$

$\int_0^2 \dfrac{dx}{(x-2)^2}$ does not exist by Definition 13.7.5. Therefore $\int_0^5 \dfrac{dx}{(x-2)^2}$ does not exist by Definition 13.7.6.

Note that since the function given by $f(x) = 1/(x-2)^2$ is not continuous on $[0,5]$, we cannot invoke the second fundamental theorem and obtain the erroneous result:

$$\int_0^5 \frac{dx}{(x-2)^2} = -\frac{1}{(x-2)}\Big|_0^5 = -\frac{1}{3} - \frac{1}{2} = -\frac{5}{6}.$$

It should be apparent that this calculation is nonsense since an integral cannot be negative when its integrand is never negative on the interval of integration. This example should therefore serve as a warning that an improper integral of the type mentioned in Definition 13.7.6 should not be evaluated by ignoring the fact that it is improper.

We sometimes encounter examples of other types of improper integrals (see Exercises 25 and 26, Exercise Set 13.7) in which the integrand is defined at all but a finite number of numbers in the interval of integration, which may be either finite or infinite. In such cases the original interval is subdivided into subintervals where the improper integral over each subinterval is one of the types covered in Definitions 13.7.1, 13.7.2, 13.7.4, or 13.7.5. The improper integral over the original interval, if it exists, is the sum of the improper integrals over each of the subintervals.

Exercise Set 13.7

In Exercises 1–26 evaluate the given improper integral, or prove that it does not exist.

1. $\displaystyle\int_{1}^{+\infty}\frac{dx}{x^{2}\sqrt{x}}$

2. $\displaystyle\int_{1}^{+\infty}\frac{dx}{x^{2/3}}$

3. $\displaystyle\int_{0}^{+\infty}\frac{x\,dx}{e^{3x}}$

4. $\displaystyle\int_{4}^{+\infty}\frac{dx}{x^{2}-4x+3}$

5. $\displaystyle\int_{-\infty}^{-1}\frac{t\,dt}{(t^{2}+1)^{3/2}}$

6. $\displaystyle\int_{-\infty}^{0}\frac{dx}{x-2}$

7. $\displaystyle\int_{-\infty}^{-5}\frac{dy}{y^{2}-9}$

8. $\displaystyle\int_{-\infty}^{0}\frac{dx}{e^{x}+e^{-x}}$

9. $\displaystyle\int_{0}^{4}\frac{dx}{(x-4)^{4/3}}$

10. $\displaystyle\int_{1}^{3}\frac{dx}{\sqrt{x-1}}$

11. $\displaystyle\int_{0}^{2}\frac{e^{-\sqrt{x}}\,dx}{\sqrt{x}}$

12. $\displaystyle\int_{-1}^{3}\frac{x\,dx}{x-3}$

13. $\displaystyle\int_{1}^{3}\frac{dx}{x^{2}\sqrt{9-x^{2}}}$

14. $\displaystyle\int_{0}^{\pi}\frac{dt}{1-\cos t}$

15. $\displaystyle\int_{-2}^{1}\frac{dx}{(3x+5)^{1/3}}$

16. $\displaystyle\int_{0}^{3}\frac{dx}{(2x-1)^{2}}$

17. $\displaystyle\int_{0}^{\pi}\tan^{2}x\,dx$

18. $\displaystyle\int_{-1}^{1}\frac{dx}{|x|^{1/3}}$

19. $\displaystyle\int_{-\infty}^{+\infty}xe^{-x^{2}}\,dx$

20. $\displaystyle\int_{e}^{+\infty}\frac{dx}{x\ln^{2}x}$

21. $\displaystyle\int_{-\infty}^{+\infty}\operatorname{csch}x\,dx$

22. $\displaystyle\int_{-\infty}^{+\infty}\frac{x^{2}\,dx}{x^{2}+4}$

23. $\displaystyle\int_0^{+\infty} e^{-st} \cos kt$

24. $\displaystyle\int_0^{+\infty} e^{-st} \sin kt$

25. $\displaystyle\int_0^3 \frac{dx}{\sqrt{3x - x^2}}$

26. $\displaystyle\int_1^{+\infty} \frac{dx}{(x - 1)^2}$

27. Prove that if $\int_{-\infty}^{+\infty} f(x)\, dx$ exists, then $\lim_{t \to +\infty} \int_{-t}^{t} f(x)\, dx$ exists. Is the converse true? Give an example to explain your answer.

28. Prove that when $\int_{-\infty}^{+\infty} f(x)\, dx$ in Definition 13.7.3 exists, the improper integral is independent of the choice of the number a. In other words, prove that if $\int_{-\infty}^{+\infty} f(x)\, dx$ is given by (13.58), and $b \neq a$, then

$$\int_{-\infty}^{+\infty} f(x)\, dx = \int_{-\infty}^{b} f(x)\, dx + \int_{b}^{+\infty} f(x)\, dx.$$

29. Show that the improper integral $\int_1^{+\infty} (1/x)\, dx$ does not exist and therefore one cannot assign an area to the infinite region $\{(x, y): y \in [0, (1/x)] \text{ and } x \geq 1\}$. However, if this region is revolved about the x axis show that it is possible to assign a volume to the solid of revolution obtained.

14. Infinite Series

14.1 Sequences

From a nontechnical point of view a sequence exists whenever there is a set whose elements can be arranged in a definite order; that is, the set contains a first element, a second element, a third element, and so on. For example, we could consider the noon temperature readings taken on successive days beginning with a particular day, or the values of 2^n obtained for $n = 1, 2, 3, \ldots$. A precise definition of a sequence is given in the following statement.

14.1.1 Definition

A *sequence* is a function whose domain is the set of all positive integers.

If f is a sequence, it can be denoted by

$$\{(n, f(n)): n = 1, 2, 3, \ldots\}$$

or simply by

$$\{f(n)\},$$

where it is understood with the latter notation that n is an arbitrary positive integer. If $f(n) = a_n$ for any n, then f could also be given in any of the following ways:

$$\{a_n\}$$
$$a_1, a_2, \ldots, a_n, \ldots$$
$$a_1, a_2, a_3, \ldots.$$

a_1 is called the *first term* of the sequence $\{a_n\}$, a_2 the *second term*, and so forth, and a_n the *nth term* or *general term* of the sequence.

For example, the sequence $\{n/(n + 1)\}$ could be expressed as

$$\frac{1}{2}, \frac{2}{3}, \frac{3}{4}, \ldots, \frac{n}{n + 1}, \ldots \tag{14.1}$$

515

The first term of this sequence is $\frac{1}{2}$, the second term is $\frac{2}{3}$, the third term is $\frac{3}{4}$, and so on, and the nth term is $n/(n + 1)$. The 25th term of the sequence (14.1), for example, is obtained by letting $n = 25$ in the nth term and is $\frac{25}{26}$. A sketch of the graph of the sequence (14.1) is shown in Figure 14–1.

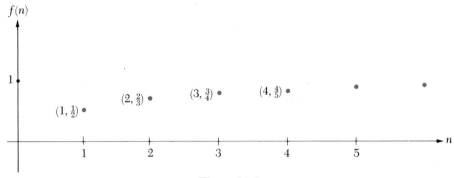

Figure 14-1

We observe that the larger we choose n, the closer to 1 is the corresponding number $n/(n + 1)$ in the sequence (14.1). Note that

$$\left| \frac{n}{n + 1} - 1 \right| = \left| -\frac{1}{n + 1} \right| = \frac{1}{n + 1} < \frac{1}{n}.$$

Since for any positive number ϵ, no matter how small, $1/n < \epsilon$ if $n > 1/\epsilon$, we have

$$\left| \frac{n}{n + 1} - 1 \right| < \epsilon \quad \text{if } n > \frac{1}{\epsilon}. \tag{14.2}$$

Another example of a sequence is $\{(-1)^{n+1}/2^{n-1}\}$, which can also be expressed as

$$1, \ -\frac{1}{2}, \ \frac{1}{4}, \ -\frac{1}{8}, \ \dots, \ \frac{(-1)^{n+1}}{2^{n-1}}, \ \dots \tag{14.3}$$

A sketch of the graph of this sequence is shown in Figure 14–2.

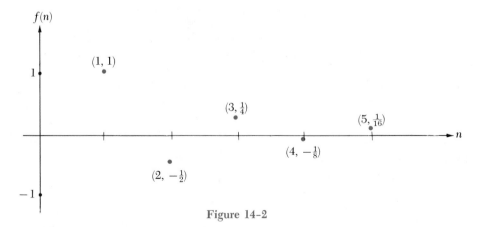

Figure 14-2

Intuition suggests that the larger n is chosen, the closer to 0 is the corresponding number $(-1)^{n+1}/2^{n-1}$ of the sequence (14.3). In fact, for any $\epsilon > 0$

$$\left| \frac{(-1)^{n+1}}{2^{n-1}} - 0 \right| = \frac{1}{2^{n-1}} < \epsilon$$

if $2^{n-1} > 1/\epsilon$—that is, if $(n-1) \ln 2 > -\ln \epsilon$, and hence if $n > (\ln 2 - \ln \epsilon)/\ln 2$. This leads us to the definition of the *limit of a sequence*.

14.1.2   Definition

The sequence $\{a_n\}$ has the limit L if and only if for every $\epsilon > 0$ there is a number $M > 0$ such that

$$|a_n - L| < \epsilon \quad \text{whenever } n > M.$$

When L is the limit of the sequence $\{a_n\}$, we express this fact by writing

$$\lim_{n \to +\infty} a_n = L.$$

Also, if L is the limit of $\{a_n\}$, one says that the sequence $\{a_n\}$ *converges* to L, and that $\{a_n\}$ is a *convergent sequence*. If there is no number L satisfying Definition 14.1.1, then $\{a_n\}$ is called a *divergent sequence* and is said to diverge. We leave as an exercise the proof that a convergent sequence has a unique limit (Exercise 29, Exercise Set 14.1).

Definition 14.1.2 is illustrated in Figure 14–3.

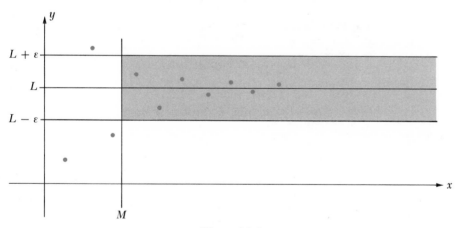

Figure 14-3

For any $\epsilon > 0$ there corresponds a band of width 2ϵ between the lines $y = L - \epsilon$ and $y = L + \epsilon$. Also corresponding to the given ϵ is a line $x = M > 0$ such that whenever $n > M$, the points (n, a_n) lie to the right of this line and within the band of width 2ϵ. This region is shaded in Figure 14–3. The smaller we choose ϵ, the narrower is the band and usually the farther to the right is the line $x = M$.

Since (14.2) is satisfied for every $\epsilon > 0$, from Definition 14.1.2

$$\lim_{n \to +\infty} \frac{n}{n+1} = 1.$$

Here $M = 1/\epsilon$. Also, from the discussion preceding Definition 14.1.2, for every $\epsilon > 0$ there is an $M > 0$ such that

$$\left| \frac{(-1)^{n+1}}{2^{n-1}} - 0 \right| < \epsilon \quad \text{if } n > M \geq \frac{\ln 2 - \ln \epsilon}{\ln 2}.$$

Hence by Definition 14.1.2

$$\lim_{n \to +\infty} \frac{(-1)^{n+1}}{2^{n-1}} = 0.$$

Before considering the next theorem, which is a consequence of Definition 14.1.2, we mention that a sequence $\{a_n\}$ is said to be *bounded* if and only if there is some $M > 0$ such that

$$|a_n| \leq M \quad \text{for every } n.$$

The reader may wish to compare this definition with that given for a bounded function on a closed interval (see Theorem 7.5.6).

14.1.3   Theorem

A convergent sequence is bounded.

PROOF If $\lim_{n \to +\infty} a_n = L$, then for $\epsilon = 1$, for example, there is an integer $M > 0$ such that

$$|a_n - L| < 1 \quad \text{if } n > M.$$

Hence

$$L - 1 < a_n < L + 1 \quad \text{if } n > M,$$

and therefore

$$-|L - 1| < a_n < |L + 1| \quad \text{if } n > M.$$

If

$$B = \text{greatest of the numbers } |a_1|, |a_2|, \ldots, |a_M|, |L - 1|, |L + 1|,$$

then certainly

$$|a_n| \leq B \quad \text{for every } n,$$

which proves the theorem.

Theorem 14.1.3 can be restated to read: *An unbounded sequence is divergent.* The theorem as restated is used in the proof of Theorem 14.1.4(b) below.

The converse of Theorem 14.1.3 is not true. For example, the sequence

$$1, 0, 1, 0, \ldots, \tfrac{1}{2}[1 + (-1)^{n+1}], \ldots$$

is certainly bounded. However, the terms of the sequence alternate between 0 and 1 and therefore cannot converge to any possible limit.

Notice the similarity between the wording of Definition 14.1.2 and Definition 5.6.1. Since a sequence $\{a_n\}$ is really a function f with $f(n) = a_n$, this similarity can be exploited to prove part (a) of the following theorem.

14.1.4   Theorem

Suppose f is defined on the interval $[1, +\infty)$ and $a_n = f(n)$ for every positive integer n.

(a) If $\lim_{x \to +\infty} f(x) = L$, then $\lim_{n \to +\infty} a_n = L$.
(b) If $\lim_{x \to +\infty} f(x) = +\infty$ or $-\infty$, then $\{a_n\}$ diverges.

PROOF OF (a)   By hypothesis, for every $\epsilon > 0$ there is an $M > 0$ such that

$$|f(x) - L| < \epsilon \quad \text{when } x > M. \tag{14.4}$$

Since (14.4) holds, in particular, for every positive integer $x = n > M$, $\lim_{n \to +\infty} f(n) = L$ by Definition 14.1.2.

The proof of (b) is left as an exercise.

Example 1   Find $\displaystyle\lim_{n \to +\infty} \frac{n^2 + 2n - 5}{2n^2 + 1}$.

SOLUTION   Since

$$\lim_{x \to +\infty} \frac{x^2 + 2x - 5}{2x^2 + 1} = \lim_{x \to +\infty} \frac{1 + \dfrac{2}{x} - \dfrac{5}{x^2}}{2 + \dfrac{1}{x^2}} = \frac{1}{2},$$

by Theorem 14.1.4(a),

$$\lim_{n \to +\infty} \frac{n^2 + 2n - 5}{2n^2 + 1} = \frac{1}{2}.$$

Example 2   Find $\lim_{n \to +\infty} \ln n$ or show that $\{\ln n\}$ diverges.

SOLUTION   Since $\lim_{x \to +\infty} \ln x = +\infty$, by Theorem 14.1.4(b) $\{\ln n\}$ diverges.

Example 3   Find $\displaystyle\lim_{n \to +\infty} n \sin \frac{1}{n}$.

SOLUTION   From Exercise 21, Exercise Set 5.6 and Theorem 10.2.1(c)

$$\lim_{x \to +\infty} x \sin \frac{1}{x} = \lim_{x \to +\infty} \frac{\sin \frac{1}{x}}{\frac{1}{x}} = \lim_{t \to 0^+} \frac{\sin t}{t} = 1.$$

Hence by Theorem 14.1.4(a)

$$\lim_{n \to +\infty} n \sin \frac{1}{n} = 1.$$

Theorems 14.1.5(a) through (d) below are analogous to limit theorems from Section 5.6 and hence their proofs are omitted.

14.1.5   Theorem

If $\lim_{n \to +\infty} a_n = a$ *and* $\lim_{n \to +\infty} b_n = b$, *then*

(a) $\lim_{n \to +\infty} (a_n + b_n) = a + b,$

(b) $\lim_{n \to +\infty} (a_n - b_n) = a - b,$

(c) $\lim_{n \to +\infty} a_n b_n = ab,$

(d) $\lim_{n \to +\infty} \dfrac{a_n}{b_n} = \dfrac{a}{b}$ *if* $b \neq 0$ *and* $b_n \neq 0$ *for every* n.

The formulas in (a) and (c) can be extended to the sum and product of any finite number of sequences.

14.1.6   Theorem

If there is an N *such that* $|a_n| \leq b_n$ *for every* $n \geq N$ *and* $\lim_{n \to +\infty} b_n = 0$, *then* $\lim_{n \to +\infty} a_n = 0$.

The proof of Theorem 14.1.6 is left as an exercise. The theorem itself is used in the following example.

Example 4   Find $\lim_{n \to +\infty} 2^n / n!$.

SOLUTION   We note that

$$0 < \frac{2^n}{n!} = \frac{2^2}{1 \cdot 2} \cdot \overbrace{\frac{2 \cdot 2 \cdots 2}{3 \cdot 4 \cdots n}}^{n - 2 \text{ factors}} < 2 \left(\frac{2}{3} \right)^{n-2} \quad \text{for every } n.$$

By Theorem 9.6.2(c) and Theorem 14.1.4(a)

$$\lim_{n \to +\infty} 2(\tfrac{2}{3})^{n-2} = \lim_{n \to +\infty} \tfrac{9}{2}(\tfrac{2}{3})^n = 0.$$

Hence by Theorem 14.1.6

$$\lim_{n \to +\infty} \frac{2^n}{n!} = 0.$$

We next define the terms *increasing sequence* and *decreasing sequence* in order to give an important property of these sequences.

14.1.7   Definition

(a) The sequence $\{a_n\}$ is *increasing* if and only if $a_{n+1} \ge a_n$ for every n.
(b) The sequence $\{a_n\}$ is *decreasing* if and only if $a_{n+1} \le a_n$ for every n.

A sequence that is either increasing or decreasing is called a *monotone sequence*. If instead of the inequality given in part (a), we have $a_{n+1} > a_n$, then $\{a_n\}$ is called a *strictly increasing* sequence. Similarly, if the inequality in (b) is replaced by $a_{n+1} < a_n$, then $\{a_n\}$ is called a *strictly decreasing* sequence.

Example 5   Show that the sequence $\left\{\dfrac{n}{n+1}\right\}$ is increasing.

SOLUTION   If $a_n = \dfrac{n}{n+1}$, then $a_{n+1} = \dfrac{n+1}{n+2}$. Since for any n

$$\frac{n+1}{n+2} - \frac{n}{n+1} = \frac{(n+1)^2 - n(n+2)}{(n+1)(n+2)} = \frac{1}{(n+1)(n+2)} \ge 0,$$

and so

$$\frac{n+1}{n+2} \ge \frac{n}{n+1};$$

hence from Definition 14.1.7 $\{n/(n+1)\}$ is increasing.

Example 6   Is the sequence $\left\{\sinh\dfrac{1}{n}\right\}$ an increasing or a decreasing sequence?

SOLUTION   Rather than consider the difference $a_{n+1} - a_n$ as we did in Example 5, we obtain the solution more directly by examining the sign of the derivative of the function f where

$$f(x) = \sinh\frac{1}{x}.$$

Then

$$f'(x) = -\frac{1}{x^2}\cosh\frac{1}{x} \quad \text{for } x \neq 0,$$

and since the function cosh is always positive-valued,

$$f'(x) < 0 \quad \text{for } x \neq 0.$$

Therefore by Theorem 5.3.2(b) the function f is certainly decreasing on the interval $[1, +\infty)$. Thus

$$\sinh\frac{1}{n+1} < \sinh\frac{1}{n} \quad \text{for } n = 1, 2, 3, \ldots,$$

and hence the sequence $\left\{\sinh\dfrac{1}{n}\right\}$ is a decreasing sequence.

14.1.8   Theorem

(a) *If $\{a_n\}$ is an increasing sequence, and $a_n \leq B$ for every n, then $\{a_n\}$ converges to a limit L where $L \leq B$, and $a_n \leq L$ for every n.*

(b) *If $\{a_n\}$ is a decreasing sequence, and $a_n \geq B$ for every n, then $\{a_n\}$ converges to a limit L where $L \geq B$, and $a_n \geq L$ for every n.*

DISCUSSION OF PROOF OF (a)   The proof that $\{a_n\}$ converges, which is found in advanced calculus texts, utilizes a property of real numbers called the completeness property. The proof that the limit of the sequence is $\leq B$ is left as an exercise.

Theorem 14.1.8(a) is illustrated in Figure 14–4.

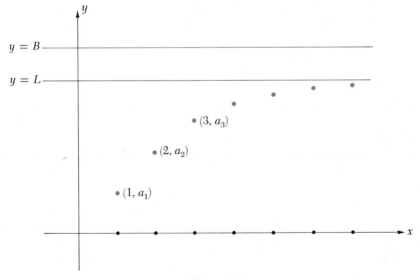

Figure 14–4

Example 7 Is the sequence $\{a_n\}$, where $a_n = \dfrac{1 \cdot 3 \cdots (2n-1)}{2 \cdot 4 \cdots 2n}$, convergent or divergent?

PROOF For every n, $a_n > 0$, and

$$\frac{a_{n+1}}{a_n} = \frac{1 \cdot 3 \cdots (2n-1)(2n+1)}{2 \cdot 4 \cdots (2n)(2n+2)} \cdot \frac{2 \cdot 4 \cdots 2n}{1 \cdot 3 \cdots (2n-1)} = \frac{2n+1}{2n+2} < 1.$$

Hence

$$a_{n+1} < a_n \quad \text{for every } n,$$

and so the sequence $\{a_n\}$ is decreasing. Also, since $a_n > 0$ for every n, $\{a_n\}$ is convergent by Theorem 14.1.8(b), which we use here letting $B = 0$.

Exercise Set 14.1

In Exercises 1–22 obtain the limit of the sequence having the given nth term or prove that the sequence diverges.

1. $\dfrac{n+2}{n^2+1}$

2. $\dfrac{n^3+1}{2n^2+n+5}$

3. $\left(\dfrac{3n+1}{2n-1}\right)^3$

4. $\left(\dfrac{2n}{4n+3}\right)^5$

5. $\left(\dfrac{5}{3}\right)^n$

6. $\dfrac{3 \cdot 4^n + 4 \cdot 3^n}{4^n}$

7. $\dfrac{3^n-1}{3^n}$

8. $\sqrt{n+2} - \sqrt{n}$

9. $\sin \dfrac{\pi n}{2n+1}$

10. $\cos^3 \left(\dfrac{1}{n}\right)$

11. $\ln(n+1) - \ln(n+2)$

12. $\dfrac{\sin n}{n}$

13. $\dfrac{\sum_{i=1}^n i}{2n^2}$

14. $\dfrac{\sum_{i=1}^n i^2}{n^3}$

15. $n^2 \tan\left(\dfrac{1}{n^2}\right)$

16. $1 + (-1)^n$

17. $n^2 e^{-3n}$

18. $\dfrac{\ln n}{n}$

19. $\sqrt[n]{n}$

20. $\left(1 + \dfrac{3}{n}\right)^{4n}$

21. $\dfrac{n!}{n^n}$

22. $\displaystyle\sum_{i=1}^n \dfrac{1}{\sqrt{n^2+i}}$

HINT: In Exercises 21 and 22 use Theorems 14.1.6 and 5.6.8, respectively.

In Exercises 23 and 24 prove that the sequence is either increasing or decreasing and also determine if it is convergent or divergent.

23. $\dfrac{1 \cdot 3 \cdot 5 \cdots (2n-1)}{n!}$ 24. $\dfrac{2 \cdot 4 \cdot 6 \cdots (2n)}{3 \cdot 5 \cdot 7 \cdots (2n+1)}$

25. Prove Theorem 14.1.4(b).

26. Prove Theorem 14.1.6.

27. Prove: If $\{a_n\}$ converges, then for every $\epsilon > 0$ there is an $M > 0$ such that if $n > M$ and $m > M$, then $|a_m - a_n| < \epsilon$.

28. If $\{a_n\}$ satisfies the hypotheses of Theorem 14.1.8(a), prove that the limit L of the sequence satisfies the inequality $a_n \leq L$, for every n, and $L \leq B$.

29. If $\lim\limits_{n \to +\infty} a_n = L_1$ and $\lim\limits_{n \to +\infty} a_n = L_2$, then $L_1 = L_2$.

14.2   Infinite Series

Speaking informally, an *infinite series* may be regarded as a sum of an infinite number of terms. The following example suggests how such a sum might arise. Suppose that a college student wishing to set a personal record for distance running begins running laps on his college track (one lap is 440 yards in length). He finishes the first lap in exactly one minute, but finds that in each succeeding minute the distance he covers is only $\frac{9}{10}$ of his distance during the preceding minute. If our student can continue running indefinitely, how far will he run? It would seem that the total distance in yards run by the student is the sum

$$440 + (0.9)(440) + (0.9)^2(440) + \cdots. \tag{14.5}$$

We will compute this sum after first attacking the more immediate problem of satisfactorily defining an "infinite sum." It should be realized that the notion of such a sum cannot be satisfactorily explained in terms of pre-calculus mathematics. Usually, in a more elementary mathematics course the existence of a unique sum of two real numbers is accepted axiomatically, and then one can define the sum of three numbers by the statement

$$a_1 + a_2 + a_3 = (a_1 + a_2) + a_3$$

and, in general, the sum of n numbers by

$$a_1 + a_2 + \cdots + a_n = (a_1 + a_2 + \cdots + a_{n-1}) + a_n.$$

Such an approach, though, does not lead us to any meaning for the term "infinite sum."

A more promising direction will be to define an infinite series in terms of a sequence.

14.2.1   Definition

Suppose $\{a_n\}$ is a sequence and $\{S_n\}$ is a second sequence given by the successive terms

$$S_n = \sum_{i=1}^{n} a_i = a_1 + a_2 + \cdots + a_n \quad \text{for } n = 1, 2, 3, \ldots. \tag{14.6}$$

The *infinite series*

$$\sum_{n=1}^{+\infty} a_n = a_1 + a_2 + a_3 + \cdots + a_n + \cdots, \tag{14.7}$$

which is read "sigma *a* sub *n* from 1 to plus infinity," is the sequence $\{S_n\}$.

The numbers a_1, a_2, a_3, and so on, are called the *1st term*, *2nd term*, *3rd term*, and so on, of the series $\sum_{n=1}^{+\infty} a_n$. The numbers S_1, S_2, S_3, $\ldots$, are called the *1st partial sum*, the *2nd partial sum*, the *3rd partial sum*, and so forth, of the series.

An example of an infinite series is

$$\sum_{n=1}^{+\infty} \frac{1}{n(n+1)} = \frac{1}{1 \cdot 2} + \frac{1}{2 \cdot 3} + \frac{1}{3 \cdot 4} + \cdots + \frac{1}{n(n+1)} + \cdots. \tag{14.8}$$

The first four terms of the series are

$$a_1 = \frac{1}{1 \cdot 2} = \frac{1}{2}, \quad a_2 = \frac{1}{2 \cdot 3} = \frac{1}{6}, \quad a_3 = \frac{1}{3 \cdot 4} = \frac{1}{12}, \quad a_4 = \frac{1}{4 \cdot 5} = \frac{1}{20}$$

and the *n*th term is

$$a_n = \frac{1}{n(n+1)}.$$

The first four partial sums of this series are

$$S_1 = \frac{1}{1 \cdot 2} = \frac{1}{2}$$

$$S_2 = \frac{1}{1 \cdot 2} + \frac{1}{2 \cdot 3} = \frac{2}{3}$$

$$S_3 = \frac{1}{1 \cdot 2} + \frac{1}{2 \cdot 3} + \frac{1}{3 \cdot 4} = \frac{3}{4}$$

$$S_4 = \frac{1}{1 \cdot 2} + \frac{1}{2 \cdot 3} + \frac{1}{3 \cdot 4} + \frac{1}{4 \cdot 5} = \frac{4}{5}$$

and the *n*th partial sum is

$$S_n = \frac{1}{1 \cdot 2} + \frac{1}{2 \cdot 3} + \cdots + \frac{1}{n(n+1)}.$$

By mathematical induction it can be proved that

$$S_n = \frac{n}{n+1} \quad \text{for } n = 1, 2, \ldots. \tag{14.9}$$

Although the notation (14.7) is the most common way of representing the infinite series with successive terms a_1, a_2, a_3, . . . there is nothing to say that 1 must be the least value assumed by the index n. In fact, (14.7) has the same meaning as $\Sigma_{n=k+1}^{+\infty} a_{n-k}$, where k is any integer. Thus, for example, the series (14.8) could be denoted by $\Sigma_{n=2}^{+\infty} 1/(n(n-1))$. This notation can be obtained from (14.8) by replacing n by $n-1$ and beginning the sum with $n=2$ instead of $n=1$.

Instead of starting with a sequence $\{a_n\}$ and obtaining the sequence of partial sums, $\{S_n\}$, one can begin with a sequence of partial sums and then calculate the terms of the corresponding series. From (14.7) it follows that

$$a_1 = S_1 \quad \text{and} \quad a_n = S_n - S_{n-1} \quad \text{if } n = 2, 3, 4, \ldots . \tag{14.10}$$

Example 1 Find the series having $\{\ln (n+1)\}$ as its sequence of partial sums.

SOLUTION From (14.10) letting $S_n = \ln (n+1)$ for $n = 1, 2, \ldots ,$

$$a_1 = \ln 2$$
$$a_2 = \ln 3 - \ln 2 = \ln \tfrac{3}{2}$$
$$a_3 = \ln 4 - \ln 3 = \ln \tfrac{4}{3}$$

and in general

$$a_n = \ln (n+1) - \ln n = \ln \frac{n+1}{n}.$$

The desired series is therefore $\displaystyle\sum_{n=1}^{+\infty} \ln \frac{n+1}{n}.$

We next define a *convergent series* and a *divergent series*.

14.2.2 Definition

The series Σa_n converges to S and is called a *convergent series* if and only if there is a number S such that $\lim_{n \to +\infty} S_n = S$, where S_n is defined by (14.6). Then S is called the *sum* of the series $\Sigma_{n=1}^{+\infty} a_n$. The series $\Sigma_{n=1}^{+\infty} a_n$ is called a *divergent series* and is said to diverge if and only if $\lim_{n \to +\infty} S_n$ fails to exist. A divergent series has no sum.

Briefly stated, *a series converges if and only if the corresponding sequence of partial sums converges*.

By Definition 14.2.2 the series (14.8) converges to the sum 1 since $S_n = n/(n+1)$ from (14.9) and $\lim_{n \to +\infty} S_n = 1$. However, the series $\Sigma_{n=1}^{+\infty}$ $\ln (n+1)/n$, which was discussed in Example 1, diverges since $S_n = \ln (n+1)$ and therefore $\lim_{n \to +\infty} S_n = +\infty$.

14.2.3   Theorem

If $\Sigma_{n=1}^{+\infty} a_n$ converges, then $\lim_{n \to +\infty} a_n = 0$.

PROOF   From (14.10)

$$a_n = S_n - S_{n-1} \quad \text{if } n > 1.$$

Then letting S be the sum of the series, which we know converges,

$$a_n = (S_n - S) - (S_{n-1} - S).$$

Hence

$$|a_n| \leq |S_n - S| + |S_{n-1} - S|. \tag{14.11}$$

Since $\lim_{n \to +\infty} S_n = S$, given $\epsilon > 0$ there is an $M > 0$ such that

$$|S_n - S| < \frac{\epsilon}{2} \quad \text{when } n > M. \tag{14.12}$$

If $n > M + 1$, then $n - 1 > M$, and hence from (14.11) and (14.12)

$$|a_n| < \frac{\epsilon}{2} + \frac{\epsilon}{2} = \epsilon.$$

Thus $\lim_{n \to +\infty} a_n = 0$.

If the sequence of the nth terms of a series does not have 0 as a limit, by Theorem 14.2.3 the *series must diverge*. For otherwise, if $\{a_n\}$ did not have the limit 0 and $\Sigma_{n=1}^{+\infty} a_n$ converged, then by Theorem 14.2.3, we would have $\lim_{n \to +\infty} a_n = 0$, a contradiction.

Example 2   Does the series

$$\frac{1}{2} + \frac{1}{\sqrt{2}} + \frac{1}{\sqrt[3]{2}} + \cdots + \frac{1}{\sqrt[n]{2}} + \cdots$$

converge or diverge?

SOLUTION   Since $\lim_{x \to +\infty} 2^{-1/x} = 2^{\lim_{x \to +\infty}(-1/x)} = 2^0 = 1$, by Theorem 14.1.4(a) $\lim_{n \to +\infty} 2^{-1/n} = 1$. Since $\lim_{n \to +\infty} 2^{-1/n} \neq 0$, the series diverges by Theorem 14.2.3.

Although the condition

$$\lim_{n \to +\infty} a_n = 0$$

is necessary for the convergence of the series $\Sigma_{n=1}^{+\infty} a_n$, the converse of Theorem 14.2.3 is *not* true. For example, the series $\Sigma_{n=1}^{+\infty} \ln(n + 1)/n$ was observed to be divergent, and yet $\lim_{n \to +\infty} \ln(n + 1)/n = \ln 1 = 0$. Thus if the nth term of a series tends to 0, the series need not converge, and Theorem 14.2.3 gives no clue as to whether it converges or not.

A useful series for our later work in determining the convergence or divergence of a series is

$$a + ar + ar^2 + \cdots + ar^{n-1} + \cdots, \tag{14.13}$$

which is called a *geometric series*. The series

$$440 + 440(\tfrac{9}{10}) + 440(\tfrac{9}{10})^2 + \cdots, \tag{14.5}$$

which was referred to at the beginning of this section, is an example of a geometric series.

14.2.4 Theorem

The geometric series (14.13) converges to $a/(1 - r)$ if $|r| < 1$ or $a = 0$, and diverges if $|r| \geq 1$ and $a \neq 0$.

PROOF If $|r| \geq 1$ and $a \neq 0$, then $|ar^{n-1}| = |a||r|^{n-1} \geq |a| > 0$ for every n. Hence the nth term cannot approach 0, and by Theorem 14.2.3 the series diverges.

If $a = 0$, then each term of (14.13) is 0, and hence the nth partial sum is $S_n = 0$. The series therefore converges to $\lim_{n \to +\infty} S_n = 0$.

Suppose $|r| < 1$. For any n,

$$S_n = a(1 + r + r^2 + \cdots + r^{n-1}).$$

Since

$$1 - r^n = (1 - r)(1 + r + r^2 + \cdots + r^{n-1}),$$

$$S_n = \frac{a(1 - r^n)}{1 - r}. \tag{14.14}$$

We can prove that

$$\lim_{n \to +\infty} r^n = 0 \quad \text{when } |r| < 1. \tag{14.15}$$

From Theorem 9.6.2(c)

$$\lim_{n \to +\infty} |r|^x = 0 \quad \text{if } 0 < |r| < 1.$$

Hence by Definition 5.6.1, for any $\epsilon > 0$ there is an $M > 0$ such that $|r|^x < \epsilon$ when $x > M$. In particular, if $x = n$ we have

$$|r^n| = |r|^n < \epsilon \quad \text{when } n > M.$$

Since this statement is also true when $r = 0$, the limit (14.15) follows by Definition 14.1.2. Then from (14.14)

$$\lim_{n \to +\infty} S_n = \frac{a}{1 - r} \quad \text{if } |r| < 1,$$

thereby proving the theorem.

Example 3 Find the sum of the series (14.5), or show that the series diverges.

SOLUTION   This series is a geometric series with $a = 440$ and $r = \frac{9}{10}$. From Theorem 14.2.4 the series converges to the sum $440/(1 - \frac{9}{10}) = 4400$. Thus the student in the example at the beginning of the section would run a total distance of 4400 yards, or $2\frac{1}{2}$ miles, if he could run forever.

The remaining theorems in this section will be useful in determining the convergence (or divergence) of a series that is obtained from one or more series that are known to be convergent (or divergent). The first of these theorems states that *the convergence or divergence of a series is not affected by the addition or deletion of a finite number of terms at the beginning of the series.*

14.2.5   Theorem

Let k be any positive integer. Then the series $\Sigma_{n=1}^{+\infty} a_n$ and $\Sigma_{n=k+1}^{+\infty} a_n$ either both converge or both diverge.

PROOF   Let S_n and T_n be the nth partial sums of $\Sigma_{n=1}^{+\infty} a_n$ and $\Sigma_{n=k+1}^{+\infty} a_n$, respectively. Then $S_n = a_1 + a_2 + \cdots + a_n$ and

$$T_n = a_{k+1} + a_{k+2} + \cdots + a_{k+n}$$
$$= S_{n+k} - S_k. \qquad (14.16)$$

We leave as an exercise the proof that

$$\lim_{n \to +\infty} S_{n+k} = \lim_{n \to +\infty} S_n \qquad (14.17)$$

if either limit exists. Hence if $\lim_{n \to +\infty} S_n = S = \lim_{n \to +\infty} S_{n+k}$, from (14.16)

$$\lim_{n \to +\infty} T_n = S - S_k.$$

Thus the convergence of $\Sigma_{n=1}^{+\infty} a_n$ implies the convergence of $\Sigma_{n=k+1}^{+\infty} a_n$. Conversely, if $\Sigma_{n=k+1}^{+\infty} a_n$ converges, then $\lim_{n \to +\infty} T_n$ exists, and from (14.16) and (14.17)

$$\lim_{n \to +\infty} S_n = \lim_{n \to +\infty} S_{n+k} = \lim_{n \to +\infty} (T_n + S_k) = \lim_{n \to +\infty} T_n + S_k.$$

Thus $\Sigma_{n=1}^{+\infty} a_n$ converges, and the part of the theorem about convergence is proved.
     To prove the part about divergence, suppose one of the series converges and the other diverges. Then from the previous discussion, if one series converges then the other must converge. Thus we have a contradiction, and hence both series diverge.

From the next theorem *the sum of two convergent series is convergent.*

14.2.6   Theorem

If $\Sigma_{n=1}^{+\infty} a_n$ and $\Sigma_{n=1}^{+\infty} b_n$ converge to S and T, respectively, then $\Sigma_{n=1}^{+\infty} (a_n \pm b_n)$ converges to $S \pm T$.

PROOF If $S_n = a_1 + a_2 + \cdots + a_n$ and $T_n = b_1 + b_2 + \cdots + b_n$, then $S_n \pm T_n$ is the nth partial sum of the series $\Sigma_{n=1}^{+\infty}(a_n \pm b_n)$. From Theorem 14.1.5

$$\lim_{n \to +\infty} (S_n \pm T_n) = \lim_{n \to +\infty} S_n \pm \lim_{n \to +\infty} T_n = S \pm T,$$

and the theorem is proved.

Example 4 Prove that the series $\Sigma_{n=1}^{+\infty}(2^{-n} + 3^{-n})$ converges.

PROOF The series $\Sigma_{n=1}^{+\infty}2^{-n}$ and $\Sigma_{n=1}^{+\infty}3^{-n}$ converge since they are each geometric series for which $|r| < 1$. Hence, by Theorem 14.2.6 $\Sigma_{n=1}^{+\infty}(2^{-n} + 3^{-n})$ converges.

The last theorem of this section states that *the convergence or divergence of a series is unchanged if each term of the series is multiplied by a nonzero constant.* The proof of this theorem is left as an exercise.

14.2.7 Theorem

If k is any number, and the series $\Sigma_{n=1}^{+\infty}a_n$ converges to S, then the series $\Sigma_{n=1}^{+\infty}ka_n$ converges to kS. If $k \neq 0$ and $\Sigma_{n=1}^{+\infty}a_n$ diverges, then $\Sigma_{n=1}^{+\infty}ka_n$ also diverges.

Exercise Set 14.2

In Exercises 1–12 either find the sum of the series or show that the series diverges.

1. $\sum_{n=1}^{+\infty} (0.99)^{-n}$

2. $\sum_{n=1}^{+\infty} 2(-\frac{5}{3})^{n-1}$

3. $\sum_{n=1}^{+\infty} \tan \pi n$

4. $\sum_{n=1}^{+\infty} \frac{3n + 2}{2n + 1}$

5. $\sum_{n=1}^{+\infty} e^{-1/n}$

6. $\sum_{n=1}^{+\infty} (3^{-n} + 3^{-2n})$

7. $\sum_{n=1}^{+\infty} \ln \frac{n + 2}{n}$

8. $\sum_{n=1}^{+\infty} (\sqrt{n + 1} - \sqrt{n})$

9. $\sum_{n=1}^{+\infty} (\cosh n - \sinh n)$

10. $\sum_{n=1}^{+\infty} \cos \pi n$

11. $\sum_{n=1}^{+\infty} \frac{1}{(2n + 1)(2n - 1)}$

12. $\sum_{n=1}^{+\infty} \frac{1}{(3n + 1)(3n - 2)}$

In Exercises 13–18 obtain the series having the given nth partial sum.

13. $\dfrac{n}{2n + 1}$

14. $\dfrac{n + 1}{n + 2}$

15. $1 - \dfrac{1}{3^n}$

16. n^2

17. $\dfrac{n}{(n + 1)!}$

18. $\ln\left(\dfrac{n + 2}{n + 1}\right)$

19. Prove that the sequence of partial sums for the series (14.8) is $\left\{\dfrac{n}{n + 1}\right\}$.

20. Obtain equation (14.17) if either limit exists.

21. Prove Theorem 14.2.7.

22. Is the sum of two divergent series always divergent? Why or why not?

23. Prove: If $\Sigma_{n=1}^{+\infty} a_n$ converges and S_n is given by (14.6), then for every $\epsilon > 0$ there is an $M > 0$ such that if $n > M$ and $m > M$, then $|S_m - S_n| < \epsilon$.

14.3   Convergence Tests for Positive Series

It is usually difficult to obtain a simple expression for the nth term of the sequence of partial sums of a series, and therefore in practice the convergence or divergence of the series is seldom determined directly from Definition 14.2.2. Instead, the question of convergence is usually answered by some test that is applicable to the series under consideration. As we will see, however, these tests will not give the sums of convergent series.

Most of the tests are applicable to *positive series*—that is, series in which each term is positive. In this section we consider three such tests. The first is one where the convergence or divergence of a series is determined by the evaluation of a corresponding improper integral.

14.3.1   Theorem (Integral Test)

Suppose f is a function that is continuous, decreasing, and positive-valued on the interval $[1, +\infty)$.

> *(a) If the improper integral $\int_1^{+\infty} f(x)\, dx$ exists, then the series $\Sigma_{n=1}^{+\infty} f(n)$ converges.*
>
> *(b) If $\int_1^{+\infty} f(x)\, dx = +\infty$, then $\Sigma_{n=1}^{+\infty} f(n)$ diverges.*

PROOF   Since f is continuous on $[1, n]$, n being any positive integer, $\int_1^n f(x)\, dx$ certainly exists. Also, since f is decreasing on $[1, +\infty)$, for every positive integer i we have

$$f(i + 1) \le f(x) \le f(i) \quad \text{for } x \in [i, i + 1].$$

Then the sums $f(2) + f(3) + \cdots + f(n)$ and $f(1) + f(2) + \cdots + f(n-1)$ are, respectively, the lower sum and upper sum for f associated with the partition $\{1, 2, \ldots, n\}$. The sum of the areas of the inscribed rectangles, which are shaded in Figure 14–5, is this lower sum.

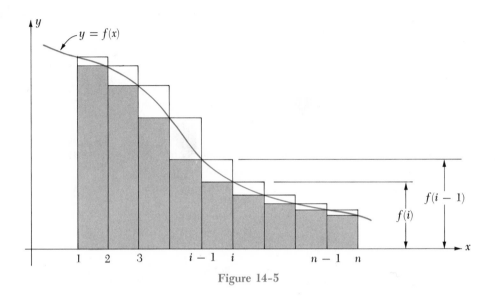

Figure 14-5

Thus by Theorem 7.5.4 and Definition 7.4.1

$$f(2) + f(3) + \cdots + f(n) \leq \int_1^n f(x)\, dx \leq f(1) + f(2) + \cdots + f(n-1). \quad (14.18)$$

The sequence $\{\int_1^n f(x)\, dx\}$ is increasing since $f(x) > 0$ on $[1, +\infty)$.

If $\int_1^{+\infty} f(x)\, dx$ exists, then by Theorem 14.1.4(a), $\lim_{n \to +\infty} \int_1^n f(x)\, dx = \int_1^{+\infty} f(x)\, dx$. Hence from the left-hand inequality in (14.18) and Theorem 14.1.8(a),

$$f(1) + f(2) + \cdots + f(n) \leq f(1) + \int_1^n f(x)\, dx \leq f(1) + \int_1^{+\infty} f(x)\, dx.$$

Since $\{f(1) + f(2) + \cdots + f(n)\}$ is also an increasing sequence, it too converges by Theorem 14.1.8(a). In other words, $\Sigma_{n=1}^{+\infty} f(n)$ converges and therefore part (a) of the theorem is proved.

If $\int_1^{+\infty} f(x)\, dx = +\infty$, then $\{\int_1^n f(x)\, dx\}$ diverges by Theorem 14.1.4(b). Since $\{\int_1^n f(x)\, dx\}$ is increasing, it must be unbounded for otherwise it would converge by Theorem 14.1.8(a). Hence from (14.18) and Theorem 14.1.3 the sequence $\{f(1) + f(2) + \cdots + f(n-1)\}$ diverges, and so the series $\Sigma_{n=1}^{+\infty} f(n)$ diverges.

The integral test may be useful in testing a given positive series $\Sigma_{n=1}^{+\infty} f(n)$ if the function f satisfies the hypotheses of the test and the definite integral $\int_1^t f(x)\, dx$ (where $t > 1$) can be readily evaluated.

Example 1 Show that the p-series

$$\sum_{n=1}^{+\infty} \frac{1}{p} = 1 + \frac{1}{2^p} + \frac{1}{3^p} + \cdots$$

converges if $p > 1$ and diverges if $p \leq 1$.

SOLUTION The p-series is of the form $\sum_{n=1}^{+\infty} f(n)$, where $f(n) = 1/n^p$. The function f given by $f(x) = 1/x^p$ is continuous, decreasing, and positive-valued for $x > 0$. Applying the integral test for $p \neq -1$, we obtain

$$\int_1^{+\infty} \frac{1}{x^p}\, dx = \lim_{t \to +\infty} \int_1^t \frac{1}{x^p}\, dx = \lim_{t \to +\infty} \frac{x^{1-p}}{1-p}\Big|_1^t$$

$$= \lim_{t \to +\infty} \frac{1}{1-p}(t^{1-p} - 1)$$

$$= \begin{cases} \dfrac{-1}{1-p} & \text{if } p > 1 \\ +\infty & \text{if } p < 1 \end{cases}$$

Hence, by (a) of the integral test, the p-series converges when $p > 1$, and by (b) of the test the series diverges when $p < 1$. If $p = 1$,

$$\int_1^{+\infty} \frac{1}{x}\, dx = \lim_{t \to +\infty} \ln t = +\infty$$

and hence by (b) of the integral test the *harmonic series*

$$1 + \frac{1}{2} + \frac{1}{3} + \cdots + \frac{1}{n} + \cdots$$

is divergent.

Example 2 Does the series $\displaystyle\sum_{n=1}^{+\infty} \frac{1}{n^2 + 1}$ converge or diverge?

SOLUTION Since this series is of the form $\sum_{n=1}^{+\infty} f(n)$ where $f(n) = 1/(n^2 + 1)$, we let $f(x) = 1/(x^2 + 1)$ and note that f satisfies the hypotheses of the integral test. Then

$$\int_1^{+\infty} \frac{dx}{x^2 + 1} = \lim_{t \to +\infty} \tan^{-1} x \Big|_1^t = \lim_{t \to +\infty} \left(\tan^{-1} t - \frac{\pi}{4} \right)$$

$$= \frac{\pi}{2} - \frac{\pi}{4} = \frac{\pi}{4}.$$

Since $\int_1^{+\infty} dx/(x^2 + 1)$ exists, the series converges by (a) of the integral test.

We next prove two *comparison tests* by which a positive series may sometimes be proved to converge (or diverge) depending upon whether a known positive series converges (or diverges).

14.3.2   Theorem (First Comparison Test)

Suppose $\Sigma_{n=1}^{+\infty} a_n$ and $\Sigma_{n=1}^{+\infty} b_n$ are two series such that $0 \leq a_n \leq b_n$ for every n.

(a) *If $\Sigma_{n=1}^{+\infty} b_n$ converges, then $\Sigma_{n=1}^{+\infty} a_n$ converges. Also, if S and T are the sums of $\Sigma_{n=1}^{+\infty} a_n$ and $\Sigma_{n=1}^{+\infty} b_n$, respectively, then $S \leq T$.*
(b) *If $\Sigma_{n=1}^{+\infty} a_n$ diverges, then $\Sigma_{n=1}^{+\infty} b_n$ diverges.*

PROOF OF (a)   Let $S_n = \Sigma_{i=1}^{n} a_i$ and $T_n = \Sigma_{i=1}^{n} b_i$. Since the terms in the given series are non-negative, $\{S_n\}$ and $\{T_n\}$ are increasing sequences, and from our hypothesis

$$S_n \leq T_n \quad \text{for every } n. \tag{14.19}$$

Since $\Sigma_{n=1}^{+\infty} b_n$ converges, $\{T_n\}$ converges to some limit T and by Theorem 14.1.8(a)

$$T_n \leq T \quad \text{for every } n. \tag{14.20}$$

From (14.19) and (14.20) $\lim_{n \to +\infty} S_n$ exists by Theorem 14.1.8(a) and hence $\Sigma_{n=1}^{+\infty} a_n$ converges to a sum S where $S \leq T$.

PROOF OF (b)   If $\Sigma_{n=1}^{+\infty} b_n$ converged, then by Theorem 14.3.2(a) $\Sigma_{n=1}^{+\infty} a_n$ would converge, which is contrary to the hypothesis. Hence $\Sigma_{n=1}^{+\infty} b_n$ must diverge.

In summary, by the first comparison test

(a) *A positive series is convergent if it is term-by-term less than or equal to the corresponding terms of a convergent series.*
(b) *A series is divergent if it is term-by-term greater than or equal to the corresponding terms of a divergent positive series.*

In applications of the comparison tests a series is often compared with a *p*-series or a geometric series.

Example 3   Does the series $\displaystyle\sum_{n=1}^{+\infty} \frac{1}{2^n + n}$ converge or diverge?

SOLUTION   We note that $1/(2^n + n) < 1/2^n$ for every n and that $\Sigma_{n=1}^{+\infty} 1/2^n$ is a convergent geometric series. Thus by Theorem 14.3.2(a) the given series converges.

Example 4   Does the series $\displaystyle\sum_{n=1}^{+\infty} \frac{1}{3\sqrt{n} - 2}$ converge or diverge?

SOLUTION We note that $1/(3\sqrt{n} - 2) > 1/(3\sqrt{n})$ for every n. Now $\sum_{n=1}^{+\infty} 1/\sqrt{n}$ is a divergent p-series $(p = \frac{1}{2})$, and hence by Theorem 14.2.7 $\sum_{n=1}^{+\infty} 1/(3\sqrt{n})$ diverges. Thus by Theorem 14.3.2(b) the given series diverges.

14.3.3 Theorem (Second Comparison Test)

Suppose $\sum_{n=1}^{+\infty} a_n$ and $\sum_{n=1}^{+\infty} b_n$ are positive series and $\lim_{n\to+\infty} a_n/b_n = L$, where L is some number. Then:

(a) *If $\sum_{n=1}^{+\infty} b_n$ converges, $\sum_{n=1}^{+\infty} a_n$ converges.*
(b) *If $\sum_{n=1}^{+\infty} a_n$ diverges, $\sum_{n=1}^{+\infty} b_n$ diverges.*

PROOF OF (a) Since $\{a_n/b_n\}$ is convergent, by Theorem 14.1.3 there is a $B > 0$ such that

$$\frac{a_n}{b_n} < B \quad \text{for every } n.$$

Hence

$$a_n < Bb_n \quad \text{for every } n. \tag{14.21}$$

Since $\sum_{n=1}^{+\infty} b_n$ converges, $\sum_{n=1}^{+\infty} Bb_n$ converges by Theorem 14.2.7. Since $\sum_{n=1}^{+\infty} Bb_n$ is a convergent positive series and (14.21) holds, $\sum_{n=1}^{+\infty} a_n$ converges by the first comparison test.

PROOF OF (b) From (14.21) $b_n > (1/B)a_n$, and since $\sum_{n=1}^{+\infty} a_n$ diverges, $\sum_{n=1}^{+\infty} (1/B)a_n$ also diverges. Hence $\sum_{n=1}^{+\infty} b_n$ diverges by the first comparison test.

Example 5 Does the series $\displaystyle\sum_{n=1}^{+\infty} \frac{1}{n^2 - \sqrt{n} + 1}$ converge or diverge?

SOLUTION Since the degree of the denominator of the nth term of the given series is 2, we shall use $\sum_{n=1}^{+\infty} 1/n^2$ as the comparing series. This series is a convergent p-series. Since

$$\lim_{n\to+\infty} \frac{\dfrac{1}{n^2 - \sqrt{n} + 1}}{\dfrac{1}{n^2}} = \lim_{n\to+\infty} \frac{n^2}{n^2 - \sqrt{n} + 1} = \lim_{n\to+\infty} \frac{1}{1 - \dfrac{1}{n^{3/2}} + \dfrac{1}{n^2}} = 1,$$

we conclude that the series $\sum_{n=1}^{+\infty} 1/(n^2 - \sqrt{n} + 1)$ converges by (a) of the second comparison test. It will be observed that in applying this test, $1/(n^2 - \sqrt{n} + 1)$ plays the role of a_n and $1/n^2$ the role of b_n in the statement of the test.

Example 6 Does the series $\displaystyle\sum_{n=1}^{+\infty} \frac{2n}{3n^2 + 1}$ converge or diverge?

SOLUTION We note that the degree of the denominator of the nth term of the given series is 1 more than the degree of the numerator, just as in the nth term of the harmonic series. Since the harmonic series diverges and

$$\lim_{n \to +\infty} \frac{\dfrac{1}{n}}{\dfrac{2n}{3n^2 + 1}} = \lim_{n \to +\infty} \frac{3n^2 + 1}{2n^2} = \lim_{n \to +\infty} \frac{3 + \dfrac{1}{n^2}}{2} = \frac{3}{2},$$

the series $\sum_{n=1}^{+\infty} 2n/(3n^2 + 1)$ diverges by (b) of the second comparison test. In this case $a_n = 1/n$ and $b_n = 2n/(3n^2 + 1)$ in the statement of the test.

Exercise Set 14.3

In Exercises 1–22 determine if the given series converges or diverges using a convergence test introduced in this section.

1. $\displaystyle\sum_{n=1}^{+\infty} \frac{1}{n^2 + 3n}$

2. $\displaystyle\sum_{n=1}^{+\infty} \frac{1}{2n - 1}$

3. $\displaystyle\sum_{n=1}^{+\infty} \frac{1}{\sqrt{2n^2 + 1}}$

4. $\displaystyle\sum_{n=1}^{+\infty} \frac{1}{n\sqrt{n^2 + 1}}$

5. $\displaystyle\sum_{n=1}^{+\infty} \frac{1}{n \cdot 3^n}$

6. $\displaystyle\sum_{n=1}^{+\infty} \frac{1}{n^3 + 5n}$

7. $\displaystyle\sum_{n=1}^{+\infty} \frac{n}{4n^2 + 1}$

8. $\displaystyle\sum_{n=1}^{+\infty} \frac{1}{(n + 1)[\ln(n + 1)]^2}$

9. $\displaystyle\sum_{n=1}^{+\infty} \frac{n}{3^{n^2}}$

10. $\displaystyle\sum_{n=1}^{+\infty} \frac{n + 4}{n^3 + 1}$

11. $\displaystyle\sum_{n=1}^{+\infty} \frac{1}{(n + 2)(n + 3)}$

12. $\displaystyle\sum_{n=1}^{+\infty} \left(3^{-n} - \frac{1}{n}\right)$

13. $\displaystyle\sum_{n=1}^{+\infty} \frac{\sqrt{n}}{n^2 + 2}$

14. $\displaystyle\sum_{n=1}^{+\infty} \frac{1}{\sqrt{3n + 1}}$

15. $\displaystyle\sum_{n=1}^{+\infty} \frac{\ln n}{n}$

16. $\displaystyle\sum_{n=1}^{+\infty} \frac{\tan^{-1} 2n}{4n^2 + 1}$

17. $\displaystyle\sum_{n=1}^{+\infty} \frac{1}{\sqrt{2n^3 - 1}}$

18. $\displaystyle\sum_{n=1}^{+\infty} \frac{1}{n!}$

19. $\displaystyle\sum_{n=1}^{+\infty} \sin \frac{\pi}{2n}$

20. $\displaystyle\sum_{n=1}^{+\infty} \frac{n}{e^n}$

21. $\displaystyle\sum_{n=1}^{+\infty} \frac{1 \cdot 3 \cdots (2n-1)}{4 \cdot 8 \cdots 4n}$

22. $\displaystyle\sum_{n=1}^{+\infty} \frac{n!}{n^n}$

23. If f satisfies the main hypotheses of the integral test, (a) prove that

$$f(1) + \int_2^{n+1} f(x)\, dx \le \sum_{i=1}^n f(i) \le f(1) + \int_1^n f(x)\, dx.$$

(b) From the inequality obtained in part (a) find bounds for the 50th partial sum of the harmonic series.

14.4 Remainders; Alternating Series

If the series $\Sigma_{n=1}^{+\infty} a_n$ converges to S, then

$$S = \lim_{n \to +\infty} \sum_{i=1}^n a_i = \lim_{n \to +\infty} \left[\sum_{i=1}^k a_i + \sum_{i=k+1}^n a_i \right]$$

where k is any positive integer. Hence

$$\sum_{n=k+1}^{+\infty} a_n = \lim_{n \to +\infty} \sum_{i=k+1}^n a_i$$

$$= S - \sum_{i=1}^k a_i. \tag{14.22}$$

We therefore have the following definition for R_k, the *remainder after k terms* for a convergent series.

14.4.1 Definition

If the series $\displaystyle\sum_{n=1}^{+\infty} a_n$ converges, then

$$R_k = \sum_{n=k+1}^{+\infty} a_n.$$

From (14.22), R_k is the error that is introduced when a convergent series is approximated by its kth partial sum. The meaning of the notation for this remainder differs slightly from the meaning of the remainder, $R_k(x)$, which was used in Taylor's formula. It will be recalled from Section 13.2 that $R_k(x)$, which is the remainder when $f(x)$ is approximated by $P_k(x)$, is actually a remainder after $k+1$ terms. However, this slight difference in meaning should present no difficulty since we will never use R_k and $R_k(x)$ in the same discussion.

The next theorem gives a useful upper bound for R_k for a positive series $\Sigma_{n=1}^{+\infty} f(n)$, when f satisfies the hypotheses of the integral test.

14.4.2   Theorem

If f is continuous, decreasing, and positive-valued on the interval $[1, +\infty)$, and if $\int_1^{+\infty} f(x)\,dx$ exists, then for the corresponding series $\Sigma_{n=1}^{+\infty} f(n)$ and any positive integer k,

$$0 \le R_k \le \int_k^{+\infty} f(x)\,dx. \tag{14.23}$$

PROOF   We first note that since $\int_1^{+\infty} f(x)\,dx$ exists, the series $\Sigma_{n=1}^{+\infty} f(n)$ converges by the integral test and hence R_k exists for every k. If the positive integer $n > k$, then $f(k+1) + f(k+2) + \cdots + f(n)$ is a lower sum for f associated with the partition $\{k, k+1, \ldots, n\}$, and

$$0 \le f(k+1) + f(k+2) + \cdots + f(n) \le \int_k^n f(x)\,dx.$$

Now $\lim_{n \to +\infty} \int_k^n f(x)\,dx = \int_k^{+\infty} f(x)\,dx$ exists since $\int_1^{+\infty} f(x)\,dx$ exists. Then since $\{\int_k^n f(x)\,dx\}$ is increasing, by Theorem 14.1.8(a)

$$f(k+1) + f(k+2) + \cdots + f(n) \le \int_k^n f(x)\,dx \le \int_k^{+\infty} f(x)\,dx.$$

Also since $\{\Sigma_{i=k+1}^n f(i)\}$ is an increasing sequence, by a second use of Theorem 14.1.8(a), $\lim_{n \to +\infty} \Sigma_{i=k+1}^n f(i) = R_k$ exists and

$$0 \le \sum_{i=k+1}^n f(i) \le R_k \le \int_k^{+\infty} f(x)\,dx.$$

Thus the theorem is proved.

Example 1   If the series $\Sigma_{n=1}^{+\infty} 1/n^2$ is approximated by its 20th partial sum, $\Sigma_{n=1}^{20} 1/n^2$, find an upper bound for the error introduced using (14.23).

SOLUTION   From (14.23), $R_{20} \le \displaystyle\int_{20}^{+\infty} \frac{1}{x^2}\,dx = \frac{1}{20}.$

We now turn our attention from positive series to another type of series, the *alternating* series.

14.4.3   Definition

Suppose $\{a_n\}$ is a sequence of *positive* terms. The series

$$\sum_{n=1}^{+\infty} (-1)^{n+1} a_n = a_1 - a_2 + a_3 - a_4 + - \cdots$$

and

$$\sum_{n=1}^{+\infty} (-1)^n a_n = -a_1 + a_2 - a_3 + a_4 - + \cdots$$

are called *alternating series*.

An example of an alternating series is the *alternating harmonic series*

$$\sum_{n=1}^{+\infty} \frac{(-1)^{n+1}}{n} = 1 - \frac{1}{2} + \frac{1}{3} - \frac{1}{4} + - \cdots$$

The following theorem gives a convergence test for alternating series and a simple upper bound for the absolute error introduced when the sum of a convergent alternating series is approximated by its kth partial sum. It will be noted this absolute error is no greater than the absolute value of the $(k + 1)$st term in the series, the first term to be omitted from this partial sum.

14.4.4   Theorem

If the sequence of positive terms $\{a_n\}$ is decreasing and $\lim_{n \to +\infty} a_n = 0$, then

(a) *The series $\sum_{n=1}^{+\infty} (-1)^{n+1} a_n$ is convergent;*
(b) *The remainder R_k for this series satisfies*

$$|R_k| \le a_{k+1}. \tag{14.24}$$

PROOF OF (a)   Let S_n be the nth partial sum of the series. Then any odd-numbered partial sum of the series, S_{2n-1}, has the form

$$S_{2n-1} = a_1 - (a_2 - a_3) - (a_4 - a_5) - \cdots - (a_{2n-2} - a_{2n-1}).$$

Since $\{a_n\}$ is a decreasing sequence, each of the expressions in parentheses can only assume non-negative values. Thus $\{S_{2n-1}\}$ is a decreasing sequence. Also, since

$$S_{2n-1} = (a_1 - a_2) + (a_3 - a_4) + \cdots + (a_{2n-3} - a_{2n-2}) + a_{2n-1},$$

and $a_{2n-1} > 0$ for every n, the sequence $\{S_{2n-1}\}$ is bounded from below by 0. Hence by Theorem 14.1.8(b) there is a number $S \ge 0$ such that

$$\lim_{n \to +\infty} S_{2n-1} = S \tag{14.25}$$

where

$$S_{2n-1} \ge S \quad \text{for every } n. \tag{14.26}$$

On the other hand, any even-numbered partial sum of the series, S_{2n}, must satisfy

$$S_{2n} = S_{2n-1} - a_{2n}.$$

Hence because of (14.25) and since $\lim_{n\to+\infty} a_n = 0$,

$$\lim_{n\to+\infty} S_{2n} = \lim_{n\to+\infty} S_{2n-1} - \lim_{n\to+\infty} a_{2n}$$
$$= S - 0 = S. \tag{14.27}$$

From (14.25) and (14.27) for every $\epsilon > 0$ there is some $M > 0$ such that

$$|S_n - S| < \epsilon$$

whenever $n > M$ regardless of whether n is even or odd. Thus $\lim_{n\to+\infty} S_n = S$, and therefore the series converges to S.

PROOF OF (b) To obtain (14.24), we first note that since

$$S_{2n} = (a_1 - a_2) + (a_3 - a_4) + \cdots + (a_{2n-1} - a_{2n})$$

and since each of the expressions in parentheses can only assume non-negative values, the sequence $\{S_{2n}\}$ is increasing. Thus we can prove that

$$S_{2n} \leq S \quad \text{for every } n, \tag{14.28}$$

for otherwise there would be a positive integer N such that $S_{2N} > S$. Then $S_{2n} \geq S_{2N} > S$ for every $n \geq N$ and S could not be the limit of $\{S_{2n}\}$, a contradiction of (14.27). Hence from (14.26) and (14.28)

$$S_k \leq S \leq S_{k+1} \quad \text{for } k \text{ even} \tag{14.29}$$

and

$$S_k \geq S \geq S_{k+1} \quad \text{for } k \text{ odd.} \tag{14.30}$$

From either (14.29) or (14.30) we can obtain

$$|S - S_k| \leq |S_{k+1} - S_k|$$
$$\leq |(-1)^{k+1} a_{k+1}| = a_{k+1}.$$

Then (14.24) follows from Definition 14.4.1 and (14.22).

Example 2 Prove that the alternating harmonic series $\sum_{n=1}^{+\infty} (-1)^{n+1}/n$ converges. Find the absolute error introduced if the sum of this series is approximated by $1 - \frac{1}{2} + \frac{1}{3} - \frac{1}{4} + \frac{1}{5}$.

SOLUTION Here $1/n$ plays the role of a_n in the statement of Theorem 14.4.4. We note that $\lim_{n\to+\infty} 1/n = 0$; also, for every positive integer n, $1/n > 1/(n+1) > 0$. Thus the hypotheses of Theorem 14.4.4 are satisfied by the sequence $\{1/n\}$. Then by Theorem 14.4.4(a) the series $\sum_{n=1}^{+\infty} (-1)^{n+1}/n$ converges. By Theorem 14.4.4(b) for any positive integer k

$$|R_k| \leq \frac{1}{k+1}. \tag{14.31}$$

Since the given series is to be approximated by its 5th partial sum, we let $k = 5$ in (14.31) and obtain

$$|R_5| \leq \tfrac{1}{6}.$$

We leave as an exercise the proof that the theorem is true which is obtained by replacing $\Sigma_{n=1}^{+\infty}(-1)^{n+1}a_n$ by $\Sigma_{n=1}^{+\infty}(-1)^n a_n$ in the statement of Theorem 14.4.4.

Exercise Set 14.4

In Exercises 1–12 determine if the given series converges or diverges.

1. $\displaystyle\sum_{n=1}^{+\infty} \frac{(-1)^{n+1}}{2n-1}$

2. $\displaystyle\sum_{n=1}^{+\infty} \frac{(-1)^{n+1}}{(2n+1)^2}$

3. $\displaystyle\sum_{n=1}^{+\infty} \frac{(-1)^n}{\sqrt{n}}$

4. $\displaystyle\sum_{n=1}^{+\infty} \frac{(-1)^{n+1}}{1+\ln n}$

5. $\displaystyle\sum_{n=1}^{+\infty} \frac{(-1)^{n+1}n}{2n-1}$

6. $\displaystyle\sum_{n=1}^{+\infty} (-1)^{n+1}\tanh n$

7. $\displaystyle\sum_{n=1}^{+\infty} \frac{(-1)^{n+1}n!}{3\cdot 5\cdots(2n+1)}$

8. $\displaystyle\sum_{n=1}^{+\infty} \frac{(-1)^n n!}{n^n}$

9. $\displaystyle\sum_{n=1}^{+\infty} (-1)^n ne^{-n^2}$

10. $\displaystyle\sum_{n=1}^{+\infty} \frac{(-1)^{n+1}\sqrt{n}}{3n+1}$

11. $\displaystyle\sum_{n=1}^{+\infty} \frac{(-1)^{n+1}\ln n}{n^2}$

12. $\displaystyle\sum_{n=1}^{+\infty} \frac{(-1)^{n+1}\ln(n+1)}{n}$

In Exercises 13–16 approximate the sum of the series correct to the third decimal place using a partial sum S_k for which $|R_k| < 0.0005$.

13. $\displaystyle\sum_{n=1}^{+\infty} \frac{(-1)^{n+1}}{n!}$

14. $\displaystyle\sum_{n=1}^{+\infty} \frac{(-1)^{n+1}}{3^{2n-2}}$

15. $\displaystyle\sum_{n=1}^{+\infty} \frac{n}{2^{n^2}}$

16. $\displaystyle\sum_{n=1}^{+\infty} \frac{1}{(7n+1)^3}$

17. Prove the theorem mentioned in the remarks following Example 2. HINT: Use Theorem 14.4.4.

14.5 Absolute Convergence; Ratio Test

The first theorem of this section is useful when investigating series having some negative terms.

14.5.1 Theorem

If the series $\Sigma_{n=1}^{+\infty}|a_n|$ is convergent, then $\Sigma_{n=1}^{+\infty}a_n$ is convergent, and

$$\left|\sum_{n=1}^{+\infty} a_n\right| \leq \sum_{n=1}^{+\infty} |a_n|. \tag{14.32}$$

PROOF We recall from Theorem 1.5.6 that for every positive integer i, $-|a_i| \leq a_i \leq |a_i|$. Hence if $|a_i|$ is added to each member of this inequality, we have

$$0 \leq a_i + |a_i| \leq 2|a_i| \quad \text{for every } i. \tag{14.33}$$

Since $\Sigma_{n=1}^{+\infty}|a_n|$ converges to a sum, which we shall denote by T, the series $\Sigma_{n=1}^{+\infty}2|a_n|$ converges to $2T$ by Theorem 14.2.7. Then from (14.33), using the first comparison test (Theorem 14.3.2), the series $\Sigma_{n=1}^{+\infty}(a_n + |a_n|)$ converges to a sum, which we denote by U, and also

$$0 \leq U \leq 2T. \tag{14.34}$$

Since $\Sigma_{n=1}^{+\infty}(a_n + |a_n|)$ converges to U and $\Sigma_{n=1}^{+\infty}|a_n|$ converges to T, $\Sigma_{n=1}^{+\infty}a_n$ converges by Theorem 14.2.6 to $U - T$. By adding $-T$ to each member of (14.34), we have

$$-T \leq U - T \leq T,$$

and hence

$$|U - T| \leq T. \tag{14.35}$$

Since $U - T$ is the sum of the series $\Sigma_{n=1}^{+\infty}a_n$, (14.32) is obtained as a restatement of (14.35).

To summarize, by Theorem 14.5.1 a series having some negative terms will converge if the series formed from the absolute values of the terms in the given series converges.

Example 1 Determine whether the series

$$\sum_{n=1}^{+\infty} \frac{\sin n\pi/4}{n^2} = \frac{\sqrt{2}}{2} + \frac{1}{4} + \frac{\sqrt{2}}{18} - \frac{\sqrt{2}}{50} - \frac{1}{36} - \frac{\sqrt{2}}{98} + \cdots$$

is convergent or divergent.

SOLUTION We consider the series $\displaystyle\sum_{n=1}^{+\infty}\left|\frac{\sin n\pi/4}{n^2}\right|$. Since $|\sin n\pi/4| \leq 1$ for every n,

$$\left|\frac{\sin n\pi/4}{n^2}\right| \leq \frac{1}{n^2}.$$

The series $\Sigma_{n=1}^{+\infty}1/n^2$ converges since it is a convergent p-series. Hence

$$\sum_{n=1}^{+\infty} \left| \frac{\sin n\pi/4}{n^2} \right|$$

converges by the first comparison test and so $\displaystyle\sum_{n=1}^{+\infty} \frac{\sin n\pi/4}{n^2}$ converges by

Theorem 14.5.1.

We next define *absolute convergence.*

14.5.2 Definition

The series $\sum_{n=1}^{+\infty} a_n$ *converges absolutely* if and only if the series $\sum_{n=1}^{+\infty} |a_n|$ is convergent.

In view of Definition 14.5.2 we can summarize the result of Theorem 14.5.1:

A series is convergent if it is absolutely convergent.

An example of an absolutely convergent series is $\displaystyle\sum_{n=1}^{+\infty} \frac{\sin n\pi/4}{n^2}$, which was just

discussed. However, the alternating harmonic series $\sum_{n=1}^{+\infty} (-1)^{n+1}/n$, which is convergent (Example 2, Section 14.4), does not converge absolutely since $\sum_{n=1}^{+\infty} |(-1)^{n+1}/n| = \sum_{n=1}^{+\infty} 1/n$, which diverges.

14.5.3 Definition

The series $\sum_{n=1}^{+\infty} a_n$ *converges conditionally* if and only if $\sum_{n=1}^{+\infty} a_n$ converges, but $\sum_{n=1}^{+\infty} |a_n|$ diverges.

$\sum_{n=1}^{+\infty} (-1)^{n+1}/n$ is an example of a conditionally convergent series.

A frequently effective test for analyzing series is the *ratio test.* This convergence test, which can be applied to series with some negative terms, as well as positive series, is simple to use because it involves the ratio of successive terms in the series.

14.5.4 Theorem (Ratio Test)

Suppose that $\sum_{n=1}^{+\infty} a_n$ is a given series and that $\lim_{n \to +\infty} |a_{n+1}/a_n| = L$.

    (a) *If $0 \leq L < 1$, then $\sum_{n=1}^{+\infty} a_n$ converges absolutely.*
    (b) *If $L > 1$, then $\sum_{n=1}^{+\infty} a_n$ diverges.*

PROOF OF (a) Since $0 \le L < 1$ there is certainly a number b such that $L < b < 1$. Then by hypothesis for $\epsilon = b - L$, there is a positive integer M such that

$$\left| \left| \frac{a_{n+1}}{a_n} \right| - L \right| < b - L \quad \text{if } n \ge M. \tag{14.36}$$

For this condition to make sense it is required that $a_n \ne 0$ if $n \ge M$ and so we may assume this is the case. Hence from (14.36)

$$|a_{n+1}| < b|a_n| \quad \text{if } n \ge M. \tag{14.37}$$

In particular from (14.37)

$$|a_{M+1}| < b|a_M|$$
$$|a_{M+2}| < b|a_{M+1}| < b^2|a_M|$$
$$|a_{M+3}| < b|a_{M+2}| < b^3|a_M|$$
$$\cdots$$

The series $\sum_{n=M}^{+\infty} |a_M| b^{n-M}$ is a convergent geometric series since $0 < b < 1$ and hence $\sum_{n=M}^{+\infty} |a_n|$ converges by the first comparison test. Then $\sum_{n=1}^{+\infty} |a_n|$ converges by Theorem 14.2.5, and hence $\sum_{n=1}^{+\infty} a_n$ converges absolutely.

PROOF OF (b) Since $L > 1$ there is certainly a positive integer M such that

$$\left| \frac{a_{n+1}}{a_n} \right| > 1 \quad \text{if } n \ge M. \tag{14.38}$$

From (14.38) we have $a_n \ne 0$ if $n \ge M$ and also

$$|a_{M+1}| > |a_M|$$
$$|a_{M+2}| > |a_{M+1}| > |a_M|$$
$$|a_{M+3}| > |a_{M+2}| > |a_M|$$
$$\cdots$$

Thus

$$|a_n| > |a_M| \quad \text{if } n \ge M,$$

and hence $\{a_n\}$ cannot converge to 0. Therefore $\sum_{n=1}^{+\infty} a_n$ diverges by Theorem 14.2.3.

If $\lim_{n \to +\infty} |a_{n+1}/a_n| = 1$, then the series may converge or diverge, and so some other test for convergence must be used. For example, we know the p-series, $\sum_{n=1}^{+\infty} 1/n^p$, converges if $p > 1$ and diverges if $p \le 1$. However, for any p,

$$\lim_{n \to +\infty} \left| \frac{a_{n+1}}{a_n} \right| = \lim_{n \to +\infty} \frac{\dfrac{1}{(n+1)^p}}{\dfrac{1}{n^p}} = \lim_{n \to +\infty} \left(\frac{n}{n+1} \right)^p = 1^p = 1.$$

Note that if $\lim_{n \to +\infty} |a_{n+1}/a_n| = 1$, and there is some positive integer M such that

$|a_{n+1}/a_n| > 1$ when $n \geq M$, the series must diverge according to the proof in part (b).

Example 2 Using the ratio test, determine if the series

$$\sum_{n=1}^{+\infty} \frac{(-1)^{n+1}n^2}{3^{2n}}$$

converges absolutely or diverges.

SOLUTION If a_n denotes the nth term of the series, then

$$a_{n+1} = \frac{(-1)^{n+2}(n+1)^2}{3^{2(n+1)}}$$

and hence

$$\left| \frac{a_{n+1}}{a_n} \right| = \left| \frac{(-1)^{n+2}(n+1)^2}{3^{2(n+1)}} \cdot \frac{3^{2n}}{(-1)^{n+1}n^2} \right| = \frac{(n+1)^2}{3^2 n^2}.$$

Then

$$\lim_{n \to +\infty} \left| \frac{a_{n+1}}{a_n} \right| = \lim_{n \to +\infty} \frac{(n+1)^2}{3^2 n^2} = \frac{1}{9}$$

and by part (a) of the ratio test, the series converges absolutely.

Example 3 Using the ratio test, determine if the series

$$\sum_{n=1}^{+\infty} \frac{1 \cdot 3 \cdot 5 \cdots (2n-1)}{n!}$$

converges or diverges.

SOLUTION Letting a_n denote the nth term of the series, we have

$$a_{n+1} = \frac{1 \cdot 3 \cdot 5 \cdots (2n-1)(2n+1)}{(n+1)!},$$

and hence

$$\left| \frac{a_{n+1}}{a_n} \right| = \frac{1 \cdot 3 \cdot 5 \cdots (2n-1)(2n+1)}{(n+1)!} \cdot \frac{n!}{1 \cdot 3 \cdot 5 \cdots (2n-1)} = \frac{2n+1}{n+1}.$$

Then

$$\lim_{n \to +\infty} \left| \frac{a_{n+1}}{a_n} \right| = \lim_{n \to +\infty} \frac{2n+1}{n+1} = 2,$$

and by part (b) of the ratio test, the series diverges.

The convergence or divergence of most series of interest can be successfully determined using the convergence tests introduced thus far in Chapter 14. For the beginner, who may be uncertain about which test to use, the following course of action is suggested in studying a series $\sum_{n=1}^{+\infty} a_n$:

1. First, obtain $\lim_{n\to+\infty} a_n$, if possible. If $\{a_n\}$ does not converge to 0, then the series diverges. If $\lim_{n\to+\infty} a_n = 0$, go to step 2.

2. Apply the ratio test. If $0 \le L < 1$, then the series converges absolutely. If $L > 1$, then the series diverges. If $L = 1$ and the series is positive, go to step 3. If $L = 1$ and the series has at least some negative terms, go to step 7.

3. Apply the integral test. If $\int_1^{+\infty} f(x)\,dx$ exists, then the series converges. If $\int_1^{+\infty} f(x)\,dx = +\infty$, the series diverges. If neither of these outcomes occurs, go to step 4.

4. Apply the second comparison test. If this test cannot be successfully used, go to step 5.

5. Apply the first comparison test. If this test cannot be successfully used, go to step 6.

6. Try to obtain a formula for the nth partial sum, S_n. Then obtain $\lim_{n\to+\infty} S_n$, if possible.

7. Apply Theorem 14.4.4, if possible. If the signs of the terms are $- + - + - +$, and so on, apply Theorem 14.4.4 to the series $\sum_{n=1}^{+\infty} - a_n$. If this procedure is unsuccessful, apply steps 3, 4, 5, and 6, if necessary, to the series $\sum_{n=1}^{+\infty} |a_n|$.

Exercise Set 14.5

In Exercises 1–29 determine if the series converges or diverges using any theory that has been introduced in this chapter.

1. $\displaystyle\sum_{n=1}^{+\infty} n\left(\frac{1}{2}\right)^n$

2. $\displaystyle\sum_{n=1}^{+\infty} \frac{n^2 + 1}{3^n}$

3. $\displaystyle\sum_{n=1}^{+\infty} \frac{3^{n+1}}{2^{2n}}$

4. $\displaystyle\sum_{n=1}^{+\infty} \frac{5^{n+1}}{n3^n}$

5. $\displaystyle\sum_{n=1}^{+\infty} \frac{2\cdot 4\cdots 2n}{5^n}$

6. $\displaystyle\sum_{n=1}^{+\infty} \frac{3^n}{n!}$

7. $\displaystyle\sum_{n=1}^{+\infty} \frac{3^{2n}}{n!}$

8. $\displaystyle\sum_{n=1}^{+\infty} n^2\left(\frac{2}{5}\right)^n$

9. $\displaystyle\sum_{n=1}^{+\infty} \frac{2^n - 1}{2^n}$

10. $\displaystyle\sum_{n=1}^{+\infty} \frac{n^3}{n^4 + 1}$

11. $\displaystyle\sum_{n=1}^{+\infty} (-1)^{n+1} \frac{n}{2n^2 + 3}$

12. $\displaystyle\sum_{n=1}^{+\infty} \frac{n + 2}{n!}$

13. $\displaystyle\sum_{n=1}^{+\infty} \frac{1}{2^n - n}$

14. $\displaystyle\sum_{n=1}^{+\infty} \frac{n + 1}{2n^2}$

15. $\displaystyle\sum_{n=1}^{+\infty} \frac{1}{n\sqrt{n+1}}$

16. $\displaystyle\sum_{n=1}^{+\infty} \frac{(-1)^{n+1}\tan^{-1}n}{n}$

17. $\displaystyle\sum_{n=1}^{+\infty} \frac{\sin n}{n^2}$

18. $\displaystyle\sum_{n=1}^{+\infty} \left(\frac{n}{n+1}\right)^n$

19. $\displaystyle\sum_{n=1}^{+\infty} \frac{2^{n^2}}{n!}$

20. $\displaystyle\sum_{n=1}^{+\infty} \frac{(n+1)!}{3\cdot6\cdot9\cdots3n}$

21. $\displaystyle\sum_{n=1}^{+\infty} \frac{1\cdot3\cdot5\cdots(2n-1)}{2\cdot4\cdot6\cdots2n}$

22. $\displaystyle\sum_{n=1}^{+\infty} \frac{(-1)^{n+1}2\cdot4\cdot6\cdots2n}{3\cdot5\cdot7\cdots(2n+1)}$

23. $\displaystyle\sum_{n=1}^{+\infty} \frac{1\cdot4\cdot7\cdots(3n-2)}{4\cdot8\cdot12\cdots4n}$

24. $\displaystyle\sum_{n=1}^{+\infty} \frac{2^n n!}{(2n)!}$

25. $\displaystyle\sum_{n=1}^{+\infty} \frac{n!}{n^n}$

26. $\displaystyle\sum_{n=1}^{+\infty} \left[\frac{\ln(n+1)}{n+1} - \frac{\ln n}{n}\right]$

27. $\displaystyle\sum_{n=1}^{+\infty} \frac{5+(-1)^{n+1}}{3^n}$

28. $\dfrac{1}{2} - \dfrac{2}{4} - \dfrac{3}{8} + \dfrac{4}{16} - \dfrac{5}{32} - \dfrac{6}{64} + - - \cdots$

29. $\dfrac{1}{3} + \dfrac{4}{9} - \dfrac{9}{27} - \dfrac{16}{81} + \dfrac{25}{243} + \dfrac{36}{729} - - - + + \cdots$

30. Prove: If $|a_{n+1}/a_n| \le b < 1$ when $n \ge k$, where k is a positive integer, then the remainder R_k for the series $\sum_{n=1}^{+\infty} a_n$ satisfies the inequality

$$|R_k| \le \frac{b|a_k|}{1-b}.$$

31. Prove the *Root Test:* Suppose that $\sum_{n=1}^{+\infty} a_n$ is a given series and let $L = \lim_{n\to+\infty} \sqrt[n]{|a_n|}$ if this limit exists.

    (a) If $0 \le L < 1$, then $\sum_{n=1}^{+\infty} a_n$ converges absolutely.
    (b) If $L > 1$, then $\sum_{n=1}^{+\infty} a_n$ diverges.

Hint: If $0 \le L < 1$, there is a b such that $L < b < 1$. Then by hypothesis there is a positive integer M such that

$$\left|\sqrt[n]{|a_n|} - L\right| < b - L \quad \text{if } n \ge M.$$

Then

$$\sqrt[n]{|a_n|} < b \quad \text{if } n \ge M.$$

14.6   Power Series

An infinite series of the form

$$\sum_{n=0}^{+\infty} c_n(x - a)^n = c_0 + c_1(x - a) + c_2(x - a)^2 + c_3(x - a)^3 + \cdots \quad (14.39)$$

where $a, c_0, c_1, c_2, \ldots, c_n, \ldots$ are given real numbers is called a *power series in* $x - a$. If $a = 0$, then (14.39) has the form $\sum_{n=0}^{+\infty} c_n x^n$ and is called a power series in x. It will be noted that in (14.39) the sum starts with $n = 0$ instead of $n = 1$ as was usually the case with series of constant terms in Sections 14.2–14.5. Therefore $c_n(x - a)^n$ is the $(n + 1)$st term and not the nth term of the power series (14.39).

When x assumes a particular value in (14.39), the resulting series may converge or diverge. Thus, it will be our purpose here to find the set of all numbers x for which a given power series converges. First of all, it should be apparent that if $x = a$ in (14.39), then the resulting series $c_0 + \sum_{n=1}^{+\infty} 0$ converges to c_0. Thus, *any power series in* $x - a$ *converges (absolutely) if* $x = a$. To extend our knowledge of the convergence of power series, we prove the following theorem.

14.6.1   Theorem

If the power series $\sum_{n=0}^{+\infty} c_n(x - a)^n$ converges when $x = x_1 \neq a$ and if $|x_2 - a| < |x_1 - a|$, then the series converges absolutely when $x = x_2$.

PROOF   If $\sum_{n=0}^{+\infty} c_n(x_1 - a)^n$ converges, then $\lim_{n \to +\infty} c_n(x_1 - a)^n = 0$. Since the sequence $\{c_n(x_1 - a)^n\}$ converges, the sequence is bounded; that is, there is an $M > 0$ such that

$$|c_n(x_1 - a)^n| \leq M \quad \text{for every } n.$$

Then

$$|c_n(x_2 - a)^n| = |c_n(x_1 - a)^n| \cdot \left| \frac{(x_2 - a)^n}{(x_1 - a)^n} \right| \leq M \left| \frac{x_2 - a}{x_1 - a} \right|^n. \quad (14.40)$$

If

$$b = \left| \frac{x_2 - a}{x_1 - a} \right|, \quad (14.41)$$

then by hypothesis, $b < 1$ and from (14.40)

$$|c_n(x_2 - a)^n| \leq M b^n.$$

Since $|b| < 1$, the series $\sum_{n=0}^{+\infty} M b^n$ is a convergent geometric series. Hence by the first comparison test the series $\sum_{n=0}^{+\infty} |c_n(x_2 - a)^n|$ converges, or saying the same thing, the series $\sum_{n=0}^{+\infty} c_n(x_2 - a)^n$ converges absolutely, thereby proving the theorem.

By Theorem 14.6.1, if the power series (14.39) converges for $x = x_1 \neq a$, then it converges absolutely for every x such that $|x - a| < |x_1 - a|$—that is, for every number x that is nearer to a then x_1.

It is possible that a power series may converge for every x. However, if the series diverges for some number x, then the following theorem is applicable.

14.6.2   Theorem

If the power series (14.39) diverges when $x = x_1 \neq a$ and if $|x_2 - a| > |x_1 - a|$, then the series diverges when $x = x_2$.

PROOF   Suppose the series converged when $x = x_2$. Then by Theorem 14.6.1 the series would converge absolutely when $x = x_1$, which is contrary to the hypothesis. Therefore the series diverges when $x = x_2$.

Because of Theorem 14.6.2 if the power series (14.39) diverges for $x = x_1 \neq a$, then the series must also diverge for every x that is farther from a than x_1. From these remarks and those following Theorem 14.6.1 it seems reasonable to suppose when a power series converges for some, but not all numbers x, that there must be some $r \geq 0$ as defined in the following statement.

14.6.3   Definition

Suppose $r \geq 0$ and the power series (14.39) converges absolutely if $|x - a| = r$ or $|x - a| < r$ and diverges if $|x - a| > r$. Then r is called the *radius of convergence* and when $r > 0$, the interval with endpoints $a - r$ and $a + r$ on which the power series converges is called the *interval of convergence* of the series (Figure 14–6). If (14.39) converges for every x, its interval of convergence is the set R.

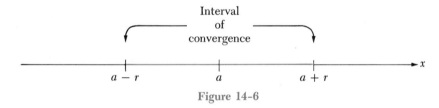

Figure 14–6

From Definition 14.6.3 a power series converges absolutely at any number in the interior of its interval of convergence, and diverges at any number outside its interval of convergence. The series may also converge at both, at neither, or at one of the endpoints of the interval of convergence.

As shown in the following examples, the ratio test can be used to obtain the interior of the interval of convergence. At endpoints of the interval of convergence, the ratio test will fail, and therefore other tests must be employed to investigate possible convergence at these numbers.

Example 1   Find the interval of convergence of the power series

$$\sum_{n=0}^{+\infty} \frac{(x - 2)^n}{n + 1}.$$

SOLUTION   Using the ratio test, we form

$$\left|\frac{a_{n+1}}{a_n}\right| = \left|\frac{(x-2)^{n+1}}{n+2}\frac{n+1}{(x-2)^n}\right| = \frac{n+1}{n+2}|x-2|$$

and obtain

$$\lim_{n\to+\infty}\left|\frac{a_{n+1}}{a_n}\right| = |x-2|.$$

Then by the ratio test the series converges absolutely if $|x-2| < 1$—that is, if $1 < x < 3$. Also by the same test, the series diverges if $|x-2| > 1$—that is, if $x > 3$ or $x < 1$.

      If $x = 3$, the series becomes $\sum_{n=0}^{+\infty}1/(n+1)$, which diverges by comparison with the harmonic series, using the second comparison test. If $x = 1$ the given series becomes $\sum_{n=0}^{+\infty}(-1)^n/(n+1)$, which converges by Theorem 14.4.4(a). The interval of convergence for the given series is therefore $[1, 3)$ (Figure 14–7).

Figure 14-7

Example 2   Find the interval of convergence of the power series

$$\sum_{n=0}^{+\infty}\frac{(x-a)^{n+1}}{(n+1)!}.$$

SOLUTION   Here

$$\left|\frac{a_{n+1}}{a_n}\right| = \left|\frac{(x-a)^{n+2}}{(n+2)!}\cdot\frac{(n+1)!}{(x-a)^{n+1}}\right| = \frac{|x-a|}{n+2},$$

and hence

$$\lim_{n\to+\infty}\left|\frac{a_{n+1}}{a_n}\right| = \lim_{n\to+\infty}\frac{|x-a|}{n+2} = 0 \quad\text{for every } x.$$

Hence the interval of convergence is the set R.

Example 3   Find the interval of convergence of the power series $\sum_{n=1}^{+\infty}(-1)^{n+1}2^n(x+1)^{2n}$.

SOLUTION   Here

$$\left|\frac{a_{n+1}}{a_n}\right| = \left|\frac{(-1)^{n+2}2^{n+1}(x+1)^{2n+2}}{(-1)^{n+1}2^n(x+1)^{2n}}\right| = 2|x+1|^2$$

and hence

$$\lim_{n\to+\infty}\left|\frac{a_{n+1}}{a_n}\right| = 2|x+1|^2.$$

Then by the ratio test the given series converges absolutely if $2|x + 1|^2 < 1$. This inequality is equivalent to each of the following inequalities:

$$|x + 1| < \frac{1}{\sqrt{2}}$$

$$-\frac{1}{\sqrt{2}} < x + 1 < \frac{1}{\sqrt{2}}$$

$$-1 - \frac{1}{\sqrt{2}} < x < -1 + \frac{1}{\sqrt{2}}.$$

Also by the ratio test, the series diverges if $2|x + 1|^2 > 1$, that is, if $x > -1 + (1/\sqrt{2})$ or $x < -1 - (1/\sqrt{2})$.

If $x = -1 \pm (1/\sqrt{2})$, then the series becomes $\sum_{n=1}^{+\infty}(-1)^{n+1}$, which diverges. Thus, the interval of convergence of the given series is $\left(-1 - \frac{1}{\sqrt{2}}, -1 + \frac{1}{\sqrt{2}}\right)$.

We conclude this section with a useful theorem about convergent power series.

14.6.4   Theorem

If $\sum_{n=0}^{+\infty} c_n(x - a)^n$ converges for every x such that $|x - a| < r$, the series converges absolutely for such numbers x.

PROOF   If x satisfies $|x - a| < r$ there is certainly a $p > a$ such that $|x - a| < p - a < r$, and hence $\sum_{n=0}^{+\infty} c_n(p - a)^n$ converges. Hence by Theorem 14.6.1 $\sum_{n=0}^{+\infty} c_n(x - a)^n$ converges absolutely.

Exercise Set 14.6

In Exercises 1–20 find the interval of convergence of the given power series.

1. $\displaystyle\sum_{n=0}^{+\infty} nx^{n+1}$

2. $\displaystyle\sum_{n=0}^{+\infty} n!x^n$

3. $\displaystyle\sum_{n=1}^{+\infty} \frac{(x + 2)^n}{2^n}$

4. $\displaystyle\sum_{n=1}^{+\infty} \frac{x^{n+1}}{n^2}$

5. $\displaystyle\sum_{n=0}^{+\infty} \frac{(-1)^n x^{2n+1}}{n + 1}$

6. $\displaystyle\sum_{n=0}^{+\infty} \frac{(x + 1)^n}{\sqrt{n + 1}}$

7. $\displaystyle\sum_{n=1}^{+\infty} \frac{(x - 1)^n}{2^n n}$

8. $\displaystyle\sum_{n=1}^{+\infty} \frac{(x - 2)^n}{n(n + 1)}$

9. $\displaystyle\sum_{n=0}^{+\infty} \frac{3^n(x-2)^n}{n+1}$

10. $\displaystyle\sum_{n=2}^{+\infty} \frac{x^n}{\ln n}$

11. $\displaystyle\sum_{n=0}^{+\infty} \frac{(-1)^n(x-1)^n}{(2n+1)2^n}$

12. $\displaystyle\sum_{n=1}^{+\infty} ne^n(x-3)^n$

13. $\displaystyle\sum_{n=0}^{+\infty} \frac{(-1)^{n+1}(x+1)^n}{3^{2n}}$

14. $\displaystyle\sum_{n=1}^{+\infty} \frac{3^n(x+2)^n}{n^2}$

15. $\displaystyle\sum_{n=0}^{+\infty} 2^n(x-1)^{2n}$

16. $\displaystyle\sum_{n=0}^{+\infty} \frac{(-1)^{n+1}(x-2)^{2n}}{4^{n+1}}$

17. $\displaystyle\sum_{n=0}^{+\infty} \frac{1\cdot 3\cdot 5\cdots(2n-1)}{2^n n!}x^n$

18. $\displaystyle\sum_{n=0}^{+\infty} \frac{(-1)^n 3\cdot 5\cdot 7\cdots(2n+1)}{2^n n!}x^{2n+1}$

19. $\displaystyle\sum_{n=0}^{+\infty} \frac{n!x^{2n}}{(2n)!}$

20. $\displaystyle\sum_{n=1}^{+\infty} \frac{n!}{n^n}x^n$

14.7 Taylor Series

Suppose a function f is continuous and has continuous derivatives through order n on a closed interval I with endpoints a and x, and $f^{(n+1)}$ exists in the interior of I. Then by Taylor's formula, (13.26),

$$f(x) = P_n(x) + R_n(x),\tag{14.42}$$

where the Taylor polynomial $P_n(x)$ is given by

$$P_n(x) = \sum_{i=0}^{n} \frac{f^{(i)}(a)(x-a)^i}{i!}\tag{14.43}$$

and the remainder $R_n(x)$ by

$$R_n(x) = \frac{f^{(n+1)}(c)(x-a)^{n+1}}{(n+1)!},\tag{14.44}$$

c being some number between a and x.

In the next theorem we state the conditions which guarantee that

$$f(x) = \lim_{n\to+\infty} P_n(x)$$

thereby justifying the use of $P_n(x)$ as an approximation for $f(x)$ when n is sufficiently large. If these conditions are fulfilled, then $f(x)$ is the sum of a convergent power series in $x-a$.

14.7.1   Theorem

Suppose f has derivatives of all orders on an interval I containing a, and $x \in I$.
Then

$$f(x) = f(a) + \frac{f'(a)(x-a)}{1!} + \frac{f''(a)(x-a)^2}{2!} + \cdots + \frac{f^{(n)}(a)(x-a)^n}{n!} + \cdots$$

$$(14.45)$$

if and only if

$$\lim_{n \to +\infty} R_n(x) = 0. \qquad (14.46)$$

PROOF   By hypothesis the expressions (14.43) and (14.44) exist for every
n. If $\lim_{n \to +\infty} R_n(x) = 0$, then from (14.42)

$$\lim_{n \to +\infty} f(x) = \lim_{n \to +\infty} [P_n(x) + R_n(x)]$$

$$f(x) = \lim_{n \to +\infty} P_n(x) + \lim_{n \to +\infty} R_n(x)$$

$$= \lim_{n \to +\infty} P_n(x)$$

and (14.45) is obtained. Conversely, if (14.45) holds, then $f(x) = \lim_{n \to +\infty} P_n(x)$ and
hence from (14.42)

$$\lim_{n \to +\infty} R_n(x) = \lim_{n \to +\infty} [f(x) - P_n(x)] = f(x) - f(x) = 0.$$

The power series in the right member of (14.45) is called a *Taylor series*
of $f(x)$ at a. In the special case where $a = 0$ the Taylor series becomes

$$f(0) + \frac{f'(0)x}{1!} + \frac{f''(0)x^2}{2!} + \cdots + \frac{f^{(n)}(0)x^n}{n!} + \cdots$$

and is called a *Maclaurin's series*.

If $f(x)$ is the sum of a power series, we say that $f(x)$ *is represented by*
the series. In Examples 1 and 2 below a Taylor series is obtained for a given $f(x)$
and it is proved using Theorem 14.7.1 that the Taylor series actually represents
the $f(x)$.

Example 1   Find the Maclaurin series for e^x. For what numbers x is
e^x the sum of this series?

SOLUTION   If $f(x) = e^x$, then $f^{(n)}(x) = e^x$ for $n = 1, 2, 3, \ldots$. We let
$a = 0$ and hence obtain $f(0) = 1$ and $f^{(n)}(0) = 1$ for $n = 1, 2, 3, \ldots$. The required
Taylor series for e^x which is of the form

$$f(0) + \frac{f'(0)x}{1!} + \frac{f''(0)x^2}{2!} + \cdots + \frac{f^{(n)}(0)x^n}{n!} + \cdots$$

is therefore

$$1 + \frac{x}{1!} + \frac{x^2}{2!} + \cdots + \frac{x^n}{n!} + \cdots. \tag{14.47}$$

From (14.44)

$$R_n(x) = \frac{e^c x^{n+1}}{(n+1)!} \quad \text{where } c \text{ is between } 0 \text{ and } x. \tag{14.48}$$

From Example 2 of Section 14.6 $\Sigma_{n=0}^{+\infty} x^{n+1}/(n+1)!$ converges for every x. Hence by Theorem 14.2.3 $\lim_{n\to+\infty} x^{n+1}/(n+1)! = 0$. Thus, since $e^c < e^{|x|}$, which is independent of n, $\lim_{n\to+\infty} R_n(x) = 0$ for every x. Therefore, for every x

$$e^x = 1 + \frac{x}{1!} + \frac{x^2}{2!} + \cdots + \frac{x^n}{n!} + \cdots. \tag{14.49}$$

In particular, we can obtain from (14.49) a power series representing $e^{-x^2/2}$. If x is replaced by $-x^2/2$ in (14.49), then

$$e^{-x^2/2} = 1 + \frac{\left(-\frac{x^2}{2}\right)}{1!} + \frac{\left(-\frac{x^2}{2}\right)^2}{2!} + \cdots + \frac{\left(-\frac{x^2}{2}\right)^n}{n!} + \cdots$$

$$= 1 - \frac{x^2}{1!2} + \frac{x^4}{2!4} - + \cdots + \frac{(-1)^n x^{2n}}{n!2^n} + \cdots.$$

Example 2 Find the Taylor series for $\sin x$ at $\frac{\pi}{6}$. For what numbers x is $\sin x$ the sum of this series?

SOLUTION   If $f(x) = \sin x$, then $f(\frac{\pi}{6}) = \frac{1}{2}$, and also:

$$f'(x) = \cos x \qquad f'\left(\frac{\pi}{6}\right) = \frac{\sqrt{3}}{2}$$

$$f''(x) = -\sin x \qquad f''\left(\frac{\pi}{6}\right) = -\frac{1}{2}$$

$$f'''(x) = -\cos x \qquad f'''\left(\frac{\pi}{6}\right) = -\frac{\sqrt{3}}{2}$$

The required Taylor series for $\sin x$ is therefore

$$\frac{1}{2} + \frac{\sqrt{3}(x - \frac{\pi}{6})}{2 \cdot 1!} - \frac{(x - \frac{\pi}{6})^2}{2 \cdot 2!} - \frac{\sqrt{3}(x - \frac{\pi}{6})^3}{2 \cdot 3!} + \cdots.$$

From (14.44)

$$|R_n(x)| = \frac{|f^{(n+1)}(c)||(x - \frac{\pi}{6})^{n+1}|}{(n+1)!}$$

and since $|f^{(n+1)}(c)|$ is either $|\sin c|$ or $|\cos c|$, $|f^{(n+1)}(c)| \leq 1$. Hence

$$|R_n(x)| \leq \frac{|x - \frac{\pi}{6}|^{n+1}}{(n+1)!}. \tag{14.50}$$

Again invoking Example 2 of Section 14.6, since the series $\displaystyle\sum_{n=0}^{+\infty} \frac{(x - \frac{\pi}{6})^{n+1}}{(n + 1)!}$ conver-

ges absolutely everywhere, we have $\displaystyle\lim_{n\to+\infty} \frac{|x - \frac{\pi}{6}|^{n+1}}{(n + 1)!} = 0$ for every x. Thus for

every x, $\lim_{n\to+\infty} R_n(x) = 0$ from (14.50), and

$$\sin x = \frac{1}{2} + \frac{\sqrt{3}(x - \frac{\pi}{6})}{2 \cdot 1!} - \frac{(x - \frac{\pi}{6})^2}{2 \cdot 2!} - \frac{\sqrt{3}(x - \frac{\pi}{6})^3}{2 \cdot 3!} + \cdots.$$

Example 3 Find the Maclaurin series for $f(x)$ if

$$f(x) = \begin{cases} e^{-1/x^2} & \text{if } x \ne 0 \\ 0 & \text{if } x = 0. \end{cases}$$

For what numbers x is $f(x)$ the sum of this series?

SOLUTION We have

$$f'(0) = \lim_{x\to0} \frac{f(x) - f(0)}{x} = \lim_{x\to0} \frac{e^{-1/x^2}}{x}.$$

The limit on the right can be evaluated by letting $x = 1/t$ and using the limit formula from Exercise 21, Exercise Set 5.6. Then by L'Hôpital's fourth rule

$$f'(0) = \lim_{t\to+\infty} \frac{t}{e^{t^2}} = \lim_{t\to+\infty} \frac{1}{2te^{t^2}} = 0.$$

We leave as an exercise the proof that if $x \ne 0$, then

$$f^{(n)}(x) = P_{3n}\left(\frac{1}{x}\right)e^{-1/x^2} \quad \text{when } n = 1, 2, 3, \ldots \tag{14.51}$$

where P_{3n} is a polynomial function of degree $3n$ (Exercise 13, Exercise Set 14.7). From (14.51) and the limit definition for the derivative the student can prove (Exercise 14, Exercise Set 14.7) that

$$f^{(n)}(0) = 0 \quad \text{when } n = 1, 2, 3, \ldots.$$

The Maclaurin series for $f(x)$ is therefore

$$0 + \frac{x \cdot 0}{1!} + \frac{x^2 \cdot 0}{2!} + \cdots$$

which converges to 0 for every x. *Thus $f(x)$ is represented by its Maclaurin series only when $x = 0$.*

It should be noted here that the Taylor polynomial for $f(x)$ is $P_n(x) = 0$, for every n, and therefore

$$f(x) = 0 + R_n(x) = R_n(x) \quad \text{for every } n \text{ and } x.$$

Here *$f(x)$ is its own remainder.*

To obtain another example of a Taylor series, we consider the power series

$$\sum_{n=0}^{+\infty} x^n = 1 + x + x^2 + \cdots.$$

Since this series is a geometric series, it converges (absolutely) when $|x| < 1$ to the sum $1/(1 - x)$, and diverges when $|x| \geq 1$. Thus

$$\frac{1}{1 - x} = 1 + x + x^2 + \cdots + x^n + \cdots \qquad \text{if } |x| < 1. \qquad (14.52)$$

It is also readily verified that this series is indeed the Maclaurin series for $1/(1 - x)$. Since one can obtain by long division

$$\frac{1}{1 - x} = 1 + x + x^2 + \cdots + x^n + \frac{x^{n+1}}{1 - x}, \qquad (14.53)$$

the remainder for this Maclaurin series is $R_n(x) = x^{n+1}/(1 - x)$.

Exercise Set 14.7

In Exercises 1–11 obtain a power series in $x - a$ for the given $f(x)$. Find the set of numbers on which $f(x)$ is represented by this power series.

1. $f(x) = \sin x, \quad a = 0$

2. $f(x) = \sin x, \quad a = \frac{\pi}{4}$

3. $f(x) = \cos x, \quad a = 0$

4. $f(x) = \cos x, \quad a = \frac{\pi}{4}$

5. $f(x) = e^{-x/3}, \quad a = 0$

6. $f(x) = e^x, \quad a = 1$

7. $f(x) = \ln x, \quad a = 1$

8. $f(x) = \dfrac{1}{1 + x^2}, \quad a = 0$

9. $f(x) = \sin 2x, \quad a = 0$

10. $f(x) = \cos 3x, \quad a = 0$

11. $f(x) = \dfrac{1}{4 - x}, \quad a = 0$

12. Prove that the series (14.52) is the Maclaurin series for $\dfrac{1}{1 - x}$.

13. Obtain (14.51) for the function f in Example 3.

14. For the function f in Example 3, prove that $f^{(n)}(0) = 0$.

15. Prove from the series obtained in Exercise 7 that

$$\ln 2 = 1 - \tfrac{1}{2} + \tfrac{1}{3} - \tfrac{1}{4} + - \cdots.$$

14.8 Properties of Power Series

In this section we will state and prove several theorems dealing with properties of power series. Using these theorems, it is sometimes easier to derive a Taylor series for a function from other Taylor series rather than from the synthesis approach in which one finds $f(a)$, $f'(a)$, $f''(a)$, and so on.

We first note that it is possible to add or subtract convergent power series term by term. If

$$f(x) = \sum_{n=0}^{+\infty} a_n(x - a)^n \qquad \text{and} \qquad g(x) = \sum_{n=0}^{+\infty} b_n(x - a)^n \quad \text{when } |x - a| < r,$$

(14.54)

then from Theorem 14.2.6

$$f(x) \pm g(x) = \sum_{n=0}^{+\infty} (a_n \pm b_n)(x - a)^n \quad \text{if } |x - a| < r.$$

Also from (14.54), using Theorem 14.2.7, for any number c

$$cf(x) = \sum_{n=0}^{+\infty} ca_n(x - a)^n \quad \text{if } |x - a| < r$$

and for any positive integer k

$$(x - a)^k f(x) = \sum_{n=0}^{+\infty} a_n(x - a)^{n+k} \quad \text{if } |x - a| < r.$$

Example 1   Find a Maclaurin series expansion for $\cosh x$.

SOLUTION   From Example 1 of Section 14.7

$$e^x = 1 + \frac{x}{1!} + \frac{x^2}{2!} + \frac{x^3}{3!} + \cdots + \frac{x^n}{n!} + \cdots \quad \text{for every } x.$$

Hence

$$e^{-x} = 1 - \frac{x}{1!} + \frac{x^2}{2!} - \frac{x^3}{3!} + \cdots + (-1)^n \frac{x^n}{n!} + \cdots \quad \text{for every } x$$

and from our remarks immediately preceding Example 1

$$\cosh x = \frac{e^x + e^{-x}}{2} = 1 + \frac{x^2}{2!} + \frac{x^4}{4!} + \cdots + \frac{x^{2n}}{(2n)!} + \cdots \quad \text{for every } x.$$

Next suppose F is defined by

$$F(x) = \sum_{n=0}^{+\infty} a_n x^n \quad \text{if } |x| < r.$$

(14.55)

If the power series in (14.55) is differentiated term by term, the series

$$\sum_{n=1}^{+\infty} na_n x^{n-1} = a_1 + 2a_2 x + 3a_3 x^2 + \cdots$$

(14.56)

is obtained. We will prove in Theorem 14.8.2 below that this power series defines the derivative of the function F given by (14.55). However, this proof requires the following preliminary theorem.

14.8.1 Theorem

If the power series $\sum_{n=0}^{+\infty} a_n x^n$ in (14.55) converges when $|x| < r$, then the series $\sum_{n=1}^{+\infty} n a_n x^{n-1}$ also converges when $|x| < r$.

PROOF Suppose x is chosen so that $|x| < |p| < r$. Then, as in the proof of Theorem 14.6.1, there is an $M > 0$ such that

$$|a_n p^n| \leq M \quad \text{for every } n. \tag{14.57}$$

If b is given by

$$b = \left| \frac{x}{p} \right|,$$

then from (14.57)

$$|n a_n x^{n-1}| = \frac{n|a_n p^n|}{|p|} b^{n-1} \leq \frac{nM}{|p|} b^{n-1}. \tag{14.58}$$

Since $0 < b < 1$, the series $\sum_{n=1}^{+\infty} nMb^{n-1}/|p|$ converges by the ratio test. Hence from (14.58), using the first comparison test, $\sum_{n=1}^{+\infty} |n a_n x^{n-1}|$ converges; that is, $\sum_{n=1}^{+\infty} n a_n x^{n-1}$ converges absolutely.

From a second application of Theorem 14.8.1 $\sum_{n=2}^{+\infty} n(n-1) a_n x^{n-2}$ converges when $|x| < r$; from a third application $\sum_{n=3}^{+\infty} n(n-1)(n-2) a_n x^{n-3}$ converges when $|x| < r$, and so on.

14.8.2 Theorem

If F is given by the power series (14.55), then

$$F'(x) = \sum_{n=1}^{+\infty} n a_n x^{n-1} \quad \text{if } |x| < r.$$

PROOF Choosing $|x| < r$, we note that there is certainly a p such that $0 \leq |x| < |p| < r$. We next choose h so that $|x + h| < |p|$. From (14.55)

$$F(x + h) - F(x) = \sum_{n=1}^{+\infty} a_n [(x + h)^n - x^n]. \tag{14.59}$$

In the right member of (14.59) the sum is started with $n = 1$ instead of $n = 0$ since the term corresponding to $n = 0$ is 0.

If Taylor's formula, (13.26), is applied to the function g where $g(t) = t^n$, and $x + h$ and x are in the roles of x and a respectively, then

$$(x + h)^n = x^n + nx^{n-1}h + \frac{n(n-1)c^{n-2}h^2}{2} \tag{14.60}$$

where c is between x and $x + h$. Hence from (14.59) and (14.60)

$$\frac{F(x+h) - F(x)}{h} = \sum_{n=1}^{+\infty} na_n x^{n-1} + \frac{h}{2} \sum_{n=2}^{+\infty} n(n-1)a_n c^{n-2}. \tag{14.61}$$

The sum on the right in (14.61) is started with $n = 2$ instead of $n = 1$ since the term corresponding to $n = 1$ is 0. As we mentioned just before starting this theorem, $\sum_{n=2}^{+\infty} n(n-1)a_n x^{n-2}$ converges whenever $|x| < r$ and hence by Theorem 14.6.4 the series $\sum_{n=2}^{+\infty} n(n-1)a_n p^{n-2}$ converges absolutely since $|p| < r$. Note that $|c| < |p|$ since c is between x and $x + h$. Hence from (14.61) and (14.32)

$$\left| \frac{F(x+h) - F(x)}{h} - \sum_{n=1}^{+\infty} na_n x^{n-1} \right| = \frac{|h|}{2} \left| \sum_{n=2}^{+\infty} n(n-1)a_n c^{n-2} \right|$$

$$\leq \frac{|h|}{2} \sum_{n=2}^{+\infty} n(n-1)|a_n||c^{n-2}|$$

$$< \frac{|h|}{2} \sum_{n=2}^{+\infty} n(n-1)|a_n||p^{n-2}|. \tag{14.62}$$

Now $\lim_{h \to 0}(|h|/2)\sum_{n=2}^{+\infty} n(n-1)|a_n||p^{n-2}| = 0$ and therefore from (14.62)

$$F'(x) = \lim_{h \to 0} \frac{F(x+h) - F(x)}{h} = \sum_{n=1}^{+\infty} na_n x^{n-1}.$$

Example 2   Find a power series for $\sinh x$ by differentiation.

SOLUTION   From Example 1

$$\cosh x = \sum_{n=0}^{+\infty} \frac{x^{2n}}{(2n)!} \quad \text{for every } x$$

and since $D_x \cosh x = \sinh x$, from Theorem 14.8.2

$$\sinh x = \sum_{n=1}^{+\infty} \frac{(2n)x^{2n-1}}{(2n)!} = \sum_{n=1}^{+\infty} \frac{x^{2n-1}}{(2n-1)!}.$$

Before considering the next example, we recall from the binomial formula that if m is a non-negative integer, then

$$(1 + x)^m = 1 + mx + \frac{m(m-1)}{2!}x^2 + \cdots + \frac{m(m-1) \cdots (m-n+1)}{n!}x^n$$

$$+ \cdots + x^m. \tag{14.63}$$

It can be proved that for any m

$$(1 + x)^m = 1 + mx + \frac{m(m - 1)}{2!}x^2 + \cdots + \frac{m(m - 1) \cdots (m - n + 1)}{n!}x^n + \cdots$$

$$\text{if } |x| < 1. \quad (14.64)$$

The series in the right member of (14.64) is called a *binomial series*. Note that the equation (14.63) is obtained as a special case of (14.64) when m is a non-negative integer, but if m is not a non-negative integer, then the right member of (14.64) is an infinite series.

Example 3 Verify the binomial series representation (14.64) of $(1 + x)^m$.

PROOF We begin by letting

$$F(x) = 1 + \sum_{n=1}^{+\infty} \frac{m(m - 1) \cdots (m - n + 1)}{n!}x^n \quad \text{if } |x| < 1 \quad (14.65)$$

and will prove that $F(x) = (1 + x)^m$.

If the ratio test is applied to the series in the right member of (14.65), then

$$\lim_{n \to +\infty} \left| \frac{a_{n+1}}{a_n} \right| = \lim_{n \to +\infty} \frac{|m - n||x|}{n + 1} = |x|.$$

Therefore the series converges absolutely when $|x| < 1$ and diverges when $|x| > 1$. It can also be proved that the series converges and represents $F(x)$ when $x = 1$ if $m > -1$ and when $x = -1$ if $m > 0$, although we will not concern ourselves with the behavior of the series at $x = \pm 1$ in this proof.

Differentiating, using Theorem 14.8.2, gives

$$F'(x) = \sum_{n=1}^{+\infty} \frac{m(m - 1) \cdots (m - n + 1)}{(n - 1)!}x^{n-1} \quad \text{if } |x| < 1, \quad (14.66)$$

and therefore from our remarks preceding Example 1

$$xF'(x) = \sum_{n=1}^{+\infty} \frac{m(m - 1) \cdots (m - n + 1)}{(n - 1)!}x^n. \quad (14.67)$$

The series on the right in (14.66) can be started with $n = 0$ if in the expression behind the summation sign n is replaced by $n + 1$. Thus

$$F'(x) = \sum_{n=0}^{+\infty} \frac{m(m - 1) \cdots (m - n)}{n!}x^n$$

$$= m + \sum_{n=1}^{+\infty} \frac{m(m - 1) \cdots (m - n)}{n!}x^n. \quad (14.68)$$

From (14.67) and (14.68)

$$(1 + x)F'(x) = m + \sum_{n=1}^{+\infty} \left[\frac{m(m-1)\cdots(m-n)}{n!} + \frac{m(m-1)\cdots(m-n+1)}{(n-1)!} \right] x^n$$

$$= m + \sum_{n=1}^{+\infty} \frac{m(m-1)\cdots(m-n+1)[(m-n)+n]}{n!} x^n$$

$$= m + \sum_{n=1}^{+\infty} \frac{m^2(m-1)\cdots(m-n+1)}{n!} x^n$$

$$= m \left[1 + \sum_{n=1}^{+\infty} \frac{m(m-1)\cdots(m-n+1)}{n!} x^n \right],$$

and from (14.65)

$$(1 + x)F'(x) = mF(x). \tag{14.69}$$

Thus from (14.69)

$$\frac{F'(x)}{F(x)} = \frac{m}{1 + x} \quad \text{for } |x| < 1.$$

Since the left member of this equation is $D_x \ln F(x)$, we therefore have

$$D_x \ln F(x) = \frac{m}{1 + x}.$$

Hence there is a constant C such that

$$\ln F(x) = m \ln (1 + x) + C \quad \text{for } |x| < 1. \tag{14.70}$$

In particular from (14.65) and (14.70)

$$0 = \ln F(0) = C,$$

and thus (14.70) simplifies to

$$\ln F(x) = m \ln (1 + x) \quad \text{for } |x| < 1.$$

This equation is equivalent to

$$F(x) = (1 + x)^m \quad \text{for } |x| < 1,$$

which verifies the binomial series representation of $(1 + x)^m$.

Because of Theorem 14.8.4 below the binomial series in (14.64) is the Maclaurin series for $(1 + x)^m$.

Example 4   Obtain the binomial series representation of $1/\sqrt{1 + x}$.

SOLUTION   Letting $m = -\frac{1}{2}$ in (14.64), we obtain

$$\frac{1}{\sqrt{1+x}} = 1 - \frac{1}{2}x + \frac{(-\frac{1}{2})(-\frac{3}{2})}{2!}x^2 + \cdots + \frac{(-\frac{1}{2})(-\frac{3}{2})\cdots(\frac{1}{2}-n)}{n!}x^n + \cdots$$

$$= 1 - \frac{1}{2}x + \frac{3}{8}x^2 + \cdots + \frac{(-1)^n 1 \cdot 3 \cdots (2n-1)}{2^n n!}x^n + \cdots.$$

From Example 3, $1/\sqrt{1+x}$ is represented by this series if $x \in (-1, 1]$.

We can easily prove from Theorem 14.8.2 that a power series in $x - a$ can be differentiated term by term. If f is defined by

$$f(x) = \sum_{n=0}^{+\infty} a_n(x-a)^n \quad \text{when } |x-a| < r, \tag{14.71}$$

then from (14.55)

$$f(x) = F(x-a) \quad \text{when } |x-a| < r.$$

By the chain rule and Theorem 14.8.2

$$f'(x) = F'(x-a) \cdot 1 \quad \text{when } |x-a| < r$$

$$= \sum_{n=1}^{+\infty} na_n(x-a)^{n-1} \quad \text{when } |x-a| < r. \tag{14.72}$$

Thus we have proved the analogous theorem to Theorem 14.8.2 for functions defined by power series in $x - a$:

14.8.3   Theorem

If f is given by (14.71), then

$$f'(x) = \sum_{n=1}^{+\infty} na_n(x-a)^{n-1} \quad \text{if } |x-a| < r.$$

From a repeated application of Theorem 14.8.3

$$f''(x) = \sum_{n=2}^{+\infty} n(n-1)a_n(x-a)^{n-2} \qquad \qquad \text{if } |x-a| < r$$

$$\cdots$$

$$f^{(k)}(x) = \sum_{n=k}^{+\infty} n(n-1)\cdots(n-k+1)a_n(x-a)^{n-k} \quad \text{if } |x-a| < r \tag{14.73}$$

when $k = 1, 2, 3, \ldots$.

In particular, if $x = a$ in (14.73), then the only nonzero contribution to the sum on the right comes from its kth term. Hence

$$f^{(k)}(a) = k(k-1)\cdots 1a_k \qquad k = 1, 2, \ldots,$$

and therefore

$$a_k = \frac{f^{(k)}(a)}{k!} \qquad k = 1, 2, \ldots .$$

We have therefore proved the following theorem.

14.8.4 Theorem

A convergent power series in $x - a$ which represents a function in the interval $(a - r, a + r)$, where $r > 0$, is the Taylor series of the function at a.

Because of Theorem 14.8.4, if a function is represented in the interval $(a - r, a + r)$ by a power series in $x - a$, then the function cannot be represented by any other power series in $x - a$ in the interval.

Also, we note from Theorem 14.8.3 that a function defined by a power series is continuous in the interior of the interval of convergence of the series.

According to the final theorem of this section, *a power series may be integrated term by term throughout the interior of its interval of convergence.*

14.8.5 Theorem

If f is given by (14.71), then

$$\int_a^x f(t)\, dt = \sum_{n=0}^{+\infty} \frac{a_n(x - a)^{n+1}}{n + 1} \qquad \text{if } |x - a| < r.$$

PROOF From Theorem 14.8.3 f is differentiable and hence continuous on the interval $(a - r, a + r)$. Thus $\int_a^x f(t)\, dt$ exists for any $x \in (a - r, a + r)$. If

$$g(x) = \sum_{n=0}^{+\infty} \frac{a_n(x - a)^{n+1}}{n + 1}, \text{ then by Theorem 14.8.3}$$

$$g'(x) = \sum_{n=0}^{+\infty} a_n(x - a)^n = f(x) \quad \text{if } |x - a| < r. \tag{14.74}$$

By the first fundamental theorem $D_x \int_a^x f(t)\, dt = f(x)$ for $x \in (a - r, a + r)$. Hence from (14.74), using Theorem 7.1.3

$$\int_a^x f(t)\, dt = g(x) + C = \sum_{n=0}^{+\infty} \frac{a_n(x - a)^{n+1}}{n + 1} \qquad \text{if } |x - a| < r. \tag{14.75}$$

From letting $x = a$ in (14.75) we obtain $C = 0$ and hence the theorem is proved.

Example 5 Find a Maclaurin series for $\tan^{-1} x$. In what interval does this series represent $\tan^{-1} x$?

SOLUTION   In (14.52) if x is replaced by $-x^2$, then

$$\frac{1}{1 + x^2} = 1 - x^2 + x^4 - x^6 + \cdots + (-1)^n x^{2n} + \cdots \quad \text{if } |x| < 1. \quad (14.76)$$

Hence by Theorem 14.8.5, term-by-term integration gives

$$\tan^{-1} x = \int_0^x \frac{1}{1 + t^2} \, dt$$

$$= x - \frac{x^3}{3} + \frac{x^5}{5} - \frac{x^7}{7} + \cdots + (-1)^n \frac{x^{2n+1}}{2n + 1} + \cdots \quad (14.77)$$

for $|x| < 1$. If $x \neq 0$, the series in the right member of (14.77) is an alternating series, and hence by Theorem 14.4.4(b) and our remarks following Example 2 of Section 14.4 the remainder, $R_n(x)$, for this series satisfies

$$|R_n(x)| \leq \left| \frac{(-1)^n x^{2n+3}}{2n + 3} \right| = \frac{|x|^{2n+3}}{2n + 3}. \quad (14.78)$$

If $x \in [-1, 1]$, then from (14.78), $\lim_{n \to +\infty} R_n(x) = 0$ and hence the representation of $\tan^{-1} x$ by the series (14.77) is valid when $|x| \leq 1$. It is notable here that $\tan^{-1} x$ is represented by its series at the numbers 1 and -1, but this series was obtained from the integration of a series which diverges at these numbers.

Letting $x = 1$ in (14.77) gives

$$\frac{\pi}{4} = \tan^{-1} x = 1 - \frac{1}{3} + \frac{1}{5} - \frac{1}{7} + - \cdots . \quad (14.79)$$

Although this series has the sum $\frac{\pi}{4}$, it converges very slowly and is therefore not particularly useful for calculating π. For example, if we wanted to approximate π correct to the 4th decimal place using (14.79), we must sum the first 10,000 terms of the series since from (14.78) $|R_n(1)| < \frac{1}{20,000} = 0.00005$ if $n \geq 10,000$.
    A more useful formula for this calculation is

$$\frac{\pi}{4} = 2 \tan^{-1} \frac{1}{3} + \tan^{-1} \frac{1}{7} \quad (14.80)$$

where $\tan^{-1} \frac{1}{3}$ and $\tan^{-1} \frac{1}{7}$ can be obtained from (14.77). We leave as exercises the problems of deriving (14.80) and approximating π correct to the 4th decimal place using (14.80).

Exercise Set 14.8

In Exercises 1–13 obtain the Maclaurin series for the given value at x and indicate the interval in which the series represents the value at x.

1. $\sqrt{1 + x}$ 2. $(1 - x)^{-1/3}$ 3. $(1 - \frac{1}{2}x)^{-2}$

4. $\sqrt{4 + 2x}$ 5. $\dfrac{x}{\sqrt{1 - x^2}}$ 6. $x \sin x$

7. $f(x)$ where $f(x) = \begin{cases} \dfrac{\sin x}{x} & \text{if } x \neq 0 \\ 1 & \text{if } x = 0 \end{cases}$

8. $\ln (x + \sqrt{x^2 + 1})$

9. $\sin^{-1} x$

10. $\cos^2 x$ HINT: $\cos 2x = 2 \cos^2 x - 1$

11. $\sin^2 x$ 12. $\displaystyle\int_0^x \frac{dt}{\sqrt{1 + t^3}}$ 13. $\displaystyle\int_0^x e^{-t^2/2}\, dt$

14. By summing the required number of terms in the series obtained in Exercise 13 approximate $\int_0^1 e^{-x^2/2}\, dx$ correct to the 3rd decimal place.

15. Find an upper bound for $|R_n(x)|$ for the series in (14.64) when $0 < x < 1$.

16. Using the result obtained in Exercise 15, approximate $(1.03)^{3/2}$ correct to the 4th decimal place by summing the required number of terms in a binomial series.

17. Derive (14.80).

18. Approximate π correct to the 4th decimal place using (14.80). Use (14.78).

19. From the Maclaurin series for $1/(1 - x)$ (14.52), obtain Maclaurin series for (a)

$(1 - x)^{-2}$ (b) $\ln (1 - x)$ (c) $\ln (1 + x)$ (d) $\ln (1 - x^2)$ (e) $\ln \dfrac{1 + x}{1 - x}$.

20. For each of the series obtained in Exercise 19 obtain $R_n(x)$ and give the interval in which the series is represented by its Maclaurin series. Use (14.53).

21. By summing the required number of terms in the series in Exercise 19(e) approximate correct to the 4th decimal place: (a) $\ln 2$ (b) $\ln 3$. Use the $R_n(x)$ obtained in part (e) of Exercise 20.

15. Vectors in the Plane; Parametric Equations

15.1   Vectors in the Plane

In physics and engineering certain physical quantities—for example, force, velocity, and acceleration—are encountered, which have the properties of *magnitude* and *direction*. Such quantities, which are called *vector quantities*, can be conveniently represented by an arrow where the length of the arrow measures the magnitude of the quantity, and the direction in which the quantity acts is indicated by the direction in which the arrow points. Suppose an arrow is drawn from an arbitrary *initial point* $P(x, y)$ to a *terminal point* $Q(x + a_1, y + a_2)$ (Figure 15–1). Then its length is

$$|PQ| = \sqrt{a_1^2 + a_2^2} \tag{15.1}$$

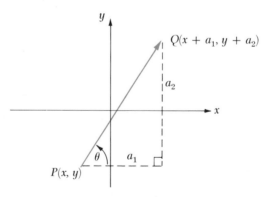

Figure 15–1

and its direction is given by θ, where $0 \le \theta < 2\pi$, the angle formed by a counterclockwise rotation from the positive x direction to the arrow (Figure 15–1). From basic trigonometry, θ satisfies the equations

$$\cos \theta = \frac{a_1}{\sqrt{a_1^2 + a_2^2}} \qquad \sin \theta = \frac{a_2}{\sqrt{a_1^2 + a_2^2}}. \tag{15.2}$$

Because of (15.1) and (15.2), the length and direction of the arrow PQ are determined by the numbers a_1 and a_2, which are the differences in the x and y coordinates, respectively, of P and Q.

The arrow PQ is a geometric representation of the *vector* $\langle a_1, a_2 \rangle$. This fact motivates the algebraic definition of this vector, which is given in the following statement.

15.1.1 Definition

A *two-dimensional vector* is an ordered pair of real numbers. The x *component* and y *component* of the vector $\langle a_1, a_2 \rangle$ are, respectively, a_1 and a_2. The *length* $|\langle a_1, a_2 \rangle|$ of the vector is given by

$$|\langle a_1, a_2 \rangle| = \sqrt{a_1^2 + a_2^2}.$$

Also by definition, $\langle a_1, a_2 \rangle = \langle b_1, b_2 \rangle$ if and only if $a_1 = b_1$ and $a_2 = b_2$. Whenever $\langle a_1, a_2 \rangle \neq \langle 0, 0 \rangle$, the vector has a *direction* θ, where $0 \leq \theta < 2\pi$, which can be obtained from the equations (15.2).

As an example of a two-dimensional vector, $\langle 3, -5 \rangle$ has an x component of 3 and a y component of -5. Its length is $\sqrt{3^2 + (-5)^2} = \sqrt{34}$, and its direction is $\theta = 2\pi - \cos^{-1} 3/\sqrt{34}$. The vector can be represented by an arrow drawn from an arbitrary point (x, y) to the point $(x + 3, y - 5)$, and, in particular, the arrow can be drawn from the origin to the point $(3, -5)$ (Figure 15-2).

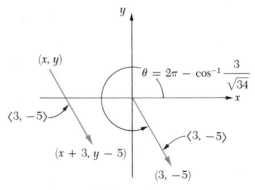

Figure 15-2

A vector is frequently denoted by a boldface letter, such as **A**, **B**, or **V**. When these letters are written longhand, an arrow is sometimes written over the letter to indicate that it denotes a vector. For example, the vector **A** could be denoted by $\vec{A}$. If a vector **A** is represented by an arrow drawn from the point P to the point Q, **A** may also be denoted by $\overrightarrow{PQ}$. We can then write $\mathbf{A} = \overrightarrow{PQ}$ and speak of $\overrightarrow{PQ}$ as a representative of **A**. The points P and Q are termed, respectively, the *initial point* and the *terminal point* of the vector $\overrightarrow{PQ}$. A representative of a vector having the origin as its initial point is called the *positional representative* of the vector.

Even though a two-dimensional vector and an element of the Cartesian plane, R^2, are each defined as an ordered pair of numbers, we shall distinguish between these concepts because of the different uses we make of them. First of all, a point (a_1, a_2) is fixed in the plane, but the representative of the vector $\langle a_1, a_2 \rangle$ may have any point in the plane as its initial point. Also, we will discuss arithmetic operations on vectors since these operations have useful applications whereas we will have no need for an arithmetic on elements of R^2.

The first arithmetic operation to be considered is vector addition. Suppose $\overrightarrow{OA}$ and $\overrightarrow{OB}$ are vectors which represent forces exerted on a particle located at the origin (Figure 15–3(a)). From physics we know that these forces have the same effect on the particle as the force represented by $\overrightarrow{OC}$, where OC is a diagonal of the parallelogram having OA and OB as adjacent sides.

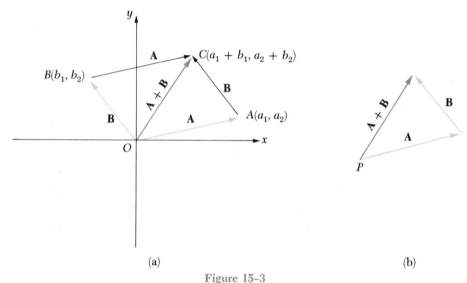

(a)   (b)

Figure 15–3

If $\overrightarrow{OA} = \langle a_1, a_2 \rangle$ and $\overrightarrow{OB} = \langle b_1, b_2 \rangle$, we shall show that $\overrightarrow{OC} = \langle a_1 + b_1, a_2 + b_2 \rangle$. Note that the points A and B are respectively (a_1, a_2) and (b_1, b_2) and suppose C is assigned coordinates (x, y). Since OA and BC are opposite sides of a parallelogram, $OA \parallel BC$, and similarly, $OB \parallel AC$. Therefore, since parallel lines have equal slopes,

$$\frac{a_2}{a_1} = \frac{y - b_2}{x - b_1} \quad \text{and} \quad \frac{b_2}{b_1} = \frac{y - a_2}{x - a_1}.$$

From these equations one obtains

$$a_2 x - a_1 y = a_2 b_1 - a_1 b_2$$

and

$$b_2 x - b_1 y = a_1 b_2 - a_2 b_1,$$

respectively. From solving these equations we obtain

$$x = a_1 + b_1 \qquad y = a_2 + b_2.$$

for the coordinates of C, and therefore $\overrightarrow{OC} = \langle a_1 + b_1, a_2 + b_2 \rangle$. If $\overrightarrow{OA} = \mathbf{A}$ and $\overrightarrow{OB} = \mathbf{B}$, then according to the following definition, $\overrightarrow{OC}$ is a representative of the *sum* of $\mathbf{A}$ and $\mathbf{B}$.

15.1.2   Definition

If $\mathbf{A} = \langle a_1, a_2 \rangle$ and $\mathbf{B} = \langle b_1, b_2 \rangle$, then

$$\mathbf{A} + \mathbf{B} = \langle a_1 + b_1, a_2 + b_2 \rangle.$$

If the points A, B, and C have coordinates as given in Figure 15–3(a), then $\overrightarrow{AC} = \langle b_1, b_2 \rangle = \mathbf{B}$. Now the arrows representing $\mathbf{A}$, $\overrightarrow{AC}$, and $\mathbf{A} + \mathbf{B}$, respectively, will continue to represent these vectors if each arrow is displaced the same distance parallel to itself. Thus, $\mathbf{A} + \mathbf{B}$ can be represented by the arrow shown in Figure 15–3(b), which is obtained by the following construction:

Draw the arrow representing $\mathbf{A}$ *from any initial point P. Next, from the tip of arrow* $\mathbf{A}$ *draw the arrow representing* $\mathbf{B}$. *Then an arrow drawn from P to the tip of arrow* $\mathbf{B}$ *will represent* $\mathbf{A} + \mathbf{B}$.

As an illustration of Definition 15.1.2,

$$\langle -6, 3 \rangle + \langle 4, 2 \rangle = \langle -6 + 4, 3 + 2 \rangle = \langle -2, 5 \rangle.$$

The usual properties for the addition of real numbers also hold for the addition of vectors. For any vectors $\mathbf{A}$, $\mathbf{B}$, and $\mathbf{C}$

$$\mathbf{A} + \mathbf{B} = \mathbf{B} + \mathbf{A} \qquad \text{\textit{(commutative law of addition)}} \qquad (15.3)$$

$$\mathbf{A} + (\mathbf{B} + \mathbf{C}) = (\mathbf{A} + \mathbf{B}) + \mathbf{C} \qquad \text{\textit{(associative law of addition)}} \qquad (15.4)$$

and denoting the *zero vector* by $\mathbf{0} = \langle 0, 0 \rangle$,

$$\mathbf{A} + \mathbf{0} = \mathbf{0} + \mathbf{A} = \mathbf{A} \qquad (15.5)$$

PROOF OF (15.3)   If $\mathbf{A} = \langle a_1, a_2 \rangle$ and $\mathbf{B} = \langle b_1, b_2 \rangle$, then from Definition 15.1.2 and the commutative law of addition for real numbers,

$$\mathbf{A} + \mathbf{B} = \langle a_1, a_2 \rangle + \langle b_1, b_2 \rangle = \langle a_1 + b_1, a_2 + b_2 \rangle = \langle b_1 + a_1, b_2 + a_2 \rangle$$
$$= \langle b_1, b_2 \rangle + \langle a_1, a_2 \rangle = \mathbf{B} + \mathbf{A}.$$

We next define the difference of two vectors and the *negative* of a vector.

15.1.3   Definition

If $\mathbf{A} = \langle a_1, a_2 \rangle$ and $\mathbf{B} = \langle b_1, b_2 \rangle$, then

(a) $\mathbf{A} - \mathbf{B} = \langle a_1 - b_1, a_2 - b_2 \rangle$
(b) $-\mathbf{A} = \langle -a_1, -a_2 \rangle$

As illustrations of these definitions,

$$\langle -1, 3 \rangle - \langle 5, 2 \rangle = \langle -1 - 5, 3 - 2 \rangle = \langle -6, 1 \rangle,$$
$$-\langle -1, 3 \rangle = \langle 1, -3 \rangle.$$

From the definitions it follows easily that

$$\mathbf{A} - \mathbf{B} = \mathbf{A} + (-\mathbf{B}) \qquad\qquad (15.6)$$
$$\mathbf{A} + (-\mathbf{A}) = 0. \qquad\qquad (15.7)$$

An arrow representing $\mathbf{A} - \mathbf{B}$ can be constructed as follows:

Draw the arrow representing $\mathbf{A}$ from some initial point P. Also with P as an initial point, draw the arrow for $\mathbf{B}$. The arrow then drawn from the tip of the arrow $\mathbf{B}$ to the tip of the arrow $\mathbf{A}$ will represent $\mathbf{A} - \mathbf{B}$ (Figure 15–4).

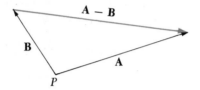

Figure 15–4

Geometrically, therefore, $\mathbf{A} - \mathbf{B}$ is the vector that when added to $\mathbf{B}$ yields $\mathbf{A}$, as is expressed algebraically in the following vector identity.

$$(\mathbf{A} - \mathbf{B}) + \mathbf{B} = \mathbf{A} \qquad\qquad (15.8)$$

Equation (15.8) is readily derived from (15.6) and (15.7).

15.1.4   Definition

If k is any number (called a *scalar*), and $\mathbf{A} = \langle a_1, a_2 \rangle$, then

$$k\mathbf{A} = \langle ka_1, ka_2 \rangle.$$

For example,

$$-2\langle -1, 3 \rangle = \langle 2, -6 \rangle.$$

The vector $k\mathbf{A}$ is called a *scalar multiple* of $\mathbf{A}$. The length of $k\mathbf{A}$ is

$$|k\mathbf{A}| = \sqrt{k^2 a_1{}^2 + k^2 a_2{}^2} = |k| \sqrt{a_1{}^2 + a_2{}^2}$$
$$= |k|\,|\mathbf{A}|.$$

If $k > 0$, an arrow representing $k\mathbf{A}$ is in the same direction as an arrow representing $\mathbf{A}$. If $k < 0$, the two arrows point in opposite directions (Figure 15–5). Thus, the vectors $k\mathbf{A}$, where $k \neq 0$, and $\mathbf{A}$ are said to be in the *same direction* if $k > 0$, and in *opposite directions* if $k < 0$.

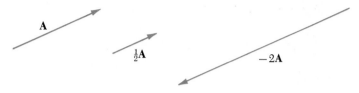

Figure 15-5

Some properties of scalar multiplication, which are easily derived, are listed below. Here **A** and **B** are any vectors, and p and q are any numbers.

$$p(\mathbf{A} + \mathbf{B}) = p\mathbf{A} + p\mathbf{B} \tag{15.9}$$
$$(p + q)\mathbf{A} = p\mathbf{A} + q\mathbf{A} \tag{15.10}$$
$$(pq)\mathbf{A} = p(q\mathbf{A}) \tag{15.11}$$
$$(-p)\mathbf{A} = -(p\mathbf{A}) \tag{15.12}$$
$$1\mathbf{A} = \mathbf{A} \tag{15.13}$$
$$0\mathbf{A} = \mathbf{0} \tag{15.14}$$
$$p\mathbf{0} = \mathbf{0} \tag{15.15}$$

We also may divide a vector by a scalar. If $k \neq 0$, then by definition

$$\frac{\mathbf{A}}{k} = \frac{1}{k}\mathbf{A}.$$

When a nonzero vector is divided by its length, it is easily shown that the quotient is a *unit vector*—that is, a vector of length 1 (Exercise 32, Exercise Set 15.1).

Using addition and scalar multiplication of vectors, an arbitrary vector $\mathbf{A} = \langle a_1, a_2 \rangle$ can be represented in terms of the unit vectors

$$\mathbf{i} = \langle 1, 0 \rangle \qquad \mathbf{j} = \langle 0, 1 \rangle. \tag{15.16}$$

To show this, we write

$$\mathbf{A} = \langle a_1, a_2 \rangle = \langle a_1, 0 \rangle + \langle 0, a_2 \rangle = a_1 \langle 1, 0 \rangle + a_2 \langle 0, 1 \rangle$$
$$= a_1 \mathbf{i} + a_2 \mathbf{j}. \tag{15.17}$$

This representation (15.17) of **A** is illustrated in Figure 15–6.

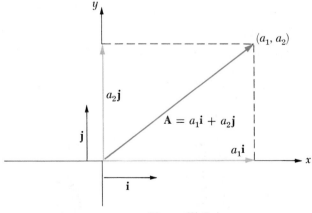

Figure 15–6

As an illustration of (15.17),

$$\langle 3, -2 \rangle = 3\mathbf{i} - 2\mathbf{j}.$$

Exercise Set 15.1

In Exercises 1–6 find the length and direction of the given vector. Draw a sketch showing the positional representative of each vector.

1. $\mathbf{A} = \langle -2, 2 \rangle$ 2. $\mathbf{A} = \langle -3, -3\sqrt{3} \rangle$

3. $\mathbf{A} = \langle 3, 4 \rangle$ 4. $\mathbf{A} = \langle 5, -12 \rangle$

5. $\mathbf{A} = \langle -2, -5 \rangle$ 6. $\mathbf{A} = \langle -1, 2\sqrt{2} \rangle$

In Exercises 7–10 find Q such that $\overrightarrow{PQ}$ is a representative of the given vector $\mathbf{A}$ when P is the given point.

7. $\mathbf{A} = \langle 3, -6 \rangle$, $P = (-1, 2)$ 8. $\mathbf{A} = \langle 4, 1 \rangle$, $P = (0, -2)$

9. $\mathbf{A} = \langle 6, 2 \rangle$, $P = (3, -4)$ 10. $\mathbf{A} = \langle -2, 7 \rangle$, $P = (-5, -3)$

11. If $P = (1, -3)$, $Q = (-1, 2)$, $S = (-2, 5)$, and $T = (-4, 10)$, prove that $\overrightarrow{PQ}$, $\overrightarrow{OS}$, and $\overrightarrow{ST}$ are representatives of the same vector.

In Exercises 12–15 find (a) $\mathbf{A} + \mathbf{B}$, (b) $\mathbf{A} - \mathbf{B}$, (c) $|\mathbf{A} + \mathbf{B}|$, and (d) $|\mathbf{A} - \mathbf{B}|$, for the given vectors $\mathbf{A}$ and $\mathbf{B}$. Draw a sketch showing the positional representatives of $\mathbf{A}$, $\mathbf{B}$, $\mathbf{A} + \mathbf{B}$, and $\mathbf{A} - \mathbf{B}$.

12. $\mathbf{A} = \langle 5, 2 \rangle$, $\mathbf{B} = \langle -2, 4 \rangle$ 13. $\mathbf{A} = \langle -3, 1 \rangle$, $\mathbf{B} = \langle 5, 9 \rangle$

14. $\mathbf{A} = \langle -2, -3 \rangle$, $\mathbf{B} = \langle -4, 5 \rangle$ 15. $\mathbf{A} = \langle 7, 8 \rangle$, $\mathbf{B} = \langle -4, 2 \rangle$

16. If $\mathbf{A} = \langle 2, -3 \rangle$, $\mathbf{B} = \langle -1, -5 \rangle$, and $\mathbf{C} = \langle -4, 2 \rangle$, obtain $\mathbf{A} + (\mathbf{B} + \mathbf{C})$ and $(\mathbf{A} + \mathbf{B}) + \mathbf{C}$. Illustrate with sketches for $\mathbf{A}$, $\mathbf{B}$, and $\mathbf{C}$ and these sums.

17. If $\mathbf{A} = \langle -3, -4 \rangle$, $\mathbf{B} = \langle 5, -2 \rangle$, and $\mathbf{C} = \langle 0, 3 \rangle$, find:

     (a) $-2\mathbf{A}$ (b) $3\mathbf{B} + \mathbf{C}$ (c) $5\mathbf{A} - 2\mathbf{B} + 3\mathbf{C}$

     (d) $\mathbf{B} - 2\mathbf{C} + 4\mathbf{A}$ (e) $|3\mathbf{C} - 2\mathbf{A}|$ (f) $|3\mathbf{C}| - |2\mathbf{A}|$

     (g) $|\mathbf{A} + 2\mathbf{B} - 4\mathbf{C}|$ (h) the unit vector in the same direction as $\mathbf{A}$

     (i) the unit vector in a direction opposite to that of $\mathbf{B}$

18. Let $P_1(x_1, y_1)$, $P_2(x_2, y_2)$, and $P_3(x_3, y_3)$ be any three points in the plane. Obtain the vector sum $\overrightarrow{P_1P_2} + \overrightarrow{P_2P_3} + \overrightarrow{P_3P_1}$.

19. Find m and n if $m\langle -1, 2 \rangle + n\langle -3, -4 \rangle = \langle 5, 6 \rangle$.

In Exercises 20–31 derive the indicated equation in Section 15.1.

20. (15.4)   21. (15.5)   22. (15.6)   23. (15.7)

24. (15.8)   25. (15.9)   26. (15.10)   27. (15.11)

28. (15.12)   29. (15.13)   30. (15.14)   31. (15.15)

32. If $\mathbf{A} \neq 0$, prove that $\dfrac{\mathbf{A}}{|\mathbf{A}|}$ is a unit vector.

33. Prove that $|\mathbf{A} + \mathbf{B}|^2 + |\mathbf{A} - \mathbf{B}|^2 = 2(|\mathbf{A}|^2 + |\mathbf{B}|^2)$.

34. Prove that $|\mathbf{A} + \mathbf{B}| \leq |\mathbf{A}| + |\mathbf{B}|$.

15.2   Dot Product

In our discussion of the dot product of two vectors it will be useful to speak of the *angle between two vectors*.

15.2.1   Definition

Suppose $\mathbf{A}$ and $\mathbf{B}$ are any two nonzero vectors. The angle ϕ between $\mathbf{A}$ and $\mathbf{B}$ is defined as follows:

 (i) $\phi = 0$ if and only if $\mathbf{A} = k\mathbf{B}$ for some $k > 0$.

 (ii) $\phi = \pi$ if and only if $\mathbf{A} = k\mathbf{B}$ for some $k < 0$.

 (iii) Otherwise, let $\overrightarrow{OP}$ and $\overrightarrow{OQ}$ be positional representations of $\mathbf{A}$ and $\mathbf{B}$, respectively. Then ϕ is the interior angle O in triangle POQ (Figure 15–7).

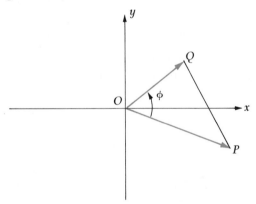

Figure 15-7

From this definition we note that $0 \leq \phi \leq \pi$.

If a constant force, represented by the vector $\mathbf{F}$, is exerted on a particle as it moves in a straight line from a point P to a point Q (Figure 15–8) and ϕ denotes

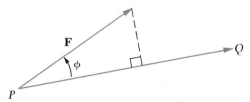

Figure 15-8

the angle between $\mathbf{F}$ and $\overrightarrow{PQ}$, then the component of $\mathbf{F}$ in the direction of $\overrightarrow{PQ}$ is defined to be $|\mathbf{F}| \cos \phi$. From physics the work W done by the force $\mathbf{F}$ in moving the particle from P to Q is given by

$$W = (|\mathbf{F}| \cos \phi)|\overrightarrow{PQ}|. \tag{15.18}$$

We will show that W is the *dot product* of the vectors $\mathbf{F}$ and $\overrightarrow{PQ}$.

15.2.2 Definition

Suppose $\mathbf{A} = \langle a_1, a_2 \rangle$ and $\mathbf{B} = \langle a_1, b_2 \rangle$. The *dot product* of $\mathbf{A}$ and $\mathbf{B}$, denoted by $\mathbf{A} \cdot \mathbf{B}$, is given by

$$\mathbf{A} \cdot \mathbf{B} = a_1 b_1 + a_2 b_2.$$

The dot product is sometimes called a *scalar product* since it is a number rather than a vector. As an illustration,

$$\langle 3, 2 \rangle \cdot \langle 1, -4 \rangle = 3(1) + 2(-4) = -5.$$

In the following properties of the dot product, $\mathbf{A}$ and $\mathbf{B}$ are arbitrary vectors, and k is any number. The proof of (15.20) is given below, and the proofs of the other properties are left as exercises.

$$\mathbf{A} \cdot \mathbf{B} = \mathbf{B} \cdot \mathbf{A} \tag{15.19}$$
$$\mathbf{A} \cdot (\mathbf{B} + \mathbf{C}) = \mathbf{A} \cdot \mathbf{B} + \mathbf{A} \cdot \mathbf{C} \tag{15.20}$$
$$k(\mathbf{A} \cdot \mathbf{B}) = (k\mathbf{A}) \cdot \mathbf{B} \tag{15.21}$$
$$\mathbf{0} \cdot \mathbf{A} = 0 \tag{15.22}$$
$$\mathbf{A} \cdot \mathbf{A} = |\mathbf{A}|^2 \tag{15.23}$$

The unit vectors $\mathbf{i}$ and $\mathbf{j}$ given in (15.16) satisfy the following properties:

$$\mathbf{i} \cdot \mathbf{i} = 1 \tag{15.24}$$
$$\mathbf{i} \cdot \mathbf{j} = 0 \tag{15.25}$$
$$\mathbf{j} \cdot \mathbf{j} = 1. \tag{15.26}$$

PROOF OF (15.20) Using Definition 15.2.2 and the distributive law for real numbers, we obtain

$$\begin{aligned}
\mathbf{A} \cdot (\mathbf{B} + \mathbf{C}) &= \langle a_1, a_2 \rangle \cdot (\langle b_1, b_2 \rangle + \langle c_1, c_2 \rangle) \\
&= \langle a_1, a_2 \rangle \cdot (\langle b_1 + c_1, b_2 + c_2 \rangle) \\
&= a_1(b_1 + c_1) + a_2(b_2 + c_2) \\
&= a_1 b_1 + a_1 c_1 + a_2 b_2 + a_2 c_2 \\
&= (a_1 b_1 + a_2 b_2) + (a_1 c_1 + a_2 c_2) \\
&= \mathbf{A} \cdot \mathbf{B} + \mathbf{A} \cdot \mathbf{C}.
\end{aligned}$$

The next theorem gives a geometric interpretation of the dot product of two vectors.

15.2.3   Theorem

If ϕ, where $0 \leq \phi \leq \pi$, denotes the angle between the nonzero vectors $\mathbf{A} = \langle a_1, a_2 \rangle$ and $\mathbf{B} = \langle b_1, b_2 \rangle$, then

$$\cos \phi = \frac{\mathbf{A} \cdot \mathbf{B}}{|\mathbf{A}| \, |\mathbf{B}|}. \tag{15.27}$$

PROOF   A proof will be given for the separate cases where $\mathbf{A}$ is or is not a scalar multiple of $\mathbf{B}$.

First, if there is a $k \neq 0$ such that $\mathbf{A} = k\mathbf{B}$, then

$$\frac{\mathbf{A} \cdot \mathbf{B}}{|\mathbf{A}| \, |\mathbf{B}|} = \frac{k\mathbf{B} \cdot \mathbf{B}}{|k\mathbf{B}| \, |\mathbf{B}|} = \frac{k(\mathbf{B} \cdot \mathbf{B})}{|k| \, |\mathbf{B}| \, |\mathbf{B}|}$$

$$= \begin{cases} 1 & \text{if } k > 0 \\ -1 & \text{if } k < 0 \end{cases}$$

$$= \begin{cases} \cos 0 & \text{if } k > 0 \\ \cos \pi & \text{if } k < 0. \end{cases}$$

Since by Definition 15.2.1 $\phi = 0$ when $k > 0$ and $\phi = \pi$ when $k < 0$, (15.27) holds if $\mathbf{A}$ is a scalar multiple of $\mathbf{B}$.

Second, if $\mathbf{A}$ is not a scalar multiple of $\mathbf{B}$, the law of cosines can be applied to the triangle with vertices $O(0, 0)$, $P(a_1, a_2)$, and $Q(b_1, b_2)$ (Figure 15–9). Then

$$|PQ|^2 = |OP|^2 + |OQ|^2 - 2|OP| \, |OQ| \cos \phi$$

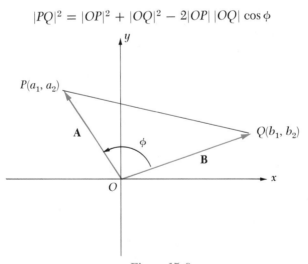

Figure 15–9

and hence

$$\cos \phi = \frac{|OP|^2 + |OQ|^2 - |PQ|^2}{2|OP| \, |OQ|}$$

$$= \frac{(a_1^2 + a_2^2) + (b_1^2 + b_2^2) - [(a_1 - b_1)^2 + (a_2 - b_2)^2]}{2|\mathbf{A}| \, |\mathbf{B}|}$$

$$= \frac{a_1 b_1 + a_2 b_2}{|\mathbf{A}| \, |\mathbf{B}|} = \frac{\mathbf{A} \cdot \mathbf{B}}{|\mathbf{A}| \, |\mathbf{B}|}.$$

Example 1   Find the angle between the vectors $\langle 3, 1 \rangle$ and $\langle -2, 4 \rangle$.

SOLUTION   From (15.27) this angle is expressed by

$$\cos \phi = \frac{\langle 3, 1 \rangle \cdot \langle -2, 4 \rangle}{|\langle 3, 1 \rangle| \, |\langle -2, 4 \rangle|} = \frac{-2}{\sqrt{10} \, \sqrt{20}} = -\frac{\sqrt{2}}{10}$$

$$\approx -.1414.$$

Hence

$$\phi \approx 180° - 81.9° = 98.1°.$$

Using (15.27), the component of the force $\mathbf{F}$ in the direction of $\overrightarrow{PQ}$, which was mentioned in the opening discussion, is

$$|\mathbf{F}| \cos \phi = \frac{\mathbf{F} \cdot \overrightarrow{PQ}}{|\overrightarrow{PQ}|},$$

and hence the work W given by (15.18) is

$$W = \mathbf{F} \cdot \overrightarrow{PQ}.$$

When $\mathbf{A}$ and $\mathbf{B}$ are nonzero vectors, by (15.27) the angle between the vectors is $\phi = \frac{\pi}{2}$, and hence $\cos \phi = 0$, if and only if $\mathbf{A} \cdot \mathbf{B} = 0$.

15.2.4   Definition

The vectors $\mathbf{A}$ and $\mathbf{B}$ are said to be *orthogonal* (*perpendicular*) if and only if $\mathbf{A} \cdot \mathbf{B} = 0$.

Example 2   Determine if the vectors $\langle -2, 3 \rangle$ and $\langle 6, 4 \rangle$ are orthogonal (Figure 15–10).

SOLUTION   Since $\langle -2, 3 \rangle \cdot \langle 6, 4 \rangle = 0$, by Definition 15.2.4 the vectors are orthogonal.

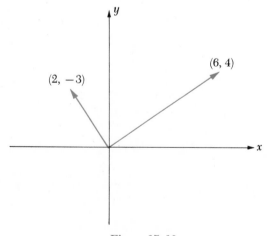

Figure 15–10

In particular, we note from Definition 15.2.4 that the zero vector, **0**, and any vector **A** are orthogonal since $\mathbf{0} \cdot \mathbf{A} = 0$.

Since nonzero vectors that are scalar multiples of each other have either the same or opposite directions, the following definition is given.

15.2.5   Definition

The vectors **A** and **B** are said to be *parallel* if and only if one of the vectors is a scalar multiple of the other.

For example, $\langle 2, -3 \rangle$ and $\langle 1, -3/2 \rangle$ are parallel since $\langle 2, -3 \rangle = 2\langle 1, -3/2 \rangle$. Also, **0** is parallel to any vector **A** since $\mathbf{0} = 0\mathbf{A}$.

A vector approach can be used to prove theorems from elementary geometry when geometric statements are expressible in vector language. For example, if AB and CD are line segments in the plane:

$$|AB| = |CD| \quad \text{if and only if} \quad |\overrightarrow{AB}| = |\overrightarrow{CD}|,$$
$$AB \perp CD \quad \text{if and only if} \quad \overrightarrow{AB} \cdot \overrightarrow{CD} = 0,$$
$$AB \parallel CD \quad \text{if and only if} \quad AB \text{ and } CD \text{ do not intersect and}$$
$$\overrightarrow{AB} \text{ is a scalar multiple of } \overrightarrow{CD}.$$

Example 3   Prove that the opposite sides of a parallelogram have the same length.

PROOF   If A, B, C, and D are consecutive vertices of the parallelogram (Figure 15–11), then

$$AB \parallel DC \qquad AD \parallel BC.$$

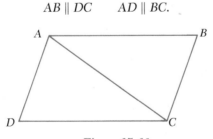

Figure 15-11

Hence there exist nonzero constants m and n such that

$$\overrightarrow{AB} = m\overrightarrow{DC} \qquad \text{and} \qquad \overrightarrow{AD} = n\overrightarrow{BC}. \tag{15.28}$$

Then

$$\overrightarrow{AC} = \overrightarrow{AB} + \overrightarrow{BC} = m\overrightarrow{DC} + \overrightarrow{BC} \tag{15.29}$$

and also

$$\overrightarrow{AC} = \overrightarrow{AD} + \overrightarrow{DC} = n\overrightarrow{BC} + \overrightarrow{DC}. \tag{15.30}$$

From (15.29) and (15.30)

$$m\overrightarrow{DC} + \overrightarrow{BC} = n\overrightarrow{BC} + \overrightarrow{DC},$$

and hence

$$(m - 1)\overrightarrow{DC} = (n - 1)\overrightarrow{BC}.$$

Since $\overrightarrow{DC} \neq \mathbf{0} \neq \overrightarrow{BC}$, either $m = n = 1$ or $m \neq 1 \neq n$. In the latter case

$$\overrightarrow{DC} = \left(\frac{n-1}{m-1}\right)\overrightarrow{BC},$$

and hence $\overrightarrow{DC}$ would be a scalar multiple of $\overrightarrow{BC}$. But this is impossible as BC and DC intersect, but do not lie on the same line. Thus $m = 1$ and $n = 1$, and from (15.28)

$$\overrightarrow{AB} = \overrightarrow{DC} \qquad \text{and} \qquad \overrightarrow{AD} = \overrightarrow{BC}.$$

Hence

$$|\overrightarrow{AB}| = |\overrightarrow{DC}| \qquad \text{and} \qquad |\overrightarrow{AD}| = |\overrightarrow{BC}|,$$

which proves the theorem.

Exercise Set 15.2

In Exercises 1–4 obtain the dot product of the given vectors.

1. $\langle 2, -3\rangle$, $\langle 5, 7\rangle$ 2. $\langle -5, -1\rangle$, $\langle 4, -6\rangle$

3. $-3\mathbf{i} + \mathbf{j}$, $2\mathbf{i} - 4\mathbf{j}$ 4. $\mathbf{i} - 5\mathbf{j}$, $4\mathbf{i} - 2\mathbf{j}$

In Exercises 5–9 find the angle between the given vectors.

5. $\langle 3, 0\rangle$, $\langle -2, 2\rangle$ 6. $\langle 1, \sqrt{3}\rangle$, $\langle -4, 0\rangle$

7. $\langle -2, 1\rangle$, $\langle 3, 4\rangle$ 8. $\langle 3, -5\rangle$, $\langle -1, -2\rangle$

9. If $\mathbf{A} = \langle -3, 2\rangle$, and $\mathbf{B} = \langle 4, k\rangle$, find k if

    (a) $\mathbf{A}$ and $\mathbf{B}$ are orthogonal   (b) $\mathbf{A}$ and $\mathbf{B}$ are parallel.

10. If $\mathbf{A} = \langle 2, 1\rangle$ and $\mathbf{B} = \langle k, -3\rangle$, find k if the angle between $\mathbf{A}$ and $\mathbf{B}$ is $\frac{\pi}{4}$.

11. If $\mathbf{A} = \langle -1, k\rangle$ and $\mathbf{B} = \langle 1, \sqrt{3}\rangle$ find k if the angle between $\mathbf{A}$ and $\mathbf{B}$ is $2\pi/3$.

In Exercises 12–18 derive the following properties of the dot product.

12. (15.19)   13. (15.21)   14. (15.22)

15. (15.23)   16. (15.24)   17. (15.25)

18. (15.26)

19. If $\mathbf{U}$ and $\mathbf{V}$ are orthogonal unit vectors, and $\mathbf{A} = p\mathbf{U} + q\mathbf{V}$, prove that $p = \mathbf{A} \cdot \mathbf{U}$ and $q = \mathbf{A} \cdot \mathbf{V}$.

20. Prove that $(\mathbf{A} + \mathbf{B}) \cdot (\mathbf{C} + \mathbf{D}) = \mathbf{A} \cdot \mathbf{C} + \mathbf{A} \cdot \mathbf{D} + \mathbf{B} \cdot \mathbf{C} + \mathbf{B} \cdot \mathbf{D}$.

21. Prove that $\mathbf{A}$ and $\mathbf{B}$ are orthogonal if and only if $|\mathbf{A} + \mathbf{B}| = |\mathbf{A} - \mathbf{B}|$.

22. If $\mathbf{A} \neq \mathbf{0}$, and $\mathbf{A} \cdot \mathbf{B} = \mathbf{A} \cdot \mathbf{C}$ does $\mathbf{B} = \mathbf{C}$? Why?

23. Prove that a parallelogram is a rectangle if and only if its diagonals have the same length.

24. Prove that the line segment connecting the midpoints of two sides of a triangle is parallel to the third side and has one-half the length of the third side.

25. Prove that a parallelogram is a rhombus if and only if its diagonals are perpendicular.

26. Prove that the medians of a triangle pass through a point that is two-thirds of the distance from any vertex of the triangle to the midpoint of the opposite side.

15.3 Parametric Equations

Before continuing our treatment of vectors in the plane, we will digress to consider the notions of a *plane curve* and the graph of *parametric equations*. Then in Section 15.5 we will unite these ideas with the material from Sections 15.1 and 15.2 by giving a vectorial description of a plane curve.

We have often considered sets of points in the plane which are graphs of equations of the form

$$y = F(x) \tag{15.31}$$

or

$$x = G(y). \tag{15.32}$$

In this section we will consider sets of points $\{(x, y)\}$ such that x and y are given in terms of a third variable t by equations of the form

$$x = f(t) \qquad y = g(t). \tag{15.33}$$

As an illustration, it will be recalled from Theorem 2.2.2, that a point (x, y) is on the line passing through the points (x_1, y_1) and (x_2, y_2) if and only if for some number t

$$x = x_1 + t(x_2 - x_1) \qquad y = y_1 + t(y_2 - y_1).$$

The variable t in (15.33) is called a *parameter* of the variables x and y and the equations (15.33) are termed *parametric equations*. The use of the symbol t as the parameter is suggested by the fact that in motion problems the parameter usually denotes time.

This discussion can be summarized with the following definition.

15.3.1 Definition

The *graph of the parametric equations* (15.33) is the set of points $\{(f(t), g(t)) : t \in \mathcal{D}_f \cap \mathcal{D}_g\}$.

If t_0 is any number in the set $\mathcal{D}_f \cap \mathcal{D}_g$, then $f(t_0)$ and $g(t_0)$ exist, and by Definition 15.3.1 the point $(f(t_0), g(t_0))$ will be in the graph of the equations (15.33) (Figure 15–12).

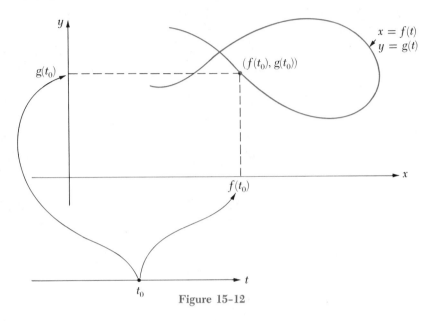

Figure 15–12

Example 1 Sketch the graph of the parametric equations

$$x = t^{1/2} + 1 \qquad y = t^{3/2}. \tag{15.34}$$

SOLUTION The graph consists of all points of the form $(t^{1/2} + 1, t^{3/2})$ and therefore contains points corresponding to every $t \geq 0$. From (15.34) the x coordinates of these points will satisfy $x \geq 1$, and the y coordinates will satisfy $y \geq 0$. By assigning certain non-negative values to t, the coordinates of points in the graph of equations (15.34) can be obtained, as noted in the following table.

t	0	$\frac{1}{4}$	1	$\frac{9}{4}$	4	$\frac{25}{4}$
x	1	$\frac{3}{2}$	2	$\frac{5}{2}$	3	$\frac{7}{2}$
y	0	$\frac{1}{8}$	1	$\frac{27}{8}$	8	$\frac{125}{8}$

A sketch of the graph of the parametric equations (15.34) is shown in Figure 15–13.

It is possible to derive a *Cartesian equation*—that is, an equation in x and y—for the graph of the equations in (15.34). From (15.34) $t^{1/2} = x - 1$, and hence a Cartesian equation for this graph is

$$y = (x - 1)^3. \tag{15.35}$$

However, it is important to realize that the graphs of (15.35) and the parametric

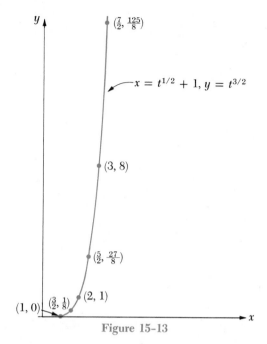

Figure 15–13

equations (15.34) are not identical. The graph of the parametric equations (15.34) consists of only those points (x, y) with coordinates satisfying (15.35) where $x \geq 1$. As this example illustrates, a graph of parametric equations is always a *subset* of the graph of the corresponding Cartesian equation; however, the two graphs may coincide as happens in the next example.

> **Example 2**   Sketch the graph of the parametric equations

$$x = 4 \cos t \qquad y = 3 \sin t. \qquad (15.36)$$

Find the corresponding Cartesian equation.

> **SOLUTION**   The graph consists of all points of the form $(4 \cos t, 3 \sin t)$ and therefore contains points associated with any number t. Since $\cos t$ and $\sin t$ assume values in the interval $[-1, 1]$, the x and y coordinates of these points will satisfy

$$-4 \leq x \leq 4 \qquad -3 \leq y \leq 3.$$

By choosing certain values for t, we obtain from (15.36) the coordinates of points in the graph as noted in the following table.

t	$-\dfrac{\pi}{2}$	$-\dfrac{\pi}{4}$	0	$\dfrac{\pi}{4}$	$\dfrac{\pi}{2}$	$\dfrac{3\pi}{4}$	π	$\dfrac{5\pi}{4}$	$\dfrac{3\pi}{2}$	$\dfrac{7\pi}{4}$	2π
x	0	$2\sqrt{2}$	4	$2\sqrt{2}$	0	$-2\sqrt{2}$	-4	$-2\sqrt{2}$	0	$2\sqrt{2}$	4
y	-3	$-\dfrac{3\sqrt{2}}{2}$	0	$\dfrac{3\sqrt{2}}{2}$	3	$\dfrac{3\sqrt{2}}{2}$	0	$-\dfrac{3\sqrt{2}}{2}$	-3	$-\dfrac{3\sqrt{2}}{2}$	0

A sketch of the graph of the parametric equations (15.36) is shown in Figure 15–14.

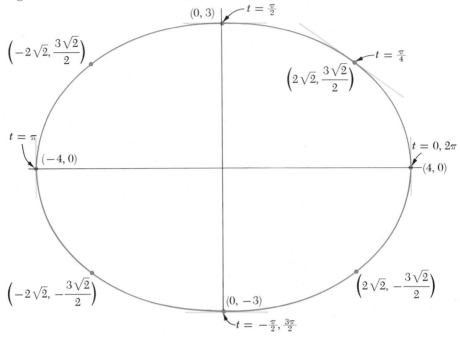

$$\left(-2\sqrt{2}, \frac{3\sqrt{2}}{2}\right)$$

$(0, 3)$ — $t = \frac{\pi}{2}$

$\left(2\sqrt{2}, \frac{3\sqrt{2}}{2}\right)$

$t = \frac{\pi}{4}$

$t = \pi$

$(-4, 0)$

$t = 0, 2\pi$

$(4, 0)$

$\left(-2\sqrt{2}, -\frac{3\sqrt{2}}{2}\right)$

$(0, -3)$

$\left(2\sqrt{2}, -\frac{3\sqrt{2}}{2}\right)$

$t = -\frac{\pi}{2}, \frac{3\pi}{2}$

Figure 15–14

To obtain a Cartesian equation for the graph, we note from (15.36) that

$$\cos t = \frac{x}{4} \qquad \sin t = \frac{y}{3}. \tag{15.37}$$

Then from (15.37), using the identity $\cos^2 t + \sin^2 t = 1$, we obtain

$$\frac{x^2}{16} + \frac{y^2}{9} = 1.$$

It should be noted that the graph of an equation of the form (15.31) has the parametric equations of the form

$$x = t \qquad y = F(t).$$

Similarly, the graph of an equation like (15.32) can be described parametrically by equations of the form

$$x = G(t) \qquad y = t.$$

The graphs obtained in Examples 1 and 2 are examples of *plane curves*.

15.3.2   Definition

Suppose f and g are continuous on an interval I. The set of points $C = \{(f(t), g(t)): t \in I\}$ is called the *plane curve* (or the *plane arc*) that is parametrized by the equations (15.33). The point $(f(a), g(a))$ is called an *endpoint* of the curve if and only if a is an endpoint of I and $a \in I$.

For economy of expression we will refer to plane curves merely as curves in this chapter.

It should be noted that a curve does not have a unique parametrization. For example, the ellipse in Example 2 also has the parametric equations

$$x = -4 \cos t \qquad y = 3 \sin t \tag{15.38}$$

among an infinite number of others. With the parametrization (15.38) the ellipse is traversed in the clockwise direction from the point $(-4, 0)$, as t increases from 0.

Also a parametrization gives the curve an *orientation;* one could, for example, take as the positive direction along a curve the direction in which t increases. In this connection, the parametrizations (15.36) and (15.38) give opposite orientations for the ellipse in Example 2.

The notion of a *tangent* to a curve defined by parametric equations can be developed through a discussion which parallels that in Section 4.1. Suppose f and g satisfy the conditions mentioned in Definition 15.3.2 and $t_0 \in I$. We shall define the tangent T at the point $P_0(f(t_0), g(t_0))$ of the curve C given in Definition 15.3.2 (Figure 15–15). If $P(f(t_0 + h), g(t_0 + h))$ is an arbitrary point in C distinct from P_0, intuition suggests that if the slope of the tangent T exists, then this slope will differ from $\dfrac{g(t_0 + h) - g(t_0)}{f(t_0 + h) - f(t_0)}$, the slope of secant P_0P, by as little as desired if h is sufficiently close to 0. Thus, the slope of T can be defined as

$$\lim_{h \to 0} \frac{g(t_0 + h) - g(t_0)}{f(t_0 + h) - f(t_0)}$$

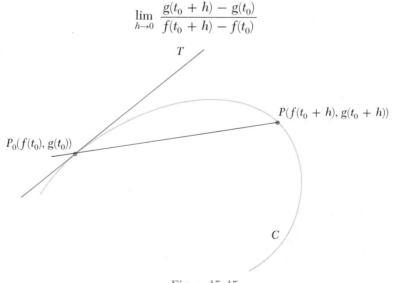

Figure 15–15

if this limit exists. If however, $\dfrac{f(t_0 + h) - f(t_0)}{g(t_0 + h) - g(t_0)}$, the reciprocal of the slope of P_0P, tends to 0 as h approaches 0, intuition suggests that P_0P tends toward a vertical line through the point P_0.

Therefore, the following definition can be stated, which is an extension of Definition 4.1.1 to parametrically defined curves.

15.3.3 Definition

Suppose f and g are continuous on an interval I, and $t_0 \in I$. The *tangent to the curve C* in Definition 15.3.2 at the point $P_0(f(t_0), g(t_0))$ is

(a) The line through P_0 with slope $\lim\limits_{h \to 0} \dfrac{g(t_0 + h) - g(t_0)}{f(t_0 + h) - f(t_0)}$

(in particular, this line is termed a *horizontal tangent* if this slope is 0),

or

(b) The line $x = f(t_0)$ if $\lim\limits_{h \to 0} \dfrac{f(t_0 + h) - f(t_0)}{g(t_0 + h) - g(t_0)} = 0$

(this line is then called a *vertical tangent* to C at P_0).

In both of the limits mentioned in parts (a) and (b) fail to exist, then C has no tangent at P_0.

The following theorem is useful in locating tangents to a curve.

15.3.4 Theorem

Suppose f and g are continuous on an interval I, $t_0 \in I$, and f and g are differentiable at t_0. Let C be the curve given in Definition 15.3.2.

(a) *If $f'(t_0) \neq 0$, then the slope of the tangent to C through $(f(t_0), g(t_0))$ is $g'(t_0)/f'(t_0)$. (In particular if $g'(t_0) = 0$ and $f'(t_0) \neq 0$, then the curve C has a horizontal tangent at P_0.)*

(b) *If $g'(t_0) \neq 0$ and $f'(t_0) = 0$, then C has a vertical tangent at P_0.*

PROOF OF (a) Since $f'(t_0) = \lim_{h \to 0}(f(t_0 + h) - f(t_0))/h \neq 0$, by Theorem 3.3.3, regardless of the sign of $f'(t_0)$, there is a $\delta > 0$ such that

$$\frac{f(t_0 + h) - f(t_0)}{h} \neq 0$$

whenever

$$h \neq 0, \ h \in (-\delta, \delta), \ \text{and} \ t_0 + h \in I. \tag{15.39}$$

If h satisfies (15.39), then

$$\frac{g(t_0 + h) - g(t_0)}{f(t_0 + h) - f(t_0)} = \frac{\dfrac{g(t_0 + h) - g(t_0)}{h}}{\dfrac{f(t_0 + h) - f(t_0)}{h}}$$

certainly is defined and hence by Theorems 3.5.1 and 3.4.3(d)

$$\lim_{h \to 0} \frac{g(t_0 + h) - g(t_0)}{f(t_0 + h) - f(t_0)} = \frac{\lim\limits_{h \to 0} \dfrac{g(t_0 + h) - g(t_0)}{h}}{\lim\limits_{h \to 0} \dfrac{f(t_0 + h) - f(t_0)}{h}} = \frac{g'(t_0)}{f'(t_0)}. \tag{15.40}$$

Thus from (15.40) and Definition 15.3.3 $g'(t_0)/f'(t_0)$ is the slope of the tangent to the curve C at P_0.

The proof of (b) is similar and is left as an exercise.

Example 3 Find an equation of the tangent to the ellipse $x = 4 \cos t$, $y = 3 \sin t$ at the point where $t = \frac{\pi}{4}$ (Figure 15–14). At what points in the ellipse is the tangent horizontal? vertical?

SOLUTIONS Since the functions f and g where

$$x = f(t) = 4 \cos t \qquad y = g(t) = 3 \sin t$$

are continuous and differentiable everywhere, by Theorem 15.3.4(a) the slope of the tangent at $t = \frac{\pi}{4}$ is

$$\frac{g'(\frac{\pi}{4})}{f'(\frac{\pi}{4})} = \frac{3 \cos \frac{\pi}{4}}{-4 \sin \frac{\pi}{4}} = -\frac{3}{4}.$$

The point in the curve where $t = \frac{\pi}{4}$ is $(2\sqrt{2}, 3\sqrt{2}/2)$. Hence, using the point-slope form for an equation of a line, an equation of the required tangent is

$$y - \frac{3\sqrt{2}}{2} = -\frac{3}{4}(x - 2\sqrt{2}),$$

or

$$3x + 4y = 12\sqrt{2}.$$

At the points $(4 \cos t, 3 \sin t)$ on the curve the tangent is horizontal by Theorem 15.3.4(a) if

$$g'(t) = \frac{dy}{dt} = 3 \cos t = 0 \qquad \text{and} \qquad f'(t) = \frac{dx}{dt} = -4 \sin t \neq 0. \quad (15.41)$$

The conditions (15.41) are fulfilled if $t = \frac{\pi}{2} + 2n\pi$ or $t = -\frac{\pi}{2} + 2n\pi$, where n is any integer. The point corresponding to $t = \frac{\pi}{2} + 2n\pi$ is $(0, 3)$ and the point corresponding to $t = -\frac{\pi}{2} + 2n\pi$ is $(0, -3)$. Thus there are horizontal tangents at $(0, 3)$ and $(0, -3)$.

At points $(4 \cos t, 3 \sin t)$ on the curve the tangent is vertical by Theorem 15.3.4(b) if

$$f'(t) = \frac{dx}{dt} = -4 \sin t = 0 \qquad \text{and} \qquad g'(t) = \frac{dy}{dt} = 3 \cos t \neq 0. \quad (15.42)$$

The conditions are fulfilled if $t = n\pi$ where n is any integer. Thus at the corresponding points in the curve, $(4, 0)$ and $(-4, 0)$, there are vertical tangents. The tangents obtained in this example are shown in Figure 15–14.

Exercise Set 15.3

In Exercises 1–12:

(a) Sketch the graph of the given parametric equations.
(b) Find a Cartesian equation associated with the graph.

(c) Find an equation of the tangent to the curve at the point corresponding to the given value of t.

(d) Find the points in the curve where the curve has a horizontal tangent.

(e) Find the points in the curve where the curve has a vertical tangent.

1. $x = 3t^2, \ y = 2t; \quad t = 1$ 2. $x = t^{2/3}, \ y = t + 2; \quad t = -1$

3. $x = 2t, \ y = \sqrt{4 - t}; \quad t = 0$ 4. $x = 2 - t^2, \ y = 1 + 2t^2; \quad t = 2$

5. $x = 3 - \cos\theta, \ y = 2 + 4\sin\theta; \quad \theta = \frac{5\pi}{6}$

6. $x = \tan t, \ y = \sec t; \quad t = \frac{3\pi}{4}$

7. $x = e^t, \ y = e^{-t}; \quad t = \ln 2$

8. $x = 1 - 2\sinh t, \ y = -2 + 4\cosh t; \quad t = \ln 3$

9. $x = 2\cos t, \ y = 1 + \cos 2t; \quad t = \frac{\pi}{3}$

10. $x = \cos 2t, \ y = \sin t; \quad t = \frac{4\pi}{3}$

11. $x = \dfrac{2t}{1 + t^2}, \ y = \dfrac{1 - t^2}{1 + t^2}; \quad t = \dfrac{1}{2}$

12. $x = 4t - t^2, \ y = 1 + t; \quad t = -2$

13. The graph of each of the following pairs of parametric equations is a subset of the graph of the Cartesian equation $y = 1 - x^2$. Describe each such subset and how, as t changes, the subset is traversed by a moving point with coordinates expressed by the given parametric equations.

(a) $x = \sqrt{1 - t}, y = t$

(b) $x = \cos t, y = \sin^2 t$

(c) $x = \dfrac{1}{t}, y = \dfrac{t^2 - 1}{t^2}$

(d) $x = \dfrac{1}{\sqrt{t + 1}}, y = \dfrac{t}{t + 1}$

(e) $x = \sec t, y = -\tan^2 t$

In Exercises 14–18 find parametric equations for the graph of the given Cartesian equation.

14. $3x - 4y = 5$ 15. $x^2 + 2y = 4$

16. $y = -1$ 17. $x^2 + y^2 - 2x + 4y - 11 = 0$

18. $x^2 + y^2 + 3x - 6y - 20 = 0$

19. Prove Theorem 15.3.4(b).

20. Suppose the functions f and g are continuous on an interval I and $f'(t) > 0$ (or $f'(t) < 0$) for every $t \in I$. Let $J = \{f(t): t \in I\}$, and suppose x and y are defined by (15.33). Then if $y = h(x) = g(f^{-1}(x))$, prove that

$$h'(x) = \frac{g'(f^{-1}(x))}{f'(f^{-1}(x))} \quad \text{for } x \in J;$$

that is,

$$\frac{dy}{dx} = \frac{\dfrac{dy}{dt}}{\dfrac{dx}{dt}}.$$

21. Suppose that in addition to the conditions described in Exercise 20, f and g have second derivatives on I. Prove that

$$h''(x) = \frac{f'(f^{-1}(x))g''(f^{-1}(x)) - g'(f^{-1}(x))f''(f^{-1}(x))}{(f'(f^{-1}(x)))^3} \quad \text{for } x \in J;$$

that is,

$$\frac{d^2y}{dx^2} = \frac{\dfrac{dx}{dt}\dfrac{d^2y}{dt} - \dfrac{dy}{dt}\dfrac{d^2x}{dt^2}}{\left(\dfrac{dx}{dt}\right)^3}.$$

22. Suppose a circle of radius r rolls without slipping on a straight line. Obtain parametric equations for the locus of a fixed point P in the circle (Figure 15–16).

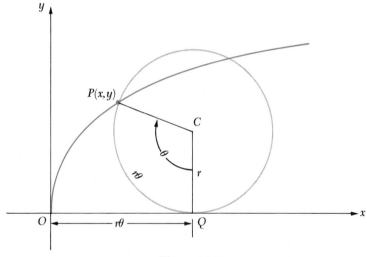

Figure 15–16

This curve is called a *cycloid*. HINT: Suppose that when the circle has its center C at $(0, r)$, then $P(x, y)$ is at the origin. Also suppose P moves counterclockwise as the circle rolls to the right along the x axis. If Q is the point of tangency of the circle with the x axis and angle PCQ has radian measure θ, then $|\overset{\frown}{QP}| = r\theta = |OQ|$.

23. Suppose that a circle of radius r rolls without slipping on the inside of a circle

of radius a where $a > r$ (Figure 15–17). If θ denotes the angle QOQ' in Figure 15–17, prove that the locus of a fixed point $P(x, y)$ on the smaller circle has parametric equations

$$x = (a - r) \cos \theta + r \cos \theta \left(\frac{a}{r} - 1 \right) \qquad y = (a - r) \sin \theta - r \sin \theta \left(\frac{a}{r} - 1 \right).$$

This curve is called a *hypocycloid*.

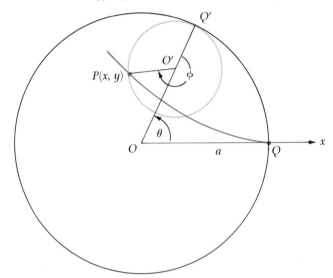

Figure 15-17

24. If a circle of radius r rolls without slipping on the outside of a circle of radius a, where $a > r$, obtain parametric equations for the curve generated by a fixed point on the smaller circle. This curve is called an *epicycloid*.

15.4   Length of a Plane Curve

In this section we will define the notion of the length of a curve and calculate these lengths using the definition that will be given. We begin by considering the curve expressed parametrically by

$$x = f(t) \qquad y = g(t) \quad \text{for } t \in [a, b]. \tag{15.43}$$

We will suppose that this curve is *smooth;* that is, the functions f and g are continuous and have continuous derivatives on $[a, b]$. Also, it will be supposed here that for any t_1 and t_2 in $[a, b)$,

$$(f(t_1), g(t_1)) \neq (f(t_2), g(t_2)) \quad \text{when } t_1 \neq t_2. \tag{15.44}$$

The restriction (15.44) requires that no portion of the curve be traversed more than once as t goes from a to b. For example, the parametrization

$$x = 4 \cos t \qquad y = 3 \sin t \quad \text{for } t \in [0, 4\pi]$$

for the ellipse in Example 2 of Section 15.3 would violate (15.44) since the ellipse is traversed once as t goes from 0 to 2π and again as t goes from 2π to 4π. However, the parametric representation

$$x = 4 \cos t \qquad y = 3 \sin t \quad \text{for } t \in [0, 2\pi]$$

for the ellipse is in conformity with (15.44) since only the point $(4, 0)$ on the graph is obtained from two different values of the parameter, but these values are $t = 0$ and 2π, the endpoints of the interval $[0, 2\pi]$.

If $\{a = t_0, t_1, t_2, \ldots, t_{n-1}, t_n = b\}$ is a partition of $[a, b]$, then for each number t_i in the partition let $P_i = (f(t_i), g(t_i))$ be the corresponding point in the arc (15.43). The sum of the lengths of the chords of the arc formed by joining the points P_i successively (Figure 15–18) is

$$\sum_{i=1}^{n} |P_{i-1}P_i| = \sum_{i=1}^{n} \sqrt{[f(t_i) - f(t_{i-1})]^2 + [g(t_i) - g(t_{i-1})]^2}. \qquad (15.45)$$

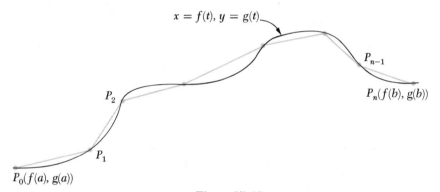

Figure 15-18

Since the functions f and g satisfy the hypotheses of the mean value theorem on each subinterval $[t_{i-1}, t_i]$, where $i = 1, 2, \ldots, n$, there exist numbers u_i and v_i in each interval (t_{i-1}, t_i) such that

$$f(t_i) - f(t_{i-1}) = f'(u_i) \, \Delta t_i \qquad g(t_i) - g(t_{i-1}) = g'(v_i) \, \Delta t_i \qquad (15.46)$$

where $\Delta t_i = t_i - t_{i-1}$.

Substitutions from (15.46) into (15.45) give

$$\sum_{i=1}^{n} |P_{i-1}P_i| = \sum_{i=1}^{n} \sqrt{(f'(u_i))^2 + (g'(v_i))^2} \, \Delta t_i. \qquad (15.47)$$

Now, the sum (15.47) is not a Riemann sum of the function $(f')^2 + (g')^2$ on $[a, b]$, since f' and g' are not evaluated at the same number in each interval (t_{i-1}, t_i). However, it can be proved, using theory beyond the scope of this text, that the sum (15.47) will differ from the definite integral $\int_a^b \sqrt{(f'(t))^2 + (g'(t))^2} \, dt$ by as little as we please provided the norm of the associated partition is sufficiently small.

In summary, the following definition of arc length can be given.

15.4.1 Definition

Suppose the functions f and g have continuous derivatives on the interval $[a, b]$ and (15.44) is satisfied. The length L of the curve (15.43) is given by

$$L = \int_a^{b} \sqrt{(f'(t))^2 + (g'(t))^2}\, dt = \int_a^{b} \sqrt{\left(\frac{dx}{dt}\right)^2 + \left(\frac{dy}{dt}\right)^2}\, dt. \qquad (15.48)$$

Example 1 Find the length of the curve (Figure 15–19) given by

$$x = \tfrac{1}{3}t^3 \qquad y = t^2 \quad \text{if } t \in [-\sqrt{5}, 0].$$

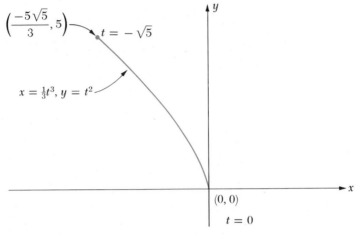

Figure 15–19

SOLUTION From the given parametric equations

$$\frac{dx}{dt} = t^2 \qquad \frac{dy}{dt} = 2t,$$

and hence from (15.48)

$$L = \int_{-\sqrt{5}}^{0} \sqrt{t^4 + 4t^2}\, dt = \int_{-\sqrt{5}}^{0} |t|\sqrt{t^2 + 4}\, dt.$$

Since $|t| = -t$ when $t < 0$,

$$L = \int_{-\sqrt{5}}^{0} -t\sqrt{t^2 + 4}\, dt$$

$$= -\tfrac{1}{2} \cdot \tfrac{2}{3}(t^2 + 4)^{3/2}\Big|_{-\sqrt{5}}^{0} = -\tfrac{1}{3}(8 - 27) = \tfrac{19}{3}.$$

Example 2 Show that the circumference of a circle of radius r is $2\pi r$.

SOLUTION Suppose the coordinate axes are chosen so that the origin is the center of the circle. Since the circle has the parametrization

$$x = r \cos t \qquad y = r \sin t \quad \text{for } t \in [0, 2\pi],$$

we have

$$\frac{dx}{dt} = -r \sin t \qquad \frac{dy}{dt} = r \cos t \quad \text{for } t \in [0, 2\pi].$$

Therefore by (15.48) the circumference of the circle is

$$C = \int_0^{2\pi} \sqrt{(-r \sin t)^2 + (r \cos t)^2}\, dt$$

$$= \int_0^{2\pi} r\, dt = 2\pi r.$$

The reader may notice that the reasoning employed in Example 1 is circular because the existence of arc length on a circle was assumed in order to define the functions sine and cosine and then these functions were used to obtain an arc length on a circle. However, as we noted in Section 10.1, these functions could have been defined without assuming in advance the existence of arc length on a circle, thereby removing this reasoning difficulty. The steps leading to these definitions are outlined below and the reader may wish to investigate them in more detail:

1. In (8.17) we gave the definition for the number π,

$$\pi = 4 \int_0^1 \sqrt{1 - u^2}\, du. \qquad (15.49)$$

2. Then, consider the function A where

$$A(t) = \int_0^t \sqrt{1 - u^2}\, du - \tfrac{1}{2} t \sqrt{1 - t^2} \quad \text{if } t \in [-1, 1]. \quad (15.50)$$

If $0 < t \leq 1$, $A(t)$ is the area of the sector POQ of the circle $x^2 + y^2 = 1$ determined by the points $O(0,0)$, $Q(1,0)$, and $P(\sqrt{1 - t^2}, t)$ (Figure 15–20). To prove this, one notes that by (8.13)

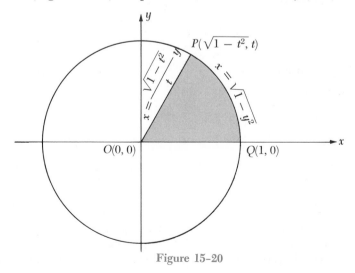

Figure 15–20

$$\text{Area of sector } POQ = \int_0^t \left(\sqrt{1 - y^2} - \frac{\sqrt{1 - t^2}}{t} y \right) dy$$

$$= \int_0^t \sqrt{1 - y^2} \, dy - \tfrac{1}{2} t \sqrt{1 - t^2} = A(t).$$

If $-1 \leq t < 0$, then $A(t)$ is the negative of the area of sector POQ.

3. From (15.50) using the first fundamental theorem

$$A'(t) = \frac{1}{2\sqrt{1 - t^2}} \quad \text{for } t \in (-1, 1),$$

and hence since $A(0) = 0$,

$$A(t) = \frac{1}{2} \int_0^t \frac{du}{\sqrt{1 - u^2}} \quad \text{for } t \in (-1, 1). \qquad (15.51)$$

4. From (15.50) A is continuous on $[-1, 1]$ and therefore from (15.49)

$$\lim_{t \to 1} A(t) = A(1) = \frac{\pi}{4}.$$

Hence from (15.51)

$$\lim_{t \to 1} \int_0^t \frac{du}{\sqrt{1 - u^2}} = \frac{\pi}{2}. \qquad (15.52)$$

5. Since $\displaystyle \int_0^{-t} \sqrt{1 - u^2} \, du = -\int_0^t \sqrt{1 - u^2} \, du$, from (15.49) and (15.50)

$$\lim_{t \to -1} A(t) = A(-1) = -\frac{\pi}{4}.$$

Hence from (15.51)

$$\lim_{t \to -1} \int_0^t \frac{du}{\sqrt{1 - u^2}} = -\frac{\pi}{2}. \qquad (15.53)$$

6. We define the function $\sin^{-1}$ by requiring that:

$$\theta = \sin^{-1} t = \begin{cases} \dfrac{\pi}{2} & \text{if } t = 1 \\[2mm] \displaystyle\int_0^t \dfrac{du}{\sqrt{1 - u^2}} & \text{if } -1 < t < 1 \\[2mm] -\dfrac{\pi}{2} & \text{if } t = -1. \end{cases} \qquad (15.54)$$

7. Because of (15.52), (15.53), and the first fundamental theorem, the function $\sin^{-1}$ is continuous and increasing on its domain, $[-1, 1]$, and has the interval $[-\frac{\pi}{2}, \frac{\pi}{2}]$ as its range. Hence by Theorem 9.4.1 $\sin^{-1}$ has an inverse function, which we denote by f, that has domain

$[-\frac{\pi}{2}, \frac{\pi}{2}]$ and range $[-1, 1]$. From (15.54) $\sin^{-1}(-t) = -\sin^{-1} t$ and therefore

$$f(-\sin^{-1} t) = -f(\sin^{-1} t).$$

8. Finally we define the sine and cosine functions by requiring that

$$\sin \theta = f(\theta) \qquad \text{if } \theta \in \left[-\frac{\pi}{2}, \frac{\pi}{2}\right]$$

$$\left.\begin{array}{l} \sin(-\theta) = -\sin \theta \\ \sin(\theta + \pi) = -\sin \theta \end{array}\right\} \quad \text{for every } \theta$$

and   (15.55)

$$\cos \theta = \sqrt{1 - (f(\theta))^2} \quad \text{if } \theta \in \left[-\frac{\pi}{2}, \frac{\pi}{2}\right]$$

$$\left.\begin{array}{l} \cos(-\theta) = \cos \theta \\ \cos(\theta + \pi) = -\cos \theta \end{array}\right\} \quad \text{for every } \theta$$

where the function f is defined in step 7.

The properties of sine and cosine that were derived in Chapter 10 can also be derived from these definitions.

We return to the notion of arc length in order to define the length of a *chain of curves*.

15.4.2   Definition

Suppose C_1, C_2, . . . , C_n are curves such that for $i = 1, 2, \ldots, n$ each C_i is given parametrically by

$$x = f_i(t) \qquad y = g_i(t) \quad \text{if } t \in [a_i, b_i]$$

and suppose also that the curves C_i are joined *end-to-end*; that is, for $i = 1, 2, \ldots,$ $n - 1$

$$f_i(b_i) = f_{i+1}(a_{i+1}) \qquad g_i(b_i) = g_{i+1}(a_{i+1}).$$

The curve $C = C_1 \cup C_2 \cup \cdots \cup C_n$ is then called a *chain* of the curves C_1, C_2, . . . , C_n (Figure 15–21).

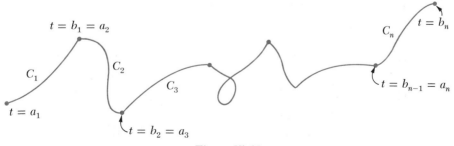

Figure 15–21

If the curves $C_1, C_2, \ldots, C_n$ in the chain C in Definition 15.4.2 are smooth curves, then C is called a *sectionally smooth curve*. If $L_1, L_2, \ldots, L_n$ are the respective lengths of $C_1, C_2, \ldots, C_n$, the length L of C is defined to be

$$L = L_1 + L_2 + \cdots + L_n,$$

as an extension of Definition 15.4.1.

Exercise Set 15.4

In Exercises 1–10 obtain the length of the given curve.

1. $\left.\begin{array}{l} x = 2t - 1 \\ y = -t + 3 \end{array}\right\} t \in [1, 5]$

2. $\left.\begin{array}{l} x = (t - 2)^{3/2} \\ y = \frac{3}{2}t - 1 \end{array}\right\} t \in [0, 2]$

3. $\left.\begin{array}{l} x = \frac{1}{2}(t - 1)^2 \\ y = \frac{4}{3}t^{3/2} \end{array}\right\} t \in [0, 4]$

4. $\left.\begin{array}{l} x = t^2 \\ y = 2t \end{array}\right\} t \in [0, 2]$

5. $\left.\begin{array}{l} x = t^2 \\ y = 4t^3 \end{array}\right\} t \in [-1, 1]$

6. $\left.\begin{array}{l} x = e^{-t}\cos t \\ y = e^{-t}\sin t \end{array}\right\} t \in \left[0, \dfrac{\pi}{2}\right]$

7. $\left.\begin{array}{l} x = a\cos^3 t \\ y = a\sin^3 t \end{array}\right\} t \in [0, 2\pi]$

8. $\left.\begin{array}{l} x = \sin^{-1}\dfrac{t}{a} \\ y = \ln\sqrt{a^2 - t^2} \end{array}\right\} t \in \left[0, \dfrac{1}{2}a\right]$

9. $\left.\begin{array}{l} x = t + \sin t \\ y = 1 - \cos t \end{array}\right\} t \in [0, 2\pi]$

10. $\left.\begin{array}{l} x = \dfrac{t^2}{4} + \dfrac{1}{2t} \\ y = 2t^{1/2} \end{array}\right\} t \in [1, 4]$

11. Calculate the length of the upper semicircle of the circle $x^2 + y^2 = 1$ using the parametrization

$$x = \frac{2t}{1 + t^2}, \quad y = \frac{1 - t^2}{1 + t^2} \text{ for } t \in [-1, 1].$$

12. If F has a continuous derivative on $[a, b]$, prove from (15.48) that the length of the arc $y = F(x)$ if $x \in [a, b]$ is

$$\int_a^b \sqrt{1 + (F'(x))^2}\ dx.$$

In Exercises 13–15 find the length of the given curve.

13. $y = \ln\sec x$ from $x = 0$ to $x = \frac{\pi}{4}$

14. $y = \cosh x$ from $x = 0$ to $x = \ln 2$

15. $y^2 = 4x$ from $y = -2$ to $y = 2$

16. It will be recalled that in the derivation of the limit $\lim_{\theta \to 0} \sin\theta/\theta = 1$ (Theorem 10.2.1(c)) the formula

$$A = \tfrac{1}{2}r^2\theta$$

for the area of a sector of a circle with radius r and central angle θ was utilized. Obtain this formula when $0 < \theta < \frac{\pi}{2}$ using properties of trigonometric functions as may be required (these properties can be developed independently of the area formula using the definitions in (15.55)).

17. Assuming no other knowledge of the trigonometric functions except what is given in steps 6, 7, and 8 in the analytic development of the sine and cosine functions, prove that

(a) $\sin 0 = 0$

(b) $\sin' (0) = 1$

(c) $\displaystyle\lim_{\theta \to 0} \frac{\sin \theta}{\theta} = 1$

15.5   Vector-Valued Functions

In order to give a vectorial treatment of plane curves, which will be useful in discussing curvilinear motion in the plane, we now consider the notion of a *vector-valued function*. In such a function the first element of each ordered pair is some number, and the second element is a vector. Vector-valued functions are usually denoted by boldface letters, and the usual conventions about functional notation are followed. For example, if $\mathbf{P} = \{(t, \mathbf{X})\}$, then one may write

$$\mathbf{X} = \mathbf{P}(t). \tag{15.56}$$

Since $\mathbf{P}(t)$ is a vector, it will be of the form

$$\mathbf{P}(t) = \langle f(t), g(t) \rangle \tag{15.57}$$

where f and g are real-valued functions of the real variable t. In contrast to the vector-valued function $\mathbf{P}$, functions like f and g, here, are sometimes called *scalar functions*. The domain of $\mathbf{P}$, $\mathcal{D}_{\mathbf{P}}$, is the set of all numbers t such that

$$t \in \mathcal{D}_f \cap \mathcal{D}_g.$$

Example 1   Find the domain of the vector-valued function given by

$$\mathbf{P}(t) = \left\langle \sqrt{1 - t}, \frac{1}{t} \right\rangle.$$

SOLUTION   Letting $f(t) = \sqrt{1 - t}$ and $g(t) = 1/t$, we note that $\mathcal{D}_f = (-\infty, 1]$ and $\mathcal{D}_g = \{t : t \neq 0\}$. Therefore the domain of $\mathbf{P}$ is

$$\begin{aligned} \mathcal{D}_{\mathbf{P}} &= (-\infty, 1] \cap \{t : t \neq 0\} \\ &= (-\infty, 0) \cup (0, 1]. \end{aligned}$$

If $\mathbf{P}(t)$ is given by (15.57), then for any $t_0 \in \mathcal{D}_{\mathbf{P}}$, the vector $\mathbf{P}(t_0)$ can be represented by an arrow drawn from the origin to the point $(f(t_0), g(t_0))$, which is in the graph of the parametric equations $x = f(t)$, $y = g(t)$. Thus $\mathbf{P}(t_0)$ can be called the *position vector* of the point $(f(t_0), g(t_0))$ (Figure 15–22).

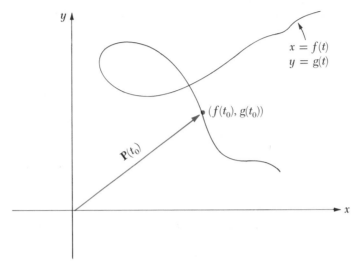

Figure 15–22

Also, the *graph* of $\mathbf{P}$ is defined to be the graph of the parametric equations $x = f(t)$, $y = g(t)$ (Definition 15.3.1).

We next consider limits of vector-valued functions.

15.5.1 Definition

Suppose $\mathbf{P}$ is defined by (15.57). Then $\lim_{t \to t_0} \mathbf{P}(t) = \langle L_1, L_2 \rangle$ if and only if $\lim_{t \to t_0} f(t) = L_1$ and $\lim_{t \to t_0} g(t) = L_2$.

Example 2   If $\mathbf{P}(t) = \langle 2 - \cos t, -2 \sin t \rangle$ find $\lim_{t \to \pi/2} \mathbf{P}(t)$.

SOLUTION   Since

$$\lim_{t \to \pi/2} (2 - \cos t) = 2 \qquad \text{and} \qquad \lim_{t \to \pi/2} (-2 \sin t) = -2,$$

from Definition 15.5.1

$$\lim_{t \to \pi/2} \mathbf{P}(t) = \lim_{t \to \pi/2} \langle 2 - \cos t, -2 \sin t \rangle = \langle 2, -2 \rangle.$$

The following definitions of continuity and differentiability for vector-valued functions are analogous to Definitions 3.7.1 and 4.3.1, respectively.

15.5.2 Definition

The vector-valued function $\mathbf{P}$ is *continuous* at a number $t_0 \in \mathcal{D}_\mathbf{P}$ if and only if

$$\lim_{t \to t_0} \mathbf{P}(t) = \mathbf{P}(t_0).$$

15.5.3   Definition

The derivative of a vector-valued function **P** is the vector-valued function **P'** defined
by

$$\mathbf{P}'(t) = \lim_{h \to 0} \frac{\mathbf{P}(t + h) - \mathbf{P}(t)}{h}.$$

The domain of **P'**, $\mathfrak{D}_{\mathbf{P}'}$, is the set of all numbers t for which this limit exists. Also,
P is said to be *differentiable* at a number t if and only if **P'**(t) exists.

The derivative of a vector-valued function is usually obtained using the
following theorem, which can be proved from Definitions 15.5.3 and 15.5.1. (See
Exercise 16, Problem Set 15.5.)

15.5.4   Theorem

Suppose **P**(t) *is given by* (15.57). *Then*

$$\mathbf{P}'(t) = \langle f'(t), g'(t) \rangle$$

if $f'(t)$ *and* $g'(t)$ *exist.*

From Theorem 15.5.4 if $f'(t_0)$ and $g'(t_0)$ exist and are not both zero, then
P'(t_0) can be represented by an arrow from the point $(f(t_0), g(t_0))$ to $(f(t_0) + f'(t_0)$,
$g(t_0) + g'(t_0))$ (Figure 15–23). Now, the line through $(f(t_0), g(t_0))$ and $(f(t_0) + f'(t_0)$,
$g(t_0) + g'(t_0))$ has slope $g'(t_0)/f'(t_0)$ and, by Theorem 15.3.4, is therefore the tangent
to the graph of **P** at $(f(t_0), g(t_0))$. Hence one is justified in calling **P'**(t_0) a *tangent
vector* to the graph of **P** at $(f(t_0), g(t_0))$.

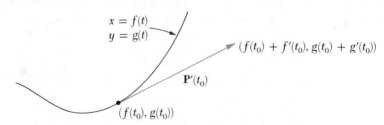

Figure 15–23

Example 3   If **P**$(t) = \langle 2 - \cos t, -2 \sin t \rangle$, find **P'**$(t)$ and **P'**$(\frac{\pi}{3})$.

SOLUTION   By Theorem 15.5.4

$$\mathbf{P}'(t) = \langle D_t(2 - \cos t), D_t(-2 \sin t) \rangle = \langle \sin t, -2 \cos t \rangle.$$

Hence

$$\mathbf{P}'\left(\frac{\pi}{3}\right) = \left\langle \frac{\sqrt{3}}{2}, -1 \right\rangle.$$

The vector $\mathbf{P}'(\frac{\pi}{3})$ obtained here is represented in Figure 15–24. Since the graph of $\mathbf{P}(t) = \langle 2 - \cos t, -2 \sin t \rangle$ is the ellipse $(x - 2)^2 + (y^2/4) = 1$, $\mathbf{P}'(\frac{\pi}{3})$ is tangent to the ellipse at the point having position vector $\mathbf{P}(\frac{\pi}{3}) = \langle \frac{3}{2}, -\sqrt{3} \rangle$.

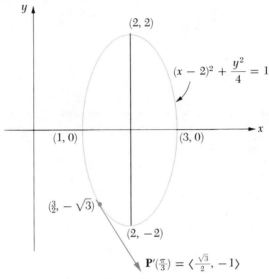

Figure 15-24

Because of Theorem 15.5.4, if f and g satisfy the conditions expressed in Definition 15.4.1, the length of the curve $\{(f(t), g(t)) : t \in [a, b]\}$ is given by

$$L = \int_a^b |\mathbf{P}'(t)| \, dt.$$

Each of the following differentiation formulas for vector-valued functions can be verified for every t at which the indicated values of the derivatives in the right members exist. Their proofs utilize the important differentiation formulas from Chapter 4. In the first two formulas h is a real-valued function of t.

$$D_t[h(t)\mathbf{P}(t)] = h(t)\mathbf{P}'(t) + h'(t)\mathbf{P}(t) \tag{15.58}$$

$$D_t[\mathbf{P}(h(t))] = h'(t)\mathbf{P}'(h(t)) \tag{15.59}$$

$$D_t[\mathbf{P}(t) + \mathbf{Q}(t)] = \mathbf{P}'(t) + \mathbf{Q}'(t) \tag{15.60}$$

$$D_t[\mathbf{P}(t) \cdot \mathbf{Q}(t)] = \mathbf{P}(t) \cdot \mathbf{Q}'(t) + \mathbf{P}'(t) \cdot \mathbf{Q}(t) \tag{15.61}$$

PROOF OF (15.58)   If $\mathbf{P}(t) = \langle f(t), g(t) \rangle$, then

$$D_t[h(t)\mathbf{P}(t)] = D_t\langle h(t)f(t), h(t)g(t) \rangle,$$

and by Theorem 15.5.4

$$D_t[h(t)\mathbf{P}(t)] = \langle D_t(h(t)f(t)), D_t(h(t)g(t)) \rangle$$
$$= \langle h(t)f'(t) + h'(t)f(t), h(t)g'(t) + h'(t)g(t) \rangle.$$

Since the right member is the vector sum $\langle h(t)f'(t), h(t)g'(t) \rangle + \langle h'(t)f(t) +$

$h'(t)g(t)\rangle$, the scalars $h(t)$ and $h'(t)$ can be factored out, and

$$D_t[h(t)\mathbf{P}(t)] = h(t)\langle f'(t), g'(t)\rangle + h'(t)\langle f(t), g(t)\rangle$$
$$= h(t)\mathbf{P}'(t) + h'(t)\mathbf{P}(t).$$

The proofs of the other properties are left as exercises.

We can define higher derivatives of vector-valued functions. The second derivative of the vector-valued function $\mathbf{P}$ is given by

$$\mathbf{P}''(t) = D_t\mathbf{P}'(t)$$

and in general

$$\mathbf{P}^{(n)}(t) = D_t\mathbf{P}^{(n-1)}(t) \quad \text{if } n = 2, 3, \ldots.$$

Example 4 If $\mathbf{P}(t) = \langle 2 - \cos t, -2 \sin t\rangle$, find $\mathbf{P}''(t)$ and $\mathbf{P}''(\frac{\pi}{3})$.

SOLUTION In Example 3 we obtained

$$\mathbf{P}'(t) = \langle \sin t, -2 \cos t\rangle$$

and hence

$$\mathbf{P}''(t) = D_t\mathbf{P}'(t) = \langle \cos t, 2 \sin t\rangle. \tag{15.62}$$

In particular from (15.62)

$$\mathbf{P}''\left(\frac{\pi}{3}\right) = \left\langle \frac{1}{2}, \sqrt{3}\right\rangle.$$

The vector $\mathbf{P}''(\frac{\pi}{3})$ is illustrated in Figure 15–25.

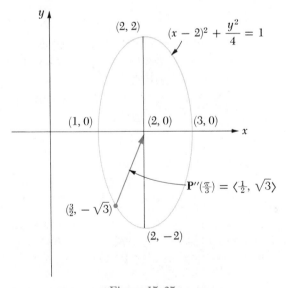

Figure 15-25

Exercise Set 15.5

In Exercises 1–6 find the domain of the vector-valued function **P**.

1. $\mathbf{P}(t) = \langle 2t, \sqrt{3 - t} \rangle$

2. $\mathbf{P}(t) = \left\langle \dfrac{1}{t}, \sqrt{4 - t^2} \right\rangle$

3. $\mathbf{P}(t) = \langle \cos^{-1} t, \sec^{-1} t \rangle$

4. $\mathbf{P}(t) = \langle \sin^{-1} t, e^{-t} \rangle$

5. $\mathbf{P}(t) = \langle \ln (t - 1), \sqrt{2 - t} \rangle$

6. $\mathbf{P}(t) = \langle \sec t, \tan 2t \rangle$

In Exercises 7–12 obtain $\mathbf{P}'(t)$ and $\mathbf{P}''(t)$ at the given number t.

7. $\mathbf{P}(t) = \langle t^2, t^3 \rangle, \quad t = \frac{3}{2}$

8. $\mathbf{P}(t) = \langle \cos^2 t, 1 + \sin t \rangle, \quad t = \frac{\pi}{3}$

9. $\mathbf{P}(t) = \langle \ln (1 + t^2), \tan^{-1} t \rangle, \quad t = 1$

10. $\mathbf{P}(t) = \langle 2 \sin t, 3 \cos t \rangle, \quad t = \frac{3\pi}{4}$

11. $\mathbf{P}(t) = \left\langle \dfrac{1}{e^t + 1}, \dfrac{e^t}{e^t + 1} \right\rangle, \quad t = 0$

12. $\mathbf{P}(t) = \langle (4 + t^2)^{3/2}, t\sqrt{4 + t^2} \rangle, \quad t = 2$

10. $\mathbf{P}(t) = \langle 2 \sin t, 3 \cos t \rangle, \quad t = \frac{3\pi}{4}$

In Exercises 13–15 derive the following differentiation formulas if the values of the derivatives in the right member of the formula exist.

13. Formula (15.59)   14. Formula (15.60)   15. Formula (15.61)

16. Prove Theorem 15.5.4.

17. Prove that if $t_0 \in \mathfrak{D}_\mathbf{P}$, and $\mathbf{P}'(t_0)$ exists, then **P** is continuous at t_0.

18. Suppose $\mathbf{P}(t) = \langle f(t), g(t) \rangle$ is defined for every t in some interval with endpoint t_0. Prove that $\lim_{t \to t_0} \mathbf{P}(t) = \langle L_1, L_2 \rangle$ if and only if for every $\epsilon > 0$ there is a $\delta > 0$ such that $|\mathbf{P}(t) - \langle L_1, L_2 \rangle| < \epsilon$ whenever $t \in \mathfrak{D}_\mathbf{P}$ and $0 < |t - t_0| < \delta$.

19. Prove: If the vector-valued function **P** is differentiable on an interval I, and $\mathbf{P}(t)$ has a constant length for every $t \in I$, then $\mathbf{P}(t)$ and $\mathbf{P}'(t)$ are orthogonal for every $t \in I$. HINT: Consider $D_t[\mathbf{P}(t) \cdot \mathbf{P}(t)]$.

20. Prove that $D_t \left[\dfrac{\mathbf{P}(t)}{|\mathbf{P}(t)|} \right] = \dfrac{(\mathbf{P}(t) \cdot \mathbf{P}(t))\mathbf{P}'(t) - (\mathbf{P}(t) \cdot \mathbf{P}'(t))\mathbf{P}(t)}{[\mathbf{P}(t) \cdot \mathbf{P}(t)]^{3/2}}$ if $\mathbf{P}'(t)$ exists.

15.6 Plane Curvilinear Motion

In Chapters 4 and 6 we discussed the velocity and acceleration of a particle in straight-line motion. In such motion the particle at any time is moving in either of two directions and hence its velocity is satisfactorily described just by giving some real number expressed in the proper units. The acceleration of the particle can also be described merely by a real number since the acceleration is the time rate of change of the velocity. However, in curvilinear motion we are concerned not only with the magnitude of the velocity and acceleration of the particle but also the directions in which these quantities act, and hence it is desirable to treat them vectorially.

Suppose at any time t the position vector of a moving particle is

$$\mathbf{P}(t) = \langle f(t), g(t) \rangle$$

where the first and second derivatives of f and g exist for every $t \in \mathfrak{D}_\mathbf{P}$. Since $dx/dt = f'(t)$ and $dy/dt = g'(t)$ are the respective rates of change of the x coordinate and the y coordinate of the particle with respect to time t, these derivatives can be considered the velocities of the particle in the x direction and y direction, respectively. Thus the vector

$$\mathbf{V} = \mathbf{P}'(t) = \left\langle \frac{dx}{dt}, \frac{dy}{dt} \right\rangle \tag{15.63}$$

is defined to be the *velocity vector* of the particle at time t. The length of this vector

$$|\mathbf{V}| = \sqrt{\left(\frac{dx}{dt}\right)^2 + \left(\frac{dy}{dt}\right)^2} \tag{15.64}$$

is defined to be the *speed* of the particle at time t. Since $d^2x/dt^2 = f''(t)$ and $d^2y/dt^2 = g''(t)$ are the respective rates of change with respect to time of the velocities of the particle in the x direction and y direction, these second derivatives are considered the accelerations of the particle in the x direction and y direction, respectively. Hence

$$\mathbf{A} = \mathbf{P}''(t) = \left\langle \frac{d^2x}{dt^2}, \frac{d^2y}{dt^2} \right\rangle \tag{15.65}$$

is defined to be the *acceleration vector* of the particle at time t.

Example 1   If the position of a particle at any time t is given by $\mathbf{P}(t) = \langle t^2 - 4t, -\frac{1}{3}t^3 \rangle$, find the vectors $\mathbf{V}$ and $\mathbf{A}$ and the speed of the particle when $t = \frac{3}{2}$.

SOLUTION   From (15.63)

$$\mathbf{V} = \mathbf{P}'(t) = \langle 2t - 4, -t^2 \rangle. \tag{15.66}$$

Hence when $t = \frac{3}{2}$, $\mathbf{V} = \langle -1, -\frac{9}{4} \rangle$, and from (15.64) the speed at $t = \frac{3}{2}$ is

$$|\mathbf{V}| = \left| \left\langle -1, -\frac{9}{4} \right\rangle \right| = \frac{\sqrt{97}}{4}.$$

After differentiating in (15.66),

$$\mathbf{A} = \mathbf{P}''(t) = \langle 2, -2t \rangle.$$

Then when $t = \frac{3}{2}$, $\mathbf{A} = \langle 2, -3 \rangle$.

Example 2   A particle moves along the parabola $y^2 = 4x$ with speed

3 units/sec and with $dy/dt > 0$. Find the velocity and acceleration vectors for the particle at the point $(2, 2\sqrt{2})$ (Figure 15–26).

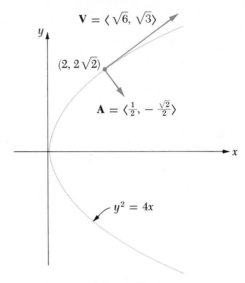

Figure 15–26

    SOLUTION Since the x and y coordinates of the particle are functions of time t, and satisfy $y^2 = 4x$, we differentiate in this equation with respect to t, obtaining

$$2y\frac{dy}{dt} = 4\frac{dx}{dt}. \tag{15.67}$$

Hence at the point $(2, 2\sqrt{2})$

$$\sqrt{2}\frac{dy}{dt} = \frac{dx}{dt}. \tag{15.68}$$

Since the speed of the particle is $|\mathbf{V}| = \sqrt{(dx/dt)^2 + (dy/dt)^2} = 3$,

$$\left(\frac{dx}{dt}\right)^2 + \left(\frac{dy}{dt}\right)^2 = 9. \tag{15.69}$$

Substitution from (15.68) into (15.69) gives

$$2\left(\frac{dy}{dt}\right)^2 + \left(\frac{dy}{dt}\right)^2 = 9.$$

Since $dy/dt > 0$, solving this equation gives

$$\frac{dy}{dt} = \sqrt{3} \text{ unit/sec}. \tag{15.70}$$

Then from (15.70) and (15.68)

$$\frac{dx}{dt} = \sqrt{6} \text{ unit/sec}. \tag{15.71}$$

Hence from (15.70) and (15.71) the velocity vector at $(2, 2\sqrt{2})$ is $\mathbf{V} = \langle \sqrt{6}, \sqrt{3} \rangle$.

To find the acceleration vector at $(2, 2\sqrt{2})$, we differentiate in (15.67) with respect to t and obtain by the product rule

$$y\frac{d^2y}{dt^2} + \left(\frac{dy}{dt}\right)^2 = 2\frac{d^2x}{dt^2}. \tag{15.72}$$

From differentiating in (15.69) with respect to t,

$$\frac{dx}{dt}\frac{d^2x}{dt^2} + \frac{dy}{dt}\frac{d^2y}{dt^2} = 0. \tag{15.73}$$

Letting $y = 2\sqrt{2}$, $dx/dt = \sqrt{6}$, and $dy/dt = \sqrt{3}$ in (15.72) and (15.73), we obtain

$$2\sqrt{2}\frac{d^2y}{dt^2} + 3 = 2\frac{d^2x}{dt^2}$$

$$\sqrt{6}\frac{d^2x}{dt^2} + \sqrt{3}\frac{d^2y}{dt^2} = 0.$$

The solution of these equations gives the acceleration vector at $(2, 2\sqrt{2})$,

$$\mathbf{A} = \left\langle \frac{d^2x}{dt^2}, \frac{d^2y}{dt^2} \right\rangle = \left\langle \frac{1}{2}, -\frac{\sqrt{2}}{2} \right\rangle.$$

Example 3   Suppose that a projectile is fired at an angle α with level ground and with an initial speed v_0. Assuming that the only force on the projectile is that due to gravity, find the velocity vector and position vector for the projectile at any time t (Figure 15–27).

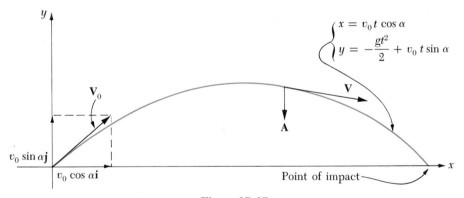

$$\begin{cases} x = v_0 t \cos\alpha \\ y = -\frac{gt^2}{2} + v_0 t \sin\alpha \end{cases}$$

Figure 15–27

SOLUTION   We shall assume that the projectile is fired from the origin, and that α is measured counterclockwise from the positive x axis. Then

$$\mathbf{A} = \left\langle \frac{d^2x}{dt^2}, \frac{d^2y}{dt^2} \right\rangle = \langle 0, -g \rangle \tag{15.74}$$

where $g \approx 32$ ft/sec. If the components of $\mathbf{A}$ are integrated, we have

$$\frac{dx}{dt} = C_1 \qquad \frac{dy}{dt} = -gt + C_2. \tag{15.75}$$

Since the initial speed is v_0, the velocity vector at $t = 0$ is

$$V_0 = \langle v_0 \cos \alpha, v_0 \sin \alpha \rangle. \tag{15.76}$$

By letting $t = 0$ in (15.75) and equating these derivatives to the corresponding components of (15.76), we obtain

$$C_1 = v_0 \cos \alpha \qquad C_2 = v_0 \sin \alpha.$$

Hence from (15.75)

$$\frac{dx}{dt} = v_0 \cos \alpha \qquad \frac{dy}{dt} = -gt + v_0 \sin \alpha \tag{15.77}$$

and the velocity vector at any time t is $V = \langle v_0 \cos \alpha, -gt + v_0 \sin \alpha \rangle$. An integration in (15.77) then gives

$$x = v_0 t \cos \alpha + C_3 \qquad y = -\frac{gt^2}{2} + v_0 t \sin \alpha + C_4 \tag{15.78}$$

Since $x = 0 = y$ when $t = 0$, $C_3 = C_4 = 0$, and from (15.78) the position vector of the projectile is

$$P(t) = \left\langle v_0 t \cos \alpha, -\frac{gt^2}{2} + v_0 t \sin \alpha \right\rangle.$$

Exercise Set 15.6

In Exercises 1–4 the motion of a particle is described by the indicated expression for $P(t)$. At the given value of t, find the vectors V, A, and the speed of the particle.

1. $P(t) = \langle t^3 - 3t^2, t^2 \rangle$, $t = 2$

2. $P(t) = \langle \cos t, \sin 2t \rangle$, $t = \frac{\pi}{3}$

3. $P(t) = \langle \sinh t, \cosh t \rangle$, $t = \ln 2$

4. $P(t) = \left\langle \sqrt{1 + t}, \frac{t}{\sqrt{1 + t}} \right\rangle$, $t = 3$.

In Exercises 5–8 find V and A at the given point for the particle having the given motion.

5. Particle moving on the parabola $4y = x^2$ with $dy/dt = 2$ (constant); $(-2, 1)$.

6. Particle moving counterclockwise on the circle $x^2 + y^2 = 25$ with $dx/dt = -3$ (constant); $(4, 3)$.

7. Particle moving on $y^2 = 9x$, and at $(1, -3)$, $dy/dt = -2$ and $d^2y/dt^2 = 4$; $(1, -3)$.

8. Particle moving clockwise on the ellipse $x^2 + 4y^2 = 40$ with speed 10; $(2\sqrt{5}, -\sqrt{5})$.

9. How fast is the particle in Example 2 moving away from the origin at the instant the particle is at $(2, 2\sqrt{2})$?

10. When a particle on the right-hand branch of the hyperbola $x^2 - y^2 = 1$ is at the point $(\sqrt{2}, -1)$, it is approaching the origin at the rate of $2\sqrt{6}$ units/sec. Find the speed of the particle at that instant.

11. For the particle in Example 3 obtain
    (a) the x coordinate of the point of impact of the projectile; its range.
    (b) the speed of the projectile at the time of impact.
    (c) the maximum altitude attained by the projectile.

12. Prove that the maximum range for the projectile in Example 3 is attained when $\alpha = 45°$.

13. If a particle moves counterclockwise around a circle of radius r with angular velocity $d\theta/dt = \omega$ (constant), prove that (a) the speed of the particle is $r\omega$, and (b) the acceleration vector at any point on the circle is directed toward its center. HINT: Let $x = r\cos\theta$, $y = r\sin\theta$ be parametric equations of the circle.

16. Polar Coordinates

16.1 Introduction to Polar Coordinates

Heretofore we have located a point in the plane using rectangular coordinates, which are "signed" distances from the point to two mutually perpendicular number lines. We will now describe another coordinate system in which the location of the point is determined from its distance along a ray from the origin through the point and the angle between the positive x axis and the ray.

Suppose we consider the point, denoted here by A (Figure 16–1), which has rectangular coordinates $(-1, \sqrt{3})$. We note that the distance from the origin to A is $|OA| = 2$. Then from basic trigonometry the ray from O through A makes

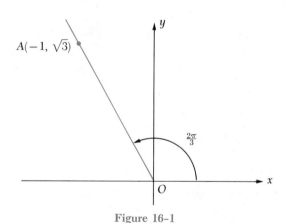

Figure 16–1

an angle of $\frac{2\pi}{3}$ rad with the positive x axis and

$$-1 = 2 \cos \frac{2\pi}{3} \qquad \sqrt{3} = 2 \sin \frac{2\pi}{3}.$$

Hence, by the following definition A has *polar coordinates* $(2, \frac{2\pi}{3})$.

16.1.1   Definition

Suppose that the rectangular coordinates of a point P are (x, y) and r and θ are any numbers such that

$$x = r \cos \theta \qquad y = r \sin \theta. \tag{16.1}$$

Then P is said to have *polar coordinates* (r, θ).

The *pole* or *origin*, O, of the polar coordinate system, is the origin of the rectangular coordinate system and can be expressed in polar coordinates by $(0, \theta)$, where θ is arbitrarily chosen. In fact, by Theorem 16.1.2 below, *every point in the plane has an infinite number of different representations in polar coordinates*. However, if we restrict ourselves to coordinates such that $r > 0$ and $0 \le \theta < 2\pi$, then any point other than the pole has a unique representation (r, θ) (Exercise 38, Exercise set 16.1).

16.1.2   Theorem

If a point in the plane has polar coordinates (r, θ), then it also has polar coordinates $((-1)^n r, \theta + n\pi)$, where n is any integer.

PROOF   Suppose the point has rectangular coordinates (x, y). Then Equations (16.1) certainly hold, and if n is *even*,

$$(-1)^n r \cos (\theta + n\pi) = r \cos \theta = x$$
$$(-1)^n r \sin (\theta + n\pi) = r \sin \theta = y.$$

Hence by Definition 16.1.1 the point has polar coordinates $((-1)^n r, \theta + n\pi)$, when n is any even integer. The remainder of the proof, which consists of showing that the point also has polar coordinates $((-1)^n r, \theta + n\pi)$ when n is odd, is left as an exercise.

By Theorem 16.1.2 the point A, with rectangular coordinates $(-1, \sqrt{3})$, above, has polar coordinates

$$\left((-1)^n 2, \frac{2\pi}{3} + n\pi\right) \quad n \text{ any integer.} \tag{16.2}$$

The coordinates $(2, \frac{2\pi}{3})$ and $(-2, \frac{5\pi}{3})$, for example, can be obtained for A by letting $n = 0$ and $n = 1$, respectively, in (16.2). NOTE: *We assume here and in the rest of this chapter, unless otherwise indicated, that whenever the coordinates of a point are given, these numbers are polar coordinates of the point.*

To locate a point $P(r, \theta)$ in the plane, the *polar axis* (the positive x axis) is first rotated through the angle θ. If $\theta > 0$, the rotation is counterclockwise, and if $\theta < 0$, the rotation is clockwise. In either case let the ray L be the terminal position of the rotated axis. Then if $r > 0$, P is located on L so that $|OP| = r$ (Figure

16–2(a)). However, if $r < 0$, P is located by reflecting L through O and measuring a distance $|r|$ from O along the reflected ray L' (Figure 16–2(b)).

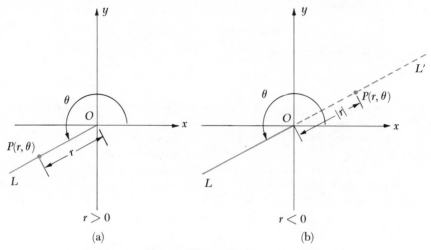

Figure 16–2

In Figure 16–3 the points $(4, \frac{5\pi}{6})$, $(2, \frac{2\pi}{3})$, $(3, \frac{\pi}{4})$, $(-3, \frac{\pi}{4})$, $(1, 0)$, and $(4, \frac{3\pi}{2})$ have been plotted. The plotting of points given by polar coordinates is facilitated by the use of printed polar coordinate paper, which is available at many college bookstores.

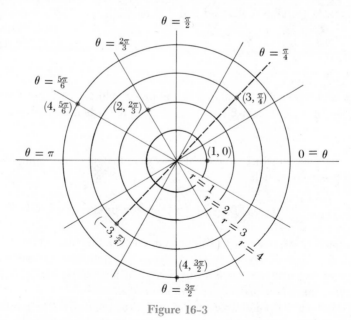

Figure 16–3

The formulas (16.1) give an easy conversion from polar coordinates back to rectangular coordinates.

***Example* 1**   Find the rectangular coordinates of the point having polar coordinates $(4, \frac{5\pi}{6})$ (Figure 16–3).

SOLUTION   From (16.1)

$$x = 4 \cos \frac{5\pi}{6} = 4\left(-\frac{\sqrt{3}}{2}\right) = -2\sqrt{3}$$

and

$$y = 4 \sin \frac{5\pi}{6} = 4 \cdot \frac{1}{2} = 2.$$

Thus, the point is denoted by $(-2\sqrt{3}, 2)$ in rectangular coordinates.

To change from rectangular coordinates to polar coordinates, we obtain from (16.1) $x^2 + y^2 = r^2$ and hence

$$r = \pm\sqrt{x^2 + y^2}. \tag{16.3}$$

Then regardless of which sign is used in (16.3) the corresponding coordinate, θ, can be computed from the following formulas, which are obtained from (16.1):

$$\cos \theta = \frac{x}{r} \qquad \sin \theta = \frac{y}{r}. \tag{16.4}$$

***Example* 2**   Find polar coordinates for the point having rectangular coordinates $(-2, -3)$ (Figure 16–4).

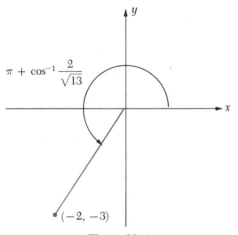

Figure 16-4

SOLUTION   If we restrict r to be positive, then from (16.3)

$$r = \sqrt{(-2)^2 + (-3)^2} = \sqrt{13}$$

and by (16.4)

$$\cos \theta = -\frac{2}{\sqrt{13}} \qquad \sin \theta = -\frac{3}{\sqrt{13}}. \tag{16.5}$$

From trigonometry, the equations in (16.5) are satisfied by $\theta = \pi + \cos^{-1}(2/\sqrt{13})$ and therefore the point has polar coordinates $(\sqrt{13}, \pi + \cos^{-1}(2/\sqrt{13}))$.

Note that if we had used $r = -\sqrt{13}$, then we would have obtained the polar coordinates $(-\sqrt{13}, \cos^{-1}(2/\sqrt{13}))$ for the point.

The graph of an equation in r and θ is the set of all points whose polar coordinates satisfy the equation. The equation is then termed a *polar equation* of the graph. If a set of points has the polar equation

$$r = F(\theta) \tag{16.6}$$

for every θ in some interval I, then from (16.1) the graph has the parametric equations

$$x = F(\theta) \cos \theta \qquad y = F(\theta) \sin \theta \tag{16.7}$$

if $\theta \in I$. If F is continuous on I, then by Definition 15.3.2 the graph is a curve, which can be termed a *polar curve*.

Many curves of interest have a simpler polar equation than Cartesian equation—an important advantage for the use of polar coordinates. For example, the graph of the polar equation

$$r = a, \quad a \text{ constant}$$

is a circle having the pole as a center and radius $|a|$. In Figure 16–3 the circles $r = 1, 2, 3,$ and 4 have been plotted. The graph of

$$\theta = \alpha, \quad \alpha \text{ constant} \tag{16.8}$$

is the line having the Cartesian equation $y = x \tan \alpha$ if α is not an odd multiple of $\frac{\pi}{2}$. If n is an odd integer, the line $\theta = \frac{n\pi}{2}$ is the y axis. Since a point with coordinates (r, θ) also has the coordinates $(-r, \theta + \pi)$, the graph of (16.8) also has the equation

$$\theta = \alpha + \pi.$$

Thus, for example, the line $\theta = \frac{5\pi}{6}$ (Figure 16–3) is the same as the line $\theta = \frac{11\pi}{6}$.

In the following example we go from a Cartesian equation of a circle to a polar equation.

Example 3 Find a polar equation for the circle $x^2 + y^2 - 2y = 0$. (Note that this equation is equivalent to $x^2 + (y - 1)^2 = 1$, which represents a circle with center $(0, 1)$ and radius 1.)

SOLUTION If P is any point in the circle, its rectangular coordinates, of course, satisfy the given equation. Then from (16.1) and (16.3) its polar coordinates satisfy

$$r^2 - 2r \sin \theta = 0. \tag{16.9}$$

Conversely, if the polar coordinates of P satisfy (16.9), from (16.1) and (16.3) its rectangular coordinates satisfy the given equation and hence (16.9) is a polar equation of the circle.

However, we can derive a simpler polar equation for the circle. From (16.9) a point (r, θ) is on the circle if and only if

$$r(r - 2 \sin \theta) = 0$$

and hence if and only if

$$r = 0 \qquad \text{or} \qquad r - 2 \sin \theta = 0.$$

The only point whose coordinates satisfy the equation $r = 0$ is the pole, but the pole is also in the graph of $r - 2 \sin \theta = 0$ since $(r, \theta) = (0, 0)$, for example, satisfies this equation. Therefore a polar equation of the given circle is

$$r = 2 \sin \theta.$$

We next proceed in the reverse direction and find a Cartesian equation from a polar equation.

Example 4 Obtain a Cartesian equation for the curve $r(1 + \cos \theta) = 2$.

SOLUTION If $P(r, \theta)$ is on the curve, then $r + r \cos \theta = 2$, and hence using (16.1) and (16.3), we have successively

$$\pm \sqrt{x^2 + y^2} + x = 2$$
$$x^2 + y^2 = (2 - x)^2 \qquad (16.10)$$
$$y^2 = 4 - 4x \qquad (16.11)$$

Conversely, if $P(x, y)$ is in the graph of (16.11), then (16.10) is true. Since $y^2 \geq 0$ for every y, from (16.11) $x \leq 1$, and hence $2 - x \geq 1$. Thus $2 - x$ is positive and from (16.10)

$$\sqrt{x^2 + y^2} = 2 - x. \qquad (16.12)$$

Since any point in the plane has a notation (r, θ) where $r \geq 0$, we shall let $r = \sqrt{x^2 + y^2}$. Then from (16.12) $r = 2 - r \cos \theta$, and hence $r(1 + \cos \theta) = 2$. Therefore (16.11) is a Cartesian equation of the curve, which is a parabola.

Exercise Set 16.1

In Exercises 1–8 the polar coordinates of a point are given. Find the rectangular coordinates of the point and plot the points in a sketch of the plane.

1. $(3, \pi)$ 　　　　　　　　2. $(4, \frac{\pi}{3})$ 　　　　　　　　3. $(-4, \frac{\pi}{3})$

4. $(2, 0)$ 　　　　　　　　5. $(4, 0)$ 　　　　　　　　6. $(2\sqrt{2}, -\frac{\pi}{4})$

7. $(\sqrt{2}, \frac{3\pi}{4})$ 　　　　　　　　8. $(-\frac{1}{2}, -\frac{4\pi}{3})$

In Exercises 9–16 the rectangular coordinates of a point are given. Find polar coordinates (r, θ) of the point where $r > 0$ and $0 \leq \theta < 2\pi$ and plot the point in a sketch of the plane.

9. $(2, 2)$ 10. $(-3, 3)$ 11. $(-1, -\sqrt{3})$

12. $(-2\sqrt{3}, -2)$ 13. $(-4, 0)$ 14. $(3, 0)$

15. $(\sqrt{6}, -\sqrt{2})$ 16. $(3\sqrt{2}, -3\sqrt{2})$

In Exercises 17–26 find a Cartesian equation of the graph of the given polar equation.

17. $r = 4$ 18. $\theta = -\frac{\pi}{6}$

19. $r = 3 \cos \theta$ 20. $r = 4 \sin \theta$

21. $r^2 \sin 2\theta = 2$ 22. $r^2 = \cos 2\theta$

23. $r = 3 \cos \theta + 2 \sin \theta$ 24. $r = \dfrac{3}{2 + \cos \theta}$

25. $r = a(1 + \cos \theta)$ 26. $r = a(1 - 2 \sin \theta)$

In Exercises 27–34 find a polar equation of the graph of the given Cartesian equation.

27. $x = 2$ 28. $y = -3$

29. $2x + 3y = 5$ 30. $x^2 - y^2 = 4$

31. $xy = a$ 32. $x^2 = 4y$

33. $x^2 + y^2 - 2x + 4y = 0$ 34. $x^2 + y^2 + 6x - 4y = 0$

35. Prove Theorem 16.1.2 for the case where n is odd.

36. Prove that the distance between the points (r_1, θ_1) and (r_2, θ_2) is
$$\sqrt{r_1^2 + r_2^2 - 2r_1 r_2 \cos (\theta_2 - \theta_1)}.$$

37. Find a polar equation for the line passing through the points (r_1, θ_1) and (r_2, θ_2).

38. Prove that any point in the plane other than the pole has a unique representation (r, θ) where $r > 0$ and $0 \leq \theta < 2\pi$. HINT: Suppose (r_1, θ_1) and (r_2, θ_2) are polar coordinates of a point having rectangular coordinates (x, y).

16.2   Graphing in Polar Coordinates

In this section we will consider some techniques involved in graphing a polar equation. In our first example the sketch of the graph is made merely by plotting a sufficient number of the points whose coordinates satisfy the given polar equation, and then sketching the curve in which these points lie. Following this example, properties of polar graphs will be discussed which facilitate the graphing process by lessening the dependence on just the plotting of points.

Example 1   Graph the *limaçon* $r = 1 + 2 \cos \theta$ (Figure 16–5).

SOLUTION We first complete a table that gives the coordinates (r, θ) of a sufficient number of points in the graph. In this table θ will be given in increments of $\frac{\pi}{6}$ radian starting with 0, and the corresponding values for r will then be obtained from the given equation. The increment of $\frac{\pi}{6}$ for θ is chosen rather arbitrarily here, and in other graphing problems a different increment may be advisable.

θ	0	$\frac{\pi}{6}$	$\frac{\pi}{3}$	$\frac{\pi}{2}$	$\frac{2\pi}{3}$	$\frac{5\pi}{6}$	π	$\frac{7\pi}{6}$	$\frac{4\pi}{3}$	$\frac{3\pi}{2}$	$\frac{5\pi}{3}$	$\frac{11\pi}{6}$	2π
r	3	$1 + \sqrt{3}$	2	1	0	$1 - \sqrt{3}$	-1	$1 - \sqrt{3}$	0	1	2	$1 + \sqrt{3}$	3

We need not extend the table beyond $\theta = 2\pi$ because when $\theta > 2\pi$ the graph will only repeat itself. This is true because the cosine function is periodic with period 2π (cf. (10.6)).

After plotting the points obtained from the table, the required curve can be drawn by connecting the points in order of increasing θ.

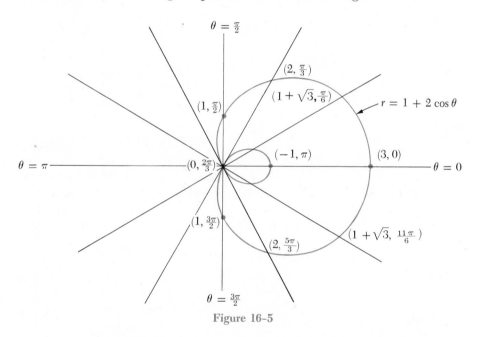

Figure 16–5

The sketch of the graph of a polar equation should indicate if the graph contains the pole. The pole will be in the graph if $r = 0$ is obtained from the equation for some value of θ. For example, the pole is in the graph of the limaçon in Example 1 since $r = 1 + 2 \cos \theta = 0$ when, for example, $\theta = \frac{2\pi}{3}$.

Also any intersection points of the graph with the *quadrantal axes*, the lines $\theta = \frac{n\pi}{2}$ where n is any integer, should be plotted. We note that the points of intersection of the limaçon in Example 1 with these axes: the points $(3, 0)$, $(1, \frac{\pi}{2})$, $(-1, \pi)$, and $(1, \frac{3\pi}{2})$, have been plotted in Figure 16–5.

The graph may be symmetric to the line $\theta = 0$, the line $\theta = \frac{\pi}{2}$, or to the pole.

(i) A graph S is symmetric to the line $\theta = 0$ if and only if whenever $(r, \theta) \in$ S, either $(r, -\theta) \in$ S or $(-r, \pi - \theta) \in$ S.

(ii) A graph S is symmetric to the line $\theta = \frac{\pi}{2}$ if and only if whenever $(r, \theta) \in$ S, either $(r, \pi - \theta) \in$ S or $(-r, -\theta) \in$ S.

(iii) A graph S is symmetric to the pole if and only if whenever $(r, \theta) \in$ S, either $(r, \pi + \theta) \in$ S or $(-r, \theta) \in$ S.

These symmetries are illustrated in Figure 16–6.

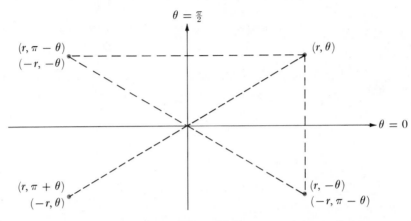

Figure 16-6

For example, from (i), above, the limaçon in Example 1 is symmetric to the line $\theta = 0$ because if (r, θ) is in the graph, that is, $r = 1 + 2 \cos \theta$, then $1 + 2 \cos(-\theta) = 1 + 2 \cos \theta = r$, that is, $(r, -\theta)$ is in the graph. However, the limaçon is not symmetric to the line $\theta = \frac{\pi}{2}$ as the point $(3, 0)$, for example, is in the graph but $(-3, 0)$ is not. Also it can be shown that the graph is not symmetric to the pole.

We recall from Section 16.1 that the graph of the polar equation

$$r = F(\theta) \tag{16.13}$$

also has the parametric equations

$$x = F(\theta) \cos \theta \qquad y = F(\theta) \sin \theta.$$

If F is differentiable on an interval I having θ_0, and $dx/d\theta \neq 0$ when $\theta = \theta_0$, then by Theorem 15.3.4(a) the slope m of the tangent at the point $(F(\theta_0), \theta_0)$ in the graph of (16.13) is

$$m = \frac{\left(\dfrac{dy}{d\theta}\right)_{\theta=\theta_0}}{\left(\dfrac{dx}{d\theta}\right)_{\theta=\theta_0}} = \frac{F'(\theta_0) \sin \theta_0 + F(\theta_0) \cos \theta_0}{F'(\theta_0) \cos \theta_0 - F(\theta_0) \sin \theta_0}. \tag{16.14}$$

Suppose the pole is in the graph of (16.13) when $\theta = \theta_0$ and $F'(\theta_0) \neq 0$. If θ_0 is not an odd multiple of $\frac{\pi}{2}$, then from (16.14) the slope of the graph at the pole is

$$m = \frac{F'(\theta_0) \sin \theta_0}{F'(\theta_0) \cos \theta_0} = \tan \theta_0, \tag{16.15}$$

and hence *the line $\theta = \theta_0$ is tangent to the graph at the pole.* The italicized statement is also true even if θ_0 is an odd multiple of $\frac{\pi}{2}$. For then, $\cos \theta_0 = 0$ and

$$\left(\frac{dx}{d\theta} \right)_{\theta = \theta_0} = F'(\theta_0) \cos \theta_0 - F(\theta_0) \sin \theta_0 = 0.$$

By Theorem 15.3.4(b) the line $\theta = \frac{\pi}{2}$ is therefore a vertical tangent to the graph at the pole.

To illustrate our statement about the tangents at the pole we first note that the graph of $r = F(\theta) = 1 + 2 \cos \theta$ (see Example 1) passes through the pole when $\theta = \frac{2\pi}{3}$ and $\frac{4\pi}{3}$ since the equation is satisfied by $(0, \frac{2\pi}{3})$ and $(0, \frac{4\pi}{3})$. Then, since $dr/d\theta = F'(\theta) = -2 \sin \theta$, we have $F'(\theta) \neq 0$ when $\theta = \frac{2\pi}{3}$ or $\frac{4\pi}{3}$. Hence the lines $\theta = \frac{2\pi}{3}$ and $\theta = \frac{4\pi}{3}$ are tangent to the limaçon $r = 1 + 2 \cos \theta$ at the pole.

In the next example the equation will be graphed after plotting specific points, as before. However, the sketch obtained will be examined in the light of our discussion about the pole, tangents at the pole, and symmetry to the quadrantal axes.

Example 2   Graph the *three-leaf rose* $r = 4 \sin 3\theta$ (Figure 16–7).

SOLUTION   As before, we complete a table giving the polar coordinates of a sufficient number of points in the graph. In this table it will be convenient to have an increment of $\frac{\pi}{12}$ for θ. The table need not be extended beyond $\theta = \pi$ because for any θ, $\sin 3(\theta + \pi) = -\sin 3\theta$, and hence the point $(4 \sin 3(\theta + \pi), \theta + \pi)$, which can be denoted by $(-4 \sin 3\theta, \theta + \pi)$, is the same as the point $(4 \sin 3\theta, \theta)$.

θ	0	$\frac{\pi}{12}$	$\frac{\pi}{6}$	$\frac{\pi}{4}$	$\frac{\pi}{3}$	$\frac{5\pi}{12}$	$\frac{\pi}{2}$	$\frac{7\pi}{12}$	$\frac{2\pi}{3}$	$\frac{3\pi}{4}$	$\frac{5\pi}{6}$	$\frac{11\pi}{12}$	π
r	0	$2\sqrt{2}$	4	$2\sqrt{2}$	0	$-2\sqrt{2}$	-4	$-2\sqrt{2}$	0	$2\sqrt{2}$	4	$2\sqrt{2}$	0

From this table the graph can be sketched. It will be noted that if $\theta = 0$, $\frac{\pi}{3}$, $\frac{2\pi}{3}$, or π, then the corresponding point in the graph is the pole. Furthermore, since $dr/d\theta = 12 \cos 3\theta \neq 0$ for these values of θ, the radial lines associated with these values are tangents to the graph at the pole (Figure 16–7). It will be seen

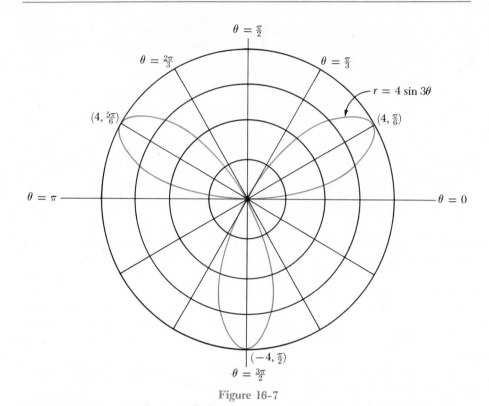

$$\theta = \frac{\pi}{2}$$

$$\theta = \frac{2\pi}{3} \qquad \theta = \frac{\pi}{3}$$

$r = 4 \sin 3\theta$

$\left(4, \frac{5\pi}{6}\right)$ $\qquad \left(4, \frac{\pi}{6}\right)$

$\theta = \pi$ $\qquad\qquad\qquad\qquad\qquad\qquad\qquad \theta = 0$

$\left(-4, \frac{\pi}{2}\right)$

$$\theta = \frac{3\pi}{2}$$

Figure 16-7

from (ii) above that the graph is symmetric to the line $\theta = \frac{\pi}{2}$. For if (r, θ) is in the graph, then $r = 4 \sin 3\theta$, and $4 \sin 3(-\theta) = -4 \sin 3\theta = -r$; that is, $(-r, -\theta)$ is in the graph.

Example 3 Graph the *lemniscate* $r^2 = a^2 \cos 2\theta$, where $a > 0$ (Figure 16–8).

SOLUTION From the given equation $r = \pm a \sqrt{\cos 2\theta}$, and hence there will be points in the graph whenever $\cos 2\theta \geq 0$—that is, if $\theta \in [-\frac{\pi}{4} + n\pi, \frac{\pi}{4} + n\pi]$ where n is any integer. Accordingly, the accompanying table can be prepared. We

θ	$-\frac{\pi}{4}$	$-\frac{\pi}{8}$	0	$\frac{\pi}{8}$	$\frac{\pi}{4}$	No points for	$\frac{3\pi}{4}$	$\frac{7\pi}{8}$	π
r	0	$\pm 0.84a$	$\pm a$	$\pm 0.84a$	0	$\theta \in \left(\frac{\pi}{4}, \frac{3\pi}{4}\right)$	0	$\pm 0.84a$	$\pm a$

shall end the table with $\theta = \pi$ because for any θ, $\cos 2(\theta + \pi) = \cos 2\theta$ and hence the points $(\pm a \sqrt{\cos 2(\theta + \pi)}, \theta + \pi)$, in the graph can be denoted by $(\pm a \sqrt{\cos 2\theta}, \theta + \pi)$, which is just another notation for the points $(\mp a \sqrt{\cos 2\theta}, \theta)$ in the graph.

We note that if $\theta = -\frac{\pi}{4}$ or $\frac{\pi}{4}$, then the corresponding point in the graph is the pole and the radial lines given by these values of θ are tangents to the graph at the pole.

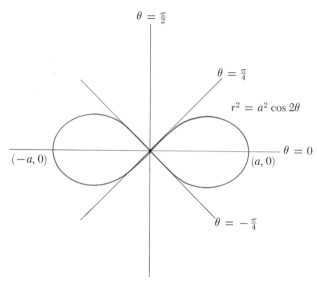

Figure 16-8

It can be proved that the graph is symmetric to the pole and to the lines $\theta = 0$ and $\theta = \frac{\pi}{2}$. To prove that the graph is symmetric to the pole, consider any point (r, θ) in the graph. Then $a^2 \cos 2(-\theta) = a^2 \cos 2\theta = r^2 = (-r)^2$ and hence $(-r, -\theta)$ is in the graph. Thus by (iii) the graph is symmetric to the pole. The proof of symmetry to the lines $\theta = 0$ and $\theta = \frac{\pi}{2}$ is left for the reader.

Exercise Set 16.2

In Exercises 1–23 sketch the graph of the given equation.

1. $r = 3$
2. $\theta = \frac{\pi}{6}$
3. $r = -2$
4. $\theta = -\frac{3\pi}{4}$
5. $r \cos \theta = -3$
6. $r \sin \theta = 1$
7. $r = 4 \cos \theta$
8. $r = 3 \sin \theta$
9. $r = 3(1 - \cos \theta)$
10. $r = 2(1 + \cos \theta)$
11. $r = 3 - 2 \sin \theta$
12. $r = 1 + 3 \sin \theta$
13. $r = 3 \cos 4\theta$
14. $r = 3 \sin 4\theta$
15. $r = 4 \cos 2\theta$
16. $r = 4 \cos 3\theta$
17. $r(2 - \sin \theta) = 3$
18. $r(1 + 2 \cos \theta) = -4$
19. $r = \dfrac{1}{\theta}$
20. $r = e^{\theta}$
21. $r^2 = 4 \sin \theta$
22. $r^2 = 4 \cos \theta$
23. $r^2 = 9 \cos 2\theta$
24. $r^2 = -4 \sin 2\theta$

25. Find the slope of the tangent to the graph of the three-leaf rose $r = 4 \sin 3\theta$ at the point $(2\sqrt{2}, \frac{3\pi}{4})$.

26. Find the slope of the tangent to the limaçon $r = 1 + 2 \cos \theta$ at the point $(2, \frac{\pi}{3})$.

16.3   Conics in Polar Coordinates

In this section we will obtain a fairly simple polar equation for a conic that has its focus at the pole and is symmetric to a quadrantal axis. To derive this equation, we begin by supposing that P is any point in the plane, O is the pole, and Q is the foot of a perpendicular from P to the line with the Cartesian equation $x = -k$. Here, k is any number except 0. If $e > 0$ is given, we shall consider the set of all points $P(x, y)$ such that

$$\frac{|OP|}{|PQ|} = e. \tag{16.16}$$

Figure 16–9 applies if $k > 0$.

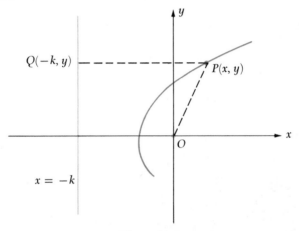

Figure 16–9

If $e = 1$, it will be recalled from Section 12.1 that this set is a parabola having the pole as its focus and the line $x = -k$ as its directrix. We shall prove that even if $e \neq 1$, the graph of (16.16) is a conic with the pole as a focus. The line $x = -k$ will be called a *directrix* of this graph as an extension of the previous meaning of this term.

If (r, θ) are the polar coordinates of a point P satisfying (16.16) then $|OP| = |r|$, $|PQ| = |x + k| = |r \cos \theta + k|$, and hence from (16.16)

$$|r| = e|r \cos \theta + k|. \tag{16.17}$$

When r and $r \cos \theta + k$ are of the same sign, we have successively from (16.17)

$$r = e(r \cos \theta + k). \tag{16.18}$$

Then if (16.18) is solved for r, we obtain

$$r = \frac{ke}{1 - e \cos \theta}. \tag{16.19}$$

However, if r and $r \cos \theta + k$ are of opposite signs, then from (16.17) $-r = e(r \cos \theta + k)$, and hence

$$r = \frac{-ke}{1 + e \cos \theta}. \tag{16.20}$$

If (r, θ) satisfies either of the equations (16.19) or (16.20), then $(-r, \theta + \pi)$ will satisfy the other. Since (r, θ) and $(-r, \theta + \pi)$ are polar representations of the same point, *the equations (16.19) and (16.20) are equivalent.*

Thus, any point P satisfying (16.16) has polar coordinates satisfying (16.19). Conversely, if the polar coordinates (r, θ) of a point P satisfy (16.19), the above steps can be retraced and P satisfies (16.16). We therefore have the following theorem.

16.3.1   Theorem

Let $e > 0$ be given. Then

$$r = \frac{ke}{1 - e \cos \theta}, \tag{16.19}$$

or equivalently

$$r = \frac{-ke}{1 + e \cos \theta}, \tag{16.20}$$

is a polar equation of the set of all points P satisfying (16.16).

It should be noted that the graph of (16.16), or equivalently (16.19), is symmetric to the polar axis by (i) in Section 16.2 since $\cos(-\theta) = \cos \theta$.

From (16.17) a Cartesian equation for the graph of (16.16) is

$$\sqrt{x^2 + y^2} = e|x + k|,$$

which, if $e \neq 1$, is equivalent to the following equations through (16.22)

$$x^2 + y^2 = e^2(x^2 + 2kx + k^2)$$

$$x^2 - \frac{2e^2k}{1 - e^2}x + \frac{y^2}{1 - e^2} = \frac{e^2k^2}{1 - e^2}. \tag{16.21}$$

Completing the square in x in (16.21) gives

$$x^2 - \frac{2e^2k}{1 - e^2}x + \frac{e^4k^2}{(1 - e^2)^2} + \frac{y^2}{1 - e^2} = \frac{e^2k^2}{1 - e^2} + \frac{e^4k^2}{(1 - e^2)^2}$$

$$\left(x - \frac{e^2k}{1 - e^2}\right)^2 + \frac{y^2}{1 - e^2} = \frac{e^2k^2}{(1 - e^2)^2},$$

and finally

$$\frac{\left(x - \dfrac{e^2k}{1 - e^2}\right)^2}{\dfrac{e^2k^2}{(1 - e^2)^2}} + \frac{y^2}{\dfrac{e^2k^2}{1 - e^2}} = 1. \tag{16.22}$$

If $0 < e < 1$, the graph of (16.22) is an ellipse with center $\left(\dfrac{e^2k}{1 - e^2}, 0\right)$ and semimajor and semiminor axes with respective lengths

$$a = \frac{e|k|}{1 - e^2} \qquad b = \frac{e|k|}{\sqrt{1 - e^2}}.$$

The distance from the center to either focus of the ellipse is

$$c = \sqrt{a^2 - b^2} = \sqrt{\frac{e^2 k^2}{(1 - e^2)^2} - \frac{e^2 k^2}{1 - e^2}}$$

$$= \frac{e^2 |k|}{1 - e^2}.$$

The foci of the ellipse are therefore $\left(\dfrac{e^2 k}{1 - e^2} \pm \dfrac{e^2 |k|}{1 - e^2}, 0\right)$, or $\left(\dfrac{2e^2 k}{1 - e^2}, 0\right)$ and $(0, 0)$, regardless of the sign of k. Thus, the pole is a focus of the ellipse.

We leave as an exercise the proof that if $e > 1$, then the graph of (16.19) is a hyperbola with center $\left(\dfrac{e^2 k}{1 - e^2}, 0\right)$ having the pole as a focus. Also the student will be asked to prove that the asymptotes, which pass through the center, have slopes $\pm\sqrt{e^2 - 1}$ and are therefore parallel to the lines $\theta = \cos^{-1}(\pm 1/e)$. This result is certainly plausible since from (16.19) and (16.20), respectively, r is undefined if $\cos\theta = 1/e$, and $\cos\theta = -1/e$.

Our results can be summarized with the following theorem:

16.3.2   Theorem

If $e > 0$ and $k \neq 0$, then the graph of the polar equation

$$r = \frac{ke}{1 - e\cos\theta} \tag{16.19}$$

or equivalently

$$r = \frac{-ke}{1 + e\cos\theta} \tag{16.20}$$

is a conic that is symmetric to the polar axis and has the pole as a focus.

   (i) *If $0 < e < 1$, the conic is an ellipse.*
  (ii) *If $e = 1$, the conic is a parabola.*
 (iii) *If $e > 1$, the conic is a hyperbola.*

The constant e is called the *eccentricity* of the conic.

Example 1   Discuss the graph of the polar equation $r = 3/(2 - \cos\theta)$ (Figure 16–10).

SOLUTION   First we rewrite the given equation in the standard form (16.19) and obtain $r = \frac{3}{2}/(1 - \frac{1}{2}\cos\theta)$. By comparison with (16.19) we note that $e = \frac{1}{2}$, $ek = \frac{3}{2}$, and therefore $k = 3$. Since $0 < e < 1$, the graph is an ellipse that has the pole as a focus. The center of the ellipse is $(e^2k/(1 - e^2), 0) = (1, 0)$. Some points on the ellipse, which are shown in Figure 16–10, are

$$(3, 0), \quad \left(\frac{3}{2}, \frac{\pi}{2}\right), \quad (1, \pi), \quad \left(\frac{3}{2}, \frac{3\pi}{2}\right).$$

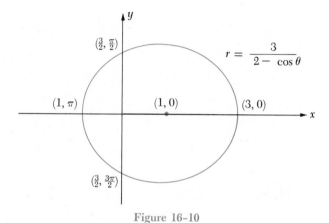

Figure 16–10

Suppose that O and P have the same meaning as in the previous discussion and Q is the foot of a perpendicular from P to the line $y = -k$. A polar equation of the set of all points P such that $|OP|/|PQ| = e$ can be shown to be

$$r = \frac{ke}{1 - e\sin\theta}, \tag{16.23}$$

or equivalently

$$r = \frac{-ke}{1 + e\sin\theta}. \tag{16.24}$$

The graph of either equation is a conic that is symmetric to the line $\theta = \frac{\pi}{2}$ and has the pole as a focus. The conic obtained is given by (i), (ii), and (iii) in Theorem 16.3.2, and if $e \neq 1$ the center of the conic is the point $(e^2k/(1 - e^2), \frac{\pi}{2})$.

Example 2   Discuss the graph of the polar equation

$$r = \frac{6}{1 + 3\sin\theta} \quad \text{(Figure 16–11)}.$$

SOLUTION   The given equation is in the standard form (16.24) for a conic. Hence, $-ke = 6$, $e = 3$, and $k = -2$. Since $e > 1$, the graph is a hyperbola having the pole as a focus and center $(e^2k/(1 - e^2), \frac{\pi}{2}) = (\frac{9}{4}, \frac{\pi}{2})$. From (16.23) and (16.24) the asymptotes, which pass through the center, are parallel to the lines $\theta =$

$\sin^{-1}(\pm\frac{1}{3})$. Some points in the hyperbola, which are shown in Figure 16–11, are

$$(6, 0), \left(\frac{3}{2}, \frac{\pi}{2}\right), (6, \pi), \left(-12, \frac{7\pi}{6}\right), \left(-3, \frac{3\pi}{2}\right), \left(-12, \frac{11\pi}{6}\right).$$

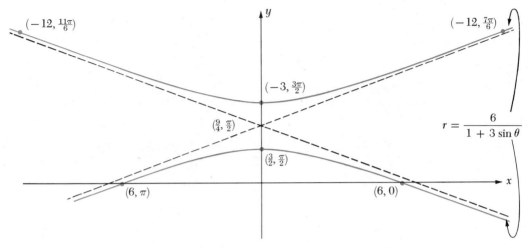

Figure 16–11

Exercise Set 16.3

In Exercises 1–8 discuss and sketch the graph of the given polar equation.

1. $r = \dfrac{4}{1 - \cos\theta}$

2. $r = \dfrac{-2}{5 + 3\cos\theta}$

3. $r = \dfrac{2}{4 - 3\sin\theta}$

4. $r = \dfrac{3}{2 - 5\sin\theta}$

5. $r = \dfrac{6}{3 + 4\cos\theta}$

6. $r = \dfrac{-3}{1 + \sin\theta}$

7. $r = \dfrac{-5}{2 + \sin\theta}$

8. $r = \dfrac{3}{1 - 2\cos\theta}$

In Exercises 9–12 find a polar equation for the conic having a focus at the pole and satisfying the given conditions.

9. $e = \frac{3}{4}$, directrix $x = 3$

10. $e = \frac{3}{2}$, directrix $x = -4$

11. $e = 3$, directrix $y = -1$

12. $e = \frac{1}{3}$, directrix $y = 4$

13. Derive the equation (16.23) for a conic having the pole as a focus, the line $y = -k$, where $k \neq 0$, as a directrix, and eccentricity e.

14. Prove that if $e > 1$, then the graph of (16.19) is a hyperbola having the pole as a focus.

15. Prove that the asymptotes of the hyperbola in Exercise 14 have slopes $\pm\sqrt{e^2 - 1}$.

16. Suppose that P has the same meaning as in connection with equation (16.16), Q is the foot of a perpendicular from P to the lines with Cartesian equation
$$x = \frac{k(1 + e^2)}{1 - e^2},$$
and F is the point $(2e^2k/(1 - e^2), 0)$. Prove that the set of all points P such that $|FP|/|PQ| = e$ has the Cartesian equation (16.21) (and hence the polar equation (16.19)).

16.4 Intersections of Graphs of Polar Equations

The student will recall that by solving simultaneously two equations in x and y, all of the intersection points of the graphs of the equations can be obtained. However, a simultaneous solution of two polar equations may not yield every intersection point of their graphs. For example, consider the graphs of the equations

$$r = a \cos \theta \qquad \text{and} \qquad r = a \sin \theta \qquad (16.25)$$

(Figure 16–12). By direct substitution in these equations it is readily verified that $(\sqrt{2}a/2, \frac{\pi}{4})$ is an intersection point of their graphs. In addition, the pole is in both graphs since $(0, \frac{\pi}{2})$ satisfies $r = a \cos \theta$, and $(0, 0)$ satisfies $r = a \sin \theta$.

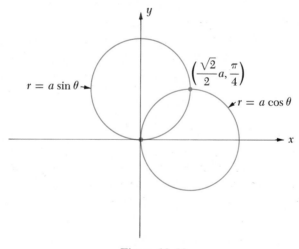

Figure 16–12

From solving the equations (16.25) we have $\tan \theta = 1$ and hence

$$\theta = \frac{\pi}{4} + n\pi \qquad n \text{ any integer.} \qquad (16.26)$$

By letting $n = 0$ and $n = 1$, respectively, in (16.26) we obtain from (16.25) the points $(\sqrt{2}a/2, \frac{\pi}{4})$ and $(-\sqrt{2}a/2, \frac{5\pi}{4})$. However, $(-\sqrt{2}a/2, \frac{5\pi}{4})$ is the same point as $(\sqrt{2}a/2, \frac{\pi}{4})$. We also obtain the same point as $(\sqrt{2}a/2, \frac{\pi}{4})$ by assigning other integral values to n. However, it is important to note that *the pole is not obtained from the simultaneous solution of equations* (16.25).

As a second example consider the graphs (Figure 16–13) of the equations

$$r = \frac{3}{2} - \cos\theta \qquad \theta = \frac{\pi}{3}. \tag{16.27}$$

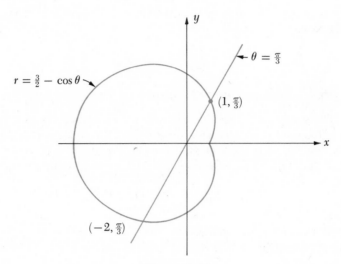

Figure 16–13

These graphs intersect at the point $(1, \frac{\pi}{3})$ since the coordinates of this point satisfy the equations in (16.27). Also, it appears from Figure 16–13 that the line $\theta = \frac{\pi}{3}$ intersects the graph of $r = \frac{3}{2} - \cos\theta$ in a second point, $(2, \frac{4\pi}{3})$, which is also denoted by $(-2, \frac{\pi}{3})$. The coordinates $(-2, \frac{\pi}{3})$ do not satisfy $r = \frac{3}{2} - \cos\theta$. However, since the point (r, θ) is the same as $(-r, \theta + \pi)$, the graph of the equation $r = \frac{3}{2} - \cos\theta$ is the same as that of $-r = \frac{3}{2} - \cos(\theta + \pi)$ or

$$-r = \frac{3}{2} + \cos\theta. \tag{16.28}$$

Thus, since the coordinates of the point $(-2, \frac{\pi}{3})$ satisfy (16.28) and the equation $\theta = \frac{\pi}{3}$, the point is in the intersection of the graphs of equations (16.27).

In summary, the following approach can be used to obtain all of the intersection points of the graphs of two polar equations.

(i) First, determine if the pole is in the intersection of the graphs. The pole will be an intersection point if it has polar coordinates that satisfy both given polar equations.

(ii) Suppose the given polar equations are denoted by I and II and m and n are arbitrary integers. Solve the different systems of polar equations that can be formed by pairing an equation obtained from I by replacing r and θ by $(-1)^m r$ and $\theta + m\pi$, respectively, with an equation obtained from II by replacing r and θ by $(-1)^n r$ and $\theta + n\pi$, respectively.

Example 1 Find the intersection points of the graphs of the equations $r = 1$ and $r = 1 + 2\cos\theta$ (Figure 16–14).

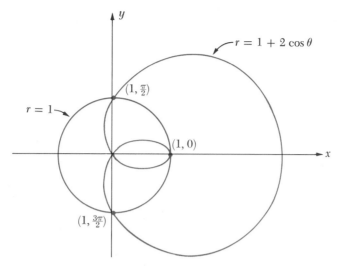

$$r = 1 + 2 \cos \theta$$

$$\left(1, \tfrac{\pi}{2}\right)$$

$$r = 1$$

$$(1, 0)$$

$$\left(1, \tfrac{3\pi}{2}\right)$$

Figure 16-14

SOLUTION   First we note that the pole is not an intersection point since it is not in the graph of $r = 1$. We then consider the systems of equations that can be formed from the given equations in accordance with step (ii) above. These systems will be of the form

$$(-1)^m r = 1 \qquad (-1)^n r = 1 + 2 \cos (\theta + n\pi).$$

For various integral values of m the first of these equations becomes either $r = 1$ or $-r = 1$, and for different integral values of n the second equation simplifies to either $r = 1 + 2 \cos \theta$ or $-r = 1 - 2 \cos \theta$. Therefore, we shall consider the systems

$r = 1 + 2 \cos \theta$	$-r = 1 - 2 \cos \theta$	$r = 1 + 2 \cos \theta$	$-r = 1 - 2 \cos \theta$
$r = 1$	$r = 1$	$-r = 1$	$-r = 1.$

In solving the first system, we obtain after equating the expressions for r

$$1 = 1 + 2 \cos \theta$$
$$\cos \theta = 0$$
$$\theta = \frac{\pi}{2}, \frac{3\pi}{2}, \text{ etc.,}$$

From these values of θ we have the intersection points $\left(1, \frac{\pi}{2}\right)$ and $\left(1, \frac{3\pi}{2}\right)$. In solving the second system, we obtain

$$-1 = 1 - 2 \cos \theta$$
$$\cos \theta = 1$$
$$\theta = 0, \text{ etc.,}$$

from which the intersection point $(1, 0)$ is found. In the third system, we obtain

$$-1 = 1 + 2 \cos \theta$$
$$\cos \theta = -1$$
$$\theta = \pi, \text{ etc.,}$$

from which the intersection point $(-1, \pi)$ is found. However, $(-1, \pi)$ is the same as $(1, 0)$. A solution of the fourth system gives again the points $(1, \frac{\pi}{2})$ and $(1, \frac{3\pi}{2})$. The intersection points of the graphs are therefore $(1, \frac{\pi}{2})$, $(1, \frac{3\pi}{2})$, and $(1, 0)$.

Exercise Set 16.4

In Exercises 1–10 find the intersection points of the graphs of the given pair of equations. Sketch the graphs of the equations.

1. $r = 1$, $r = 2 \sin \theta$

2. $r = \dfrac{5}{3 - \cos \theta}$, $r = 2$

3. $r = 2$, $r^2 = 4 \cos 2\theta$

4. $r = a(1 + \sin \theta)$, $r = 2a \sin \theta$

5. $r = 2 - 2 \cos \theta$, $r = \dfrac{1}{1 + \cos \theta}$

6. $r = 2\theta$, $\theta = \dfrac{\pi}{6}$

7. $r = \sin \theta - 1$, $r = \cos 2\theta$

8. $r = 1 - \cos \theta$, $r = \sin \theta$

9. $r = \sin 2\theta$, $r = \cos 2\theta$

10. $r = \csc \theta$, $r = \cot \theta$

16.5 Areas Using Polar Coordinates

In this section we will find the area of the region in the plane which is bounded by a polar curve and two rays that have the pole as an endpoint. To obtain this area, we will use the formula from Euclidean geometry for the area of a sector of a circle,

$$A = \tfrac{1}{2}\theta r^2 \qquad (16.29)$$

where r is the radius of the sector and θ is the radian measure of its central angle.

Suppose the function f is continuous and non-negative valued on the interval $[\alpha, \beta]$, and that T is the region given in polar coordinates by

$$T = \{(r, \theta): 0 \le r \le f(\theta) \text{ and } \theta \in [\alpha, \beta]\}. \qquad (16.30)$$

This region is shaded in Figure 16–15. Next we let $P = \{\alpha = \theta_0, \theta_1, \theta_2, \ldots, \theta_{n-1},$

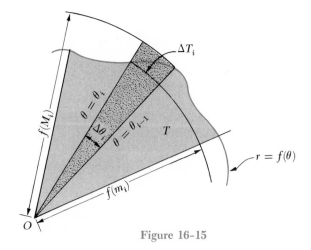

Figure 16–15

$\theta_n = \beta\}$ be a partition of $[\alpha, \beta]$ and

$$\Delta\theta_i = \theta_i - \theta_{i-1} \quad \text{if } i = 1, 2, \ldots, n.$$

Also, we let

$$f(m_i) = \text{minimum of } f \text{ on } [\theta_{i-1}, \theta_i]$$
$$f(M_i) = \text{maximum of } f \text{ on } [\theta_{i-1}, \theta_i].$$

Then the subregion ΔT_i of T, which is bounded by the rays $\theta = \theta_{i-1}$ and $\theta = \theta_i$ and the graph of $r = f(\theta)$ is contained in a "circumscribed" sector which has central angle $\Delta\theta_i$ and radius $f(M_i)$ and contains an "inscribed" sector having central angle $\Delta\theta_i$ and radius $f(m_i)$. If ΔA_i, the area of ΔT_i, exists, then its area will be sandwiched between the areas of these sectors; that is,

$$\tfrac{1}{2}(f(m_i))^2 \,\Delta\theta_i \leq \Delta A_i \leq \tfrac{1}{2}(f(M_i))^2 \,\Delta\theta_i. \tag{16.31}$$

Thus if A, the area of T, is defined, summing in (16.31) from 1 to n will give

$$\sum_{i=1}^{n} \tfrac{1}{2}(f(m_i))^2 \,\Delta\theta_i \leq A \leq \sum_{i=1}^{n} \tfrac{1}{2}(f(M_i))^2 \,\Delta\theta_i. \tag{16.32}$$

Since f is continuous on $[\alpha, \beta]$, so is the function g where $g(\theta) = \tfrac{1}{2}(f(\theta))^2$ and hence the definite integral $\int_\alpha^\beta \tfrac{1}{2}(f(\theta))^2 \, d\theta$ certainly exists. Now the left and right members of (16.32) are respectively the lower sum and upper sum of the function g. Hence by Definition 7.5.3 if the norm of the associated partition P is sufficiently small, these Riemann sums can be made to differ from $\int_\alpha^\beta \tfrac{1}{2}(f(\theta))^2 \, d\theta$ by as little as we please.

Therefore, we have the following polar coordinate definition of area:

16.5.1 Definition

The area A of the region T given by (16.30) is

$$A = \int_\alpha^\beta \tfrac{1}{2}(f(\theta))^2 \, d\theta. \tag{16.33}$$

Example 1 Find the area of one leaf of the three-leaf rose $r = 4 \sin 3\theta$ (Figure 16–16).

SOLUTION We note that a leaf of the graph is traced as θ varies from 0 to $\frac{\pi}{3}$. Hence from (16.33) the area of the leaf is

$$A = \int_0^{\pi/3} \tfrac{1}{2}(4 \sin 3\theta)^2 \, d\theta = \int_0^{\pi/3} 8 \sin^2 3\theta \, d\theta.$$

Letting

$$\sin^2 3\theta = \frac{1 + \cos 6\theta}{2}$$

(compare (10.19)), we have

$$A = \int_0^{\pi/3} 4(1 + \cos 6\theta) \, d\theta = 4(\theta + \tfrac{1}{6} \sin 6\theta) \Big|_0^{\pi/3}$$

$$= \frac{4\pi}{3}.$$

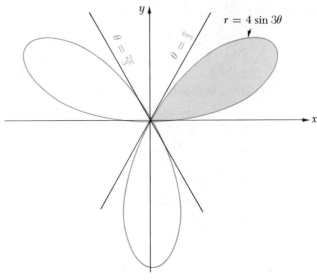

Figure 16-16

Example 2 Find the area of the region that is inside the cardioid $r = a(1 + \cos\theta)$ and outside the circle $r = 2a\cos\theta$.

SOLUTION The region is sketched in Figure 16–17. The upper half of the cardioid is traced as θ increases from 0 to π, and the upper half of the circle is described as θ increases from 0 to $\frac{\pi}{2}$. Since the region is symmetric to the polar axis, we will obtain the area of the upper half of the region and double this number.

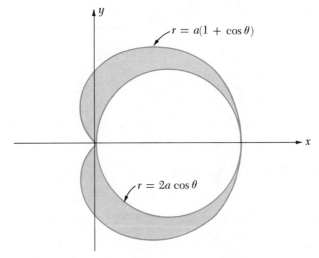

Figure 16-17

The area A of the upper half of the region is obtained by subtracting the area of the portion under the upper semicircle of $r = 2a\cos\theta$ from the area of the portion under the upper half of the cardioid. Thus

$$A = 2\left(\int_0^\pi \tfrac{1}{2}[a(1 + \cos\theta)]^2\, d\theta - \int_0^{\pi/2} \tfrac{1}{2}(2a\cos\theta)^2\, d\theta \right)$$

$$= \int_0^\pi a^2(1 + 2\cos\theta + \cos^2\theta)\, d\theta - \int_0^{\pi/2} 4a^2\cos^2\theta\, d\theta,$$

and letting $\cos^2\theta = (1 + \cos 2\theta)/2$ (see equation (10.20)), we have

$$A = \int_0^\pi a^2(\tfrac{3}{2} + 2\cos\theta + \tfrac{1}{2}\cos 2\theta)\, d\theta - \int_0^{\pi/2} 2a^2(1 + \cos 2\theta)\, d\theta$$

$$= a^2(\tfrac{3}{2}\theta + 2\sin\theta + \tfrac{1}{4}\sin 2\theta)\,\Big|_0^\pi - 2a^2(\theta + \tfrac{1}{2}\sin 2\theta)\,\Big|_0^{\pi/2}$$

$$= \frac{3\pi a^2}{2} - \pi a^2$$

$$= \frac{\pi a^2}{2}.$$

The only other application of the definite integral involving polar coordinates that we will mention in this section is arc length. The derivation of the arc length formula is left as an exercise (Exercise 26, Exercise Set 16.5).

Exercise Set 16.5

In Exercises 1–6 find the area of the region enclosed by the graphs of the given polar equations.

1. $r = \dfrac{\theta}{2}$, $\theta = 0$, $\theta = \pi$

2. $r = e^{a\theta}$, $\theta = 0$, $\theta = \pi$

3. $r = a\sin 2\theta$, $\theta = 0$, $\theta = \dfrac{\pi}{4}$

4. $r = \tan\theta$, $\theta = 0$, $\theta = \dfrac{\pi}{4}$

5. $r = \dfrac{2}{1 - \cos\theta}$, $\theta = \dfrac{\pi}{2}$, $\theta = \dfrac{3\pi}{2}$

6. $r = \dfrac{2\sqrt{2}}{\sin\theta - \cos\theta}$, $\theta = \dfrac{\pi}{2}$, $\theta = \pi$

In Exercises 7–16 find the area of the region enclosed by the given polar curve.

7. $r = 4\sin\theta$

8. $r = \cos\theta + \sin\theta$

9. $r = 2 + 2\cos\theta$

10. $r = \sqrt{3} - \sin\theta$

11. $r = 3\cos 2\theta$

12. $r = 4\sin 5\theta$

13. $r^2 = a^2\sin 2\theta$

14. $r^2 = 9\cos 2\theta$

15. $r = 2\cos\theta\sin^2\theta$

16. $r = 2 + \sin^2\theta$

In Exercises 17–25 find the area of the given region.

17. The smaller loop of the limaçon $r = 1 - 2\sin\theta$.

18. The smaller loop of the limaçon $r = 1 + \sqrt{3}\cos\theta$.

19. The region bounded by the graphs of the equations $r = a \sin \theta$, $r = a(1 + \sin \theta)$, $\theta = 0$, and $\theta = \pi$.

20. The region that is inside the cardioid $r = a(1 + \cos \theta)$ and outside the circle $r = a$.

21. The region that is inside both the cardioid $r = a(1 + \cos \theta)$ and the circle $r = a$.

22. The region that is inside both the cardioids $r = a(1 + \cos \theta)$ and $r = a(1 - \cos \theta)$.

23. The region inside both the circles $r = a \cos \theta$ and $r = a \sin \theta$.

24. The region inside the lemniscate $r^2 = a^2 \cos 2\theta$ and outside the circle $r = a \sin \theta$.

25. A smaller loop of the curve $r = 1 + 2 \sin 2\theta$.

26. Prove: If f has a continuous derivative on the interval $[\alpha, \beta]$, then the length L of the polar curve $\{(r, \theta): r = f(\theta) \text{ and } \theta \in [\alpha, \beta]\}$ is given by

$$L = \int_{\alpha}^{\beta} \sqrt{(f(\theta))^2 + (f'(\theta))^2}\, d\theta.$$

HINT: The curve can be expressed parametrically by equations (16.7).

In Exercises 27–30 obtain the length of the given polar curves using the formula derived in Exercise 26.

27. $r = 2(1 + \cos \theta)$

28. $r = 3 \cos \theta + 4 \sin \theta$

29. $r = \sin^3 \dfrac{\theta}{3}$

30. $r = \dfrac{2}{1 + \cos \theta}$

31. Suppose a particle moves along the polar curve $r = f(\theta)$, where f is twice differentiable, in such a way that a line segment from the pole to the particle sweeps out equal areas in equal intervals of time. Then from (16.33)

$$\frac{dA}{dt} = \frac{dA}{d\theta} \frac{d\theta}{dt} = \frac{1}{2} r^2 \frac{d\theta}{dt}$$

is constant. Prove that the acceleration vector of the particle is given by

$$\frac{d^2\mathbf{P}}{dt^2} = \left[\frac{d^2 r}{dt^2} - r \left(\frac{d\theta}{dt} \right)^2 \right] \langle \cos \theta, \sin \theta \rangle.$$

17. Vectors in Space; Solid Analytical Geometry

17.1 Three-Dimensional Coordinates

In order to solve problems in solid geometry, a three-dimensional coordinate system is utilized just as a two-dimensional system is employed for discussions involving plane geometry. To introduce the notion of three-dimensional Cartesian coordinates, we consider three mutually perpendicular number lines, the *x axis*, *y axis*, and *z axis*, which intersect at a point O that is the origin of each axis. These *coordinate axes* will be oriented so that the *x* axis is horizontal with its positive direction directly out from the plane of the paper, the *y* axis is horizontal with its positive direction to the right, and the *z* axis is vertical with its positive direction upward (Figure 17–1(a)). We then have a *right-handed* coordinate system, which is so named because of the following *right-hand rule:* if the right hand is held so that the fingers curl in the direction of rotation from the positive *x* axis to the positive *y* axis, the thumb will point in the direction of the positive *z* axis (Figure 17–1(b)).

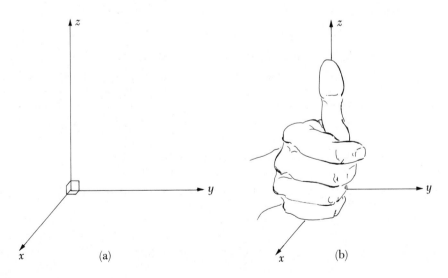

Figure 17–1

As before, the point O is called the *origin* of this three-dimensional coordinate system. Each pair of coordinate axes determine a plane called a *coordinate plane*. The coordinate plane determined by the x and y axes is called the xy plane and an analogous naming scheme also identifies the xz and yz planes. The coordinate planes divide the three-dimensional space into eight subregions called *octants*.

Through any point P in space we can construct planes that are perpendicular to the coordinate axes. If these planes intersect the x axis, the y axis, and the z axis in points having one-dimensional coordinates a, b, and c, respectively, then the *ordered triple* (a, b, c) is the *Cartesian coordinate* representation of P (Figure 17–2).

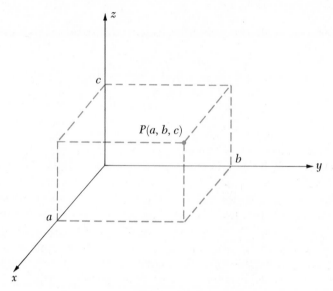

Figure 17–2

In particular, the numbers a, b, and c are respectively *the x coordinate, the y coordinate,* and *the z coordinate* of P. Conversely, for any triple of numbers (a, b, c) there is a unique point P in space having the Cartesian coordinate representation (a, b, c). Hence there is a one-to-one correspondence between points in space and ordered triples of numbers. Because of this one-to-one correspondence, the point P with coordinates (a, b, c) can be denoted by $P(a, b, c)$. We note that this procedure for determining the coordinates of a point in space is analogous to that used in Section 2.1 to find the two-dimensional coordinates of a point in the plane, in which perpendiculars are drawn from the point to the coordinate axes.

In Figure 17–3 the points $(3, -2, 4)$ and $(2, 5, -3)$, $(0, 0, 4)$, $(0, 5, 0)$, $(3, -2, 0)$, and $(2, 0, -3)$ have been located. To locate the point $(3, -2, 4)$, for example, one could begin at the origin, proceed 3 units along the positive x axis, then 2 units to the left (in the negative y direction), and finally 4 units upward (in the positive z direction).

We note that all points with the same x coordinate, a, are in the plane perpendicular to the x axis at the point $(a, 0, 0)$. Similarly, all points with the same y coordinate b are in the plane perpendicular to the y axis at the point $(0, b, 0)$,

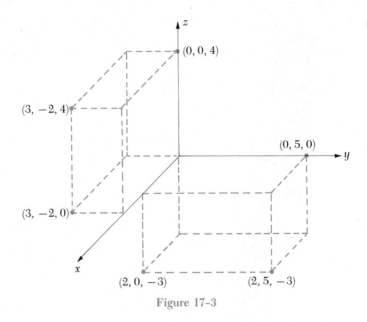

Figure 17–3

and those points with the same z coordinate c are in the plane perpendicular to the z axis at the point $(0, 0, c)$. The set of all points $\{(a, b, z) : z \in R\}$ is a line that is perpendicular to the xy plane at the point $(a, b, 0)$. Analogous remarks could be made about a set of points in which any two of the coordinates are constant but the remaining coordinate is not fixed.

Because of the one-to-one correspondence between the set of points in space and the set of all ordered triples of real numbers, the set

$$R^3 = \{(x, y, z) : x, y, \text{ and } z \in R\}$$

is called the *Cartesian 3-space* and the elements (x, y, z) of R^3 are called "points" in R^3. A frequently mentioned subset of R^3 is its *first octant*, which is the set

$$\{(x, y, z) : x, y, \text{ and } z \text{ are positive}\}.$$

In the next theorem the two-point distance formula for points in R^3 is derived. This formula is a natural extension of the two-point distance formula in Theorem 2.1.1.

17.1.1   Theorem

The distance between the points $P_1(x_1, y_1, z_1)$ and $P_2(x_2, y_2, z_2)$ is given by

$$|P_1P_2| = \sqrt{(x_2 - x_1)^2 + (y_2 - y_1)^2 + (z_2 - z_1)^2}. \tag{17.1}$$

PROOF   We first construct planes through P_1 and P_2 which are perpendicular to the z axis in the points $A(0, 0, z_1)$ and $B(0, 0, z_2)$, respectively (Figure 17–4).

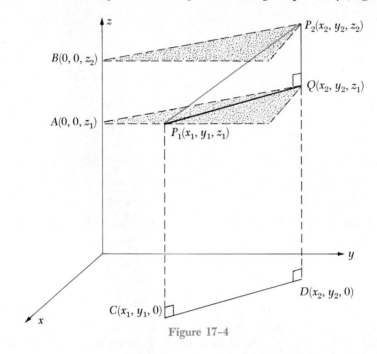

Figure 17-4

Also, we construct lines through P_1 and P_2 which are perpendicular to the xy plane and intersect the xy plane in points $C(x_1, y_1, 0)$ and $D(x_2, y_2, 0)$, respectively. If Q denotes the point (x_2, y_2, z_1), then

$$|QP_2| = |AB| = |z_2 - z_1|.$$

Also, since P_1Q and CD are opposite sides of a rectangle,

$$|P_1Q| = |CD| = \sqrt{(x_2 - x_1)^2 + (y_2 - y_1)^2}.$$

Hence by the Pythagorean Theorem

$$
\begin{aligned}
|P_1P_2|^2 &= |P_1Q|^2 + |QP_2|^2 \\
&= (x_2 - x_1)^2 + (y_2 - y_1)^2 + (z_2 - z_1)^2,
\end{aligned}
$$

and (17.1) is obtained.

Example 1   Find the distance between the points $P_1(1, -2, 3)$ and $P_2(4, -1, -2)$.

SOLUTION   From (17.1)

$$|P_1P_2| = \sqrt{(4 - 1)^2 + (-1 + 2)^2 + (-2 - 3)^2} = \sqrt{35}.$$

Formula (17.1) can be used to obtain an equation of a *sphere*. First, however, we shall require some definitions.

17.1.2 Definition

The *graph* of a set of ordered triples of numbers is the set of all points in space that are associated with these ordered triples. In particular, *the graph of an equation in x, y, and z* is the graph of the set of all solutions (x, y, z) of the equation. As in Definition 2.5.1, the equation is then called an *equation of the graph*.

17.1.3 Definition

Let r be a positive number and C be a point in three-dimensional space. The *sphere* with center C and radius r is the set of all points P in three-dimensional space such that $|PC| = r$.

17.1.4 Theorem

An equation of the sphere with center (a, b, c) and radius r is

$$(x - a)^2 + (y - b)^2 + (z - c)^2 = r^2. \tag{17.2}$$

PROOF From Definition 17.1.3, a point $P(x, y, z)$ is on the sphere if and only if its distance from the center is r—that is, if and only if

$$\sqrt{(x - a)^2 + (y - b)^2 + (z - c)^2} = r,$$

which is equivalent to equation (17.2).

Example 2 Find the center and radius of the sphere having the equation $x^2 + y^2 + z^2 - 4x + 6y - 3z - 5 = 0$.

SOLUTION Upon completing the square in x, y, and z,

$$(x^2 - 4x + 4) + (y^2 + 6y + 9) + (z^2 - 3z + \tfrac{9}{4}) = 5 + 4 + 9 + \tfrac{9}{4}$$

or

$$(x - 2)^2 + (y + 3)^2 + (z - \tfrac{3}{2})^2 = \tfrac{81}{4}. \tag{17.3}$$

Then by comparison of (17.3) with (17.2), the center of the sphere is $(2, -3, \tfrac{3}{2})$ and the radius is $\tfrac{9}{2}$.

We will mention two sets of points associated with a sphere which will be useful in our work in this text. The set

$$\{(x, y, z) : (x - a)^2 + (y - b)^2 + (z - c)^2 < r^2\},$$

which includes all points inside the sphere (17.2), is called the *open ball* with center (a, b, c) and radius r. Also, the set

$$\{(x, y, z) : (x - a)^2 + (y - b)^2 + (z - c)^2 \le r\},$$

which includes all points that are either inside the sphere (17.2) or on the sphere, is called the *closed ball* with center (a, b, c) and radius r.

A point of division formula that is an extension of Theorem 2.2.1 can be easily derived for line segments in space.

17.1.5 Theorem

The point $P(x, y, z)$ is in the line segment with endpoints $P_1(x_1, y_1, z_1)$ and $P_2(x_2, y_2, z_2)$ and satisfies

$$\frac{|P_1P|}{|P_1P_2|} = t \quad \text{where } 0 \le t \le 1 \tag{17.4}$$

if and only if

$$x = x_1 + t(x_2 - x_1) \qquad y = y_1 + t(y_2 - y_1) \qquad z = z_1 + t(z_2 - z_1). \tag{17.5}$$

PROOF We shall prove the "only if" part and leave the proof of the converse as an exercise. Suppose $P(x, y, z)$ is in the line segment and (17.4) is satisfied. If lines perpendicular to the xy plane are constructed through P_1, P, and P_2, these lines intersect the xy plane in points $Q_1(x_1, y_1, 0)$, $Q(x, y, 0)$, and $Q_2(x_2, y_2, 0)$, respectively (Figure 17–5). Since the line segments P_1Q_1, PQ, and P_2Q_2 are perpen-

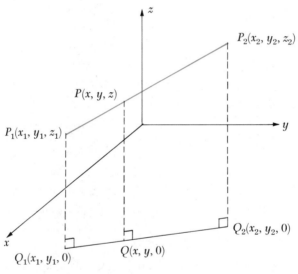

Figure 17–5

dicular to the xy plane, either they are parallel or else $Q_1 = Q = Q_2$. If P_1Q_1, PQ, and P_2Q_2 are parallel, from elementary geometry

$$\frac{|Q_1Q|}{|Q_1Q_2|} = \frac{|P_1P|}{|P_1P_2|} = t.$$

Then by Theorem 2.2.1

$$x = x_1 + t(x_2 - x_1) \qquad y = y_1 + t(y_2 - y_1). \tag{17.6}$$

Equations (17.6) also hold if $Q_1 = Q = Q_2$. The equation for z in (17.5) can then be obtained by repeating the same construction with the xz or yz planes.

Letting $t = \frac{1}{2}$ in (17.5), we obtain

$$x = \frac{x_1 + x_2}{2} \qquad y = \frac{y_1 + y_2}{2} \qquad z = \frac{z_1 + z_2}{2}$$

for the coordinates of the midpoint of $P_1 P_2$.

Exercise Set 17.1

In Exercises 1–4 suppose the given points are opposite vertices of a rectangular parallelepiped whose faces are parallel to the coordinate axes; (a) sketch the rectangular parallelepiped and give the coordinates of the other vertices, (b) find the length of the diagonal connecting the given vertices, and (c) find the midpoint of the diagonal in (b).

1. $(2, -1, 0), (-3, 4, 2)$ 2. $(-1, 4, 3), (2, -1, -2)$

3. $(-5, 2, 6), (3, -1, 2)$ 4. $(4, 0, -3), (1, 1, -5)$

In Exercises 5–8 find the center and radius of the sphere having the given equation.

5. $x^2 + y^2 + z^2 + 6x - 10y - 12z - 11 = 0$

6. $x^2 + y^2 + z^2 - 8x + 4y + 10z + 9 = 0$

7. $x^2 + y^2 + z^2 - x - 5y + 6z + 1 = 0$

8. $3x^2 + 3y^2 + 3z^2 + 2x - 5y + 9z + 5 = 0$

In Exercises 9 and 10 prove that the triangle with the given vertices is a right triangle.

9. $(-2, 3, 4), (1, -1, 5), (-4, 2, 6)$ 10. $(5, 0, -3), (2, -1, 4), (6, 4, -2)$

In Exercises 11 and 12 prove that the given points are collinear.

11. $(1, -1, 5), (-2, 4, 3), (4, -6, 7)$ 12. $(-2, 6, 1), (0, -2, 3), (6, -26, 9)$

In Exercises 13–18 describe the set of all points whose coordinates satisfy the given conditions.

13. $x = 4, \quad z = 3$ 14. $y = -3, \quad z = 2$

15. $-2 \le y < 4$ 16. $0 < z < 3$

17. $x = 3y, \quad z = 1$ 18. $x^2 + z^2 = 9, \quad y = -1$

19. Find an equation satisfied by all points that are equidistant from $(3, 2, -1)$ and $(0, -2, 4)$.

20. Find an equation of the sphere having a diameter with endpoints $(2, 5, -1)$ and $(6, -3, 5)$.

21. Prove the "if" part of Theorem 17.1.5.

22. Find an equation of the sphere passing through the points $(-1, 3, 5), (5, -6, 4), (7, -1, 5),$ and $(7, 2, -4)$.

23. Prove that the three line segments which join the midpoints of the opposite edges of a tetrahedron bisect each other.

24. Prove that the medians of a triangle are concurrent in a point that is two-thirds of the distance from any vertex to the midpoint of the opposite side.

17.2   Three-Dimensional Vectors

In this section we shall discuss three-dimensional vectors in some detail in order to facilitate the presentation of topics from solid analytic geometry. It will be noted that the definitions and theorems presented here are either identical with or are natural extensions of statements given in Section 15.1.

17.2.1   Definition

A *three-dimensional vector* is an ordered triple of real numbers. The *x component*, *y component*, and *z component* of the three-dimensional vector $\mathbf{A} = \langle a_1, a_2, a_3 \rangle$ are, respectively, a_1, a_2, and a_3. The *zero vector*, denoted by $\mathbf{0}$, is given by

$$\mathbf{0} = \langle 0, 0, 0 \rangle.$$

The *length* of the vector $\mathbf{A} = \langle a_1, a_2, a_3 \rangle$ is given by

$$|\mathbf{A}| = \sqrt{a_1{}^2 + a_2{}^2 + a_3{}^2}.$$

A *unit vector* is a vector having length 1. Two vectors $\mathbf{A} = \langle a_1, a_2, a_3 \rangle$ and $\mathbf{B} = \langle b_1, b_2, b_3 \rangle$ are *equal* if and only if $a_1 = b_1$, $a_2 = b_2$, and $a_3 = b_3$.

The vector $\mathbf{A} = \langle a_1, a_2, a_3 \rangle$ may be represented by an arrow drawn from an arbitrary point (x, y, z), called the *initial point*, to the point $(x + a_1, y + a_2, z + a_3)$, which is called the *terminal point*. If (x, y, z) and $(x + a_1, y + a_2, z + a_3)$ are denoted by P and Q respectively, we call the vector $\overrightarrow{PQ}$ a *representative* of $\mathbf{A}$ and also write $\overrightarrow{PQ} = \mathbf{A}$. An arrow from the origin to the point (a_1, a_2, a_3) is called the *positional representative* of $\mathbf{A}$ (Figure 17–6).

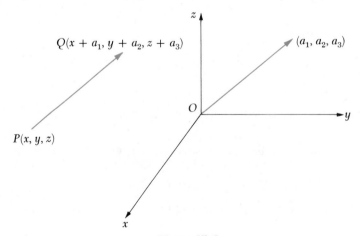

Figure 17–6

17.2.2   Definition

Suppose $\mathbf{A} = \langle a_1, a_2, a_3 \rangle$ and $\mathbf{B} = \langle b_1, b_2, b_3 \rangle$ are arbitrary three-dimensional vectors and k is any number.

(a) (Sum of two vectors) $\mathbf{A} + \mathbf{B} = \langle a_1 + b_1, a_2 + b_2, a_3 + b_3 \rangle$

(b) (Difference of two vectors) $\mathbf{A} - \mathbf{B} = \langle a_1 - b_1, a_2 - b_2, a_3 - b_3 \rangle$

(c) (Negative of a vector) $-\mathbf{A} = \langle -a_1, -a_2, -a_3 \rangle$

(d) (Multiplication by a scalar) $k\mathbf{A} = \langle ka_1, ka_2, ka_3 \rangle$

(e) (Dot product) $\mathbf{A} \cdot \mathbf{B} = a_1 b_1 + a_2 b_2 + a_3 b_3$

The properties (15.3) through (15.15) from Section 15.1 and (15.19) through (15.23) from Section 15.2 can also be proved for three-dimensional vectors. The proofs here are identical with those of the corresponding properties in Section 15.1 and 15.2 except for the presence of the z component.

Example 1   If $\mathbf{A} = \langle 3, -1, 4 \rangle$ and $\mathbf{B} = \langle -2, 4, -5 \rangle$, find: (a) $\mathbf{A} + \mathbf{B}$, (b) $\mathbf{A} - \mathbf{B}$, (c) $-\mathbf{A}$, (d) $3\mathbf{B}$, (e) $\mathbf{A} \cdot \mathbf{B}$, (f) $|\mathbf{A}|$.

SOLUTION   From Definition 17.2.2

$$\mathbf{A} + \mathbf{B} = \langle 3, -1, 4 \rangle + \langle -2, 4, -5 \rangle = \langle 3 - 2, -1 + 4, 4 - 5 \rangle = \langle 1, 3, -1 \rangle$$
$$\mathbf{A} - \mathbf{B} = \langle 3, -1, 4 \rangle - \langle -2, 4, -5 \rangle = \langle 3 - (-2), -1 - 4, 4 - (-5) \rangle$$
$$= \langle 5, -5, 9 \rangle$$
$$-\mathbf{A} = -\langle 3, -1, 4 \rangle = \langle -3, 1, -4 \rangle$$
$$3\mathbf{B} = 3\langle -2, 4, -5 \rangle = \langle 3(-2), 3(4), 3(-5) \rangle = \langle -6, 12, -15 \rangle$$
$$\mathbf{A} \cdot \mathbf{B} = 3(-2) + (-1)(4) + (4)(-5) = -30$$

Also, from Definition 17.2.1

$$|\mathbf{A}| = \sqrt{3^2 + (-1)^2 + 4^2} = \sqrt{26}.$$

A three-dimensional vector $\mathbf{A} = \langle a_1, a_2, a_3 \rangle$ can be expressed in terms of the unit vectors (Figure 17–7(a) and (b)):

$$\mathbf{i} = \langle 1, 0, 0 \rangle \qquad \mathbf{j} = \langle 0, 1, 0 \rangle \qquad \mathbf{k} = \langle 0, 0, 1 \rangle. \tag{17.7}$$

We then obtain

$$\mathbf{A} = \langle a_1, a_2, a_3 \rangle = \langle a_1, 0, 0 \rangle + \langle 0, a_2, 0 \rangle + \langle 0, 0, a_3 \rangle$$
$$= a_1 \langle 1, 0, 0 \rangle + a_2 \langle 0, 1, 0 \rangle + a_3 \langle 0, 0, 1 \rangle$$
$$= a_1 \mathbf{i} + a_2 \mathbf{j} + a_3 \mathbf{k}.$$

We also define the angle between two three-dimensional vectors as in Definition 15.2.1, and can prove as before that the angle between the nonzero vectors $\mathbf{A} = \langle a_1, a_2, a_3 \rangle$ and $\mathbf{B} = \langle b_1, b_2, b_3 \rangle$ satisfies

$$\cos \phi = \frac{\mathbf{A} \cdot \mathbf{B}}{|\mathbf{A}| \, |\mathbf{B}|} \qquad \text{where } 0 \le \phi \le \pi. \tag{17.8}$$

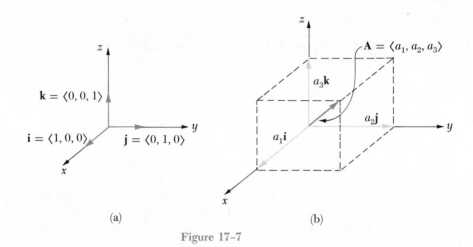

Figure 17-7

Example 2 Find the angle between the vectors $\mathbf{P} = \langle 1, 0, -3 \rangle$ and $\mathbf{Q} = \langle -2, 1, 0 \rangle$ (Figure 17-8).

SOLUTION From (17.8) the angle ϕ between $\mathbf{P}$ and $\mathbf{Q}$ satisfies

$$\cos \phi = \frac{\langle 1, 0, -3 \rangle \cdot \langle -2, 1, 0 \rangle}{|\langle 1, 0, -3 \rangle| \, |\langle -2, 1, 0 \rangle|} = \frac{-2}{\sqrt{10} \, \sqrt{5}} = -\frac{\sqrt{2}}{5}.$$

Hence

$$\phi = \cos^{-1}\left(-\frac{\sqrt{2}}{5}\right) \approx 180° - 73.6° = 106.4°.$$

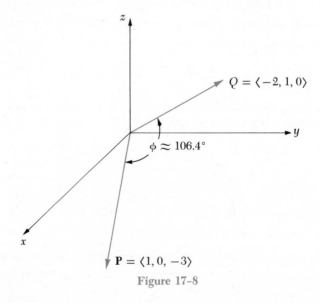

Figure 17-8

The direction of a nonzero three-dimensional vector is determined by its *direction angles*.

17.2.3 Definition

The *direction angles* α, β, and γ of a nonzero vector $\mathbf{A} = \langle a_1, a_2, a_3 \rangle$ are the angles between the vector and the unit vectors $\mathbf{i}$, $\mathbf{j}$, and $\mathbf{k}$, respectively (Figure 17–9).

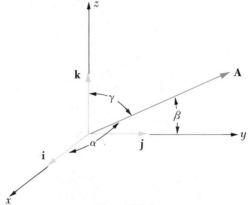

Figure 17-9

From (17.8) and Definition 17.2.3

$$\cos \alpha = \frac{\mathbf{A} \cdot \mathbf{i}}{|\mathbf{A}| \, |\mathbf{i}|} = \frac{a_1}{\sqrt{a_1^2 + a_2^2 + a_3^2}}$$

$$\cos \beta = \frac{\mathbf{A} \cdot \mathbf{j}}{|\mathbf{A}| \, |\mathbf{j}|} = \frac{a_2}{\sqrt{a_1^2 + a_2^2 + a_3^2}} \tag{17.9}$$

$$\cos \gamma = \frac{\mathbf{A} \cdot \mathbf{k}}{|\mathbf{A}| \, |\mathbf{k}|} = \frac{a_3}{\sqrt{a_1^2 + a_2^2 + a_3^2}}$$

The numbers $\cos \alpha$, $\cos \beta$, and $\cos \gamma$ in (17.9) are called the *direction cosines* of the vector $\langle a_1, a_2, a_3 \rangle$. From (17.9) the reader can easily derive the identity

$$\cos^2 \alpha + \cos^2 \beta + \cos^2 \gamma = 1. \tag{17.10}$$

Example 3 Find the direction cosines and direction angles of the vector $\mathbf{A} = \langle 1, -2, 3 \rangle$.

SOLUTION From (17.9), the direction cosines of $\mathbf{A}$ are

$$\cos \alpha = \frac{1}{\sqrt{14}} \qquad \cos \beta = -\frac{2}{\sqrt{14}} \qquad \cos \gamma = \frac{3}{\sqrt{14}},$$

and hence the direction angles of $\mathbf{A}$ are

$$\alpha = \cos^{-1} \frac{1}{\sqrt{14}} \approx 74.5°$$

$$\beta = \cos^{-1} \left(-\frac{2}{\sqrt{14}} \right) \approx 180° - 57.7° = 122.3°$$

$$\gamma = \cos^{-1} \frac{3}{\sqrt{14}} \approx 36.7°$$

If ϕ is the angle between **A** and **B**, then from (17.8) the *scalar component of* **A** *in the direction of* **B** is given by

$$|\mathbf{A}| \cos \phi = \frac{\mathbf{A} \cdot \mathbf{B}}{|\mathbf{B}|}. \qquad (17.11)$$

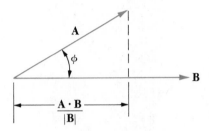

Figure 17–10

The number given in (17.11) is also called the *scalar projection of* **A** *onto* **B**. (Figure 17–10 applies if ϕ is a positive acute angle.) The components of $\mathbf{A} = \langle a_1, a_2, a_3 \rangle$ in the directions of **i**, **j**, and **k** are respectively

$$\frac{\mathbf{A} \cdot \mathbf{i}}{|\mathbf{i}|} = a_1 \qquad \frac{\mathbf{A} \cdot \mathbf{j}}{|\mathbf{j}|} = a_2 \qquad \frac{\mathbf{A} \cdot \mathbf{k}}{|\mathbf{k}|} = a_3.$$

We shall repeat for three-dimensional vectors the following definitions from Section 15.2.

17.2.4   Definition

The vectors **A** and **B** are *orthogonal* if and only if $\mathbf{A} \cdot \mathbf{B} = 0$.

17.2.5   Definition

The vectors **A** and **B** are *parallel* if and only if one of the vectors is a scalar multiple of the other.

Example 4   Show that the vectors $\mathbf{A} = \langle -2, -3, 5 \rangle$ and $\mathbf{B} = \langle 2, -3, -1 \rangle$ are orthogonal.

SOLUTION   Since $\mathbf{A} \cdot \mathbf{B} = -2(2) + (-3)(-3) + 5(-1) = 0$, **A** and **B** are orthogonal by Theorem 17.2.4.

Exercise Set 17.2

In Exercises 1–4 find the (a) length, (b) direction cosines, (c) direction angles of the given vector. Sketch the positional representative of the vector.

1. $\langle 1, -5, 0 \rangle$ 2. $\langle -2, 3, 0 \rangle$

3. $\langle 4, 2, -3 \rangle$ 4. $\langle -1, -5, 2 \rangle$

In Exercises 5–8 find Q if $\overrightarrow{PQ}$ is a representative of the given vector when P is the given point.

5. $\langle 3, 1, -4 \rangle$, $P = (-2, 0, 5)$ 6. $\langle -1, -2, 4 \rangle$, $P = (4, -5, 0)$

7. $\langle -1, 6, -2 \rangle$, $P = (3, -2, -1)$ 8. $\langle 3, 3, -7 \rangle$, $P = (-4, 1, -1)$

In Exercises 9–16 obtain the required vector if $\mathbf{A} = \langle 3, -5, 4 \rangle$, $\mathbf{B} = \langle -2, 1, 7 \rangle$, and $\mathbf{C} = \langle 1, -3, -6 \rangle$.

9. $3\mathbf{A} + \mathbf{C}$ 10. $\mathbf{B} - 2\mathbf{A}$

11. $\mathbf{A} - 4\mathbf{B} + 2\mathbf{C}$ 12. $2\mathbf{A} + 3\mathbf{B} - \mathbf{C}$

13. $(\mathbf{A} \cdot \mathbf{B})\mathbf{C} + (\mathbf{B} \cdot \mathbf{C})\mathbf{A}$ 14. $|\mathbf{A} + \mathbf{B} + \mathbf{C}|\mathbf{A}$

15. The unit vector in the same direction as $\mathbf{A}$.
16. The unit vector in the opposite direction to $\mathbf{C}$.

17. Find the component of $\langle -1, 3, 2 \rangle$ in the direction of $\langle 2, 0, -5 \rangle$.

18. Find the component of $\langle 2, -1, 4 \rangle$ in the direction of $\langle -3, 2, 1 \rangle$.

19. Find the angle between the vectors $\langle -2, 1, 3 \rangle$ and $\langle 1, 1, -2 \rangle$.

20. Find the angle between the vectors $\langle 3, -1, -2 \rangle$ and $(2, 1, -1)$.

In Exercises 21 and 22 suppose $\mathbf{A} = \langle a_1, a_2, a_3 \rangle$, $\mathbf{B} = \langle b_1, b_2, b_3 \rangle$, and $\mathbf{C} = \langle c_1, c_2, c_3 \rangle$. Verify the given equations.

21. $\mathbf{A} + (\mathbf{B} + \mathbf{C}) = (\mathbf{A} + \mathbf{B}) + \mathbf{C}$

22. $\mathbf{A} \cdot (\mathbf{B} + \mathbf{C}) = \mathbf{A} \cdot \mathbf{B} + \mathbf{A} \cdot \mathbf{C}$

23. Prove: If $\mathbf{A}$, $\mathbf{B}$, and $\mathbf{C}$ are mutually orthogonal nonzero three-dimensional vectors and α, β, and γ are numbers such that $\alpha \mathbf{A} + \beta \mathbf{B} + \gamma \mathbf{C} = 0$, then $\alpha = \beta = \gamma = 0$.

24. Show that the vectors $\langle 1, 4, -6 \rangle$ and $\langle 0, 3, 2 \rangle$ are orthogonal. Also find two unit vectors that are orthogonal to each of the given vectors.

25. Suppose P_1, P_2, P_3, and P_4 are any four points in space no three of which are collinear. If Q_1, Q_2, Q_3, and Q_4 are the midpoints of P_1P_2, P_2P_3, P_3P_4, and P_4P_1, respectively, prove that $Q_1Q_2Q_3Q_4$ is a parallelogram.

17.3   Cross Product

In this section we will consider another product of vectors which is called the *cross product*. As will be seen from the following discussion, this product arises in finding the vectors that are orthogonal to two given nonparallel vectors.

Suppose $\mathbf{A} = \langle a_1, a_2, a_3 \rangle$ and $\mathbf{B} = \langle b_1, b_2, b_3 \rangle$ are the given nonparallel vectors and k_1, k_2, and k_3 are numbers such that

$$a_1 = k_1 b_1 \qquad a_2 = k_2 b_2 \qquad a_3 = k_3 b_3. \tag{17.12}$$

At least two of the k's, say k_1 and k_2, must be unequal, for otherwise, if $k_1 = k_2 = k_3$, then $\mathbf{A}$ would be a scalar multiple of $\mathbf{B}$; that is, $\mathbf{A}$ and $\mathbf{B}$ would be parallel. Thus, supposing that $k_1 \neq k_2$, we have from (17.12),

$$a_1 b_2 - a_2 b_1 = k_1 b_1 b_2 - k_2 b_1 b_2 \neq 0. \tag{17.13}$$

By Definition 17.2.4 a vector $\mathbf{X} = \langle x, y, z \rangle$ is orthogonal to $\mathbf{A}$ and $\mathbf{B}$ if and only if

$$\mathbf{A} \cdot \mathbf{X} = 0 \qquad \text{and} \qquad \mathbf{B} \cdot \mathbf{X} = 0,$$

and hence if and only if the following equations hold:

$$\left. \begin{array}{c} a_1 x + a_2 y + a_3 z = 0 \\ b_1 x + b_2 y + b_3 z = 0 \end{array} \right\} \tag{17.14}$$

Because of (17.13) we can solve these equations for x and y in terms of z, obtaining

$$x = \frac{a_2 b_3 - a_3 b_2}{a_1 b_2 - a_2 b_1} z \qquad y = \frac{a_3 b_1 - a_1 b_3}{a_1 b_2 - a_2 b_1} z. \tag{17.15}$$

As can be shown by a substitution from (17.15) into (17.14), any vector of the form

$$\mathbf{X} = \left\langle \frac{a_2 b_3 - a_3 b_2}{a_1 b_2 - a_2 b_1} z, \; \frac{a_3 b_1 - a_1 b_3}{a_1 b_2 - a_2 b_1} z, \; z \right\rangle,$$

z being arbitrary, is orthogonal to $\mathbf{A}$ and $\mathbf{B}$ including the vector obtained by letting $z = a_1 b_2 - a_2 b_1$. Using this value for z, we obtain from $\mathbf{X}$ the vector

$$\langle a_2 b_3 - a_3 b_2, \; a_3 b_1 - a_1 b_3, \; a_1 b_2 - a_2 b_1 \rangle,$$

which is the subject of the following definition.

17.3.1   Definition

Suppose $\mathbf{A} = \langle a_1, a_2, a_3 \rangle$ and $\mathbf{B} = \langle b_1, b_2, b_3 \rangle$. The *cross product* $\mathbf{A} \times \mathbf{B}$ is the vector given by

$$\mathbf{A} \times \mathbf{B} = \langle a_2 b_3 - a_3 b_2, \; a_3 b_1 - a_1 b_3, \; a_1 b_2 - a_2 b_1 \rangle. \tag{17.16}$$

Unlike the dot product, the cross product of two vectors is a vector rather than a number and thus is sometimes called the *vector product* of the vectors. A useful scheme for remembering the cross product of $\mathbf{A} = \langle a_1, a_2, a_3 \rangle$ and $\mathbf{B} = \langle b_1, b_2, b_3 \rangle$ is the determinant notation

$$\begin{vmatrix} \mathbf{i} & \mathbf{j} & \mathbf{k} \\ a_1 & a_2 & a_3 \\ b_1 & b_2 & b_3 \end{vmatrix} \tag{17.17}$$

The symbol (17.17) is not actually a determinant since $\mathbf{i}$, $\mathbf{j}$, and $\mathbf{k}$ are vectors while the other elements are numbers. However, if one applies the usual properties of determinants to this expression, then an expansion by minors along the first row gives

$$\mathbf{A} \times \mathbf{B} = \begin{vmatrix} a_2 & a_3 \\ b_2 & b_3 \end{vmatrix} \mathbf{i} - \begin{vmatrix} a_1 & a_3 \\ b_1 & b_3 \end{vmatrix} \mathbf{j} + \begin{vmatrix} a_1 & a_2 \\ b_1 & b_2 \end{vmatrix} \mathbf{k}, \qquad (17.18)$$

which is the expression for $\mathbf{A} \times \mathbf{B}$ in terms of $\mathbf{i}$, $\mathbf{j}$, and $\mathbf{k}$.

Example 1   Find $\mathbf{A} \times \mathbf{B}$ if $\mathbf{A} = \langle 1, -1, 2 \rangle$ and $\mathbf{B} = \langle 0, 1, -3 \rangle$ (Figure 17–11).

SOLUTION   From (17.16) using (17.17),

$$\mathbf{A} \times \mathbf{B} = \begin{vmatrix} \mathbf{i} & \mathbf{j} & \mathbf{k} \\ 1 & -1 & 2 \\ 0 & 1 & -3 \end{vmatrix} = \mathbf{i} + 3\mathbf{j} + \mathbf{k} = \langle 1, 3, 1 \rangle.$$

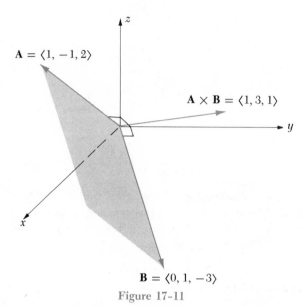

$$\mathbf{A} = \langle 1, -1, 2 \rangle$$
$$\mathbf{A} \times \mathbf{B} = \langle 1, 3, 1 \rangle$$
$$\mathbf{B} = \langle 0, 1, -3 \rangle$$

Figure 17-11

A basic property of the cross product (given in Theorem 17.32) follows immediately from Definition 17.3.1 and the discussion preceding this definition.

17.3.2   Theorem

The vector $\mathbf{A} \times \mathbf{B}$ is orthogonal to both $\mathbf{A}$ and $\mathbf{B}$; that is,

$$\mathbf{A} \cdot (\mathbf{A} \times \mathbf{B}) = 0 \qquad \text{and} \qquad \mathbf{B} \cdot (\mathbf{A} \times \mathbf{B}) = 0. \qquad (17.19)$$

This theorem should be used in checking the computation of a cross product. From Example 1, as an illustration,

$$\mathbf{A} \cdot (\mathbf{A} \times \mathbf{B}) = \langle 1, -1, 2 \rangle \cdot \langle 1, 3, 1 \rangle = 0$$
$$\mathbf{B} \cdot (\mathbf{A} \times \mathbf{B}) = \langle 0, 1, -3 \rangle \cdot \langle 1, 3, 1 \rangle = 0.$$

Theorem 17.3.2 does not, of course, give any clue as to which of two possible directions the cross product of two vectors assumes. However, since our coordinate system for three-dimensional space is right-handed, it can be shown (Exercise 35, Exercise Set 17.10) that this direction is given by the following right-hand rule: if the fingers of the right hand curl in the direction of rotation (through an angle $< \pi$) from **A** to **B**, the thumb will point in the direction of **A** × **B** (Figure 17–12). Note that this right-hand rule gives the direction of the vector **A** × **B** in Figure 17–11.

Figure 17–12

A number of properties of the cross product can be derived directly from its definition. The proof of (17.20) below is left as an exercise and the proof of (17.21) is outlined. In each of these properties k is any number.

$$k\mathbf{A} \times \mathbf{B} = k(\mathbf{A} \times \mathbf{B}). \tag{17.20}$$

$$\mathbf{A} \times k\mathbf{A} = k\mathbf{A} \times \mathbf{A} = \mathbf{0}. \tag{17.21}$$

PROOF OF (17.21)   From (17.16)

$$\mathbf{A} \times k\mathbf{A} = \langle a_2ka_3 - a_3ka_2, a_3ka_1 - a_1ka_3, a_1ka_2 - a_2ka_1 \rangle$$
$$= \mathbf{0}.$$

Similarly, it may be proved from (17.16) that $k\mathbf{A} \times \mathbf{A} = \mathbf{0}$.

From (17.21) the cross product of two parallel vectors is the zero vector. Also, since $0\mathbf{A} = \mathbf{0}$, we obtain as a special case of (17.21) when $k = 0$,

$$\mathbf{A} \times \mathbf{0} = \mathbf{0} \times \mathbf{A} = \mathbf{0}. \tag{17.22}$$

The following cross products involving the unit vectors **i**, **j**, and **k** are also derived from (17.16):

$$\mathbf{i} \times \mathbf{j} = \mathbf{k} \qquad \mathbf{j} \times \mathbf{k} = \mathbf{i} \qquad \mathbf{k} \times \mathbf{i} = \mathbf{j} \tag{17.23}$$

and also

$$\mathbf{j} \times \mathbf{i} = -\mathbf{k} \qquad \mathbf{k} \times \mathbf{j} = -\mathbf{i} \qquad \mathbf{i} \times \mathbf{k} = -\mathbf{j}. \tag{17.24}$$

Note that each of these cross products is formed in accordance with the right-hand rule mentioned above. The particular cross products $\mathbf{i} \times \mathbf{j} = \mathbf{k}$ and $\mathbf{i} \times \mathbf{k} = -\mathbf{j}$ are illustrated in Figure 17–13.

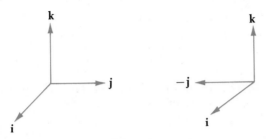

Figure 17–13

We note from (17.23) and (17.24) that the cross product of vectors is not commutative. In fact, it can be proved (Exercise 17, Exercise Set 17.3) that the product is *anti-commutative*; that is,

$$\mathbf{A} \times \mathbf{B} = -(\mathbf{B} \times \mathbf{A}). \tag{17.25}$$

Also, the cross product is not associative since from (17.23) and (17.24)

$$\mathbf{i} \times (\mathbf{i} \times \mathbf{j}) = \mathbf{i} \times \mathbf{k} = -\mathbf{j},$$

but from (17.21) and (17.22)

$$(\mathbf{i} \times \mathbf{i}) \times \mathbf{j} = \mathbf{0} \times \mathbf{j} = \mathbf{0}.$$

Yet, the distributive law is retained for cross products since

$$\mathbf{A} \times (\mathbf{B} + \mathbf{C}) = \mathbf{A} \times \mathbf{B} + \mathbf{A} \times \mathbf{C} \tag{17.26}$$

(Exercise 18, Exercise Set 17.3).

A useful formula involving the cross product is the identity

$$|\mathbf{A} \times \mathbf{B}|^2 = |\mathbf{A}|^2 |\mathbf{B}|^2 - (\mathbf{A} \cdot \mathbf{B})^2. \tag{17.27}$$

PROOF   We note that

$$
\begin{aligned}
|\mathbf{A}|^2 |\mathbf{B}|^2 - (\mathbf{A} \cdot \mathbf{B})^2 &= (a_1{}^2 + a_2{}^2 + a_3{}^2)(b_1{}^2 + b_2{}^2 + b_3{}^2) - (a_1 b_1 + a_2 b_2 + a_3 b_3)^2 \\
&= (a_2{}^2 b_3{}^2 - 2a_2 a_3 b_2 b_3 + a_3{}^2 b_2{}^2) + (a_3{}^2 b_1{}^2 - 2a_1 a_3 b_1 b_3 + a_1{}^2 b_3{}^2) \\
&\qquad + (a_1{}^2 b_2{}^2 - 2a_1 a_2 b_1 b_2 + a_2{}^2 b_1{}^2) \\
&= (a_2 b_3 - a_3 b_2)^2 + (a_3 b_1 - a_1 b_3)^2 + (a_1 b_2 - a_2 b_1)^2 \\
&= |\mathbf{A} \times \mathbf{B}|^2.
\end{aligned}
$$

The next theorem about the length of the cross product can be proved from (17.27).

17.3.3   Theorem

If ϕ is the angle between the three-dimensional vectors $\mathbf{A}$ and $\mathbf{B}$, then

$$|\mathbf{A} \times \mathbf{B}| = |\mathbf{A}|\,|\mathbf{B}|\sin\phi. \tag{17.28}$$

PROOF From (17.27) and (17.8)

$$|\mathbf{A} \times \mathbf{B}|^2 = |\mathbf{A}|^2|\mathbf{B}|^2 - |\mathbf{A}|^2|\mathbf{B}|^2 \cos^2 \phi$$
$$= |\mathbf{A}|^2|\mathbf{B}|^2(1 - \cos^2 \phi)$$
$$= |\mathbf{A}|^2|\mathbf{B}|^2 \sin^2 \phi.$$

Since $0 \leq \phi \leq \pi$, $\sin \phi \geq 0$, and therefore (17.28) is obtained.

Formula (17.28) is illustrated in Figure 17–14. The parallelogram formed by the representations of **A** and **B** has a base length $|\mathbf{A}|$ and altitude $|\mathbf{B}| \sin \phi$ and hence the area of the parallelogram is $|\mathbf{A}| \, |\mathbf{B}| \sin \phi = |\mathbf{A} \times \mathbf{B}|$.

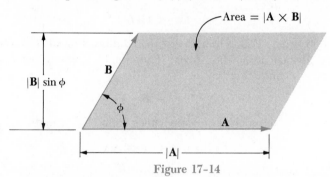

Figure 17–14

We will also discuss the geometric interpretation of the *triple scalar product* $\mathbf{C} \cdot (\mathbf{A} \times \mathbf{B})$. Consider the parallelepiped formed by the representations of **A**, **B**, and **C** (Figure 17–15). The area of the base of the parallelepiped is $|\mathbf{A} \times \mathbf{B}|$

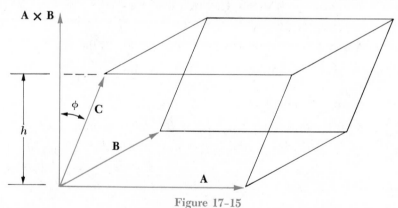

Figure 17–15

and its altitude is $h = |\mathbf{C}| \, |\cos \phi|$ where ϕ is the angle between **C** and $\mathbf{A} \times \mathbf{B}$. (In Figure 17–15 ϕ is shown as a positive acute angle.) Hence the volume of the parallelepiped is

$$V = h|\mathbf{A} \times \mathbf{B}| = |\mathbf{C}| \, |\mathbf{A} \times \mathbf{B}| \, |\cos \phi|$$
$$= |\mathbf{C} \cdot (\mathbf{A} \times \mathbf{B})|. \tag{17.29}$$

Example 2 Find the volume of the parallelepiped with vertices

$M(2, 0, 1)$, $N(4, 3, -1)$, $P(3, 2, 3)$, and $Q(-2, 3, 0)$ (Figure 17–16) and sides MN, MP, and MQ.

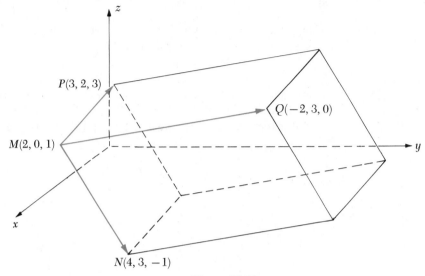

Figure 17-16

SOLUTION   From (17.29) the volume of the parallelepiped is

$$V = |\overrightarrow{MQ} \cdot (\overrightarrow{MN} \times \overrightarrow{MP})| = |\langle -4, 3, -1 \rangle \cdot (\langle 2, 3, -2 \rangle \times \langle 1, 2, 2 \rangle)|$$
$$= |\langle -4, 3, -1 \rangle \cdot \langle 10, -6, 1 \rangle| = |-59|$$
$$= 59.$$

In our work in solid analytic geometry the following theorems are useful.

17.3.4   Theorem

A *and* **B** *are parallel if and only if* **A** × **B** = **0**.

   PROOF   To prove the "if" part, we note that if **A** × **B** = **0** and **A** ≠ **0** ≠ **B**, then from (17.28), $\sin \phi = 0$ and hence $\phi = 0$ or π. Then by the extension of Definition 15.2.1 to three-dimensional vectors and Definition 17.2.5, **A** and **B** are parallel vectors. If either **A** or **B** is the zero vector, then **A** × **B** = **0** from (17.22). The proof of the "only if" part is left as an exercise.

17.3.5   Theorem

$$\mathbf{A} \times (\mathbf{B} \times \mathbf{C}) = (\mathbf{A} \cdot \mathbf{C})\mathbf{B} - (\mathbf{A} \cdot \mathbf{B})\mathbf{C}. \tag{17.30}$$

   The proof of Theorem 17.3.5, which is also left as an exercise, involves finding the components of the vector in each member of (17.30).

17.3.6 Theorem

If **A** *is orthogonal to* **B** *and* **C**, *then* **A** *is a parallel to* **B × C**.

PROOF By hypothesis $\mathbf{A} \cdot \mathbf{B} = 0$ and $\mathbf{A} \cdot \mathbf{C} = 0$. Hence, from (17.30) $\mathbf{A} \times (\mathbf{B} \times \mathbf{C}) = \mathbf{0}$. Thus **A** is parallel to **B × C** by Theorem 17.3.4.

The cross product is found in physical applications of vectors. For example, the moment M about a point O of a force **F** at a point A is defined as $M = |\mathbf{F}|\, d$ where d is the distance from O to the line of action of the force **F** (Figure 17–17). Thus, since $d = |\overrightarrow{OA}| \sin \phi$,

$$M = |\overrightarrow{OA}|\,|\mathbf{F}| \sin \phi = |\overrightarrow{OA} \times \mathbf{F}|.$$

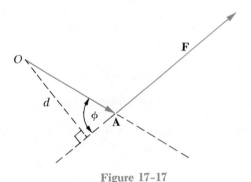

Figure 17-17

We can then define **M**, the *moment vector* of **F** about O, as

$$\mathbf{M} = \overrightarrow{OA} \times \mathbf{F}.$$

For a second example of a physical application of a cross product we consider a particle P rotating about a fixed line with angular velocity ω. If O is a fixed point on the axis of rotation, then the linear velocity (speed) of the particle P (Exercise 13, Exercise Set 15.6) is

$$v = \omega |\overrightarrow{OP}| \sin \phi \tag{17.31}$$

where ϕ is the angle between $\overrightarrow{OP}$ and the axis of rotation.

The velocity vector **V** of the particle P is tangent to the circle of rotation at P, being in the direction of motion of the particle, and $|\mathbf{V}| = v$. If the vector ω has length ω and is directed along the axis from O as shown in Figure 17–18, then from (17.31) and our description of ω and **V**,

$$\mathbf{V} = \omega \times \overrightarrow{OP}.$$

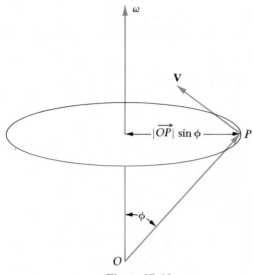

Figure 17–18

Exercise Set 17.3

In Exercises 1–10, $\mathbf{A} = \langle 1, 0, 2 \rangle$, $\mathbf{B} = \langle -2, 3, 1 \rangle$, and $\mathbf{C} = \langle -1, 4, 0 \rangle$. Find

1. $\mathbf{A} \times \mathbf{B}$ 2. $\mathbf{B} \times \mathbf{A}$

3. $|2\mathbf{A} \times \mathbf{C}|$ 4. $|\mathbf{A} \times 3\mathbf{B}|$

5. $\mathbf{A} \cdot (\mathbf{B} \times \mathbf{C})$ 6. $\mathbf{C} \cdot (\mathbf{A} \times \mathbf{B})$

7. $(\mathbf{A} - \mathbf{B}) \times (\mathbf{B} + \mathbf{C})$ 8. $(\mathbf{A} \times \mathbf{B}) \cdot (\mathbf{B} \times \mathbf{C})$

9. $\mathbf{A} \times (\mathbf{B} \times \mathbf{C})$ 10. $\mathbf{C} \times (\mathbf{A} \times \mathbf{B})$

11. Find a unit vector that is orthogonal to the vectors $\langle -1, 1, 3 \rangle$ and $\langle 0, -2, 1 \rangle$.

12. Find the area of the triangle with vertices $(2, 1, 3)$, $(-4, 3, -1)$, and $(5, 0, 2)$.

13. Find the area of the triangle with vertices $(1, 4, 0)$, $(5, -2, 2)$, and $(3, 1, 6)$.

14. Find the volume of the parallelepiped with vertices $M(1, 0, 3)$, $N(-2, 1, 4)$, $P(3, -4, 1)$, and $Q(-1, 2, 3)$ and sides MN, MP, and MQ. Check your answer with a different calculation.

15. Find the volume of the tetrahedron with vertices $(1, 2, 1)$, $(0, 2, 3)$, $(2, 4, 5)$, and $(3, 7, 4)$.

16. Prove (17.20).

17. Prove (17.25).

18. Prove (17.26).

19. Prove the "only if" part of Theorem 17.3.4.

20. If $\mathbf{A} = \langle a_1, a_2, a_3 \rangle$, $\mathbf{B} = \langle b_1, b_2, b_3 \rangle$, and $\mathbf{C} = \langle c_1, c_2, c_3 \rangle$, prove that

$$\mathbf{A} \cdot (\mathbf{B} \times \mathbf{C}) = \begin{vmatrix} a_1 & a_2 & a_3 \\ b_1 & b_2 & b_3 \\ c_1 & c_2 & c_3 \end{vmatrix}.$$

21. Prove that $\mathbf{A} \cdot (\mathbf{B} \times \mathbf{C}) = \mathbf{C} \cdot (\mathbf{A} \times \mathbf{B}) = \mathbf{B} \cdot (\mathbf{C} \times \mathbf{A})$.

22. Prove Theorem 17.3.5.

23. Prove the *Jacobi identity*

$$\mathbf{A} \times (\mathbf{B} \times \mathbf{C}) + \mathbf{B} \times (\mathbf{C} \times \mathbf{A}) + \mathbf{C} \times (\mathbf{A} \times \mathbf{B}) = 0.$$

24. Prove that $\mathbf{A} \times (\mathbf{B} \times \mathbf{C}) = (\mathbf{A} \times \mathbf{B}) \times \mathbf{C}$ if and only if $(\mathbf{A} \times \mathbf{C}) \times \mathbf{B} = \mathbf{0}$.

17.4   Lines

We begin with the following vector definition for a line in three-dimensional space.

17.4.1   Definition

Suppose $P_1 \in R^3$ and $\mathbf{N}$ is a nonzero three-dimensional vector. The line L through P_1 in the direction of $\mathbf{N}$ is the set of all points $P \in R^3$ such that $\overrightarrow{P_1P}$ is parallel to $\mathbf{N}$ (Figure 17–19).

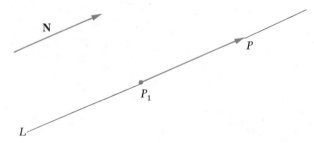

Figure 17-19

If $\mathbf{N} = \langle a, b, c \rangle$ and $P_1 = (x_1, y_1, z_1)$, then by Definition 17.4.1 $P(x, y, z) \in L$ if and only if there is some number t such that

$$\langle x - x_1, y - y_1, z - z_1 \rangle = t \langle a, b, c \rangle$$

or saying the same thing

$$x = x_1 + at \qquad y = y_1 + bt \qquad z = z_1 + ct. \tag{17.32}$$

The equations (17.32) are called the *parametric equations* of the line L.

The components a, b, and c of the vector $\mathbf{N}$ determine the direction of L and are therefore called *direction numbers* of L. We can also obtain different parametric equations for L using any vector that is parallel to $\mathbf{N}$. Thus any numbers that are proportional to direction numbers for a line are themselves direction numbers for the line.

Example 1 Find parametric equations for the line through the points $P_1(1, -2, 4)$ and $P_2(0, 3, 2)$.

SOLUTION We can think of the line as the set of all points $P(x, y, z)$ such that $\overrightarrow{P_1P}$ is parallel to $\overrightarrow{P_1P_2} = \langle -1, 5, -2 \rangle$. Hence from (17.32) the line has parametric equations

$$x = 1 - t \qquad y = -2 + 5t \qquad z = 4 - 2t. \qquad (17.33)$$

Example 2 Is the point $(2, -7, 6)$ in the line in Example 1?

SOLUTION The point $(2, -7, 6)$ is in the line if and only if the equations obtained from (17.33) by replacing x, y, and z by 2, -7, and 6, respectively, are satisfied by the same number t. Since these equations

$$2 = 1 - t \qquad -7 = -2 + 5t \qquad 6 = 4 - 2t$$

are satisfied by $t = -1$, the point is in the line.

If a, b, and c are not zero, the equations in (17.32) for the line L can be solved for t. When these solutions are equated, we have

$$\frac{x - x_1}{a} = \frac{y - y_1}{b} = \frac{z - z_1}{c}, \qquad (17.34)$$

which are called *symmetric equations* of L. The equations (17.34) could not be stated if at least one of the numbers a, b, and c is 0. If, for example, $c = 0$ but $a \neq 0 \neq b$, then instead of (17.34) L would have the equations

$$\frac{x - x_1}{a} = \frac{y - y_1}{b}, \qquad z = z_1.$$

Two lines in three-dimensional space are called *skew lines* if and only if they are not parallel and do not intersect. The distance between two skew lines can be obtained by a vector method.

Example 3 Find the distance between the skew lines

$$\frac{x - 1}{2} = \frac{y}{-1} = \frac{z + 4}{1}$$

and $x = -1 + t, y = 3 - 2t$, and $z = 2$.

SOLUTION For convenience let L_1 and L_2 denote the lines with the first equations and second equations, respectively, above. Also let $\mathbf{N}_1$ and $\mathbf{N}_2$ be vectors denoting the directions of L_1 and L_2, respectively. Then $\mathbf{N}_1 \times \mathbf{N}_2$ is a vector that is orthogonal to $\mathbf{N}_1$ and $\mathbf{N}_2$ and is therefore directed parallel to the shortest line segment connecting L_1 and L_2. The length d of this line segment is the required distance between the lines (Figure 17–20). If P_1 and P_2 are points in L_1 and L_2,

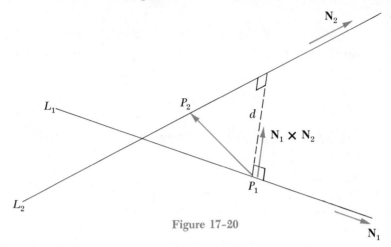

Figure 17-20

respectively, then d is the absolute value of the scalar projection of $\overrightarrow{P_1P_2}$ on $\mathbf{N}_1 \times \mathbf{N}_2$. Thus by (17.11),

$$d = \frac{|\overrightarrow{P_1P_2} \cdot (\mathbf{N}_1 \times \mathbf{N}_2)|}{|\mathbf{N}_1 \times \mathbf{N}_2|}. \tag{17.35}$$

In this example

$$\mathbf{N}_1 = \langle 2, -1, 1 \rangle \qquad \mathbf{N}_2 = \langle 1, -2, 0 \rangle,$$

and therefore

$$\mathbf{N}_1 \times \mathbf{N}_2 = \begin{vmatrix} \mathbf{i} & \mathbf{j} & \mathbf{k} \\ 2 & -1 & 1 \\ 1 & -2 & 0 \end{vmatrix} = \langle 2, 1, -3 \rangle.$$

If the points $P_1(1, 0, -4)$ and $P_2(-1, 3, 2)$ are chosen arbitrarily in L_1 and L_2, respectively, then

$$\overrightarrow{P_1P_2} = \langle -2, 3, 6 \rangle.$$

Then from (17.35) the distance d between the lines is given by

$$d = \frac{|\langle -2, 3, 6 \rangle \cdot \langle 2, 1, -3 \rangle|}{|\langle 2, 1, -3 \rangle|} = \frac{19}{\sqrt{14}}.$$

Exercise Set 17.4

1. Find equations for the line through $(-3, 1, 5)$ with direction numbers given by $\langle -1, 2, 4 \rangle$.

2. Find equations for the line through the points $(2, 0, -4)$ and $(1, -1, 2)$.

3. Find equations for the line through the points $(3, 5, 1)$ and $(-2, 6, 4)$.

4. Find equations for the line through $(4, 1, -3)$ which is parallel to the line $x = 2$, $y = 5 - 3t$, $z = 2t$.

5. Find equations for the line through $(1, 2, -1)$ which is parallel to the line $\dfrac{2x - 1}{2} = \dfrac{y}{-1} = \dfrac{3z + 2}{4}$.

6. Find the distance between the skew lines $x = 3 - t$, $y = 1 + 4t$, $z = t$ and $\dfrac{x + 2}{3} = \dfrac{y - 1}{1} = \dfrac{z + 1}{-2}$.

7. Find the distance between the skew lines $x = 1 + 2t$, $y = 2 + t$, $z = 3 - 4t$ and $\dfrac{x - 2}{4} = \dfrac{z - 3}{3}$, $y = -1$.

8. Find the intersection point of the lines $x = -2 + t$, $y = 3t$, $z = 5 - t$ and $\dfrac{x + 3}{2} = \dfrac{y + 4}{5} = \dfrac{z - 5}{-3}$.

9. Do the lines $x = 1 - t$, $y = 2 + 3t$, $z = 4 + t$ and $\dfrac{x - 1}{-2} = \dfrac{y - 5}{3} = \dfrac{z + 1}{1}$ intersect? If so, find their intersection point.

10. Show that the lines $x = 2 + t$, $y = 3 - t$, $z = 4 + 3t$, and $\dfrac{x + 1}{2} = \dfrac{y - 6}{-2} = \dfrac{z + 5}{6}$ coincide.

11. Write equations for the line through $(-3, 1, 2)$ which is perpendicular to the line $x = -t$, $y = 1 + t$, $z = 3 - 2t$.

12. Find the distance between the parallel lines

$$\frac{x}{3} = \frac{y + 2}{2} = \frac{z - 3}{-1} \quad \text{and} \quad \frac{x + 1}{-3} = \frac{y}{-2} = \frac{z + 4}{1}.$$

13. Find the distance between the point $(2, -1, 4)$ and the line $x = 4 + t$, $y = 2$, $z = 1 + 2t$.

17.5 Planes

We begin with the following vector definition for a plane.

17.5.1 Definition

Suppose $P_1 \in R^3$ and $\mathbf{N}$ is a nonzero three-dimensional vector. The set of all points $P \in R^3$ such that $\overrightarrow{P_1P}$ is orthogonal to $\mathbf{N}$ is called the plane through P_1 with *normal vector* $\mathbf{N}$ (Figure 17–21).

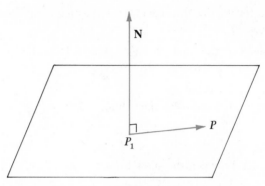

Figure 17-21

In stating this definition, we recognize that any nonzero scalar multiple of a normal vector to a plane is also a normal vector of the plane defined here.

Using Definition 17.5.1, a standard form for an equation of a plane, is easily derived.

17.5.2   Theorem

An equation for the plane through $P_1(x_1, y_1, z_1)$ with normal vector $\langle a, b, c \rangle$ is

$$a(x - x_1) + b(y - y_1) + c(z - z_1) = 0. \tag{17.36}$$

PROOF   From Definition 17.5.1 $P(x, y, z)$ is in the plane if and only if $\langle a, b, c \rangle \cdot \overrightarrow{P_1P} = 0$—that is, if and only if (17.36) holds.

Example 1   Find an equation of the plane through $(3, 1, -1)$ having a normal vector $\langle 3, -2, 1 \rangle$. Sketch the plane.

SOLUTION   An equation for the plane in the form (17.36) is

$$3(x - 3) - 2(y - 1) + (z + 1) = 0,$$

or in simplest form

$$3x - 2y + z = 6.$$

The plane intersects the coordinate axes in the points $(2, 0, 0), (0, -3, 0)$, and $(0, 0, 6)$ (Figure 17–22), and therefore we say that 2, -3, and 6 are, respectively, the x intercept, y intercept, and z intercept of the plane.

A plane perpendicular to the x axis at the point $(a, 0, 0)$ has the normal vector $\langle 1, 0, 0 \rangle$, and hence from (17.36) the plane has the equation

$$x = a.$$

Similarly, an equation of the plane perpendicular to the y axis at $(0, b, 0)$ is

$$y = b,$$

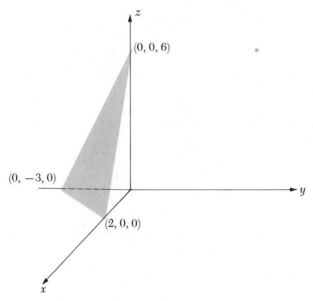

Figure 17-22

and an equation of the plane perpendicular to the z axis at $(0, 0, c)$ is

$$z = c.$$

By letting $d = ax_1 + by_1 + cz_1$ in (17.36), we obtain

$$ax + by + cz = d \qquad (17.37)$$

for an equation of the plane. Note that the final equation obtained for the plane in Example 1 is of this form. An equation of the form (17.37) in which a, b, and c are not all zero is called a *linear equation* in x, y, and z.

17.5.3   Theorem

The graph of every linear equation in x, y, and z is a plane.

PROOF   If (17.37) is the linear equation, then at least one of the numbers a, b, and c is not zero. If $a \neq 0$, then (17.37) is satisfied by $(d/a, 0, 0)$. The plane through $(d/a, 0, 0)$ with normal vector $\langle a, b, c \rangle$ then has equation (17.37), or saying the same thing, the graph of (17.37) is this plane.

If P_1, P_2, and P_3 are three noncollinear points, then $\overrightarrow{P_1 P_2}$ and $\overrightarrow{P_1 P_3}$ are not parallel vectors and hence by Theorem 17.3.4 $\overrightarrow{P_1 P_2} \times \overrightarrow{P_1 P_3} \neq 0$. By Theorem 17.3.2 $\overrightarrow{P_1 P_2} \times \overrightarrow{P_1 P_3}$ is orthogonal to $\overrightarrow{P_1 P_2}$ and $\overrightarrow{P_1 P_3}$. Therefore, by Definition 17.5.1 *the plane through P_1 with normal vector $\overrightarrow{P_1 P_2} \times \overrightarrow{P_1 P_3}$ also passes through P_2 and P_3.* From the remarks following Definition 17.5.1 this plane is uniquely determined by the three points since any other plane through these points would have a normal vector

N that is orthogonal to $\overrightarrow{P_1P_2}$ and $\overrightarrow{P_1P_3}$, and by Theorem 17.3.6 **N** would be parallel to $\overrightarrow{P_1P_2} \times \overrightarrow{P_1P_3}$.

Example 2 Find an equation of the plane through the points $(1, 3, -2)$, $(2, 1, 4)$, and $(5, -2, 1)$.

SOLUTION Denoting these points by P_1, P_2, and P_3, respectively, we have $\overrightarrow{P_1P_2} = \langle 1, -2, 6 \rangle$, $\overrightarrow{P_1P_3} = \langle 4, -5, 3 \rangle$ and

$$\overrightarrow{P_1P_2} \times \overrightarrow{P_1P_3} = \begin{vmatrix} \mathbf{i} & \mathbf{j} & \mathbf{k} \\ 1 & -2 & 6 \\ 4 & -5 & 3 \end{vmatrix} = \langle 24, 21, 3 \rangle \cdot$$

For the normal vector to the plane we can select $\frac{1}{3}(\overrightarrow{P_1P_2} \times \overrightarrow{P_1P_3}) = \langle 8, 7, 1 \rangle$. From (17.36) the equation for the plane is therefore

$$8(x - 1) + 7(y - 3) + (z + 2) = 0$$

or

$$8x + 7y + z = 27.$$

We will discuss a condition which when satisfied guarantees that two planes are intersecting. Suppose the equations of the planes are

$$a_1x + b_1y + c_1z = d_1 \quad \text{and} \quad a_2x + b_2y + c_2z = d_2. \quad (17.38)$$

If their normal vectors satisfy the condition

$$\langle a_1, b_1, c_1 \rangle \times \langle a_2, b_2, c_2 \rangle \neq \mathbf{0}$$

then by Theorem 17.3.4 $\langle a_1, b_1, c_1 \rangle$ and $\langle a_2, b_2, c_2 \rangle$ are not scalar multiples of each other. Hence the respective coefficients of at least two of the three variables x, y, and z in the equations (17.38) are not in the same ratio. If, for example, the respective coefficients of y and z are not in the same ratio, we can assign an arbitrary value t to x and then solve equations (17.38) for y and z as linear functions of t. The graph of these equations giving x, y, and z in terms of t is the intersection line of the planes.

Example 3 Find the intersection of the planes $4x + y + z = 5$ and $-y + 3z = 7$.

SOLUTION We could let $x = t$ in the given equations of the planes and then solve the equations

$$y + z = 5 - 4t \quad \text{and} \quad -y + 3z = 7$$

for y and z. We then have the parametric equations

$$x = t \quad y = 2 - 3t \quad z = 3 - t$$

for the intersection line of the planes.

Parallel planes are defined as planes having no intersection. If $k \neq 0$ and $d_2 \neq kd_1$, the planes $ax + by + cz = d_1$ and $kax + kby + kcz = d_2$ are parallel since any point satisfying the first of these equations cannot satisfy the second equation. To summarize our discussion of parallel and intersecting planes, we have the following theorem.

17.5.4   Theorem

Two distinct planes are parallel if and only if their normal vectors are parallel.

For example, the planes

$$3x - 2y + z = 5 \quad \text{and} \quad 6x - 4y + 2z = -3$$

are parallel since their respective normal vectors $\langle 3, -2, 1 \rangle$ and $\langle 6, -4, 2 \rangle$ satisfy $\langle 6, -4, 2 \rangle = 2\langle 3, -2, 1 \rangle$.

Two planes are *perpendicular* if and only if their normal vectors are orthogonal. Thus the planes

$$3x - 2y + z = 5 \quad \text{and} \quad x + y - z = 2$$

are perpendicular since their respective normal vectors, $\langle 3, -2, 1 \rangle$ and $\langle 1, 1, -1 \rangle$, are orthogonal. Similarly, the plane $ax + by = c$ is perpendicular to a plane $z = $ constant since the normal vectors $\langle a, b, 0 \rangle$ and $\langle 0, 0, 1 \rangle$ are orthogonal.

Example 4   Find an equation of the plane S that contains the line $x = 1 + t, y = 1, z = -1 + t$ and is perpendicular to the plane $3x - y + z = 7$.

SOLUTION   We note that the direction numbers of the line are given by the vector $\langle 1, 0, 1 \rangle$ and also that a normal vector to the plane $3x - y + z = 7$ is $\langle 3, -1, 1 \rangle$. A normal vector $\mathbf{N}$ to plane S is therefore orthogonal to $\langle 1, 0, 1 \rangle$ and $\langle 3, -1, 1 \rangle$, and by Theorem 17.3.6 $\mathbf{N}$ is parallel to $\langle 1, 0, 1 \rangle \times \langle 3, -1, 1 \rangle$. We therefore let

$$\mathbf{N} = \langle 1, 0, 1 \rangle \times \langle 3, -1, 1 \rangle = \langle 1, 2, -1 \rangle.$$

By letting $t = 0$ in the equations for the line, the point $(1, 1, -1)$ in the plane is obtained. Hence from (17.36) an equation of the plane is

$$(x - 1) + 2(y - 1) - (z + 1) = 0$$

or

$$x + 2y - z = 4.$$

A line is said to be parallel to a plane if and only if the line does not intersect the plane. Also, a line is perpendicular to a plane if and only if the components of a normal vector for the plane are direction numbers of the line.

The last example of the section illustrates a vector method for finding the distance from a point to the plane.

Example 5 Find the distance from the point $(1, -2, 3)$ to the plane $x + 2y + z = 5$.

SOLUTION To find the distance d between a point P_1 and a plane, we select any point P_2 in the plane and obtain the absolute value of the scalar projection of $\overrightarrow{P_2P_1}$ on **N**, a normal vector to the plane (Figure 17–23).

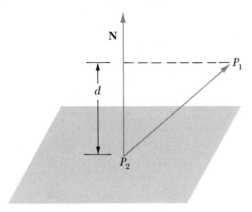

Figure 17–23

Thus by (17.11)

$$d = \frac{|\mathbf{N} \cdot \overrightarrow{P_2P_1}|}{|\mathbf{N}|}. \tag{17.39}$$

Here P_1 is the point $(1, -2, 3)$ and the point P_2 in the plane could be chosen to be $(5, 0, 0)$. Thus $\overrightarrow{P_2P_1} = \langle -4, -2, 3 \rangle$. Letting $\mathbf{N} = \langle 1, 2, 1 \rangle$, we then have from (17.39).

$$d = \frac{|\langle 1, 2, 1 \rangle \cdot \langle -4, -2, 3 \rangle|}{|\langle 1, 2, 1 \rangle|} = \frac{5}{\sqrt{6}}.$$

Exercise Set 17.5

In Exercises 1–10 obtain the equation of the plane satisfying the given conditions.

1. Has the point $(-2, 1, 3)$ and the normal vector $\langle 5, -2, 1 \rangle$.
2. Has the point $(4, 0, -1)$ and the normal vector $\langle -3, 1, 6 \rangle$.
3. Contains the points $(1, -2, 3)$, $(2, -1, 5)$, and $(0, 1, -2)$.
4. Contains the points $(-1, 0, 4)$, $(1, 3, -2)$, and $(2, -1, -3)$.
5. Contains the point $(1, 3, 2)$ and the line $x = -1 + 3t$, $y = -t$, $z = 2 + t$.
6. Contains the intersecting lines

$$\frac{x - 1}{2} = \frac{y + 2}{-3} = \frac{z}{1}$$

and $x = 2 - t$, $y = -1 + 4t$, $z = 3 + 2t$.

7. Contains the parallel lines

$$\frac{x}{2} = \frac{y-3}{1} = \frac{z+1}{-3} \quad \text{and} \quad \frac{x-2}{-2} = \frac{y}{-1} = \frac{z-1}{3}.$$

8. Through $(2, 1, 4)$ and parallel to the plane $x - 2y + 3z = 4$.

9. Through $(-1, 2, 5)$ and $(2, 1, 0)$ and perpendicular to the plane $2x + y + z = 3$.

10. Through $(1, -3, 1)$ and perpendicular to the plane $3x - 2y + 4z = 5$ and the xz plane.

11. Find the intersection of the line $x = 3 + t$, $y = 2t$, $z = -1 - t$ and the plane $x + 5y - 3z = 9$.

12. Find the intersection of the line

$$\frac{x}{4} = \frac{y-1}{-1} = \frac{z+2}{2}$$

and the plane $3x - y + 2z = 6$.

13. Find the intersection line of the planes $x + y + 2z = 6$ and $2x - y - z = 4$.

14. Find the intersection line of the planes $2x + 3z = 5$ and $3x + y - 2z = 6$.

15. Find equations of the projection of the line $x = 1 + 2t$, $y = 2 - t$, $z = 3t$ on the plane $x + y + z = 4$.

16. Prove that the line

$$\frac{x-1}{2} = \frac{y+2}{3} = \frac{z-4}{-1}$$

is in the plane $x + 2y + 8z = 29$.

17. Prove that the plane $x + 2y - z = 5$ is parallel to the line

$$\frac{x-3}{1} = \frac{y-1}{-1} = \frac{z+2}{-1}.$$

18. Find the distance between the point $(2, 3, -1)$ and the plane $x - 4y + 5z = 10$.

19. Prove that the distance between the point (x_1, y_1, z_1) and the plane $ax + by + cz = d$ is

$$\frac{|ax_1 + by_1 + cz_1 - d|}{\sqrt{a^2 + b^2 + c^2}}.$$

20. Find the distance between the parallel planes $3x - 4y + z = 5$ and $3x - 4y + z = -10$.

17.6   Surfaces

In this section and in Section 17.7 we will discuss the graph of an equation in three variables. Such a graph is called a *surface*. We have already encountered two examples of surfaces in this chapter, the plane and the sphere. Another example of a surface is a *cylinder*.

17.6.1   Definition

Let C be a plane curve and suppose that passing through every point P in C there is a line L_p and that all such lines L_p are parallel. The set of all points on these lines L_p is called a *cylinder*. The curve C here is termed the *directrix* of the cylinder and each of the lines L_p is called a *ruling* of the cylinder.

If the directrix of the cylinder is a curve C in the xy plane, and each ruling of the surface is perpendicular to the xy plane, then the surface is the set of all points (x, y, z) such that $(x, y, 0) \in C$. Hence an equation in x and y for C is also an equation of the cylinder. For example, the *right circular cylinder*

$$x^2 + y^2 = 9 \tag{17.40}$$

(Figure 17–24) has the circle $x^2 + y^2 = 9$ in the xy plane as its directrix. The rulings of the cylinder are the lines parallel to the z axis which intersect the circle.

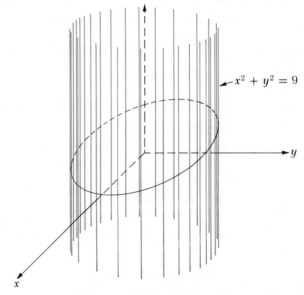

Figure 17–24

In general, *if a surface has an equation in only two of the three variables x, y, and z, then the surface is a cylinder whose rulings are parallel to the axis associated with the missing variable.* The directrix of the cylinder is in the coordinate plane associated with the two variables that are present in the equation. For example, the parabolic cylinder $y = z^2$ has the parabola $y = z^2$ in the yz plane as its directrix. Its rulings are the lines parallel to the x axis which intersect the parabola (Figure 17–25).

The right circular cylinder (17.40) that we just discussed is also an example of a *surface of revolution*. A·surface of revolution is a surface that is generated

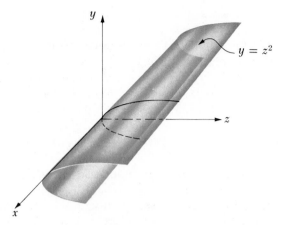

Figure 17-25

by revolving a plane curve about a line in the plane of the curve, called the *axis* of the surface of revolution. The cylinder (17.40) can be generated by revolving the line $x = 3$ in the xz plane about the z axis. Another example of a surface of revolution is a sphere, which is generated by revolving a semicircle about its diameter. It will be noted that the intersection of a surface of revolution and a plane perpendicular to its axis is always a *circle*.

Suppose a surface of revolution S is generated by revolving a curve in the xy plane with equation

$$y = f(x) \tag{17.41}$$

about the x axis. If $P(x, y, z)$ is an arbitrary point in S and $P_0(x, f(x), 0)$ and $Q(x, 0, 0)$ are points having the same x coordinate as P, then $|QP|$ and $|QP_0|$ are equal radii of a circular cross-section of the surface of revolution (Figure 17–26). Then $|QP|^2 = |QP_0|^2$ or

$$y^2 + z^2 = (f(x))^2. \tag{17.42}$$

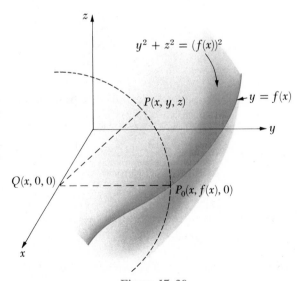

Figure 17-26

Conversely, if the coordinates of a point P satisfy (17.42), then $|QP| = |QP_0|$ and hence P is in S. Thus (17.42) is an equation of the surface of revolution S.

Note that the equation (17.42) for S can be formed by squaring each side of (17.41) and then replacing y^2 by $y^2 + z^2$. If any of the roles of x, y, and z in the foregoing discussion are interchanged, then equations analogous to (17.42) are obtained for the surfaces of revolution generated. For example, if the curve $z = f(y)$ in the yz plane is revolved about the y axis, the resulting surface of revolution has an equation

$$x^2 + z^2 = (f(y))^2. \tag{17.43}$$

Thus, in general we have a surface of revolution S that is formed by revolving a curve C about a *coordinate axis*. If the equation of C is in a form where (i) its right member is a function of the variable associated with this coordinate axis, and (ii) its left member is one of the other two coordinate variables, one obtains an equation for S by squaring each member of the equation for C, and then replacing the square on the left by the sum of the squares of the coordinate variables that are not associated with the axis of S.

Example 1   Find an equation for the surface of revolution obtained by revolving the parabola $z^2 = 4y$ in the yz plane about the y axis (Figure 17–27).

SOLUTION   The surface of revolution is generated by revolving the curve

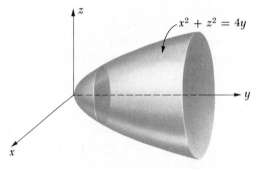

Figure 17–27

$z = f(y) = 2\sqrt{y}$ in the yz plane about the y axis. Hence from our previous discussion the surface has an equation of the form (17.43). Since $f(y) = 2\sqrt{y}$, this equation is

$$x^2 + z^2 = 4y.$$

The surface in this example, which is obtained by revolving a parabola about its axis, is called a *paraboloid of revolution*.

Example 2   Find an equation for the surface of revolution obtained by revolving the parabola in Example 1 about the z axis (Figure 17–28).

SOLUTION   The surface of revolution is generated by revolving the parabola $y = f(z) = z^2/4$ about the z axis. Thus our surface has an equation of the form

$$x^2 + y^2 = (f(z))^2.$$

Since $f(z) = z^2/4$, an equation for the surface is

$$x^2 + y^2 = \frac{z^4}{16}.$$

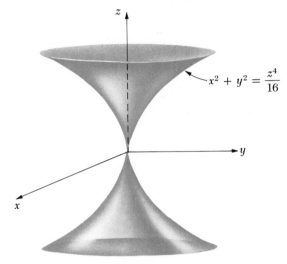

$$x^2 + y^2 = \frac{z^4}{16}$$

Figure 17-28

Example 3   Sketch the graph of the equation $4x^2 + y^2 + z^2 = 16$ (Figure 17-29).

SOLUTION   Since the equation is of the form

$$y^2 + z^2 = 16 - 4x^2 = (f(x))^2$$

the graph of the equation is a surface of revolution. From our above discussion this surface can be obtained by revolving the curve

$$y = f(x) = 2\sqrt{4 - x^2} \tag{17.44}$$

in the xy plane about the x axis (or by revolving the curve $z = f(x) = 2\sqrt{4 - x^2}$ in the xz plane about the x axis). The graph of (17.44) is a semiellipse and hence the surface of revolution is called an *ellipsoid of revolution.*

In sketching this surface a knowledge of its intersection with planes perpendicular to the coordinate axes is helpful. We note that if $x = x_0$ in the given equation, then

$$y^2 + z^2 = 16 - 4x_0^2. \tag{17.45}$$

From (17.45) if $-2 < x_0 < 2$, the intersection of the plane $x = x_0$ with the ellipsoid of revolution is a circle with center on the x axis and radius $\sqrt{16 - 4x_0^2} = 2\sqrt{4 - x_0^2}$. If we let $y = y_0$ in the given equation, we obtain

$$4x^2 + z^2 = 16 - y_0^2. \tag{17.46}$$

From (17.46) if $-4 < y_0 < 4$, the intersection of the plane $y = y_0$ with the surface is an ellipse with center on the y axis. If $-4 < z_0 < 4$, it can also be shown that the intersection of the surface with a plane $z = z_0$ is an ellipse.

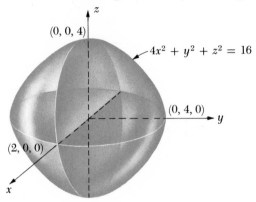

Figure 17-29

Exercise Set 17.6

In Exercises 1–12 discuss the graph of the given equation. Sketch the surface.

1. $4x^2 + y^2 = 16$

2. $y^2 + z^2 = 16$

3. $x = |z|$

4. $x^2 - z^2 = 4$

5. $yz = 12$

6. $y^2 = 4x$

7. $x^2 + y^2 = 4z$

8. $x^2 + z^2 = y^2$

9. $y^2 + z^2 - e^{-2x} = 0.$

10. $y^2 + z^2 - [\sin^{-1}(x - 1)]^2 = 0$

11. $4x^2 + 4z^2 = y - 4$

12. $9x^2 + 9y^2 + 4z^2 = 36$

In Exercises 13–18 obtain the equation for the surface of revolution generated by revolving the given curve about the given axis.

13. $y = \ln x$ in the xy plane, about the x axis

14. $y = \ln x$ in the xy plane about the y axis

15. $y^2 + 4z^2 = 36$ in the yz plane about the y axis

16. $y^2 + 4z^2 = 36$ in the yz plane about the z axis

17. $x^2 - 9z^2 = 9$ in the xz plane about the x axis

18. $x^2 - 9z^2 = 9$ in the xz plane about the z axis

17.7 Quadric Surfaces

It will be recalled from Chapter 12 that the graph of an equation of second degree in x and y is a conic or a degenerate form of a conic. As an extension into three variables, we now consider the general equation of second degree in x, y, and z,

$$a_{11}x^2 + a_{22}y^2 + a_{33}z^2 + a_{12}xy + a_{13}xz + a_{23}yz + b_1x + b_2y$$
$$+ b_3z + c = 0, \quad (17.47)$$

where the constants a_{ij}, b_k, and c can be any numbers except that not all the numbers a_{ij} are zero. The graph of an equation of the form (17.47) is called a *quadric surface*. Examples of quadric surfaces that have already been discussed are (i) cylinders in which the directrix is a conic and (ii) surfaces of revolution obtained by revolving a conic in a coordinate plane about a coordinate axis in that plane.

We will now consider some common examples of quadric surfaces, and in order to familiarize ourselves with these surfaces, we will discuss, in particular, their intercepts and their intersections with planes perpendicular to the coordinate axes. It will be noted that these intersections are either conics or degenerate forms of conics. In the equations for these quadric surfaces it will be assumed that a constant a, b, or c is positive if it appears in a^2, b^2, or c^2, respectively.

The first quadric surface to be considered is the *ellipsoid* (Figure 17–30)

$$\frac{x^2}{a^2} + \frac{y^2}{b^2} + \frac{z^2}{c^2} = 1. \quad (17.48)$$

Intercepts: $x = \pm a$, $y = \pm b$, $z = \pm c$.

Intersection with plane $z = z_0$: The ellipse

$$\frac{x^2}{a^2\left(1 - \dfrac{z_0^2}{c^2}\right)} + \frac{y^2}{b^2\left(1 - \dfrac{z_0^2}{c^2}\right)} = 1 \quad \text{if } |z_0| < c.$$

The largest ellipse is the graph of $x^2/a^2 + y^2/b^2 = 1$, which is in the plane $z = 0$. The intersection is empty if $|z_0| > c$.

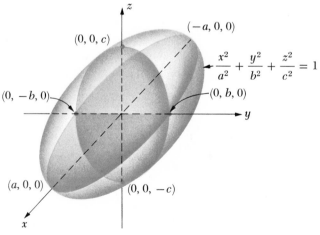

Figure 17–30

The intersection of the ellipsoid with a plane $x = x_0$ where $|x_0| < a$ is an ellipse, as is the intersection with a plane $y = y_0$ where $|y_0| < b$.

If $a = b$, then the surface (17.48) is an ellipsoid of revolution that can be obtained by revolving the ellipse $x^2/a^2 + z^2/c^2 = 1$ in the xz plane about the z axis. If $a = b = c$, the ellipsoid of revolution is a sphere.

The graph of an equation of the form

$$\frac{x^2}{a^2} + \frac{y^2}{b^2} - \frac{z^2}{c^2} = 1 \tag{17.49}$$

is called a *hyperboloid of one sheet* (Figure 17–31).

Intercepts: $x = \pm a$, $y = \pm b$.

Intersection with the plane $z = z_0$: The ellipse

$$\frac{x^2}{a^2\left(1 + \dfrac{z_0^2}{c^2}\right)} + \frac{y^2}{b^2\left(1 + \dfrac{z_0^2}{c^2}\right)} = 1.$$

The smallest ellipse is the graph of $x^2/a^2 + y^2/b^2 = 1$, which is in the plane $z = 0$.

Intersection with the plane $x = x_0$: The hyperbola

$$\frac{y^2}{b^2\left(1 - \dfrac{x_0^2}{a^2}\right)} - \frac{z^2}{c^2\left(1 - \dfrac{x_0^2}{a^2}\right)} = 1 \quad \text{if } |x_0| \neq a.$$

If $|x_0| < a$, the transverse axis is parallel to the y axis; if $|x_0| > a$, the transverse axis is parallel to the z axis. If $|x_0| = a$, the intersection consists of the lines $y/b = \pm z/c$. An analogous description can be given of the intersection with a plane $y = y_0$.

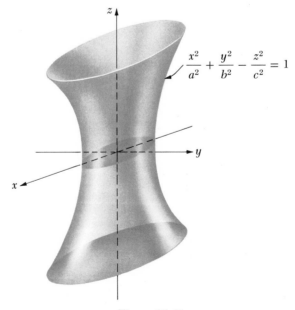

$$\frac{x^2}{a^2} + \frac{y^2}{b^2} - \frac{z^2}{c^2} = 1$$

Figure 17–31

If $a = b$, the graph of (17.49) is a solid of revolution which can be obtained by revolving the hyperbola $x^2/a^2 - z^2/c^2 = 1$ in the xz plane about the z axis. It should also be noted that any equation that is formed from (17.49) by interchanging the variables x, y, and z is also a hyperboloid of one sheet. For example, the graph of the equation $x^2/a^2 + z^2/b^2 - y^2/c^2 = 1$ is a hyperboloid of one sheet whose intersections with the planes $y = y_0$ are ellipses. Also, the graph of $y^2/a^2 + z^2/b^2 - x^2/c^2 = 1$ is a hyperboloid of one sheet whose intersections with the planes $x = x_0$ are ellipses. Analogous remarks can be made about the graphs of equations that are obtained from (17.50), (17.51), (17.52) and (17.53), below, by interchanges of x, y, and z.

The graph of an equation of the form

$$-\frac{x^2}{a^2} - \frac{y^2}{b^2} + \frac{z^2}{c^2} = 1 \tag{17.50}$$

is termed a *hyperboloid of two sheets* (Figure 17–32).

Intercepts: $z = \pm c$.

Intersection with plane $z = z_0$: The ellipse

$$\frac{x^2}{a^2\left(\dfrac{z_0^{\,2}}{c^2} - 1\right)} + \frac{y^2}{b^2\left(\dfrac{z_0^{\,2}}{c^2} - 1\right)} = 1 \quad \text{if } |z_0| > c.$$

The ellipse increases in size as $|z_0|$ increases. The intersection is empty if $|z_0| < c$.

Intersection with plane $x = x_0$: The hyperbola

$$\frac{z^2}{c^2\left(1 + \dfrac{x_0^{\,2}}{a^2}\right)} - \frac{y^2}{b^2\left(1 + \dfrac{x_0^{\,2}}{a^2}\right)} = 1,$$

which has its transverse axis parallel to the z axis.

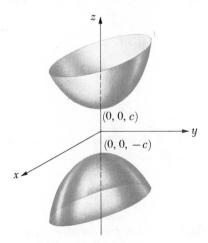

Figure 17-32

If $a = b$, the graph of (17.50) is a solid of revolution which can be generated by revolving the hyperbola $z^2/c^2 - x^2/a^2 = 1$ in the xz plane about the z axis.

The graph of the surface

$$\frac{x^2}{a^2} + \frac{y^2}{b^2} = \frac{z^2}{c^2} \tag{17.51}$$

is called an *elliptic cone* (Figure 17–33). This surface can be regarded as an asymptotic surface to the hyperboloids (17.49) and (17.50) just as in the xy plane the lines $y/b = \pm x/a$ are asymptotes to the hyperbolas $x^2/a^2 - y^2/b^2 = 1$ and $y^2/b^2 - x^2/a^2 = 1$.

Intercepts: $x = 0$, $y = 0$, $z = 0$.

Intersection with plane $z = z_0$: The ellipse

$$\frac{x^2}{\dfrac{a^2 z_0^2}{c^2}} + \frac{y^2}{\dfrac{b^2 z_0^2}{c^2}} = 1 \quad \text{if } z_0 \neq 0.$$

The ellipse increases in size as $|z_0|$ increases.

Intersection with plane $x = x_0$: The hyperbola

$$\frac{z^2}{\dfrac{c^2 x_0^2}{a^2}} - \frac{y^2}{\dfrac{b^2 x_0^2}{a^2}} = 1 \quad \text{if } x_0 \neq 0,$$

which has its transverse axis parallel to the z axis.

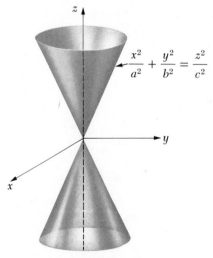

$$\frac{x^2}{a^2} + \frac{y^2}{b^2} = \frac{z^2}{c^2}$$

Figure 17–33

If $a = b$, the graph of (17.51) is a *right circular cone*, which can be obtained by revolving the line $x/a = z/c$ about the z axis.

We next consider the *elliptic paraboloid*.

$$\frac{x^2}{a^2} + \frac{y^2}{b^2} = cz \quad \text{where } c \neq 0. \tag{17.52}$$

A sketch of this surface when $c > 0$ is shown in Figure 17–34.

Intercepts: $x = 0$, $y = 0$, $z = 0$.

Intersection with plane $z = z_0$: The ellipse

$$\frac{x^2}{a^2 c z_0} + \frac{y^2}{b^2 c z_0} = 1 \quad \text{if } c z_0 > 0.$$

The intersection is empty if $c z_0 < 0$.

Intersection with plane $x = x_0$: The parabola

$$y^2 = b^2 \left(cz - \frac{x_0}{a^2} \right),$$

which is concave upward if $c > 0$ and concave downward if $c < 0$.

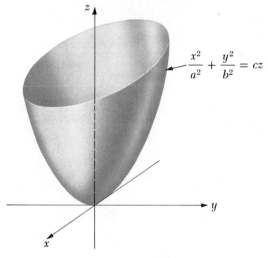

$$-\frac{x^2}{a^2} + \frac{y^2}{b^2} = cz$$

Figure 17-34

If $a = b$, the surface (17.52) is called a *paraboloid of revolution*.

The graph of an equation of the form

$$\frac{y^2}{b^2} - \frac{x^2}{a^2} = cz \quad \text{where } c \neq 0 \tag{17.53}$$

is called a *hyperbolic paraboloid* (Figure 17–35).

Intercepts: $x = 0$, $y = 0$, $z = 0$.

Intersection with plane $z = z_0$: The hyperbola

$$\frac{y^2}{b^2 cz_0} - \frac{x^2}{a^2 cz_0} = 1 \quad \text{if } cz_0 \neq 0.$$

If $cz_0 > 0$, the transverse axis is parallel to the y axis; if $cz_0 < 0$, the transverse axis is parallel to the x axis. The intersection of the surface (17.53) with the plane $z = 0$ consists of the lines $y/b = \pm x/a$.

Intersection with plane $x = x_0$: A parabola that is concave upward if $c > 0$ and concave downward if $c < 0$.

Intersection with plane $y = y_0$: A parabola that is concave downward if $c > 0$ and concave upward if $c < 0$.

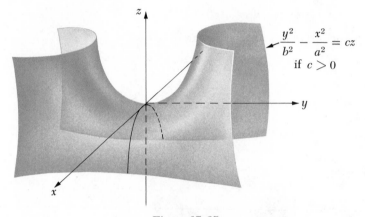

$$\frac{y^2}{b^2} - \frac{x^2}{a^2} = cz$$
$$\text{if } c > 0$$

Figure 17-35

Example 1 Discuss and sketch the graph of the equation $4x^2 - y^2 - 4z^2 = 9$.

SOLUTION   As before, we shall discuss the intercepts of the surface and its intersection with planes perpendicular to the coordinate axes.

Intercepts: $x = \pm\frac{3}{2}$.

Intersection with a plane $x = x_0$: The ellipse

$$\frac{y^2}{4x_0{}^2 - 9} + \frac{z^2}{\dfrac{4x_0{}^2 - 9}{4}} = 1 \quad \text{if } |x_0| > \tfrac{3}{2}.$$

In particular for $x_0 = \pm\frac{5}{2}$ the ellipse $y^2/16 + z^2/4 = 1$. No intersection if $|x_0| < \frac{3}{2}$.

Intersection with a plane $y = y_0$: The hyperbola

$$\frac{x^2}{\dfrac{y_0{}^2 + 9}{4}} - \frac{z^2}{\dfrac{y_0{}^2 + 9}{4}} = 1.$$

In particular the hyperbola $x^2/\frac{9}{4} - z^2/\frac{9}{4} = 1$ if $y_0 = 0$.

Intersection with a plane $z = z_0$: The hyperbola

$$\frac{x^2}{\dfrac{4z_0{}^2 + 9}{4}} - \frac{y^2}{4z_0{}^2 + 9} = 1.$$

In particular the hyperbola $x^2/\frac{9}{4} - y^2/9 = 1$ if $z_0 = 0$.

A sketch of the graph of the given equation, a hyperboloid of two sheets, is shown in Figure 17–36.

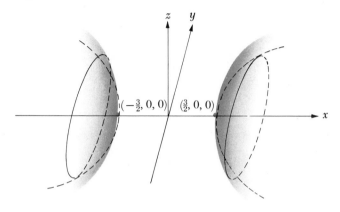

Figure 17–36

Exercise Set 17.7

Discuss and sketch the graph of the given equation.

1. $x^2 + 4y^2 + 9z^2 = 36$ 2. $x^2 + 4y^2 + z^2 = 25$

3. $4y^2 + z^2 = x$ 4. $z^2 - 4x^2 = 4y^2$

5. $x^2 - 4y^2 + z^2 = 16$ 6. $x^2 + 4z^2 = 36y$

7. $4y^2 - z^2 = x^2$ 8. $x^2 - 4y^2 - z^2 = 36$

9. $-9x^2 + 16y^2 - z^2 = 144$ 10. $-4x^2 + y^2 + 9z^2 = 36$

11. $x^2 - 4z^2 = 2y$ 12. $9z^2 - y^2 = 3x$

17.8   Vector-Valued Functions, Space Curves

We recall from Section 15.5 that plane curves can be described analytically by vector-valued functions in which the value of the function at a number in its domain is a two-dimensional vector. Similarly, vector-valued functions that are defined by three-dimensional vectors can be used to give an analytic characterization of curves in three-dimensional space, and will be discussed in this section. It will be noted that the ideas presented here are analogous to those already covered in Chapter 15.

Suppose $\mathbf{P}$ is the vector-valued function given by

$$\mathbf{P}(t) = \langle f_1(t), f_2(t), f_3(t) \rangle. \tag{17.54}$$

The domain of **P**, denoted by $\mathfrak{D}_{\mathbf{P}}$, is the intersection of the domains of the functions f_1, f_2, and f_3. The graph of **P**, or, saying the same thing, the graph of the parametric equations

$$x = f_1(t) \qquad y = f_2(t) \qquad z = f_3(t) \tag{17.55}$$

is the set of all points $(f_1(t), f_2(t), f_3(t))$ in space. If $t_0 \in \mathfrak{D}_{\mathbf{P}}$, then $\mathbf{P}(t_0)$ is called the *position vector* of the point $(f_1(t_0), f_2(t_0), f_3(t_0))$.

The limit definition for **P** in (17.54) is analogous to Definition 15.5.1.

17.8.1   Definition

Suppose **P** is given by (17.54). Then $\lim_{t \to t_0} \mathbf{P}(t) = \langle L_1, L_2, L_3 \rangle$ if and only if $\lim_{t \to t_0} f_1(t) = L_1$, $\lim_{t \to t_0} f_2(t) = L_2$, and $\lim_{t \to t_0} f_3(t) = L_3$.

Continuity for the function **P** here is defined as in Definition 15.5.2. Thus **P** is continuous at t_0 if and only if the functions f_1, f_2, and f_3 are continuous at t_0. If **P** is continuous on an interval I, its graph on I—that is, the set of points $\{(f_1(t), f_2(t), f_3(t)): t \in I\}$—is called the *space curve* defined by (17.54) (or by the parametric equations (17.55)) when $t \in I$.

Example 1   Sketch the graph of the space curve given by $\mathbf{P}(t) = \langle 4 \cos t, 3 \sin t, t/2 \rangle$.

SOLUTION   We note that the curve has parametric equations

$$x = 4 \cos t \qquad y = 3 \sin t \qquad z = \frac{t}{2}. \tag{17.56}$$

Since x and y in (17.56) satisfy the equation

$$\frac{x^2}{16} + \frac{y^2}{9} = 1, \tag{17.57}$$

a point is in the curve if and only if (x, y, z) is in the elliptic cylinder with equation (17.57). As t increases from some value t_0 to $t_0 + 2\pi$ the corresponding point $(4 \cos t, 3 \sin t, t/2)$ in the curve ascends the cylinder while moving in a counterclockwise direction from the point $P_0(4 \cos t_0, 3 \sin t_0, t_0/2)$ to the point $(4 \cos t_0, 3 \sin t_0, t_0/2 + \pi)$, which is π units above P_0. Some representative points in the curve can be plotted with the aid of the accompanying table of values.

t	$-\dfrac{\pi}{4}$	0	$\dfrac{\pi}{4}$	$\dfrac{\pi}{2}$	π	$\dfrac{3\pi}{2}$	2π	$\dfrac{5\pi}{2}$	3π	$\dfrac{7\pi}{2}$	4π
x	$2\sqrt{2}$	4	$2\sqrt{2}$	0	-4	0	4	0	-4	0	4
y	$-\dfrac{3\sqrt{2}}{2}$	0	$\dfrac{3\sqrt{2}}{2}$	3	0	-3	0	3	0	-3	0
z	$-\dfrac{\pi}{8}$	0	$\dfrac{\pi}{8}$	$\dfrac{\pi}{4}$	$\dfrac{\pi}{2}$	$\dfrac{3\pi}{4}$	π	$\dfrac{5\pi}{4}$	$\dfrac{3\pi}{2}$	$\dfrac{7\pi}{4}$	2π

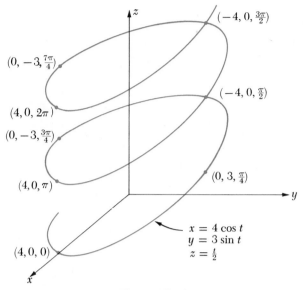

Figure 17–37

The curve, which is called a helix, is sketched in Figure 17–37.

Example 2   Obtain parametric equations for the portion of the inter-section of the cylinders $x^2 + 4z^2 = 4$ and $y = x^2$ where x, y, and z are non-negative. Obtain several points in the intersection curve and plot the curve.

SOLUTION   If we let $x = t$, then from the second given equation $y = t^2$. Hence, from the first given equation $z = (\sqrt{4 - t^2})/2$ since we require z to be non-negative on the curve. The parametric equations of the curve are therefore

$$x = t \qquad y = t^2 \qquad z = \frac{\sqrt{4 - t^2}}{2} \quad \text{if } t \in [0, 2].$$

If values are assigned to t as noted in the following table, the corresponding x, y, and z coordinates of the points given in the table are then obtained.

t	0	$\dfrac{\sqrt{2}}{2}$	1	$\sqrt{2}$	$\sqrt{3}$	2
x	0	$\dfrac{\sqrt{2}}{2}$	1	$\sqrt{2}$	$\sqrt{3}$	2
y	0	$\dfrac{1}{2}$	1	2	3	4
z	1	$\dfrac{\sqrt{14}}{4}$	$\dfrac{\sqrt{3}}{2}$	$\dfrac{\sqrt{2}}{2}$	$\dfrac{1}{2}$	0

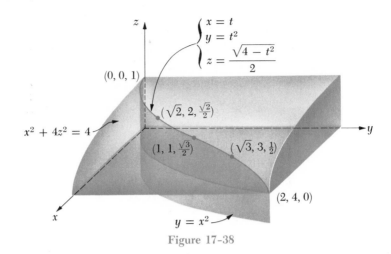

Figure 17–38

The curve is sketched in Figure 17–38.

Differentiability for $\mathbf{P}$ in (17.54) can be defined as in Definition 15.5.3. Then by analogy with Theorem 15.5.4 the following theorem can be stated. Its proof follows by the same argument used for Theorem 15.5.4.

17.8.2   Theorem

Suppose $\mathbf{P}$ is given by (17.54). Then

$$\mathbf{P}'(t) = \langle f_1'(t), f_2'(t), f_3'(t) \rangle$$

if and only if $f_1'(t)$, $f_2'(t)$, and $f_3'(t)$ exist.

Example 3   If $\mathbf{P}(t) = \langle 4 \cos t, 3 \sin t, t/2 \rangle$, find $\mathbf{P}'(t)$ and $\mathbf{P}'(2\pi/3)$.

SOLUTION   By Theorem 17.8.2

$$\mathbf{P}'(t) = \langle -4 \sin t, 3 \cos t, \tfrac{1}{2} \rangle$$

and therefore

$$\mathbf{P}'\left(\frac{2\pi}{3}\right) = \left\langle -4 \sin \frac{2\pi}{3}, 3 \cos \frac{2\pi}{3}, \frac{1}{2} \right\rangle = \left\langle -2\sqrt{3}, -\frac{3}{2}, \frac{1}{2} \right\rangle.$$

The following differentiation formulas can be proved when the derivatives on the right exist:

$$D_t[h(t)\mathbf{P}(t)] = h(t)\mathbf{P}'(t) + h'(t)\mathbf{P}(t) \tag{17.58}$$

$$D_t[\mathbf{P}(h(t))] = h'(t)\mathbf{P}'(h(t)) \tag{17.59}$$

$$D_t[\mathbf{P}(t) + \mathbf{Q}(t)] = \mathbf{P}'(t) + \mathbf{Q}'(t) \tag{17.60}$$

$$D_t[\mathbf{P}(t) \cdot \mathbf{Q}(t)] = \mathbf{P}(t) \cdot \mathbf{Q}'(t) + \mathbf{P}'(t) \cdot \mathbf{Q}(t) \tag{17.61}$$

$$D_t[\mathbf{P}(t) \times \mathbf{Q}(t)] = \mathbf{P}(t) \times \mathbf{Q}'(t) + \mathbf{P}'(t) \times \mathbf{Q}(t) \tag{17.62}$$

Note that formulas (17.58) through (17.61) are the same as formulas (15.58) through

(15.61), which were given for vector-valued functions defined by two-dimensional vectors.

Suppose a space curve C is given by (17.54) for every t in some interval I and $t_0 \in I$; that is, $Q_0(f_1(t_0), f_2(t_0), f_3(t_0))$ is a point in C. Also suppose $\mathbf{P}'(t_0)$ exists and is not the zero vector. If $h \neq 0$ and $t_0 + h \in I$, then the secant through the points Q_0 and $Q(f_1(t_0 + h), f_2(t_0 + h), f_3(t + h))$ (Figure 17–39) has direction

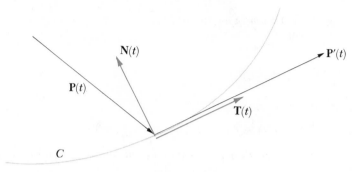

Figure 17–39

numbers that are given by the components of $\mathbf{P}(t_0 + h) - \mathbf{P}(t_0)$ and hence also of $(\mathbf{P}(t_0 + h) - \mathbf{P}(t_0))/h$. Since

$$\lim_{h \to 0} \frac{\mathbf{P}(t_0 + h) - \mathbf{P}(t_0)}{h} = \mathbf{P}'(t_0),$$

from intuitive considerations it is reasonable to call $\mathbf{P}'(t_0)$ a *tangent vector* to C at Q_0, just as in Section 15.5 where $\mathbf{P}'(t_0)$ was a two-dimensional vector. We note that the arrow representing $\mathbf{P}'(t_0)$ points in the direction in which t is increasing when $t = t_0$.

We are thus led to the following definition of a tangent to a space curve.

17.8.3   Definition

Suppose a curve C is defined vectorially by (17.54) for every t in some interval I. Also suppose P is differentiable at some $t_0 \in I$ and $\mathbf{P}'(t_0) \neq 0$. The *tangent* to C at $(f_1(t_0), f_2(t_0), f_3(t_0))$ is the line through this point which is parallel to $\mathbf{P}'(t_0)$.

Example 4   Find the tangent to the curve $\mathbf{P}(t) = \langle 4 \cos t, 3 \sin t, t/2 \rangle$ at the point where $t = \frac{2\pi}{3}$.

SOLUTION   In Example 3, we obtained $\mathbf{P}'(\frac{2\pi}{3}) = \langle -2\sqrt{3}, -\frac{3}{2}, \frac{1}{2} \rangle$. The components of this vector are direction numbers of the required tangent. The point in the curve where $t = \frac{2\pi}{3}$ is $(-2, 3\sqrt{3}/2, \frac{\pi}{3})$. Hence the tangent has the parametric equations

$$x = -2 - 2\sqrt{3}t \qquad y = \frac{3\sqrt{3}}{2} - \frac{3}{2}t \qquad z = \frac{\pi}{3} + \frac{1}{2}t.$$

If t denotes time, and the position vector of the particle at any time t is given by (17.54), then since $dx/dt = f'_1(t)$, $dy/dt = f'_2(t)$, and $dz/dt = f'_3(t)$ are the velocities of the particle in the x, y, and z directions, respectively, the vector $\mathbf{P}'(t)$ is called the *velocity vector* of the particle. Also since $d^2x/dt^2 = f''_1(t)$, $d^2y/dt^2 = f''_2(t)$, and $d^2z/dt^2 = f''_3(t)$ are the accelerations of the particle in the x, y, and z directions, respectively, $\mathbf{P}''(t)$ is called the *acceleration vector* of the particle.

By the same reasoning used in Section 15.4 we are led to the following definition for arc length:

17.8.4 Definition

Suppose the curve C is given by

$$\mathbf{P}(t) = \langle f_1(t), f_2(t), f_3(t) \rangle \quad \text{if } t \in [a, b]$$

where $\mathbf{P}$ has a continuous derivative on $[a, b]$ and for any t_1 and t_2 in $[a, b)$

$$\mathbf{P}(t_1) \neq \mathbf{P}(t_2) \quad \text{when } t_1 \neq t_2. \tag{17.63}$$

The *length* L of C is given by

$$L = \int_a^b |\mathbf{P}'(t)|\, dt = \int_a^b \sqrt{(f'_1(t))^2 + (f'_2(t))^2 + (f'_3(t))^2}\, dt.$$

As with plane curves, the restriction (17.63) requires that no portion of the curve C be traversed more than once as t goes from a to b. Thus from Definition 17.8.4 a unique arc length is obtained for C.

Example 5 Find the length of the curve $\mathbf{P}(t) = \langle t^2 + 1, t^3/3, 2t \rangle$ from $t = 0$ to $t = 3$.

SOLUTION Here

$$f_1(t) = t^2 + 1 \qquad f_2(t) = \frac{t^3}{3} \qquad f_3(t) = 2t,$$

and hence from Definition 17.8.4 the length of the curve is

$$L = \int_0^3 \sqrt{(2t)^2 + (t^2)^2 + 2^2}\, dt = \int_0^3 \sqrt{t^4 + 4t^2 + 4}\, dt$$

$$= \int_0^3 (t^2 + 2)\, dt = \left(\frac{t^3}{3} + 2t \right) \Big]_0^3$$

$$= 15.$$

The notion of a chain of curves can also be extended to space curves. Then suppose a curve C is the chain of curves $C_1, C_2, \ldots, C_n$ which satisfy the requirements given in Definition 17.8.4 and therefore have respective lengths $L_1, L_2, \ldots, L_n$. The length L of C is defined as

$$L = L_1 + L_2 + \cdots + L_n.$$

Exercise Set 17.8

In Exercises 1–4, (a) find $\mathbf{P}'(t)$ and $\mathbf{P}''(t)$ and their values at the given number t; (b) find equations of the tangent to the curve at the point associated with the given number t.

1. $\mathbf{P}(t) = \langle 2t + 1, 3 - 4t, t^2 \rangle$; $t = 2$
2. $\mathbf{P}(t) = \langle t^3 + 2t, -2t^3, 4t^2 \rangle$; $t = 1$
3. $\mathbf{P}(t) = \langle e^{-2t} \sin t, e^{-2t} \cos t, e^{-2t} \rangle$; $t = \frac{\pi}{2}$
4. $\mathbf{P}(t) = \langle t \cos 2t, t \sin 2t, te^{-t} \rangle$; $t = 0$

In Exercises 5–10 obtain at least six points in the curve. Plot the curve.

5. $\mathbf{P}(t) = \langle 2t, \frac{1}{3}t^3, t^2 \rangle$; $t \in [0, 2]$
6. $\mathbf{P}(t) = \langle 2 - 3t, 4t, 2t^{3/2} \rangle$; $t \in [1, \frac{9}{4}]$
7. $\mathbf{P}(t) = \langle 3 \cos 2t, 3 \sin 2t, 2t \rangle$; $t \in [0, \pi]$
8. $\mathbf{P}(t) = \langle t \cos t, t \sin t, t \rangle$; $t \in [0, 4\pi]$
9. $\mathbf{P}(t) = \langle 2\sqrt{1 + t^2}, \ln(1 + t^2), 2 \tan^{-1} t \rangle$; $t \in [0, \sqrt{3}]$
10. $\mathbf{P}(t) = \langle 2t - \sin 2t, \cos 2t, 4 \sin t \rangle$; $t \in [0, 2\pi]$

In Exercises 11–16 find the length of the curve in the given exercise.

11. Exercise 5   12. Exercise 6
13. Exercise 7   14. Exercise 8
15. Exercise 9   16. Exercise 10

In Exercises 17–20, (a) obtain parametric equations for the given curve obtained from the intersection of the surfaces having the given equations as noted, (b) find at least six new points in the curve, and (c) sketch the intersecting surfaces and the curve.

17. $z = 4 - x^2 - y^2$, $y = 2x$ between the points $(0, 0, 4)$ and $(1, 2, -1)$
18. $z = \sqrt{x^2 + y^2}$, $y + z = 3$
19. $x^2 + y^2 = 4$, $x^2 + z^2 = 4$ between the points $(2, 0, 0)$ and $(0, 2, 2)$
20. $x^2 - 2x + y^2 = 0$, $z = \sqrt{9 - x^2 - y^2}$
21. Derive (17.62).

17.9   Unit Tangent and Unit Principal Normal Vectors; Curvature

Suppose C is a space curve having a position vector $\mathbf{P}(t)$ for every t in some interval I where $\mathbf{P}$ is twice differentiable on I. We noted in Section 17.8 that when $\mathbf{P}'(t) \neq 0$, this vector is a tangent vector at the point in C associated with the number t. The *unit tangent vector* $\mathbf{T}(t)$ is then defined as follows:

$$\mathbf{T}(t) = \frac{\mathbf{P}'(t)}{|\mathbf{P}'(t)|} \tag{17.64}$$

(Figure 17–40). From (17.64) it is seen that $\mathbf{T}(t)$ has length one and is in the same direction as $\mathbf{P}'(t)$.

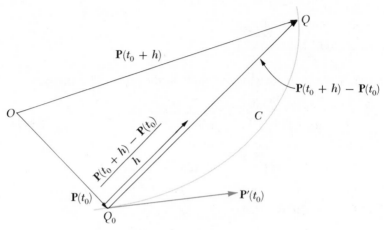

Figure 17–40

If $s(t)$ is defined by

$$s(t) = \int_{t_0}^{t} |\mathbf{P}'(u)| \, du \tag{17.65}$$

then when $t > t_0$, $s(t)$ is the distance traversed along the subarc of C whose endpoints correspond to the numbers t_0 and t. By the first fundamental theorem of calculus (Theorem 7.7.1),

$$s'(t) = |\mathbf{P}'(t)|. \tag{17.66}$$

If t is time, then $s'(t)$ is the speed of a particle at the instant its position vector is $\mathbf{P}(t)$.

Example 1 Find the vector $\mathbf{T}(1)$ for the curve $\mathbf{P}(t) = \langle \frac{2}{3}t^3, -t^2, t \rangle$.

SOLUTION Since $\mathbf{P}'(t) = \langle 2t^2, -2t, 1 \rangle$, from (17.64)

$$\mathbf{T}(t) = \frac{\langle 2t^2, -2t, 1 \rangle}{|\langle 2t^2, -2t, 1 \rangle|} = \frac{\langle 2t^2, -2t, 1 \rangle}{2t^2 + 1}$$

and therefore when $t = 1$,

$$\mathbf{T}(1) = \frac{\langle 2, -2, 1 \rangle}{3} = \left\langle \frac{2}{3}, -\frac{2}{3}, \frac{1}{3} \right\rangle.$$

For the unit tangent vector $\mathbf{T}(t)$ we have

$$\mathbf{T}(t) \cdot \mathbf{T}(t) = |\mathbf{T}(t)|^2 = 1.$$

Hence

$$D_t[\mathbf{T}(t) \cdot \mathbf{T}(t)] = D_t 1 = 0.$$

If (17.61) is applied to the left member of this equation, we obtain $2\mathbf{T}(t) \cdot \mathbf{T}'(t) = 0$ and

$$\mathbf{T}(t) \cdot \mathbf{T}'(t) = 0. \tag{17.67}$$

Thus, in words, $\mathbf{T}'(t)$ is orthogonal to $\mathbf{T}(t)$.

The *unit principal normal vector* $\mathbf{N}(t)$ at an arbitrary point on the curve C is defined by the equation

$$\mathbf{N}(t) = \frac{\mathbf{T}'(t)}{|\mathbf{T}'(t)|} \quad \text{if } \mathbf{T}'(t) \neq 0. \tag{17.68}$$

Since by (17.68) and (17.67)

$$\mathbf{N}(t) \cdot \mathbf{T}(t) = \frac{1}{|\mathbf{T}'(t)|}(\mathbf{T}'(t) \cdot \mathbf{T}(t)) = 0,$$

$\mathbf{N}(t)$ is orthogonal to $\mathbf{T}(t)$ (Figure 17–40).

Example 2   Find the vector $\mathbf{N}(1)$ for the curve $\mathbf{P}(t) = \langle \frac{2}{3}t^3, -t^2, t \rangle$.

SOLUTION   In Example 1 we calculated

$$\mathbf{T}(t) = \frac{1}{2t^2 + 1}\langle 2t^2, -2t, 1 \rangle.$$

Then after differentiation using (17.58)

$$\begin{aligned}
\mathbf{T}'(t) &= \frac{1}{2t^2 + 1}\langle 4t, -2, 0 \rangle - \frac{4t}{(2t^2 + 1)^2}\langle 2t^2, -2t, 1 \rangle \\
&= \frac{(2t^2 + 1)\langle 4t, -2, 0 \rangle - 4t\langle 2t^2, -2t, 1 \rangle}{(2t^2 + 1)^2} \\
&= \frac{\langle 4t, 4t^2 - 2, -4t \rangle}{(2t^2 + 1)^2}.
\end{aligned}$$

In particular when $t = 1$,

$$\mathbf{T}'(1) = \langle \tfrac{4}{9}, \tfrac{2}{9}, -\tfrac{4}{9} \rangle.$$

Hence by (17.68)

$$\begin{aligned}
\mathbf{N}(1) &= \frac{\mathbf{T}'(1)}{|\mathbf{T}'(1)|} = \frac{\langle \tfrac{4}{9}, \tfrac{2}{9}, -\tfrac{4}{9} \rangle}{\tfrac{2}{3}} \\
&= \langle \tfrac{2}{3}, \tfrac{1}{3}, -\tfrac{2}{3} \rangle.
\end{aligned}$$

Since the component of the acceleration vector in the direction of $\mathbf{N}(t)$, the *normal component* of the acceleration vector, is often expressed in terms of the *curvature* of the curve of motion, we pause to define this notion. The *curvature* $K(t)$ of a curve at a point with position vector $\mathbf{P}(t)$ is given by

$$K(t) = \frac{|\mathbf{T}'(t)|}{s'(t)} \qquad\qquad (17.69)$$

where $\mathbf{T}(t)$ and $s'(t)$ are given by (17.64) and (17.66), respectively. The right side of (17.69) expresses the change in $\mathbf{T}(t)$ for a given change in $s(t)$ since

$$\frac{|\mathbf{T}'(t)|}{s'(t)} \approx \frac{\left|\dfrac{\mathbf{T}(t + h) - \mathbf{T}(t)}{h}\right|}{\left|\dfrac{s(t + h) - s(t)}{h}\right|} = \frac{|\mathbf{T}(t + h) - \mathbf{T}(t)|}{|s(t + h) - s(t)|}$$

when $h \neq 0$ but close to 0. The greater the length $|\mathbf{T}(t + h) - \mathbf{T}(t)|$ for a given $|s(t + h) - s(t)|$, the more bending there is to the curve at the point associated with t, and by (17.69) the greater is the number $K(t)$.

We will work an example illustrating the definition (17.69) and then prove an important property about the curvature of a circle.

Example 3 Find the curvature of the curve $\mathbf{P}(t) = \langle \frac{2}{3}t^3, -t^2, t \rangle$ where $t = 1$.

SOLUTION   We recall from the solution of Example 1 that

$$s'(t) = |\mathbf{P}'(t)| = 2t^2 + 1$$

and therefore

$$s'(1) = 3.$$

Also, from the solution of Example 2

$$\mathbf{T}'(1) = \langle \tfrac{4}{9}, \tfrac{2}{9}, -\tfrac{4}{9} \rangle.$$

Hence by (17.69)

$$K(1) = \frac{|\mathbf{T}'(1)|}{s'(1)} = \frac{\tfrac{2}{3}}{3}$$

$$= \tfrac{2}{9}.$$

Example 4 Prove that the curvature of a circle of radius a at any point is $1/a$.

PROOF   The coordinate system in space can be chosen so that its origin is the center of the circle and the circle is in the xy plane. Hence a position vector at any point on the circle is of the form

$$\mathbf{P}(t) = \langle a \cos t, a \sin t, 0 \rangle.$$

Then

$$\mathbf{P}'(t) = \langle -a \sin t, a \cos t, 0 \rangle,$$

and by (17.66)

$$s'(t) = |\mathbf{P}'(t)| = a. \qquad\qquad (17.70)$$

Also, by (17.64)

$$\mathbf{T}(t) = \frac{\mathbf{P}'(t)}{|\mathbf{P}'(t)|} = \frac{\langle -a \sin t, \, a \cos t, \, 0 \rangle}{a}$$

$$= \langle -\sin t, \, \cos t, \, 0 \rangle.$$

Then

$$\mathbf{T}'(t) = \langle -\cos t, \, -\sin t, \, 0 \rangle$$

and from (17.69) and (17.70)

$$K(t) = \frac{|\mathbf{T}'(t)|}{s'(t)} = \frac{1}{a}.$$

We leave as an exercise the proof that the curvature at any point on a line in space is 0.

To obtain the tangential and normal components of the velocity and acceleration vectors of a particle in curvilinear motion, suppose that the location of the particle at any time t is expressed by the position vector $\mathbf{P}(t)$. From (17.64) and (17.66) the velocity vector for the particle can be given in the form

$$\mathbf{V}(t) = \mathbf{P}'(t) = s'(t)\mathbf{T}(t). \tag{17.71}$$

Thus the tangential and normal components of the velocity vector are, respectively,

$$v_T(t) = s'(t) \qquad \text{and} \qquad v_N(t) = 0.$$

Differentiating in (17.71), we next obtain

$$\mathbf{A}(t) = \mathbf{P}''(t) = s''(t)\mathbf{T}(t) + s'(t)\mathbf{T}'(t)$$

and after substituting from (17.69) and (17.68)

$$\mathbf{A}(t) = s''(t)\mathbf{T}(t) + (s'(t))^2 K(t)\mathbf{N}(t). \tag{17.72}$$

From (17.72) the tangential and normal components of the acceleration vector (Figure 17–41) are

$$a_T(t) = s''(t) \qquad \text{and} \qquad a_N(t) = (s'(t))^2 K(t). \tag{17.73}$$

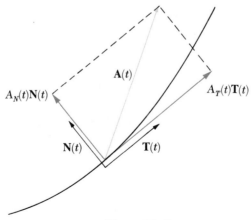

Figure 17–41

Exercise Set 17.9

In Exercises 1–8 find the following for the given curve at the point denoted by the given value of t.

(a) $\mathbf{T}(t)$ (b) $\mathbf{N}(t)$ (c) $K(t)$ (d) $a_T(t)$ (e) $a_N(t)$

1. $\mathbf{P}(t) = \langle 2t, \frac{1}{3}t^3, t^2 \rangle$; $t = 2$

2. $\mathbf{P}(t) = \langle t^3, 0, 3t \rangle$; $t = 2$

3. $\mathbf{P}(t) = \langle 1 - 2t, 3t + 2, (t - 1)^2 \rangle$; $t = 0$

4. $\mathbf{P}(t) = \langle 2 - 3t, 4t, 2t^{3/2} \rangle$; $t = 1$

5. $\mathbf{P}(t) = \langle 3 \cos t, 4 \sin t, 0 \rangle$; $t = \frac{3\pi}{4}$

6. $\mathbf{P}(t) = \langle 3 \cos 2t, 3 \sin 2t, 2t \rangle$; $t = \frac{3\pi}{8}$

7. $\mathbf{P}(t) = \langle e^{-t} \cos t, e^{-t} \sin t, t \rangle$; $t = \pi$

8. $\mathbf{P}(t) = \langle \ln (\csc t + \cot t), \ln \csc t, t \rangle$; $t = \frac{1\pi}{4}$

9. Prove that the curvature of any line in space is 0.

10. If a particle moves along a space curve with a constant speed, what can you say about its acceleration vector? Why?

11. At what points on the hypocycloid $x = a \cos^3 t$, $y = a \sin^3 t$, $z = 0$ (where $a > 0$) is the curvature a minimum?

12. If the function f is twice differentiable, prove that the curvature at any point in the graph of $y = f(x)$ in the xy plane is

$$\frac{|f''(x)|}{[1 + (f'(x))^2]^{3/2}}.$$

13. Find the curvature of the graph of $y = e^{2x}$ at the point $(0, 1)$.

14. Find the curvature of the graph of $y^2 = 3x$ at the point $(3, 3)$.

15. Find the curvature of the graph of $y = a \cosh (x/a)$ at any point (x, y).

16. If the curvature $K(t)$ exists for a plane curve, prove that

$$K(t) = \frac{|\theta'(t)|}{s'(t)}$$

where $\theta(t)$ is the angle between the unit vectors $\mathbf{i}$ and $\mathbf{T}(t)$.

17. A car weighing 3000 lb drives at a constant speed of 60 miles per hour along a parabolic track having the equation $y = x^2$ (where x and y are in miles). Find the force exerted on the car in a direction normal to the track when $x = \frac{1}{2}$.

18. Prove that the curvature $K(t)$ for the curve defined by $\mathbf{P}(t)$ is given by

$$K(t) = \frac{|\mathbf{P}'(t) \times \mathbf{P}''(t)|}{|\mathbf{P}'(t)|^3}.$$

HINT: Use (17.71) with $\mathbf{P}'(t)$ in place of $\mathbf{V}(t)$ and (17.72) with $\mathbf{P}''(t)$ in place of $\mathbf{A}(t)$.

17.10 Cylindrical and Spherical Coordinates

Instead of locating a point or a set of points in space using the Cartesian coordinates (x, y, z), it is sometimes more convenient to use *cylindrical coordinates* (r, θ, z). The r and θ here are the polar coordinate variables and the z is the same as in Cartesian coordinates. The transformation equations from cylindrical coordinates to Cartesian coordinates are therefore

$$x = r \cos \theta \qquad y = r \sin \theta \qquad z = z. \tag{17.74}$$

The cylindrical coordinates of a point P in the first octant are illustrated in Figure 17–42.

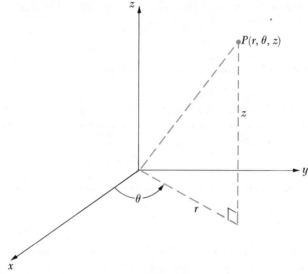

Figure 17–42

Example 1 Find the Cartesian coordinates of the point having cylindrical coordinates $(3, \frac{2\pi}{3}, -4)$.

SOLUTION From (17.74)

$$x = 3 \cos \frac{2\pi}{3} = -\frac{3}{2} \qquad y = 3 \sin \frac{2\pi}{3} = \frac{3\sqrt{3}}{2} \qquad z = -4,$$

and hence the point has Cartesian coordinates $\left(-\dfrac{3}{2}, \dfrac{3\sqrt{3}}{2}, -4\right)$.

Since a point in the xy plane has an infinite number of different representations in polar coordinates, a point in space has an infinite number of different representations in cylindrical coordinates. However, if the restriction $0 \leq \theta < 2\pi$ is imposed, all points except those on the z axis have a unique cylindrical coordinate representation. To obtain cylindrical coordinates for a point with given Cartesian coordinates, we use the following equations, which are derived from (17.74),

$$r^2 = x^2 + y^2 \qquad \tan \theta = \frac{y}{x} \qquad z = z. \tag{17.75}$$

Example 2 Find cylindrical coordinates for the point having Cartesian coordinates $(-1, 2, -3)$.

SOLUTION If the coordinate r is to be positive, then using (17.75)
$$r = \sqrt{x^2 + y^2} = \sqrt{(-1)^2 + 2^2} = \sqrt{5}.$$

Also, using (17.75),
$$\tan \theta = \frac{2}{-1} = -2.$$

Here θ is a second quadrant angle since the point has coordinates $x < 0$ and $y > 0$. Since $\tan^{-1} 2$ is between 0 and $\frac{\pi}{2}$, we can select
$$\theta = \pi - \tan^{-1} 2.$$

Since $z = -3$, the cylindrical coordinates of the point are $(\sqrt{5}, \pi - \tan^{-1} 2, -3)$.

We can give simple equations using the cylindrical coordinate variables for some commonly encountered surfaces that are symmetric to the z axis. The graph of the *cylindrical equation*
$$r = a$$

is the right circular cylinder having the z axis as its axis and radius $|a|$. The graph of
$$\theta = k$$

is the plane that is perpendicular to the xy plane and intersects the xy plane in the line with polar equation $\theta = k$. The sphere $x^2 + y^2 + z^2 = a^2$ has the cylindrical equation
$$r^2 + z^2 = a^2.$$

A point P in space is sometimes located using its *spherical coordinates* (ρ, ϕ, θ). As is illustrated in Figure 17–43,

$\rho = |OP|$ where $\rho \geq 0$.
ϕ = angle between the vectors $\mathbf{k} = \langle 0, 0, 1 \rangle$ and $\overrightarrow{OP}$ where $0 \leq \phi \leq \pi$.
θ is any θ in cylindrical coordinates which is associated with the point P when $r \geq 0$.

The x and y coordinates of P satisfy the relations
$$x = |OQ| \cos \theta \qquad y = |OQ| \sin \theta \tag{17.76}$$

where Q is the projection of P into the xy plane. Since $|OQ| = \rho \sin \phi$, using (17.76), we obtain for the transformation equations from spherical coordinates to Cartesian coordinates
$$x = \rho \sin \phi \cos \theta \qquad y = \rho \sin \phi \sin \theta \qquad z = \rho \cos \phi. \tag{17.77}$$

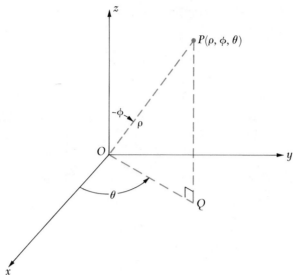

Figure 17-43

Example 3   Find the Cartesian coordinates of the point having spherical coordinates $(5, \frac{3\pi}{4}, \frac{5\pi}{3})$.

SOLUTION   From (17.77)

$$x = 5 \sin \frac{3\pi}{4} \cos \frac{5\pi}{3} = 5\left(\frac{\sqrt{2}}{2}\right)\left(\frac{1}{2}\right) = \frac{5\sqrt{2}}{4}$$

$$y = 5 \sin \frac{3\pi}{4} \sin \frac{5\pi}{3} = 5\left(\frac{\sqrt{2}}{2}\right)\left(-\frac{\sqrt{3}}{2}\right) = -\frac{5\sqrt{6}}{4}$$

$$z = 5 \cos \frac{3\pi}{4} = 5\left(-\frac{\sqrt{2}}{2}\right) = -\frac{5\sqrt{2}}{2},$$

and thus the point has Cartesian coordinates $(5\sqrt{2}/4, \, -5\sqrt{6}/4, \, -5\sqrt{2}/2)$.

To obtain spherical coordinates for a point with given Cartesian coordinates, we will utilize the following equations, which can be derived from (17.77):

$$\rho = \sqrt{x^2 + y^2 + z^2} \qquad \phi = \cos^{-1}\frac{z}{\rho} \qquad \tan \theta = \frac{y}{x}. \qquad (17.78)$$

Example 4   Find the spherical coordinates of the point having Cartesian coordinates $(2, -3, 1)$.

SOLUTION   From (17.78)

$$\rho = \sqrt{2^2 + (-3)^2 + 1^2} = \sqrt{14}$$

$$\phi = \cos^{-1}\frac{1}{\sqrt{14}}$$

$$\tan \theta = -\tfrac{3}{2}.$$

Since the point $(2, -3, 1)$ projects into the point $(2, -3, 0)$ which is in the fourth quadrant of the xy plane,

$$\theta = 2\pi - \tan^{-1} \tfrac{3}{2}.$$

Thus the spherical coordinates of the point are $(\sqrt{14}, \cos^{-1} 1/\sqrt{14}, 2\pi - \tan^{-1} \tfrac{3}{2})$.

Some surfaces that are symmetric to the z axis or origin have simple equations in spherical coordinates. For example, the graph of

$$\rho = a \quad \text{where } a > 0$$

is the sphere with center at the origin and radius a. The graph of

$$\phi = \alpha$$

is a nappe of a cone with vertex at the origin and vertex angle 2α if $\alpha \in (0, \tfrac{\pi}{2})$, or $2(\pi - \alpha)$ if $\alpha \in (\tfrac{\pi}{2}, \pi)$. The graph of $\theta = k$ in spherical coordinates is the half-plane given in cylindrical coordinates by $\theta = k$, $r \geq 0$.

Example 5 Find a *spherical equation* for the sphere $x^2 + y^2 + z^2 - 4y = 0$.

SOLUTION From (17.77) the sphere has the equation

$$\rho^2 - 4\rho \sin \phi \sin \theta = 0. \tag{17.79}$$

For any point in the sphere other than the origin, $\rho \neq 0$. Then the coordinates of the point satisfy $\rho - 4 \sin \phi \sin \theta = 0$ and hence

$$\rho = 4 \sin \phi \sin \theta. \tag{17.80}$$

The origin also satisfies (17.80) since $\rho = 0$, for example, when $\phi = 0$. Conversely, if the coordinates of a point satisfy (17.80), they also satisfy (17.79), and hence the point is on the sphere. Thus (17.80) is a spherical equation of the sphere.

Exercise Set 17.10

1. Give the Cartesian coordinates for the points having cylindrical coordinates as follows:

   (a) $\left(2, \tfrac{\pi}{6}, -4\right)$ (b) $\left(4, \tfrac{4\pi}{3}, 2\right)$ (c) $\left(-3, \cos^{-1}\left(\tfrac{-2}{5}\right), 1\right)$

2. Give cylindrical coordinates for the points having Cartesian coordinates as follows:

   (a) $(3, -3, 5)$ (b) $(-2\sqrt{3}, 2, -3)$ (c) $(-4, 2, -3)$

3. Give the Cartesian coordinates for the points having spherical coordinates as follows:

   (a) $\left(2, \tfrac{\pi}{3}, \tfrac{3\pi}{4}\right)$ (b) $\left(4, \tfrac{3\pi}{4}, \tfrac{7\pi}{6}\right)$ (c) $\left(3, \cos^{-1}\left(-\tfrac{1}{3}\right), \sin^{-1}\left(-\tfrac{2}{3}\right)\right)$

4. Give spherical coordinates for the points having Cartesian coordinates as follows:

(a) $(1, -\sqrt{3}, -2\sqrt{3})$ (b) $(2, -2, 5)$ (c) $(-3, -4, 5)$

5. Give spherical coordinates of the point having cylindrical coordinates as follows:

(a) $\left(2, \frac{\pi}{6}, -4\right)$ (b) $\left(4, \sin^{-1}\left(-\frac{2}{3}\right), 1\right)$ (c) $\left(-3, -\frac{\pi}{3}, -4\right)$

6. Give cylindrical coordinates of the point having spherical coordinates as follows:

(a) $\left(2, \frac{\pi}{3}, \frac{3\pi}{4}\right)$ (b) $\left(5, \frac{5\pi}{6}, \frac{7\pi}{4}\right)$ (c) $\left(3, \pi - \tan^{-1}\frac{4}{3}, \frac{\pi}{3}\right)$

In Exercises 7–14 obtain an equation for the given surface in (a) cylindrical coordinates, (b) spherical coordinates. Sketch the surface.

7. $x + 2y - 4z = 6$ 8. $x^2 + y^2 = 9$

9. $x^2 + y^2 = 3z^2$ 10. $x^2 + y^2 = 4z$

11. $(x - 1)^2 + y^2 + z = 1$ 12. $x + y = 0$

13. $x^2 + 2y^2 + 4z^2 = 16$ 14. $x^2 + y^2 + (z - 3)^2 = 9$

In Exercises 15–19 an equation of a surface is given in cylindrical coordinates. Find an equation for the surface in (a) Cartesian coordinates, (b) spherical coordinates. Sketch the surface.

15. $r = 3$ 16. $\theta = \frac{2}{3}\pi$

17. $r = 2\sin\theta$ 18. $3r = 2z$

19. $z = 2r^2\sin 2\theta$

In Exercises 20–24 an equation of a surface is given in spherical coordinates. Find an equation for the surface in (a) Cartesian coordinates (b) cylindrical coordinates.

20. $\rho = 3$ 21. $\theta = \frac{2}{3}\pi$

22. $\phi = \frac{3}{4}\pi$ 23. $\rho = 4\cos\phi$

24. $\rho = 4\cot\phi\csc\phi\cos\theta$

In Exercises 25–28 sketch the space curve described by the given cylindrical equations.

25. $z = 2, \theta = \frac{\pi}{4}$ 26. $r = 3, z = 2\theta$

27. $z = 2r, \theta = r$ 28. $r = \theta, z = 2$

In Exercises 29–32 sketch the space curve described by the given spherical equations.

29. $\rho = 3, \phi = \frac{3\pi}{4}$ 30. $\theta = \frac{\pi}{3}, \phi = \frac{\pi}{4}$

31. $\rho = 4\sec\phi, \phi = \tan^{-1}2$ 32. $\rho = 4\csc\phi, \theta = \frac{\pi}{2}$

33. A space curve is given by the parametric equations

$$r = f_1(t) \qquad \theta = f_2(t) \qquad z = f_3(t) \quad \text{if } t \in [a, b]$$

using cylindrical coordinates. If the functions f_1, f_2, and f_3 have continuous derivatives on $[a, b]$, prove that the length of the curve is

$$\int_a^b \sqrt{\left(\frac{dr}{dt}\right)^2 + r^2\left(\frac{d\theta}{dt}\right)^2 + \left(\frac{dz}{dt}\right)^2}\, dt.$$

34. A space curve is given by the parametric equations

$$\rho = f_1(t) \qquad \phi = f_2(t) \qquad \theta = f_3(t) \quad \text{if } t \in [a, b]$$

using spherical coordinates. If the functions f_1, f_2, and f_3 have continuous derivatives on $[a, b]$, prove that the length of the curve is

$$\int_a^b \sqrt{\left(\frac{d\rho}{dt}\right)^2 + \rho^2 \sin^2\phi \left(\frac{d\theta}{dt}\right)^2 + \rho^2\left(\frac{d\phi}{dt}\right)^2}\, dt.$$

35. Show that the direction of $\mathbf{A} \times \mathbf{B}$ is determined by the right-hand rule given in Section 17.3. HINT: Express the x, y, and z components of $\mathbf{A}$ and $\mathbf{B}$ in terms of spherical coordinates. Consider the direction of $\mathbf{A} \times \mathbf{B}$ for the separate cases where (i) the rotation from $\mathbf{A}$ to $\mathbf{B}$ is counterclockwise and (ii) the rotation is clockwise.

17.11   Cartesian n-Space, n-Dimensional Vectors

In the next chapter we will discuss the differential calculus of functions of more than one variable, and so it will be appropriate now to introduce the *Cartesian n-space*, R^n. This set, which is a generalization of the Cartesian spaces R, R^2, and R^3, is the set of all *ordered n-tuples* $(x_1, x_2, \ldots, x_n)$ of real numbers. We cannot attempt any pictorial representation of R^n because of the impossibility of representing an n-dimensional space where $n > 3$ on a sheet of paper. However, this inadequacy does not prevent our discussion of extensions to R^n of ideas about the earlier Cartesian spaces. For example, as an extension of the notion of distance, the distance between any two points $X = (x_1, x_2, \ldots, x_n)$ and $Y = (y_1, y_2, \ldots, y_n)$ in R^n is defined as

$$|XY| = \sqrt{\sum_{i=1}^{n} (x_i - y_i)^2}. \tag{17.81}$$

To illustrate (17.81), the distance between the points $X = (2, 3, -1, 4)$ and $Y = (-4, 5, 0, 3)$ in R^4 is

$$|XY| = \sqrt{(2 - (-4))^2 + (3 - 5)^2 + (-1 - 0)^2 + (4 - 3)^2}$$
$$= \sqrt{42}.$$

We will also discuss some types of sets in R^n. Suppose $A = (a_1, a_2, \ldots, a_n)$ and $X = (x_1, x_2, \ldots, x_n)$ are points in R^n and ϵ is a positive number. The set S given by

$$S = \{X \in R^n : |AX| \in \epsilon\}$$

is called the *open ball* with center A and radius ϵ. (However, an open ball in R^2 is customarily called an *open disc*). For example, the open ball in R^3 with center $(1, 2, -3)$ and radius 5 is the set

$$\{(x, y, z) \in R^3 : \sqrt{(x - 1)^2 + (y - 2)^2 + (z + 3)^2} < 5\}.$$

A point A in a set $E \subset R^n$ is called an *interior point* of E if and only if there is some open ball with center A which contains only points of E (Figure 17–44(a)). A set is called an *open set* if and only if all its points are interior points.

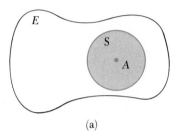
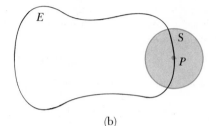

(a)   (b)

A, an interior point of E P, a boundary point of E

Figure 17–44

If a and b are positive numbers, the *open rectangle*

$$\{(x, y) : 0 < x < a \text{ and } 0 < y < b\}$$

is an open set in R^2.

Example 1   Prove that the union of two open sets is an open set.

PROOF   Suppose E_1 and E_2 are open sets and P is any point in $E_1 \cup E_2$. To prove that $E_1 \cup E_2$ is an open set, we must show that P is an interior point of $E_1 \cup E_2$. From the defining property of P, the point is either in E_1 or E_2. We shall suppose that $P \in E_1$ as the proof is analogous when $P \in E_2$. Since $P \in E_1$, which is an open set, P is an interior point of E_1 and there is some open ball with center P which contains only points in E_1. Hence P is also an interior point of $E_1 \cup E_2$ and therefore this set is an open set.

A point P in a set $E \subset R^n$ is called a *boundary point* of E if and only if every open ball with center P contains at least one point in E and at least one point not in E (Figure 17–44(b)). A boundary point of a set may or may not be an element of the set. For example, the set

$$T = \{(x, y) : 0 \le x < a \text{ and } 0 \le y < b\}$$

in R^2 has both $(0, 0)$ and (a, b), for example, as boundary points, and $(0, 0) \in T$ but $(a, b) \notin T$. The *boundary* of a set is the set of all of its boundary points.

A set that contains its boundary is called a *closed set*. The closed ball

$$\{(x, y, z) : (x - 1)^2 + (y - 2)^2 + (z + 3)^2 \le 5\} \tag{17.82}$$

is a closed set since its boundary, the sphere $(x - 1)^2 + (y - 2)^2 + (z + 3)^2 = 5$, is a subset of (17.82).

A set $E \subset R^n$ is called a *bounded set* if and only if E is a subset of some open ball in R^n with center at the origin. The set (17.82), for example, is a bounded set in R^3; however, the first quadrant

$$\{(x, y) : x > 0 \text{ and } y > 0\}$$

is *unbounded* in R^2.

Just as we considered the set R^n as a generalization of the Cartesian spaces R^2 and R^3, we will treat n-dimensional vectors as generalizations of two- and three-dimensional vectors. First of all, an n-dimensional vector is simply an ordered n-*tuple* of real numbers. Then the following definitions are given, and in their statements

$$\mathbf{A} = \langle a_1, a_2, \ldots, a_n \rangle \qquad \mathbf{B} = \langle b_1, b_2, \ldots, b_n \rangle \tag{17.83}$$

denote arbitrary n-dimensional vectors.

(i) *Length:* $|\mathbf{A}| = $ length of $\mathbf{A} = \sqrt{a_1{}^2 + a_2{}^2 + \cdots + a_n{}^2}$

(ii) *Equality:* $\mathbf{A} = \mathbf{B}$ if and only if $a_i = b_i$ for $i = 1, 2, \ldots, n$

(iii) *Addition:* $\mathbf{A} + \mathbf{B} = \langle a_1 + b_1, a_2 + b_2, \ldots, a_n + b_n \rangle$

(iv) *Multiplication by a scalar:* For any number k,

$$k\mathbf{A} = \langle ka_1, ka_2, \ldots, ka_n \rangle$$

(v) *Subtraction:* $\mathbf{A} - \mathbf{B} = \langle a_1 - b_1, a_2 - b_2, \ldots, a_n - b_n \rangle$

(vi) *Negative of a vector:* $-\mathbf{A} = \langle -a_1, -a_2, \ldots, -a_n \rangle$

(vii) *Zero vector:* $\mathbf{0} = \underbrace{\langle 0, 0, \ldots, 0 \rangle}_{n \text{ components}}$

(viii) *Dot product:* $\mathbf{A} \cdot \mathbf{B} = \displaystyle\sum_{i=1}^{n} a_i b_i$

However, the cross product $\mathbf{A} \times \mathbf{B}$ is defined only when $n = 3$. The student can readily prove that the properties (15.3) through (15.15) and (15.19) through (15.23) for two-dimensional vectors are also valid for n-dimensional vectors. Also, we will prove that the vectors $\mathbf{A}$ and $\mathbf{B}$ expressed in (17.83) satisfy the *Cauchy-Schwarz inequality*

$$|\mathbf{A} \cdot \mathbf{B}| \leq |\mathbf{A}| \, |\mathbf{B}|. \tag{17.84}$$

PROOF We note that for any numbers p and q

$$(p\mathbf{A} - q\mathbf{B}) \cdot (p\mathbf{A} - q\mathbf{B}) = |p\mathbf{A} - q\mathbf{B}|^2 \geq 0.$$

An expansion of the left member then gives

$$p^2 \mathbf{A} \cdot \mathbf{A} - 2pq\mathbf{A} \cdot \mathbf{B} + q^2 \mathbf{B} \cdot \mathbf{B} \geq 0.$$

Hence

$$2pq\mathbf{A} \cdot \mathbf{B} \leq p^2 \mathbf{A} \cdot \mathbf{A} + q^2 \mathbf{B} \cdot \mathbf{B}. \tag{17.85}$$

Now (17.85) is true, in particular if $p = |\mathbf{B}|$ and $q = |\mathbf{A}|$. Then from (17.85)

$$2|\mathbf{B}|\,|\mathbf{A}|\mathbf{A}\cdot\mathbf{B}\le 2|\mathbf{A}|^2|\mathbf{B}|^2,$$

and hence

$$\mathbf{A}\cdot\mathbf{B}\le|\mathbf{A}|\,|\mathbf{B}|, \tag{17.86}$$

even if $\mathbf{A}$ or $\mathbf{B}$ is the zero vector. Since the argument used to obtain (17.86) is also valid if $\mathbf{A}$ is replaced by $-\mathbf{A}$,

$$-(\mathbf{A}\cdot\mathbf{B})=-\mathbf{A}\cdot\mathbf{B}\le|-\mathbf{A}|\,|\mathbf{B}|=|\mathbf{A}|\,|\mathbf{B}|. \tag{17.87}$$

Hence from (17.86) and (17.87), the inequality (17.84) is proved.

We can also discuss the calculus of a vector-valued function $\mathbf{P}$ of one variable of the form

$$\mathbf{P}(t)=\langle f_1(t),\,f_2(t),\dots,f_n(t)\rangle. \tag{17.88}$$

The domain of $\mathbf{P}$ here is the intersection of the domains of the functions $f_1, f_2,\dots,f_n$. If $\mathbf{L}=\langle L_1, L_2,\dots,L_n\rangle$, and $\mathbf{P}$ is given by (17.88), then by definition

$$\lim_{t\to t_0}\mathbf{P}(t)=L \quad\text{if and only if}\quad \lim_{t\to t_0}f_i(t)=L_i \quad\text{for } i=1,2,\dots,n.$$

The same definitions hold for the continuity and differentiability of vector-valued functions of the type (17.88) as were given in Section 15.5.

If $\mathbf{P}$, here, is continuous on an interval I, the set $\{(f_1(t), f_2(t),\dots,f_n(t)): t\in I\}$ is called a *curve in R^n*. This curve can also be defined by the parametric equations

$$x_1=f_1(t),\quad x_2=f_2(t),\quad\dots,\quad x_n=f_n(t)\quad\text{if } t\in I. \tag{17.89}$$

In particular, the *line* in R^n through the point $(p_1, p_2,\dots,p_n)$ parallel to the nonzero vector $\mathbf{A}=\langle a_1, a_2,\dots,a_n\rangle$ is defined to be the curve with parametric equations

$$x_1=p_1+a_1 t,\quad x_2=p_2+a_2 t,\quad\dots,\quad x_n=p_n+a_n t. \tag{17.90}$$

If t in (17.90) is restricted to the interval $[a, b]$, then the parametric equations in (17.90) define the line segment in R^n joining the points $(p_1+a_1 a,\dots,p_n+a_n a)$ and $(p_1+a_1 b,\dots,p_n+a_n b)$.

A set $E\subset R^n$ is then said to be a *convex set* if and only if the line segment joining any two points in E is also in E. Any open ball, for example, is a convex set (Exercise 4 below). However, the set E sketched in Figure 17–45 is not convex since the line segment joining the points A and B in E is not in E.

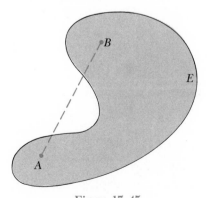

Figure 17–45

Exercise Set 17.11

In Exercises 1–5 prove the given statement.

1. The intersection of two open sets is an open set.

2. The union of two closed sets is a closed set.

3. The intersection of two closed sets is a closed set.

4. An open ball is a convex set.

5. If $E \subset R^n$ is an open set, the *complement of E* (the set of all points in R^n that are not points of E) is a closed set.

6. If $\mathbf{A}$ and $\mathbf{B}$ are n-dimensional vectors, prove that $|\mathbf{A} + \mathbf{B}| \leq |\mathbf{A}| + |\mathbf{B}|$. HINT: Prove that $|\mathbf{A} + \mathbf{B}|^2 \leq (|\mathbf{A}| + |\mathbf{B}|)^2$.

18. Multivariable Differential Calculus

18.1 Functions of n Variables

Heretofore in this text we have been concerned exclusively with real-valued functions of a real variable—that is, functions whose domain and range are subsets of R, the set of all real numbers. In this chapter, however, we will study functions whose domains are subsets of R^n, the n-dimensional Cartesian space, and whose ranges are subsets of R. If f is such a function, which is called a *real-valued function of n real variables*, then f is a set of ordered pairs of the form

$$f = \{((x_1, x_2, \ldots, x_n), u)\}. \tag{18.1}$$

Here, u is the element in the range of f corresponding to the ordered n-tuple $(x_1, x_2, \ldots, x_n)$ in the domain of f. Therefore, using functional notation, we can write

$$u = f(x_1, x_2, \ldots, x_n).$$

The graph of the function f in (18.1) is the set of all points $(x_1, x_2, \ldots, x_n, f(x_1, x_2, \ldots, x_n)) \in R^{n+1}$ such that $(x_1, x_2, \ldots, x_n) \in \mathcal{D}_f$. In this chapter as in Chapter 17 we shall find it convenient to use the symbols X and A where

$$X = (x_1, x_2, \ldots, x_n) \qquad A = (a_1, a_2, \ldots, a_n)$$

to denote arbitrary n-tuples in R^n.

Functions of more than one variable have many applications. For example, the volume V of a solid right circular cylinder of radius r and altitude h can be expressed by

$$V = f(r, h) = \pi r^2 h.$$

Also, the distance d in R^n from the origin to any point $X = (x_1, x_2, \ldots, x_n)$ is given by

$$d = f(X) = \sqrt{x_1{}^2 + x_2{}^2 + \cdots + x_n{}^2}.$$

Like functions of one variable, a function of n variables can be defined by specifying the element $f(X)$ in the range of f which corresponds to an arbitrary element X in the domain of f. For example, the function $f = \{((x, y), \sqrt{9 - x^2 - y^2})\}$ can be given by

$$f(x, y) = \sqrt{9 - x^2 - y^2}. \tag{18.2}$$

The domain of this function f is the *closed circular disc* given by

$$\mathcal{D}_f = \{(x, y): x^2 + y^2 \leq 9\}$$

(Figure 18–1) and the range is the set

$$\mathcal{R}_f = \{z: 0 \leq z \leq 3\}.$$

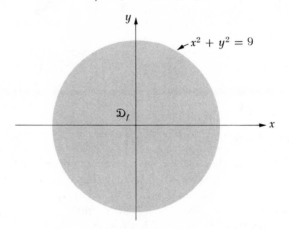

Figure 18-1

The graph of the function given by (18.2) is the set of all points $(x, y, \sqrt{9 - x^2 - y^2}) \in R^3$ such that $(x, y) \in \mathcal{D}_f$, and is the upper hemisphere $x^2 + y^2 + z^2 = 9$ (Figure 18–2).

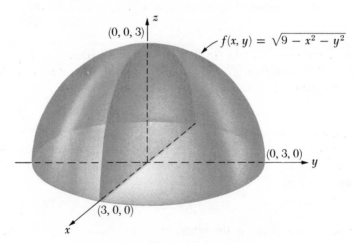

Figure 18-2

To illustrate the use of functional notation for functions of two variables, we shall obtain from (18.2)

$$f(2, -1) = \sqrt{9 - 2^2 - (-1)^2} = 2 \tag{18.3}$$

$$f(0, -2) = \sqrt{9 - 0^2 - (-2)^2} = \sqrt{5} \tag{18.4}$$

$$f\left(x + h, \frac{y}{2}\right) = \sqrt{9 - (x + h)^2 - \frac{y^2}{4}}.$$

From (18.3) and (18.4) the points $(2, -1, 2)$ and $(0, -2, \sqrt{5})$ are in the graph of f. Another example of a function of two variables is g where

$$g(x, y) = \frac{2xy}{x^2 + y^2} \quad \text{if } x \geq 0 \text{ and } y \geq 0. \tag{18.5}$$

This function has domain

$$\mathfrak{D}_g = \{(x, y): x \geq 0 \text{ and } y \geq 0, \quad \text{but } (x, y) \neq (0, 0)\}.$$

This domain, which is illustrated in Figure 18–3, includes the first quadrant in R^2 and the positive x and y axes, but excludes the point $(0, 0)$.

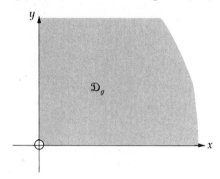

Figure 18-3

Since $(x - y)^2 = x^2 - 2xy + y^2 \geq 0$ for every x and y, $x^2 + y^2 \geq 2xy$ with equality holding when $x = y$. Thus if $x \geq 0$ and $y \geq 0$, but $(x, y) \neq (0, 0)$, then

$$0 \leq \frac{2xy}{x^2 + y^2} \leq 1.$$

Hence the range of g is

$$\mathfrak{R}_g = \{z: 0 \leq z \leq 1\}.$$

The graph of g is the set of all points $\left(x, y, \dfrac{2xy}{x^2 + y^2}\right) \in R^3$ such that $(x, y) \in \mathfrak{D}_g$ (Figure 18–4).

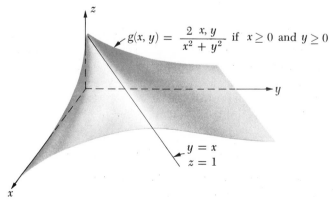

Figure 18-4

It will be noted in Figure 18–4 that the graph of g is a "ridge" in the first octant, the top of which lies in the line $y = x$, $z = 1$. The sloping sides of the ridge become steeper as the z axis is approached, although no part of the z axis is in the graph of g. Some of the curves of intersection of the surface with planes of the form $x = k$ are sketched in Figure 18–4.

A surface obtained as the graph of a function f of two variables may also be described by the *level curves* of the function, which are the graphs of the family of equations

$$f(x, y) = c, \quad c \text{ any constant.}$$

A level curve $f(x, y) = c$ is the projection into the xy plane of the intersection of the surface $z = f(x, y)$ and the plane $z = c$ (Figure 18–5).

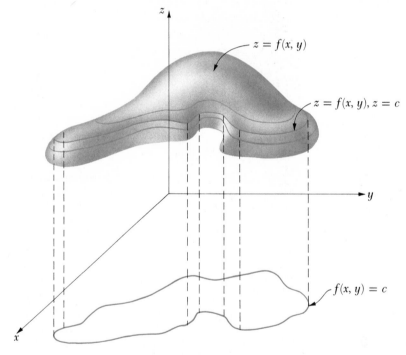

Figure 18-5

Since a level curve results from the projection into the xy plane of points on the surface with the same elevation (same z coordinate), a level curve is actually a "contour line" for the surface just like a contour line on a United States Geological Survey topographic map. Thus a sketch of the level curves of a surface can be considered a topographic map of the surface.

The level curves of the function f given by (18.2) have equations of the form $\sqrt{9 - x^2 - y^2} = c$ or

$$x^2 + y^2 = 9 - c^2 \quad \text{where } 0 \le c \le 3. \tag{18.6}$$

Thus, a typical level curve here is a circle with center at the origin and radius $\sqrt{9 - c^2}$. Those circles corresponding to $c = 0, 1, 2,$ and 3 in (18.6) have been graphed in Figure 18–6.

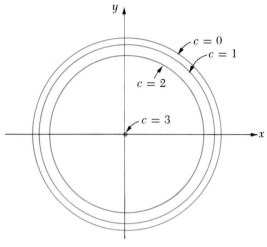

Figure 18-6

The level curves of the function g given by (18.5) have equations of the form

$$\frac{2xy}{x^2 + y^2} = c \quad \text{where } 0 \le c \le 1 \text{ and } (x, y) \in \mathfrak{D}_g. \qquad (18.7)$$

If $c = 0$, then from (18.7), the corresponding level curve is given by

$$xy = 0 \quad \text{if } (x, y) \in \mathfrak{D}_g$$

and is therefore the union of the positive x axis and the positive y axis. If $c \ne 0$, $2xy/(x^2 + y^2) = c$ can be written as $y^2 - (2xy/c) + x^2 = 0$. Then solving for y, we have

$$y = \frac{1 \pm \sqrt{1 - c^2}}{c} x \quad \text{if } (x, y) \in \mathfrak{D}_g. \qquad (18.8)$$

As is seen from (18.8), any level curve associated with $c \in [0, 1)$ is the union of two half-lines drawn from, but not including, the origin. The level curve corresponding to $c = 1$ is the half-line $y = x$ where $x > 0$.

The level curves associated with $c = 0, \frac{1}{4}, \frac{1}{2}, \frac{3}{4}$, and 1 have been plotted in Figure 18–7.

We will also consider functions of three variables. Suppose F is given by

$$F(x, y, z) = \sin^{-1}(x - 2y + 3z). \qquad (18.9)$$

Since $\sin^{-1}$, a function of one variable, has domain $[-1, 1]$ and range $[-\frac{\pi}{2}, \frac{\pi}{2}]$, the domain and range of F, here, are respectively:

$$\mathfrak{D}_F = \{(x, y, z): -1 \le x - 2y + 3z \le 1\}$$
$$\mathfrak{R}_F = [-\frac{\pi}{2}, \frac{\pi}{2}].$$

The set $\mathfrak{D}_F$ is the set of all points in R^3 that are in either of the parallel planes $x - 2y + 3z = 1$ or $x - 2y + 3x = -1$ or lie between the planes.

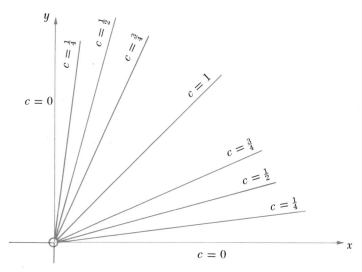

Figure 18-7

To illustrate the use of functional notation with functions of three variables, we obtain from (18.9)

$$F(\tfrac{1}{2}, \tfrac{1}{2}, \tfrac{1}{2}) = \sin^{-1}(\tfrac{1}{2} - 2 \cdot \tfrac{1}{2} + 3 \cdot \tfrac{1}{2}) = \sin^{-1} 1$$
$$= \tfrac{\pi}{2}$$

and

$$F(2, 0, -\tfrac{2}{3}) = \sin^{-1}(2 - 2 \cdot 0 + 3(-\tfrac{2}{3})) = \sin^{-1} 0.$$
$$= 0.$$

If $g_1, g_2, \ldots, g_m$ are functions of n variables and f is a function of m variables, the function F where

$$F(X) = f(g_1(X), g_2(X), \ldots, g_m(X))$$

is a *composite function* of n variables. For example, if

$$f(x, y, z) = x^2 y \cos z$$

and

$$g_1(t) = t^2 \qquad g_2(t) = e^{-t} \qquad g_3(t) = \pi t,$$

then F where $F(t) = f(g_1(t), g_2(t), g_3(t))$ is given by

$$F(t) = f(t^2, e^{-t}, \pi t) = t^4 e^{-t} \cos \pi t$$

and is a composite function of one variable. Also if

$$f(x, y) = x^2 + 4y^2$$

and

$$g_1(s, t) = s^2 \ln t \qquad g_2(s, t) = e^s t,$$

then F where $F(s, t) = f(g_1(s, t), g_2(s, t))$ is given by

$$F(s, t) = f(s^2 \ln t, e^s t) = s^4 \ln^2 t + 4e^{2s} t^2$$

and is a composite function of two variables.

Definitions can also be given for the sum, difference, product, and quotient of two functions of n variables and for the sum and product of more than two such functions. These definitions are analogous to those given in Section 3.2 for functions of one variable. If f is a function of the form

$$f(x_1, x_2, \ldots, x_n) = kx_1^{m_1}x_2^{m_2} \cdots x_n^{m_n} \qquad (18.10)$$

where k is any number and $m_1, m_2, \ldots, m_n$ are non-negative integers, f is called a *monomial function* (of n variables). The sum of a finite number of monomial functions is called a *polynomial function*. For example, P is a polynomial function if

$$P(x, y) = 3x^5y^4 - xy^2 + 2x - 5.$$

As before, the quotient of two polynomial functions is termed a *rational function*. The function f where

$$f(x, y, z) = \frac{x^2 + 3xy^3z - z^4}{xy + yz}$$

is an example of a rational function.

In the ensuing sections of this chapter we shall discuss the differential calculus of functions of more than one variable. It will be noted that the definitions of limit, continuity, and derivative which were previously given for functions of one variable are in reality special cases of corresponding definitions for functions of n variables.

Exercise Set 18.1

1. If $f(x, y) = \dfrac{3x}{2}$, find

   (a) $f(-1, 2)$

   (b) $f(3, -4)$

   (c) $\dfrac{f(x + h, y) - f(x, y)}{h}$

   (d) $\dfrac{f(-1, 2 + h) - f(-1, 2)}{h}$

2. If $f(x, y) = e^x \sin y$, find

   (a) $f(0, \frac{\pi}{3})$

   (b) $f(\ln \frac{1}{2}, \frac{5\pi}{6})$

   (c) $\dfrac{f(x, y + h) - f(x, y)}{h}$

   (d) $\dfrac{f(h, \frac{\pi}{3}) - f(0, \frac{\pi}{3})}{h}$

3. If $g(x, y, z) = \dfrac{x^2 + 4\sqrt{y}}{z}$, find

   (a) $g(2, 4, -3)$

   (b) $g(-1, 9, 5)$

   (c) $\dfrac{g(x + h, y, z) - g(x, y, z)}{h}$

   (d) $\dfrac{g(x, y + h, z) - g(x, y, z)}{h}$

   (e) $\dfrac{g(2, 4, -3 + h) - g(2, 4, -3)}{h}$

4. If $F(x, y, z) = \dfrac{\ln x^2 z}{y^2}$, find

   (a) $F(-1, 2, 1)$ (b) $F(e^{-2}, 2, e)$

   (c) $\dfrac{F(-1 + h, 2, 1) - F(-1, 2, 1)}{h}$ (d) $\dfrac{F(x, y + h, z) - F(x, y, z)}{h}$

   (e) $\dfrac{F(x, y, z + h) - F(x, y, z)}{h}$

5. Suppose $f(x, y) = x^2 + 3xy$, $x = s + 2t$, and $y = 3s - t$. If $F(s, t) = f(s + 2t, 3s - t)$, find

   (a) $F(-1, 2)$ (b) $\dfrac{F(-1, 2 + h) - F(-1, 2)}{h}$

6. Suppose $g(x, y, z) = \sqrt{x^2 + y^2 + z^2}$, $x = e^{-s} \cos t$, $y = e^{-s} \sin t$, and $z = e^{-s}$. If $f(s, t) = g(e^{-s} \cos t, e^{-s} \sin t, e^{-s})$, find

   (a) $f\left(\ln 2, \dfrac{\pi}{6}\right)$ (b) $\dfrac{f(\ln 2 + h, \frac{\pi}{6}) - f(\ln 2, \frac{\pi}{6})}{h}$

In Exercises 7–16 determine the domain and range of the given function. In Exercises 7–14 sketch the subset of R^2 that is the domain of the given function.

7. $f(x, y) = 6 - 2x - 3y$ 8. $f(x, y) = x^2 + 4y^2$

9. $f(x, y) = \sqrt{25 - 9x^2 - 16y^2}$ 10. $f(x, y) = \sqrt{4x^2 + 9y^2 - 9}$

11. $f(x, y) = \dfrac{\ln (x - 2y)}{x + y}$ 12. $f(x, y) = \dfrac{x^2 + 4y^2}{x^2 - 4y^2}$

13. $f(x, y) = \sqrt{\dfrac{x}{x - 3y}}$ 14. $f(x, y) = \sqrt{\dfrac{x + y + 2}{2x - y + 4}}$

15. $f(x, y, z) = \cos^{-1} \sqrt{\dfrac{y - z}{y + z}}$ 16. $f(x, y, z) = \sec^{-1} \sqrt{xy}$

In Exercises 17–20 (a) sketch the graph of the given function; (b) sketch four level curves for the given function.

17. The function in Exercise 7.

18. The function in Exercise 8.

19. The function in Exercise 9.

20. The function in Exercise 10.

18.2   Limit of a Function of n Variables

      Suppose L is some number, (a, b) is a point in R^2, and f is a function of two variables. In ordinary terms, L is called *the limit of f at (a, b)* if and only if $f(x, y)$ can be made to differ from L by as little as we please whenever (x, y)

is in the domain of f and is sufficiently close to, but unequal to, (a, b). If L is this limit, we express this fact by writing the equation

$$\lim_{(x,y)\to(a,b)} f(x, y) = L.$$

The precise statement of this limit is an extension of Definition 3.3.1.

18.2.1   Definition

Suppose f is defined on some open set in R^2 having (a, b) in its boundary. Then $\lim_{(x,y)\to(a,b)} f(x, y) = L$ if and only if for every $\epsilon > 0$ there is some $\delta > 0$ such that

$$|f(x, y) - L| < \epsilon \quad \text{if } (x, y) \in \mathcal{D}_f \text{ and } 0 < \sqrt{(x - a)^2 + (y - b)^2} < \delta. \quad (18.11)$$

Statement (18.11) is illustrated in Figure 18–8. If $(x, y) \neq (a, b)$ is chosen

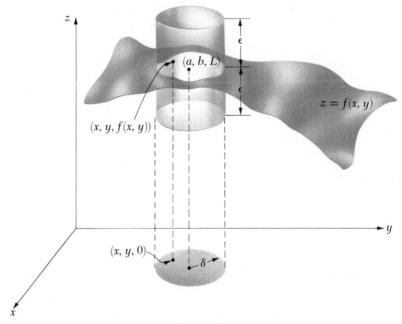

Figure 18–8

inside the open disc in the xy plane with center (a, b) and radius δ, then the corresponding point $(x, y, f(x, y)) \in R^3$ is inside the three-dimensional region bounded by the cylinder $(x - a)^2 + (y - b)^2 = \delta^2$ and the planes $z = L - \epsilon$ and $z = L + \epsilon$. Note however that the existence or nonexistence of the limit has nothing to do with the value of f at (a, b) or whether f is even defined at (a, b).

To verify a limit of a function of two variables using Definition 18.2.1, we can use analogous approaches to those suggested in Section 3.3. Suppose, for example, that there exist positive numbers M and n such that

$$|f(x, y) - L| \le M(\sqrt{(x - a)^2 + (y - b)^2})^n \quad (18.12)$$

for every $(x, y) \neq (a, b)$. Then if $\epsilon > 0$ is given and

$$0 < \sqrt{(x - a)^2 + (y - b)^2} < \left(\frac{\epsilon}{M}\right)^{1/n} = \delta, \tag{18.13}$$

from (18.12) and (18.13)

$$|f(x, y) - L| < M\left[\left(\frac{\epsilon}{M}\right)^{1/n}\right]^n = \epsilon. \tag{18.14}$$

Hence from Definition 18.2.1 $\lim_{(x,y)\to(a,b)} f(x, y) = L$. This approach is used in the next two examples.

Example 1 Prove that $\lim_{(x,y)\to(1,-2)} (3x - 4y) = 11$.

    SOLUTION Since the limit of the function is considered at $(1, -2)$, we first express $|f(x, y) - L| = |(3x - 4y) - 11|$ in terms of $x - 1$ and $y + 2$. Then

$$|(3x - 4y) - 11| = |3(x - 1) - 4(y + 2)| \leq 3|x - 1| + 4|y + 2|$$

and since $|x - 1|$ and $|y + 2|$ are both $\leq \sqrt{(x - 1)^2 + (y + 2)^2}$,

$$|(3x - 4y) - 11| \leq 7\sqrt{(x - 1)^2 + (y + 2)^2}. \tag{18.15}$$

We note that (18.15) is an inequality of the form (18.12) above. Thus if $\epsilon > 0$ is given, and $0 < \sqrt{(x - 1)^2 + (y + 2)^2} < \epsilon/7 = \delta$, from (18.15)

$$|(3x - 4y) - 11| < 7 \cdot \frac{\epsilon}{7} = \epsilon,$$

and the desired limit is verified.

Example 2 Prove that $\displaystyle\lim_{(x,y)\to(0,0)} \frac{2xy^2}{x^2 + y^2} = 0$.

    SOLUTION Since $\dfrac{y^2}{x^2 + y^2} \leq 1$ when $(x, y) \neq (0, 0)$,

$$\left|\frac{2xy^2}{x^2 + y^2} - 0\right| = \frac{2|x|y^2}{x^2 + y^2} \leq 2|x|$$

$$\leq 2\sqrt{x^2 + y^2}. \tag{18.16}$$

If $\epsilon > 0$ is given, and $0 < \sqrt{x^2 + y^2} < \epsilon/2 = \delta$, then from (18.16)

$$\left|\frac{2xy^2}{x^2 + y^2} - 0\right| < 2 \cdot \frac{\epsilon}{2} = \epsilon,$$

and the desired limit is verified.

    As an extension of Definition 18.2.1 we shall define the limit of a function of n variables.

18.2.2   Definition

Suppose f, a function of n variables, is defined on some open set having A in its boundary. Then L is the limit of f at A and we write

$$\lim_{X \to A} f(X) = L$$

if and only if for every $\epsilon > 0$ there is some $\delta > 0$ such that $|f(X) - L| < \epsilon$ if $X \in \mathcal{D}_f$ and

$$0 < \sqrt{(x_1 - a_1)^2 + (x_2 - a_2)^2 + \cdots + (x_n - a_n)^2} < \delta.$$

The following theorem for functions of n variables is analogous to Theorem 3.5.1 and follows from Definition 18.2.2.

18.2.3   Theorem

Let $f(X) = g(X)$ for every $X \neq A$ in some open ball with center A. Then

(a) $\lim\limits_{X \to A} f(X) = L$ *if* $\lim\limits_{X \to A} g(X) = L$

(b) $\lim\limits_{X \to A} f(X)$ *fails to exist if* $\lim\limits_{X \to A} g(X)$ *fails to exist.*

The proof of Theorem 18.2.3 is left as an exercise.

Exercise Set 18.2

In Exercises 1–6 verify the given limit using Definition 18.2.1.

1. $\lim\limits_{(x,y) \to (2,3)} (2x + 5y) = 19$

2. $\lim\limits_{(x,y) \to (4,-1)} (3x - 2y + 1) = 15$

3. $\lim\limits_{(x,y) \to (0,0)} (2x^2 + 3y^2) = 0$

4. $\lim\limits_{(x,y) \to (0,0)} \dfrac{2xy}{\sqrt{x^2 + y^2}} = 0$

5. $\lim\limits_{(x,y) \to (2,3)} (x^2 + y^2) = 13$.   HINT: Let $x^2 + y^2 - 13 = (x - 2)^2 + (y - 3)^2 + 4(x - 2) + 6(y - 3)$.

6. $\lim\limits_{(x,y) \to (2,3)} xy = 6$

7. Define $\lim\limits_{(x,y,z) \to (a,b,c)} f(x, y, z) = L$.

8. Prove Theorem 18.2.3.

18.3   Limit Theorems

As with functions of one variable, it is usually easier to obtain a limit of a function of more than one variable using a suitable limit theorem than by invoking the ϵ-δ definition. However, Definitions 18.2.1 and 18.2.2 are still useful in verifying a limit that is intuitively apparent and yet no limit theorem is available.

The first of these limit theorems concerns the limit of a function of n variables which is defined using only one variable.

18.3.1   Theorem

Suppose f is a function of n variables and g is a function of one variable. If k is a positive integer $\leq n$, suppose also that

$$f(X) = g(x_k)$$

for any $X = (x_1, x_2, \ldots, x_k, \ldots, x_n)$ such that $x_k \in \mathcal{D}_g$. If $A = (a_1, a_2, \ldots, a_k, \ldots, a_n)$ and g is continuous at a_k, then

$$\lim_{X \to A} f(X) = \lim_{x_k \to a_k} g(x_k) = g(a_k).$$

PROOF   Since g is continuous at a_k, for every $\epsilon > 0$ there is a $\delta > 0$ such that

$$|g(x_k) - g(a_k)| < \epsilon \quad \text{when } x_k \in \mathcal{D}_g \text{ and } |x_k - a_k| < \delta. \tag{18.17}$$

Suppose $X \in \mathcal{D}_f$ and $0 < \sqrt{(x_1 - a_1)^2 + \cdots + (x_n - a_n)^2} < \delta$. Then $x_k \in \mathcal{D}_g$ and since $|x_k - a_k| \leq \sqrt{\sum_{i=1}^{n} (x_i - a_i)^2} < \delta$, we have

$$|f(X) - g(a_k)| = |g(x_k) - g(a_k)| < \epsilon.$$

The conclusion then follows from Definition 18.2.2.

In particular from Theorem 18.3.1 if g is continuous at a and

$$f(x, y) = g(x) \quad \text{for every } x \in \mathcal{D}_g,$$

then for any b

$$\lim_{(x,y) \to (a,b)} f(x, y) = \lim_{x \to a} g(x) = g(a).$$

Thus if $f(x, y) = 2x^2$, then

$$\lim_{(x,y) \to (-1,3)} 2x^2 = \lim_{x \to -1} 2x^2 = 2.$$

Also from Theorem 18.3.1 if $f(x, y) = e^{-y^2}$, then

$$\lim_{(x,y) \to (3,0)} e^{-y^2} = \lim_{y \to 0} e^{-y^2} = 1.$$

The limit of the sum, difference, product, and quotient of two functions of n variables follows from a theorem that is analogous to Theorem 3.4.3. In this theorem and in the remaining theorems of this chapter we will make this tacit assumption: whenever the existence of the limit of a function at some point, which we might call A, in R^n is to be proved, then the function is actually defined on some open set having A in its boundary (recall Definition 18.2.2).

18.3.2   Theorem

Suppose f and g are functions of n variables such that $\lim\limits_{X \to A} f(X) = L_1$ *and* $\lim\limits_{X \to A} g(X) = L_2$. *Then*

(a) $\lim\limits_{X \to A} [f(X) + g(X)] = L_1 + L_2$

(b) $\lim\limits_{X \to A} [f(X) - g(X)] = L_1 - L_2$

(c) $\lim\limits_{X \to A} f(X)g(X) = L_1 L_2$

(d) $\lim\limits_{X \to A} \dfrac{f(X)}{g(X)} = \dfrac{L_1}{L_2}$ *if* $L_2 \neq 0$.

The proof of this theorem is omitted since it is analogous to the proof of Theorem 3.4.3.

   Theorems 18.3.2(a) and 18.3.2(c) can be generalized and hence the limit of the sum (or product) of a finite number of functions is the sum (or product) of their limits provided these limits exist.

Example 1   Find $\lim\limits_{(x,y) \to (-1,2)} (3x^2 y + xy^2 - y^4)$.

SOLUTION   From Theorem 18.3.1 and extensions of Theorems 18.3.2(a) and 18.3.2(c),

$\lim\limits_{(x,y) \to (-1,2)} (3x^2 y + xy^2 - y^4)$

$$= \lim\limits_{(x,y) \to (-1,2)} 3x^2 y + \lim\limits_{(x,y) \to (-1,2)} xy^2 + \lim\limits_{(x,y) \to (-1,2)} (-y^4)$$

$$= \lim\limits_{(x,y) \to (-1,2)} 3x^2 \cdot \lim\limits_{(x,y) \to (-1,2)} y$$

$$+ \lim\limits_{(x,y) \to (-1,2)} x \cdot \lim\limits_{(x,y) \to (-1,2)} y^2 + \lim\limits_{(x,y) \to (-1,2)} (-y^4)$$

$$= 3 \cdot 2 + (-1) \cdot 2^2 - 16 = -14.$$

   If P is a polynomial function, then for any $A \in R^n$ it can be proved, using the same method as in Example 1 that

$$\lim\limits_{X \to A} P(X) = P(A). \tag{18.18}$$

If Q is also a polynomial function, then by Theorem 18.3.2(d)

$$\lim\limits_{X \to A} \frac{P(X)}{Q(X)} = \frac{\lim\limits_{X \to A} P(X)}{\lim\limits_{X \to A} Q(X)} = \frac{P(A)}{Q(A)} \quad \text{if } Q(A) \neq 0. \tag{18.19}$$

   The next theorem is useful in evaluating the limits of certain types of composite functions.

18.3.3   Theorem

Suppose that

> *(i) g is a function of n variables and $\lim_{X \to A} g(X) = L$, and*
> *(ii) f, a function of one variable, is continuous at L.*

Then the function F that is defined by $F(X) = f(g(X))$ has the limit

$$\lim_{X \to A} F(X) = f\left(\lim_{X \to A} g(X)\right) = f(L).$$

The proof of Theorem 18.3.3, which is analogous to the proof of Theorem 3.8.1, is left as an exercise.

Example 2   Find $\displaystyle\lim_{(x,y,z) \to (2,-1,\pi/3)} \sin x^2 yz$.

SOLUTION   Here f is a function of one variable such that

$$f(s) = \sin s$$

and g is a function of three variables given by

$$g(x, y, z) = x^2 yz.$$

Then by Theorem 18.3.3

$$\lim_{(x,y,z) \to (2,-1,\pi/3)} \sin x^2 yz = \sin\left(\lim_{(x,y,z) \to (2,-1,\pi/3)} x^2 yz\right)$$

$$= \sin\left(-\frac{4\pi}{3}\right) = \frac{\sqrt{3}}{2}.$$

Suppose f is a function of two variables which is defined at all points near (a, b) and $\lim_{(x,y) \to (a,b)} f(x, y) = L$. Then as (x, y) approaches (a, b) along any curve $C : x = \phi(t)$, $y = \psi(t)$ through the point (a, b), intuition suggests that the numbers $f(\phi(t), \psi(t))$ also approach L (Figure 18–9). This idea is expressed in the next theorem.

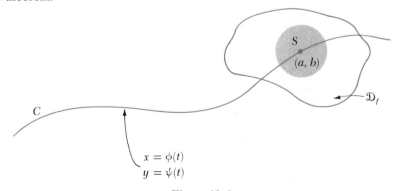

$$x = \phi(t)$$
$$y = \psi(t)$$

Figure 18–9

18.3.4   Theorem

Suppose the following conditions are satisfied:

> (i) *f, a function of two variables is defined for every* $(x, y) \neq (a, b)$ *in some open disc S with center* (a, b).
>
> (ii) $\lim_{(x,y) \to (a,b)} f(x, y) = L$.
>
> (iii) *C is a curve with parametrization* $x = \phi(t)$, $y = \psi(t)$ *for every t in some interval I and* $(\phi(t_1), \psi(t_1)) \neq (\phi(t_2), \psi(t_2))$ *whenever* $t_1 \neq t_2$ *in I.*
>
> (iv) *There is a* $t_0 \in I$ *such that* $\phi(t_0) = a$ *and* $\psi(t_0) = b$.

Then

$$\lim_{t \to t_0} f(\phi(t), \psi(t)) = L.$$

PROOF   From (ii) and Definition 18.2.1 given $\epsilon > 0$ there is a δ, where $0 < \delta < $ radius of S, such that

$$|f(x, y) - L| < \epsilon \quad \text{if } 0 < \sqrt{(x - a)^2 + (y - b)^2} < \delta. \qquad (18.20)$$

Then from (iii) the functions ϕ and ψ are continuous at t_0 and hence there exist positive numbers γ_1 and γ_2 such that

$$|\phi(t) - a| < \frac{\delta}{\sqrt{2}} \quad \text{if } t \in I \text{ and } |t - t_0| < \gamma_1 \qquad (18.21)$$

and

$$|\psi(t) - b| < \frac{\delta}{\sqrt{2}} \quad \text{if } t \in I \text{ and } |t - t_0| < \gamma_2. \qquad (18.22)$$

Also from (iii) for $t \in I$

$$\sqrt{[\phi(t) - a]^2 + [\psi(t) - b]^2} > 0 \quad \text{if } t \neq t_0. \qquad (18.23)$$

Every open interval containing t_0 certainly contains a $t \in I$ where $t \neq t_0$. Hence if $0 < |t - t_0| < \gamma = $ smaller of γ_1 and γ_2, then from (18.23), (18.21), and (18.22)

$$0 < \sqrt{[\phi(t) - a]^2 + [\psi(t) - b]^2} < \sqrt{\frac{\delta^2}{2} + \frac{\delta^2}{2}} = \delta$$

and thus from (18.20)

$$|f(\phi(t), \psi(t)) - L| < \epsilon.$$

Theorem 18.3.4 is useful in proving that a limit of a function of two variables *fails to exist*.

Example 3   If $f(x, y) = \begin{cases} \dfrac{2xy}{x^2 + y^2} & \text{if } (x, y) \neq (0, 0) \\ 0 & \text{if } (x, y) = (0, 0) \end{cases}$

prove that $\lim_{(x,y) \to (0,0)} f(x, y)$ fails to exist.

PROOF If $\lim_{(x,y)\to(0,0)} f(x, y)$ did exist, then by Theorem 18.3.4, the limit for $f(\phi(t), \psi(t))$ would be the same as (x, y) approaches $(0, 0)$ along any curve having a parametrization that satisfies (iii) in Theorem 18.3.4. If (x, y) approaches $(0, 0)$ along the x axis, which is given parametrically by

$$x = t \qquad y = 0,$$

then

$$\lim_{t\to 0} f(t, 0) = \lim_{t\to 0} 0 = 0.$$

However, if (x, y) approaches $(0, 0)$ along the line $y = x$, which has the parametrization

$$x = t \qquad y = t,$$

then

$$\lim_{t\to 0} f(t, t) = \lim_{t\to 0} \frac{2t^2}{t^2 + t^2} = 1.$$

Since different limits were obtained over different paths through $(0, 0)$, $\lim_{(x,y)\to(0,0)} f(x, y)$ fails to exist.

Exercise Set 18.3

In Exercises 1–11 evaluate the following limits or show that they do not exist. Give the limit theorems used.

1. $\displaystyle\lim_{(x,y)\to(-2,3)} (3x^2 + 2y^2)$

2. $\displaystyle\lim_{(x,y)\to(3,1)} \frac{3x^2 - 2xy}{x^2 + xy + y^2}$

3. $\displaystyle\lim_{(x,y)\to(-1,3)} \frac{\sqrt{2xy + y^2}}{x^2}$

4. $\displaystyle\lim_{(x,y)\to(\pi/2,\pi/4)} \cos(2x + 3y)$

5. $\displaystyle\lim_{(x,y)\to(0,0)} \frac{x^2 - y^2}{x^2 + y^2}$

6. $\displaystyle\lim_{(x,y)\to(0,0)} \frac{x^3 + y^3}{x + y}$

7. $\displaystyle\lim_{(x,y)\to(0,0)} \frac{\sin(x + y)}{x + y}$

8. $\displaystyle\lim_{(x,y)\to(0,0)} \frac{|xy|}{x^2 + y^2}$

9. $\displaystyle\lim_{(x,y)\to(0,0)} \frac{x^2 y}{x^4 + y^2}$

10. $\displaystyle\lim_{(x,y)\to(0,0)} y^2 \ln(x^2 + y^2)$

11. $\displaystyle\lim_{(x,y)\to(2,1)} |x - 2|^{y-1}$

12. State the generalization of Theorem 18.3.4 to functions of n variables.

13. Prove Theorem 18.3.3.

18.4 Continuous Functions

The same intuitive description of continuity can be given for functions of n variables as for functions of one variable. A function f of n variables is continuous at an ordered n-tuple $A \in \mathcal{D}_f$ if $f(X)$ can be made to differ from $f(A)$

by as little as desired when X is sufficiently close to A. Thus the following definition is given.

18.4.1   Definition

Let f be a function of n variables. Then f is continuous at $A \in R^n$ if and only if the following conditions are satisfied:

> (i) $f(A)$ is defined,
> (ii) $\lim_{X \to A} f(X)$ exists,
> (iii) $\lim_{X \to A} f(X) = f(A)$.

The function f is said to be *continuous on a set* $S \subset R^n$ if and only if f is continuous at every element $A \in S$. Also, f is called a *continuous function* if and only if f is continuous on $\mathfrak{D}_f$.

As an illustration of Definition 18.4.1 any polynomial function P of n variables is a continuous function because $\lim_{X \to A} P(X) = P(A)$ for any $A \in R^n$ (Equation (18.18)). Also any rational function is continuous on its domain except where its denominator assumes the value zero because of (18.19).

Three examples are now presented to illustrate Definition 18.4.1.

Example 1   Is the function f where

$$f(x, y) = \begin{cases} \dfrac{2xy}{x^2 + y^2} & \text{if } (x, y) \neq (0, 0) \\ 0 & \text{if } (x, y) = (0, 0) \end{cases}$$

continuous at $(0, 0)$? Why?

SOLUTION   Condition (i) of Definition 18.4.1 is satisfied since $f(0, 0) = 0$. However, from Example 3 of Section 18.3 $\lim_{(x, y) \to (0, 0)} f(x, y)$ fails to exist and therefore (ii) of Definition 18.4.1 does not hold. Thus f is discontinuous at $(0, 0)$.

The function f in Example 1 merits more than passing notice since the functions g and h which are associated with f by the equations

$$g(x) = f(x, 0) = 0 \qquad h(y) = f(0, y) = 0$$

are continuous at $(0, 0)$. Thus a function of two variables may be discontinuous at a point in its domain and yet be a continuous function of each variable separately.

Example 2   Is the function f where

$$f(x, y) = \begin{cases} x^2 + y^2 & \text{if } (x, y) \neq (2, 3) \\ -1 & \text{if } (x, y) = (2, 3) \end{cases}$$

continuous at $(2, 3)$? Why?

SOLUTION   Since $f(2, 3) = -1$, condition (i) in Definition 18.4.1 is satisfied. Also since

$$\lim_{(x,y)\to(2,3)} f(x, y) = \lim_{(x,y)\to(2,3)} (x^2 + y^2) = 13,$$

condition (ii) in the definition is fulfilled. However, since $\lim_{(x,y)\to(2,3)} f(x, y) \neq f(2, 3)$, condition (iii) is not satisfied and therefore f is discontinuous at $(2, 3)$.

Example 3   Is the function f where

$$f(x, y) = \begin{cases} \dfrac{2xy^2}{x^2 + y^2} & \text{if } (x, y) \neq (0, 0) \\ 0 & \text{if } (x, y) = (0, 0) \end{cases}$$

continuous at $(0, 0)$? Why?

SOLUTION   Since $f(0, 0) = 0$, condition (i) in Definition 18.4.1 is satisfied. From Example 2 of Section 18.2, using Theorem 18.2.3,

$$\lim_{(x,y)\to(0,0)} f(x, y) = \lim_{(x,y)\to(0,0)} \frac{2xy^2}{x^2 + y^2} = 0$$

and therefore condition (ii) of Definition 18.4.1 is satisfied. Also condition (iii) of the definition holds since $\lim_{(x,y)\to(0,0)} f(x, y) = 0 = f(0, 0)$ and therefore by the definition, f is continuous at $(0, 0)$.

The definition of continuity at a point can also be stated in ϵ-δ language:

Let f be a function of n variables which is defined at a point $A = (a_1, a_2, \ldots, a_n) \in R^n$. Then f is continuous at A if and only if for every $\epsilon > 0$ there is a $\delta > 0$ such that

$$|f(X) - f(A)| < \epsilon \quad \text{if } x \in \mathcal{D}_f \text{ and } \sqrt{\sum_{i=1}^{n}(x_i - a_i)^2} < \delta. \qquad (18.24)$$

Also, the following theorem states a condition for continuity which is equivalent to that given in Definition 18.4.1.

18.4.2   Theorem

Suppose f, a function of n variables, is defined at $A = (a_1, a_2, \ldots, a_n) \in R^n$. Then f is continuous at A if and only if

$$\lim_{(h_1,h_2,\ldots,h_n)\to(0,0,\ldots,0)} f(a_1 + h_1, a_2 + h_2, \ldots, a_n + h_n) = f(A).$$

The proof of Theorem 18.4.2 is left as an exercise.

We next mention some theorems about continuity of functions of n variables which are analogous to those for functions of one variable.

18.4.3   Theorem

If f and g are functions of n variables which are continuous at A, then

(a) $f + g$ *is continuous at A.*
(b) $f - g$ *is continuous at A.*
(c) fg *is continuous at A.*
(d) $\dfrac{f}{g}$ *is continuous at A if $g(A) \neq 0$.*

PROOF OF (a)   By Definition 18.4.1 $\lim_{X \to A} f(X) = f(A)$ and $\lim_{X \to A} g(X) = g(A)$. Therefore from Theorem 18.3.2(a)

$$\lim_{X \to A} (f + g)(X) = \lim_{X \to A} [f(X) + g(X)] = \lim_{X \to A} f(X) + \lim_{X \to A} g(X)$$
$$= f(A) + g(A) = (f + g)(A).$$

Hence by Definition 18.4.1 $f + g$ is continuous at A.

The proofs of the other parts of Theorem 18.4.3 are similar. Parts (a) and (c) of the theorem can be extended to prove that the sum or product of a finite number of functions that are continuous at a point is also continuous at the point.

In contrast to these extensions the following generalization of Theorem 18.3.3 is of basic importance and requires a detailed proof.

18.4.4   Theorem (Composite Function Limit Theorem)

Suppose that

(i) $g_1, g_2, \ldots, g_m$ *are each function of n variables and*
$$\lim_{X \to A} g_1(X) = L_1, \lim_{X \to A} g_2(X) = L_2, \ldots, \lim_{X \to A} g_m(X) = L_m,$$

(ii) *f, a function of m variables, is continuous at $L = (L_1, L_2, \ldots, L_m) \in R^m$,*
(iii) $F(X) = f(g_1(X), g_2(X), \ldots, g_m(X))$.

Then
$$\lim_{X \to A} F(X) = f\left(\lim_{X \to A} g_1(X), \ldots, \lim_{X \to A} g_m(X) \right) = f(L).$$

PROOF   Since f is continuous at L, for every $\epsilon > 0$ there is a $\delta > 0$ such that

$$|f(s_1, s_2, \ldots, s_m) - f(L)| < \epsilon \quad \text{if } (s_1, \ldots, s_m) \in \mathfrak{D}_f \text{ and } \sqrt{\sum_{k=1}^{m} (s_k - L_k)^2} < \delta.$$
$$(18.25)$$

From (i) for $k = 1, 2, \ldots, m$ there is a $\gamma_k > 0$ such that

$$|g_k(X) - L_k| < \frac{\delta}{\sqrt{m}} \quad \text{if } X \in \mathcal{D}_{g_k} \text{ and } 0 < \sqrt{\sum_{i=1}^{n}(x_i - a_i)^2} < \gamma_k. \quad (18.26)$$

If $X \in \mathcal{D}_F$, then $X \in \mathcal{D}_{g_k}$ for every k. Also if $0 < \sqrt{\sum_{i=1}^{n}(x_i - a_i)^2} < \gamma = $ smallest of $\gamma_1, \gamma_2, \ldots, \gamma_k$, from (18.26)

$$\sqrt{\sum_{k=1}^{m}(g_k(X) - L_k)^2} < \sqrt{\sum_{k=1}^{m}\frac{\delta^2}{m}} = \delta,$$

and hence from (18.25)

$$|f(g_1(X), \ldots, g_m(X)) - f(L)| < \epsilon.$$

Thus the conclusion is obtained.

As an illustration of Theorem 18.4.4 we have the following example.

Example 4 Suppose f is a continuous function of two variables and g_1 and g_2 are functions of two variables given by

$$g_1(s, t) = s - 2 \sin t \qquad g_2(s, t) = s + 3 \cos t.$$

Then if

$$F(s, t) = f(s - 2 \sin t, s + 3 \cos t),$$

we have by Theorem 18.4.4

$$\lim_{(s,t) \to (1, -\pi/2)} f(s - 2 \sin t, s + 3 \cos t)$$

$$= f\left(\lim_{(s,t) \to (1, -\pi/2)} (s - 2 \sin t), \lim_{(s,t) \to (1, -\pi/2)} (s + 3 \cos t)\right) = f(3, 1).$$

The proof of the following corollary of Theorem 18.4.4 is left as an exercise.

18.4.5   Theorem

Suppose that

(i) $g_1, g_2, \ldots, g_m$ *are each functions of n variables which are continuous at A.*

(ii) f, *a function of m variables, is continuous at* $(g_1(A), g_2(A), \ldots, g_m(A)) \in R^m$.

(iii) $F(X) = f(g_1(X), g_2(X), \ldots, g_m(X))$.

Then F is continuous at A.

Example 5   At what points (x, y) is the function G continuous where

$$G(x, y) = \frac{\sqrt{x - 4y - 4}}{x + y}?$$

Illustrate with a sketch.

SOLUTION   The function g where $g(x, y) = x - 4y - 4$ is a polynomial function and is therefore everywhere continuous. The function given by $f(s) = \sqrt{s}$ is continuous if $s \geq 0$, and hence by Theorem 18.4.5 the function F where $F(x, y) = \sqrt{x - 4y - 4}$ is continuous wherever $x - 4y - 4 \geq 0$. The function h where $h(x, y) = x + y$ is also everywhere continuous. Hence, by Theorem 18.4.3(d) the function G is continuous wherever

$$x - 4y - 4 \geq 0 \qquad \text{and} \qquad x + y \neq 0. \qquad (18.27)$$

A sketch of the graph of all points (x, y) satisfying the inequalities in (18.27) is shown shaded in Figure 18–10. This graph is the set of all points in the plane which are on or below the line $x - 4y = 4$, but not on the line $x + y = 0$.

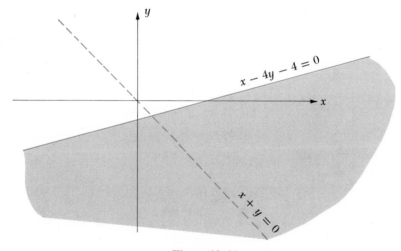

Figure 18–10

From the next theorem a continuous function of n variables is also *a continuous function of each variable separately.*

18.4.6   Theorem

If f, a function of n variables, is continuous at $A = (a_1, a_2, \ldots, a_n)$ and g, a function of one variable, is given by

$$g(x_i) = f(a_1, a_2, \ldots, a_{i-1}, x_i, a_{i+1}, \ldots, a_n),$$

then g is continuous at a_i.

The proof of this theorem is left as an exercise.

However, a function of n variables which is continuous in each variable separately need not be a continuous function, as was noted in our remarks following Example 1.

Exercise Set 18.4

In Exercises 1–8 determine the region in the xy plane on which the given function f is continuous. Sketch this region.

1. $f(x, y) = \sqrt{x - 2y}$

2. $f(x, y) = \ln(2x - 3y + 6)$

3. $f(x, y) = \ln(x^2 - y^2)$

4. $f(x, y) = \sqrt{9x^2 + y^2 - 16}$

5. $f(x, y) = \sqrt{\dfrac{x + 2y + 2}{x - 4y}}$

6. $f(x, y) = \dfrac{x + 3y}{\sqrt{4 - x^2 - y^2}}$

7. $f(x, y) = \dfrac{\sqrt{2x - y}}{\sin^{-1}(x + y)}$

8. $f(x, y) = \begin{cases} \dfrac{x^2 + y^3}{x + y^2} & \text{if } (x, y) \neq (0, 0) \\ 0 & \text{if } (x, y) = (0, 0) \end{cases}$

In Exercises 9–16 determine if the function f is continuous at $(0, 0)$.

9. $f(x, y) = \begin{cases} \dfrac{x^2 - 4xy + 3y^2}{x - y} & \text{if } (x, y) \neq (0, 0) \\ 0 & \text{if } (x, y) = (0, 0) \end{cases}$

10. $f(x, y) = \begin{cases} \dfrac{x^3 y + xy^3}{x^2 + y^2} & \text{if } (x, y) \neq (0, 0) \\ 0 & \text{if } (x, y) = (0, 0) \end{cases}$

11. $f(x, y) = \begin{cases} \dfrac{x^2 y^2}{x^2 + y^2} & \text{if } (x, y) \neq (0, 0) \\ 0 & \text{if } (x, y) = (0, 0) \end{cases}$

12. $f(x, y) = \begin{cases} \dfrac{x^2 + xy + y^2}{x^2 + 3xy + y^2} & \text{if } (x, y) = (0, 0) \\ 1 & \text{if } (x, y) = (0, 0) \end{cases}$

13. $f(x, y) = \begin{cases} \dfrac{x^2 y^2}{x^4 + y^4} & \text{if } (x, y) \neq (0, 0) \\ 0 & \text{if } (x, y) = (0, 0) \end{cases}$

14. $f(x, y) = \begin{cases} \dfrac{x^2 + 2xy + y^2}{|x| + |y|} & \text{if } (x, y) \neq (0, 0) \\ 0 & \text{if } (x, y) \neq (0, 0) \end{cases}$

15. $f(x, y) = \begin{cases} |x|^y & \text{if } (x, y) \neq (0, 0) \\ 0 & \text{if } (x, y) = (0, 0) \end{cases}$

16. $f(x, y) = \begin{cases} \dfrac{\sin xy}{xy} & \text{if } (x, y) \neq (0, 0) \\ 1 & \text{if } (x, y) = (0, 0) \end{cases}$

17. Is the function f where

$$f(x, y, z) = \begin{cases} \dfrac{xyz}{x^2 + y^2 + z^2} & \text{if } (x, y, z) \neq (0, 0, 0) \\ 0 & \text{if } (x, y, z) = (0, 0, 0) \end{cases}$$

continuous at $(0, 0, 0)$? Why?

18. Prove Theorem 18.4.3(d).

19. Prove Theorem 18.4.5.

20. Prove Theorem 18.4.2.

21. Prove Theorem 18.4.6.

18.5 Partial Derivatives

It will be recalled that the derivative of a function f of one variable is the function f' defined by

$$f'(x) = \lim_{h \to 0} \frac{f(x + h) - f(x)}{h}$$

when this limit exists. A function f of two variables, however, has two *partial derivatives* $D_1 f$ and $D_2 f$, which are defined as follows:

$$D_1 f(x, y) = \lim_{h \to 0} \frac{f(x + h, y) - f(x, y)}{h} \tag{18.28}$$

$$D_2 f(x, y) = \lim_{h \to 0} \frac{f(x, y + h) - f(x, y)}{h}. \tag{18.29}$$

The domain of each partial derivative consists of all ordered pairs (x, y) for which the corresponding limit exists.

Example 1 If $f(x, y) = 3x^2/y$ find (a) $D_1 f(x, y)$ (b) $D_1 f(-1, 2)$ (c) $D_2 f(x, y)$ (d) $D_2 f(-1, 2)$.

SOLUTION To obtain $D_1 f(x, y)$, we first form

$$\frac{f(x + h, y) - f(x, y)}{h} = \frac{\dfrac{3(x + h)^2}{y} - \dfrac{3x^2}{y}}{h} = \frac{6hx + 3h^2}{hy}$$

$$= \frac{6x + 3h}{y}. \tag{18.30}$$

Hence from (18.28) and (18.30)

$$D_1 f(x, y) = \lim_{h \to 0} \frac{6x + 3h}{y} = \frac{6x}{y}. \tag{18.31}$$

Then from (18.31)

$$D_1(-1, 2) = \frac{6(-1)}{2} = -3.$$

In obtaining $D_2 f(x, y)$, we utilize

$$\frac{f(x, y + h) - f(x, y)}{h} = \frac{\dfrac{3x^2}{y + h} - \dfrac{3x^2}{y}}{h}$$

$$= \frac{-3x^2}{y(y + h)}. \tag{18.32}$$

Then from (18.29) and (18.32)

$$D_2 f(x, y) = \lim_{h \to 0} \frac{-3x^2}{y(y + h)} = -\frac{3x^2}{y^2} \tag{18.33}$$

and from (18.33)

$$D_2 f(-1, 2) = -\frac{3(-1)^2}{2^2} = -\frac{3}{4}.$$

It should be noted that in forming $D_1 f(x, y)$, we differentiate $f(x, y)$ with respect to x while holding y constant; in forming $D_2 f(x, y)$ we differentiate $f(x, y)$ with respect to y while holding x constant. Thus, since the values of $D_1 f$ and $D_2 f$ at any (x, y) are obtained by differentiation with respect to only one of the variables while the other is kept constant, it is appropriate to call these functions partial derivatives.

Example 2   Find $D_1 f(x, y)$ and $D_2 f(x, y)$ if $f(x, y) = e^{-2x} \cos 3y + y^2$.

SOLUTION   From our preceding remarks, differentiating $f(x, y)$ with respect to x while holding y constant gives

$$D_1 f(x, y) = -2e^{-2x} \cos 3y.$$

Also, differentiating $f(x, y)$ with respect to y while keeping x constant yields

$$D_2 f(x, y) = -3e^{-2x} \sin 3y + 2y.$$

The geometric significance of the partial derivative of a function of two variables is illustrated in Figure 18–11. In this figure the surface $z = f(x, y)$ intersects the plane $x = a$ in the curve C_a and the plane $y = b$ in the curve C_b. Since C_b has the equation $z = f(x, b) = \phi(x)$, the slope of C_b at the point $(a, b, f(a, b))$ is

$$\phi'(a) = \lim_{h \to 0} \frac{\phi(a + h) - \phi(a)}{h} = \lim_{h \to 0} \frac{f(a + h, b) - f(a, b)}{h}$$

$$= D_1 f(a, b).$$

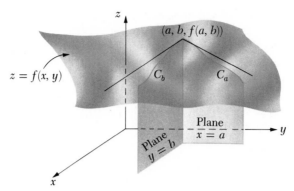

Figure 18–11

Also, since C_a has the equation $z = f(a, y) = \psi(y)$, the slope of C_a at the point $(a, b, f(a, b))$ is

$$\psi'(b) = \lim_{h \to 0} \frac{\psi(b + h) - \psi(b)}{h} = \lim_{h \to 0} \frac{f(a, b + h) - f(a, b)}{h}$$

$$= D_2 f(a, b).$$

As a generalization of (18.28) and (18.29) we next define the partial derivative of a function of n variables.

18.5.1 Definition

Let f be a function of n variables. The *partial derivative of f with respect to the ith variable*, denoted by $D_i f$ is given by

$$D_i f(X) = \lim_{h \to 0} \frac{f(x_1, x_2, \ldots, x_{i-1}, x_i + h, x_{i+1}, \ldots, x_n) - f(x_1, x_2, \ldots, x_n)}{h}$$

if this limit exists.

If f is a function of one variable, then the partial derivative $D_1 f$ is the same as the ordinary derivative f'.

The function value $D_1 f(X)$ is called, variously, *the partial derivative of f with respect to x_i at X* and "the partial derivative of $f(X)$ with respect to x_i." However, for brevity of expression the word "derivative" is often omitted from the verbal description of $D_i f$ and $D_i f(X)$, and thus $D_i f$ is called simply the "partial of f with respect to the ith variable."

In mathematical literature the notations

$$f_{x_i}(X), \quad \frac{\partial f}{\partial x_i}, \quad \frac{\partial f(X)}{\partial x_i}, \quad \frac{\partial}{\partial x_i} f(X)$$

are also used in place of $D_i f(X)$. Further, if $f(X)$ is expressed by the equation, say, $u = f(X)$, then $D_1 f(X)$ can be denoted by

$$\frac{\partial u}{\partial x_i} \quad \text{or} \quad u_{x_i}.$$

$\partial u / \partial x_i$, or u_{x_i}, is termed the *rate of change of u with respect to* x_i *when* x_1, x_2, . . . , x_{i-1}, x_{i+1}, . . . , x_n *are held constant.*

Although there are a variety of possible notations for the value of a given partial derivative, any symbol actually used should be as simple as possible without sacrificing clarity.

For example, if f is a function of three variables and one has occasion to refer to $D_1 f(x, y, z)$ repeatedly, it might be simpler to let $u = f(x, y, z)$ and then denote $D_1 f(x, y, z)$ by u_x. But if $u = f(x, y, g(x, y))$, then u_x would refer to

$$\frac{\partial}{\partial x} f(x, y, g(x, y)) = \lim_{h \to 0} \frac{f(x + h, y, g(x + h, y)) - f(x, y, g(x, y))}{h},$$

whereas

$$D_1 f(x, y, g(x, y)) = \lim_{h \to 0} \frac{f(x + h, y, g(x, y)) - f(x, y, g(x, y))}{h}.$$

The process of obtaining $D_i f(X)$ from a given $f(X)$ for a certain variable x_i is called *partial differentiation*. In computing $D_i f(X)$, the variables x_1, x_2, . . . , x_{i-1}, x_{i+1}, . . . , x_n are held constant and $f(X)$ is differentiated with respect to x_i. Example 2 illustrates the technique of partial differentiation for functions of two variables, while in the following example the process is applied to a function of three variables.

Example 3 If $f(x, y, z) = e^{-y} \tan^{-1} x^2 yz$, find $\partial f / \partial x$, $\partial f / \partial y$, and $\partial f / \partial z$.

SOLUTION To find $\partial f / \partial x$, y and z are held constant and $f(x, y, z)$ is differentiated with respect to x. Then

$$\frac{\partial f}{\partial x} = e^{-y} \frac{2xyz}{1 + x^4 y^2 z^2} = \frac{2e^{-y} xyz}{1 + x^4 y^2 z^2}.$$

Analogously, in obtaining $\partial f / \partial y$, x and z are kept constant and $f(x, y, z)$ is differentiated with respect to y.

$$\frac{\partial f}{\partial y} = e^{-y} \frac{x^2 z}{1 + x^4 y^2 z^2} - e^{-y} \tan^{-1} x^2 yz$$

$$= e^{-y} \left(\frac{x^2 z}{1 + x^4 y^2 z^2} - \tan^{-1} x^2 yz \right).$$

Also

$$\frac{\partial f}{\partial z} = e^{-y} \frac{x^2 y}{1 + x^4 y^2 z^2} = \frac{e^{-y} x^2 y}{1 + x^4 y^2 z^2}.$$

To obtain the partial derivatives of "pathological" functions, the following theorem, which is immediately derived from Definition 18.5.1, is useful.

18.5.2   Theorem

If $f(X) = g(X)$ *for every* $X \neq A$ *in some open ball with center* A, *and* $D_i g(A)$ *exists, then* $D_i f(A) = D_i g(A)$.

Example 4   Let

$$f(x, y) = \begin{cases} \dfrac{2xy}{x^2 + y^2} & \text{if } (x, y) \neq (0, 0) \\ 0 & \text{if } (x, y) = (0, 0). \end{cases}$$

Find $D_1 f(x, y)$ and $D_2 f(x, y)$ for any $(x, y) \in \mathcal{D}_f$.

SOLUTION   If $(x, y) \neq (0, 0)$, there is certainly an open disc S with center (x, y) which does not contain $(0, 0)$. Then since $f(s, t) = 2st/(s^2 + t^2)$ for every $(s, t) \in S$, by Theorem 18.5.2

$$D_1 f(x, y) = \frac{\partial}{\partial x}\left(\frac{2xy}{x^2 + y^2}\right) = \frac{(x^2 + y^2)2y - 2xy \cdot 2x}{(x^2 + y^2)^2}$$

$$= \frac{2y^3 - 2x^2 y}{(x^2 + y^2)^2} \quad \text{if } (x, y) \neq (0, 0).$$

$$D_2 f(x, y) = \frac{\partial}{\partial y}\left(\frac{2xy}{x^2 + y^2}\right) = \frac{(x^2 + y^2)2x - 2xy \cdot 2y}{(x^2 + y^2)^2}$$

$$= \frac{2x^3 - 2xy^2}{(x^2 + y^2)^2} \quad \text{if } (x, y) \neq (0, 0).$$

To compute the partial derivatives of f at $(0, 0)$, we use (18.28) and (18.29). Then

$$D_1 f(0, 0) = \lim_{h \to 0} \frac{f(h, 0) - f(0, 0)}{h} = \lim_{h \to 0} \frac{0 - 0}{h}$$

$$= 0.$$

$$D_2 f(0, 0) = \lim_{h \to 0} \frac{f(0, h) - f(0, 0)}{h} = \lim_{h \to 0} \frac{0 - 0}{h}$$

$$= 0.$$

It will be recalled from Example 3, Section 18.4 that the function f here is discontinuous at $(0, 0)$. Thus the existence of the partial derivatives of a function of two variables at a point does not guarantee the continuity of the function at the point.

We next introduce an important n-dimensional vector whose components are obtained from the partial derivatives of a function.

18.5.3   Definition

Suppose f is a function of n variables and $X \in \mathcal{D}_f$. Then the *gradient of* f *at* X, denoted by $\nabla f(X)$ is given by

$$\nabla f(X) = \langle D_1 f(X), D_2 f(X), \dots, D_n f(X) \rangle.$$

The symbol ∇ used here is read "del," and hence $\nabla f(X)$ can be read "del f at X." If $u = f(X)$, then the gradient of f at X is often denoted by ∇u.

Example 5   If $f(x, y, z) = x^2 + y^2 + xyz$, find $\nabla f(1, 2, -3)$.

SOLUTION   Since

$$D_1 f(x, y, z) = 2x + yz$$
$$D_2 f(x, y, z) = 2y + xz$$
$$D_3 f(x, y, z) = xy$$

we have

$$D_1 f(1, 2, -3) = -4 \qquad D_2 f(1, 2, -3) = 1 \qquad D_3 f(1, 2, -3) = 2$$

and hence $\nabla f(1, 2, -3) = \langle -4, 1, 2 \rangle$.

The usefulness of the gradient will be discussed in succeeding sections of this chapter and a geometric illustration of the vector will be given in Section 18.9. For the present we will only concern ourselves with the calculation of the vector.

Exercise Set 18.5

In Exercises 1–10 obtain $D_1 f(x, y)$ and $D_2 f(x, y)$

1. $f(x, y) = 3x^2 y^4 - 5xy^3$

2. $f(x, y) = \tan^{-1} \dfrac{y}{x}$

3. $f(x, y) = \sin^3 (e^{-xy})$

4. $f(x, y) = \sec^2 (2x + 3y)$

5. $f(x, y) = \ln \dfrac{x - y}{x + y}$

6. $f(x, y) = x^y$

7. $f(x, y) = \sin^{-1} \dfrac{y}{\sqrt{x^2 + y^2}}$

8. $f(x, y) = e^{x/y} \cos xy$

9. $f(x, y) = \displaystyle\int_x^y \sqrt{1 + t^3}\, dt$

10. $f(x, y) = \displaystyle\int_x^{2x+y} e^{t^2}\, dt$

In Exercises 11–16 find the partial derivatives $D_1 f$, $D_2 f$, and $D_3 f$.

11. $f(x, y, z) = x^3 y + 2xy^2 z - 3yz^3$

12. $f(\rho, \theta, \phi) = \rho \sin \phi \cos \theta$

13. $f(u, v, w) = uv \tan^{-1} \dfrac{w}{uv}$

14. $f(x, y, z) = \ln \sqrt{x^2 + y^2 + z^2}$

15. $f(x, y, z) = \dfrac{x}{\sqrt{x^2 + y^2 + z^2}}$

16. $f(r, s, t) = \dfrac{\sin (r - s + 2t)}{e^{st^2}}$

17. If $f(x, y) = \sqrt{x^2 + 2y^2}$ find the vectors $\nabla f(x, y)$ and $\nabla f(1, -2)$.

18. If $f(x, y) = y \sin xy$, find the vectors $\nabla f(x, y)$ and $\nabla f(\frac{\pi}{3}, 4)$.

19. If $f(x, y, z) = xy^2 + 2yz^2 - x^2 z$, find $\nabla f(x, y, z)$ and $\nabla f(-1, 1, 2)$.

20. If $g(x, y, z) = (x^2 + y^2 + z^2)^{1/2}$, find $\nabla g(x, y, z)$ and $\nabla g(-3, 4, 5)$.

21. Find the slope of the tangent to the intersection of the surface $z = x^2 + 2y^2$ and the plane $x = 2$ at the point $(2, -1, 6)$.

22. Find the slope of the tangent to the intersection of the surface $z = 16 - x^2 - y^2$ and the plane $y = 3$ at the point $(-2, 3, 3)$.

In Exercises 23–26 evaluate $D_1 f$ and $D_2 f$ at the given point using the definitions (18.28) and (18.29).

23. $f(x, y) = \dfrac{2x}{y^2}$, at $(2, -3)$

24. $f(x, y) = \ln(x + 2y)$, at $(1, 2)$

25. $f(x, y) = \begin{cases} \dfrac{x^2 y}{x^2 + y^2} & \text{if } (x, y) \neq (0, 0) \\ 0 & \text{if } (x, y) = (0, 0) \end{cases}$, at $(0, 0)$

26. $f(x, y) = \begin{cases} \dfrac{x^3 + y^3}{|x| + 2|y|} & \text{if } (x, y) \neq (0, 0) \\ 0 & \text{if } (x, y) = (0, 0) \end{cases}$, at $(0, 0)$

27. If $z = \tan^{-1} \dfrac{y}{x}$, prove that $x \dfrac{\partial z}{\partial x} + y \dfrac{\partial z}{\partial y} = 0$.

28. If $u = x^3 y - 2y^2 z^2 + xz^3$, prove that $x \dfrac{\partial u}{\partial x} + y \dfrac{\partial u}{\partial y} + z \dfrac{\partial u}{\partial z} = 4u$.

29. Are the partial derivatives for the function f in Exercise 25 continuous at $(0, 0)$? Why?

30. Prove Theorem 18.5.2.

18.6 Differentiability

In this section we will derive a formula for a function of two variables that is analogous to formula (6.40) for a function of one variable. This formula and its extension to function of n variables (Theorem 18.6.2) will prove useful in a variety of applications including the derivation of an approximation formula and a chain rule for partial differentiation.

We begin by assuming that S is an open disc with center (a, b) and that f, a function of two variables, has partial derivatives that exist on S and are continuous at (a, b). If $(a + h, b + k)$, where $h \neq 0 \neq k$, is some other point in S (Figure 18–12), then $(a, b + k)$ is also in S since its distance from (a, b) is no greater than

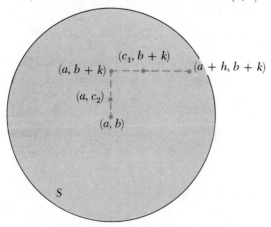

Figure 18–12

that of $(a + h, b + k)$. The difference $f(a + h, b + k) - f(a, b)$ will next be expressed in the form

$$f(a + h, b + k) - f(a, b) = [f(a + h, b + k) - f(a, b + k)]$$
$$+ [f(a, b + k) - f(a, b)]. \qquad (18.34)$$

If $y = b + k$ is fixed, then f is a function of x alone that satisfies the hypothesis of the mean value theorem on the closed interval with endpoints a and $a + h$. Hence, by this theorem there is a c_1 between a and $a + h$ such that

$$\frac{f(a + h, b + k) - f(a, b + k)}{(a + h) - a} = D_1 f(c_1, b + k). \qquad (18.35)$$

Also if $x = a$ is fixed, f is a function of y alone that satisfies the hypothesis of the mean value theorem on the closed interval with endpoints b and $b + k$. Thus there is a c_2 between b and $b + k$ such that

$$\frac{f(a, b + k) - f(a, b)}{(b + k) - b} = D_2 f(a, c_2). \qquad (18.36)$$

The differences in brackets in the right member of (18.34) can be rewritten using substitutions from (18.35) and (18.36), and therefore

$$f(a + h, b + k) - f(a, b) = D_1 f(c_1, b + k) \cdot h + D_2 f(a, c_2) \cdot k. \qquad (18.37)$$

The equation (18.37) will be modified if $h = 0$ or $k = 0$. If $h = 0$, then in place of (18.37) we obtain the equation $f(a, b + k) - f(a, b) = D_2 f(a, c_2) \cdot k$ while if $k = 0$ we have $f(a + h, b) - f(a, b) = D_1 f(c_1, b) \cdot h$.

Since the partials $D_1 f$ and $D_2 f$ are continuous at (a, b), intuition suggests that $D_1 f(c_1, b + k)$ and $D_2 f(a, c_2)$ are close to $D_1 f(a, b)$ and $D_2 f(a, b)$, respectively, when (h, k) is close to $(0, 0)$. Then for the functions η_1 and η_2 given by

$$\eta_1(h, k) = \begin{cases} D_1 f(c_1, b + k) - D_1 f(a, b) & \text{if } (h, k) \neq (0, 0) \\ 0 & \text{if } (h, k) = (0, 0) \end{cases}$$
$$\eta_2(h, k) = \begin{cases} D_2 f(a, c_2) - D_2 f(a, b) & \text{if } k \neq 0 \\ 0 & \text{if } k = 0 \end{cases} \qquad (18.38)$$

we shall obtain the limits

$$\lim_{(h,k) \to (0,0)} \eta_1(h, k) = 0 = \eta_1(0, 0) \qquad \lim_{(h,k) \to (0,0)} \eta_2(h, k) = 0 = \eta_2(0, 0). \qquad (18.39)$$

Since $D_1 f$ is continuous at (a, b), the first of the limits in (18.39) can be obtained by noting that for $\epsilon > 0$ there is a $\delta > 0$ such that

$$|D_1 f(x, y) - D_1 f(a, b)| < \epsilon \qquad (18.40)$$

when $\sqrt{(x - a)^2 + (y - b)^2} < \delta$. Suppose $\sqrt{h^2 + k^2} < \delta$. Then since c_1 is between a and $a + h$,

$$\sqrt{(c_1 - a)^2 + k^2} < \sqrt{h^2 + k^2} < \delta,$$

and therefore (18.40) is true when $(x, y) = (c_1 - a, b + k)$; that is,

$$|\eta_1(h, k)| = |D_1 f(c_1, b + k) - D_1 f(a, b)| < \epsilon.$$

Thus $\lim_{(h,k)\to(0,0)} \eta_1(h, k) = 0$. The proof of the other part of statement (18.39) is similar and is left as an exercise.

From (18.38) we can rewrite (18.37) in the following form for any h and k:

$$f(a + h, b + k) - f(a, b) = D_1 f(a, b)h + D_2 f(a, b)k + \eta_1(h, k)h + \eta_2(h, k)k.$$
$$(18.41)$$

Equation (18.41) is sometimes called an *increment formula* since h and k are spoken of as *increments* in a and b, respectively.

Therefore, we have proved the following theorem.

18.6.1   Theorem

Suppose f, a function of two variables, has partial derivatives that exist on some open disc S with center (a, b) and are continuous at (a, b). Then there exist functions η_1 and η_2 of two variables such that if $(a + h, b + k)$ is any point in S, (18.41) holds where η_1 and η_2 satisfy the equations (18.39).

Example 1   If $f(x, y) = x^2 y$, obtain functions η_1 and η_2 that satisfy (18.39) and (18.41).

SOLUTION   We note here that

$$D_1 f(x, y) = 2xy \qquad D_2 f(x, y) = x^2.$$

Then if (a, b) is any point in R^2,

$$
\begin{aligned}
f(a + h, b + k) - f(a, b) &- D_1 f(a, b)h - D_2 f(a, b)k \\
&= (a + h)^2(b + k) - a^2 b - 2abh - a^2 k \\
&= bh^2 + 2ahk + h^2 k.
\end{aligned}
$$
$$(18.42)$$

The right member of (18.42) can be expressed in the form

$$\underbrace{(bh + 2ak)h}_{\eta_1(h, k)} + \underbrace{h^2 k}_{\eta_2(h, k)}$$

where

$$\lim_{(h,k)\to(0,0)} \eta_1(h, k) = \lim_{(h,k)\to(0,0)} (bh + 2ak) = 0$$

and

$$\lim_{(h,k)\to(0,0)} \eta_2(h, k) = \lim_{(h,k)\to(0,0)} h^2 = 0.$$

It should be noted that there is no unique choice here for the functions η_1 and η_2. From (18.42), for example, we could alternatively have selected

$$\eta_1(h, k) = bh + ak \qquad \text{and} \qquad \eta_2(h, k) = ah + h^2.$$

Theorem 18.6.1 is extended in the next theorem. Its proof, being analogous to that for Theorem 18.6.1, is omitted.

18.6.2   Theorem

Suppose f, a function of n variables, has partial derivatives that are defined on some open ball S with center $A = (a_1, a_2, \ldots, a_n)$ and are continuous at A. Then there exist functions $\eta_1, \eta_2, \ldots, \eta_n$ of n variables such that if $(a_1 + h_1, a_2 + h_2, \ldots, a_n + h_n)$ is any point in S,

$$f(a_1 + h_1, \ldots, a_n + h_n) - f(A) = \sum_{i=1}^{n} D_i f(A) h_i + \sum_{i=1}^{n} \eta_i(h_1, h_2, \ldots, h_n) h_i \quad (18.43)$$

where

$$\lim_{(h_1, h_2, \ldots, h_n) \to (0, 0, \ldots, 0)} \eta_i(h_1, h_2, \ldots, h_n) = 0 = \eta_i(0, 0, \ldots, 0) \quad (18.44)$$

for $i = 1, 2, \ldots, n$.

We note that the first sum on the right in (18.43) can be rewritten using the notion of the gradient of f (Definition 18.5.3). Thus

$$f(a_1 + h_1, \ldots, a_n + h_n) - f(A) = \nabla f(A) \cdot \langle h_1, h_2, \ldots, h_n \rangle + \sum_{i=1}^{n} \eta_i(h_1, h_2, \ldots, h_n) h_i.$$

The conclusion of Theorem 18.6.2 leads us to an important definition.

18.6.3   Definition

Suppose f is a function of n variables and $A \in R^n$. Then f is *differentiable* at A if and only if there exist functions $\eta_1, \eta_2, \ldots, \eta_n$ such that (18.43) and (18.44) hold for any point $(a_1 + h_1, \ldots, a_n + h_n) \in \mathcal{D}_f$.

In Example 1 we showed that the function f given by $f(x, y) = x^2 y$ is differentiable at any point (a, b) in R^2. In fact, by a restatement of Theorem 18.6.2, *if a function f has partial derivatives on an open ball which are continuous at its center, then f is differentiable at the center.* Thus the usual function that we encounter in this text is differentiable at most points, if not all, in its domain.

The reader will be asked to prove below that if f is a function of one variable, then Definition 18.6.3 is equivalent to the definition of differentiability given just after Definition 4.3.1. This connection between functions of n variables and functions of one variable is further emphasized by the following extension of Theorem 4.3.2.

18.6.4   Theorem

If f, a function of n variables is differentiable at A, then f is continuous at A.

PROOF   By hypothesis there exist functions $\eta_1, \eta_2, \ldots, \eta_n$ of n variables such that (18.43) holds for any point $(a_1 + h_1, a_2 + h_2, \ldots, a_n + h_n) \in \mathcal{D}_f$ and

(18.44) is satisfied. Then from taking the limits in each member of (18.43) at $(h_1, h_2, \ldots, h_n) = (0, 0, \ldots, 0)$, we have

$$\lim_{(h_1, h_2, \ldots, h_n) \to (0, 0, \ldots, 0)} [f(a_1 + h_1, \ldots, a_n + h_n) - f(A)] = 0.$$

Hence

$$\lim_{(h_1, h_2, \ldots, h_n) \to (0, 0, \ldots, 0)} f(a_1 + h_1, \ldots, a_n + h_n) = f(A),$$

and the conclusion follows by Theorem 18.4.2.

Exercise Set 18.6

In Exercises 1 and 2 obtain functions η_1 and η_2 that satisfy (18.41) and (18.39).

1. $f(x, y) = x^2 + xy$

2. $f(x, y) = x^2/y$ if $(a, b) \neq (0, 0)$

3. Give the proof that $\lim_{(h, k) \to (0, 0)} \eta_2(h, k) = 0 = \eta_2(0, 0)$, which is needed in proving Theorem 18.6.1.

4. Prove the assertion made after Definition 18.6.3 about the equivalence of two definitions of differentiability for functions of one variable. HINT: Show that either definition implies the other.

18.7 Approximations; Differentials

It will be recalled from Section 6.6 that if f is a function of one variable and $c \in \mathcal{D}_f$, then

$$f(c + h) \approx f(c) + f'(c)h. \tag{18.45}$$

This approximation formula was justified by (6.41) and (6.42), which together are equivalent to the statement

$$\lim_{h \to 0} \frac{f(c + h) - f(c) - f'(c)h}{h} = 0. \tag{18.46}$$

In this section we shall obtain an extension of (18.45) for functions of n variables. Suppose f is such a function and is differentiable at $A = (a_1, a_2, \ldots, a_n) \in R^n$. Then for $(a_1 + h_1, \ldots, a_n + h_n) \in \mathcal{D}_f$

$$f(a_1 + h_1, \ldots, a_n + h_n) = f(A) + \sum_{i=1}^{n} D_i f(A) h_i + \sum_{i=1}^{n} \eta_1(h_1, \ldots, h_n) h_i \tag{18.47}$$

where

$$\lim_{(h_1, \ldots, h_n) \to (0, \ldots, 0)} \eta_i(h_1, \ldots, h_n) = \eta_i(0, \ldots, 0) = 0. \tag{18.48}$$

From (18.47) if we divide by $\sqrt{h_1^2 + \cdots + h_n^2}$,

$$\frac{f(a_1 + h_1, \ldots, a_n + h_n) - f(A) - \sum_{i=1}^{n} D_i f(A) h_i}{\sqrt{h_1^2 + \cdots + h_n^2}}$$

$$= \sum_{i=1}^{n} \eta_i(h_1, \ldots, h_n) \frac{h_i}{\sqrt{h_1^2 + \cdots + h_n^2}}. \qquad (18.49)$$

Since

$$\frac{|h_i|}{\sqrt{h_1^2 + \cdots + h_n^2}} \le 1 \quad \text{for } i = 1, 2, \ldots, n,$$

we have for every i

$$\lim_{(h_1, \ldots, h_n) \to (0, \ldots, 0)} \eta_i(h_1, \ldots, h_n) \frac{h_i}{\sqrt{h_1^2 + \cdots + h_n^2}} = 0.$$

Hence from (18.49)

$$\lim_{(h_1, \ldots, h_n) \to (0, \ldots, 0)} \frac{f(a_1 + h_1, \ldots, a_n + h_n) - f(A) - \sum_{i=1}^{n} D_i f(A) h_i}{\sqrt{h_1^2 + \cdots + h_n^2}} = 0.$$
$$(18.50)$$

Because of the limit (18.50) $f(a_1 + h_1, \ldots, a_n + h_n) - f(A) - \sum_{i=1}^{n} D_i f(A) h_i$ approaches 0 much more rapidly than $\sqrt{h_1^2 + \cdots + h_n^2}$ as $(h_1, \ldots, h_n)$ approaches $(0, \ldots, 0)$. Therefore $f(A) + \sum_{i=1}^{n} D_i f(A) h_i$ is considered a "good approximation" of $f(a_1 + h_1, \ldots, a_n + h_n)$ when $(h_1, \ldots, h_n)$ is sufficiently close to $(0, \ldots, 0)$, and we shall write

$$f(a_1 + h_1, \ldots, a_n + h_n) \approx f(A) + \sum_{i=1}^{n} D_i f(A) h_i. \qquad (18.51)$$

Using the notion of the gradient of f (see Definition 18.5.3) and the dot product of two vectors, we may write

$$f(a_1 + h_1, \ldots, a_n + h_n) \approx f(A) + \nabla f(A) \cdot \langle h_1, h_2, \ldots, h_n \rangle.$$

The approximation formula (18.51) is the extension of (18.45) which we were seeking.

For a function f of two variables which is differentiable at (a, b) the approximation formula (18.51) becomes

$$f(a + h, b + k) \approx f(a, b) + D_1 f(a, b) h + D_2 f(a, b) k. \qquad (18.52)$$

Formula (18.52) can be used to approximate the value of a function at a point if the function is differentiable at a nearby point.

Example 1 If $f(x, y) = x^2/y$, approximate $f(-5.03, 2.02)$.

SOLUTION Here f is differentiable at $(-5, 2)$ and

$$D_1 f(x, y) = \frac{2x}{y} \qquad D_2 f(x, y) = -\frac{x^2}{y^2} \quad \text{for } y \ne 0.$$

Since the point $(-5.03, 2.02)$ is near $(-5, 2)$, we let

$$a = -5 \qquad b = 2 \qquad h = -0.03 \qquad k = 0.02.$$

Then from (18.52)

$$f(-5.03, 2.02) \approx f(-5, 2) + D_1 f(-5, 2)(-0.03) + D_2 f(-5, 2)(0.02);$$

that is,

$$f(-5.03, 2.02) \approx 12.5 + (-5)(-0.03) + (-6.25)(0.02)$$
$$\approx 12.525.$$

The next example utilizes (18.51) for a function of three variables.

Example 2 The force of attraction F between two particles with masses m_1 and m_2 which are a distance r apart is given by $F = Gm_1m_2/r^2$ where G is constant. If the maximum errors are 1% in the measurements of each mass and 2% in the measurement of the distance between the particles, what is the approximate maximum percentage error in measuring the force F?

SOLUTION Let $F = f(m_1, m_2, r)$ and h_1, h_2, and h_3 be the increments in the variables m_1, m_2, and r. Since f is differentiable at any point (m_1, m_2, r) except where $r = 0$, by (18.51):

$$f(m_1 + h_1, m_2 + h_2, r + h_3) - f(m_1, m_2, r) \approx \frac{\partial f}{\partial m_1}h_1 + \frac{\partial f}{\partial m_2}h_2 + \frac{\partial f}{\partial m_3}h_3$$

$$\approx \frac{Gm_2}{r^2}h_1 + \frac{Gm_1}{r^2}h_2 - \frac{2Gm_1m_2}{r^3}h_3.$$

$$(18.53)$$

The given maximum errors in m_1, m_2, and r are, respectively,

$$h_1 = \pm 0.01m_1 \qquad h_2 = \pm 0.01m_2 \qquad h_3 = \pm 0.02r. \qquad (18.54)$$

Because of the signs of the terms in the right member of (18.53), the error in F, which is given by the left member of (18.53), is approximately a maximum when we choose from (18.54)

$$h_1 = 0.01m_1 \qquad h_2 = 0.01m_2 \qquad h_3 = -0.02r. \qquad (18.55)$$

Then from (18.55) and (18.53)

$$f(m_1 + h_1, m_2 + h_2, r + h_3) - f(m_1, m_2, r)$$

$$\approx \frac{0.01Gm_1m_2}{r^2} + \frac{0.01Gm_1m_2}{r^2} + \frac{0.04Gm_1m_2}{r^2}$$

$$\approx \frac{0.06Gm_1m_2}{r^2} = 0.06F.$$

Thus the approximate maximum percentage error in measuring F is 6%.

We also mention another, somewhat old-fashioned, notation that is sometimes used in approximations of this type. If the function f of n variables is given by

$$w = f(x_1, x_2, \ldots, x_n) \tag{18.56}$$

and $h_1, h_2, \ldots, h_n$ are arbitrary numbers, then dw, the *differential* of w, is defined at any point common to the domains of $D_1 f, D_2 f, \ldots, D_n f$ by the equation

$$dw = D_1 f(X) h_1 + D_2 f(X) h_2 + \cdots + D_n f(X) h_n. \tag{18.57}$$

From (18.57) it is seen that dw is a function of the $2n$ variables $x_1, x_2, \ldots, x_n$, $h_1, h_2, \ldots, h_n$. The notion of the differential dw here is the extension of the idea of the differential of a dependent variable which was presented in Section 6.7. When w is defined by (18.56), we also define Δw as a function of $x_1, x_2, \ldots, x_n, h_1, h_2, \ldots, h_n$ by the equation

$$\Delta w = f(x_1 + h_1, x_2 + h_2, \ldots, x_n + h_n) - f(x_1, x_2, \ldots, x_n). \tag{18.58}$$

Thus from (18.51) with $x_1, x_2, \ldots, x_n$ in place of $a_1, a_2, \ldots, a_n$,

$$\Delta w \approx dw. \tag{18.59}$$

It is also convenient to define the differentials of the variables $x_1, x_2, \ldots, x_n$ by letting

$$h_i = dx_i \quad \text{for } i = 1, 2, \ldots, n. \tag{18.60}$$

Then (18.57) can be rewritten

$$dw = \sum_{i=1}^{n} D_i f(X) \, dx_i \tag{18.61}$$

or

$$dw = \nabla f(X) \cdot \langle dx_1, dx_2, \ldots, dx_n \rangle.$$

The following example where f is a function of two variables illustrates our remarks here.

Example 3 If $w = f(x, y) = x^2 - xy$, (a) express dw and Δw in terms of x, y, dx, and dy, and (b) obtain dw and Δw when $x = 2$, $y = 3$, $dx = 0.02$, and $dy = -0.01$.

SOLUTION From (18.58) and (18.60)

$$\begin{aligned}
\Delta w &= f(x + dx, y + dy) - f(x, y) \\
&= (x + dx)^2 - (x + dx)(y + dy) - x^2 + xy \\
&= (2x - y) \, dx - x \, dy + (dx)^2 - dx \, dy.
\end{aligned}$$

Then if $x = 2$, $y = 3$, $dx = 0.02$, and $dy = -0.01$,

$$\begin{aligned}
\Delta w &= (2 \cdot 2 - 3)(0.02) - 2(-0.01) + (0.02)^2 - (0.02)(-0.01) \\
&= 0.0406.
\end{aligned}$$

From (18.61)

$$dw = \frac{\partial f}{\partial x}\, dx + \frac{\partial f}{\partial y}\, dy$$

$$= (2x - y)\, dx - x\, dy.$$

Then if $x = 2$, $y = 3$, $dx = 0.02$, and $dy = -0.01$,

$$dw = (2 \cdot 2 - 3)(0.02) - 2(-0.01)$$

$$= 0.04.$$

Exercise Set 18.7

In Exercises 1–4 (a) express Δw and dw in terms of x, y, dx, and dy, and (b) obtain Δw and dw for the given values of x, y, dx, and dy.

1. $w = 2xy^2$; $x = 3, y = -2, dx = -0.01, dy = 0.03$

2. $w = \dfrac{x - y}{x + y}$; $x = 5, y = 2, dx = 0.02, dy = 0.01$

3. $w = e^{-xy}$; $x = 1.5, y = 0.6, dx = 0.03, dy = -0.02$

4. $w = \ln \dfrac{y}{x}$; $x = 3, y = 4, dx = -0.03, dy = 0.02$

5. Approximate $\sqrt{(2.97)^2 + (4.02)^2}$.

6. Approximate $\sin 46° \cos 43°$.

7. Find the approximate maximum error in the volume of a right circular cone if its altitude and base are respectively 15 in. and 20 in. with a possible error of 0.04 in.

8. The effective resistance R of two resistances R_1 and R_2 that are connected in parallel is given by $R = R_1 R_2/(R_1 + R_2)$. If R_1 and R_2 are, respectively, 10 ohms and 20 ohms and there are possible errors of 0.02 ohm and 0.05 ohm, respectively, in measuring these resistances, find the approximate maximum error in R.

9. Find the approximate maximum error in obtaining the volume of a rectangular box with dimensions 15 in. $\times$ 12 in. $\times$ 10 in. if the maximum error in the dimensions is $\frac{1}{4}$ in.

10. The possible errors in measuring the sides and angles of a certain triangle are 0.1 in. and $0.5°$, respectively. Find the approximate maximum error in the area of the triangle if the measurements of the lengths of two sides are 12 in. and 18 in. and the angle between the sides is $30°$.

18.8   General Chain Rule

If f and g are functions of one variable which are related by the equations

$$y = f(x) \qquad x = g(t)$$

and $g'(t)$ and $f'(g(t))$ exist, then by the chain rule (Theorem 4.5.1) the derivative of the composite function $F = f \circ g$ is given by

$$F'(t) = f'(g(t))g'(t), \tag{18.62}$$

or equivalently

$$\frac{dy}{dt} = \frac{dy}{dx}\frac{dx}{dt}.$$

Suppose next that $g_1, g_2, \ldots, g_m$ are each functions of one variable defined by the equations

$$x_i = g_i(t) \quad i = 1, 2, \ldots, m \tag{18.63}$$

and f is a function of m variables given by

$$w = f(x_1, x_2, \ldots, x_m). \tag{18.64}$$

The derivative of the composite function F where

$$F(t) = f(g_1(t), g_2(t), \ldots, g_m(t))$$

is given by the following extension of the chain rule.

18.8.1   Theorem

Suppose that

(i) $g_1, g_2, \ldots, g_m$, which are each functions of one variable, are differentiable at t;

(ii) f, a function of m variables, is differentiable at

$$G(t) = (g_1(t), g_2(t), \ldots, g_m(t));$$

(iii) the composite function F is defined by

$$F(t) = f(g_1(t), g_2(t), \ldots, g_m(t)).$$

Then the derivative of F at t is given by

$$F'(t) = \sum_{k=1}^{m} D_k f(g_1(t), g_2(t), \ldots, g_m(t))g_k'(t) \tag{18.65}$$

or in vector notation

$$F'(t) = \nabla f(g_1(t), g_2(t), \ldots, g_m(t)) \cdot \langle g_1'(t), g_2'(t), \ldots, g_m'(t) \rangle. \tag{18.66}$$

PROOF   From hypothesis (ii) there exist functions $\eta_1, \eta_2, \ldots, \eta_m$ of n variables such that if

$$G(t + h) = (g_1(t + h), g_2(t + h), \ldots, g_m(t + h)) \in \mathcal{D}_f,$$

then

$$f(G(t + h)) - f(G(t)) = \sum_{k=1}^{m} D_k f(G(t))[g_k(t + h) - g_k(t)]$$

$$+ \sum_{k=1}^{m} \eta_k(g_1(t + h) - g_1(t), \ldots, g_m(t + h) - g_m(t)) \cdot [g_k(t + h) - g_k(t)] \quad (18.67)$$

where for $k = 1, 2, \ldots, m$

$$\lim_{(h_1, h_2, \ldots, h_m) \to (0,0,\ldots,0)} \eta_k(h_1, h_2, \ldots, h_m) = 0 = \eta_k(0, 0, \ldots, 0). \quad (18.68)$$

Since each function g_k is differentiable at t, by Theorem 4.3.2 each g_k is also continuous at t and therefore

$$\lim_{h \to 0} [g_k(t + h) - g_k(t)] = 0 \quad k = 1, 2, \ldots, m. \quad (18.69)$$

From (18.68) each η_k is continuous at $(0, 0, \ldots, 0) \in R^m$, and therefore from (18.68) and (18.69) using Theorem 18.4.4

$$\lim_{h \to 0} \eta_k(g_1(t + h) - g_1(t), \ldots, g_m(t + h) - g_m(t))$$

$$= \eta_k \left(\lim_{h \to 0} [g_1(t + h) - g_1(t)], \ldots, \lim_{h \to 0} [g_m(t + h) - g_m(t)] \right)$$

$$= \eta_k(0, 0, \ldots, 0) = 0 \quad \text{for } k = 1, 2, \ldots, m. \quad (18.70)$$

Dividing each member of (18.67) by h gives

$$\frac{f(G(t + h)) - f(G(t))}{h} = \sum_{k=1}^{m} D_k f(G(t)) \frac{g_k(t + h) - g_k(t)}{h}$$

$$+ \sum_{k=1}^{m} \eta_k(g_1(t + h) - g_1(t), \ldots, g_m(t + h) - g_m(t)) \frac{g_k(t + h) - g_k(t)}{h}. \quad (18.71)$$

Invoking (18.70) to obtain the limit in (18.71) as $h \to 0$, we have

$$F'(t) = \lim_{h \to 0} \frac{f(G(t + h)) - f(G(t))}{h} = \sum_{k=1}^{m} D_k f(G(t)) g_k'(t) + \sum_{k=1}^{m} 0 \cdot g_k'(t)$$

$$= \sum_{k=1}^{m} D_k f(g_1(t), g_2(t), \ldots, g_m(t)) g_k'(t).$$

Example 1   Suppose $F(t) = f(t^2, t^3, 2t)$. Find $F'(t)$ if f is differentiable at $(t^2, t^3, 2t)$.

SOLUTION   Using the notation in the statement of Theorem 18.8.1

$$g_1(t) = t^2 \qquad g_2(t) = t^3 \qquad g_3(t) = 2t,$$

and hence

$$g_1'(t) = 2t \qquad g_2'(t) = 3t^2 \qquad g_3'(t) = 2.$$

Then from (18.66)

$$F'(t) = \nabla f(t^2, t^3, 2t) \cdot \langle 2t, 3t^2, 2 \rangle$$
$$= 2t\, D_1 f(t^2, t^3, 2t) + 3t^2\, D_2 f(t^2, t^3, 2t) + 2\, D_3 f(t^2, t^3, 2t).$$

If the functions f, g_1, g_2, . . . , and g_m which satisfy the hypotheses of Theorem 18.8.1 are related by the equations (18.63) and (18.64), then $F'(t) = dw/dt$ is expressed by

$$\frac{dw}{dt} = \frac{\partial w}{\partial x_1}\frac{dx_1}{dt} + \frac{\partial w}{\partial x_2}\frac{dx_2}{dt} + \cdots + \frac{\partial w}{\partial x_n}\frac{dx_n}{dt}. \tag{18.72}$$

If f in (ii) of Theorem 18.8.1 is a function of one variable, then $\partial w/\partial x_1$ in (18.72) is replaced by dw/dx_1, $D_k f(g_1(t), g_2(t), \ldots, g_n(t))$ becomes merely $f'(g_1(t))$, and we have the familiar chain rule equations

$$F'(t) = f'(g_1(t))g_1'(t) \qquad \text{or} \qquad \frac{dw}{dt} = \frac{dw}{dx_1}\frac{dx_1}{dt}.$$

Thus the chain rule formula for functions of one variable can be viewed as a special case of (18.65) or (18.72).

In the next example the same notation as in (18.72) is used to express the derivative of a composite function.

Example 2 The altitude of a right circular cylinder is increasing at the rate of 4 in./sec while the radius is decreasing at 3 in./sec. At a certain instant of time the altitude and radius of the cylinder are 12 ft and 15 ft respectively. Find the rate of change of the volume of the cylinder with respect to time at that instant.

SOLUTION The volume of the cylinder is given by $V = \pi r^2 h$ where r and h denote the base radius and altitude, respectively. Since r and h are differentiable functions of time t, V is also a differentiable function of t, and from (18.72)

$$\frac{dV}{dt} = \frac{\partial V}{\partial r}\frac{dr}{dt} + \frac{\partial V}{\partial h}\frac{dh}{dt}$$
$$= 2\pi r h\frac{dr}{dt} + \pi r^2\frac{dh}{dt}. \tag{18.73}$$

Since $dh/dt = 4$ in./sec $= \frac{1}{3}$ ft/sec and $dr/dt = -3$ in./sec $= -\frac{1}{4}$ ft/sec, from (18.73) when $h = 12$ ft and $r = 15$ ft

$$\frac{dV}{dt} = 2\pi \cdot 12 \cdot 15\left(-\frac{1}{4}\right) + \pi \cdot 15^2 \cdot \frac{1}{3} = -15\pi \ \text{ft}^3/\text{sec}.$$

Thus the volume of the cylinder is decreasing at 15π ft³/sec.

If g_1, g_2, . . . , g_m are each functions of n variables and f is a function of m variables, we shall obtain the partial derivatives $D_k F$ of the composite function F defined by

$$F(t_1, t_2, \ldots, t_n) = f(g_1(t_1, t_2, \ldots, t_n), g_2(t_1, t_2, \ldots, t_n), \ldots, g_m(t_1, t_2, \ldots, t_n))$$
$$(18.74)$$

using the following *general chain rule*.

18.8.2   Theorem (General Chain Rule)

Suppose that

(i) $g_1, g_2, \ldots, g_m$ *are each functions of n variables which are differentiable at* $(t_1, t_2, \ldots, t_n)$;

(ii) f *is a function of m variables which is differentiable at* $(g_1(t_1, \ldots, t_n), g_2(t_1, \ldots, t_n), \ldots, g_m(t_1, \ldots, t_n))$.

Then the partial derivatives at $(t_1, t_2, \ldots, t_n)$ *of the function F in (18.74) are given by*

$$D_i F(t_1, t_2, \ldots, t_n) = \sum_{k=1}^{m} D_k f(g_1(t_1, \ldots, t_n), \ldots, g_m(t_1, \ldots, t_n)) D_i g_k(t_1, \ldots, t_n)$$
$$for\ i = 1, 2, \ldots, n. \qquad (18.75)$$

The same steps can be used in the proof of this theorem as were given in the proof of Theorem 18.8.1 except that in place of t and $t + h$, we now have

$$T = (t_1, t_2, \ldots, t_n)$$

and

$$P = (t_1, t_2, \ldots, t_{i-1}, t_i + h, t_{i+1}, \ldots, t_n),$$

respectively. Then in place of $G(t)$ and $G(t + h)$, we will have

$$G(T) = (g_1(T), g_2(T), \ldots, g_m(T))$$

and

$$G(P) = (g_1(P), g_2(P), \ldots, g_m(P)),$$

respectively. The details of the proof are left for the interested reader.

Suppose the functions $f, g_1, g_2, \ldots,$ and g_m that satisfy the hypotheses of Theorem 18.8.2 are related by the equations

$$w = f(x_1, x_2, \ldots, x_m) \qquad (18.76)$$

and

$$x_k = g_k(t_1, t_2, \ldots, t_n) \quad \text{when } k = 1, 2, \ldots, m. \qquad (18.77)$$

Then (18.75) is equivalent to the system of equations

$$\frac{\partial w}{\partial t_1} = \frac{\partial w}{\partial x_1}\frac{\partial x_1}{\partial t_1} + \frac{\partial w}{\partial x_2}\frac{\partial x_2}{\partial t_1} + \cdots + \frac{\partial w}{\partial x_m}\frac{\partial x_m}{\partial t_1}$$

$$\frac{\partial w}{\partial t_2} = \frac{\partial w}{\partial x_1}\frac{\partial x_1}{\partial t_2} + \frac{\partial w}{\partial x_1}\frac{\partial x_2}{\partial t_2} + \cdots + \frac{\partial w}{\partial x_m}\frac{\partial x_m}{\partial t_2}$$

$$\cdots$$

$$\frac{\partial w}{\partial t_n} = \frac{\partial w}{\partial x_1}\frac{\partial x_1}{\partial t_n} + \frac{\partial w}{\partial x_2}\frac{\partial x_2}{\partial t_n} + \cdots + \frac{\partial w}{\partial x_m}\frac{\partial x_m}{\partial t_n}$$

that can be expressed in the condensed form

$$\frac{\partial w}{\partial t_i} = \sum_{k=1}^{m} \frac{\partial w}{\partial x_k}\frac{\partial x_k}{\partial t_i} \quad \text{for } i = 1, 2, \ldots, n. \tag{18.78}$$

It will be noted that for each of the *intermediate* variables $x_1, x_2, \ldots, x_m$ there is a corresponding term $\dfrac{\partial w}{\partial x_k}\dfrac{\partial x_k}{\partial t_i}$ in the sum (18.78).

If f is a function of one variable, (18.74) simplifies to

$$F(t_1, t_2, \ldots, t_n) = f(g_1(t_1, t_2, \ldots, t_n)).$$

Then in place of (18.75) and (18.78), respectively, we have

$$D_i F(t_1, t_2, \ldots, t_n) = f'(g_1(t_1, t_2, \ldots, t_n)) D_i g_1(t_1, t_2, \ldots, t_n)$$

and

$$\frac{\partial w}{\partial t_i} = \frac{dw}{dx_1}\frac{\partial x_1}{\partial t_i}.$$

The next example illustrates the equations (18.78).

Example 3 If $w = x^2 + 2y^2 - 3z$, and $x = 2s + \cos t$, $y = s - \sin t$, and $z = 2t$, find $\partial w/\partial s$ and $\partial w/\partial t$.

SOLUTION By the general chain rule

$$\frac{\partial w}{\partial s} = \frac{\partial w}{\partial x}\frac{\partial x}{\partial s} + \frac{\partial w}{\partial y}\frac{\partial y}{\partial s} + \frac{\partial w}{\partial z}\frac{\partial z}{\partial s}$$

$$= 2x \cdot 2 + 4y \cdot 1 + (-3) \cdot 0$$

$$= 4x + 4y = 4(2s + \cos t) + 4(s - \sin t)$$

$$= 12s + 4\cos t - 4\sin t$$

$$\frac{\partial w}{\partial t} = \frac{\partial w}{\partial x}\frac{\partial x}{\partial t} + \frac{\partial w}{\partial y}\frac{\partial y}{\partial t} + \frac{\partial w}{\partial z}\frac{\partial z}{\partial t}$$

$$= 2x(-\sin t) + 4y(-\cos t) + (-3)(2)$$

$$= -2x\sin t - 4y\cos t - 6$$

$$= -2(2s + \cos t)\sin t - 4(s - \sin t)\cos t - 6$$

$$= -4s\sin t - 4s\cos t + 2\sin t\cos t - 6$$

The results obtained here will agree with those obtained if w is first expressed in terms of s and t and $\partial w/\partial s$ and $\partial w/\partial t$ are obtained directly by partial differentiation. To show this, we first write

$$w = (2s + \cos t)^2 + 2(s - \sin t)^2 - 6t.$$

Then

$$\frac{\partial w}{\partial s} = 2(2s + \cos t) \cdot 2 + 4(s - \sin t)$$

$$= 12s + 4 \cos t - 4 \sin t$$

and

$$\frac{\partial w}{\partial t} = 2(2s + \cos t)(-\sin t) + 4(s - \sin t)(-\cos t) - 6$$

$$= -4s \sin t - 4s \cos t + 2 \sin t \cos t - 6.$$

Example 4 If $F(x, y) = f\left(\dfrac{x}{y}, \dfrac{x - y}{x}\right)$ where f is differentiable on its domain, prove that $x\dfrac{\partial F}{\partial x} + y\dfrac{\partial F}{\partial y} = 0$.

SOLUTION For convenience we introduce the variables u and v by letting $u = x/y$ and $v = (x - y)/x$. Then by the general chain rule

$$\frac{\partial F}{\partial x} = \frac{\partial f}{\partial u}\frac{\partial u}{\partial x} + \frac{\partial f}{\partial v}\frac{\partial v}{\partial x} = \frac{1}{y}\frac{\partial f}{\partial u} + \frac{y}{x^2}\frac{\partial f}{\partial v}$$

$$\frac{\partial F}{\partial y} = \frac{\partial f}{\partial u}\frac{\partial u}{\partial y} + \frac{\partial f}{\partial v}\frac{\partial v}{\partial y} = -\frac{x}{y^2}\frac{\partial f}{\partial u} - \frac{1}{x}\frac{\partial f}{\partial v}.$$

Hence

$$x\frac{\partial F}{\partial x} + y\frac{\partial F}{\partial y} = x\left(\frac{1}{y}\frac{\partial f}{\partial u} + \frac{y}{x^2}\frac{\partial f}{\partial v}\right) + y\left(-\frac{x}{y^2}\frac{\partial f}{\partial u} - \frac{1}{x}\frac{\partial f}{\partial v}\right)$$

$$= 0.$$

Exercise Set 18.8

In Exercises 1–10 obtain the indicated partial derivatives using a chain rule.

1. $w = x^2 - 2xy - y^2$, $x = 2t$, $y = -t^2$; $\dfrac{dw}{dt}$

2. $w = \sqrt{x^2 + y^2}$, $x = 3 \cos t$, $y = 3 \sin t$; $\dfrac{dw}{dt}$

3. $w = (x^3 - y^2z)^3$, $x = t$, $y = \frac{1}{2}t$, $z = 2t$; $\dfrac{dw}{dt}$

4. $w = \dfrac{b + y}{z^2}$, $x = t$, $y = t^2$, $z = t^3$; $\dfrac{dw}{dt}$

5. $w = \ln x^2 y$, $\quad x = 2s + t$, $\quad y = 3s - 2t$; $\quad \dfrac{\partial w}{\partial s}$, $\dfrac{\partial w}{\partial t}$

6. $z = e^{x^2 - 2y^2}$, $\quad x = 2uv$, $\quad y = u^2 - v^2$; $\quad \dfrac{\partial z}{\partial u}$, $\dfrac{\partial z}{\partial v}$

7. $z = \dfrac{2xy}{x^2 + y^2}$, $\quad x = \rho \sin \phi \cos \theta$, $\quad y = \rho \sin \phi \sin \theta$; $\quad \dfrac{\partial z}{\partial \rho}$, $\dfrac{\partial z}{\partial \phi}$, $\dfrac{\partial z}{\partial \theta}$

8. $w = e^{xy} \cos (x + y)$, $\quad x = s + 2t - u$, $\quad y = 3s - t + 2u$; $\quad \dfrac{\partial w}{\partial s}$, $\dfrac{\partial w}{\partial t}$, $\dfrac{\partial w}{\partial u}$

9. $u = \sqrt{x^2 + y^2 + z^2}$, $\quad x = e^{-s} \cos t$, $\quad y = e^{-s} \sin t$, $\quad z = e^{-s}$; $\quad \dfrac{\partial u}{\partial s}$, $\dfrac{\partial u}{\partial t}$

10. $w = \sin^{-1} \dfrac{xy}{z}$, $\quad x = u + v$, $\quad y = 2v$, $\quad z = u - v$; $\quad \dfrac{\partial w}{\partial u}$, $\dfrac{\partial w}{\partial v}$

11. From Exercise 2 obtain dw/dt when $t = \frac{\pi}{3}$.

12. From Exercise 5 obtain $\partial w/\partial s$ and $\partial w/\partial t$ when $s = 2$ and $t = 1$.

13. From Exercise 7 obtain $\partial z/\partial \rho$, $\partial z/\partial \phi$, and $\partial z/\partial \theta$ when $\rho = 2$, $\phi = \frac{\pi}{4}$ and $\theta = \frac{2\pi}{3}$.

14. From Exercise 9, obtain $\partial u/\partial s$ and $\partial u/\partial t$ when $s = 0$ and $t = \frac{3\pi}{4}$.

15. Find $D_2 f(s + \sin t, s - \cos t)$ if $f(x, y) = xy \sqrt{x^2 + y^2}$.

16. Find $D_1 g(e^u - e^v, e^u + e^v)$ if $g(x, y) = \dfrac{x^2 - y^2}{x^2 + y^2}$.

17. Find $D_1 f(u, v)$, $D_2 f(u, v)$, $D_1 g(u, v)$, and $D_2 g(u, v)$ when these partials exist if

$$3u - v^2 + 2f(u, v) - 3g(u, v) = 0$$
$$2 + v^2 - u - f(u, v) + 2g(u, v) = 0.$$

18. If $w = f(x^2 + y^2)$ where f is a differentiable function, prove that

$$y \dfrac{\partial w}{\partial x} - x \dfrac{\partial w}{\partial y} = 0.$$

19. If $w = f(y/x)$ where f is a differentiable function, prove that

$$x \dfrac{\partial w}{\partial x} + y \dfrac{\partial w}{\partial y} = 0.$$

20. If $u = f(x, y)$ where f is a differentiable function and $x = r \cos \theta$, $y = r \sin \theta$, prove that

$$\left(\dfrac{\partial u}{\partial r} \right)^2 + \dfrac{1}{r^2} \left(\dfrac{\partial u}{\partial \theta} \right)^2 = \left(\dfrac{\partial u}{\partial x} \right)^2 + \left(\dfrac{\partial u}{\partial y} \right)^2.$$

21. A particle moves in a plane in accordance with the equations $x = t^2$, $y = 1 - t$ while a second particle moves in accordance with the equations $x = 2t$, $y = t + 2$, where x and y are in feet and t is in seconds. Find the rate of change of the distance between the particles with respect to t when $t = 4$.

22. A particle moves in space in accordance with the equations $x = t^2/2$, $y = 2t$, $z = t$, where x, y, and z are in feet and t is in seconds. If the temperature in

space is given by $T = x^2 + y^2 + z^2 - xz - yz$, find dT/dt for the particle when $t = 3$.

23. At a certain instant the radius of the base of a right circular cone is 12 in. and is increasing at $\frac{1}{2}$ in./sec while the altitude of the cone is 16 in. and is decreasing at $\frac{2}{3}$ in./sec. Find the rate of change of the lateral area of the cone with respect to time at that instant.

24. The power P in watts expended by a resistor of R ohms with a current of I amps is given by $P = I^2R$. At a certain instant of time $I = 40$ amps and is increasing at 0.5 amp/sec while $R = 30$ ohms and is decreasing at 0.2 ohm/sec. Find the rate of change of P with respect to time at that instant.

25. Prove the mean value theorem for functions of two variables: Suppose that $A(a_1, a_2)$ and $B(b_1, b_2)$ are any points in R^2 and S is an open convex set in R^2 which contains these points. If f, a function of two variables, is differentiable on S, then there is some point (c_1, c_2) in the line segment AB such that

$$f(b_1, b_2) - f(a_1, a_2) = \nabla f(c_1, c_2) \cdot \langle b_1 - a_1, b_2 - a_2 \rangle.$$

HINT: Let $g(t) = f(a_1 + (b_1 - a_1)t, a_2 + (b_2 - a_2)t)$ and show that g satisfies the hypotheses of the mean value theorem on $[0, 1]$.

18.9   Directional Derivative

Suppose a mountain climber is ascending a mountain whose surface has an equation of the form

$$z = f(x, y). \tag{18.79}$$

Here, (x, y) might represent a point in the plane at sea level and z the elevation above sea level of the corresponding point $(x, y, f(x, y))$ on the surface of the mountain. The climber is aware that at any point in his ascent the steepness of the mountain in a given horizontal direction depends upon the vertical distance traversed per unit distance traversed in that horizontal direction. Thus the steepness at the point $(a, b, f(a, b))$ is the *rate of change* of z when $x = a$ and $y = b$ along an arbitrary line L in the xy plane through the point (a, b). To define this rate of change, we first note that line L has parametric equations of the form

$$x = a + t \cos \theta \qquad y = b + t \sin \theta \tag{18.80}$$

where θ is an angle measured counterclockwise from the positive x axis to the line L (Figure 18–13). Since the distance between the points (a, b) and $(a + t \cos \theta, b + t \sin \theta)$ is

$$\sqrt{(a + t \cos \theta - a)^2 + (b + t \sin \theta - b)^2} = |t|,$$

we can set up a one-dimensional coordinate system on L with the point (a, b) as the origin and t as the coordinate of the corresponding point $(a + t \cos \theta, b + t \sin \theta)$. Thus, if

$$z = g(t) = f(a + t \cos \theta, b + t \sin \theta) \tag{18.81}$$

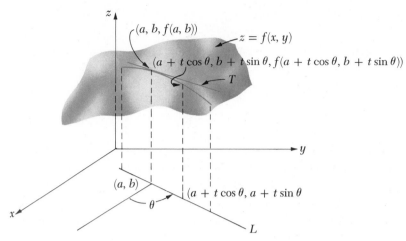

Figure 18-13

we are seeking the rate of change of z with respect to t when $t = 0$. This rate of change is simply

$$g'(0) = \lim_{t \to 0} \frac{g(t) - g(0)}{t} = \lim_{t \to 0} \frac{f(a + t \cos \theta, b + t \sin \theta) - f(a, b)}{t}$$

From the following definition this rate of change is a *directional derivative* of f at the point (a, b).

18.9.1   Definition

Suppose f, a function of two variables, is defined in an open disc with center (a, b). The *directional derivative* of f at (a, b) in the direction of the unit vector $\mathbf{u} = \langle \cos \theta, \sin \theta \rangle$, denoted by $D_{\mathbf{u}} f(a, b)$, is given by

$$D_{\mathbf{u}} f(a, b) = \lim_{t \to 0} \frac{f(a + t \cos \theta, b + t \sin \theta) - f(a, b)}{t} \tag{18.82}$$

if this limit exists.

Geometrically, this directional derivative is the slope of the tangent T at the point $(a, b, f(a, b))$ in the curve of intersection of the surface (18.79) and the vertical plane (18.80) (Figure 18–13).

Example 1   Find the directional derivative of the function f where $f(x, y) = x^2 y$ at $(2, 5)$ in the direction of the vector $\langle 3, -4 \rangle$.

SOLUTION   The graph of the function is represented by level curves $f(x, y) = c$ in Figure 18–14. We note that the unit vector in the same direction as $\langle 3, -4 \rangle$ is $\mathbf{u} = \langle \frac{3}{5}, -\frac{4}{5} \rangle$. Then since

$$\frac{f(2 + \frac{3}{5}t, 5 - \frac{4}{5}t) - f(2, 5)}{t} = \frac{(2 + \frac{3}{5}t)^2(5 - \frac{4}{5}t) - 20}{t}$$

$$= \tfrac{44}{5} - \tfrac{3}{25}t - \tfrac{36}{125}t^2 \qquad (18.83)$$

from (18.82) and (18.83)

$$D_{\mathbf{u}}f(2, 5) = \lim_{t \to 0} (\tfrac{44}{5} - \tfrac{3}{25}t - \tfrac{36}{125}t^2) = \tfrac{44}{5}.$$

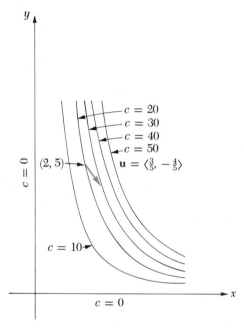

Figure 18–14

If in (18.82) we let $\theta = 0$, then $\mathbf{u} = \langle 1, 0 \rangle = \mathbf{i}$, and hence

$$D_{\mathbf{i}}f(a, b) = \lim_{t \to 0} \frac{f(a + t, b) - f(a, b)}{t} = D_1 f(a, b).$$

Also in (18.82) if $\theta = \frac{\pi}{2}$, then $\mathbf{u} = \langle 0, 1 \rangle = \mathbf{j}$, and

$$D_{\mathbf{j}}f(a, b) = \lim_{t \to 0} \frac{f(a, b + t) - f(a, b)}{t} = D_2 f(a, b).$$

Thus the partial derivatives of f at (a, b) occur as special cases of the directional derivative of f at (a, b) when $\theta = 0$ or $\frac{\pi}{2}$.

The directional derivative for a function may not always exist at a point in the domain of the function. For example, consider the now-familiar function given by:

$$f(x, y) = \begin{cases} \dfrac{2xy}{x^2 + y^2} & \text{if } (x, y) \neq (0, 0) \\ 0 & \text{if } (x, y) = (0, 0) \end{cases} \qquad (18.84)$$

Here

$$\frac{f(t\cos\theta, t\sin\theta) - f(0,0)}{t} = \frac{2\cos\theta\sin\theta}{t}$$

$$= \frac{\sin 2\theta}{t}. \tag{18.85}$$

If θ is an integral multiple of $\frac{\pi}{2}$, then $\sin 2\theta = 0$, and from (18.85)

$$D_{\mathbf{u}}f(0,0) = \lim_{t\to 0}\frac{f(t\cos\theta, t\sin\theta) - f(0,0)}{t} = 0.$$

However, if θ is not an integral multiple of $\frac{\pi}{2}$, $\sin 2\theta \neq 0$, and from (18.85) $D_{\mathbf{u}}f(0,0)$ does not exist.

The notion of a directional derivative is easily extended to functions of n variables.

18.9.2   Definition

Suppose f, a function of n variables, is defined on an open ball with center $A = (a_1, a_2, \ldots, a_n)$. The directional derivative of f at A in the direction of the unit vector $\mathbf{u} = \langle u_1, u_2, \ldots, u_n \rangle$, denoted by $D_{\mathbf{u}}f(A)$, is given by

$$D_{\mathbf{u}}f(A) = \lim_{t\to 0}\frac{f(a_1 + tu_1, a_2 + tu_2, \ldots, a_n + tu_n) - f(A)}{t} \tag{18.86}$$

if this limit exists.

If $w = f(X)$, then $D_{\mathbf{u}}f(A)$ is also called the rate of change of w in the direction of $\mathbf{u}$ at A.

It is usually easier to calculate directional derivatives using the following theorem rather than the limits (18.86) or (18.82).

18.9.3   Theorem

If f, a function of n variables, is defined on an open ball with center $A = (a_1, \ldots, a_n)$ and is differentiable at A, then for any unit vector $\mathbf{u} = \langle u_1, \ldots, u_n \rangle$

$$D_{\mathbf{u}}f(A) = \nabla f(A) \cdot \mathbf{u}. \tag{18.87}$$

PROOF   We first define the function F where

$$F(t) = f(a_1 + tu_1, a_2 + tu_2, \ldots, a_n + tu_n).$$

From the chain rule equation (18.66)

$$F'(t) = \nabla f(a_1 + tu_1, \ldots, a_n + tu_n) \cdot$$

$$\langle D_t(a_1 + tu_1), D_t(a_2 + tu_2), \ldots, D_t(a_n + tu_n) \rangle$$

$$= \nabla f(a_1 + tu_1, \ldots, a_n + tu_n) \cdot \langle u_1, u_2, \ldots, u_n \rangle$$

$$= \nabla f(a_1 + tu_1, \ldots, a_n + tu_n) \cdot \mathbf{u}$$

and in particular

$$F'(0) = \nabla f(A) \cdot \mathbf{u}. \tag{18.88}$$

Since $F'(0)$ exists, by the definition of the derivative

$$F'(0) = \lim_{t \to 0} \frac{F(t) - F(0)}{t}$$

$$= \lim_{t \to 0} \frac{f(a_1 + tu_1, \ldots, a_n + tu_n) - f(A)}{t}$$

and from (18.86)

$$F'(0) = D_\mathbf{u} f(A). \tag{18.89}$$

The conclusion is then obtained by equating the expressions for $F'(0)$ from (18.88) and (18.89).

It will be instructive to rework Example 1 using (18.87). Since

$$\nabla f(x, y) = \langle 2xy, x^2 \rangle$$

we have

$$\nabla f(2, 5) = \langle 20, 4 \rangle.$$

Also since $\mathbf{u} = \langle \frac{3}{5}, -\frac{4}{5} \rangle$, by (18.87)

$$D_\mathbf{u} f(2, 5) = \langle 20, 4 \rangle \cdot \langle \tfrac{3}{5}, -\tfrac{4}{5} \rangle = 20(\tfrac{3}{5}) + 4(-\tfrac{4}{5})$$

$$= \tfrac{44}{5}.$$

Example 2   If $f(x, y, z) = 4x^2 + 9y^2 - 18z$, find the directional derivative of f at $(3, 2, 1)$ along the line $(x - 3)/2 = (y - 2)/-3 = z - 1$ in the direction of decreasing x.

SOLUTION   Since $\nabla f(x, y, z) = \langle 8x, 18y, -18 \rangle$, we have

$$\nabla f(3, 2, 1) = \langle 24, 36, -18 \rangle.$$

From the discussion of the symmetric equations of a line in Section 17.4, the direction numbers of this line are 2, -3, and 1. Hence the direction along the line in which x is decreasing is given by vector $\langle -2, 3, -1 \rangle$, or

$$\mathbf{u} = \left\langle -\frac{2}{\sqrt{14}}, \frac{3}{\sqrt{14}}, -\frac{1}{\sqrt{14}} \right\rangle.$$

Then by (18.87)

$$D_\mathbf{u} f(3, 2, 1) = \nabla f(3, 2, 1) \cdot \mathbf{u} = \langle 24, 36, -18 \rangle \cdot \left\langle -\frac{2}{\sqrt{14}}, \frac{3}{\sqrt{14}}, -\frac{1}{\sqrt{14}} \right\rangle$$

$$= \frac{39\sqrt{14}}{7}.$$

Our mountain climber mentioned at the beginning of this section will be ascending in the direction of maximum steepness at a point if in that direction the maximum number of contour lines are traversed per unit of horizontal distance. Analytically speaking, this is the direction in which the directional derivative at the point is a maximum.

To find this direction in general, we obtain from (18.87), the Cauchy-Schwarz inequality (17.84), and the fact that $\mathbf{u}$ is a unit vector

$$|D_{\mathbf{u}}f(A)| = |\nabla f(A) \cdot \mathbf{u}| \le |\nabla f(A)|\,|\mathbf{u}|$$
$$\le |\nabla f(A)|\,|\mathbf{u}| = |\nabla f(A)|. \tag{18.90}$$

If in particular we let $\mathbf{u} = \nabla f(A)/|\nabla f(A)|$, then $\mathbf{u}$ is in the same direction as $\nabla f(A)$ and

$$D_{\mathbf{u}}f(A) = \nabla f(A) \cdot \frac{\nabla f(A)}{|\nabla f(A)|} = \frac{|\nabla f(A)|^2}{|\nabla f(A)|}$$
$$= |\nabla f(A)|.$$

Thus:

    (i) The maximum value of the directional derivative $D_{\mathbf{u}}f(A)$ is $|\nabla f(A)|$.
    (ii) This maximum value is obtained when $\mathbf{u}$ is in the same direction as $\nabla f(A)$. In other words, *the vector $\nabla f(A)$ is in the direction in which the function f is increasing most rapidly at A.*

Example 3 If $f(x, y) = x^2 y$, find the maximum value of $D_{\mathbf{u}}f(2, 5)$ and the direction in which this maximum is attained.

    SOLUTION Since $\nabla f(x, y) = \langle 2xy, x^2 \rangle$ here, the maximum value of $D_{\mathbf{u}}f(2, 5)$ from the above discussion is

$$|\nabla f(2, 5)| = |\langle 20, 4 \rangle| = 4\sqrt{26}$$

and is attained in the direction of the unit vector

$$\mathbf{u} = \frac{\nabla f(2, 5)}{|\nabla f(2, 5)|} = \left\langle \frac{5}{\sqrt{26}}, \frac{1}{\sqrt{26}} \right\rangle.$$

It would appear from Figure 18–15 that the unit vector $\langle 5/\sqrt{26}, 1/\sqrt{26} \rangle$ is in the direction of the maximum value of $D_{\mathbf{u}}f(2, 5)$ since the vector cuts more level curves per unit distance than any other unit vector drawn from $(2, 5)$.

Exercise Set 18.9

In Exercises 1–6 find the directional derivative to the given function in the direction of the given vector at the given point.

1. $f(x, y) = x^3 + x^2 y - 2y^3$, $\left\langle \dfrac{1}{\sqrt{2}}, -\dfrac{1}{\sqrt{2}} \right\rangle$, $(-1, 2)$

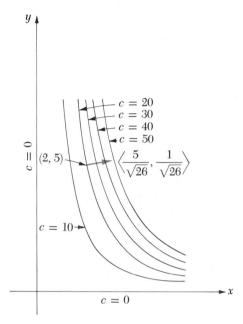

Figure 18–15

2. $f(x, y) = \dfrac{x + 2y}{2x - 3y}$, $\left\langle \dfrac{1}{\sqrt{5}}, \dfrac{2}{\sqrt{5}} \right\rangle$, $(2, -3)$

3. $f(x, y) = \sqrt{x^2 - 2y^2}$, $\langle -3, \sqrt{7} \rangle$, $(3, 1)$

4. $f(x, y, z) = x^2 + y^2 - xz + 2yz + z^2$, $\langle \frac{6}{11}, -\frac{2}{11}, \frac{9}{11} \rangle$, $(2, 0, 4)$

5. $f(x, y, z) = \dfrac{1}{\sqrt{x^2 + y^2 + z^2}}$, $\left\langle \dfrac{1}{3}, \dfrac{2}{3}, -\dfrac{2}{3} \right\rangle$, $(3, 4, -1)$

6. $f(x, y, z) = \sin xy + \cos xz - \sin yz$, $\langle -3, 2, 5 \rangle$, $\left(\dfrac{\sqrt{\pi}}{2}, \dfrac{\sqrt{\pi}}{3}, \sqrt{\pi} \right)$

7. If $f(x, y) = \tan^{-1} y/x$, in what direction at $(-4, 3)$ is $D_{\mathbf{u}}f(-4, 3) = 0$?

8. If $f(x, y) = e^{-x} \cos y$, in what direction at $(0, \frac{\pi}{6})$ is $D_{\mathbf{u}}f(0, \frac{\pi}{6}) = 0$?

9. If f is differentiable at a point A, what is the minimum value of $D_{\mathbf{u}}f(A)$? In what direction is this minimum value attained?

In Exercises 10–13 obtain the following information for the given function f at the given point.

(a) The directional derivative in the direction of the given unit vector obtained from the limit definition (18.82)

(b) The same directional derivative as in part (a) obtained from (18.87)

(c) The maximum value of the directional derivative

(d) The direction in which the maximum directional derivative obtained in part (c) is obtained

(e) A sketch showing at least four level curves associated with f in the vicinity of the given point and the gradient associated with the function at the point

10. $f(x, y) = 3x + 2y$, $(1, 2)$, $\langle -\frac{5}{13}, \frac{12}{13} \rangle$

11. $f(x, y) = \sqrt{x/y}$, $(9, 4)$, $\left\langle -\dfrac{1}{\sqrt{5}}, \dfrac{2}{\sqrt{5}} \right\rangle$

12. $f(x, y) = 4x^2 - 9y^2$, $(4, -2)$, $\langle \frac{3}{5}, \frac{4}{5} \rangle$

13. $f(x, y) = x^2 + 4y^2$, $(-2, 3)$, $\left\langle \dfrac{2}{\sqrt{13}}, -\dfrac{3}{\sqrt{13}} \right\rangle$

14. If the temperature T at any point (x, y, z) is given by $T(x, y, z) = e^{-(x^2+y^2+z^2)}$, find the rate of change of T on the sphere $x^2 + y^2 + z^2 = 4$, (a) in the direction of the origin, and (b) in any direction tangent to the sphere.

15. If the electrostatic potential at a point (x, y, z) in space is given by $f(x, y, z) = 1/(\sqrt{x^2 + y^2 + z^2})$, find the rate of change of this potential at the point $(1, 2, -2)$ in the direction of the point $(5, 2, 0)$.

18.10   Tangent Plane

As an extension of the notion of a tangent to a line, we now consider the problem of defining the *tangent plane* to a surface when this plane exists. Suppose that the point $P_0(x_0, y_0, z_0)$ is on the surface

$$F(x, y, z) = 0 \tag{18.91}$$

and that the function F is differentiable at P_0. Also, let C

$$x = f(t) \qquad y = g(t) \qquad z = h(t) \tag{18.92}$$

be any curve in the surface (18.91) which passes through P_0 and suppose that $t = t_0$ at P_0 (Figure 18–16). We shall require that the functions f, g, and h in (18.92) be

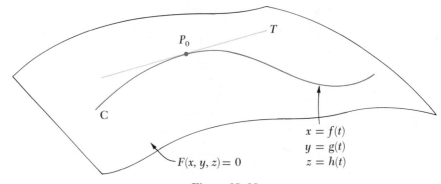

Figure 18–16

differentiable at t_0 and that not all of the numbers $f'(t_0)$, $g'(t_0)$, $h'(t_0)$ be zero. Then by Definition 17.8.3 and Theorem 17.8.2 there is a tangent T to C at P_0, which has the parametric equations (u being the parameter)

$$x - x_0 = uf'(t_0) \qquad y - y_0 = ug'(t_0) \qquad z - z_0 = uh'(t_0). \tag{18.93}$$

At an arbitrary point in C the equation

$$F(f(t), g(t), h(t)) = 0 \tag{18.94}$$

is certainly satisfied. Thus if the left member of (18.94) is denoted by $G(t)$, we obtain from the chain rule

$$G'(t_0) = D_1F(P_0)f'(t_0) + D_2F(P_0)g'(t_0) + D_3F(P_0)h'(t_0) = 0. \qquad (18.95)$$

From multiplying each member of (18.95) by u and invoking (18.93), we have

$$D_1F(x_0, y_0, z_0)(x - x_0) + D_2F(x_0, y_0, z_0)(y - y_0) + D_3F(x_0, y_0, z_0)(z - z_0) = 0. \qquad (18.96)$$

If the numbers $D_1F(x_0, y_0, z_0)$, $D_2F(x_0, y_0, z_0)$, and $D_3F(x_0, y_0, z_0)$ are not all zero, then (18.96) is an equation of a plane through P_0 having $\nabla F(x_0, y_0, z_0)$ as its normal vector. From (18.95) this plane contains the tangent T. In fact, since C was an arbitrary curve through P_0 in S, *the plane* (18.96) *contains all tangents at P_0 of curves in S which pass through P_0.*

18.10.1   Definition

Suppose that the point $P_0(x_0, y_0, z_0)$ is in the surface $F(x, y, z) = 0$ where F is differentiable at P_0 and $\nabla F(P_0) \neq 0$. The plane with equation (18.96) is then called the *tangent plane* to the surface at P_0 (Figure 18–17). The line through P_0 having the components of $\nabla F(P_0)$ as direction numbers is the *normal line* to the surface at P_0, and the vector $\nabla F(P_0)$ is a *normal vector* to the surface at P_0.

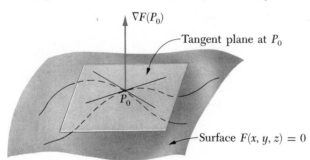

Figure 18–17

In vector form the equation (18.96) for the tangent plane can be given as

$$\nabla F(x_0, y_0, z_0) \cdot \langle x - x_0, y - y_0, z - z_0 \rangle = 0.$$

Example 1   Find the tangent plane and normal line to the surface $y = x^2 + 2xz$ at $(-1, 1, 0)$.

SOLUTION   Here we may let $F(x, y, z) = x^2 + 2xz - y$, and then

$$D_1F(x, y, z) = 2x + 2z \qquad D_2F(x, y, z) = -1 \qquad D_3F(x, y, z) = 2x.$$

Hence

$$D_1F(-1, 1, 0) = -2 \qquad D_2F(-1, 1, 0) = -1 \qquad D_3F(-1, 1, 0) = -2.$$

Thus $\nabla F(-1, 1, 0) = \langle -2, -1, -2 \rangle$ is a normal vector to the surface at $(-1, 1, 0)$. From (18.96) an equation for the tangent plane to the surface at $(-1, 1, 0)$ is

$$-2(x + 1) - (y - 1) - 2z = 0$$

or

$$2x + y + 2z + 1 = 0.$$

Since the components of $\nabla F(-1, 1, 0)$ are direction numbers for the normal line, this line has parametric equations

$$x = -2t - 1 \qquad y = -t + 1 \qquad z = -2t.$$

The intersection of two surfaces can be discussed in terms of their tangent planes (or normal vectors). Two surfaces which intersect in a point P are said to be *tangent surfaces* at P if and only if they have the same tangent plane at the point. Also, two surfaces are said to be *orthogonal* if and only if their tangent planes at each intersection point of the surfaces are perpendicular.

Next, suppose that

$$F(x, y, z) = 0 \qquad \text{and} \qquad G(x, y, z) = 0 \tag{18.97}$$

are equations of two surfaces which intersect in a curve C. The tangent to C at a point P_0 can be determined (when it exists) if the normal vectors to each surface, $\nabla F(P_0)$ and $\nabla G(P_0)$, exist and are not parallel. Since the tangent (denoted by T in Figure 18–18) is in the tangent planes to both of the surfaces (18.97) at P_0, it has as direction numbers the components of any vector that is orthogonal to both normal vectors $\nabla F(P_0)$ and $\nabla G(P_0)$—for example, $\nabla F(P_0) \times \nabla G(P_0)$.

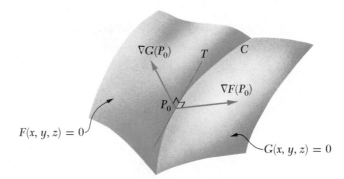

Figure 18-18

Example 2 Find the tangent line at $(1, -2, 3)$ of the curve of intersection of the surfaces $x^2 + y^2 + yz + 1 = 0$ and $xy^2 - 5z + 11 = 0$.

SOLUTION   We let

$$F(x, y, z) = x^2 + y^2 + yz + 1 = 0 \qquad G(x, y, z) = xy^2 - 5z + 11 = 0.$$

Then

$$\nabla F(x, y, z) = \langle 2x, 2y + z, y \rangle \qquad \nabla G(x, y, z) = \langle y^2, 2xy, -5 \rangle$$

and in particular

$$\nabla F(1, -2, 3) = \langle 2, -1, -2 \rangle \qquad \nabla G(1, -2, 3) = \langle 4, -4, -5 \rangle.$$

From our previous discussion the direction numbers of the tangent at $(1, -2, 3)$ are the components of

$$\nabla F(1, -2, 3) \times \nabla G(1, -2, 3) = \langle -3, 2, -4 \rangle.$$

Thus the tangent has the parametric equations

$$x = 1 - 3t \qquad y = -2 + 2t \qquad z = 3 - 4t.$$

Exercise Set 18.10

In Exercises 1–6 obtain equations for the tangent plane and normal line at the given point in the given surface.

1. $z = x^2 + 3xy$, $\quad (-1, 2, -5)$
2. $z = \sqrt{x^2 + 4y^2}$, $\quad (3, -2, 5)$
3. $x^2 + 2y^2 + 4z^2 = 15$, $\quad (-3, 1, -1)$
4. $x^2y + y^2z - xz^2 = 1$, $\quad (1, -2, 3)$
5. $2e^{-xy} \cos z = 1$, $\quad (1, 0, \frac{5\pi}{3})$
6. $x^2 \ln \dfrac{y}{z} + 4y = 0$, $\quad (2, 1, e)$

7. Suppose f, a function of two variables, has partial derivatives at (x_0, y_0). Derive the equation

$$z = z_0 + D_1 f(x_0, y_0)(x - x_0) + D_2 f(x_0, y_0)(y - y_0)$$

   for the tangent plane to the surface $z = f(x, y)$ at the point $(x_0, y_0, f(x_0, y_0))$.

8. Prove that the tangent plane at any point (x_1, y_1, z_1) in the paraboloid $z = x^2 + y^2$ passes through the point $(0, 0, -z_1)$.

9. Prove that a normal line at any point (x_1, y_1, z_1) in the sphere $x^2 + y^2 + z^2 = a^2$ also passes through the point $(-x_1, -y_1, -z_1)$.

10. Find the point(s) in the ellipsoid $x^2 + 4y^2 + 9z^2 = 36$ where the tangent plane is parallel to the plane $x + 2y - 3z = 6$.

11. Prove that the sum of the intercepts of the tangent plane at any point in the surface $x^{1/2} + y^{1/2} + z^{1/2} = a^{1/2}$ is constant.

12. Prove that the tangent plane at any point in the surface $xyz = 1$ forms a tetrahedron of constant volume with the coordinate axes.

13. The surfaces $F(x, y, z) = 0$ and $G(x, y, z) = 0$ are orthogonal if and only if what equation involving the partial derivatives of F and G is satisfied at every intersection point of the surfaces?

In Exercises 14–17 obtain equations of the tangent to the curve of intersection of the given surfaces at the given point.

14. $x^2 + y^2 + z^2 = 25$, $\quad 4x^2 - y^2 + 2z^2 = 23$; $\quad (0, 3, 4)$
15. $z = x^2 + 1$, $\quad x^2 + y^2 + z^2 = 9$; $\quad (1, -2, 2)$

16. $y \cos xz = 1$, $ze^{x^2-y^2} = \pi$; $(-1, -1, \pi)$

17. $x^2y = 4$, $y^2 + z^2 = 10$; $(-2, 1, 3)$

18. Prove that the surfaces $(x - 2)^2 + y^2 + z^2 = 30$ and $x^2 - 2(y - 1)^2 - z^2 + 26 = 0$ are tangent to each other at the point $(1, 2, 5)$.

18.11   Higher-Order Partial Derivatives

Just as we were concerned earlier with second derivatives of functions of one variable, we will now consider partial derivatives of *order two*.

18.11.1   Definition

Suppose f is a function of n variables. The *second partial derivative*, $D_{ij}f$, where i and j are arbitrary positive integers chosen from the set $\{1, 2, \ldots, n\}$ is the function defined by

$$D_{ij}f(X) = D_j(D_i f(X)) \qquad (18.98)$$

when this function exists.

In obtaining $D_{ij}f(X)$, we first perform the partial differentiation with respect to the ith variable x_i, and then with respect to the jth variable, x_j.
The function $D_{ij}f$ is also given by

$$D_{ij}f(X) = \lim_{h \to 0} \frac{D_i f(x_1, x_2, \ldots, x_{j-1}, x_j + h, x_{j+1}, \ldots, x_n) - D_i f(x_1, \ldots, x_n)}{h} \qquad (18.99)$$

when this limit exists. Other notations that are used to represent $D_{ij}f(X)$ are

$$f_{x_i x_j}(X), \qquad \frac{\partial}{\partial x_j}\left(\frac{\partial f}{\partial x_i}\right), \qquad \text{and} \qquad \frac{\partial^2 f}{\partial x_j\, \partial x_i}.$$

Also if $u = f(X)$,

$$\frac{\partial^2 u}{\partial x_j\, \partial x_i}, \qquad \frac{\partial}{\partial x_j}\left(\frac{\partial u}{\partial x_i}\right), \qquad \text{and} \qquad u_{x_i x_j}$$

are used to denote $D_{ij}f(X)$.
For a function f of two variables, the second partial derivatives of f are given by

$$\frac{\partial^2 f}{\partial x^2} = D_{11}f(x, y) = D_1(D_1 f(x, y)) \qquad \frac{\partial^2 f}{\partial y\, \partial x} = D_{12}f(x, y) = D_2(D_1 f(x, y))$$

$$\frac{\partial^2 f}{\partial x\, \partial y} = D_{21}f(x, y) = D_1(D_2 f(x, y)) \qquad \frac{\partial^2 f}{\partial y^2} = D_{22}f(x, y) = D_2(D_2 f(x, y))$$

Example 1   Obtain the second partial derivatives associated with the function f where

$$f(x, y) = x^2y + x \sin y^2.$$

SOLUTION   First we obtain

$$\frac{\partial f}{\partial x} = D_1 f(x, y) = 2xy + \sin y^2 \qquad \frac{\partial f}{\partial y} = D_2 f(x, y) = x^2 + 2xy \cos y^2.$$

Then

$$\frac{\partial^2 f}{\partial x^2} = D_{11} f(x, y) = 2y \qquad\qquad \frac{\partial^2 f}{\partial x \, \partial y} = D_{21} f(x, y) = 2x + 2y \cos y^2$$

$$\frac{\partial^2 f}{\partial y \, \partial x} = D_{12} f(x, y) = 2x + 2y \cos y^2 \qquad \frac{\partial^2 f}{\partial y^2} = D_{22} f(x, y) = 2x \cos y^2 - 4xy^2 \sin y^2.$$

As is illustrated in Example 1, when using the subscript notation $D_{ij} f(X)$ or $f_{x_i x_j}(X)$ we differentiate with respect to the variables denoted by subscripts going from left to right, but when using the fractional notation $\dfrac{\partial^2 f}{\partial x_j \, \partial x_i}$ or $\dfrac{\partial}{\partial x_j}\left(\dfrac{\partial f}{\partial x_i}\right)$ the order of differentiation is from right to left.

Example 2   If $F(x, y, z) = xe^{2y-z}$, find the second partial derivatives associated with the function F.

SOLUTION   First we calculate

$$\frac{\partial F}{\partial x} = e^{2y-z} \qquad \frac{\partial F}{\partial y} = 2xe^{2y-z} \qquad \frac{\partial F}{\partial z} = -xe^{2y-z}.$$

The second partials of F are then given by

$$\frac{\partial^2 F}{\partial x^2} = 0 \qquad\qquad \frac{\partial^2 F}{\partial x \, \partial y} = 2e^{2y-z} \qquad\qquad \frac{\partial^2 F}{\partial x \, \partial z} = -e^{2y-z}$$

$$\frac{\partial^2 F}{\partial y \, \partial x} = 2e^{2y-z} \qquad \frac{\partial^2 F}{\partial y^2} = 4xe^{2y-z} \qquad\quad \frac{\partial^2 F}{\partial y \, \partial z} = -2xe^{2y-z}$$

$$\frac{\partial^2 F}{\partial z \, \partial x} = -e^{2y-z} \qquad \frac{\partial^2 F}{\partial z \, \partial y} = -2xe^{2y-z} \qquad \frac{\partial^2 F}{\partial z^2} = xe^{2y-z}$$

We note from Example 1 that $D_{12} f(x, y) = D_{21} f(x, y)$ for every (x, y) and from Example 2 that

$$\frac{\partial^2 F}{\partial x \, \partial y} = \frac{\partial^2 F}{\partial y \, \partial x} \qquad \frac{\partial^2 F}{\partial x \, \partial z} = \frac{\partial^2 F}{\partial z \, \partial x} \qquad \frac{\partial^2 F}{\partial y \, \partial z} = \frac{\partial^2 F}{\partial z \, \partial y}$$

for every (x, y, z). The next theorem gives a condition that guarantees the equality of the *mixed* second partials for a function of two variables.

18.11.2   Theorem

Suppose a function f of two variables has continuous first and second partial derivatives on an open disc S with center (a, b). Then

$$D_{12} f(a, b) = D_{21} f(a, b). \tag{18.100}$$

OUTLINE OF PROOF   Let $(a + h, b + k)$ where $h \neq 0 \neq k$ be an arbitrarily chosen point in S, and consider the function A where

$$A(h, k) = f(a + h, b + k) - f(a + h, b) - f(a, b + k) + f(a, b). \quad (18.101)$$

If F is given by

$$F(x) = f(x, b + k) - f(x, b), \quad (18.102)$$

then from (18.101) and (18.102)

$$A(h, k) = F(a + h) - F(a). \quad (18.103)$$

Since F satisfies the hypothesis of the mean value theorem on the closed interval with endpoints a and $a + h$, from (18.103) there is a c_1 between a and $a + h$ such that

$$A(h, k) = hF'(c_1),$$

and hence from (18.102)

$$A(h, k) = h[D_1 f(c_1, b + k) - D_1 f(c_1, b)].$$

Again using the mean value theorem, we obtain

$$A(h, k) = hkD_{12} f(c_1, c_2) \quad (18.104)$$

where c_2 is between b and $b + k$.

By considering the function G where

$$G(y) = f(a + h, y) - f(a, y)$$

we can obtain, analogously to the derivation of (18.104),

$$A(h, k) = hkD_{21} f(c_3, c_4) \quad (18.105)$$

where c_3 is between a and $a + h$ and c_4 is between b and $b + k$. The details in obtaining (18.105) are left as an exercise. From (18.104) and (18.105)

$$D_{12} f(c_1, c_2) = D_{21} f(c_3, c_4). \quad (18.106)$$

Since $D_{12} f$ and $D_{21} f$ are continuous at (a, b) and c_1 and c_3 are between a and $a + h$ and c_2 and c_4 are between b and $b + k$, it is intuitively apparent that (18.106) implies (18.100). The details in the proof of this implication are omitted although the interested reader may wish to supply the necessary steps.

An extension of Theorem 18.11.2 can be proved if f is a function of n variables with continuous first and second partial derivatives on an open ball S with center $A = (a_1, a_2, \ldots, a_n)$. Suppose i and j are arbitrary elements of the set $\{1, 2, \ldots, n\}$. Then after a proof, which is analogous to that given for Theorem 18.11.2, one obtains

$$D_{ij} f(A) = D_{ji} f(A)$$

Example 3   If $F(x, y) = f(x^2 + y^2, xy)$ where f has continuous first and second partial derivatives on its domain, obtain $\partial^2 F / \partial x \, \partial y$ in terms of partial derivatives of f.

SOLUTION   For convenience let $u = x^2 + y^2$ and $v = xy$. Then by the general chain rule

$$\frac{\partial F}{\partial y} = \frac{\partial f}{\partial u}\frac{\partial u}{\partial y} + \frac{\partial f}{\partial v}\frac{\partial v}{\partial y} = 2y\frac{\partial f}{\partial u} + x\frac{\partial f}{\partial v}.$$

Since $\partial f/\partial u$ and $\partial f/\partial v$ are functions of u and v it will be helpful to let

$$P(u, v) = \frac{\partial f}{\partial u} \qquad Q(u, v) = \frac{\partial f}{\partial v} \qquad\qquad (18.107)$$

and then express $\partial F/\partial y$ in the form

$$\frac{\partial F}{\partial y} = 2yP(u, v) + xQ(u, v). \qquad\qquad (18.108)$$

Differentiating in (18.108) with respect to x using the product and sum rule, we obtain

$$\frac{\partial^2 F}{\partial x\,\partial y} = 2y\frac{\partial}{\partial x}[P(u, v)] + P(u, v)\frac{\partial}{\partial x}(2y) + x\frac{\partial}{\partial x}[Q(u, v)] + Q(u, v)\frac{\partial}{\partial x}(x),$$

and after computing $\dfrac{\partial}{\partial x}[P(u, v)]$ and $\dfrac{\partial}{\partial x}[Q(u, v)]$ by the general chain rule

$$\frac{\partial^2 F}{\partial x\,\partial y} = 2y\left(\frac{\partial P}{\partial u}\frac{\partial u}{\partial x} + \frac{\partial P}{\partial v}\frac{\partial v}{\partial x}\right) + P(u, v)\cdot 0 + x\left(\frac{\partial Q}{\partial u}\frac{\partial u}{\partial x} + \frac{\partial Q}{\partial v}\frac{\partial v}{\partial x}\right) + Q(u, v)\cdot 1.$$

$$\qquad\qquad (18.109)$$

Since from (18.107)

$$\frac{\partial P}{\partial u} = \frac{\partial^2 f}{\partial u^2} \qquad\qquad \frac{\partial Q}{\partial u} = \frac{\partial^2 f}{\partial u\,\partial v}$$

$$\frac{\partial P}{\partial v} = \frac{\partial^2 f}{\partial v\,\partial u} \qquad\qquad \frac{\partial Q}{\partial v} = \frac{\partial^2 f}{\partial v^2}$$

equation (18.109) can be rewritten

$$\frac{\partial^2 F}{\partial x\,\partial y} = 2y\left(\frac{\partial^2 f}{\partial u^2}2x + \frac{\partial^2 f}{\partial v\,\partial u}y\right) + x\left(\frac{\partial^2 f}{\partial u\,\partial v}2x + \frac{\partial^2 f}{\partial v^2}y\right) + \frac{\partial f}{\partial v}. \quad (18.110)$$

From Theorem 18.11.2 $\partial^2 f/\partial v\,\partial u = \partial^2 f/\partial u\,\partial v$ and hence (18.110) can be simplified to

$$\frac{\partial^2 F}{\partial x\,\partial y} = 4xy\frac{\partial^2 f}{\partial u^2} + 2(x^2 + y^2)\frac{\partial^2 f}{\partial u\,\partial v} + xy\frac{\partial^2 f}{\partial v^2} + \frac{\partial f}{\partial v}.$$

We shall briefly mention partial derivatives of order greater than two. If f is a function of n variables and i, j, and k are any numbers chosen from the set $\{1, 2, \ldots, n\}$, then the *third partial derivative* $D_{ijk}f$ is given by

$$D_{ijk}f(X) = D_k(D_{ij}f(X)),$$

or saying the same thing,

$$\frac{\partial^3 f}{\partial x_k \, \partial x_j \, \partial x_i} = \frac{\partial}{\partial x_k}\left(\frac{\partial^2 f}{\partial x_j \, \partial x_i}\right).$$

Thus from Example 1 where $f(x, y) = x^2 y + x \sin y^2$, we can obtain

$$\frac{\partial^3 f}{\partial x \, \partial y \, \partial x} = \frac{\partial}{\partial x}\left(\frac{\partial^2 f}{\partial y \, \partial x}\right) = \frac{\partial}{\partial x}(2x + 2y \cos y^2)$$

$$= 2.$$

Exercise Set 18.11

In Exercises 1–4 obtain $D_{11}f(x, y)$, $D_{21}f(x, y)$, $D_{12}f(x, y)$, $D_{22}f(x, y)$.

1. $f(x, y) = x^3 y^2 + 3xy^4 - y^3$

2. $f(x, y) = \dfrac{x^2 y}{\sqrt{x^2 + y^2}}$

3. $f(x, y) = \sin^{-1}\dfrac{x^4}{y^4}$

4. $f(x, y) = x^2 \sin xy + y^2 \cos xy$

In Exercises 5–8 obtain the indicated second partial derivatives of the given function.

5. $f(x, y, z) = (x^2 - y^2 + 2z^2)^3$; $\quad \dfrac{\partial^2 f}{\partial y \, \partial x}, \ \dfrac{\partial^2 f}{\partial z \, \partial x}, \ \dfrac{\partial^2 f}{\partial y^2}, \ \dfrac{\partial^2 f}{\partial z \, \partial y}$

6. $g(x, y, z) = \ln(x^2 + y^2 + 4z)$; $\quad \dfrac{\partial^2 g}{\partial x \, \partial z}, \ \dfrac{\partial^2 g}{\partial z^2}, \ \dfrac{\partial^2 g}{\partial y \, \partial x}, \ \dfrac{\partial^2 g}{\partial x^2}$

7. $h(x, y, z) = \dfrac{\sin xz}{\cos yz}$; $\quad \dfrac{\partial^2 h}{\partial x^2}, \ \dfrac{\partial^2 h}{\partial z \, \partial x}, \ \dfrac{\partial^2 h}{\partial z^2}, \ \dfrac{\partial^2 h}{\partial y \, \partial z}$

8. $F(x, y, z) = \dfrac{z}{\sqrt{x^2 + y^2 + z^2}}$; $\quad \dfrac{\partial^2 F}{\partial y \, \partial z}, \ \dfrac{\partial^2 F}{\partial z^2}, \ \dfrac{\partial^2 F}{\partial x \, \partial z}, \ \dfrac{\partial^2 F}{\partial x^2}$

9. If $F(x, y, z) = x^2 y^3 + 2xz^4 - 3y^3 z^2$, find $\dfrac{\partial^3 F}{\partial y \, \partial x^2}, \ \dfrac{\partial^3 F}{\partial z \, \partial y \, \partial x}, \ \dfrac{\partial^3 F}{\partial x \, \partial y \, \partial z}$.

10. If $G(x, y, z) = \sin xy + \sin yz + \sin xz$, find $\dfrac{\partial^3 G}{\partial z \, \partial y \, \partial z}, \ \dfrac{\partial^3 G}{\partial z^2 \, \partial y}, \ \dfrac{\partial^3 G}{\partial y^3}$.

The partial differential equation $(\partial^2 u / \partial x^2) + (\partial^2 u / \partial y^2) = 0$, which is called *Laplace's*† *equation in two variables*, is important in mathematical physics. In Exercises 11–14 show that u satisfies this equation.

11. $u = \tan^{-1}\dfrac{x}{y}$

12. $u = e^{-2x}\cos 2y$

13. $u = \dfrac{y}{x^2 + y^2}$

14. $u = \ln(x^2 + y^2)^2$

---

† Named after Pierre Simon Laplace (1749–1827) a French mathematician, physicist, and astronomer who has been called "The Newton of France." Laplace is remembered for his work in celestial mechanics and particularly for his "nebular hypothesis" concerning the origin of the earth.

In Exercises 15 and 16 show that u satisfies the partial differential equation $\dfrac{\partial^2 u}{\partial x^2} + \dfrac{\partial^2 u}{\partial y^2} + \dfrac{\partial^2 u}{\partial z^2} = 0$, *Laplace's equation in three variables.*

15. $u = e^{3x+4y} \sin 5z$

16. $u = \dfrac{1}{\sqrt{x^2 + y^2 + z^2}}$

17. Show that the one-dimensional heat flow equation $\partial u/\partial t = a^2\, \partial^2 u/\partial x^2$ is satisfied by $u = (c_1 \cos Kx + c_2 \sin Kx)e^{-a^2 K^2 t}$ where c_1, c_2, and K are constants.

18. If

$$f(x, y) = \begin{cases} xy\dfrac{x^2 - y^2}{x^2 + y^2} & \text{if } (x, y) \neq (0, 0) \\ 0 & \text{if } (x, y) = (0, 0) \end{cases}$$

prove that $D_{12}f(0, 0) = -1$ and $D_{21}f(0, 0) = 1$. Why is this result not a contradiction of Theorem 18.11.2?

In Exercises 19–24 assume that all functions mentioned have continuous first and second (partial) derivatives.

19. If $x = t^2$, $y = t^3$, and $w = F(x, y)$, find d^2w/dt^2.

20. If $w = F(x, y)$ where $x = f(t)$ and $y = g(t)$, prove that

$$\frac{d^2w}{dt^2} = \frac{\partial^2 F}{\partial x^2}\left(\frac{dx}{dt}\right)^2 + 2\frac{\partial F}{\partial x\,\partial y}\frac{dx}{dt}\frac{dy}{dt} + \frac{\partial^2 F}{\partial y^2}\left(\frac{dy}{dt}\right)^2 + \frac{\partial F}{\partial x}\frac{d^2x}{dt^2} + \frac{\partial F}{\partial y}\frac{d^2y}{dt^2}.$$

21. If $x = u^2 - v^2$, $y = 2uv$, and $w = F(x, y)$, find (a) $\partial^2 w/\partial u^2$ (b) $\partial^2 w/\partial u\,\partial v$ (c) $\partial^2 w/\partial v^2$ in terms of u and v and partial derivatives of F.

22. If $x = e^{-s} \cos t$, $y = e^{-s} \sin t$, and $w = F(x, y)$, find (a) $\partial^2 w/\partial s^2$ (b) $\partial^2 w/\partial s\,\partial t$ (c) $\partial^2 w/\partial t^2$ in terms of s and t and partial derivatives of F.

23. If f and g are functions of one variable and $u = F(x, t) = f(x + ct) + g(x - ct)$, where c is constant, prove that u satisfies the *one-dimensional wave equation* $\partial^2 u/\partial t^2 = c^2\, \partial^2 u/\partial x^2$.

24. If $u = f(x, y)$, $x = r \cos \theta$, and $y = r \sin \theta$, prove that

$$\frac{\partial^2 u}{\partial x^2} + \frac{\partial^2 u}{\partial y^2} = \frac{\partial^2 u}{\partial r^2} + \frac{1}{r}\frac{\partial u}{\partial r} + \frac{1}{r^2}\frac{\partial^2 u}{\partial \theta^2}.$$

25. Derive equation (18.105) in the proof of Theorem 18.11.2.

18.12 Extrema for a Function of Two Variables

In this section we will consider maximum and minimum values for a function of two variables. It will be noted that the definitions and theorems considered here are analogous to statements that are given in Chapters 5 and 6 for functions of one variable.

First we give the basic definition.

18.12.1   Definition

Suppose S is a subset of the domain $\mathcal{D}_f$ of a function f of two variables and $(a, b) \in S$. Then:

- (a) $f(a, b)$ is a *maximum of f on S* if and only if $f(x, y) \leq f(a, b)$ for every $(x, y) \in S$.
- (b) $f(a, b)$ is a *minimum of f on S* if and only if $f(x, y) \geq f(a, b)$ for every $(x, y) \in S$.

In particular, a maximum of f on $\mathcal{D}_f$ is called the *absolute maximum* of f, and a minimum of f on $\mathcal{D}_f$ is called the *absolute minimum* of f.

For example, if

$$f(x, y) = x^2 + y^2 + 1, \tag{18.111}$$

then on the closed rectangle

$$S = \{(x, y): x \in [0, 1] \text{ and } y \in [0, 1]\}$$

f has a minimum $f(0, 0) = 1$ since from (18.111)

$$f(x, y) \geq 1 \quad \text{for every } (x, y) \in S.$$

Also, f has a maximum $f(1, 1) = 3$ since from (18.111)

$$f(x, y) \leq 3 \quad \text{for every } (x, y) \in S.$$

Since $\mathcal{D}_f$ is the whole xy plane, f given by (18.111) has no absolute maximum. However, $f(0, 0) = 1$ is an absolute minimum of the function since $f(x, y) \geq 1$ for all (x, y).

We could also define the *relative extrema* for a function of two variables by statements that are analogous to Definition 5.3.3. However, these definitions will be omitted since we will not be explicitly treating such extrema in this section.

The following theorem is analogous to Theorem 5.1.2, which is proved in advanced calculus.

18.12.2   Theorem

If a function f of two variables is continuous on a bounded closed set $S \subset R^2$, then f has a maximum and a minimum on S.

Points where a function has extrema are frequently located with the following theorem, which is analogous to Fermat's Theorem (Theorem 5.1.3).

18.12.3   Theorem

If $f(a, b)$ is an extremum of f on an open set $S \subset R^2$, and the first partial derivatives of f exist at (a, b), then

$$D_1 f(a, b) = 0 = D_2 f(a, b). \tag{18.112}$$

PROOF   By hypothesis if ϕ is given by

$$\phi(x) = f(x, b)$$

then $\phi(a) = f(a, b)$ is also a relative extremum of ϕ. Hence by Fermat's Theorem

$$\phi'(a) = \lim_{h \to 0} \frac{f(a + h, b) - f(a, b)}{h} = D_1 f(a, b)$$

$$= 0.$$

Similarly, if ψ is given by $\psi(y) = f(a, y)$, we can prove that $\psi'(b) = D_2 f(a, b) = 0$, and thus, the conclusion is obtained.

The point (a, b) in (18.112) is called a *critical point* of the function f. Because of Theorem 18.12.3 such points should always be investigated when searching for extrema of a function on a set.

Example 1   Find the extrema of the function f where $f(x, y) = -x^2 + xy - y^2 + 3x + 4$ on the closed triangle $S = \{(x, y): x \in [0, 4] \text{ and } 0 \le y \le 4 - x\}$ (Figure 18–19).

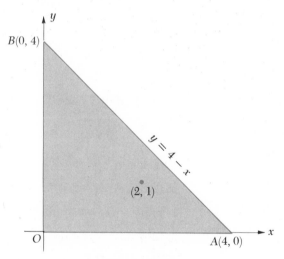

Figure 18–19

SOLUTION   Since f is everywhere continuous, the maximum and minimum of f on S exist by Theorem 18.12.2. These extrema may be assumed at either interior points or boundary points of S. If an extremum is assumed at an interior point (a, b) of S, then this value is also an extremum of f on some open disc T where $(a, b) \in T \subset S$. Then by Theorem 18.12.3 the first partial derivatives of f will vanish at (a, b). Therefore, the extrema of f on S will be assumed at either

(a) the critical points of f

or

(b) the boundary points of S.

The greatest of the values obtained at the points (a) or (b) is the maximum of f on S and the least value of f at one of these points is the minimum of f on S. This is analogous to having extrema of functions of one variable at critical points and endpoints of an interval.

First we obtain the critical points of f by solving the system of equations

$$D_1 f(x, y) = -2x + y + 3 = 0$$
$$D_2 f(x, y) = x - 2y = 0.$$

The solution to this system is $(2, 1)$. Then we calculate

$$f(2, 1) = 7. \tag{18.113}$$

To investigate the boundary points, we first consider the boundary OA. The function values of f on this boundary are given by

$$g(x) = f(x, 0) = -x^2 + 3x + 4 \quad \text{if } x \in [0, 4].$$

Since $g'(x) = -2x + 3$, $g'(x) = 0$ when $x = \frac{3}{2}$. Thus, using the methods of Section 5.1, we examine the function values $f(x, 0)$ at the critical number $\frac{3}{2}$ for g and the endpoints 0 and 4 of the interval. Thus the maximum and minimum values of f on OA are, respectively,

$$f(\tfrac{3}{2}, 0) = \tfrac{25}{4} \quad \text{and} \quad f(4, 0) = 0. \tag{18.114}$$

Similarly the function values on the boundary OB are given by

$$f(0, y) = 4 - y^2 \quad \text{if } y \in [0, 4].$$

Again by the methods of Section 5.1, the maximum and minimum values of f on OB are, respectively

$$f(0, 0) = 4 \quad \text{and} \quad f(0, 4) = -12. \tag{18.115}$$

Also, the function values on AB are given by

$$f(x, 4 - x) = -x^2 + x(4 - x) - (4 - x)^2 + 3x + 4$$
$$= -3x^2 + 15x - 12 \quad \text{if } x \in [0, 4].$$

The maximum and minimum values of f on AB are, respectively

$$f(\tfrac{5}{2}, \tfrac{3}{2}) = \tfrac{27}{4} \quad \text{and} \quad f(0, 4) = -12. \tag{18.116}$$

From our remarks at the beginning of the solution, the maximum of f on S is the greatest of the numbers obtained from (18.113), (18.114), (18.115), and (18.116) and is therefore $f(2, 1) = 7$. Similarly, the minimum of f on S, which is the least of the numbers obtained from (18.113), (18.114), (18.115), and (18.116), is $f(0, 4) = -12$.

It should be recognized that a function of two variables need not assume an extremum at a critical point of the function. For example, if

$$f(x, y) = x^2 - y^2 \quad \text{(Figure 18–20)} \tag{18.117}$$

then $D_1 f(x, y) = 2x$ and $D_2 f(x, y) = -2y$ and hence $D_1 f(0, 0) = 0 = D_2 f(0, 0)$.

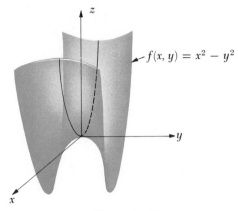

Figure 18-20

However,

$$f(x, 0) = x^2 > f(0, 0) = 0 \quad \text{for every } x \neq 0$$

and

$$f(0, y) = -y^2 < f(0, 0) = 0 \quad \text{for every } y \neq 0,$$

so $f(0, 0) = 0$ cannot be either a maximum or a minimum of f on any set containing the point $(0, 0)$. In particular, $f(0, 0) = 0$ is not an absolute extremum of f. We will show below that $(0, 0)$ is a *saddle point* of f.

To obtain extrema for a function of two variables on an open convex set, the following theorem, which is analogous to the second derivative test, is utilized.

18.12.4   Theorem

Suppose the following conditions hold:

(*i*) *f, a function of two variables, has continuous first and second partial derivatives on some open convex set $S \subset R^2$.*
(*ii*) *$(a, b) \in S$.*
(*iii*) *$D_1 f(a, b) = 0 = D_2 f(a, b)$.*
(*iv*) *The function Δ is given by*

$$\Delta(x, y) = D_{11} f(x, y) \, D_{22} f(x, y) - (D_{12} f(x, y))^2. \quad (18.118)$$

Then:

(*a*) *If $\Delta(x, y) > 0$ and $D_{11} f(x, y) > 0$ for every $(x, y) \in S$, $f(a, b)$ is a minimum of f on S.*
(*b*) *If $\Delta(x, y) > 0$ and $D_{11} f(x, y) < 0$ for every $(x, y) \in S$, $f(a, b)$ is a maximum of f on S.*
(*c*) *If $\Delta(a, b) < 0$, $f(a, b)$ is not an extremum of f on S.*

PROOF OF (a)   Let $(a + h, b + k)$ be any point distinct from (a, b) in S (Figure 18-21), and consider an arbitrary point $P(a + ht, b + kt)$ where $0 \leq t$

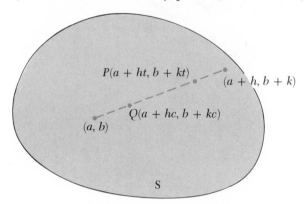

Figure 18-21

≤ 1. P is on the line segment connecting (a, b) and $(a + h, b + k)$ and therefore $P \in S$. Suppose g is defined by

$$g(t) = f(a + ht, b + kt) \quad \text{if } t \in [0, 1]. \tag{18.119}$$

By Taylor's formula there is a c between 0 and 1 such that

$$g(t) = g(0) + g'(0)t + \frac{g''(c)t^2}{2!}$$

and in particular, if $t = 1$, then

$$g(1) = g(0) + g'(0) + \tfrac{1}{2}g''(c). \tag{18.120}$$

From (18.119)

$$g(1) = f(a + h, b + k) \quad \text{and} \quad g(0) = f(a, b). \tag{18.121}$$

If the chain rule is applied in (18.119)

$$g'(t) = h\, D_1 f(a + ht, b + kt) + k\, D_2 f(a + ht, b + kt), \tag{18.122}$$

and therefore from hypothesis (iii) and (18.122)

$$g'(0) = h\, D_1 f(a, b) + k\, D_2 f(a, b) = 0. \tag{18.123}$$

Rewriting (18.120) using (18.121) and (18.123), we have

$$f(a + h, b + k) = f(a, b) + \tfrac{1}{2}g''(c). \tag{18.124}$$

From (18.124) the conclusion in part (a) can be obtained by proving that $g''(c) \geq 0$. To show this we apply the chain rule to (18.122), and obtain

$$g''(t) = h[h\, D_{11} f(P) + k\, D_{12} f(P)] + k[h\, D_{21} f(P) + k\, D_{22} f(P)]. \tag{18.125}$$

Since f has continuous first and second partial derivatives on some open disc containing P, $D_{12} f(P) = D_{21} f(P)$ by Theorem 18.11.2, and therefore from (18.125)

$$g''(t) = h^2 D_{11}f(P) + 2hk\, D_{12}f(P) + k^2 D_{22}f(P). \qquad (18.126)$$

In expressing $g''(c)$ in (18.124), it will be convenient to let the values of the second partials of f at the point $Q = (a + hc, b + kc)$ corresponding to $t = c$ be denoted by

$$A = D_{11}f(Q) \qquad B = D_{12}f(Q) \qquad C = D_{22}f(Q),$$

and thus $\Delta(Q) = AC - B^2$. Then from (18.126)

$$\begin{aligned} g''(c) &= h^2 A + 2hkB + k^2 C \\ &= \frac{1}{A}(A^2 h^2 + 2ABhk) + k^2 C, \end{aligned} \qquad (18.127)$$

and upon completing the square in $A^2 h^2 + 2ABhk$, we have

$$g''(c) = \frac{1}{A}[(Ah + Bk)^2 + k^2(AC - B^2)]. \qquad (18.128)$$

Since $D_{11}f$ and Δ can assume only positive values on S, $AC - B^2$ and A are positive. Then from (18.128), since $(Ah + Bk)^2$ and k^2 are non-negative,

$$g''(c) \ge 0.$$

The conclusion in part (a) then follows from our remark after (18.124).

The proof of Theorem 18.12.4(b) parallels the above proof and is therefore left as an exercise. To prove (c) of the theorem, one first notes that since the second partials of f are continuous at (a, b), Δ is also continuous at (a, b). Thus there is an open disc $S_1 \subset S$ with center (a, b) such that Δ is negative-valued on S_1. By properly choosing h and k for the point $P(a + h, b + k) \in S_1$, it is possible to obtain either $g''(c) > 0$ or $g''(c) < 0$. The details are also left as an exercise.

A point (a, b) in the domain of f for which $D_1 f(a, b) = 0 = D_2 f(a, b)$ and $\Delta(a, b) < 0$ (which is referred to in Theorem 18.12.4(c)) is called a *saddle point* of f. In particular the function f where $f(x, y) = x^2 - y^2$ (compare (18.117) et seq.) has a saddle point at $(0, 0)$ since $D_1 f(0, 0) = 0 = D_2 f(0, 0)$ and $\Delta(0, 0) = -4 < 0$.

Example 2   Find the absolute extrema of the function f where $f(x, y) = -x^2 + xy - y^2 + 3x + 4$.

SOLUTION   This function, which was discussed in Example 1, has the open convex set R^2 as its domain. We also recall from Example 1 that $(2, 1)$ is the only critical point of the function. Since for every $(x, y) \in \mathcal{D}_f$

$$D_{11}f(x, y) = -2 \qquad D_{12}f(x, y) = 1 \qquad D_{22}f(x, y) = -2,$$

we have

$$\Delta(x, y) = 3,$$

and therefore by Theorem 18.12.4(b) $f(2, 1) = 7$ is an absolute maximum of f. The function f here has no absolute minimum since, for example, $f(0, y)$ can be obtained less than any negative number if $|y|$ is sufficiently large.

A frequently encountered extremal problem is one in which we seek an absolute extremum of a function f of three variables given by

$$w = f(x, y, z) \tag{18.129}$$

where x, y, and z satisfy a side condition given by

$$z = g(x, y). \tag{18.130}$$

From (18.129) and (18.130)

$$w = F(x, y) = f(x, y, g(x, y))$$

and the absolute extremum is usually obtained by applying Theorem 18.12.4 to the function F.

Example 3   Find the maximum volume for a rectangular box that can be inscribed in the region bounded by the plane $x + 2y + z = 4$ and the coordinate planes as shown in Figure 18–22.

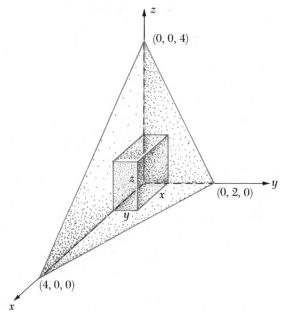

Figure 18-22

SOLUTION   If the dimensions of the box are x, y, and z as shown in Figure 18–22, the volume of the box is

$$V = xyz \qquad\qquad (18.131)$$

where

$$x + 2y + z = 4. \qquad\qquad (18.132)$$

Thus from (18.131) and (18.132) this volume is

$$V = F(x, y) = xy(4 - x - 2y)$$
$$= 4xy - x^2y - 2xy^2.$$

The domain of F is the open triangle in the xy plane which is bounded by the x and y axes and the line $x + 2y = 4$. To obtain the critical points of F, we shall solve the equations

$$\frac{\partial F}{\partial x} = 4y - 2xy - 2y^2 = 0 \qquad\qquad (18.133)$$

$$\frac{\partial F}{\partial y} = 4x - x^2 - 4xy = 0. \qquad\qquad (18.134)$$

In these equations we shall divide each term by $-2y$ and $-x$, respectively, since $x > 0$ and $y > 0$ for any $(x, y) \in \mathcal{D}_f$. Then after transposing terms, we have the system

$$x + y = 2$$
$$x + 4y = 4$$

which when solved gives the critical point $(x, y) = (\frac{4}{3}, \frac{2}{3})$.

To show that $F(\frac{4}{3}, \frac{2}{3}) = \frac{32}{27}$ is the maximum volume we will use Theorem 18.12.4(b). Therefore we obtain from (18.133) and (18.134)

$$\frac{\partial^2 F}{\partial x^2} = -2y \qquad \frac{\partial^2 F}{\partial y^2} = -4x \qquad \frac{\partial^2 F}{\partial y\, \partial x} = 4 - 2x - 4y.$$

The values of these second partials at $(\frac{4}{3}, \frac{2}{3})$ are, respectively,

$$\frac{\partial^2 F}{\partial x^2} = -\frac{4}{3} \qquad \frac{\partial^2 F}{\partial y^2} = -\frac{16}{3} \qquad \frac{\partial^2 F}{\partial y\, \partial x} = -\frac{4}{3},$$

and hence

$$\Delta(\tfrac{4}{3}, \tfrac{2}{3}) = (-\tfrac{4}{3})(-\tfrac{16}{3}) - (-\tfrac{4}{3})^2 > 0.$$

Thus by Theorem 18.12.4(b) $F(\frac{4}{3}, \frac{2}{3}) = \frac{32}{27}$ is the required maximum volume of the box.

Exercise Set 18.12

In Exercises 1–6 discuss the absolute extrema of the given functions.

1. $f(x, y) = x^2 + xy + y^2 - 4x + 4y - 2$

2. $f(x, y) = 3xy - x^2 + 4x - 6y + 5$

3. $f(x, y) = 4xy - x^2 - y^2 + 6x - 2$

4. $f(x, y) = x^2 - xy + 2y^2 - 2x - 5y - 3$

5. $f(x, y) = 2xy - x^2 - 3y^2 + 8x - 2$

6. $f(x, y) = -x^2 + xy - y^2 + 2x - 3y + 1$

In Exercises 7 and 8 find the maximum and minimum values of the given function on the closed square with vertices $(0, 0)$, $(4, 0)$, $(4, 4)$, and $(0, 4)$.

7. $f(x, y) = 3x^2 - 6xy + y^3 - 9y + 2$

8. $f(x, y) = (x - 2y - 2)^2 + (2x + y)^2$

In Exercises 9 and 10 find the maximum and minimum values of the given function on the closed unit disc $x^2 + y^2 \leq 1$.

9. $f(x, y) = 2x^2 + 4y^2 + 2x + 1$

10. $f(x, y) = e^{x-3y}$

Solve Exercises 11–20 using theory which has been discussed in Section 18.12.

11. Find three numbers whose sum is N and the sum of whose squares is a minimum.

12. Find the distance from the origin to the plane $3x + 4y - z = 12$.

13. Find the distance between the lines $x = 2 - t$, $y = -1 + 3t$, $z = 2t$, and $\dfrac{x - 1}{-2} = \dfrac{y - 3}{1} = \dfrac{z}{-1}$.

14. If $f(x, y) = x^3 + 2x^2y + y^3 - 2x^2$, at what point is the minimum value attained for the directional derivatives of f in the direction of the vector $\mathbf{u} = \left\langle \dfrac{1}{\sqrt{5}}, \dfrac{2}{\sqrt{5}} \right\rangle$?

15. If a department store chain charges p dollars per television set of a certain kind and q dollars per service policy for this kind of set, then it can sell $27{,}500 - 30p - 5q$ television sets and $6500 - 5p - 20q$ service policies. For maximum revenue, what should the chain charge per television set? per service policy?

16. A company selling a certain product spends amounts x and y for advertising on television and in a newspaper respectively. The sales s resulting from this advertising are given by

$$s = \frac{200x}{4 + x} + \frac{100y}{9 + y}.$$

The net profit resulting in these sales is given by $\frac{1}{4}s - x - y$. Find x and y for the maximum net profit.

17. Find the maximum volume for a rectangular box which can be inscribed in the ellipsoid

$$\frac{x^2}{a^2} + \frac{y^2}{b^2} + \frac{z^2}{c^2} = 1$$

so that its faces are parallel to the coordinate axes.

18. Find the dimensions of the rectangular box with no top which has the maximum volume for a surface area of 30 ft^2.

19. Find the plane through the point $(1, 2, 3)$ which forms a tetrahedron of minimum volume in the first octant with the coordinate planes.

20. In statistics the line $y = mx + b$ which gives the "best fit" for the points (x_1, y_1), (x_2, y_2), . . . , (x_n, y_n) in the sense that the sum of the squares of the vertical distances from the points to the line is a minimum is called a *least-squares line* for the points (Figure 18–23). Prove that the slope and y intercept of this line are given by

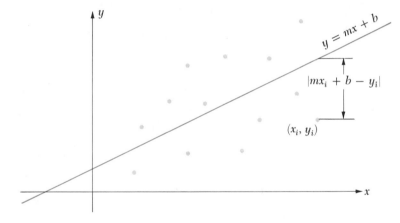

Figure 18–23

$$m = \frac{\begin{vmatrix} \sum\limits_{i=1}^{n} y_i & n \\ \sum\limits_{i=1}^{n} x_i y_i & \sum\limits_{i=1}^{n} x_i \end{vmatrix}}{\begin{vmatrix} \sum\limits_{i=1}^{n} x_i & n \\ \sum\limits_{i=1}^{n} x_i^2 & \sum\limits_{i=1}^{n} x_i \end{vmatrix}} \qquad b = \frac{\begin{vmatrix} \sum\limits_{i=1}^{n} x_i & \sum\limits_{i=1}^{n} y_i \\ \sum\limits_{i=1}^{n} x_i^2 & \sum\limits_{i=1}^{n} x_i y_i \end{vmatrix}}{\begin{vmatrix} \sum\limits_{i-1}^{n} x_i & n \\ \sum\limits_{i=1}^{n} x_i^2 & \sum\limits_{i=1}^{n} x_i \end{vmatrix}}$$

HINT: The vertical distance from a typical point (x_i, y_i) to the line $y = mx + b$ is $|(mx_i + b) - y_i|$ and hence the sum of the squares of the distances from the points $(x_1, y_1), \ldots, (x_n, y_n)$ to the line can be given by $f(m, b) = \sum_{i=1}^{n}(mx_i + b - y_i)^2$.

21. Find the least-squares line for the points $(1, 0)$, $(2, 3)$, $(3, 2)$, $(5, 4)$.

22. Prove Theorem 18.12.4(b).

23. Prove Theorem 18.12.4(c).

19. Multiple Integrals; Line Integrals

19.1   Iterated Double Integral

In this section the iterated double integral, an extension of the definite integral, will be introduced. Then in Section 19.2 this iterated integral will be utilized to obtain volumes of certain types of solid regions. In the subsequent sections of this chapter the *double integral* and *triple integral* will be defined and physical and geometric applications of these multiple integrals discussed.

In this chapter we will consider sets in R^2 of the type given by

$$T = \{(x, y): x \in [a, b] \text{ and } y \in [\phi_1(x), \phi_2(x)]\} \tag{19.1}$$

where the functions ϕ_1 and ϕ_2 are continuous on $[a, b]$ and $\phi_2(x) \geq \phi_1(x)$ for every $x \in [a, b]$ (Figure 19–1(a)). A set of this kind is called a *type I plane region*. Also, we will discuss sets in R^2 of the type given by

$$T = \{(x, y): y \in [c, d] \text{ and } x \in \{\psi_1(y), \psi_2(y)]\} \tag{19.2}$$

where the functions ψ_1 and ψ_2 are continuous on $[c, d]$ and $\psi_2(y) \geq \psi_1(y)$ for every $y \in [c, d]$ (Figure 19-1(b)). A set of this kind is called a *type II plane region*.

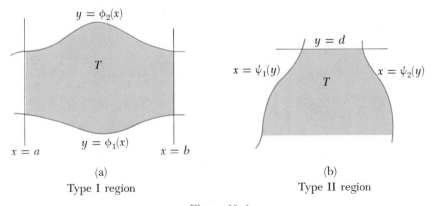

(a)
Type I region

(b)
Type II region

Figure 19-1

In the type I region y ranges over an interval whose endpoints depend upon x, and x is chosen in an interval whose endpoints are constants. For the type

II region these roles for x and y are reversed. A type I or II region will be called an *elementary region* (in the xy plane). An elementary region is clearly an example of a bounded closed set.

Suppose the function F of two variables is continuous on the set T given by (19.1). It can be proved that the function given by

$$G(x) = \int_{\phi_1(x)}^{\phi_2(x)} F(x, y)\, dy \tag{19.3}$$

is continuous on $[a, b]$ although the proof is beyond our scope here. Hence

$$\int_a^b G(x)\, dx = \int_a^b \left[\int_{\phi_1(x)}^{\phi_2(x)} F(x, y)\, dy \right] dx \tag{19.4}$$

exists. The integral on the right in (19.4) is called an *iterated (double) integral*. It is customarily written without brackets so

$$\int_a^b \int_{\phi_1(x)}^{\phi_2(x)} F(x, y)\, dy\, dx = \int_a^b \left[\int_{\phi_1(x)}^{\phi_2(x)} F(x, y)\, dy \right] dx. \tag{19.5}$$

Example 1   Evaluate the iterated double integral $\displaystyle\int_1^3 \int_{x^{1/3}}^x \frac{y^2}{x}\, dy\, dx.$

SOLUTION   From (19.5) it is seen that we first integrate with respect to y while holding x constant and then use the limits with the inside integral sign, which are in general functions of x. Thus integrating with respect to y, we obtain

$$\int_{x=1}^{x=3} \int_{y=x^{1/3}}^{y=x} \frac{y^2}{x}\, dy\, dx = \int_{x=1}^{x=3} \frac{1}{x}\left(\frac{y^3}{3} \right) \Big]_{y=x^{1/3}}^{y=x} dx$$

$$= \int_1^3 \frac{1}{x}\left(\frac{x^3}{3} - \frac{x}{3} \right) dx$$

$$= \int_1^3 \left(\frac{x^2}{3} - \frac{1}{3} \right) dx.$$

Next integrate with respect to x and use the limits with the outside integral sign. Then

$$\int_1^3 \int_{x^{1/3}}^x \frac{y^2}{x}\, dy\, dx = \left(\frac{x^3}{9} - \frac{x}{3} \right) \Big]_1^3 = \frac{20}{9}.$$

From this example it is seen that the order in which the variables of an iterated integral are integrated is indicated by the order in which their differentials appear, reading from *left to right*. The corresponding limits of integration are associated with the integral signs, reading from *right to left*. Thus we also have

$$\int_c^d \int_{\psi_1(y)}^{\psi_2(y)} F(x, y) \, dx \, dy = \int_c^d \left[\int_{\psi_1(y)}^{\psi_2(y)} F(x, y) \, dx \right] dy$$

if the integrals on the right exist.

Example 2   Evaluate the iterated integral $\int_0^4 \int_0^y \sqrt{3x + y} \, dx \, dy$.

SOLUTION   We first integrate with respect to x while keeping y constant. Thus

$$\int_0^4 \int_0^y \sqrt{3x + y} \, dx \, dy = \int_0^4 \tfrac{2}{9}(3x + y)^{3/2} \Big]_0^y dy.$$

After substituting the limits for x, we have successively

$$\int_0^4 \int_0^y \sqrt{3x + y} \, dx \, dy = \int_0^4 \tfrac{2}{9}[(4y)^{3/2} - y^{3/2}] \, dy$$

$$= \int_0^4 \tfrac{14}{9} y^{3/2} \, dy$$

$$= \tfrac{28}{45} y^{5/2} \Big]_0^4 = \tfrac{896}{45}.$$

Exercise Set 19.1

In Exercises 1–12 evaluate the given iterated integral.

1. $\displaystyle \int_1^4 \int_{-1}^3 xy^2 \, dy \, dx$

2. $\displaystyle \int_{-1}^2 \int_0^3 x^3 y^2 \, dx \, dy$

3. $\displaystyle \int_2^5 \int_{2y-1}^{y^2} dx \, dy$

4. $\displaystyle \int_1^3 \int_0^x (x^3 y - 2y^4) \, dy \, dx$

5. $\displaystyle \int_1^4 \int_x^{2x} \sqrt{\frac{x}{2y}} \, dy \, dx$

6. $\displaystyle \int_{-1}^1 \int_0^y \frac{x \, dx \, dy}{(4 - x^2)^{3/2}}$

7. $\displaystyle \int_1^4 \int_0^{4y} \frac{dx \, dy}{(2x + y)^{3/2}}$

8. $\displaystyle \int_0^{\pi/4} \int_{2a}^{2a\sqrt{2}\cos\theta} r^2 \, dr \, d\theta$

9. $\displaystyle \int_0^{\pi/2} \int_0^{a\cos\theta} r^3 \cos\theta \, dr \, d\theta$

10. $\displaystyle \int_{\pi/6}^{\pi/3} \int_0^{\csc\theta\cot\theta} r^3 \sin^2\theta \, dr \, d\theta$

11. $\displaystyle \int_0^{\pi} \int_{\tan^{-1}a/h}^{\pi/2} \sin^3\phi \sin\theta \, d\phi \, d\theta$

12. $\displaystyle \int_0^{\sqrt[3]{\ln 2}} \int_0^x x^2 y e^{xy^2} \, dy \, dx$

19.2   Volumes

We shall first prove a theorem for expressing the volume of a solid region in terms of an iterated integral.

19.2.1　Theorem

Suppose the following conditions are fulfilled:

    (*i*) *T is the type I region given by*

$$T = \{(x, y): x \in [a, b] \text{ and } y \in [\phi_1(x), \phi_2(x)]\}. \tag{19.1}$$

    (*ii*) *The functions f and g are continuous on T.*
    (*iii*) $f(x, y) \geq g(x, y)$ *for every* $(x, y) \in T.$

Then V, the volume of the solid region

$$\{(x, y, z): (x, y) \in T \text{ and } z \in [g(x, y), f(x, y)]\} \tag{19.6}$$

is given by

$$V = \int_a^b \int_{\phi_1(x)}^{\phi_2(x)} [f(x, y) - g(x, y)] \, dy \, dx. \tag{19.7}$$

PROOF　The proof utilizes the "volume by slicing" formula (Theorem 8.2.2). The area of any cross section of the solid region (19.6) in a plane perpendicular to the *x* axis will be expressed as a function of *x*. Since the cross section of (19.6) in the plane $x = c$ has the upper boundary $z = f(c, y)$, the lower boundary $z = g(c, y)$, and extends from $y = \phi_1(c)$ to $y = \phi_2(c)$ (Figure 19–2), the area of this cross section is

$$A(c) = \int_{y=\phi_1(c)}^{y=\phi_2(c)} [f(c, y) - g(c, y)] \, dy. \tag{19.8}$$

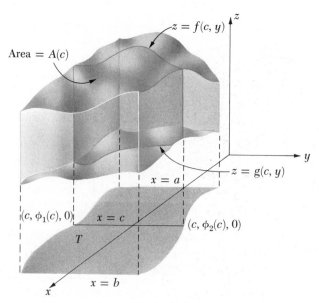

Figure 19–2

The function A defined by (19.8) is continuous at any number c in $[a, b]$ like the function G given by (19.3). Hence, as in (19.4),

$$V = \int_a^b A(x)\, dx = \int_a^b \int_{\phi_1(x)}^{\phi_2(x)} [f(x, y) - g(x, y)]\, dy\, dx.$$

In applying (19.7) it should be remembered that the surfaces $z = f(x, y)$ and $z = g(x, y)$ are respectively the upper and lower boundaries of the solid region. The limits of integration are determined from the plane region T that is obtained by projecting the points of the solid region (19.6) into the xy plane. In particular, from (19.7) if $g(x, y) = 0$ for every $(x, y) \in T$ (that is, the xy plane is the lower boundary of the solid region (19.6)), then (19.7) simplifies to

$$V = \int_a^b \int_{\phi_1(x)}^{\phi_2(x)} f(x, y)\, dy\, dx.$$

Example 1   Find the volume of the solid in the first octant which is bounded from above by the cylinder $x^2 + z^2 = 16$ and on the right by the plane $y = \frac{3}{2}x$ (Figure 19-3).

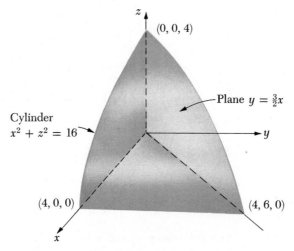

Figure 19-3

SOLUTION   Along the upper boundary surface $z = \sqrt{16 - x^2}$ and on the lower boundary surface $z = 0$. The region T in the xy plane is between the graphs of the lines $y = 0$ and $y = \frac{3}{2}x$ and extends from $x = 0$ to $x = 4$. Therefore from (19.7) the volume V of the solid is

$$V = \int_0^4 \int_0^{3x/2} \sqrt{16 - x^2}\, dy\, dx$$

$$= \int_0^4 \sqrt{16 - x^2}\, y \Big|_0^{3x/2}\, dx = \int_0^4 \tfrac{3}{2}x\sqrt{16 - x^2}\, dx$$

$$= -\tfrac{1}{2}(16 - x^2)^{3/2} \Big|_0^4 = 32.$$

If the statement of Theorem 19.2.1 is modified so that T is the type II region given by

$$T = \{(x, y): y \in [c, d] \text{ and } x \in [\psi_1(y), \psi_2(y)],$$

then the volume of the solid region (19.6) is

$$V = \int_c^d \int_{\psi_1(y)}^{\psi_2(y)} [f(x, y) - g(x, y)] \, dx \, dy. \qquad (19.9)$$

The region T in Example 1 is also a type II region since it is between the graphs of $x = \frac{2}{3}y$ and $x = 4$ and extends from $y = 0$ to $y = 6$. Hence by (19.9) the volume of the solid in Example 1 is also given by

$$V = \int_0^6 \int_{2y/3}^4 \sqrt{16 - x^2} \, dx \, dy.$$

Integrating with respect to x, using (26) from the Table of Integrals, we obtain

$$V = \int_0^6 \left(\frac{x\sqrt{16 - x^2}}{2} + 8 \sin^{-1} \frac{x}{4} \right) \Bigg|_{2y/3}^4 \, dy,$$

and hence

$$V = \int_0^6 \left[4\pi - \frac{1}{3} y \sqrt{16 - \frac{4}{9}y^2} - 8 \sin^{-1} \frac{y}{6} \right] dy.$$

After integrating with respect to y, using (34) from the Table of Integrals,

$$V = \left[4\pi y + \frac{1}{4} \left(16 - \frac{4}{9}y^2 \right)^{3/2} - 8y \sin^{-1} \frac{y}{6} - 8\sqrt{36 - y^2} \right]_0^6,$$

$$= 32$$

which agrees with the volume obtained in Example 1.

In volume problems such as this where the plane region T is both of type I and type II, the student should naturally use either formula (19.7) or (19.9), whichever is easier to apply.

Example 2   Find the volume of the solid bounded from above by the plane $2x + y + z = 4$, from below by the plane $z = y$, and on the left by the plane $y = 0$ (Figure 19–4(a)).

      SOLUTION   We first find the plane region T into which the solid region can be projected. On the intersection line of the upper and lower planes,

$$z = 4 - 2x - y = y$$

and hence if z is eliminated, the x and y coordinates of any point on this line satisfy

$$x + y = 2. \qquad (19.10)$$

Therefore (19.10) is also the equation of the projection of the intersection line into the xy plane. Thus our region T in the xy plane is bounded by the x and y axes

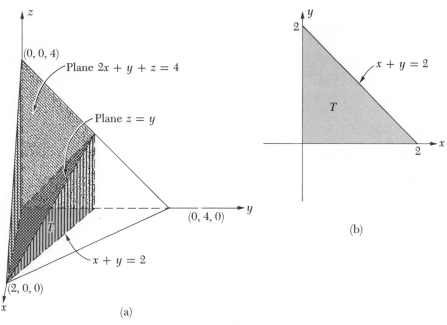

Figure 19–4

and the line (19.10) (Figure 19–4(b)). From (19.7) the volume of the solid region is given by

$$V = \int_0^2 \int_0^{2-x} [(4 - 2x - y) - y]\, dy\, dx = \int_0^2 \int_0^{2-x} 2(2 - x - y)\, dy\, dx$$

$$= \int_0^2 \left[2(2 - x)y - y^2 \right]_0^{2-x} dx = \int_0^2 (2 - x)^2\, dx$$

$$= -\frac{(2 - x)^3}{3} \Big]_0^2 = \frac{8}{3}.$$

Formulas analogous to (19.7) are obtained by rearranging the roles of the variables x, y, and z in the statement of Theorem 19.2.1. For example, if the solid region S is bounded in front by the surface $x = f(y, z)$, in back by the surface $x = g(y, z)$, and S projects into a plane region in the yz plane given by

$$\{(y, z): y \in [a, b] \text{ and } z \in [\phi_1(y), \phi_2(y)]\},$$

then the volume of S is given by

$$V = \int_a^b \int_{\phi_1(y)}^{\phi_2(y)} [f(y, z) - g(y, z)]\, dz\, dy. \qquad (19.11)$$

Example 3   Find the volume of the solid in Example 2 using formula (19.11). See Figure 19–5(a).

SOLUTION   We first find the plane region T in the yz plane into which the solid region is projected. Since the planes $y = 0$ and $y = z$ are perpendicular to the yz plane, and since the plane $2x + y + z = 4$ intersects the yz plane in the line $y + z = 4$, this region T is bounded by the lines $y = 0$, $y = z$, and $y + z = 4$ (Figure 19–5(b)). Thus from (19.11) the volume of the solid region is

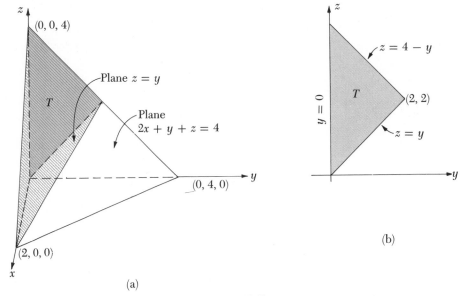

Figure 19–5

$$V = \int_0^2 \int_y^{4-y} \tfrac{1}{2}(4 - y - z)\, dz\, dy = \int_0^2 \left[\tfrac{1}{2}(4 - y)z - \tfrac{1}{4}z^2 \right]_y^{4-y}$$

$$= \int_0^2 \left[\tfrac{1}{4}(4 - y)^2 - 2y + \tfrac{3}{4}y^2 \right] dy = \left[-\tfrac{1}{12}(4 - y)^3 - y^2 + \tfrac{1}{4}y^3 \right]_0^2$$

$$= \tfrac{8}{3}.$$

Exercise Set 19.2

In Exercises 1–10 obtain the volume of the solid region described in the given statement using an iterated double integral. Sketch the solid region.

1. Bounded by the parabolic cylinder $z = 9 - x^2$ and the planes $y = -1$, $y = 4$, and $z = 0$

2. Bounded from above by the plane $z = y + 3$, from below by the plane $z = 0$, and inside the cylinder $x^2 + y^2 = 9$

3. In the first octant and bounded by the cylinders $x^2 + y^2 = a^2$ and $x^2 + z^2 = a^2$, where $a > 0$

4. Bounded by the planes $z = 4 - x - y$, $x = 0$, $y = x$, and $z = 0$

5. Bounded by the planes $x + y + z = 6$, $x + 2y = z$, $x = 0$, and $y = 0$

6. Bounded by the planes $z = 2$ and $x + y + z = 2$, and the parabolic cylinders $y^2 = x$ and $x^2 = y$

7. Bounded from above by the elliptic paraboloid $z = x^2 + 9y^2$, from below by the plane $z = 0$, and inside the elliptic cylinder $x^2 + 9y^2 = 9$

8. Bounded from above by the paraboloid $z = 4 - x^2 - y^2$ and from below by the plane $z = 2x$

9. Between the plane $x = 5$ and parabolic cylinder $x = 9 - y^2$ and inside the cylinder $y^2 + z^2 = 9$

10. Bounded by the graphs of $y = 9 - z^2$ and $y = x^2 + 2z^2$

19.3   Double Integral

In this section we will define the *double integral* of a function of two variables on a bounded closed set in R^2. If such a function is continuous on an elementary region, then its double integral on the region will be expressed in terms of an iterated double integral. To begin the discussion, let T be a bounded closed set in R^2. Suppose T is the union of the closed sets $\Delta T_1, \Delta T_2, \ldots, \Delta T_n$ (Figure 19–6) having respective areas $\Delta A_1, \Delta A_2, \ldots, \Delta A_n$ where no two of these subsets have any points in common except possibly boundary points. The set $P = \{\Delta T_1, \Delta T_2, \ldots,$

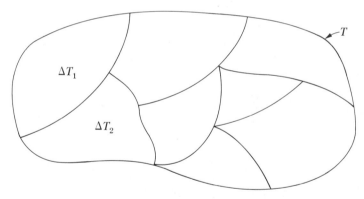

Figure 19-6

$\Delta T_n\}$ will be called a *partition* of T. Each subset ΔT_i has a *diameter*—that is, a maximum distance between any two points in the set. This is true by Theorem 18.12.2 since the distance between two points in R^2 is a continuous function of the coordinates of the points. The greatest diameter of any of the subsets ΔT_i is called the *norm* of the partition P and will be denoted, as before, by Δ_P. If f, a function of two variables, is defined on T and an arbitrary point (x_i, y_i) is chosen in each ΔT_i, then the sum

$$S_P = \sum_{i=1}^{h} f(x_i, y_i)\,\Delta A_i \tag{19.12}$$

is called a *Riemann sum of f on T associated with the partition P.* The reader should note the analogy between the ideas expressed here and those given in Sections 7.4 and 7.5.

If $f(x, y) \geq 0$ for every $(x, y) \in T$, then a typical term $f(x_i, y_i) \Delta A_i$ in the sum (19.12) is the volume of a solid cylinder having a base of area ΔA_i and altitude $f(x_i, y_i)$ (Figure 19–7).

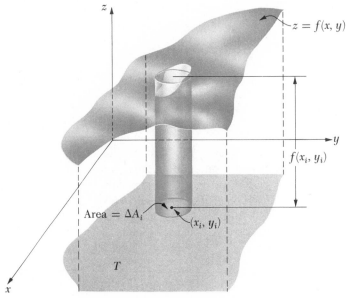

Figure 19–7

The *double integral* of a function will now be defined in terms of Riemann sums.

19.3.1 Definition

Let f, a function of two variables, be defined on a bounded closed region T in R^2. A number L is called the *double integral of f on T* if and only if for every $\epsilon > 0$ there is a $\delta > 0$ such that any Riemann sum S_P of f on T satisfies

$$|S_P - L| < \epsilon$$

whenever $\Delta_P < \delta$. The function f is said to be *integrable* on T if and only if such a number L exists.

Henceforth the double integral of f on T will be denoted by

$$\iint\limits_{T} f(x, y)\, dA$$

when it exists. We leave as an exercise the proof that the double integral is unique when it exists.

We now state some basic theorems about double integrals which parallel the properties of definite integrals that were given in Theorems 7.5.5, 7.6.1–7.6.5, and 7.6.7. In each of these theorems T is assumed to be bounded and closed. The proofs of Theorems 19.3.3–19.3.8 below are omitted since they parallel the proofs of the earlier theorems. The proof of Theorem 19.3.2 is also omitted since it is more properly discussed in advanced calculus.

19.3.2   Theorem

If f, a function of two variables, is continuous on T, then $\iint_T f(x, y)\, dA$ exists.

19.3.3   Theorem

If $f(x, y) = k$, a constant, for every $(x, y) \in T$, then

$$\iint_T k\, dA = kA$$

where A is the area of T.

19.3.4   Theorem

If $f(x, y) \geq 0$ for every $(x, y) \in T$ and $\iint_T f(x, y)\, dA$ exists, then $\iint_T f(x, y)\, dA \geq 0$.

19.3.5   Theorem

If k is constant and $\iint_T f(x, y)\, dA$ exists, then

$$\iint_T kf(x, y)\, dA = k \iint_T f(x, y)\, dA.$$

19.3.6   Theorem

If f and g are integrable on T, then

$$\iint_T [f(x, y) \pm g(x, y)]\, dA = \iint_T f(x, y)\, dA \pm \iint_T g(x, y)\, dA.$$

19.3.7   Theorem

If $f(x, y) \geq g(x, y)$ for every $(x, y) \in T$ and f and g are integrable on T, then

$$\iint_T f(x, y)\, dA \geq \iint_T g(x, y)\, dA.$$

19.3.8 Theorem

Suppose $T = T_1 \cup T_2$ where T_1 and T_2 are bounded closed sets having no points in common except possibly boundary points. If f is integrable on T_1 and T_2, then

$$\iint_T f(x, y)\, dA = \iint_{T_1} f(x, y)\, dA + \iint_{T_2} f(x, y)\, dA.$$

If T is the type **I** region given by

$$T = \{(x, y) : x \in [a, b] \text{ and } y \in [\phi_1(x), \phi_2(x)]\} \tag{19.13}$$

and f is continuous and non-negative-valued on T, we will show that the volume V of the solid region

$$\{(x, y, z) : (x, y) \in T \text{ and } z \in [0, f(x, y)]\} \tag{19.14}$$

is given by $\iint_T f(x, y)\, dA$. First recall from Theorem 19.2.1 that

$$V = \int_a^b \int_{\phi_1(x)}^{\phi_2(x)} f(x, y)\, dy\, dx \tag{19.15}$$

and also by Theorem 19.3.2 $\iint_T f(x, y)\, dA$ exists. Then for any partition of T let m_i and M_i, respectively, be the minimum and maximum values of f on the subregion ΔT_i of T. Each solid subregion of (19.14) of the form

$$\{(x, y, z) : (x, y) \in \Delta T_i \text{ and } z \in [0, f(x, y)]\} \tag{19.16}$$

contains a solid cylinder with base ΔT_i and altitude m_i and is contained in another solid cylinder with base ΔT_i and altitude M_i (Figure 19–8). Let ΔV_i be the volume

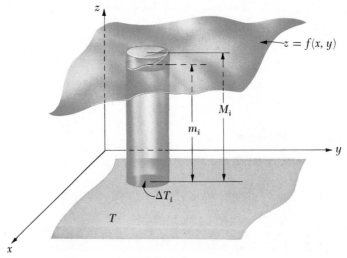

Figure 19–8

of the subregion (19.16). Then for every i, since ΔA_i is the area of ΔT_i,

$$m_i\, \Delta A_i \leq \Delta V_i \leq M_i\, \Delta A_i.$$

Upon summing over every i, V, the volume of (19.14), satisfies

$$\sum_i m_i \, \Delta A_i \le V \le \sum_i M_i \, \Delta A_i. \qquad (19.17)$$

The left and right members of (19.17), are Riemann sums of the function f on T. Since these sums can be made to differ from $\iint_T f(x, y) \, dA$ by as little as desired when the norm of the associated partition is sufficiently small,

$$V = \iint_T f(x, y) \, dA. \qquad (19.18)$$

From (19.15) and (19.18) we then obtain

$$\iint_T f(x, y) \, dA = \int_a^b \int_{\phi_1(x)}^{\phi_2(x)} f(x, y) \, dy \, dx. \qquad (19.19)$$

In the following theorem, which is proved in advanced calculus, the formula (19.19) is obtained after relaxing the condition that $f(x, y) \ge 0$ on T.

19.3.9   Theorem

If f, a function of two variables, is continuous on the type I region (19.13), then (19.19) holds.

Example 1   Evaluate $\iint_T 6xy \, dA$ where T is the region in the first quadrant bounded by the graphs of $y = x^2$ and $y = 2x$ (Figure 19-9).

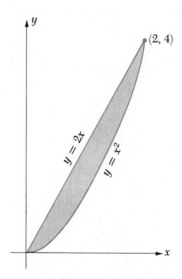

Figure 19-9

SOLUTION   From (19.19)

$$\iint_T 6xy \, dA = \int_0^2 \int_{x^2}^{2x} 6xy \, dy \, dx$$

$$= \int_0^2 3xy^2 \Big]_{x^2}^{2x} dx = \int_0^2 (12x^3 - 3x^5) \, dx$$

$$= (3x^4 - \tfrac{1}{2}x^6) \Big]_0^2 = 16.$$

If in the statement of Theorem 19.3.9 the function f is continuous on the type II region

$$\{(x, y): y \in [c, d] \text{ and } x \in [\psi_1(y), \psi_2(y)]\} \tag{19.20}$$

instead of the set (19.13), then in place of (19.19), we have the conclusion

$$\iint_T f(x, y) \, dA = \int_c^d \int_{\psi_1(y)}^{\psi_2(y)} f(x, y) \, dx \, dy. \tag{19.21}$$

Thus if T is simultaneously defined by (19.13) and (19.20) and f is continuous on T, we have from (19.19) and (19.21)

$$\int_a^b \int_{\phi_1(x)}^{\phi_2(x)} f(x, y) \, dy \, dx = \int_c^d \int_{\psi_1(y)}^{\psi_2(y)} f(x, y) \, dx \, dy. \tag{19.22}$$

Hence when the plane region associated with a given iterated integral is both type I and type II, equation (19.22) enables us to reverse the order of integration of the variables. Such a change is frequently useful when the given iterated integral is difficult to evaluate, as in the next example.

Example 2   Evaluate the iterated double integral $\int_0^2 \int_{x^2}^4 xe^{-y^2} \, dy \, dx$.

SOLUTION   This iterated double integral cannot be evaluated directly since, as we had noted at the beginning of Chapter 11, $\int e^{-y^2} \, dy$ is not expressible in terms of elementary functions. However, since the type I region associated with the given iterated double integral,

$$T = \{(x, y): y \in [x^2, 4] \text{ and } x \in [0, 2]\} \quad \text{(Figure 19–10)}$$

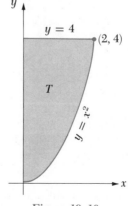

Figure 19–10

is also the type II region defined by

$$T = \{(x, y): x \in [0, \sqrt{y}] \text{ and } y \in [0, 4]\},$$

by (19.22)

$$\int_0^2 \int_{x^2}^4 xe^{-y^2} \, dy \, dx = \int_0^4 \int_0^{\sqrt{y}} xe^{-y^2} \, dx \, dy$$

$$= \int_0^4 e^{-y^2} \cdot \tfrac{1}{2}x^2 \Big]_0^{\sqrt{y}} \, dy = \int_0^4 \tfrac{1}{2}ye^{-y^2} \, dy$$

$$= -\tfrac{1}{4}e^{-y^2} \Big]_0^4 = \tfrac{1}{4}(1 - e^{-16}).$$

The next two theorems are corollaries of Theorem 19.3.9.

19.3.10   Theorem

If T is an elementary region, then

$$\iint_T dA = \text{Area of } T.   \tag{19.23}$$

PROOF   We can suppose that T is the type I region given by (19.13) as the proof is analogous when T is a type II region. If $f(x, y) = 1$ for every $(x, y) \in T$, then by (19.19)

$$\iint_T dA = \int_a^b \int_{\phi_1(x)}^{\phi_2(x)} dy \, dx = \int_a^b [\phi_2(x) - \phi_1(x)] \, dx$$

and hence by Definition 8.1.1

$$\iint_T dA = \text{Area of } T.$$

19.3.11   Theorem

If f and g are continuous functions of two variables on the elementary region T, and $f(x, y) \geq g(x, y)$ for every $(x, y) \in T$, then V, the volume of the solid region

$$\{(x, y, z): (x, y) \in T \text{ and } z \in [g(x, y), f(x, y)]\},$$

is given by

$$V = \iint_T [f(x, y) - g(x, y)] \, dA.$$

This result follows immediately from Theorems 19.3.9 and 19.2.1 (or the modifications of these theorems for **type II** regions).

19.3.12 Theorem

If f is a continuous function of two variables on the elementary region T, and A is the area of T, then there exist numbers m and M such that

$$mA \leq \iint_T f(x, y)\, dA \leq MA.$$

The proof of this theorem, which utilizes Theorem 19.3.7, is left as an exercise.

Exercise Set 19.3

In Exercises 1–6 evaluate the given iterated integral after changing the order of integration of the variables

1. $\displaystyle\int_0^1 \int_0^y e^{x-y}\, dx\, dy$

2. $\displaystyle\int_0^3 \int_0^{2-2x/3} x\, dy\, dx$

3. $\displaystyle\int_{-1}^1 \int_{x+1}^2 xy\, dy\, dx$

4. $\displaystyle\int_1^e \int_{\ln y}^1 y\, dx\, dy$

5. $\displaystyle\int_0^4 \int_{\sqrt{x}}^2 (x + y^2)^{1/2}\, dy\, dx$

6. $\displaystyle\int_0^{\sqrt{3}/2} \int_0^{\sin^{-1} x} x\, dy\, dx$

In Exercises 7–14 evaluate the given double integrals.

7. $\displaystyle\iint_T (x + 2y - 3)\, dA$ where $T = \{(x, y): x \in [-1, 2]$ and $y \in [1, 3]\}$

8. $\displaystyle\iint_T (x^2 + 2xy)\, dA$ where $T = \{(x, y): x \in [0, 2]$ and $y \in [1, 4]\}$

9. $\displaystyle\iint_T (4 - x^2 - y^2)\, dA$ where $T = \{(x, y): x \in [0, 1]$ and $y \in [x^2, x]\}$

10. $\displaystyle\iint_T (xy + 2)\, dA$ where $T = \{(x, y): y \in [0, 2]$ and $x \in [y/2, 6 - y]\}$

11. $\displaystyle\iint_T y^2 \sin xy\, dA$ where $T = \left\{(x, y): y \in \left[\sqrt{\frac{\pi}{3}}, \sqrt{\frac{\pi}{2}}\right]$ and $x \in [y, 2y]\right\}$

12. $\displaystyle\iint_T x^2 y e^{-xy^2}\, dA$ where $T = \{(x, y): x \in [0, \sqrt{\ln 3}]$ and $y \in [0, \sqrt{x}]\}$

13. $\displaystyle\iint_T e^{\cos x}\, dA$ where $T = \{(x, y): y \in [0, 1]$ and $x \in [\sin^{-1} y, \frac{\pi}{2}]\}$

14. $\displaystyle\iint_T \frac{x}{1 + y^4}\, dA$ where $T = \{(x, y): x \in [0, 1]$ and $y \in [x^2, x^{2/3}]\}$

15. Prove that $\iint_T f(x, y)\, dA$ is unique when it exists.

16. Prove Theorem 19.3.12.

17. Prove: If a function f is continuous and positive-valued on an elementary region T, then $\iint_T f(x, y)\, dA > 0$.

18. Give the details of the proof for obtaining (19.18) from (19.17).

19.4   Center of Mass; Centroid

We recall from physics that if a particle of mass m is located at a point (x, y) in the plane, the (*first*) *moment of the particle* about the y axis is defined as mx and the (first) moment of the particle about the x axis as my. Then if a system consists of n discrete particles with masses $m_1, m_2, \ldots, m_n$ which are located at points $(x_1, y_1)(x_2, y_2), \ldots, (x_n, y_n)$, respectively, by definition, the (*first*) *moment of the system* of particles about the y axis is

$$M_y = \sum_{i=1}^{n} m_i x_i, \tag{19.24}$$

and the (first) moment of the system about the x axis is

$$M_x = \sum_{i=1}^{n} m_i y_i. \tag{19.25}$$

The *center of mass* $(\bar{x}, \bar{y})$ of the system of particles is then given by the equations

$$\bar{x} = \frac{\sum_{i=1}^{n} m_i x_i}{\sum_{i=1}^{n} m_i} \qquad \bar{y} = \frac{\sum_{i=1}^{n} m_i y_i}{\sum_{i=1}^{n} m_i}. \tag{19.26}$$

Since from (19.26)

$$M_y = \left(\sum_{i=1}^{n} m_i\right)\bar{x} \qquad M_x = \left(\sum_{i=1}^{n} m_i\right)\bar{y},$$

the moment of the system about a coordinate axis is the moment about the axis obtained when the total mass of the system is considered to be concentrated at the center of mass.

We will extend the notion of the center of mass to plane regions. Suppose T is a bounded closed region having mass m that is given by

$$m = \iint_T \sigma(x, y)\, dA \tag{19.27}$$

where the function σ is continuous and positive-valued on T (from Exercise 17, Exercise Set 19.3, m must be positive). The function σ here is called the *density function* for T, and when $(x, y) \in T$, $\sigma(x, y)$ is termed the *density* of T at (x, y). A plane closed region having a density function is called a *lamina*, and for our purposes a lamina may be visualized as a thin plate. Suppose we form a partition of T into n subregions and consider an arbitrary subregion ΔT_i. Intuition suggests that if the partition is fine enough and an arbitrary point (x_i, y_i) is selected in ΔT_i, then the mass of ΔT_i is approximately $\sigma(x_i, y_i) \Delta A_i$. By analogy with the definitions for systems of particles the moments of the subregion about the y axis and x axis would seem to be approximately $x_i \sigma(x_i, y_i) \Delta A_i$ and $y_i \sigma(x_i, y_i) \Delta A_i$, respectively. Also, by analogy with (19.24) and (19.25), M_y and M_x, the moments of T about the y axis and x axis, respectively, would seem to be given by the approximations

$$M_y \approx \sum_{i=1}^{n} x_i \sigma(x_i, y_i) \Delta A_i \qquad M_x \approx \sum_{i=1}^{n} y_i \sigma(x_i, y_i) \Delta A_i. \tag{19.28}$$

Intuition suggests that the smaller the norm of the partition, the better are the approximations expressed by (19.28). Since the sums in (19.28) are Riemann sums associated with the double integrals $\iint_T x\sigma(x, y) \, dA$ and $\iint_T y\sigma(x, y) \, dA$, respectively, the following definitions are given.

19.4.1   Definition

Suppose T is a bounded closed region with mass given by (19.28). The *(first) moment* of T about the y axis is given by

$$M_y = \iint_T x\sigma(x, y) \, dA \tag{19.29}$$

and about the x axis is

$$M_x = \iint_T y\sigma(x, y) \, dA. \tag{19.30}$$

The *center of mass* of T, $(\bar{x}, \bar{y})$ is given by

$$\bar{x} = \frac{M_y}{m} \qquad \bar{y} = \frac{M_x}{m}. \tag{19.31}$$

Example 1   Find the center of mass of the lamina T bounded by the graphs of the equations $y = \sqrt{x}$, $y = 0$, and $x = 4$ if the density at any point is proportional to the distance from the y axis (Figure 19–11).

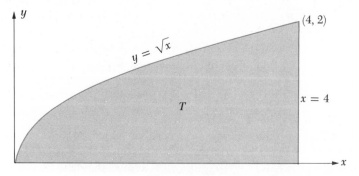

Figure 19-11

SOLUTION   Here the density is given by $\sigma(x) = kx$ where k is constant. From (19.27) the mass of the lamina is given by

$$m = \iint_T \sigma(x, y) \, dA = \iint_T kx \, dA$$

$$= \int_0^4 \int_0^{\sqrt{x}} kx \, dy \, dx = \int_0^4 kx^{3/2} \, dx = \tfrac{2}{5}kx^{5/2} \Big|_0^4$$

$$= \frac{64k}{5}.$$

The moment of T about the y axis is, by (19.29),

$$M_y = \iint_T x\sigma(x, y) \, dA = \iint_T kx^2 \, dA$$

$$= \int_0^4 \int_0^{\sqrt{x}} kx^2 \, dy \, dx = \int_0^4 kx^{5/2} \, dx = \tfrac{2}{7}kx^{7/2} \Big|_0^4$$

$$= \frac{256k}{7}.$$

From (19.30) the moment of T about the x axis is

$$M_x = \iint_T y\sigma(x, y) \, dA = \iint_T kxy \, dA$$

$$= \int_0^4 \int_0^{\sqrt{x}} kxy \, dy \, dx = \int_0^4 \tfrac{1}{2}kx^2 \, dx = \tfrac{1}{6}kx^3 \Big|_0^4$$

$$= \frac{32k}{3}.$$

Hence from (19.31) the coordinates of the center of mass of T are

$$\bar{x} = \frac{\dfrac{256k}{7}}{\dfrac{64k}{5}} = \frac{20}{7} \quad \text{and} \quad \bar{y} = \frac{\dfrac{32k}{3}}{\dfrac{64k}{5}} = \frac{5}{6}$$

and therefore the center of mass of the lamina is $(\tfrac{20}{7}, \tfrac{5}{6})$.

A lamina is said to be *homogeneous* if and only if its density is constant. The center of mass of a homogeneous region is then called the *centroid* or *center of gravity* of the region.

Example 2   Find the centroid of the homogeneous lamina bounded by the graphs of $y = 4 - 3x^2$ and $y = x^2$ (Figure 19–12).

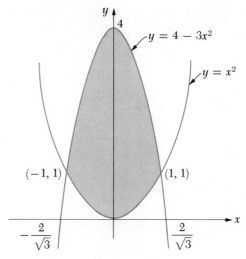

Figure 19-12

SOLUTION   Solving simultaneously the equations of the boundary curves, we obtain their intersection points $(-1, 1)$ and $(1, 1)$. Then denoting the density by $\sigma(x, y) = k$ (constant),

$$m = \iint_T k \, dA = \int_{-1}^{1} \int_{x^2}^{4-3x^2} k \, dy \, dx$$

$$= k \int_{-1}^{1} (4 - 4x^2) \, dx.$$

Since the graph of $y = 4 - 4x^2$ is symmetric to the y axis,

$$m = 2k \int_{0}^{1} (4 - 4x^2) = 2k \left(4x - \frac{4x^3}{3} \right) \Big|_{0}^{1}$$

$$= \frac{16k}{3}.$$

Also

$$M_x = \iint_T yk \, dA = \int_{-1}^{1} \int_{x^2}^{4-3x^2} ky \, dy \, dx$$

$$= \int_{-1}^{1} k \frac{y^2}{2} \Big|_{x^2}^{4-3x^2} dx = k \int_{-1}^{1} \frac{[(4 - 3x^2)^2 - x^4]}{2} \, dx$$

$$= 4k \int_{-1}^{1} (2 - 3x^2 + x^4) \, dx.$$

Since the graph of $y = 2 - 3x^2 + x^4$ is symmetric to the y axis,

$$M_x = 8k \int_0^1 (2 - 3x^2 + x^4)\, dx = 8k\left(2x - x^3 + \frac{x^5}{5}\right)\Big|_0^1$$

$$= \frac{48k}{5}.$$

Hence

$$\bar{y} = \frac{M_x}{m} = \frac{\dfrac{48k}{5}}{\dfrac{16k}{3}} = \frac{9}{5}.$$

Since the homogeneous lamina is symmetric to the y axis, it is intuitively apparent that $\bar{x} = 0$. (This is verified in Exercise 17, Exercise Set 19.4.) The centroid of the lamina is therefore $(0, \frac{9}{5})$.

We note in Example 2 that the coordinates of the centroid of the homogeneous lamina are independent of the density of the lamina.

Suppose a bounded lamina T with density function σ is partitioned into sublaminas $T_1, T_2, \ldots, T_n$. Also suppose the mass of T_i is m_i and the moments of T_i with respect to the x and y axes are, respectively, M_{x_i} and M_{y_i}. Then from (19.27)

$$m = \iint_T \sigma(x, y)\, dA = \sum_{i=1}^n \iint_{T_i} \sigma(x, y)\, dA$$

$$= \sum_{i=1}^n m_i, \tag{19.32}$$

and from (19.29) and (19.30)

$$M_x = \sum_{i=1}^n M_{x_i} \qquad M_y = \sum_{i=1}^n M_{y_i}. \tag{19.33}$$

Hence from (19.31), (19.32), and (19.33), the center of mass of T is given by

$$\bar{x} = \frac{\sum_{i=1}^n m_i \bar{x}_i}{\sum_{i=1}^n m_i} \qquad \bar{y} = \frac{\sum_{i=1}^n m_i \bar{y}_i}{\sum_{i=1}^n m_i}. \tag{19.34}$$

The equations (19.34) are extensions of (19.26) for a system of particles.

Exercise Set 19.4

In Exercises 1–8 obtain the centroid for the homogeneous lamina determined by the given conditions. Sketch each lamina.

1. Bounded by the graphs of $y = 2x + 3$, $y = 0$, $x = 0$, and $x = 3$

2. Formed by the triangle with vertices $(a, 0)$, $(0, 0)$, and $(0, b)$

3. Bounded by the graphs of $y = \sqrt{r^2 - x^2}$, and the x axis

4. Formed by the first quadrant portion of the arc of the ellipse $b^2x^2 + a^2y^2 = a^2b^2$ and the coordinate axes

5. Formed by one loop of the graph of $y^2 = x^2(4 - x^2)$

6. Bounded by the graphs of the parabola $ay = x^2$ and the line $y = a$

7. Bounded by the graphs of $y^2 = 4x$ and $x + y = 3$

8. Bounded by the graphs of $x = (y - 2)^2$ and $x = y$

In Exercises 9–14 obtain the mass and the center of mass for the lamina having the given density and bounded by the graphs of the given equations.

9. $3x + 4y = 12$, $x = 0$, and $y = 0$; density proportional to the product of the distances from the coordinate axes

10. $y = x^2 - 3x - 4$ and $y = 0$; density proportional to the distance from the y axis

11. $x^2 = 4y$ and $y^2 = 4x$; density proportional to the distance from the x axis

12. $2x = y^2$ and $x + y = 4$; density proportional to the distance from the y axis

13. $x^2 + y^2 = a^2$ and $x + y = a$; density proportional to the sum of the distances from the coordinate axes

14. $y = \sin x$, $y = \cos x$, and $x = 0$ in the first quadrant; density proportional to the distance from the line $x = \frac{\pi}{2}$

15. If the functions f and g are continuous on $[a, b]$ and $f(x) \geq g(x)$ for every $x \in [a, b]$, prove that the centroid $(\bar{x}, \bar{y})$ of the homogeneous lamina

$$\{(x, y): x \in [a, b] \text{ and } y \in [g(x), f(x)]\} \tag{19.35}$$

is given by

$$\bar{x} = \frac{\int_a^b x[f(x) - g(x)]\, dx}{A} \qquad \bar{y} = \frac{\int_a^b \frac{1}{2}[f(x))^2 - (g(x))^2]\, dx}{A}$$

where A is the area of the lamina.

16. Find the centroid of the homogeneous lamina shown in Figure 19–13. HINT: Use (19.34).

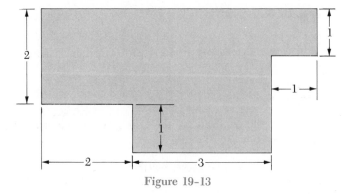

Figure 19-13

17. If the even functions f and g are continuous on $[-a, a]$ and $f(x) \geq g(x)$ for every $x \in [-a, a]$, prove that the centroid of the homogeneous lamina

$$\{(x, y): x \in [-a, a] \text{ and } y \in [g(x), f(x)]\}$$

has an x coordinate of 0.

18. If f and g satisfy the hypotheses of Exercise 15, $g(x) \geq 0$ for every $x \in [a, b]$, and the plane region (19.35) in that exercise is revolved about the x axis, prove that the volume V of the solid generated is given by

$$V = 2\pi \bar{y} A$$

where A is the area of the region (19.35). This theorem is a special case of Pappus's first theorem: *If a plane region is revolved about a line that does not intersect the region except possibly on its boundary, then the volume of the solid generated is the product of the area of the region and the circumference of a circle traversed by the centroid of the region.*

Solve Exercises 19 and 20 using Pappus's first theorem.

19. Find the volume of the solid generated by revolving the rectangle with vertices $(-1, 5)$, $(5, 5)$, $(-1, 1)$, and $(5, 1)$ about the line $x - 3y = 6$.

20. Find the volume of the solid generated by revolving the circle $(x - 4)^2 + (y - 6)^2 = 9$ about the line $x + y = 3$.

21. Define the centroid of the smooth curve $\{(f(t), g(t)): t \in [a, b]\}$ after an analysis involving a partition of $[a, b]$ which is analogous to that used to obtain Definition 19.4.1.

22. Using the definition obtained in Exercise 21, find the centroid of the curve given by

$$x = t^2 \qquad y = t \quad \text{if } t \in [0, 2].$$

23. Find the centroid of the semicircle $y = \sqrt{a^2 - x^2}$.

19.5   Double Integrals in Polar Coordinates

In this section we will consider double integrals over bounded closed regions in R^2 which are defined using polar coordinates. If these regions are not of type I or II, then the familiar formulas (19.19) and (19.21) do not apply. Also, even when these regions are elementary regions, the use of these formulas is often unduly tedious. We therefore require a formula for transforming a double integral to polar coordinates.

Suppose T is the set of points given in Cartesian coordinates by

$$T = \{(r \cos \theta, r \sin \theta): \theta \in [\alpha, \beta] \text{ and } r \in [u_1(\theta), u_2(\theta)]\} \qquad (19.36)$$

where $\beta - \alpha \leq 2\pi$, the functions u_1 and u_2 are continuous on $[\alpha, \beta]$, and $u_2(\theta) \geq u_1(\theta) \geq 0$ for every $\theta \in [\alpha, \beta]$ (Figure 19–14(b)). We shall call the set T a "type III plane region."

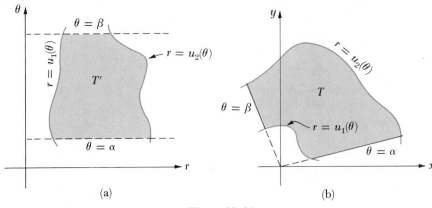

Figure 19-14

The region T can be obtained by a *transformation* of the region

$$T' = \{(r, \theta): \theta \in [\alpha, \beta] \text{ and } r \in [u_1(\theta), u_2(\theta)]\} \tag{19.37}$$

(Figure 19–14(a)) in the $r\theta$ plane in which any point $(r, \theta) \in T'$ *maps into* the point $(x, y) \in T$ given by the polar coordinate equations

$$x = r \cos \theta \qquad\qquad y = r \sin \theta. \tag{19.38}$$

Note also that because of the restrictions imposed upon $u_1(\theta)$, $u_2(\theta)$, α, and β in (19.36), two different points of T' cannot map into the same point in T using equations (19.38). As an illustration of this transformation, using (19.38), the point $(2, \frac{\pi}{3})$, for example, in the $r\theta$ plane maps into $(1, \sqrt{3})$ in the xy plane.

Suppose f is a function of two variables which is continuous on the region T given by (19.36). Then

$$\iint_{T} f(x, y)\, dA$$

exists by Theorem 19.3.2. By constructing lines and circles with polar equations $r = r_i$ and $\theta = \theta_j$, respectively, where $r_i > r_{i-1}$ and $\theta_j > \theta_{j-1}$, we can obtain a family of "truncated sectors" that cover T. Here the truncated sector ΔT_{ij} is given by

$$\Delta T_{ij} = \{(r \cos \theta, r \sin \theta): r \in [r_{i-1}, r_i] \text{ and } \theta \in [\theta_{j-1}, \theta_j]\} \tag{19.39}$$

(Figure 19–15). For $i \geq 1$ and $j \geq 1$, we also let

$$\Delta r_i = r_i - r_{i-1} \qquad \Delta \theta_j = \theta_j - \theta_{j-1}. \tag{19.40}$$

Although ΔT_{ij} may or may not intersect T, we will consider the partition P of T which consists of all possible nonempty sets $\Delta T_{ij} \cap T$. The area of each such set can be computed (by applying the polar coordinate area formula in Theorem 16.5.1 if necessary), so we shall let

$$\Delta A_{ij} = \text{Area of } \Delta T_{ij} \cap T. \tag{19.41}$$

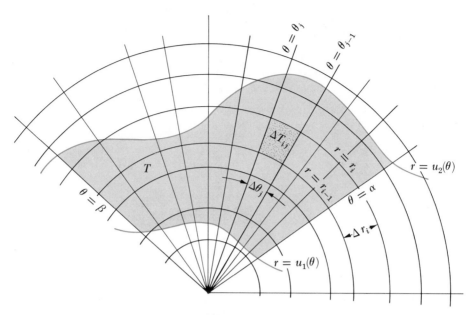

Figure 19-15

If $\Delta T_{ij} \subset T$, then from the area formula for a sector

$$\Delta A_{ij} = \tfrac{1}{2}r_i^2\, \Delta\theta_j - \tfrac{1}{2}r_{i-1}^2\, \Delta\theta_j = \tfrac{1}{2}(r_i^2 - r_{i-1}^2)\, \Delta\theta_j$$

$$= \frac{r_i + r_{i-1}}{2}\, \Delta r_i\, \Delta\theta_j.$$

Letting

$$\overline{r_i} = \frac{r_i + r_{i-1}}{2} \tag{19.42}$$

we then have

$$\Delta A_{ij} = \overline{r_i}\, \Delta r_i\, \Delta\theta_j. \tag{19.43}$$

In each nonempty $\Delta T_{ij} \cap T$ we shall choose a point (x_{ij}, y_{ij}). If $\Delta T_{ij} \subset T$, then $(x_{ij}, y_{ij}) = (\overline{r_i} \cos \overline{\theta}_j, \overline{r_i} \sin \overline{\theta}_j)$ where $\overline{r_i}$ is given by (19.42) and $\overline{\theta}_j$ is an arbitrary number in $[\theta_{j-1}, \theta_j]$. If ΔT_{ij} is not a subset of T and yet intersects T, then (x_{ij}, y_{ij}) is arbitrarily chosen in $\Delta T_{ij} \cap T$. We then form the Riemann sum of f on T:

$$S = \sum_{i,j} f(x_{ij}, y_{ij})\, \Delta A_{ij}$$

$$= \underbrace{\sum_{i,j} f(\overline{r_i} \cos \overline{\theta}_j, \overline{r_i} \sin \overline{\theta}_j)\overline{r_i}\, \Delta r_i\, \Delta\theta_j}_{\text{where } \Delta T_{ij} \subset T} + \underbrace{\sum_{i,j} f(x_{ij}, y_{ij})\, \Delta A_{ij}}_{\text{where } \Delta T_{ij} \not\subset T} \tag{19.44}$$

By choosing the norm of the partition P sufficiently small, (i) S itself, can be made to differ from $\iint_T f(x, y)\, dA$ by as little as is desired, and it can be proved that (ii) the second sum on the right in (19.44) can be made as close to 0 as desired and (iii) the first sum on the right can be made as close as desired to

$$\iint_{T'} f(r \cos \theta, r \sin \theta) \, r \, dA$$

where T' is defined by (19.37). However, the details for the proofs of (ii) and (iii) are omitted. Since T' is a type II region in the $r\theta$ plane,

$$\iint_{T'} f(r \cos \theta, r \sin \theta) \, r \, dA = \int_{\alpha}^{\beta} \int_{u_1(\theta)}^{u_2(\theta)} f(r \cos \theta, r \sin \theta) \, r \, dr \, d\theta.$$

Thus we have an informal proof of the following theorem.

19.5.1   Theorem

Suppose u_1 and u_2 are continuous on the interval $[\alpha, \beta]$ where $\beta - \alpha \le 2\pi$ and $0 \le u_1(\theta) \le u_2(\theta)$. Suppose also that f, a function of two variables, is continuous on the set

$$T = \{(r \cos \theta, r \sin \theta): \theta \in [\alpha, \beta] \text{ and } r \in [u_1(\theta), u_2(\theta)]\}.$$

Then
$$\iint_T f(x, y) \, dA = \int_{\alpha}^{\beta} \int_{u_1(\theta)}^{u_2(\theta)} f(r \cos \theta, r \sin \theta) \, r \, dr \, d\theta. \qquad (19.45)$$

We note that in applying (19.45) $f(x, y)$ is replaced by $f(r \cos \theta, r \sin \theta)$ and dA by $r \, dr \, d\theta$. Also, if $f(x, y) = 1$ for every $(x, y) \in T$, then from (19.45)

$$\text{Area of } T = \iint_T dA = \int_{\alpha}^{\beta} \int_{u_1(\theta)}^{u_2(\theta)} r \, dr \, d\theta. \qquad (19.46)$$

Example 1   Evaluate

$$\int_0^{3/\sqrt{2}} \int_x^{\sqrt{9-x^2}} e^{-(x^2+y^2)} \, dy \, dx.$$

SOLUTION   From Theorem 19.3.9

$$\int_0^{3/\sqrt{2}} \int_x^{\sqrt{9-x^2}} e^{-(x^2+y^2)} \, dy \, dx = \iint_T e^{-(x^2+y^2)} \, dy \, dx \qquad (19.47)$$

where T is the region between the graphs of $y = x$ and $y = \sqrt{9 - x^2}$ extending from $x = 0$ to $x = 3/\sqrt{2}$. In terms of polar coordinates T is between the graphs of $r = 0$ and $r = 3$ and extends from $\theta = \frac{\pi}{4}$ to $\theta = \frac{\pi}{2}$ (Figure 19–16). Hence from (19.45) and (19.47)

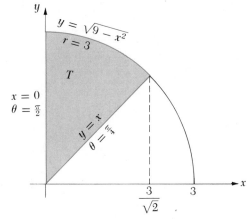

Figure 19-16

$$\int_0^{3/\sqrt{2}} \int_x^{\sqrt{9-x^2}} e^{-(x^2+y^2)} \, dy \, dx = \int_{\pi/4}^{\pi/2} \int_0^3 e^{-r^2} r \, dr \, d\theta$$

$$= \int_{\pi/4}^{\pi/2} -\tfrac{1}{2}e^{-r^2} \Big]_0^3 \, d\theta$$

$$= \int_{\pi/4}^{\pi/2} \tfrac{1}{2}(1 - e^{-9}) \, d\theta$$

$$= \frac{\pi}{8}(1 - e^{-9}).$$

Example 2   Find the mass of the first quadrant portion of the lamina bounded by the cardioid $r = 1 + \cos\theta$ and the circle $r = 1$ (Figure 19–17) if the density at any point is proportional to the distance from the line $\theta = 0$.

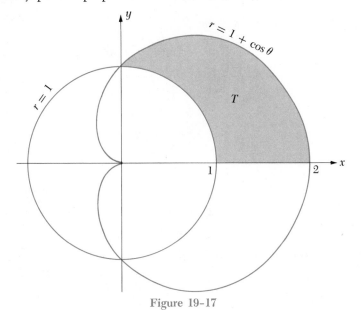

Figure 19-17

SOLUTION   The mass of the lamina, denoted by T, is given by

$$m = \iint_T \sigma(x, y)\, dA = \iint_T ky\, dA.$$

Since in polar coordinates

$$T = \left\{ (r, \theta) : \theta \in \left[0, \frac{\pi}{2}\right] \text{ and } r \in [1, 1 + \cos \theta] \right\},$$

from (19.45)

$$m = \int_0^{\pi/2} \int_1^{1+\cos\theta} kr \sin\theta \cdot r\, dr\, d\theta = \int_0^{\pi/2} \frac{k}{3}[(1 + \cos\theta)^3 - 1] \sin\theta\, d\theta$$

$$= \int_0^{\pi/2} \frac{k}{3}(3 \cos\theta + 3 \cos^2\theta + \cos^3\theta) \sin\theta\, d\theta$$

$$= \frac{k}{3}\left(-\frac{3}{2} \cos^2\theta - \cos^3\theta - \frac{1}{4} \cos^4\theta \right) \Bigg]_0^{\pi/2}$$

$$= \frac{11k}{12}.$$

Example 3   Find the volume of the solid region bounded by the paraboloid $z = x^2 + y^2$ and the plane $z = 2y$ (Figure 19–18).

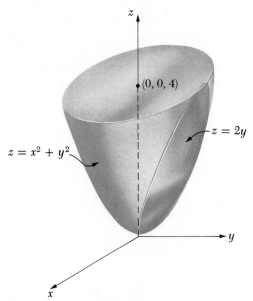

$\bullet (0, 0, 4)$

$z = 2y$

$z = x^2 + y^2$

Figure 19-18

SOLUTION   By Theorem 19.3.11 the volume V of the solid is

$$V = \iint_T [2y - (x^2 + y^2)]\, dA$$

where T is the plane region obtained by projecting the solid into the xy plane. Here T is the plane region (Figure 19–19) bounded by the graph of

$$2y - (x^2 + y^2) = 0 \tag{19.48}$$

or

$$x^2 + (y - 1)^2 = 1.$$

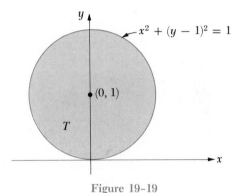

Figure 19-19

From (19.9)

$$V = \int_0^2 \int_{-\sqrt{2y-y^2}}^{\sqrt{2y-y^2}} [2y - (x^2 + y^2)] \, dx \, dy$$

but the evaluation of this iterated double integral is quite difficult. However, from (19.48) the boundary of T has the polar equation

$$2r \sin \theta - r^2 = 0$$

or

$$r = 2 \sin \theta.$$

Thus in polar coordinates

$$T = \{(r, \theta): \theta \in [0, \pi] \text{ and } r \in [0, 2 \sin \theta]\}.$$

Hence from (19.45)

$$V = \int_0^\pi \int_0^{2 \sin \theta} [2r \sin \theta - r^2] \, r \, dr \, d\theta$$

$$= \int_0^\pi \frac{2r^3}{3} \sin \theta - \frac{r^4}{4} \Big|_0^{2 \sin \theta} d\theta$$

$$= \int_0^\pi \tfrac{4}{3} \sin^4 \theta \, d\theta$$

and from (30) in the Table of Integrals

$$V = \frac{4}{3} \left[\frac{3}{8}\theta - \frac{\sin 2\theta}{4} + \frac{\sin 4\theta}{32} \right]_0^\pi.$$

$$= \frac{\pi}{2}.$$

If T is the set of points $(x, y) = (r \cos \theta, r \sin \theta)$ given in Cartesian coordinates by

$$T = \{(r \cos \theta, r \sin \theta): r \in [\gamma, \delta] \text{ and } \theta \in [v_1(r), v_2(r)]\} \tag{19.49}$$

where v_1 and v_2 are continuous on $[\gamma, \delta]$, $\gamma \geq 0$, and $v_2(r) - v_1(r) \leq 2\pi$, T is called a "type IV plane region" (Figure 19–20). If f, a function of two variables, is

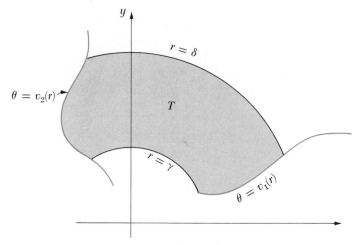

Figure 19-20

continuous on T, then we obtain

$$\iint_T f(x, y) \, dA = \int_\gamma^\delta \int_{v_1(\theta)}^{v_2(\theta)} f(r \cos \theta, r \sin \theta) \, r \, d\theta \, dr \tag{19.50}$$

after a discussion that is analogous to the development of Theorem 19.5.1.

In general, if f, a function of two variables, is continuous on any bounded closed set T in R^2 and T' is the set that is transformed into T by equations (19.38), it can be proved that

$$\iint_T f(x, y) \, dA = \iint_{T'} f(r \cos \theta, r \sin \theta) \, r \, dA \tag{19.51}$$

although the details will be omitted. We note that equation (19.45) or (19.50) can be obtained from (19.51) if T is defined by (19.36) or (19.49), respectively.

Exercise Set 19.5

In Exercises 1–4 evaluate the given iterated integral after changing to polar coordinates

1. $\displaystyle\int_0^a \int_0^{\sqrt{a^2 - x^2}} x^2 \sqrt{x^2 + y^2} \, dy \, dx$ 2. $\displaystyle\int_0^{a/\sqrt{2}} \int_y^{\sqrt{a^2 - y^2}} y^2 e^{-(x^2 + y^2)^2} \, dx \, dy$

3. $\displaystyle\int_0^2 \int_{-\sqrt{4-y^2}}^{\sqrt{4-y^2}} \frac{dx\,dy}{\sqrt{9-x^2-y^2}}$ 4. $\displaystyle\int_0^2 \int_{-\sqrt{2x-x^2}}^{\sqrt{2x-x^2}} (x^2+y^2)^{3/2}\,dy\,dx$

In Exercises 5–7 find the area of the given plane region described using an iterated double integral in polar coordinates.

5. Enclosed by one loop of the graph of $r = 3\sin 4\theta$

6. Enclosed by one loop of the graph of $(x^2+y^2)^2 = 2a^2xy$

7. Bounded by the graphs of $r = a(1+\cos\theta)$ and $r = a\cos\theta$

In Exercises 8–12 find the volume of the given solid region. Sketch the region.

8. Bounded by the upper nappe of the cone $h^2(x^2+y^2) = a^2z^2$ and the plane $z = h$

9. Inside both the sphere $x^2+y^2+z^2 = 4a^2$ and the cylinder $x^2+y^2 = a^2$

10. Bounded from above and below by the cone $z^2 = x^2+y^2$ and inside the cylinder $x^2 - 4x + y^2 = 0$

11. Bounded by the paraboloid $2z = x^2+y^2$ and the upper nappe of the cone $z^2 = x^2+y^2$

12. Bounded from above by the paraboloid $z = 3 - x^2 - y^2$ and from below by the sphere $x^2+y^2+z^2 = 3+6z$

13. Find the centroid of the homogeneous lamina bounded by the cardioid $r = 1 + \cos\theta$.

14. Find the center of mass of the lamina bounded by circle $r = 2\sin\theta$ if the density at any point in the lamina is proportional to the distance from the line $\theta = \frac{\pi}{2}$.

15. Find the center of mass of the lamina bounded by the graphs of $r = a\sin\theta$ and $r = a\cos\theta$ if the density at any point is proportional to the distance from the pole.

19.6 Triple Integral in Cartesian Coordinates

The definition of the triple integral is analogous to that given for a double integral (Definition 19.3.1). We suppose first that F, a function of three variables, is defined on a bounded closed set S in R^3. Then let S be the union of the closed subsets $\Delta S_1, \Delta S_2, \ldots, \Delta S_n$ where no two of the subsets have any points in common except possibly boundary points, and every ΔS_i has a volume ΔV_i. The set $P = \{\Delta S_1, \Delta S_2, \ldots, \Delta S_n\}$ is then called a *partition* of S. Like the elements of a partition of a set in R^2, each ΔS_i has a diameter and the greatest diameter for any ΔS_i is called the norm of the partition. The norm of the partition P will be denoted, as before, by Δ_P. If an arbitrary point (x_i, y_i, z_i) is chosen in each ΔS_i, then the sum

$$S_P = \sum_{i=1}^n F(x_i, y_i, z_i)\,\Delta V_i$$

is called a *Riemann sum of F on S associated with the partition P.*

19.6.1 Definition

Let F, a function of three variables, be defined on a bounded closed region S in R^3. A number L is called the *triple integral* of F on S if and only if for every $\epsilon > 0$ there is a $\delta > 0$ such that a Riemann sum S_P of F on S satisfies

$$|S_P - L| < \epsilon$$

whenever $\Delta_P < \delta$. The function F is said to be integrable on S if and only if such a number L exists.

The triple integral of F on S will henceforth be denoted by

$$\iiint\limits_S F(x, y, z)\, dV$$

when it exists.

A triple integral is usually computed by evaluating an *iterated triple integral* (Definition 19.3.1). We suppose first that F, a function of three variables,

$$\int_a^b \int_{\phi_1(x)}^{\phi_2(x)} \int_{g(x,y)}^{f(x,y)} F(x, y, z)\, dz\, dy\, dx$$

is evaluated by integrating first with respect to z holding x and y constant and then substituting the limits associated with the right-hand integral sign, which are in general functions of x and y. Then we integrate with respect to y holding x constant and substitute the limits associated with the middle integral sign, which are in general functions of x. Finally we integrate with respect to x using the limits connected with the left-hand integral sign, which are constants.

Example 1 Evaluate

$$\int_0^1 \int_0^{2-2x} \int_x^{2-x-y} (x + z)\, dz\, dy\, dx.$$

SOLUTION Integrating first with respect to z while holding x and y constant, we have

$$\int_0^1 \int_0^{2-2x} \int_x^{2-x-y} (x + z)\, dz\, dy\, dx = \int_0^1 \int_1^{2-2x} \tfrac{1}{2}(x + z)^2 \Big]_x^{2-x-y} dy\, dx$$

$$= \int_0^1 \int_1^{2-2x} \tfrac{1}{2}[(2 - y)^2 - 4x^2]\, dy\, dx.$$

Next we integrate with respect to y, holding x constant and obtain successively

$$\int_0^1 \int_0^{2-2x} \int_x^{2-x-y} (x + z)\, dz\, dy\, dx = \int_0^1 \frac{1}{2}\left[-\frac{(2 - y)^3}{3} - 4x^2 y \right]_0^{2-2x} dx$$

$$= \int_0^1 \left(\frac{8x^3}{3} - 4x^2 + \frac{4}{3}\right) dx$$

$$= \left(\frac{2}{3}x^4 - \frac{4}{3}x^3 + \frac{4}{3}x\right)\Big]_0^1$$

$$= \frac{2}{3}.$$

We shall now state without proof the theorem for evaluating a triple integral, which is analogous to Theorem 19.3.9 for double integrals.

19.6.2   Theorem

Suppose

$$T = \{(x, y): x \in [a, b] \text{ and } y \in [\phi_1(x), \phi_2(x)]\} \qquad (19.1)$$

is an elementary region in R^2 and $S \subset R^3$ is given by

$$S = \{(x, y, z): (x, y) \in T \text{ and } z \in [g(x, y), f(x, y)]\}. \qquad (19.52)$$

Here the functions f and g are continuous on T and $f(x, y) \geq g(x, y)$ for every $(x, y) \in T$. If F, a function of three variables, is continuous on S, then

$$\iiint_S F(x, y, z)\, dV = \int_a^b \int_{\phi_1(x)}^{\phi_2(x)} \int_{g(x,y)}^{f(x,y)} F(x, y, z)\, dz\, dy\, dx. \qquad (19.53)$$

The solid region S in (19.52) is illustrated in Figure 19–21. The upper boundary of S is the surface $z = f(x, y)$ and the lower boundary is the surface $z = g(x, y)$. The solid region can be projected into the region T in the xy plane which is defined by (19.1).

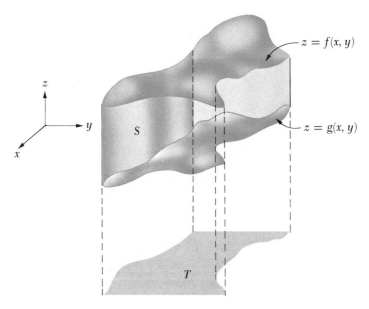

Figure 19-21

If $F(x, y, z) = 1$ on S, then from (19.53)

$$\iiint_S dV = \int_a^b \int_{\phi_1(x)}^{\phi_2(x)} [f(x, y) - g(x, y)]\, dy\, dx,$$

and hence from Theorem 19.2.1

$$\iiint\limits_S dV = \text{Volume of } S.$$

If the roles of x, y, and z in Theorem 19.6.2 are interchanged, then analogous formulas to (19.53) can be obtained. For example, if this theorem is restated with the variables z, x, and y in place of x, y, and z, respectively, then instead of (19.53) the formula

$$\iiint\limits_S F(x, y, z)\, dV = \int_a^b \int_{\phi_1(z)}^{\phi_2(z)} \int_{g(x,z)}^{f(x,z)} F(x, y, z)\, dy\, dx\, dz$$

is obtained.

Triple integrals are useful in computing the mass and moments of a solid region. If a density function σ is defined on a bounded closed subset of R^3, then, as extensions of concepts defined in Section 19.4, we have the following definitions. The mass m of S is

$$m = \iiint\limits_S \sigma(x, y, z)\, dV. \tag{19.54}$$

The (first) moments of S about the xy plane, the xz plane, and the yz plane are, respectively,

$$M_{xy} = \iiint\limits_S z\sigma(x, y, z)\, dV \tag{19.55}$$

$$M_{xz} = \iiint\limits_S y\sigma(x, y, z)\, dV \tag{19.56}$$

$$M_{yz} = \iiint\limits_S x\sigma(x, y, z)\, dV. \tag{19.57}$$

Also, the center of mass of S is

$$(\bar{x}, \bar{y}, \bar{z}) = \left(\frac{M_{yz}}{m}, \frac{M_{xz}}{m}, \frac{M_{xy}}{m} \right). \tag{19.58}$$

Example 2 Find the centroid of the homogeneous solid bounded from above by the plane $z = y$, from below by the plane $z = 0$, and inside the parabolic cylinder $y = 1 - x^2$ (Figure 19–22).

SOLUTION Since the density at any point in the solid region is $\sigma(x, y, z) = k$, where k is constant, from (19.54)

$$m = \iiint\limits_S k\, dV.$$

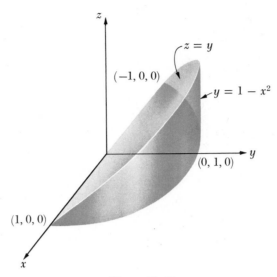

Figure 19-22

Since the solid region is the set

$$\{(x, y, z): x \in [-1, 1], \ y \in [0, 1 - x^2], \text{ and } z \in [0, y]\}$$

by Theorem 19.6.2

$$m = \int_{-1}^{1} \int_{0}^{1-x^2} \int_{0}^{y} k \, dz \, dy \, dx$$

$$= \int_{-1}^{1} \int_{0}^{1-x^2} ky \, dy \, dx = \frac{k}{2} \int_{-1}^{1} (1 - x^2)^2 \, dx.$$

Since the graph of $y = (1 - x^2)^2$ is symmetric to the y axis,

$$m = k \int_{0}^{1} (1 - x^2)^2 \, dx = k \int_{0}^{1} (1 - 2x^2 + x^4) \, dx$$

$$= k \left(x - \frac{2x^3}{3} + \frac{x^5}{5} \right) \Big]_{0}^{1}$$

$$= \frac{8k}{15}. \tag{19.59}$$

From (19.55)

$$M_{xy} = \iiint_{S} zk \, dV = \int_{-1}^{1} \int_{0}^{1-x^2} \int_{0}^{y} kz \, dz \, dy \, dx$$

$$= \int_{-1}^{1} \int_{0}^{1-x^2} \frac{ky^2}{2} \, dy \, dx = \frac{k}{6} \int_{-1}^{1} (1 - x^2)^3 \, dx. \tag{19.60}$$

Since the graph of $y = (1 - x^2)^3$ is symmetric to the y axis

$$M_{xy} = 2 \cdot \frac{k}{6} \int_0^1 (1 - x^2)^3 \, dx = \frac{k}{3} \int_0^1 (1 - 3x^2 + 3x^4 - x^6) \, dx$$

$$= \frac{k}{3} \left(x - x^3 + \frac{3x^5}{5} - \frac{x^7}{7} \right) \Big]_0^1$$

$$= \frac{16k}{105}. \tag{19.61}$$

Thus from (19.59) and (19.61)

$$\bar{z} = \frac{M_{xy}}{m} = \frac{\dfrac{16k}{105}}{\dfrac{8k}{15}} = \frac{2}{7}.$$

From (19.56)

$$M_{xz} = \iiint_S yk \, dV = \int_{-1}^1 \int_0^{1-x^2} \int_0^y ky \, dz \, dy \, dx$$

$$= \int_{-1}^1 \int_0^{1-x^2} ky^2 \, dy \, dx \tag{19.62}$$

From a comparison of the iterated double integrals in (19.62) and (19.60)

$$M_{xz} = 2M_{xy} = \frac{32k}{105}. \tag{19.63}$$

Hence from (19.59) and (19.63)

$$\bar{y} = \frac{M_{xz}}{m} = \frac{\dfrac{32k}{105}}{\dfrac{8k}{15}} = \frac{4}{7}.$$

Because of the symmetry of the solid region S with respect to the yz plane, it is intuitively apparent that $\bar{x} = 0$. Thus the centroid of S is $(0, \frac{4}{7}, \frac{2}{7})$.

We will also define another physical concept using a triple integral, the *moment of inertia* (second moment) of a solid mass with respect to an axis. In physics, a particle of mass m that is located a distance r from a line is said to have a moment of inertia I about the line given by

$$I = mr^2. \tag{19.64}$$

The number mr^2 is a measure of the resistance of the particle to changes in its angular velocity (its *rotational inertia*) about the line just as the mass m is a measure of the inertia of the particle in straight line motion. The idea of a moment of inertia is also extended to a system of particles. If n distinct particles with respective masses $m_1, m_2, \ldots, m_r$ are located at distances $r_1, r_2, \ldots, r_n$, respectively, from the line, then I, the *moment of inertia of the system* of particles about the line, is defined by

$$I = \sum_{i=1}^n m_i r_i^2. \tag{19.65}$$

We shall define the moment of inertia of a solid mass S about a line L in space. Suppose again that S is a bounded closed solid region with density function σ and a partition of S forms the subregions $\Delta S_1, \Delta S_2, \ldots, \Delta S_n$ of S. If the partition is sufficiently fine and (x_i, y_i, z_i) is an arbitrary point in ΔS_i, then the mass of ΔS_i is approximately $\sigma(x_i, y_i, z_i) \Delta V_i$ where ΔV_i is the volume of ΔS_i. Suppose the distance from (x_i, y_i, z_i) to the line L is denoted by $\delta(x_i, y_i, z_i)$ (Figure 19–23).

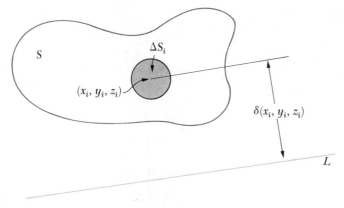

Figure 19–23

From (19.64) intuition suggests that the moment of inertia of ΔS_i about L is approximately

$$\underbrace{\sigma(x_i, y_i, z_i)\, \Delta V_i}_{\text{Mass}} \cdot \underbrace{(\delta(x_i, y_i, z_i))^2}_{\text{(Distance)}^2}.$$

Then by analogy with (19.65) the moment of inertia I_L of S about L would seem to be expressed by

$$I_L \approx \sum_{i=1}^{n} (\delta(x_i, y_i, z_i))^2 \sigma(x_i, y_i, z_i)\, \Delta V_i. \tag{19.66}$$

Intuition also suggests that the smaller the norm of the partition, the better is the approximation expressed by (19.66). Since the sum in (19.66) can be made to differ from the triple integral $\iiint_S (\delta(x, y, z))^2 \sigma(x, y, z)\, dV$ by as little as we please when the norm of the partition of S is sufficiently small, we have the following definition.

19.6.3   Definition

Suppose S is a bounded closed solid region with density function σ and suppose $\delta(x, y, z)$ is the distance from the point $(x, y, z) \in S$ to the line L in R^3. Then I_L, the *moment of inertia (second moment)* of S about L is given by

$$I_L = \iiint_S (\delta(x, y, z))^2 \sigma(x, y, z)\, dV. \tag{19.67}$$

If m is the mass of S, then

$$k_L = \sqrt{\frac{I_L}{m}} \qquad\qquad (19.68)$$

is called the *radius of gyration* of S with respect to L.

The number k_L is the distance from the *axis* L to a particle of mass m having the same moment of inertia about L as the solid region S.

Example 3   Find the moment of inertia and radius of gyration with respect to the z axis of the solid cube S which is bounded by the planes $x = 0$, $x = a$, $y = 0$, $y = a$, $z = 0$, and $z = a$ (Figure 19–24) if the density at any point in the cube varies as the square of its distance from the z axis.

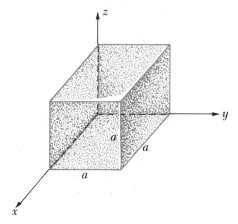

Figure 19-24

SOLUTION   Since a perpendicular from any point (x, y, z) in S to the z axis has the point $(0, 0, z)$ as its foot, the distance between these points is

$$\delta(x, y, z) = \sqrt{x^2 + y^2}.$$

Also, the density at any point in S is given by

$$\sigma(x, y, z) = k(x^2 + y^2).$$

From (19.67) the moment of inertia of S about the z axis is

$$I_z = \iiint_S (x^2 + y^2)k(x^2 + y^2)\, dV = \iiint_S k(x^2 + y^2)^2\, dV.$$

Then by Theorem 19.6.3

$$I_z = \int_0^a \int_0^a \int_0^a k(x^4 + 2x^2y^2 + y^4)\, dz\, dx\, dy$$

$$= \int_0^a \int_0^a ka(x^4 + 2x^2y^2 + y^4)\, dx\, dy$$

$$= \int_0^a ka\left(\frac{x^5}{5} + \frac{2x^3y^2}{3} + xy^4\right)\Big]_0^a dy$$

$$= \int_0^a ka\left(\frac{a^5}{5} + \frac{2a^3y^2}{3} + ay^4\right) dy$$

$$= ka\left(\frac{a^5}{5}y + \frac{2a^3y^3}{9} + \frac{ay^5}{5}\right)\Big]_0^a$$

$$= \frac{28ka^7}{45}.$$

The mass of the cube S is

$$m = \iiint_S k(x^2 + y^2)\, dV$$

$$= \int_0^a \int_0^a \int_0^a k(x^2 + y^2)\, dz\, dx\, dy = \int_0^a \int_0^a ka(x^2 + y^2)\, dx\, dy$$

$$= \int_0^a ka\left(\frac{x^3}{3} + xy^2\right)\Big]_0^a dy = \int_0^a ka\left(\frac{a^3}{3} + ay^2\right) dy$$

$$= ka\left(\frac{a^3y}{3} + \frac{ay^3}{3}\right)\Big]_0^a$$

$$= \tfrac{2}{3}ka^5$$

Hence by (19.68) the radius of gyration of S with respect to the z axis is

$$k_z = \sqrt{\frac{28ka^7/45}{\tfrac{2}{3}ka^5}} = \frac{a\sqrt{210}}{15}.$$

Exercise Set 19.6

In Exercises 1–4 evaluate the given iterated triple integral.

1. $\displaystyle \int_0^2 \int_0^{2-x} \int_{x+y}^2 3\, dz\, dy\, dx$

2. $\displaystyle \int_0^\pi \int_0^y \int_x^{x+y} \sin(y+z)\, dz\, dx\, dy$

3. $\displaystyle \int_0^1 \int_y^{\sqrt{4-y^2}} \int_1^{e^z} \frac{y}{x}\, dx\, dz\, dy$

4. $\displaystyle \int_1^2 \int_{z/2}^{\sqrt{3z}/2} \int_0^{\ln xz} \frac{e^y}{\sqrt{z^2 - x^2}}\, dy\, dx\, dz$

In Exercises 5–10 evaluate the given triple integrals. Sketch the region S involved.

5. $\iiint_S xyz \, dV$ where S is the solid region bounded on the left by the plane $y = 0$, on the right by the cylinder $x^2 + y^2 = 9$ from above by the plane $z = 2$ and from below by the plane $z = 0$

6. $\iiint_S z \, dV$ where S is the solid region bounded from above by the graph of $x^2 + y^2 + z^2 = 4$ and from below by the plane $z = 1$

7. $\iiint_S x \, dV$ where S is bounded by the planes $z = a$, $x + y = z$, $x = 0$, and $y = 0$

8. $\iiint_S y \, dV$ where S is the solid region under the plane $z = y$ above the xy plane and bounded on the sides by the graphs of $y = x$ and $x = 4y - y^2$

9. $\iiint_S y^2 \, dV$ where S is solid region for which $x \geq 0$ which is bounded by parabolic cylinder $z = 3 - y^2$ and the planes $z = x$ and $z = 2x$

10. $\iiint_S (x^2 + z^2) \, dV$ where S is bounded on the right by the parabolic cylinder $y = 2 - z^2$, on the left by the parabolic cylinder $y = z^2$, in front by the plane $x = 1$, and in back by the plane $x = -1$

In each part of Exercises 11 and 12 rewrite the given iterated integral as an equivalent iterated integral having the indicated order of integration of the variables and the correct limits. Sketch the solid region associated with the given iterated integral.

11. $\displaystyle\int_0^2 \int_{-\sqrt{4-y^2}}^{\sqrt{4-y^2}} \int_0^{2y} f(x, y, z) \, dz \, dx \, dy$

(a) $\displaystyle\int_?^? \int_?^? \int_?^? f(x, y, z) \, dy \, dx \, dz$

(b) $\displaystyle\int_?^? \int_?^? \int_?^? f(x, y, z) \, dx \, dz \, dy$

12. $\displaystyle\int_0^3 \int_0^{(3-x)/2} \int_1^{4-x-2y} f(x, y, z) \, dz \, dy \, dx$

(a) $\displaystyle\int_?^? \int_?^? \int_?^? f(x, y, z) \, dx \, dy \, dz$

(b) $\displaystyle\int_?^? \int_?^? \int_?^? f(x, y, z) \, dy \, dz \, dx$

In Exercises 13–21 solve the given problem using a triple integral. Sketch the given solid region.

13. Find the volume of the solid region bounded by the graphs of $2x + 3y + z = 6$, $2x + y - z = 0$, $x = 0$, and $y = 0$.

14. Find the volume of the solid region bounded by the graphs of $y = x - z^2$, $2y = x$, and $x = 6$.

15. Find the centroid of the homogeneous tetrahedron having vertices $(0, 0, 0)$, $(a, 0, 0)$, $(0, b, 0)$, and $(0, 0, c)$.

16. Find the moment of inertia and radius of gyration about the y axis of the solid region bounded by the planes $x = 0$, $x = a$, $y = 0$, $y = b$, $z = 0$, and $z = c$ if a, b, and c are positive and the density varies as the distance from the yz plane.

17. Find the moment of inertia and radius of gyration of the homogeneous solid region in Example 2 (Figure 19–22) about the z axis.

18. Find the center of mass of the solid region in Exercise 16.

19. Find the center of mass of the solid region bounded by paraboloid $z = 4 - x^2 - y^2$ and the plane $z = 0$ if the density at any point is proportional to the distance from the xz plane.

20. Find the centroid of the homogeneous solid region bounded by the parabolic cylinder $z^2 = 4y$, the plane $x = 0$, and the plane $x + y = 4$.

21. Find the moment of inertia of the solid region in Exercise 20 about the x axis.

19.7   Triple Integrals in Cylindrical and Spherical Coordinates

As we saw in Example 1, Section 19.5, a double integral given in Cartesian variables is sometimes more easily evaluated following a transformation to polar variables. Analogously, the evaluation of some triple integrals is facilitated by changing from Cartesian coordinates to either cylindrical or spherical coordinates. We recall from Section 17.10 that the Cartesian coordinates of a point in R^3 are related to its cylindrical coordinates (r, θ, z) by the equations

$$x = r \cos \theta \qquad y = r \sin \theta \qquad z = z \tag{19.69}$$

and to its spherical coordinates (ρ, φ, θ) by the equations

$$\left.\begin{array}{l} x = \rho \sin \phi \cos \theta \\ y = \rho \sin \phi \sin \theta \\ z = \rho \cos \phi. \end{array}\right\} \text{where } \rho \geq 0 \text{ and } \phi \in [0, \pi] \tag{19.70}$$

In this section we will assume that F, a function of three variables, is continuous on a bounded closed set S in R^3. To obtain a formula for transforming

$$\iiint_S F(x, y, z)\, dV \tag{19.71}$$

to cylindrical coordinates, a discussion can be employed which is essentially a three-dimensional analogue of that used to justify Theorem 19.5.1. Thus if a region S' in $r\theta z$ space for which $r \geq 0$ and θ is chosen from an interval of length $\leq 2\pi$ is transformed by equations (19.69) into S, we obtain

$$\iiint_S F(x, y, z)\, dV = \iiint_{S'} F(r \cos \theta,\ r \sin \theta,\ z)r\, dV. \tag{19.72}$$

Note that (19.72) is an extension of the formula (19.51) for transforming a double integral from Cartesian coordinates to polar coordinates. In deriving (19.72) S is subdivided into regions ΔS_{ijk} by constructing circular cylinders $r = r_i$, and planes $\theta = \theta_j$ and $z = z_k$. The volume ΔV_{ijk} of each ΔS_{ijk} that is contained in S is given by

$$\Delta V_{ijk} = \bar{r}_i \, \Delta r_i \, \Delta \theta_j \, \Delta z_k$$

where $\bar{r}_i$ is given by (19.42), Δr_i and $\Delta \theta_j$ are defined in (19.40), and $\Delta z_k = z_k - z_{k-1}$. In the usual problem of interest S is of the form

$$S = \{(x, y, z) : (x, y) \in T \text{ and } z \in [f_1(x, y), f_2(x, y)]\}$$

where T is of the form (19.36), f_1 and f_2 are continuous on T, and $f_2(x, y) \geq f_1(x, y)$ for every $(x, y) \in T$ (Figure 19–25(b)). Thus S is the transformation under equations

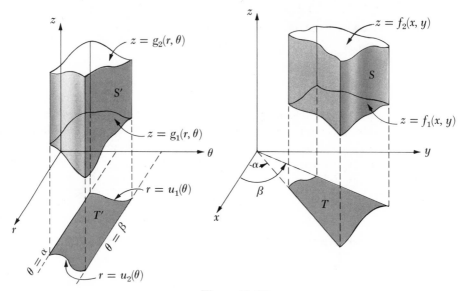

Figure 19-25

(19.69) of the region S' in the $r\theta z$ space given by

$$S' = \{(r, \theta, z) : \theta \in [\alpha, \beta], \ r \in [u_1(\theta), u_2(\theta)], \text{ and } z \in [g_1(r, \theta), g_2(r, \theta)]\}$$

(Figure 19–25(a)) where

$$g_1(r, \theta) = f_1(r \cos \theta, r \sin \theta) \qquad \text{and} \qquad g_2(r, \theta) = f_2(r \cos \theta, r \sin \theta).$$

Hence from (19.72) and the application of Theorem 19.6.2 to the function G on S' given by $G(r, \theta, z) = F(r \cos \theta, r \sin \theta, z)r$,

$$\iiint_S F(x, y, z) \, dV = \int_\alpha^\beta \int_{u_1(\theta)}^{u_2(\theta)} \int_{g_1(r,\theta)}^{g_2(r,\theta)} F(r \cos \theta, r \sin \theta, z)r \, dz \, dr \, d\theta. \quad (19.73)$$

Note that in applying (19.73), x and y in $\iiint_S F(x, y, z) \, dV$ are replaced by $r \cos \theta$ and $r \sin \theta$, respectively, and dV is replaced by $r \, dz \, dr \, d\theta$. The limits for the iterated triple integral on the right side of (19.73) correspond to the conditions satisfied by the coordinates of any point in S', or equivalently, the cylindrical coordinates of any point in S.

The use of formula (19.73) may be advisable when the graphs of $z = f(x, y)$ and $z = g(x, y)$, the upper and lower boundary surfaces of S, are *symmetric to the z axis*, as in the following example.

Example 1   Find the mass of the solid region S bounded from above by the plane $z = 5$ and from below by the paraboloid $z = x^2 + y^2$ if the density at any point is proportional to the distance from the z axis (Figure 19–26).

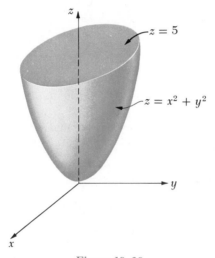

Figure 19-26

SOLUTION   The density for S is given by $\sigma(x, y, z) = k\sqrt{x^2 + y^2}$ and hence the mass of S is

$$m = \iiint \sigma(x, y, z)\, dV = \iiint k\sqrt{x^2 + y^2}\, dV.$$

Since the paraboloid has the cylindrical equation $z = r^2$ and since S can be projected into the region in the xy plane which is bounded by the circle with polar equation $r = \sqrt{5}$, in cylindrical coordinates

$$S = \{(r, \theta, z): \theta \in [0, 2\pi],\ r \in [0, \sqrt{5}],\ \text{and}\ z \in [r^2, 5]\}.$$

Thus from (19.73)

$$m = \int_0^{2\pi} \int_0^{\sqrt{5}} \int_{r^2}^{5} kr \cdot r\, dz\, dr\, d\theta$$
$$= \int_0^{2\pi} \int_0^{\sqrt{5}} k(5 - r^2)r^2\, dr\, d\theta = \int_0^{2\pi} k\left(\frac{5r^3}{3} - \frac{r^5}{5}\right)\Bigg]_0^{\sqrt{5}} d\theta$$
$$= \frac{20k\pi \sqrt{5}}{3}.$$

To derive the formula for transforming $\iiint_S F(x, y, z)\, dV$ to spherical coordinates, we will again utilize (19.72). From this equation

$$\iiint_S F(x, y, z)\, dV = \iiint_{S'} G(r, \theta, z)\, dV \tag{19.74}$$

where

$$G(r, \theta, z) = F(r \cos \theta,\ r \sin \theta,\ z)r, \tag{19.75}$$

and it is understood that for any point in S', $r \geq 0$ and θ is chosen from an interval of length $\leq 2\pi$. From (19.70)

$$x^2 + y^2 = \rho^2 \sin^2 \phi$$

and hence since $r \geq 0$

$$r = \sqrt{x^2 + y^2} = \rho \sin \phi.$$

Also from (19.70)

$$z = \rho \cos \phi.$$

Thus a region S'' in the $\rho\phi\theta$ space is transformed into S' by the transformation equations

$$z = \rho \cos \phi$$
$$r = \rho \sin \phi$$
$$\theta = \theta.$$

Since $\rho \geq 0$ and $\phi \in [0, \pi]$, an interval of length $< 2\pi$, by the same reasoning used to derive (19.72), we can obtain

$$\iiint_{S'} G(r, \theta, z)\, dV = \iiint_{S''} H(\rho, \phi, \theta)\, dV \tag{19.76}$$

where

$$H(\rho, \phi, \theta) = G(\rho \sin \phi,\ \theta,\ \rho \cos \phi)\rho.$$

From (19.75)

$$H(\rho, \phi, \theta) = F(\rho \sin \phi \cos \theta,\ \rho \sin \phi \sin \theta,\ \rho \cos \phi)\rho \sin \phi \cdot \rho$$
$$= F(\rho \sin \phi \cos \theta,\ \rho \sin \phi \sin \theta,\ \rho \cos \phi)\rho^2 \sin \phi \tag{19.77}$$

and thus from (19.74), (19.76), and (19.77),

$$\iiint_S F(x, y, z)\, dV = \iiint_{S''} F(\rho \sin \phi \cos \theta,\ \rho \sin \phi \sin \theta,\ \rho \cos \phi)\rho^2 \sin \phi\, dV. \tag{19.78}$$

In the usual situation the solid region S in R^3 is given by

$$S = \{(\rho \sin \phi \cos \theta,\ \rho \sin \phi \sin \theta,\ \rho \cos \phi): \theta \in [\alpha, \beta],$$
$$\phi \in [u_1(\theta), u_2(\theta)],\ \text{and}\ \rho \in [g_1(\phi, \theta), g_2(\phi, \theta)]\}$$

(Figure 19–27) and is therefore the transformation under equations (19.70) of the region

$$S'' = \{(\rho, \phi, \theta): \theta \in [\alpha, \beta], \phi \in [u_1(\theta), u_2(\theta)], \text{ and } \rho \in [g_1(\phi, \theta), g_2(\phi, \theta)]\}.$$

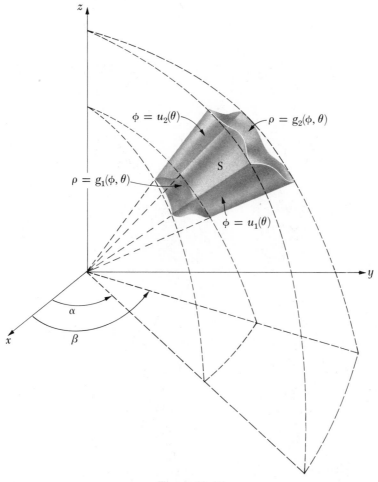

Figure 19-27

Then from (19.78) and Theorem 19.6.2,

$$\iiint\limits_{S} F(x, y, z)\, dV$$

$$= \int_{\alpha}^{\beta} \int_{u_1(\theta)}^{u_2(\theta)} \int_{g(\phi, \theta)}^{f(\phi, \theta)} F(\rho \sin \phi \cos \theta, \rho \sin \phi \sin \theta, \rho \cos \phi) \rho^2 \sin \phi\, d\rho\, d\phi\, d\theta \quad (19.79)$$

Note that in applying (19.79) x, y, and z are replaced by $\rho \sin \phi \cos \theta$, $\rho \sin \phi \sin \theta$, and $\rho \cos \phi$, respectively, and dV is replaced by $\rho^2 \sin \phi\, d\rho\, d\phi\, d\theta$. The limits for the iterated triple integral on the right side of (19.79) correspond to the conditions satisfied by the coordinates of any point in S'', or equivalently, the spherical coordinates of any point in S.

The use of formula (19.79) may be advisable when the boundary surfaces of the region S are symmetric to the z axis or to the origin.

Example 2 Find the volume of a sphere of radius a using spherical coordinates.

SOLUTION It will be convenient to locate the sphere so that its center is the origin (Figure 19–28). Then the sphere, which has the Cartesian equation

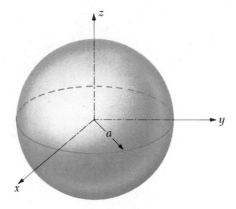

Figure 19-28

$x^2 + y^2 + z^2 = a^2$, has the spherical equation

$$\rho = a.$$

The sphere, which shall be denoted by S, is given in spherical coordinates by

$$S = \{(\rho, \phi, \theta): \theta \in [0, 2\pi], \pi \in [0, \pi], \text{ and } \rho \in [0, a]\}.$$

Hence from (19.79) its volume is

$$\iiint_S dV = \int_0^{2\pi} \int_0^\pi \int_0^a \rho^2 \sin \phi \, d\rho \, d\phi \, d\theta$$

$$= \int_0^{2\pi} \int_0^\pi \frac{\rho^3}{3} \Big]_0^a \sin \phi \, d\phi \, d\theta = \int_0^{2\pi} \int_0^\pi \frac{a^3}{3} \sin \phi \, d\phi \, d\theta$$

$$= \int_0^{2\pi} -\frac{a^3}{3} \cos \phi \Big|_0^\pi d\theta = \int_0^{2\pi} \frac{2a^3}{3} d\theta$$

$$= \frac{4\pi a^3}{3}.$$

Example 3 Find the moment of inertia about the z axis of the homogeneous solid region S bounded from below by the upper nappe of the cone

$z^2 = \dfrac{h^2}{a^2}(x^2 + y^2)$ and from above by the plane $z = h$ (Figure 19–29).

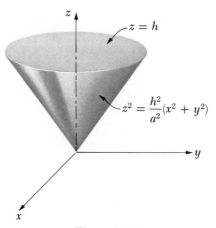

Figure 19-29

SOLUTION Since the distance from any point (x, y, z) in the region to the z axis is

$$\delta(x, y, z) = \sqrt{x^2 + y^2}$$

and since the density at any point can be given by

$$\sigma(x, y, z) = k \quad k \text{ constant,}$$

we have by (19.67)

$$I_z = \iiint_S (\delta(x, y, z))^2 \sigma(x, y, z)\, dV = \iiint_S k(x^2 + y^2)\, dV.$$

A spherical equation for the cone is

$$\phi = \tan^{-1}\frac{a}{h}$$

and for the plane is $\rho \cos \phi = h$ or

$$\rho = h \sec \phi.$$

In spherical coordinates S is therefore given by

$$S = \left\{(\rho, \phi, \theta): \theta \in [0, 2\pi],\, \phi \in \left[0, \tan^{-1}\frac{a}{h}\right],\text{ and } \rho \in [0, h \sec \phi]\right\}.$$

Then from (19.79)

$$I_z = \int_0^{2\pi} \int_0^{\tan^{-1}(a/h)} \int_0^{h\sec\phi} k\rho^2 \sin^2\phi \cdot \rho^2 \sin\phi \, d\rho \, d\phi \, d\theta$$

$$= \int_0^{2\pi} \int_0^{\tan^{-1}(a/h)} \frac{k\rho^5}{5} \Big|_0^{h\sec\phi} \sin^3\phi \, d\phi \, d\theta$$

$$= \int_0^{2\pi} \int_0^{\tan^{-1}(a/h)} \frac{kh^5}{5} \sec^5\phi \sin^3\phi \, d\phi \, d\theta$$

$$= \int_0^{2\pi} \int_0^{\tan^{-1}(a/h)} \frac{kh^5}{5} \tan^3\phi \sec^2\phi \, d\phi \, d\theta$$

$$= \int_0^{2\pi} \frac{kh^5}{5} \frac{\tan^4\phi}{4} \Big|_0^{\tan^{-1}(a/h)} d\theta = \int_0^{2\pi} \frac{a^4 hk}{20} d\theta$$

$$= \frac{a^4 hk\pi}{10}$$

Exercise Set 19.7

1. Find the center of mass of the solid region in Example 1.

2. Find the moment of inertia of the solid region in Example 1 about the z axis.

3. Find the volume of a solid sphere of radius a using cylindrical coordinates.

4. Find the volume of a solid cone with radius a and altitude h using spherical coordinates.

In Exercises 5–10 use either cylindrical or spherical coordinates.

5. Find the moment of inertia of a homogeneous solid right circular cylinder of radius a about   (a) its axis   (b) a line in its curved surface. Express the answers in terms of the mass of the solid.

6. Find the moment of inertia of a homogeneous solid sphere about its diameter.

7. Find the center of mass of a solid hemisphere of radius a if the density at any point is proportional to the distance from the center.

8. Find the moment of inertia of the solid hemisphere in Exercise 7 about its axis.

9. Find the center of mass of a solid cone with radius a and altitude h if the density at any point varies as the distance from the vertex of the cone.

10. Find the centroid of the homogeneous solid region above the xy plane which is bounded by the sphere $x^2 + y^2 + z^2 = 2a^2$ and the cylinder $x^2 + y^2 = a^2$.

11. Find the mass of the solid region bounded from above by the sphere $x^2 + y^2 + z^2 = 4a^2$ and from below by the plane $z = a$ if the density at any point is proportional to the distance from the z axis.

12. A particle of mass m_1 at a point (x, y, z) exerts a force of attraction $\mathbf{F}(x, y, z)$ on a particle of mass m_2 at a point (a, b, c) given by

$$\mathbf{F}(x, y, z) = \frac{Gm_1m_2}{(x-a)^2 + (y-b)^2 + (z-c)^2} \frac{\langle x-a, \ y-b, \ z-c \rangle}{\sqrt{(x-a)^2 + (y-b)^2 + (z-c)^2}}$$

where G is the gravitational constant. Define the force of attraction exerted by a solid region S with density $\sigma(x, y, z)$ on a particle of mass μ at a point $(a, b, c) \notin S$ after first defining the x, y, and z components of this force.

13. From the definitions obtained in Exercise 12, find the force of attraction exerted by the solid region between the hemispheres $z = \sqrt{a^2 - x^2 - y^2}$ and $z = \sqrt{b^2 - x^2 - y^2}$ where $b > a > 0$ on a particle of mass μ at the origin. Assume that the density at any point in the solid varies inversely as the distance from the origin.

14. Find the force of attraction exerted by the solid homogeneous sphere of radius a on a particle of unit mass a distance $b > a$ from the center of the sphere.

15. Prove that the moment of inertia of a homogeneous spherical shell of mass m and radius a about an axis through its center is $\frac{2}{3}ma^2$.

19.8   Line Integrals

In this section and in Section 19.9 we will discuss another kind of integral that is associated with functions of more than one variable—the *line integral*. To introduce this integral, suppose that a force $\mathbf{F}(x, y)$ that is given by

$$\mathbf{F}(x, y) = \langle M(x, y), \ N(x, y) \rangle \tag{19.80}$$

is exerted at every point (x, y) in some open set $S \subset R^2$. The function $\mathbf{F}$, here, is an example of a vector-valued function of two variables and is termed a *force field* on S.

Suppose also that the functions M and N in (19.80) are continuous on S and C is a smooth curve in S with parametric equations

$$x = f(t) \qquad y = g(t) \quad \text{if } t \in [a, b].$$

We shall define the work done by the force $\mathbf{F}$ in moving the particle along C from the point $(f(a), g(a))$ to $(f(b), g(b))$. See Figure 19–30.

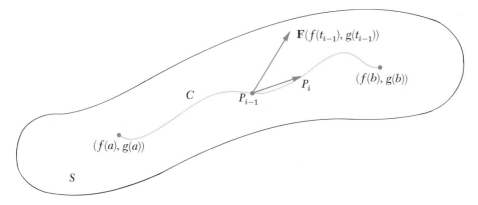

Figure 19–30

Let $T = \{a = t_0, t_1, \ldots, t_n = b\}$ be a partition of $[a, b]$ and let P_i denote the point $(f(t_i), g(t_i))$ for $i = 1, 2, \ldots, n$. It will be recalled from Section 15.2 that the work W done by a constant force $\mathbf{F}$ in moving a particle along a straight line from a point P to Q is given by

$$W = \mathbf{F} \cdot \overrightarrow{PQ}.$$

Thus, intuition suggests that the work done by the force $\mathbf{F}(x, y)$ in moving the particle along the curve from P_{i-1} to P_i can be approximated by

$$\Delta W_i = \mathbf{F}(f(t_{i-1}), g(t_{i-1})) \cdot \overrightarrow{P_{i-1}P_i}. \tag{19.81}$$

Since

$$\overrightarrow{P_{i-1}P_i} = \langle f(t_i) - f(t_{i-1}), g(t_i) - g(t_{i-1}) \rangle,$$

by the mean value theorem there exist numbers c_i and c_i' in (t_{i-1}, t_i) such that

$$\overrightarrow{P_{i-1}P_i} = (t_i - t_{i-1})\langle f'(c_i), g'(c_i') \rangle. \tag{19.82}$$

Letting $\Delta t_i = t_i - t_{i-1}$, we have from (19.81) and (19.82)

$$\Delta W_i = \mathbf{F}(f(t_{i-1}), g(t_{i-1})) \cdot \langle f'(c_i), g'(c_i') \rangle \Delta t_i.$$

Thus if W is the total work required to move the particle from $(f(a), g(a))$ to $(f(b), g(b))$, it would seem that W is approximated by the sum of the increments ΔW_i, that is,

$$W \approx \sum_{i=1}^n \mathbf{F}(f(t_{i-1}), g(t_{i-1})) \cdot \langle f'(c_i), g'(c_i') \rangle \Delta t_i,$$

and from (19.80)

$$W \approx \sum_{i=1}^n [M(f(t_{i-1}), g(t_{i-1}))f'(c_i) + N(f(t_{i-1}), g(t_{i-1}))g'(c_i')] \Delta t_i. \tag{19.83}$$

By the chain rule the functions ϕ and ψ where $\phi(t) = M(f(t), g(t))$ and $\psi(t) = N(f(t), g(t))$ are differentiable on $[a, b]$ and hence are continuous on this interval. Since the functions f' and g' are also continuous on $[a, b]$, the definite integral

$$\int_a^b [M(f(t), g(t))f'(t) + N(f(t), g(t))g'(t)] \, dt$$

certainly exists. Intuition suggests that the sum in (19.83) can be made to differ from this integral by as little as we please if the norm of the partition T is sufficiently small. Therefore the work W done by the force $\mathbf{F}(x, y)$ in moving the particle along the curve C from $(f(a), g(a))$ to $(f(b), g(b))$ *will be defined to be*

$$W = \int_a^b [M(f(t), g(t))f'(t) + N(f(t), g(t))g'(t)] \, dt \tag{19.84}$$

or in vector notation

$$W = \int_a^b \langle M(f(t), g(t)), N(f(t), g(t)) \rangle \cdot \langle f'(t), g'(t) \rangle \, dt.$$

Example 1   The force of repulsion between an electrical charge k located at the origin and a unit electrical charge of like polarity at the point (x, y) is given by

$$\mathbf{F}(x, y) = \left\langle \frac{kx}{(x^2 + y^2)^{3/2}}, \frac{ky}{(x^2 + y^2)^{3/2}} \right\rangle.$$

Find the work done by the force $\mathbf{F}$ on the unit charge as it moves along the line segment from $(1, 0)$ to $(-1, 2)$.

SOLUTION   Here $M(x, y) = kx/(x^2 + y^2)^{3/2}$ and $N(x, y) = ky/(x^2 + y^2)^{3/2}$ and the line segment can be given parametrically by

$$x = f(t) = 1 - t \qquad y = g(t) = t \quad \text{if } t \in [0, 2]. \tag{19.85}$$

Thus

$$M(f(t), g(t)) = \frac{k(1 - t)}{[(1 - t)^2 + t^2]^{3/2}} \qquad N(f(t), g(t)) = \frac{kt}{[(1 - t)^2 + t^2]^{3/2}}$$

if $t \in [0, 2]$. Then from (19.84) the work W is

$$W = \int_0^2 \left\{ \frac{k(1 - t)}{[(1 - t)^2 + t^2]^{3/2}}(-1) + \frac{kt}{[(1 - t)^2 + t^2]^{3/2}} \cdot 1 \right\} dt$$

$$= \int_0^2 \frac{k(2t - 1) \, dt}{(2t^2 - 2t + 1)^{3/2}}$$

$$= \left[\frac{k}{2} \frac{(2t^2 - 2t + 1)^{-1/2}}{-\frac{1}{2}} \right]_0^2 = -\left[\frac{k}{(2t^2 - 2t + 1)^{1/2}} \right]_0^2$$

$$= k\left(\frac{5 - \sqrt{5}}{5} \right).$$

The notion of the work done by a force in moving a particle along a curve leads us to the concept of a *line integral* in the plane.

19.8.1   Definition

Suppose the functions M and N of two variables are continuous on an open set S containing the smooth curve C, given by

$$x = f(t) \qquad y = g(t) \quad \text{if } t \in [a, b]. \tag{19.86}$$

Then the *line integral* $\int_C M \, dx + N \, dy$ over C is given by

$$\int_C M \, dx + N \, dy = \int_a^b [M(f(t), g(t))f'(t) + N(f(t), g(t))g'(t)] \, dt, \tag{19.87}$$

or in vector notation

$$\int_C M\,dx + N\,dy = \int_a^b \langle M(f(t), g(t)), N(f(t), g(t))\rangle \cdot \langle f'(t), g'(t)\rangle\,dt.$$

Thus the work W in Example 1 is the line integral

$$\int_C M\,dx + N\,dy = \int_C \frac{kx}{(x^2 + y^2)^{3/2}}\,dx + \frac{ky}{(x^2 + y^2)^{3/2}}\,dy$$

where C is the line segment from $(1, 0)$ to $(-1, 2)$.

It should be noted from Definition 19.8.1 that the line integral (19.87) is associated with the direction in which the parameter t of the curve C is *increasing* since t goes from a to b where $a < b$. Further, it may be proved that if t is replaced by a new parameter τ where $d\tau/dt > 0$ when $t \in [a, b]$, then $\int_C M\,dx + N\,dy$ is unchanged in value (Exercise 8, Exercise Set 19.8). Here, τ is increasing in the same direction along the curve C as t. However, if t is replaced by a parameter τ where $d\tau/dt < 0$ when $t \in [a, b]$, then the sign of the line integral is changed (Exercise 8, Exercise Set 19.8). Here, τ is increasing in the same direction in which t is decreasing. In particular, suppose the curve C is given by (19.86). Then with the parametrization

$$x = f(a + b - t) \qquad y = g(a + b - t) \quad \text{if } t \in [a, b] \qquad (19.88)$$

the curve C is traversed in the *reverse direction* going from $(f(b), g(b))$ to $(f(a), g(a))$. Hence we denote the curve with parametrization (19.88) by $-C$. Since $d\tau/dt < 0$ when $\tau = a + b - t$, from the above discussion

$$\int_{-C} M\,dx + N\,dy = -\int_C M\,dx + N\,dy. \qquad (19.89)$$

Next, suppose C is the smooth curve given by (19.86). If $a < \bar{t} < b$ and C_1 and C_2 are the subarcs of C corresponding to the intervals $[a, \bar{t}]$ and $[\bar{t}, b]$, respectively, then

$$\int_C M\,dx + N\,dy = \int_{C_1} M\,dx + N\,dy + \int_{C_2} M\,dx + N\,dy. \qquad (19.90)$$

More generally, if $\{a = t_0, t_1, \ldots, t_n = b\}$ is any partition of $[a, b]$, and C_i is the subarc of C given by

$$x = f(t) \qquad y = g(t) \quad \text{if } t \in [t_{i-1}, t_i]$$

for $i = 1, 2, \ldots, n$, then from (19.90)

$$\int_C M\,dx + N\,dy = \sum_{i=1}^n \left(\int_{C_i} M\,dx + N\,dy\right). \qquad (19.91)$$

The notion of a line integral is also extended to a sectionally smooth curve. Suppose the plane curve C is a chain consisting of the smooth curves C_1, $C_2, \ldots, C_n$ (recall Definition 15.4.2). Then if the functions M and N are continuous on an open set S containing C, the line integral $\int_C M\,dx + N\,dy$ is defined to be

$$\int_C M\,dx + N\,dy = \sum_{i=1}^{n}\left(\int_{C_i} M\,dx + N\,dy\right),$$ (19.92)

which is consistent with (19.91).

In the next example we evaluate a line integral over a sectionally smooth curve.

Example 2 Evaluate the line integral $\int_C (x^2 - y^2)\,dx + xy\,dy$ over the curve C in which the x axis is traversed from $(-2, 0)$ to $(1, 0)$ and then the first quadrant arc of the unit circle from $(1, 0)$ to $(0, 1)$ (Figure 19–31).

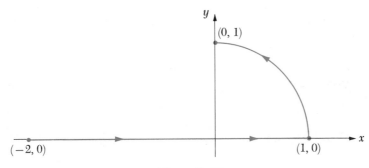

Figure 19-31

SOLUTION C_1, the part of the curve C on the x axis is given parametrically by

$$x = t \qquad y = 0 \quad \text{if } t \in [-2, 1].$$

Hence from Definition 19.8.1

$$\int_{C_1} (x^2 - y^2)\,dx + xy\,dy = \int_{-2}^{1} t^2\,dt$$
$$= 3.$$

C_2, the arc of the unit circle has the parametric equations

$$x = t \qquad y = \sqrt{1 - t^2} \quad \text{if } t \in [0, 1],$$

but if this parametrization is used, C_2 is traversed in the counterclockwise direction as t increases from 0 to 1. However, we recall that the curve with parametric equations (19.86) can also be given by equations (19.88) in which the direction of increasing parameter is reversed. Thus C_2 is also given by

$$x = 1 - t \qquad y = \sqrt{1 - (1 - t)^2} \quad \text{if } t \in [0, 1]$$ (19.93)

and with this parametrization C_2 is traversed in the clockwise direction as t goes from 0 to 1. Then by Definition 19.8.1

$$\int_{C_2} (x^2 - y^2)\, dx + xy\, dy = \int_0^1 [(1 - t)^2 - 1 + (1 - t)^2](-1)$$

$$+ (1 - t)\sqrt{1 - (1 - t)^2} \cdot \frac{2(1 - t)}{2\sqrt{1 - (1 - t)^2}}\Bigg]\, dt$$

$$= \int_0^1 [1 - (1 - t)^2]\, dt$$

$$= \left[t + \frac{(1 - t)^3}{3}\right]_0^1 = \frac{2}{3}.$$

Hence, from (19.92)

$$\int_C (x^2 - y^2)\, dx + xy\, dy = \int_{C_1} + \int_{C_2} = 3 + \tfrac{2}{3} = \tfrac{11}{3}.$$

Exercise Set 19.8

In Exercises 1–4 evaluate the given line integral over the indicated curve C.

1. $\int_C (x + 2y)\, dx + (y - 3x)\, dy$; $C: x = t^2,\, y = t$ if $t \in [0, 1]$

2. $\int_C xy\, dx - y^2\, dy$; $C: x = 1 - t,\, y = 2 + t$ if $t \in [-1, 1]$

3. $\int_C \cos y\, dx - x \sin y\, dy$; C: The line segment from $(0, \frac{\pi}{2})$ to $(1, \pi)$

4. $\int_C (x^2 + 2y^2)\, dx - x^2\, dy$; C; The arc of the ellipse $4x^2 + y^2 = 16$ in the counterclockwise direction from $(2, 0)$ to $(-2, 0)$

5. Evaluate $\int_C y^2\, dx + 2xy\, dy$ over the triangle with vertices $(-2, 0)$, $(2, 0)$, and $(0, 1)$ proceeding in the counterclockwise direction from $(0, 1)$.

6. Evaluate $\int_C (x - y)\, dx + (2x + y)\, dy$ over the path in which a line segment is traversed from $(3, -2)$ to $(1, 1)$ and then the graph of $y = x^3$ from $(1, 1)$ to $(0, 0)$.

7. Evaluate $\int_C (4x - y^2)\, dx + (x^2 - y)\, dy$ over the closed path in which the line $x = 2$ is traversed from $(2, -1)$ to $(2, 1)$ and then the parabola $y^2 = \frac{1}{2}x$ is traversed in the counterclockwise direction from $(2, 1)$ to $(2, -1)$.

8. Suppose that the curve (19.86) is parametrized by $x = F(\tau)$ and $y = G(\tau)$ where $\tau \in [\alpha, \beta]$. Assume that F and G have continuous derivatives on $[\alpha, \beta]$. Prove that:

   (a) The line integral $\int_C M\, dx + N\, dy$ is unchanged when t is replaced by τ if $d\tau/dt > 0$ for every $t \in [a, b]$.

   (b) The line integral $\int_C M\, dx + N\, dy$ is changed in sign when t is replaced by τ if $d\tau/dt < 0$ for every $t \in [a, b]$.

HINT: Let $\tau = \psi(t)$ for $t \in [a, b]$ and let $\alpha = \psi(a)$ and $\beta = \psi(b)$. Then rewrite $\int_a^b [M(f(t), g(t))f'(t) + N(f(t), g(t))g'(t)]\, dt$ so that the integrand is in terms of τ. This can be done by going from right to left in the formula in Theorem 7.8.2.

9. Find the work done by the force of repulsion in Example 1 in moving the unit charge along a polygonal path from $(1, 0)$ to $(1, 1)$ and then from $(1, 1)$ to $(-1, 2)$. How does the work calculated here compare with that obtained in Example 1?

10. Suppose M, N, and P are functions of three variables which are continuous on an open set $S \subset R^3$. Give a suitable definition for the line integral $\int_C M \, dx + N \, dy + P \, dz$ which is analogous to Definition 19.8.1.

In Exercises 11 and 12 evaluate the given line integrals over the given curves, using the definition obtained in Exercise 10.

11. $\int_C xy \, dx - yz \, dy + xz \, dz$; $C: x = t,\ y = t^2,\ z = t^4$ if $t \in [0, 1]$
12. $\int_C e^y \, dx + e^x \, dy - 2e^y \, dz$; $C: x = t,\ y = 2t - 1,\ z = t^2$ if $t \in [0, 1]$

19.9   Independence of the Path

If n is a positive integer and C is the curve from $(0, 0)$ to $(1, 1)$ given by

$$x = t \qquad y = t^n \quad \text{if } t \in [0, 1],$$

then by Definition 19.8.1 the line integral

$$\int_C 2xy \, dx + (x^2 + 3y) \, dy = \int_0^1 [2t^{n+1} \cdot 1 + (t^2 + 3t^n) \cdot nt^{n-1}] \, dt$$

$$= \int_0^1 [(2 + n)t^{n+1} + 3nt^{2n-1}] \, dt$$

$$= \left. (t^{n+2} + \tfrac{3}{2}t^{2n}) \right]_0^1 = \tfrac{5}{2}.$$

Since the value of the line integral is a constant independent of n, we might then ask the question: "Is $\int_C 2xy \, dx + (x^2 + 3y) \, dy$ the same over any sectionally smooth curve from $(0, 0)$ to $(1, 1)$?" The answer to this query involves the notion of *independence of the path* of a line integral.

19.9.1   Definition

Suppose P and Q are any two points in an open set $S \subset R^2$ and C_1 and C_2 are arbitrarily chosen sectionally smooth curves in S from P to Q (Figure 19–32). The line integral $\int_C M \, dx + N \, dy$ is *independent of the path in S* if and only if

$$\int_{C_1} M \, dx + N \, dy = \int_{C_2} M \, dx + N \, dy.$$

The following theorem states the additional conditions that must be imposed on the functions M and N in order to guarantee that the line integral is independent of the path in S. It will be noted that the "only if" part of the theorem is the analogue of the first fundamental theorem (Theorem 7.7.1) for vector-valued functions of the form (19.80).

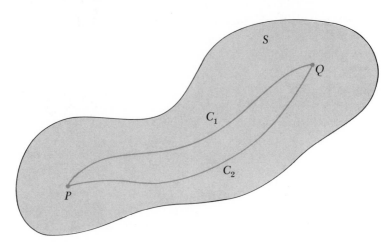

Figure 19-32

19.9.2 Theorem

Suppose the functions M and N of two variables are continuous on an open set $S \subset R^2$. The line integral $\int_C M\,dx + N\,dy$ is independent of the path in S if and only if there is a function ϕ defined on S such that

$$D_1\phi(x, y) = M(x, y) \quad \text{and} \quad D_2\phi(x, y) = N(x, y) \quad \text{if } (x, y) \in S, \qquad (19.94)$$

or saying the same thing,

$$\nabla\phi(x, y) = \langle M(x, y), N(x, y)\rangle \quad \text{if } (x, y) \in S. \qquad (19.95)$$

PROOF We shall prove the "if" part first. Suppose P and Q are distinct points in S and C is an arbitrarily chosen sectionally smooth curve from P to Q. Then by definition, C is a chain of smooth curves $C_1, C_2, \ldots, C_n$. Suppose an arbitrary curve C_i from this chain has the parametrization

$$x = f(t) \qquad y = g(t) \quad \text{if } t \in [t_{i-1}, t_i] \qquad (19.96)$$

where it is understood that $t_0 = a$ and $t_n = b$. Then let $P_{i-1} = (f(t_{i-1}), g(t_{i-1}))$ and $P_i = (f(t_i), g(t_i))$ be the endpoints of C_i. Here it is assumed that $P_0 = P$ and $P_n = Q$. Then from (19.94) and Definition 19.8.1

$$\int_{C_i} M\,dx + N\,dy = \int_{t_{i-1}}^{t_i} [D_1\phi(f(t), g(t))f'(t) + D_2\phi(f(t), g(t))g'(t)]\,dt. \qquad (19.97)$$

By the chain rule the integrand in the right member of (19.97) is $D_t\varphi(f(t), g(t))$. Hence

$$\int_{C_i} M\,dx + N\,dy = \int_{t_{i-1}}^{t_i} D_t\phi(f(t), g(t))\,dt$$

$$= \phi(f(t_i), g(t_i)) - \phi(f(t_{i-1}), g(t_{i-1}))$$

$$= \phi(P_i) - \phi(P_{i-1}). \qquad (19.98)$$

By summing the line integrals over the curves $C_1, C_2, \ldots, C_n$ we have from (19.92) and (19.98)

$$\int_C M\,dx + N\,dy = \sum_{i=1}^{n} \phi(P_i) - \phi(P_{i-1}). \qquad (19.99)$$

For $i = 1, 2, \ldots, n - 1$ each term $\phi(P_i)$ in the sum on the right in (19.99) cancels with the term $-\phi(P_i)$ also in the sum. Hence this sum collapses and we obtain

$$\int_C M\,dx + N\,dy = \phi(Q) - \phi(P). \qquad (19.100)$$

We note from (19.100) that the line integral depends on the function ϕ and the endpoints P and Q, but is independent of the choice of the curve C from P to Q.

   To prove the converse, let (a, b) be fixed in S and let C be an arbitrary sectionally smooth curve in S from (a, b) to some point (ξ, η) (Figure 19–33). Since

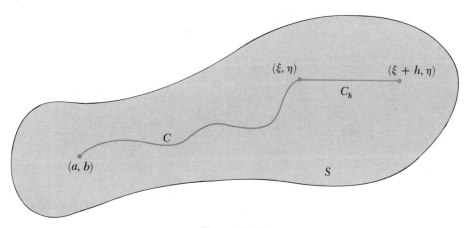

Figure 19–33

$\int_C M\,dx + N\,dy$ is, by hypothesis, independent of the choice of the curve C here, we shall let ϕ be the function given by

$$\phi(\xi, \eta) = \int_C M\,dx + N\,dy \quad \text{if } (\xi, \eta) \in S. \qquad (19.101)$$

It will be shown in the ensuing proof that this function ϕ satisfies (19.94).
   Since S is an open set, (ξ, η) is an interior point in S. Hence for h sufficiently close to 0 the point $(\xi + h, \eta)$ and the line segment C_h from (ξ, η) to $(\xi + h, \eta)$ are also in S. Then since the line integral over any sectionally smooth curve in S from (a, b) to $(\xi + h, \eta)$ is independent of the choice of the curve, we can evaluate this line integral over the chain consisting of C and C_h. Thus

$$\varphi(\xi + h, \eta) = \int_C M\,dx + N\,dy + \int_{C_h} M\,dx + N\,dy. \qquad (19.102)$$

From (19.101) and (19.102)

$$\frac{\phi(\xi + h, \eta) - \phi(\xi, \eta)}{h} = \frac{1}{h}\int_{C_h} M \, dx + N \, dy.$$

If $h > 0$, C_h is given by

$$x = t \qquad y = \eta \quad \text{if } t \in [\xi, \xi + h],$$

but if $h < 0$, C_h has the parametrization

$$x = 2\xi + h - \tau \qquad y = \eta \quad \text{if } \tau \in [\xi + h, \xi].$$

Thus, regardless of whether $h > 0$ or $h < 0$

$$\frac{\phi(\xi + h, \eta) - \phi(\xi, \eta)}{h} = \frac{1}{h}\int_{\xi}^{\xi + h} M(t, \eta) \, dt$$

and from the mean value theorem of integral calculus there is a c in the closed interval with endpoints ξ and $\xi + h$ such that

$$\frac{\phi(\xi + h, \eta) - \phi(\xi, \eta)}{h} = M(c, \eta). \tag{19.103}$$

Because M is continuous at (ξ, η), from (19.103)

$$\lim_{h \to 0} \frac{\phi(\xi + h, \eta) - \phi(\xi, \eta)}{h} = D_1\phi(\xi, \eta) = M(\xi, \eta) \quad \text{if } (\xi, \eta) \in S.$$

The proof that $D_2\phi(\xi, \eta) = N(\xi, \eta)$ if $(\xi, \eta) \in S$ is similar and is left as an exercise.

In this paragraph $\int_C M \, dx + N \, dy$ is independent of the path in an open set S and φ is the function satisfying (19.94), which is guaranteed by Theorem 19.9.2. If C is any sectionally smooth curve in S from the point (a, b) to (c, d), then $\int_C M \, dx + N \, dy$ can be denoted by $\int_{(a,b)}^{(c,d)} M \, dx + N \, dy$ since the line integral is independent of the choice of C. Thus from (19.100)

$$\int_{(a,b)}^{(c,d)} M \, dx + N \, dy = \phi(c, d) - \phi(a, b). \tag{19.104}$$

Note that (19.104) is the analogue for vector-valued functions of the type (19.80) of the integration formula in the second fundamental theorem (Theorem 7.8.1).

If $\mathbf{F}$ is the vector-valued function such that

$$\mathbf{F}(x, y) = \langle M(x, y), N(x, y)\rangle = \nabla\phi(x, y) \quad \text{if } (x, y) \in S,$$

then $\mathbf{F}$ is called a *gradient field* on S, ϕ is termed a *potential function* of $\mathbf{F}$ and $\phi(x, y)$ is termed the *potential* of $\mathbf{F}$ at (x, y). If $\mathbf{F}$ is a force field on S, then $\mathbf{F}$ is called a *conservative force field,* and a restatement of (19.104) asserts that the work done by $\mathbf{F}$ in moving a particle from (a, b) to (c, d) is *the difference in potential at the endpoints of the path.*

We return to the question posed in the early part of Section 19.9 with the next example.

Example 1   Show that the line integral $\int_C 2xy \, dx + (x^2 + 3y) \, dy$ is independent of the path in R^2 by obtaining a potential function associated with the line integral. Utilize this function to evaluate the line integral $\int_{(0,0)}^{(1,1)} 2xy \, dx + (x^2 + 3y) \, dy$.

SOLUTION   Because of Theorem 19.9.2 the line integral $\int_C 2xy \, dx + (x^2 + 3y) \, dy$ is the same over any sectionally smooth curve from $(0,0)$ to $(1,1)$ if there is a function ϕ such that

$$\frac{\partial \phi}{\partial x} = 2xy \quad \text{and} \quad \frac{\partial \phi}{\partial y} = x^2 + 3y \qquad (19.105)$$

for every (x, y). If such a ϕ exists, then upon integrating in the first of the equations in (19.105) with respect to x while keeping y constant, we obtain

$$\phi(x, y) = x^2 y + g(y). \qquad (19.106)$$

Note that if $\partial \phi / \partial x = 2xy$, then the most general form for $\phi(x, y)$ is (19.106) rather than merely $x^2 y + C$, where C is constant. Hence from (19.106) and (19.105)

$$\frac{\partial \phi}{\partial y} = x^2 + g'(y) = x^2 + 3y. \qquad (19.107)$$

From the equality of the middle and right members in (19.107)

$$g'(y) = 3y$$

and therefore

$$g(y) = \frac{3y^2}{2} + K, \qquad (19.108)$$

where K is the constant of integration. Thus from (19.106) and (19.108) we obtain

$$\phi(x, y) = x^2 + \frac{3y^2}{2} + K. \qquad (19.109)$$

It can be readily verified that any function ϕ of the form (19.109) satisfies (19.105), and thus the line integral $\int_C 2xy \, dx + (x^2 + 3y) \, dy$ is independent of the path in R^2. If we let $K = 0$ in (19.109) so that ϕ is given by

$$\phi(x, y) = x^2 + \frac{3y^2}{2},$$

then from (19.104)

$$\int_{(0,0)}^{(1,1)} 2xy \, dx + (x^2 + 3y) \, dy = \phi(1, 1) - \phi(0, 0)$$

$$= \tfrac{5}{2} - 0 = \tfrac{5}{2}.$$

It would be convenient at this point to have some criterion by which we can readily determine in advance whether or not a line integral is independent of the path without having to waste time seeking a potential function ϕ when no such function exists. The following theorem is useful because it provides a necessary condition for the existence of a potential function.

19.9.3   Theorem

Suppose M and N have continuous partial derivatives on an open set $S \subset R^2$. If there is a function ϕ on S such that

$$D_1\phi(x, y) = M(x, y) \quad and \quad D_2\phi(x, y) = N(x, y) \quad for \ (x, y) \in S, \quad (19.94)$$

then

$$D_2M(x, y) = D_1N(x, y) \quad if \ (x, y) \in S. \tag{19.110}$$

PROOF   From partial differentiation in (19.94)

$$D_{12}\phi(x, y) = D_2M(x, y) \quad and \quad D_{21}\phi(x, y) = D_1N(x, y) \quad if \ (x, y) \in S.$$
$$\tag{19.111}$$

Since the functions D_2M and D_1N are continuous on S, by Theorem 18.11.2 $D_{12}\varphi = D_{21}\varphi$ on S, and the conclusion follows from (19.111).

Example 2   Show that the line integral $\int_C xy \, dx + x^2 \, dy$ is not independent of the path in R^2.

SOLUTION   Here $M(x, y) = xy$ and $N(x, y) = x^2$. Since

$$D_2M(x, y) = \frac{\partial}{\partial y}(xy) = x \quad and \quad D_1N(x, y) = \frac{\partial}{\partial x}(x^2) = 2x,$$

$D_2M(x, y) \neq D_1N(x, y)$ for every $(x, y) \in R^2$, and hence there is no potential function φ for this line integral. Thus by Theorem 19.9.3 the given line integral is not independent of the path in R^2.

In advanced calculus it is proved that the line integral $\int_C M \, dx + N \, dy$ is independent of the path in the open set S when (19.110) is satisfied provided S is *simply connected;* in ordinary language this means that S has no "holes" (Figure 19–34). However if S is not simply connected, then (19.110) does not guarantee that $\int_C M \, dx + N \, dy$ is independent of the path in S (see Exercise 17, Exercise Set 19.9).

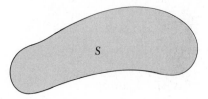

Simply-connected set

Multiply-connected set

Figure 19–34

Exercise Set 19.9

In Exercises 1–8 evaluate the given line integrals using (19.104) or prove that the integral does not exist.

1. $\displaystyle\int_{(0,1)}^{(-1,2)} y^2\,dx + 2xy\,dy$

2. $\displaystyle\int_{(-2,1)}^{(1,3)} (2xy^3 - 1)\,dx + 3x^2y^2\,dy$

3. $\displaystyle\int_{(1,\sqrt{3})}^{(-\sqrt{3},1)} \frac{y\,dx}{x^2 + y^2} - \frac{x\,dy}{x^2 + y^2}$

4. $\displaystyle\int_{(1,0)}^{(4,3)} \frac{y}{(x + y)^2}\,dx + \left[\frac{-x}{(x + y)^2} + 2\right]dy$

5. $\displaystyle\int_{(-1,0)}^{(2,\pi/3)} \cos y\,dx - (\cos y + x \sin y)\,dy$

6. $\displaystyle\int_{(1,1)}^{(3,\ln 2)} -y^2 e^{-xy}\,dx + (1 - xy)e^{-xy}\,dy$

7. $\displaystyle\int_{(0,-2)}^{(1,1)} \frac{1}{y}\,dx + \frac{2x}{y^2}\,dy$

8. $\displaystyle\int_{(0,0)}^{(\pi/4,\pi/3)} \sec x \tan x \tan^2 y\,dx + 2 \sec x \tan y \sec^2 y\,dy.$

9. Prove: If $\int_C M\,dx + N\,dy$ is independent of the path in an open rectangle S whose sides are parallel to the coordinate axes and (a, b) and (c, d) are any two points in S, then

$$\int_{(a,b)}^{(c,d)} M\,dx + N\,dy = \int_a^c M(x, b)\,dx + \int_b^d N(c, y)\,dy.$$

HINT: Evaluate $\int_{(a,b)}^{(c,d)} M\,dx + N\,dy$ over a particular path from (a, b) to (c, d).

In Exercises 10–13 evaluate the line integrals in the following exercises using the formula derived in Exercise 9.

10. Exercise 2   11. Exercise 3

12. Exercise 4   13. Exercise 6

14. Prove that the force field in Example 1 of Section 19.8 is conservative in any open set not containing the origin.

15. Complete the proof of Theorem 19.9.2 by proving that $D_2\phi(\xi, \eta) = N(\xi, \eta)$ if $(\xi, \eta) \in$ S. HINT: Obtain $(\phi(\xi, \eta + h) - \phi(\xi, \eta))/h$ when h is sufficiently close to 0 so that $(\xi, \eta + h) \in$ S.

16. Prove that if $\int_C M\,dx + N\,dy$ is independent of the path in any open set S, then $\int_K M\,dx + N\,dy = 0$ for any *closed curve* K in S. NOTE: A curve is said to be closed if and only if its endpoints coincide.

17. Show that the line integral

$$\int_C \frac{-y}{x^2 + y^2}\,dx + \frac{x}{x^2 + y^2}\,dy$$

is not independent of the path in the set $S = \{(x, y): (x, y) \neq (0, 0)\}$ even though

$$\frac{\partial}{\partial y}\left(-\frac{y}{x^2 + y^2}\right) = \frac{\partial}{\partial x}\left(\frac{x}{x^2 + y^2}\right)$$

for every $(x, y) \in S$. (Note that S is not simply connected because it has a "hole" at $(0, 0)$.) HINT: If the line integral were independent of the path in S, then by Exercise 16 the line integral would be 0 over the circle

$$x = \cos t \qquad y = \sin t \quad \text{for } t \in [0, 2\pi].$$

18. Prove the converse of the statement in Exercise 16.

19. State the theorem which is analogous to Theorem 19.9.2 for line integrals involving functions of three variables.

20. Suppose that M, N, and P are functions of three variables which have continuous partial derivatives on some open set $S \subset R^3$. Prove that the equations

$$\frac{\partial M}{\partial y} = \frac{\partial N}{\partial x} \qquad \frac{\partial N}{\partial z} = \frac{\partial P}{\partial y} \qquad \frac{\partial P}{\partial x} = \frac{\partial M}{\partial z}$$

hold for every $(x, y, z) \in S$ if $\int_C M\, dx + N\, dy + P\, dz$ is independent of the path in S.

In Exercises 21 and 22 evaluate the given integrals if they exist.

21. $\displaystyle\int_{(-1,0,1)}^{(2,1,-2)} (6x^2y^2 - 3yz^2)\, dx + (4x^3y - 3xz^2)\, dy - 6xyz\, dz$

22. $\displaystyle\int_{(0,0,\pi/6)}^{(1,-1,\pi/2)} (-e^{-x}\sin z - y^2)\, dx - 2xy\, dy + e^{-x}\cos z\, dz.$

20. Differential Equations

20.1 Introduction

The reader has already had some contact with the subject of differential equations through the consideration of problems arising from certain topics in physics. For example, it will be recalled that in Section 7.2 simple differential equations dealing with the motion of a particle were solved by antidifferentiation. Also, the solution of certain types of differential equations was discussed in Section 9.7 (exponential growth and decay) and Section 10.4 (simple harmonic motion). In this chapter a more general treatment of the subject will be undertaken and some of the important applications will be discussed.

The term *(ordinary) differential equation of order n* is applied to an equation in the variables x, y (considered as a function of x), dy/dx, d^2y/dx^2, . . . , d^ny/dx^n. For example,

$$\frac{dy}{dx} = -\frac{x}{y} \tag{20.1}$$

is a *first order* differential equation and

$$\frac{d^2y}{dx^2} + x\frac{dy}{dx} - 3y = 2 - x^2 \tag{20.2}$$

is of *second order*. The term *ordinary*, as used here, means that the equation does not involve partial derivatives. The solution of *partial differential equations* is a completely separate topic and is not treated in any depth in this text.

An important class of differential equations consists of those which are called *linear differential equations*. If P_1, P_2, . . . , P_n, and Q are functions of one variable, then

$$\frac{d^ny}{dx^n} + P_1(x)\frac{d^{n-1}y}{dx^{n-1}} + \cdots + P_{n-1}(x)\frac{dy}{dx} + P_n(x)y = Q(x) \tag{20.3}$$

is called a *linear differential equation of order n*. The term linear is used to describe these differential equations since they arise from certain kinds of functions that in linear algebra are called "linear transformations." The equation (20.2) above, for example, is a second order linear differential equation. If $Q(x) = 0$ for every x, (20.3) has the form

$$\frac{d^n y}{dx^n} + P_1(x)\frac{d^{n-1}y}{dx^{n-1}} + \cdots + P_{n-1}(x)\frac{dy}{dx} + P_n(x)y = 0. \qquad (20.4)$$

Equation (20.4) is called a *homogeneous linear differential equation of order n.*

A function of one variable, f, is said to be a *solution* of a differential equation in x, y, dy/dx, . . . , and $d^n y/dx^n$ on an interval I if and only if the equation obtained by replacing y by $f(x)$, dy/dx by $f'(x)$, . . . , and $d^n y/dx^n$ by $f^{(n)}(x)$ is an identity whenever $x \in I$. For example, if

$$f(x) = e^{2x},$$

then f is a solution of $dy/dx = 2y$ on the interval $(-\infty, +\infty)$ since

$$f'(x) = 2e^{2x} = 2f(x)$$

for every x. In fact, if C is an arbitrary constant, the equation

$$y = Ce^{2x}$$

defines infinitely many solutions of $dy/dx = 2y$. Also, the function u where

$$u(x) = x^2$$

is a solution of (20.2) on $(-\infty, +\infty)$ since

$$u''(x) + xu'(x) - 3u(x) = 2 + x \cdot 2x - 3x^2$$
$$= 2 - x^2$$

for every x. However, the equation

$$\left(\frac{dy}{dx}\right)^2 + 1 = 0$$

has no real-valued solutions, because there is no real-valued function f such that $(f'(x))^2 + 1 = 0$.

The solutions of a differential equation may be expressed implicitly. For example, the equation

$$x^2 + y^2 = C^2 \quad \text{where } C > 0 \qquad (20.5)$$

defines implicitly the functions f_1 and f_2 given by

$$f_1(x) = \sqrt{C^2 - x^2} \qquad \text{if } x \in [-C, C]$$
$$f_2(x) = -\sqrt{C^2 - x^2} \quad \text{if } x \in [-C, C],$$

both of which satisfy the equation (20.1) on the interval $(-C, C)$. In fact, if we differentiate implicitly in (20.5) with respect to x, we obtain

$$2x + 2y\frac{dy}{dx} = 0,$$

and hence

$$\frac{dy}{dx} = -\frac{x}{y}.$$

Often the explicit form of a solution is not easily obtained when the solution is given implicitly. In such situations implicit differentiation can be used to verify that the solution actually satisfies the differential equation. The following example illustrates the technique involved.

Example 1 Verify that the solutions expressed implicitly by

$$x^2 - y^2 = Cx \qquad (20.6)$$

are solutions of the differential equation

$$\frac{dy}{dx} = \frac{x^2 + y^2}{2xy}. \qquad (20.7)$$

SOLUTION Since (20.7) does not contain the constant C, we shall rewrite (20.6) in the form

$$\frac{x^2 - y^2}{x} = C \qquad (20.8)$$

when $x \neq 0$ so that the constant will be lost upon differentiation. Then after implicit differentiation in (20.8) we have

$$\frac{x(2x - 2yy') - (x^2 - y^2)}{x^2} = 0 \quad \text{where } y' = \frac{dy}{dx}$$

$$x^2 - 2xyy' + y^2 = 0$$

and finally

$$y' = \frac{x^2 + y^2}{2xy}.$$

Exercise Set 20.1

In Exercises 1–10 verify that the given function f is a solution of the given differential equation.

1. $f(x) = \dfrac{3}{1 - x^3}$, $\dfrac{dy}{dx} = x^2 y^2$

2. $f(x) = 1 + Ce^{-x^2/2}$, $\dfrac{dy}{dx} + xy = x$

3. $f(x) = C_1 e^{ax} + C_2 e^{-ax}$, $\dfrac{d^2 y}{dx^2} - a^2 y = 0$

4. $f(x) = a \sinh\left(\dfrac{x}{a} + b\right)$, $\dfrac{dy}{dx} = \sqrt{\dfrac{y^2}{a^2} + 1}$

5. $f(t) = \dfrac{a^2 kt}{1 + akt}$, $\dfrac{dx}{dt} = k(a - x)^2$

6. $f(x) = \ln(C - e^{-x})$, $\dfrac{dy}{dx} = e^{-(x+y)}$

7. $f(x) = 2 \tan^{-1} x - x$, $\dfrac{dy}{dx} = \cos(x + y)$

8. $f(t) = 3t \cos 2t$, $\dfrac{d^2x}{dt^2} + 4x = -12 \sin 2t$

9. $f(x) = x^2 \sin x$, $x^2 \dfrac{d^2y}{dx^2} - 4x \dfrac{dy}{dx} + 6y = -x^4 \sin x$

10. $f(x) = \dfrac{\cos kx}{x}$, $\dfrac{d^2y}{dx^2} + \dfrac{2}{x} \dfrac{dy}{dx} + k^2y = 0$

In Exercises 11–15 verify that the function defined implicitly by the first equation satisfies the given differential equation.

11. $r(1 + e \cos \theta) = C$, $\dfrac{dr}{d\theta} = \dfrac{er \sin \theta}{1 + e \cos \theta}$

12. $\ln(x^2 + y^2) - \tan^{-1} \dfrac{y}{x} = C$, $(x - 2y)\dfrac{dy}{dx} = 2x + y$

13. $y^2 = \dfrac{x^3}{C - x}$, $2x^3 \dfrac{dy}{dx} = y(y^2 + 3x^2)$

14. $y^2 = x(x - C)^3$, $\left(2x\dfrac{dy}{dx} - y\right)^3 = 27x^4y$

15. $(y - C_1)^2 = 4(x - C_2)$, $\left(\dfrac{dy}{dx}\right)^3 + 2\dfrac{d^2y}{dx^2} = 0$

20.2   First Order Differential Equations

Many of the first order differential equations that we will solve in this text can be expressed in the form

$$\frac{dy}{dx} = G(x, y) \tag{20.9}$$

We shall seek a solution f of (20.9) such that if x_0 and y_0 are preassigned constants, then f satisfies the *initial condition* (also called the *boundary condition*)

$$y_0 = f(x_0). \tag{20.10}$$

The problem of finding such a solution f (see Figure 20–1) is often called the *initial value problem* (or *boundary value problem*):

$$\frac{dy}{dx} = G(x, y) \qquad y(x_0) = y_0.$$

If certain conditions are satisfied by the function G for every (x, y) in the vicinity of (x_0, y_0), then it will be possible to find a sequence of functions $\{f_n\}$ that satisfy

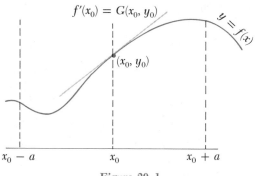

Figure 20-1

$$\left.\begin{array}{lll} f_1(x_0) = y_0 & \text{and} & f_1'(x) = G(x, y_0) \\ f_2(x_0) = y_0 & \text{and} & f_2'(x) = G(x, f_1(x)) \\ & \cdots & \\ f_n(x_0) = y_0 & \text{and} & f_n'(x) = G(x, f_{n-1}(x)) \end{array}\right\} \tag{20.11}$$

and converge to a solution f of (20.9). These conditions for G are given in the following existence and uniqueness theorem.

20.2.1   Theorem

Suppose that G, a function of two variables, is continuous and has the continuous partial derivative D_2G on some closed rectangle having (x_0, y_0) as an interior point. Then

> (*i*) *on some interval $[x_0 - a, x_0 + a]$ there is a unique solution f of (20.9) which also satisfies (20.10),*
>
> (*ii*) *there is a sequence of functions $\{f_n\}$ such that for every $x \in [x_0 - a, x_0 + a]$*

$$f_0(x) = y_0$$
$$f_n(x) = y_0 + \int_{x_0}^{x} G(t, f_{n-1}(t))\, dt \quad \text{for } n = 1, 2, \ldots, \tag{20.12}$$

and

> (*iii*) $\lim\limits_{n \to +\infty} f_n(x) = f(x).$

The proof of Theorem 20.2.1 is omitted here since it is somewhat long, although it is found in many texts on differential equations. Note that the equations (20.11) for the sequence $f_n(x)$ follow from equations (20.12). The use of the sequence (20.12) for approximating the solution f is called *Picard's method of successive approximations* after the French mathematician Émile Picard (1856–1941).

Example 1   Prove that the differential equation

$$\frac{dy}{dx} = xy \tag{20.13}$$

has a unique solution f on some interval containing 0 which satisfies the initial condition $f(0) = 1$. Then find the first few terms of a sequence $\{f_n(x)\}$ that converges to $f(x)$ for every x in that interval.

SOLUTION If G is given by $G(x, y) = xy$, then since $D_2G(x, y) = x$, G and D_2G are everywhere continuous. Hence by Theorem 20.2.1, on some interval $[-a, a]$, where $a > 0$, there is a unique solution f of (20.13) such that $f(0) = 1$. Also from Theorem 20.2.1 for $x \in [-a, a]$ there is a sequence of functions such that

$$f_0(x) = 1$$

$$f_1(x) = 1 + \int_1^x t \cdot 1 \, dt$$

$$= 1 + \frac{x^2}{2}$$

$$f_2(x) = 1 + \int_0^x t\left(1 + \frac{t^2}{2}\right) dt$$

$$= 1 + \frac{x^2}{2} + \frac{x^4}{8}$$

$$f_3(x) = 1 + \int_0^x t\left(1 + \frac{t^2}{2} + \frac{t^4}{8}\right) dt$$

$$= 1 + \frac{x^2}{2} + \frac{x^4}{8} + \frac{x^6}{48}$$

$$\cdots$$

which converge to f on $[-a, a]$.

The terms $f_0(x)$, $f_1(x)$, $f_2(x)$, and $f_3(x)$ in this sequence are actually the first four terms in the Maclaurin series for $e^{x^2/2}$. In fact, it can be easily verified that if

$$f(x) = e^{x^2/2}, \qquad\qquad (20.14)$$

then f is a solution of (20.13) satisfying $f(0) = 1$. Hence by Theorem 20.2.1 the unique solution f obtained in Example 1 is given by (20.14).

Often a first order differential equation will have a family of solutions, which can be expressed by an equation in x, y, and C where C is an arbitrary constant. Such an equation is often called a *general solution* of the differential equation. For example, it will be recalled from Section 20.1 that

$$\frac{dy}{dx} = -\frac{x}{y}$$

has the general solutions

$$x^2 + y^2 = C^2$$

or

$$y = \sqrt{C^2 - x^2}$$

and
$$y = -\sqrt{C^2 - x^2}.$$
Also
$$\frac{dy}{dx} = y^{2/3} \tag{20.15}$$
has a general solution
$$y = \tfrac{1}{27}(x + C)^3 \tag{20.16}$$
since
$$\frac{dy}{dx} = \frac{1}{9}(x + C)^2 = \left(\frac{1}{27}(x + C)^3\right)^{2/3} = y^{2/3}.$$

If a particular value is assigned to the arbitrary constant in the general solution of a first order differential equation, then a *particular solution* of the differential equation is obtained. For example, if $C = 0$ in (20.16), the particular solution expressed by
$$y = \tfrac{1}{27}x^3 \tag{20.17}$$
is obtained. It will also be noted that another solution of (20.15) is the trivial solution given by
$$y = 0 \tag{20.18}$$
but that this solution cannot be obtained from the general solution (20.16) by substituting any particular number for C. The reader will also observe that both of the solutions (20.17) and (20.18) of (20.15) satisfy $y = 0$ when $x = 0$. In other words, there are two solutions of the initial value problem
$$\frac{dy}{dx} = y^{2/3} \qquad y(0) = 0$$
and so there is no uniqueness. Why is this possible in view of Theorem 20.2.1?

Exercise Set 20.2

In Exercises 1–6 use Picard's method to find the first four terms in a sequence of functions which converges to the solution of the given initial value problem. If possible, find this solution as an elementary function.

1. $\dfrac{dy}{dx} = x + y, \quad y(0) = 1$

2. $\dfrac{dy}{dx} = 2y, \quad y(0) = 2$

3. $\dfrac{dy}{dx} = xy + 1, \quad y(1) = 0$

4. $\dfrac{dy}{dx} = e^x - y, \quad y(0) = 0$

5. $\dfrac{dy}{dx} = 1 + y^2, \quad y(0) = 0$

6. $\dfrac{dy}{dx} = 1 + xy^2, \quad y(0) = 0$

7. Suppose g, a function of one variable, is everywhere continuous. Prove that the Picard's method approximations obtained for the solution of the initial value problem $dy/dx = g(x), y(x_0) = y_0$ are the same as the solution.

20.3   Separable Differential Equations

In this section and also in Section 20.4 general solutions will be obtained for some of the common types of first order equations. The first type we shall consider consists of equations of the form

$$P(y)\frac{dy}{dx} = Q(x). \qquad (20.19)$$

Such equations are called *separable differential equations* since from (20.19) one can obtain

$$P(y)\,dy = Q(x)\,dx$$

in which the variables x and y are separated. It will be recalled that the first order equations that were discussed in Section 7.2 and Section 9.7 are of the form (20.19).

Suppose the functions P and Q here are continuous on the intervals J and I, respectively. Also suppose f is a function of one variable that has a continuous derivative on I and, further, that $f(x) \in J$ when $x \in I$. Then f is a solution of (20.19) on I if and only if

$$P(f(x))f'(x) = Q(x) \quad \text{for } x \in I,$$

and hence if and only if

$$\int P(f(x))f'(x)\,dx = \int Q(x)\,dx \quad \text{for } x \in I.$$

Therefore to solve (20.19), one can multiply each member of (20.19) by dx and integrate to obtain

$$\int P(y)\,dy = \int Q(x)\,dx. \qquad (20.20)$$

The equation (20.20) will implicitly express the solutions of (20.19).

Example 1   Solve $y^3\dfrac{dy}{dx} = x^2$.

SOLUTION   We can write the given equation in the form

$$y^3\,dy = x^2\,dx.$$

Integration then gives

$$\frac{y^4}{4} = \frac{x^3}{3} + C_1.$$

If one lets $C = 12C_1$, the solutions can be implicitly obtained in the form

$$3y^4 = 4x^3 + C.$$

Example 2   Solve $dy/dx = 2y/(x + 1)$. Also, find the particular solution for which $y = 2$ when $x = 1$.

SOLUTION If $y \neq 0$, then by separating variables

$$\frac{dy}{y} = \frac{2\,dx}{x + 1}$$

and after integrating,

$$\ln |y| = C_1 + 2 \ln |x + 1|.$$

Then by utilizing properties of the natural logarithm function

$$\ln |y| = \ln e^{C_1}(x + 1)^2,$$

Hence

$$|y| = e^{C_1}(x - 1)^2. \tag{20.21}$$

Since (20.21) defines y as a differentiable function of x on some interval, certainly y is a continuous function of x on the interval. Therefore exactly one of the following situations holds:

$$y = e^{C_1}(x + 1)^2 \qquad \text{or} \qquad y = -e^{C_1}(x + 1)^2.$$

In either case there is a constant C such that $C = e^{C_1}$ or $C = -e^{C_1}$, whichever is applicable, and we conclude that a general solution for the given equation is

$$y = C(x + 1)^2. \tag{20.22}$$

Even though this equation was derived with the proviso that $y \neq 0$, note that the function defined by $y = 0$ satisfies the given differential equation and can be obtained from (20.22) by letting $C = 0$. To find the desired particular solution, let $y = 2$ and $x = 1$ in (20.22), and obtain $C = \frac{1}{2}$. Hence this particular solution is given by

$$y = \tfrac{1}{2}(x + 1)^2.$$

Example 3 A body of mass m which falls from rest near the earth's surface is subject to a downward force of mg due to gravity and a resisting force that is proportional to the square of the velocity of the body. Then from Newton's second law, the net downward force on the body at any time t is

$$F = m\frac{dv}{dt} = mg - kv^2 \tag{20.23}$$

where v is the velocity of the body at time t and k is a positive constant. (Note here that the downward direction is taken to be positive.) If we let

$$b = \frac{k}{m},$$

then (20.23) can be rewritten

$$\frac{dv}{dt} = g - bv^2. \tag{20.24}$$

Obtain the velocity of the body as a function of time t.

SOLUTION   From (20.24)

$$\frac{dv}{g - bv^2} = dt$$

$$\frac{\frac{1}{b} dv}{\frac{g}{b} - v^2} = dt.$$

Integrating using (23) from the Table of Integrals

$$\frac{1}{2\sqrt{bg}} \ln \frac{\sqrt{\frac{g}{b}} + v}{\sqrt{\frac{g}{b}} - v} = \frac{1}{2\sqrt{bg}} \ln C + t.$$

Then as in Example 2 we can write this as

$$\frac{\sqrt{\frac{g}{b}} + v}{\sqrt{\frac{g}{b}} - v} = Ce^{2\sqrt{bg}\,t},$$

and solving for v gives

$$v = \sqrt{\frac{g}{b}} \frac{Ce^{2\sqrt{bg}\,t} - 1}{Ce^{2\sqrt{bg}\,t} + 1}. \tag{20.25}$$

Since $v = 0$ when $t = 0$, we obtain $C = 1$ from (20.25). Thus we have

$$v = \sqrt{\frac{g}{b}} \frac{e^{2\sqrt{bg}\,t} - 1}{e^{2\sqrt{bg}\,t} + 1}. \tag{20.26}$$

It is interesting to note from (20.26) that v does not increase indefinitely, but instead approaches the limiting velocity

$$\lim_{t \to +\infty} v = \lim_{t \to +\infty} \sqrt{\frac{g}{b}} \frac{e^{2\sqrt{bg}\,t} - 1}{e^{2\sqrt{bg}\,t} + 1} = \sqrt{\frac{g}{b}}$$

$$= \sqrt{\frac{mg}{k}}.$$

A differential equation of the form

$$\frac{dy}{dx} = g\left(\frac{y}{x}\right) \tag{20.27}$$

can frequently be solved by using the substitution

$$y = vx. \tag{20.28}$$

After differentiating with respect to x in (20.28)

$$\frac{dy}{dx} = v + x\frac{dv}{dx},$$ (20.29)

and hence from (20.28) and (20.29), equation (20.27) becomes

$$v + x\frac{dv}{dx} = g(v).$$

This equation is separable since it can be rewritten as

$$\frac{dv}{g(v) - v} = \frac{dx}{x}.$$ (20.30)

Example 4   Solve the differential equation $\dfrac{dy}{dx} = \dfrac{x^2 + y^2}{2xy}$.

SOLUTION   If $x \neq 0$, then from the given equation

$$\frac{dy}{dx} = \frac{1 + \dfrac{y^2}{x^2}}{2\dfrac{y}{x}}.$$

Then by letting $y = vx$ we obtain successively,

$$v + x\frac{dv}{dx} = \frac{1 + v^2}{2v}$$

$$x\frac{dv}{dx} = \frac{1 - v^2}{2v}$$

$$\frac{2v\, dv}{1 - v^2} = \frac{dx}{x}.$$ (20.31)

From integrating each member in (20.31)

$$\ln |C_1| - \ln |1 - v^2| = \ln |x|$$

or

$$|C_1| = |x(1 - v^2)|.$$ (20.32)

Using the same argument as in Example 2, we obtain from (20.32)

$$x(1 - v^2) = C,$$

and recalling the substitution $y = vx$, the general solution is

$$x\left(1 - \frac{y^2}{x^2}\right) = C$$

or

$$x^2 - y^2 = Cx.$$

Exercise Set 20.3

In Exercises 1–18 solve the given equation.

1. $\dfrac{dy}{dx} = xy^{1/2}$

2. $\dfrac{dy}{dx} = \dfrac{x^2}{1 + y^2}$

3. $\dfrac{dy}{dx} = \dfrac{\sin^2 y}{\sqrt{1 - x^2}}$

4. $\cos y \dfrac{dy}{dx} = \dfrac{\sin y}{x}$

5. $\dfrac{dy}{dx} = e^{y-x}$

6. $\dfrac{dy}{dx} = \left(\dfrac{y - 1}{x - 1}\right)^2$

7. $x\dfrac{dy}{dx} + 2y = 5$

8. $\dfrac{dy}{dx} - xy^2 = x$

9. $\dfrac{dy}{dx} = \cos x \dfrac{dy}{dx} - y \sin x$

10. $\dfrac{dy}{dx} = y + \sec x \dfrac{dy}{dx}$

11. $2x\dfrac{d^2y}{dx^2} = 1 - \left(\dfrac{dy}{dx}\right)^2$. HINT: Let $p = \dfrac{dy}{dx}$ and then $\dfrac{dp}{dx} = \dfrac{d^2y}{dx^2}$. Integrate twice to obtain y in terms of x.

12. $\dfrac{dy}{dx} = \dfrac{x + 2y}{2x - y}$

13. $x^2\dfrac{dy}{dx} = x^2 - xy - 2y^2$

14. $\dfrac{dy}{dx} = \dfrac{2xy}{x^2 + y^2}$

15. $x(x^2 - y^2)\dfrac{dy}{dx} - y(x^2 + y^2) = 0$

16. $\dfrac{dy}{dx} = \dfrac{y - \sqrt{x^2 + y^2}}{x}$

17. $\dfrac{dy}{dx} = \dfrac{x + 2y - 3}{2x - y + 1}$. HINT: Let $x = u + h$ and $y = v + k$ where h and k are constants chosen so as to eliminate the constant terms on the right side.

18. $\dfrac{dy}{dx} = \dfrac{3x + y - 1}{x - 2y + 4}$

In Exercises 19–34 obtain the given particular solution of the differential equation in the given exercise above.

19. Exercise 1; $y = 1$ when $x = 2$

20. Exercise 2; $y = -1$ when $x = 0$

21. Exercise 3; $y = \frac{\pi}{4}$ when $x = \frac{1}{2}$

22. Exercise 4; $y = \frac{\pi}{3}$ when $x = -1$

23. Exercise 5; $y = 0$ when $x = 0$

24. Exercise 6; $y = 2$ when $x = 0$

25. Exercise 7; $y = 1$ when $x = -1$

26. Exercise 8; $y = \dfrac{\sqrt{3}}{3}$ when $x = 2$

27. Exercise 9; $y = \frac{1}{2}$ when $x = \pi$

28. Exercise 10; $y = 2$ when $x = \frac{\pi}{4}$

29. Exercise 13; $y = 1$ when $x = -1$

30. Exercise 14; $y = 0$ when $x = 2$

31. Exercise 15; $y = -2$ when $x = -1$

32. Exercise 16; $y = -1$ when $x = 1$

33. Exercise 17; $y = 4$ when $x = 3$

34. Exercise 18; $y = 1$ when $x = 1$

35. The electrical circuit shown in Figure 20–2 contains a battery delivering E volts, a resistor of R ohms, an inductor of L henries, and a switch S. By Kirchhoff's second law from physics

$$E = \text{Total voltage drop in the circuit.} \tag{20.33}$$

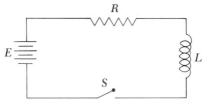

Figure 20–2

The voltage drop across the resistor is RI and across the inductor $L(dI/dt)$ where I, the current in the circuit, is a function of time t. Hence from (20.33)

$$L\frac{dI}{dt} + RI = E.$$

If the switch S is closed at time $t = 0$, obtain I as a function of t.

Two families of curves are the *orthogonal trajectories* of each other if each member of one family is orthogonal to each member of the other. Thus if $dy/dx = G(x, y)$ is the differential equation of each curve in one family, then $dy/dx = -1/G(x, y)$ is the differential equation for each curve in the other family. In Exercises 36–39 find the orthogonal trajectories of the given family of curves.

36. $x^2 - y^2 = C^2$

37. $y^2 = Cx$

38. $y = C \tan x$

39. $x^2 + (y - C)^2 = C^2$

40. From physics it is known that the rate of flow of water from a tank through an orifice of area A ft^2 is given approximately by $kA\sqrt{2gh}$ where $k = 0.6$, $g = 32$ ft/sec^2, and h is the height of the water level in feet above the orifice. Suppose a cylindrical tank with altitude 6 ft and diameter 3 ft is filled with water. At time $t = 0$ a valve is opened and the water begins draining through a hole of diameter 1 in. in the bottom of the tank. (a) Express the height of the water in the tank as a function of time t. (b) How much time is required to empty the tank?

41. Assuming a simplified mathematical model, the rate of growth of a fad at any time t among a population is proportional to the product xy where x is the number who have adopted the fad and y is the number who have not adopted the fad at time t (assuming that all the members of the population are in contact with each other). Suppose that on a certain day $(t = 0)$ 2 girls from a club of 30 girls begin wearing a new style of clothing. The next day one more girl adopts the clothing style. (a) Express the number of girls x in the club who have adopted the style, as a function of t days. (b) In approximately how many days will half of the club members have adopted the style?

20.4   Exact Differential Equation; Integrating Factors

We recall from Section 18.7 that if ϕ, a function of two variables, has first partial derivatives at an arbitrary point $(x, y) \in R^2$ and $w = \phi(x, y)$, then the differential of w is

$$dw = \frac{\partial \phi}{\partial x}\, dx + \frac{\partial \phi}{\partial y}\, dy.$$

Hence an expression of the form

$$M(x, y)\, dx + N(x, y)\, dy$$

is called an *exact differential* on some open set $S \subset R^2$ if and only if there is some function ϕ such that

$$\frac{\partial \phi}{\partial x} = M(x, y) \qquad \text{and} \qquad \frac{\partial \phi}{\partial y} = N(x, y) \quad \text{if } (x, y) \in S. \tag{20.34}$$

If M and N are continuous on S and (20.34) is satisfied, recall from Section 19.9 that ϕ is a potential function associated with a path-independent line integral $\int_C M\, dx + N\, dy$ in S.

A differential equation of the form

$$M(x, y) + N(x, y)\frac{dy}{dx} = 0, \tag{20.35}$$

or equivalently,

$$M(x, y)\, dx + N(x, y)\, dy = 0 \tag{20.36}$$

is called an *exact differential equation*, or, more briefly, is said to be *exact*, if and only if the left member of (20.36) is an exact differential on some open set $S \subset R^2$.

An exact differential equation is easily identified by the condition stated in the next theorem.

20.4.1   Theorem

Suppose M and N are functions of two variables which have continuous first partial derivatives on an open connected set $S \subset R^2$. Then (20.35) is exact if and only if

$$\frac{\partial M}{\partial y} = \frac{\partial N}{\partial x} \quad \text{for every } (x, y) \in S. \tag{20.37}$$

PROOF   By Theorem 19.9.3 the line integral $\int_C M\, dx + N\, dy$ is independent of the path in S if and only if (20.37) holds and hence by Theorem 19.9.2 if and only if (20.34) is satisfied. By definition, (20.35) is exact if and only if (20.34) is satisfied and the theorem is therefore proved.

Example 1   Is the differential equation

$$xy^2\, dx + (x^2 y - \cos y)\, dy = 0 \tag{20.38}$$

exact?

SOLUTION   Since

$$\frac{\partial}{\partial y}(xy^2) = 2xy = \frac{\partial}{\partial x}(x^2 y - \cos y) \tag{20.39}$$

the equation is exact by Theorem 20.4.1.

We next concern ourselves with the solutions of an exact differential equation. Suppose (20.35) is exact by virtue of condition (20.34). If f is a solution of (20.35) on some interval I where $\{(x, y): x \in I \text{ and } y = f(x)\} \subset S$, then

$$D_1\phi(x, f(x)) + D_2\phi(x, f(x))f'(x) = 0 \quad \text{if } x \in I. \tag{20.40}$$

By the chain rule the left member of (20.40) can be rewritten, and thus

$$D_x\phi(x, f(x)) = 0 \quad \text{if } x \in I.$$

Hence by integration there is a constant C such that

$$\phi(x, f(x)) = C \quad \text{if } x \in I. \tag{20.41}$$

Conversely, if (20.41) is fulfilled where $f'(x)$ exists for each $x \in I$, (20.40) is valid; that is, f is a solution of (20.35) on I.

Thus the following theorem has been proved.

20.4.2   Theorem

If there exists a function $\phi(x, y)$ satisfying (20.34) on some open set S, then a general solution of the exact differential equation $M(x, y) + N(x, y)\, dy/dx = 0$ is given by

$$\phi(x, y) = C$$

where C is an arbitrary constant.

The procedure for obtaining the function ϕ in Theorem 20.4.2 is the same as in Example 1, Section 19.9.

Example 2   Solve the equation $xy^2\, dx + (x^2 y - \cos y)\, dy = 0$.

SOLUTION   From (20.39) the equation is exact, and hence there is a function ϕ such that

$$\frac{\partial \phi}{\partial x} = xy^2.$$

Then, integrating with respect to x while holding y constant

$$\phi(x, y) = \frac{x^2 y^2}{2} + g(y).$$

Since

$$\frac{\partial \phi}{\partial y} = x^2 y + g'(y) = x^2 y - \cos y,$$

we have

$$g'(y) = -\cos y$$

and therefore we could select

$$g(y) = -\sin y$$

Hence $\phi(x, y) = x^2 y^2 / 2 - \sin y$ and by Theorem 20.4.2 a general solution of the given equation is

$$\frac{x^2 y^2}{2} - \sin y = C.$$

If a differential equation of the form (20.35) is not exact, it is sometimes possible to multiply each member of the equation by an *integrating factor* $H(x, y)$ that is not identically 0 and thus obtain an equation of the form

$$H(x, y)M(x, y) + H(x, y)N(x, y)\frac{dy}{dx} = 0 \qquad (20.42)$$

which is exact. For example

$$-\frac{y}{x} + \frac{dy}{dx} = 0 \qquad (20.43)$$

is not exact since by Theorem 20.4.1

$$\frac{\partial}{\partial y}\left(-\frac{y}{x}\right) = -\frac{1}{x} \quad \text{and} \quad \frac{\partial}{\partial x}(1) = 0.$$

However, if each member of (20.43) is multiplied by $1/x$ the resulting equation

$$-\frac{y}{x^2} + \frac{1}{x}\frac{dy}{dx} = 0 \qquad (20.44)$$

is exact. The left member of (20.44) is $D_x\left(\frac{1}{x}y\right)$; that is, there is a potential function given by $\phi(x, y) = y/x$ which is associated with (20.44). (This potential function can also be obtained by the method of Example 2.) Thus by Theorem 20.4.2 a general solution of (20.44) and hence of (20.43) is $y/x = C$ or

$$y = Cx. \tag{20.45}$$

The choice of the integrating factor here is not unique. In place of $1/x$ we could also have used any constant multiple of $1/x$ or, for example, $x/(x^2 + y^2)$. A multiplication in (20.43) by the latter factor gives the exact equation

$$-\frac{y}{x^2 + y^2} + \frac{x}{x^2 + y^2}\frac{dy}{dx} = 0. \tag{20.46}$$

An associated potential function for (20.46) has the form $\phi(x, y) = \tan^{-1} y/x$, or saying the same thing, the left member of (20.46) is $D_x \tan^{-1} y/x$. Hence a general solution of (20.43) has the form

$$\tan^{-1}\frac{y}{x} = C_1.$$

But here,

$$\frac{y}{x} = \tan\tan^{-1}\frac{y}{x} = \tan C_1 = C$$

and the general solution (20.45) is again obtained.

Suppose the equation (20.42) is exact on some open connected set S where the functions H, M, and N have continuous partial derivatives. Then from Theorem 20.4.1

$$\frac{\partial(HM)}{\partial y} = \frac{\partial(HN)}{\partial x} \quad \text{if } (x, y) \in S. \tag{20.47}$$

If the indicated partial differentiations are performed in (20.47),

$$H(x, y)\frac{\partial M}{\partial y} + M(x, y)\frac{\partial H}{\partial y} = H(x, y)\frac{\partial N}{\partial x} + N(x, y)\frac{\partial H}{\partial x},$$

and after transposing terms, we obtain

$$H(x, y)\left(\frac{\partial M}{\partial y} - \frac{\partial N}{\partial x}\right) = N(x, y)\frac{\partial H}{\partial x} - M(x, y)\frac{\partial H}{\partial y}. \tag{20.48}$$

We would like to be able to solve (20.48) in general for $H(x, y)$, and thereby obtain an integrating factor for any equation of the form (20.35). However, because of the difficulty involved, the following problem will be investigated: *Can conditions be imposed on M and N which will guarantee that* (20.35) *has an integrating factor of a particular form?* Specifically, suppose (20.35) has an integrating factor of the form

$$H(x, y) = g(x) > 0 \quad \text{if } (x, y) \in S.$$

If $N(x, y) \neq 0$ when $(x, y) \in S$, then $\partial H/\partial y = 0$ and from (20.48)

$$\frac{g'(x)}{g(x)} = \frac{1}{N(x, y)}\left(\frac{\partial M}{\partial y} - \frac{\partial N}{\partial x}\right). \tag{20.49}$$

Since the left member of (20.49) is a function of x alone, so is the right member, and accordingly we denote it by $\mu(x)$. Hence from (20.49)

$$\ln g(x) = \int \mu(x)\, dx$$

$$g(x) = e^{\int \mu(x)\, dx}. \tag{20.50}$$

Because of the question in italics posed earlier in the paragraph, it is reasonable to ask if an integrating factor of the form (20.50) does exist for (20.35) when $\dfrac{1}{N(x,y)}\left(\dfrac{\partial M}{\partial y} - \dfrac{\partial N}{\partial x}\right)$ is a function of x alone on S. We leave as an exercise (Exercise 12, Exercise Set 20.4) the proof that this query has an affirmative answer.

Example 3   Solve the *first order linear differential equation*

$$\frac{dy}{dx} + P(x)y = Q(x) \tag{20.51}$$

where P and Q are continuous on some interval I.

SOLUTION   If this equation is compared with (20.35), then

$$M(x, y) = P(x)y - Q(x) \qquad \text{and} \qquad N(x, y) = 1.$$

Since

$$\frac{1}{N(x, y)}\left(\frac{\partial M}{\partial y} - \frac{\partial N}{\partial x}\right) = P(x),$$

from our previous discussion $e^{\int P(x)\, dx}$, where $\int P(x)\, dx$ denotes any integral of P on I, is an integrating factor of (20.51). Thus from (20.51)

$$e^{\int P(x)\, dx}\frac{dy}{dx} + e^{\int P(x)\, dx}P(x)y = e^{\int P(x)\, dx}Q(x). \tag{20.52}$$

Since the left member of (20.52) is $D_x(e^{\int P(x)\, dx}y)$, integration in (20.52) yields

$$e^{\int P(x)\, dx}y = \int e^{\int P(x)\, dx}Q(x)\, dx,$$

or equivalently

$$y = e^{-\int P(x)\, dx}\int e^{\int P(x)\, dx}Q(x)\, dx,$$

which expresses a general solution of (20.51).

Example 4   Solve the linear differential equation

$$\frac{dy}{dx} + \frac{3y}{x} = x^2. \tag{20.53}$$

SOLUTION   From Example 3 an integrating factor for equation (20.53) is of the form

$$e^{\int P(x)\, dx} = e^{\int (3/x)\, dx} = e^{3\ln x + C_1}$$

$$= e^{C_1}x^3. \tag{20.54}$$

For simplicity we shall let $C_1 = 0$ in (20.54) and choose x^3 as the integrating factor. Multiplication of each member of (20.53) by x^3 then yields

$$x^3 \frac{dy}{dx} + 3x^2 y = x^5$$

$$D_x(x^3 y) = x^5$$

$$x^3 y = \int x^5 \, dx = \frac{x^6}{6} + C.$$

Hence a general solution of (20.53) is

$$y = \frac{x^3}{6} + \frac{C}{x^3}.$$

Exercise Set 20.4

In Exercises 1–10 show that the given equation is exact and solve.

1. $2xy + x^2 \dfrac{dy}{dx} = 0$

2. $y \, dx + (x + 2y) \, dy = 0$

3. $(3x - 2y - 4) + (5y - 2x + 1)\dfrac{dy}{dx} = 0$

4. $(3y - 2xy - 1) + (3x - x^2 - 1)\dfrac{dy}{dx} = 0$

5. $(x^2 + \ln y) \, dx + \dfrac{x}{y} \, dy = 0$

6. $(\cos y - 2x) \, dx - (x \sin y + y^2) \, dy = 0$

7. $(2x + \tan y - y) \, dx + (x \sec^2 y - x + y) \, dx = 0$

8. $(e^y + x) + (xe^y - y)\dfrac{dy}{dx} = 0$

9. $(x \tan^{-1} y - x) + \left(\dfrac{x^2}{2 + 2y^2} + e^y \right)\dfrac{dy}{dx} = 0$

10. $y\left(1 + y + \dfrac{1}{x^2}\right) dx + \left(x + 2xy - \dfrac{1}{x}\right) dy = 0$

11. Prove that a separable equation is exact.

12. Prove that if $\dfrac{1}{N(x, y)}\left(\dfrac{\partial M}{\partial y} - \dfrac{\partial N}{\partial x}\right)$ is a continuous function of x alone on the open connected set S and can therefore be denoted by $\mu(x)$, then $g(x) = e^{\int \mu(x) \, dx}$ is an integrating factor of (20.35).

In Exercises 13–17 solve the given equation after finding the required integrating factor.

13. $(2x + y) \, dx - x \, dy = 0$

14. $(x^2 + 2y^2) \, dx + xy \, dy = 0$

---

15. $(x^2y - 2x)\dfrac{dy}{dx} + x^2 + 2y = 0$ 16. $(3x^2y^2 - e^y x^4)\dfrac{dy}{dx} = 2xy^3$

17. $\dfrac{dy}{dx} = \dfrac{y}{x - y^2}$

18. Obtain a condition which when satisfied by the functions M and N guarantees that the equation (20.35) has an integrating factor that is a function of y alone. Give the form of this integrating factor and verify that it is, in fact, an integrating factor of (20.35).

In Exercises 19–26 solve the given first order linear equation.

19. $\dfrac{dy}{dx} + 4y = 3$ 20. $\dfrac{dy}{dx} - 2y = e^{-x}$

21. $\dfrac{dy}{dx} - \dfrac{y}{x} = x^{3/2}$ 22. $\dfrac{dy}{dx} + \dfrac{2y}{x} = 2$

23. $\dfrac{dy}{dx} + y \tan x = \cos^2 x$ 24. $\dfrac{dy}{dx} - y \csc x = \sin 2x$

25. $(1 + x^2)\dfrac{dy}{dx} + xy = x^3$ 26. $\cos x \dfrac{dy}{dx} + y = \sec x$

27. Suppose in the electric circuit in Exercise 35, Exercise Set 20.3, that instead of a battery there is an AC generator delivering a voltage $E(t) = E_0 \sin \omega t$. Then by Kirchhoff's second law

$$L\dfrac{dI}{dt} + RI = E_0 \sin \omega t.$$

Find I as a function of t if $I = I_0$ when $t = 0$.

28. A *Bernoulli equation* is an equation of the form

$$\dfrac{dy}{dx} + P(x)y = Q(x)y^n \quad n \neq 0 \text{ and } 1. \tag{20.55}$$

If the functions P and Q are continuous on an interval I, prove that a general solution of (20.55) on I is given by

$$e^{(1-n)\int P(x)\,dx}y^{1-n} = \int (1 - n)Q(x)e^{(1-n)\int P(x)\,dx}\,dx$$

where $\int P(x)\,dx$ is any integral of P on I. HINT: Let $v = y^{1-n}$ in (20.55). Then obtain a first order linear differential equation in v and x.

20.5   Homogeneous Second Order Linear Equations

It will be recalled from (20.4) that a second order homogeneous linear differential equation is an equation of the form

$$\dfrac{d^2y}{dx^2} + P_1(x)\dfrac{dy}{dx} + P_2(x)y = 0.$$

Then a homogeneous second order linear differential equation with constant co-efficients has the form

$$\frac{d^2y}{dx^2} + 2a\frac{dy}{dx} + by = 0 \quad a, b \text{ constant.} \tag{20.56}$$

Our purpose, therefore, in this section is to solve equations of the form (20.56). Note first that if $a = 0 = b$, (20.56) simplifies to

$$\frac{d^2y}{dx^2} = 0.$$

From integrating twice, we observe that this equation has solutions of the form

$$y = C_1 + C_2 x.$$

If $a = 0$ and $b > 0$, then letting $\omega = \sqrt{b}$, (20.56) becomes

$$\frac{d^2y}{dx^2} + \omega^2 y = 0. \tag{20.57}$$

In Section 10.4 this equation was derived in a discussion of the oscillatory motion of a weighted spring. From what was said following (10.61), if f is a solution of (20.57), there exist constants C_1 and C_2 such that

$$f(x) = C_1 \cos \omega x + C_2 \sin \omega x.$$

In solving equations of the form (20.56) two theorems are useful. The first theorem states that a linear combination of solutions of (20.56) is also a solution of this equation.

20.5.1 Theorem

If v_1 and v_2 are solutions of (20.56) on an interval I, then for any constants C_1 and C_2 the function f, where

$$f(x) = C_1 v_1(x) + C_2 v_2(x) \quad \text{for every } x \in I,$$

is also a solution of (20.56) on I.

SOLUTION Since v_1 and v_2 are solutions of (20.56) on I

$$v_1''(x) + 2av_1'(x) + bv_1(x) = 0 \quad \text{and} \quad v_2''(x) + 2av_2'(x) + bv_2(x) = 0$$

for every $x \in I$. Hence if $x \in I$,

$f''(x) + 2af'(x) + bf(x)$

$$= D_x^2[C_1 v_1(x) + C_2 v_2(x)] + 2aD_x[C_1 v_1(x) + C_2 v_2(x)]$$
$$+ b[C_1 v_1(x) + C_2 v_2(x)]$$
$$= C_1 v_1''(x) + C_2 v_2''(x) + 2a[C_1 v_1'(x) + C_2 v_2'(x)]$$
$$+ b[C_1 v_1(x) + C_2 v_2(x)]$$
$$= C_1[v_1''(x) + 2av_1'(x) + bv_1(x)] + C_2[v_2''(x) + 2av_2'(x) + bv_2(x)]$$
$$= C_1 0 + C_2 0 = 0.$$

Thus f is a solution of (20.56) on I.

Not only is every linear combination of solutions of (20.56) itself a solution of this equation, but also from the following theorem, every solution of (20.56) is a linear combination of *certain* solutions of this equation. This theorem will be proved at the end of the section after we have discussed some examples.

20.5.2   Theorem

Suppose v_1 *and* v_2 *are solutions of* (20.56) *on interval I such that*

$$\begin{vmatrix} v_1(x) & v_2(x) \\ v_1'(x) & v_2'(x) \end{vmatrix} \neq 0 \quad \text{for every } x \in I. \tag{20.58}$$

Then if f is any solution of (20.56) *on I, there exist constants* C_1 *and* C_2 *such that*

$$f(x) = C_1 v_1(x) + C_2 v_2(x) \quad \text{for every } x \in I.$$

Any two functions v_1 and v_2 that satisfy the hypothesis of Theorem 20.5.2 are said to form a *fundamental solution set* of (20.56) on the interval I. Also, if C_1 and C_2 are arbitrary constants,

$$y = C_1 v_1(x) + C_2 v_2(x) \tag{20.59}$$

is said to be the *general solution* of (20.56) on I.

It is apparent that if $f(x) = 0$ for every x, then f is certainly a solution of (20.56). Also, it is conceivable that if r is a properly chosen constant and f is given by

$$f(x) = e^{rx}, \tag{20.60}$$

then f might also be a solution of (20.56) since $f'(x)$ and $f''(x)$ are constant multiples of e^{rx}. If (20.60) does indeed define such a solution, then

$$r^2 e^{rx} + 2are^{rx} + be^{rx} = 0,$$

and if the nonzero factor e^{rx} is divided out,

$$r^2 + 2ar + b = 0. \tag{20.61}$$

Equation (20.61) is called the *auxiliary equation* or *characteristic equation* of (20.56). By the quadratic formula the roots of (20.61) are

$$r = -a \pm \sqrt{a^2 - b}.$$

The following possibilities exist for these roots:

  (i) Two distinct real roots if $a^2 > b$
 (ii) A double root of $-a$ if $a^2 = b$
(iii) Two conjugate imaginary roots if $a^2 < b$

Case (i): Suppose r_1 and r_2 are the distinct roots of (20.61). By direct substitution it can be verified that if

$$v_1(x) = e^{r_1 x} \quad \text{and} \quad v_2(x) = e^{r_2 x},$$

then v_1 and v_2 satisfy (20.56). Also, note that

$$\begin{vmatrix} v_1(x) & v_2(x) \\ v_1'(x) & v_2'(x) \end{vmatrix} = \begin{vmatrix} e^{r_1 x} & e^{r_2 x} \\ r_1 e^{r_1 x} & r_2 e^{r_2 x} \end{vmatrix} = (r_2 - r_1)e^{(r_1 + r_2)x}$$

$$\neq 0$$

since $r_1 \neq r_2$. Hence from our remarks following Theorem 20.5.2 the general solution of (20.56) in Case (i) is

$$y = C_1 e^{r_1 x} + C_2 e^{r_2 x}. \tag{20.62}$$

Example 1 Find the general solution of

$$y'' + 2y' - 3y = 0. \tag{20.63}$$

SOLUTION From (20.61) the auxiliary equation for (20.63) is

$$r^2 + 2r - 3 = 0.$$

Since the roots of this equation are -3 and 1, solutions of (20.63) are given by $v_1(x) = e^{-3x}$ and $v_2(x) = e^x$. From our preceding remarks the general solution of (20.63), expressed in the form (20.59), is

$$y = C_1 e^{-3x} + C_2 e^x.$$

Case (ii): Suppose $r = -a$ is the double root of (20.61). As before it can be verified that $v_1(x) = e^{-ax}$ gives a solution of (20.56). Now, we can certainly assume that any solution of (20.56) can be expressed in the form

$$y = h(x)e^{-ax} \tag{20.64}$$

for some properly chosen function h. (In fact, $h(x) = e^{ax}y$.) Then from (20.64)

$$\left. \begin{array}{l} y' = e^{-ax}[h'(x) - ah(x)] \\ y'' = e^{-ax}[h''(x) - 2ah'(x) + a^2h(x)]. \end{array} \right\} \tag{20.65}$$

Hence after a substitution from (20.64) and (20.65) into (20.56), dividing out e^{-ax}, and simplifying, we see that $h(x)$ must satisfy

$$h''(x) + h(x)(b - a^2) = 0. \tag{20.66}$$

Since $b = a^2$ in Case (ii), equation (20.66) simplifies to

$$h''(x) = 0.$$

Then, as noted at the beginning of this section,

$$h(x) = C_1 + C_2 x.$$

Therefore from (20.64) any solution of (20.56) must be of the form

$$y = (C_1 + C_2 x)e^{-ax}. \tag{20.67}$$

It can be readily shown that if $v_2(x) = xe^{-ax}$, then v_2 is a solution of (20.56), and also

$$\begin{vmatrix} v_1(x) & v_2(x) \\ v_1'(x) & v_2'(x) \end{vmatrix} = \begin{vmatrix} e^{-ax} & xe^{-ax} \\ -ae^{-ax} & e^{-ax} - xae^{-ax} \end{vmatrix} = e^{-2ax}$$

$$\neq 0.$$

Hence by Theorem 20.5.2, (20.67) is the general solution of (20.56).

Example 2 Solve the equation

$$y'' - 4y' + 4y = 0. \tag{20.68}$$

SOLUTION The auxiliary equation for (20.68) is

$$r^2 - 4r + 4 = 0,$$

which has the double root 2. Hence from the preceding discussion the general solution of (20.68) is

$$y = (C_1 + C_2 x)e^{2x}.$$

Case (iii): Suppose (20.61) has the conjugate imaginary roots $-a \pm \omega i$ where $\omega = \sqrt{b - a^2} > 0$. From Case (i) one might claim that

$$y = e^{(-a+\omega i)x} \qquad \text{and} \qquad y = e^{(-a-\omega i)x}$$

give solutions of (20.56). However, we have not discussed imaginary powers of numbers, a topic which is considered in a more advanced course in complex variable theory. Therefore we will utilize a different approach in obtaining the solutions of (20.56) in Case (iii).

As in Case (ii) we can assume that any solution of this equation can be expressed in the form

$$y = h(x)e^{-ax} \tag{20.69}$$

for some properly chosen $h(x)$. Since $\omega^2 = b - a^2$, from (20.66) h must satisfy

$$h''(x) + \omega^2 h(x) = 0.$$

Therefore from our discussion following (20.57), $h(x)$ must have the form

$$h(x) = C_1 \cos \omega x + C_2 \sin \omega x. \tag{20.70}$$

Thus from (20.69) and (20.70) any solution of (20.56) in Case (iii) will be of the form

$$y = e^{-ax}(C_1 \cos \omega x + C_2 \sin \omega x). \tag{20.71}$$

It can also be proved that (20.71) is the general solution of (20.56) in Case (iii). First one proves that if

$$v_1(x) = e^{-ax} \cos \omega x \qquad v_2(x) = e^{-ax} \sin \omega x, \tag{20.72}$$

then v_1 and v_2 are solutions of (20.56) (Exercise 23, Exercise Set 20.5). Then from (20.72)

$$\begin{vmatrix} v_1(x) & v_2(x) \\ v_1'(x) & v_2'(x) \end{vmatrix} \neq 0 \quad \text{for every } x$$

(Exercise 24, Exercise Set 20.5). Hence by Theorem 20.5.2 the general solution of (20.56) is (20.71).

Example 3   Solve the equation

$$y'' + y' + 2y = 0. \tag{20.73}$$

SOLUTION   The auxiliary equation for (20.73) is

$$r^2 + r + 2 = 0,$$

which has the imaginary roots $-\dfrac{1}{2} \pm \dfrac{\sqrt{7}}{2} i$. Hence from (20.71) the general solution of (20.73) is of the form

$$y = e^{-x/2}\left(C_1 \cos \frac{\sqrt{7}}{2} x + C_2 \sin \frac{\sqrt{7}}{2} x \right).$$

We shall now give the deferred proof of Theorem 20.5.2.

PROOF   Suppose $x_0 \in I$. Since v_1 and v_2 satisfy (20.58), the system

$$\left. \begin{array}{c} C_1 v_1(x_0) + C_2 v_2(x_0) = f(x_0) \\ C_1 v_1'(x_0) + C_2 v_2'(x_0) = f'(x_0) \end{array} \right\} \tag{20.74}$$

has a unique solution given by $C_1 = \gamma_1$, $C_2 = \gamma_2$, which can be computed by Cramer's rule. Thus, for the function g given by

$$g(x) = \gamma_1 v_1(x) + \gamma_2 v_2(x)$$

we obtain from (20.74)

$$g(x_0) = f(x_0) \quad \text{and} \quad g'(x_0) = f'(x_0). \tag{20.75}$$

If it can be proved that $f(x) = g(x)$ for every $x \in I$, then the desired conclusion is obtained. To this end we consider the system of equations

$$\left. \begin{array}{c} k_1 f(x) + k_2 g(x) = 0 \\ k_1 f'(x) + k_2 g'(x) = 0. \end{array} \right\} \tag{20.76}$$

The system (20.76) has a solution for k_1 and k_2 other than $(k_1, k_2) = (0, 0)$ if and only if

$$W(x) = \begin{vmatrix} f(x) & g(x) \\ f'(x) & g'(x) \end{vmatrix} \tag{20.77}$$

is 0 for every $x \in I$. We leave as an exercise (Exercise 27, Exercise Set 20.5) the proof that

$$W'(x) = -2aW(x).$$

Thus for some constant C

$$W(x) = Ce^{-2ax} \qquad \text{for } x \in I. \tag{20.78}$$

Note from (20.77) and (20.75) that

$$W(x_0) = f(x_0)g'(x_0) - f'(x_0)g(x_0) = 0$$

and therefore from (20.78) $C = 0$. Thus

$$W(x) = 0 \qquad \text{for } x \in I$$

and from our previous discussion there exist numbers k_1 and k_2 which are not both 0 such that

$$k_1 f(x) + k_2 g(x) = 0 \qquad \text{for } x \in I.$$

We can assume without loss of generality that $k_1 \neq 0$ and therefore

$$f(x) = -\frac{k_2}{k_1} g(x) \qquad \text{for } x \in I. \tag{20.79}$$

In particular then,

$$f(x_0) = -\frac{k_2}{k_1} g(x_0).$$

But from (20.75) — $k_2/k_1 = 1$, and hence from (20.79)

$$f(x) = g(x) \qquad \text{for } x \in I,$$

which proves the theorem.

We are also concerned with the initial value problem for equations of the form (20.56): Given any numbers x_0, y_0, and y_1, find a solution f of (20.56) such that

$$f(x_0) = y_0 \qquad \text{and} \qquad f'(x_0) = y_1. \tag{20.80}$$

This problem has a unique solution because of the following theorem for nth order linear equations.

20.5.3   Theorem

Suppose the functions of one variable $P_1, P_2, \ldots, P_n$, and Q are continuous on an interval I and $x_0 \in I$. Then if $y_0, y_1, \ldots, y_{n-1}$ are any numbers, there is a unique solution f of (20.3) on I such that

$$f(x_0) = y_0 \qquad \text{and} \qquad f^{(i)}(x_0) = y_i \quad \text{for } i = 1, 2, \ldots, n - 1. \tag{20.81}$$

The proof of Theorem 20.5.3 is beyond our scope here. However, the reader can, with some direction, prove that there is a unique solution of (20.56) which satisfies (20.80) (Exercise 25, Exercise Set 20.5). The theorem given in this exercise is sufficient for our use in this section.

Using the terminology introduced with first order equations, a solution of an initial value problem for an equation of the form (20.56) is often called a *particular solution* since it is derived from the general solution of the equation by assigning specific values to the constants involved.

Example 4   Find the particular solution of

$$y'' + 2y' - 3y = 0$$

such that $y' = 2$ and $y = 6$ when $x = 0$.

SOLUTION   From Example 1 the general solution of this equation is

$$y = C_1 e^{-3x} + C_2 e^x. \tag{20.82}$$

To obtain the desired particular solution f, the constants C_1 and C_2 in (20.82) must satisfy

$$\left. \begin{aligned} f(0) &= C_1 e^{-3 \cdot 0} + C_2 e^0 = C_1 + C_2 = 2 \\ f'(0) &= -3C_1 e^{-3 \cdot 0} + C_2 e^0 = -3C_1 + C_2 = 6. \end{aligned} \right\} \tag{20.83}$$

Since the solution of the system (20.83) is

$$C_1 = -1, \qquad C_2 = 3,$$

the desired particular solution is given by

$$y = f(x) = -e^{-3x} + 3e^x.$$

Exercise Set 20.5

In Exercises 1–16 solve the given differential equation.

1. $y'' - 2y' - 8y = 0$ 2. $y'' + 6y' + 9y = 0$

3. $y'' - 2y' + 4y = 0$ 4. $y'' - y' - 20y = 0$

5. $y'' + 8y' + 16y = 0$ 6. $y'' - 3y' = 0$

7. $y'' - 5y = 0$ 8. $y'' - 2y' - 5y = 0$

9. $y'' + 4y' = 0$ 10. $y'' + 2y = 0$

11. $y'' + y' - 3y = 0$ 12. $y'' + y' + 3y = 0$

13. $y'' - 2\sqrt{5}y' + 6y = 0$ 14. $y'' - \sqrt{3}y' - 2y = 0$

15. $y'' + 4\sqrt{3}y' + 12y = 0$ 16. $y'' - (2 - \sqrt{3})y' - \sqrt{3}y = 0$

In Exercises 17–22 obtain the particular solution of the differential equation in the given exercise above which satisfies the given initial conditions.

17. Exercise 1; $y' = 1$ and $y = -2$ when $x = 0$

18. Exercise 4; $y' = -1$ and $y = 3$ when $x = 0$

19. Exercise 13; $y' = 2$ and $y = 1$ when $x = 0$

20. Exercise 12; $y' = 2$ and $y = 0$ when $x = 0$

21. Exercise 5; $y' = -1$ and $y = 1$ when $x = 0$

22. Exercise 2; $y' = 0$ and $y = 1$ when $x = -1$

23. Prove that the functions v_1 and v_2 given by (20.72) in Case (iii) are solutions of (20.56) on $(-\infty, +\infty)$.

24. Prove that the solutions v_1 and v_2 in Exercise 23 form a fundamental solution set of (20.56) on the interval $(-\infty, +\infty)$.

25. Prove: If x_0, y_0, and y_1 are any numbers, there is a unique solution f of (20.56) on the interval $(-\infty, +\infty)$ such that (20.80) is valid. HINT: From the discussion of Cases (i), (ii), and (iii) there exist functions v_1 and v_2 that form a fundamental solution set of (20.56) on $(-\infty, +\infty)$. Prove that if the constants C_1 and C_2 are properly chosen, and $f(x) = C_1 v_1(x) + C_2 v_2(x)$, then f is the required solution.

26. A pendulum that consists of a mass m at the end of a weightless rod of length L oscillates in a vertical plane. For small displacements θ from its vertical position the pendulum has the equation of motion $L\, d^2\theta/dt^2 = -g\theta$ where g is the gravitational constant. Show that the pendulum executes simple harmonic motion with period $2\pi\sqrt{L/g}$.

27. Prove: If $W(x)$ is given by (20.77), then $W'(x) = -2aW(x)$.

20.6   Nonhomogeneous Second Order Linear Equations

We next consider the solutions of the nonhomogeneous second order linear differential equation with constant coefficients

$$\frac{d^2y}{dx^2} + 2a\frac{dy}{dx} + by = Q(x). \tag{20.84}$$

To solve this equation, we will utilize the associated homogeneous equation

$$\frac{d^2y}{dx^2} + 2a\frac{dy}{dx} + by = 0. \tag{20.85}$$

The general solution of (20.85), the so-called *complementary solution* for (20.84), will be denoted by

$$y_c = C_1 v_1(x) + C_2 v_2(x). \tag{20.86}$$

The following theorem will be used to obtain the solutions of (20.84).

20.6.1   Theorem

Suppose the following conditions are fulfilled on an interval I:

  (i) v_1 and v_2 form a fundamental solution set of (20.85)
  (ii) g is some solution of (20.84)

Then for any solution f of (20.84) on I, there exist unique constants C_1 and C_2 such that

$$f(x) = g(x) + C_1 v_1(x) + C_2 v_2(x) \quad \text{whenever } x \in I. \tag{20.87}$$

PROOF   Let ϕ be defined by

$$\phi(x) = f(x) - g(x). \tag{20.88}$$

Then

$$\phi''(x) + 2a\phi'(x) + b\phi(x) = [f''(x) + 2af'(x) + bg(x)] - [g''(x) + 2ag'(x) + bg(x)],$$

and since f and g are solutions of (20.84) on I,

$$\phi''(x) + 2a\phi'(x) + b\phi(x) = Q(x) - Q(x) = 0 \quad \text{if } x \in I.$$

Therefore ϕ is a solution of (20.85) on I. From (i) and Theorem 20.5.2 there exist unique constants C_1 and C_2 such that

$$\phi(x) = C_1 v_1(x) + C_2 v_2(x) \quad \text{for all } x \in I. \tag{20.89}$$

Thus (20.87) is obtained from (20.88) and (20.89).

By Theorem 20.6.1 *any solution of (20.84) is obtained by adding a linear combination of functions forming a fundamental solution set of (20.85) and some solution of (20.84).* Thus the *general solution* of (20.84) is given by

$$y = y_c + g(x) \tag{20.90}$$

where y_c defined in (20.86) and g is a *particular* solution of (20.84).

The complementary solution, y_c, can be obtained by methods that were discussed in Section 20.5. The solution g in (20.90) could conceivably be discovered by inspection although the use of a systematic method for obtaining this function is usually more rewarding. One such method, that of *undetermined coefficients*, is illustrated in the next two examples.

Example 1   Find the general solution of the linear equation

$$y'' - 3y' + 2y = \sin 2x. \tag{20.91}$$

SOLUTION   We first obtain the complementary solution for (20.91). The corresponding homogeneous equation

$$y'' - 3y' + 2y = 0$$

has the auxiliary equation

$$r^2 - 3r + 2 = 0,$$

whose roots are $r = 1$ and 2. The complementary solution for (20.91) is therefore

$$y_c = C_1 e^x + C_2 e^{2x}. \tag{20.92}$$

If (20.91) has a solution g, it is quite possible that $g(x)$ is of the form

$$g(x) = A \cos 2x + B \sin 2x \qquad A, B \text{ constant} \tag{20.93}$$

since $g'(x)$ and $g''(x)$ would then be in terms of $\cos 2x$ and $\sin 2x$, and hence $\sin 2x$, the right member of (20.91), could be expressed as the linear combination of $\cos 2x$ and $\sin 2x$ obtained by replacing y by $g(x)$ in the left member of (20.91). Thus from (20.93)

$$\left. \begin{array}{l} g'(x) = 2B \cos 2x - 2A \sin 2x \\ g''(x) = -4A \cos 2x - 4B \sin 2x. \end{array} \right\} \tag{20.94}$$

If $g(x)$ is substituted for y in (20.91), then from (20.93) and (20.94), we obtain

$$-4A \cos 2x - 4B \sin 2x - 3(2B \cos 2x - 2A \sin 2x)$$
$$+ 2(A \cos 2x + B \sin 2x) = \sin 2x.$$

This equation simplifies to

$$(-2A - 6B) \cos 2x + (6A - 2B) \sin 2x = \sin 2x. \tag{20.95}$$

By equating the coefficients for $\cos 2x$ and $\sin 2x$ in each member of (20.95), we obtain the system

$$-2A - 6B = 0$$
$$6A - 2B = 1$$

which has the solution

$$A = \tfrac{3}{20} \qquad B = -\tfrac{1}{20}.$$

Hence from (20.93) $g(x)$ is given by

$$g(x) = \tfrac{3}{20} \cos 2x - \tfrac{1}{20} \sin 2x. \tag{20.96}$$

It can be verified that the function g defined by (20.96) is a solution of (20.91), and hence the general solution of (20.91), expressed in the form (20.90), is

$$y = C_1 e^x + C_2 e^{2x} + \tfrac{3}{20} \cos 2x - \tfrac{1}{20} \sin 2x.$$

Example 2 Find the general solution of

$$y'' + 4y' + 7y = x - 3. \tag{20.97}$$

SOLUTION The corresponding homogeneous equation

$$y'' + 4y' + 7y = 0$$

has the auxiliary equation

$$r^2 + 4r + 7 = 0,$$

which has roots $r = -2 \pm \sqrt{3}i$. The complementary solution for (20.97) is therefore

$$y_c = e^{-2x}(C_1 \cos \sqrt{3}x + C_2 \sin \sqrt{3}x). \tag{20.98}$$

Suppose (20.98) has a solution g. We suspect that $g(x)$ is a polynomial of degree ≤ 3 since $x - 3$, the right member of (20.97), could then be obtained as a linear combination of $g(x)$, $g'(x)$, and $g''(x)$. Thus we let

$$g(x) = Ax^3 + Bx^2 + Cx + D, \tag{20.99}$$

and hence

$$g'(x) = 3Ax^2 + 2Bx + C \tag{20.100}$$
$$g''(x) = 6Ax + 2B. \tag{20.101}$$

Then from (20.97)

$$6Ax + 2B + 4(3Ax^2 + 2Bx + C) + 7(Ax^3 + Bx^2 + Cx + D) = x - 3,$$

and if the terms on the left are regrouped, we have

$$7Ax^3 + (12A + 7B)x^2 + (6A + 8B + 7C)x + (2B + 4C + 7D) = x - 3. \tag{20.102}$$

From equating coefficients in (20.102) the equations

$$\begin{aligned} 7A &= 0 \\ 12A + 7B &= 0 \\ 6A + 8B + 7C &= 1 \\ 2B + 4C + 7D &= -3 \end{aligned}$$

are obtained. Since the solution of this system of equations is

$$A = 0 \qquad B = 0 \qquad C = \tfrac{1}{7} \qquad D = -\tfrac{25}{49},$$

we obtain from (20.99)

$$g(x) = \tfrac{1}{7}x - \tfrac{25}{49}. \tag{20.103}$$

Since it can be verified that (20.103) defines a solution of (20.97), the general solution of (20.97), expressed in the form (20.90), is

$$y = e^{-2x}(C_1 \cos \sqrt{3}x + C_2 \sin \sqrt{3}x) + \tfrac{1}{7}x - \tfrac{25}{49}.$$

In most applications of differential equations of the type (20.84), $Q(x)$ will have one of the forms given in the first column of the following table. Then the corresponding $g(x)$ given in the second column of the table will often be a particular solution of the differential equation, and can therefore be used in connection with the method of undetermined coefficients.

$Q(x)$	$g(x)$
Polynomial in x of degree n	$A_0 x^n + A_1 x^{n-1} + \cdots + A_n$
$ke^{\alpha x}$	$Ae^{\alpha x}$
(Polynomial in x of degree n)$e^{\alpha x}$	$e^{\alpha x}(A_0 x^n + A_1 x^{n-1} + \cdots + A_n)$
Linear combination of $\cos \beta x$ and $\sin \beta x$	$A \cos \beta x + B \sin \beta x$
Linear combination of $e^{\alpha x} \cos \beta x$ and $e^{\alpha x} \sin \beta x$	$e^{\alpha x}(A \cos \beta x + B \sin \beta x)$

If the $g(x)$ obtained from the table does not satisfy the equation, then a particular solution may sometimes be discovered by multiplying this $g(x)$ by x, x^2, and so on. A general method for obtaining these particular solutions, which is called *variation of parameters*, is described in texts on differential equations. However, this method will not be included in this text since it is not as efficient as the method of undetermined coefficients for the usual problems that are encountered.

As an application of a nonhomogeneous second order linear differential equation, we consider the *LRC* electrical circuit shown in Figure 20–3. This circuit

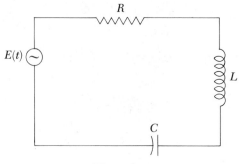

Figure 20–3

is so named because it contains a resistor of R ohms, an inductor of L henries, and a capacitor of C farads. If an AC generator supplies a voltage $E(t)$ to the circuit, then by Kirchhoff's second law from physics

$$E(t) = \text{Total voltage drop in the circuit}$$

$$= L\frac{dI}{dt} + RI + \frac{1}{C}\int_0^t I(u)\,du \qquad (20.104)$$

where I is the current in the circuit at time t. Then after differentiating in (20.104)

$$L\frac{d^2I}{dt^2} + R\frac{dI}{dt} + \frac{1}{C}I = E'(t). \qquad (20.105)$$

The foregoing application also has a mechanical analogue. Suppose a weightless spring, which hangs vertically from a fixed point of attachment, has a weight of mass m at its other end (Figure 20–4(a)). We recall from Section 10.4 that if the weight is pulled down below its equilibrium position and released, then by Newton's second law of motion (compare (10.59))

$$m\frac{d^2x}{dt^2} = -kx \quad \text{where } k > 0. \qquad (20.106)$$

Here the positive direction is downward and x is the vertical coordinate of weight with respect to an origin at the equilibrium position of the weight. Suppose also a piston which is attached to the weight is free to move in a dashpot filled with oil as shown in Figure 20–4(b). The spring system is then subject to a resisting force that is proportional to the velocity of the weight and is in a direction opposite to the motion of the weight. This resisting force must therefore have the form

$$-c\frac{dx}{dt} \quad \text{where } c > 0. \qquad (20.107)$$

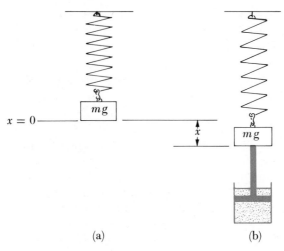

(a)   (b)

Figure 20–4

(Note from (20.107) that this force is downward when $dx/dt < 0$ and upward when $dx/dt > 0$.) Then instead of (20.106), the equation of motion

$$m \frac{d^2x}{dt^2} = -kx - c\frac{dx}{dt} \tag{20.108}$$

is obtained using Newton's second law. The constant c here is called a *damping constant*. Another modification will be made in the equation of the motion if a *driving force* $F(t)$ is applied to the weight. Then instead of (20.108) we have

$$m \frac{d^2x}{dt^2} = -kx - c\frac{dx}{dt} + F(t)$$

or

$$m \frac{d^2x}{dt^2} + c\frac{dx}{dt} + kx = F(t). \tag{20.109}$$

The motion of the spring governed by equation (20.109) is of particular interest when $c = 0$ and $F(t)$ has the form

$$F(t) = F_0 \sin \omega t.$$

Then letting $\beta = \sqrt{k/m}$, we can rewrite (20.109) in the form

$$\frac{d^2x}{dt^2} + \beta^2 x = \frac{F_0}{m} \sin \omega t. \tag{20.110}$$

The complementary solution of (20.110) is given by

$$x = C_1 \cos \beta t + C_2 \sin \beta t$$

and represents the "free motion" of the spring—that is, the motion that occurs if $F(t) = 0$ for every t. It can easily be verified that a particular solution of (20.110) is given by

$$g(t) = \frac{F_0}{m(\beta^2 - \omega^2)} \sin \omega t,$$

and therefore the general solution of (20.110) is

$$x = C_1 \cos \beta t + C_2 \sin \beta t + \frac{F_0}{m(\beta^2 - \omega^2)} \sin \omega t. \qquad (20.111)$$

Note that the motion given by (20.111) represents the combined effect of two different periodic motions. Note also from the last term on the right in (20.111) that by choosing ω sufficiently close to β the amplitude of the motion (the maximum value of $|x|$) can be made as large as we please. This phenomenon in which an arbitrarily large amplitude can be developed by having the period of the driving force, $2\pi/\omega$, sufficiently close to the period of the free motion, $2\pi/\beta$, is called *resonance*. In a spring resonance can produce oscillations of such amplitude as to exceed the elastic limit of the spring, thereby causing its permanent deformation.

   The weighted springs considered here are very simple illustrations of vibrating systems. Some examples of mechanical systems that experience more complex vibrations are bridges, airplane wings, operating machinery, and buildings in an earthquake. Therefore vibrational analysis can be an important consideration in engineering design.

Exercise Set 20.6

In Exercises 1–10 find the general solutions for the given differential equations.

1. $y'' + y' - 2y = 3$

2. $y'' - 2y' - 8y = 2x$

3. $y'' + 6y' + 9y = e^x$

4. $y'' - 4y' + 4y = x^2$

5. $y'' + y' - 3y = x - 1$

6. $y'' + 2y' - y = \sin 3x$

7. $y'' - 2y' + 5y = \cos 3x$

8. $y'' + 2y' - 3y = xe^{-x}$

9. $y'' + 4y = 3 \cos 2x$

10. $y'' - 4y = e^{2x}$

In Exercises 11–15 obtain the particular solution of the differential equation in the given exercise above which satisfies the given initial conditions.

11. Exercise 1; $y' = 2$ and $y = 1$ when $x = 0$

12. Exercise 2; $y' = -1$ and $y = 2$ when $x = 0$

13. Exercise 3; $y' = 3$ and $y = -1$ when $x = 0$

14. Exercise 6; $y' = -2$ and $y = 1$ when $x = 0$

15. Exercise 7; $y' = 1$ and $y = 0$ when $x = 0$

Exercises 16 and 17 refer to the *LRC* electrical circuit described above.

16. Suppose $L = 5$, $R = 20$, $C = 0.005$, and $E(t) = 50 \sin 20t$. If $dI/dt = 0$ and $I = 0$ when $t = 0$, obtain I as a function of t.

17. Suppose $L = 2$, $R = 10$, $C = 0.01$, and $E(t) = 100 \sin 40t$. If $dI/dt = 0$ and $I = 0$ when $t = 0$, obtain I as a function of t. What is $\lim_{t \to +\infty} I(t)$, the *steady-state current* in the circuit?

Exercises 18 and 19 refer to a weighted spring system having the differential equation (20.109).

18. Let $m = 2$ slugs, $c = 12$ lb per ft/sec, $k = 26$ lb/ft, and $F(t) = 10 \sin 10t$ lb. If the spring is stretched a distance 4 ft below its equilibrium and released at time $t = 0$ sec, find an equation of motion for the weight.

19. Suppose that in an undamped spring system $(c = 0)$, $m = 4$ slugs, $k = 36$ lb/ft, $F(t) = 8 \sin 3t$ lb. If the weight is imparted a downward velocity of 2 ft/sec at time $t = 0$ sec from its equilibrium position, find an equation of motion for the weight. Note that ultimately the spring must stretch beyond its elastic limit.

20. A cubical block of wood with edge 1 ft weighs 45 lb and floats in water having density 62.5 lb/ft^3. Neglecting the resistance of the water, there are two forces exerted on the block: a downward force due to the weight of the block and an upward force that equals the weight of the water displaced by the block. If the block is submerged until its upper face is in the surface of the water and then released, find (a) the period of the motion executed by the block and (b) the depth of the submerged portion of the block 2 sec after being released.

20.7   Power Series Solutions

Frequently one encounters differential equations having solutions that are not expressible in terms of elementary functions. In such situations it may be possible to define these solutions by power series. The technique for obtaining such series solutions is illustrated in the following example.

Example 1   Solve the differential equation

$$y'' - xy' + 2y = 0 \tag{20.112}$$

SOLUTION   Suppose that the function f where

$$f(x) = \sum_{n=0}^{+\infty} a_n x^n \quad \text{if } |x| < r \tag{20.113}$$

is a solution of (20.112) on some interval $(-r, r)$. Then by Theorem 14.8.2

$$f'(x) = \sum_{n=1}^{+\infty} n a_n x^{n-1} \tag{20.114}$$

and

$$f''(x) = \sum_{n=2}^{+\infty} n(n-1) a_n x^{n-2} \tag{20.115}$$

whenever $|x| < r$. Upon substituting in (20.112) for y, y' and y'', from (20.113), (20.114), and (20.115), respectively, one obtains

$$\sum_{n=2}^{+\infty} n(n-1)a_n x^{n-2} - \sum_{n=1}^{+\infty} na_n x^n + \sum_{n=0}^{+\infty} 2a_n x^n = 0 \quad \text{if } |x| < r. \quad (20.116)$$

We shall rewrite the left member of (20.116) as a power series, but before doing so, it is necessary to express the first sum in the equation in a form in which the nth term (where $n = 0, 1, 2, \ldots$) has the factor x^n. Because of the identity

$$\sum_{n=2}^{+\infty} n(n-1)a_n x^{n-2} = \sum_{n=0}^{+\infty} (n+2)(n+1)a_{n+2} x^n,$$

obtained by shifting the summation index, (20.116) can be rewritten as

$$\sum_{n=0}^{+\infty} (n+2)(n+1)a_{n+2} x^n - \sum_{n=1}^{+\infty} na_n x^n + \sum_{n=0}^{+\infty} 2a_n x^n = 0$$

or

$$2a_2 + 2a_0 + \sum_{n=1}^{+\infty} [(n+2)(n+1)a_{n+2} - (n-2)a_n]x^n = 0 \quad \text{if } |x| < r. \quad (20.117)$$

The left member of (20.117) is a power series for the function g where $g(x) = 0$ for every $x \in (-r, r)$. Hence, by Theorem 14.8.4, the left member of (20.117) is actually a Taylor series with $a = 0$ for this function g. Since each of the coefficients of this Taylor series is 0,

$$2a_2 + 2a_0 = 0 \quad (20.118)$$

$$(n+2)(n+1)a_{n+2} - (n-2)a_n = 0 \quad \text{for } n = 1, 2, 3, \ldots.$$

Then

$$a_{n+2} = \frac{n-2}{(n+2)(n+1)}a_n \quad \text{for } n = 1, 2, 3, \ldots. \quad (20.119)$$

Hence from (20.118) and (20.119)

$$\left.\begin{array}{c} a_2 = -a_0 \\ a_4 = 0 \\ a_{2n} = 0 \quad \text{for } n = 2, 3, 4, \ldots \end{array}\right\} \quad (20.120)$$

and

$$\left.\begin{array}{c} a_3 = -\dfrac{1}{3 \cdot 2}a_1 \\[2mm] a_5 = \dfrac{1}{5 \cdot 4}a_3 = -\dfrac{1}{5!}a_1 \\[2mm] a_7 = \dfrac{3}{7 \cdot 6}a_5 = -\dfrac{3}{7!}a_1 \\[2mm] a_9 = \dfrac{5}{9 \cdot 8}a_7 = -\dfrac{5 \cdot 3}{9!}a_1 \\[2mm] \cdots \end{array}\right\} \quad (20.121)$$

$$a_{2n+1} = -\frac{(2n-3)(2n-5)\cdots 1}{(2n+1)!}a_1 \quad \text{for } n = 2, 3, 4, \ldots$$

Substituting from (20.120) and (20.121) into (20.113),

$$f(x) = a_0(1 - x^2) + a_1\left[x - \frac{1}{3 \cdot 2}x^3 - \sum_{n=2}^{+\infty}\frac{(2n-3)(2n-5)\cdots 1}{(2n+1)!}x^{2n+1}\right] \quad (20.122)$$

if $|x| < r$. By the ratio test, the series in (20.122) converges for every x. Hence, by reversing the above steps, we find that any function f given by (20.122) is a solution of (20.112) on the interval $(-\infty, +\infty)$.

Exercise Set 20.7

In Exercises 1–12 solve the given differential equation using the power series method. When possible, express the power series obtained in terms of elementary functions.

1. $y' + 2y = 0$

2. $y' - 3y = 0$

3. $y'' + y = 0$

4. $y'' - 4y = 0$

5. $y'' - y' - 2y = 0$

6. $y'' + 2y' + 4y = 0$

7. $xy'' + 2y = 0$

8. $xy'' - y' + xy = 0$

9. $x^2y'' + 2xy' - 2y = 0$

10. $x^2y'' + xy' + (x^2 - 1)y = 0$

11. $xy'' + (x - 1)y' + y = 2$

12. $x^2y'' + 2y' - 2y = x$

Answers to Selected Exercises

Exercise Set 1.1

1. (a) $\{1, 2, 3, 4, 5, 6, 7, 9\}$ (c) $\{1, 2, 3, 4, 5, 6, 7, 9\}$ (e) $\{3\}$
2. (a) yes   (c) no   (e) yes
3. (a) $3x^2 + 3hx + x^2$
4. (a) $3 - 5\sqrt{3},\ 1,\ 4 + \sqrt{75},\ \sqrt{3}$

Exercise Set 1.4

1. $(-\infty, 5)$
3. $(-\frac{32}{25}, +\infty)$
5. $[-\frac{7}{2}, -2)$
7. $(-\frac{7}{3}, 2)$
9. $(-1, 2)$
11. $[-\sqrt{3}, \sqrt{3}]$
13. $(-\infty, +\infty)$
15. $[0, \frac{3}{2}]$
17. $(\frac{4}{3}, 2) \cup (3, +\infty)$
19. $(\frac{1}{7}, 7)$
21. $(-\infty, -\frac{5}{2}] \cup [\frac{5}{2}, +\infty)$
23. $[-\frac{5}{2}, 3]$
25. $(-2, \frac{3}{4}]$

Exercise Set 1.5

1. $x = 2, -7$
3. $x = \frac{16}{3}, \frac{2}{9}$
5. $x = \frac{5}{4}, -\frac{1}{2}$
7. $x = -2$
9. $x = -\frac{4}{3}, 2$
11. $(-4, 1)$
13. $(-\infty, -\frac{4}{5}] \cup [\frac{12}{5}, +\infty)$
15. $(-\infty, \frac{11}{9}] \cup [9, +\infty)$
17. $(-3, -\frac{3}{4}) \cup (-\frac{3}{4}, -\frac{3}{7})$
19. $[\frac{3}{8}, \frac{1}{2}) \cup (\frac{1}{2}, \frac{7}{12}]$
21. $M = 17$
23. $M = \frac{139}{9}$

Exercise Set 2.1

1. (a) 5   (c) $\frac{1}{3}\sqrt{185}$
2. (a) points not collinear
3. (a) right triangle   (c) not a right triangle or an isosceles triangle
   (e) isosceles right triangle
4. (a) parallelogram   (c) rhombus
5. $(3, 0)$

Exercise Set 2.2

1. (a) $(-\frac{3}{5}, -\frac{19}{5})$ (c) $(-2, -\frac{11}{4})$ (e) $(-\frac{9}{7}, -\frac{23}{7})$
2. (a) $(-\frac{1}{2}, -5)$
3. (a) $(1, -20)$
5. Point $(18, -4)$ is on the line.
7. $y = \frac{7}{4}$
9. Median from $A(0, 3)$ has length $\sqrt{17}$.
   Median from $B(5, 6)$ has length $\sqrt{458}/2$.
   Median from $C(-3, -8)$ has length $\sqrt{746}/2$.

Exercise Set 2.4

1. (a) 2   (c) $-\frac{62}{81}$
2. (a) Points are collinear.   (c) Points are not collinear.
3. (a) parallelogram   (c) rectangle   (e) trapezoid   (g) square
5. $x = \frac{9}{2}$ 7. $(8, -4), (-16, 6)$

Exercise Set 2.5

1. $5x - 4y = -2$ 3. $6x + 5y = 37$
5. $3x - 2y = -6$ 7. $3x + 2y = 4$
9. $7x + 2y = -11$ 11. $16x - 6y = -13$
13. $5x + 2y = -7$

Exercise Set 2.6

1. $(x + 1)^2 + (y - 1)^2 = 16$ 3. $(x - 4)^2 + (y - 5)^2 = 58$
5. $(x - 2)^2 + (y + 3)^2 = 20$ 7. $(x - 1)^2 + (y + 1)^2 = 17$
9. $(x - 5)^2 + (y - 3)^2 = 52$ 11. $(x + 5)^2 + (y + 4)^2 = 25$,
    $(x - 1)^2 + (y - 4)^2 = 25$
13. circle with center $(-4, 1)$ 15. circle with center $(-\frac{1}{4}, -\frac{3}{4})$
    and radius 3   and radius $3\sqrt{10}/4$
17. point $(2, 5)$ 19. Equation has no graph.
21. $(\frac{7}{10}, -\frac{11}{10}), (-2, -2)$ 23. $4x - 5y = 32$
25. $y = mx \pm r\sqrt{m^2 + 1}$

Exercise Set 3.1

1. (a) function   (c) function   (e) function
    (g) not a function   (i) function
3. $\mathcal{D}_f = (-\infty, +\infty)$ 5. $\mathcal{D}_h = (-\infty, 1) \cup (1, +\infty)$
7. $\mathcal{D}_F = [-1, 1]$ 9. $\mathcal{D}_F = (-3, 6]$
11. $\mathcal{D}_f = (-\infty, +\infty)$ 13. $\mathcal{D}_\phi = (-\infty, -1) \cup (-1, +\infty)$
15. $\mathcal{D}_f = (-\infty, 1) \cup [2, +\infty)$ 17. $\mathcal{D}_h = R$
19. (a) $A = C^2/4\pi$ (c) $C = 5x(7 - 0.10x)$

Exercise Set 3.2

1. (a) -5 (b) $-\frac{11}{5}$ (c) $4x - 3$ (d) $6x - 7$ (e) $\dfrac{2 - 3x}{x}$
    (f) 2
3. (a) 6   (b) $\frac{66}{25}$ (c) $4 - 6x - 4x^2$ (d) $-9x^2 + 3x + 6$
    (e) $\dfrac{4x^2 - 3x - 1}{x^2}$ (f) $-2x - 3 - h$

5. (a) 1 (b) $\frac{5}{12}$ (c) $\dfrac{1}{2x+2}$ (d) $\dfrac{1}{3x}$ (e) $\dfrac{x}{1+2x}$

(f) $\dfrac{-1}{(x+2)(x+h+2)}$

7. (a) does not exist (b) $\sqrt{\frac{1}{5}}$ (c) $\sqrt{6x-1}$ (d) $\sqrt{9x-7}$

(e) $\sqrt{\dfrac{3-x}{x}}$ (f) $\dfrac{\sqrt{3x+3h-1}-\sqrt{3x-1}}{h}$

9. (a) $\frac{19}{4}$ (b) $\frac{7}{2}$

11. (a) $4/\sqrt{19}$ (c) $\sqrt{\frac{3}{5}}$

13. (a) $f+g = \{(1,5),(2,6),(3,1),(4,6)\}$ $\mathcal{D}_{f+g} = \{1,2,3,4\}$

(b) $fg = \{(1,6),(2,8),(3,0),(4,9)\}$ $\mathcal{D}_{fg} = \{1,2,3,4\}$

(c) $f/g = \{(1,\frac{3}{2}),(2,2),(4,1)\}$ $\mathcal{D}_{f/g} = \{1,2,4\}$

(d) $g/f = \{(1,\frac{2}{3}),(2,\frac{1}{2}),(3,0),(4,1)\}$ $\mathcal{D}_{g/f} = \{1,2,3,4\}$

(e) $f \circ g = \{(1,4),(2,4),(4,1)\}$ $\mathcal{D}_{f \circ g} = \{1,2,4\}$

(f) $f \circ f = \{(1,1),(2,3),(3,3),(4,1)\}$ $\mathcal{D}_{f \circ f} = \{1,2,3,4\}$

(g) $g \circ f = \{(1,0),(2,3),(3,2),(4,0)\}$ $\mathcal{D}_{g \circ f} = \{1,2,3,4\}$

15. (a) $(f+g)(x) = \dfrac{2x^2-6x+5}{(x-2)(x-1)}$ $\mathcal{D}_{f+g} = \{x : x \neq 1 \text{ and } 2\}$

(b) $(fg)(x) = 1$ if $x \neq 1$ and 2 $\mathcal{D}_{fg} = \{x : x \neq 1 \text{ and } 2\}$

(c) $\left(\dfrac{f}{g}\right)(x) = \dfrac{(x-1)^2}{(x-2)^2}$ if $x \neq 1$ $\mathcal{D}_{f/g} = \{x : x \neq 1 \text{ and } 2\}$

(d) $\left(\dfrac{g}{f}\right)(x) = \dfrac{(x-2)^2}{(x-1)^2}$ if $x \neq 2$ $\mathcal{D}_{g/f} = \{x : x \neq 1 \text{ and } 2\}$

(e) $(f \circ g)(x) = 1/x$ if $x \neq 1$ $\mathcal{D}_{f \circ g} = \{x : x \neq 0 \text{ and } 1\}$

(f) $(f \circ f)(x) = \dfrac{1}{3-x}$ if $x \neq 2$ $\mathcal{D}_{f \circ f} = \{x : x \neq 2 \text{ and } 3\}$

(g) $(g \circ f)(x) = 3-x$ if $x \neq 2$ $\mathcal{D}_{g \circ f} = \{x : x \neq 2\}$

17. (a) $(f+g)(x) = \sqrt{x+2}+x-2$ $\mathcal{D}_{f+g} = [-2,+\infty)$

(b) $(fg)(x) = \sqrt{x-2}\,(x-2)$ $\mathcal{D}_{fg} = [-2,+\infty)$

(c) $\left(\dfrac{f}{g}\right)(x) = \dfrac{\sqrt{x+2}}{x-2}$ $\mathcal{D}_{f/g} = [-2,2) \cup (2,+\infty)$

(d) $\left(\dfrac{g}{f}\right)(x) = \dfrac{x-2}{\sqrt{x+2}}$ $\mathcal{D}_{g/f} = (-2,+\infty)$

(e) $(f \circ g)(x) = \sqrt{x}$ $\mathcal{D}_{f \circ g} = [0,+\infty)$

(f) $(f \circ f)(x) = \sqrt{\sqrt{x+2}+2}$ $\mathcal{D}_{f \circ f} = [-2,+\infty)$

(g) $(g \circ f)(x) = \sqrt{x+2}-2$ $\mathcal{D}_{g \circ f} = [-2,+\infty)$

19. (a) $(f+g)(x) = \sqrt{2x-1}+\sqrt{x+2}$ $\mathcal{D}_{f+g} = [\frac{1}{2},+\infty)$

(b) $(fg)(x) = \sqrt{2x-1}\,\sqrt{x+2}$ $\mathcal{D}_{fg} = [\frac{1}{2},+\infty)$

(c) $\left(\dfrac{f}{g}\right)(x) = \dfrac{\sqrt{2x-1}}{\sqrt{x-2}}$ $\mathcal{D}_{f/g} = \left[\dfrac{1}{2},+\infty\right)$

(d) $\left(\dfrac{g}{f}\right)(x) = \dfrac{\sqrt{x+2}}{\sqrt{2x-1}}$ $\quad \mathfrak{D}_{g/f} = \left(\dfrac{1}{2}, +\infty\right)$

(e) $(f \circ g)(x) = \sqrt{2\sqrt{x+2}-1}$ $\quad \mathfrak{D}_{f \circ g} = [-\tfrac{7}{4}, +\infty)$

(f) $(f \circ f)(x) = \sqrt{2\sqrt{2x-1}-1}$ $\quad \mathfrak{D}_{f \circ f} = [\tfrac{5}{8}, +\infty)$

(g) $(g \circ f)(x) = \sqrt{\sqrt{2x-1}+2}$ $\quad \mathfrak{D}_{g \circ f} = [\tfrac{1}{2}, +\infty)$

21. $g(x) = \dfrac{1-2x}{x-1}$; yes

24. (a) odd $\qquad$ (c) even

Exercise Set 3.3

1. (a) $\delta = \tfrac{1}{20}$ $\qquad$ (b) $\delta = 0.00025$ $\qquad$ (c) $\delta = \epsilon/4$

3. $\delta = \epsilon/2$ $\hspace{5cm}$ **5.** $\delta =$ smaller of 1 and $\epsilon/11$

7. $\delta =$ smaller of 9 and 3ϵ $\hspace{3cm}$ **9.** $\delta =$ smaller of 1 and 3ϵ

11. $\delta =$ smaller of $\tfrac{1}{2}$ and $\epsilon/6$ $\hspace{3cm}$ **13.** $\delta = \epsilon^2$

Exercise Set 3.4

1. 11; Theorem 3.4.5 $\hspace{4cm}$ **3.** 625; Theorems 3.4.5, 3.4.3(c)

5. $\tfrac{5}{32}$; Theorems 3.4.1, 3.4.3(c), 3.4.7

7. $-\tfrac{4}{9}$; Theorems 3.4.3(d), 3.4.7

Exercise Set 3.5

1. 1 $\hspace{6cm}$ **3.** doesn't exist

5. $\tfrac{2}{3}$ $\hspace{6cm}$ **7.** $-\tfrac{4}{9}$

9. $-2/x^2$ $\hspace{5cm}$ **11.** $-1/3a\sqrt[3]{a}$

13. $\tfrac{3}{4}$

Exercise Set 3.6

1. $\lim_{x \to 0^-} f(x) = 0$; $\lim_{x \to 0^+} f(x) = 4$; $\lim_{x \to 0} f(x)$ doesn't exist.

3. $\lim_{x \to 2^-} F(x) = \lim_{x \to 2^+} F(x) = \lim_{x \to 2} F(x) = 4$

5. $\lim_{x \to -3^-/2} f(x) = \lim_{x \to -3^+/2} f(x) = \lim_{x \to -3/2} f(x) = 0$

7. $\lim_{x \to 2^-} f(x) = -1$; $\lim_{x \to 2^+} f(x) = 1$; $\lim_{x \to 2} f(x)$ doesn't exist.

9. $\lim_{x \to -2^-} f(x) = 1$; $\lim_{x \to -2^+} f(x) = 0$; $\lim_{x \to -2} f(x)$ doesn't exist.

Exercise Set 3.7

1. all x except 3 and -2 $\hspace{3cm}$ **3.** no

5. no $\hspace{6cm}$ **7.** no

9. yes $\hspace{6cm}$ **11.** no

Exercise Set 3.8

1. (a) $\sqrt{3}$ $\qquad$ (c) $\sqrt[3]{5/3}$ $\qquad$ (e) doesn't exist $\qquad$ (g) 0 $\qquad$ (i) $\tfrac{1}{3}$

2. (a) all $x \neq 2$ and -4 $\qquad$ (c) all $x \in (-3, -\tfrac{2}{3}]$

Exercise Set 3.9

    1. (a) $+\infty$ (b) $-\infty$ (c) doesn't exist as an infinite limit

    3. (a) $+\infty$ (b) $+\infty$ (c) $+\infty$

    5. (a) $+\infty$ (b) doesn't exist as an infinite limit   (c) $+\infty$

Exercise Set 4.1

    1. $y = -2x$ **3.** $4x - 12y = -9$

    5. $4x + 25y = 12$ **7.** $x - 4y = -5$

    9. $y = 3x - 4$ **11.** $(0, 0)$ and $(4, -32)$

Exercise Set 4.2

    1. 3 **3.** $\dfrac{1}{2\sqrt{2}}$

    5. (a) 128 ft/sec   (b) $\frac{5}{2}$ sec   (c) $48\sqrt{3}$ ft/sec   (d) 44 ft

Exercise Set 4.3

    1. $2, 2$ **3.** $2x + 2, 4$

    5. $-2x, 4$ **7.** $4x^3, \frac{1}{2}$

    9. $-1/(x + 1)^2, -\frac{1}{9}$ **11.** $1/\sqrt{2x}, \frac{1}{4}$

    13. $-2/3(x + 2)^{5/3}, -\frac{1}{48}$

    15. (a) $f'(x) = \begin{cases} 1 & \text{if } x > 0 \\ -1 & \text{if } x < 0 \end{cases}$ (b) f continuous at 0

        (c) $f'(0)$ does not exist

    17. (a) $f'(x) = \begin{cases} 3 & \text{if } x > -1 \\ 0 & \text{if } x < -1 \end{cases}$ (b) discontinuous   (c) fails to exist

    19. (a) $f'(x) = \begin{cases} 1 & \text{if } x > 2 \\ -1 & \text{if } -1 < x < 2 \\ x & \text{if } x < -1 \end{cases}$ (b) continuous   (c) $f'(-1) = -1$

Exercise Set 4.4

    1. $-6x^2$ **3.** $4x + 5$

    5. $200x - \frac{2}{5}x^3$ **7.** $\dfrac{6 - x}{2x^3}$

    9. $\dfrac{-6(x + 1)}{(x^2 + 2x)^2}$ **11.** $\dfrac{x^3 - 6x^2 + 4x + 4}{(x - 2)^3}$

    13. $f'(x) = \begin{cases} 20/x^5 & \text{if } x > 0 \\ 2x - 2 & \text{if } x < 0 \end{cases}$

    15. $\frac{2}{3}\pi h(4a - 3h)$ **17.** $6(2s + 1)^2$

    19. $6x^2 - 10x + 5$ **25.** $2\pi(r - x)$

Exercise Set 4.5

1. $5(x^3 - 4x + 5)^4(3x^2 - 4)$

3. $-4(2x - 1)/3(x^2 - x + 1)^3$

5. $3(3x^2 - 2x + 7)^9(63x^2 - 22x + 7)$

7. $(x^2 - 3x)^2(x^2 + 4)^4(16x^3 - 39x^2 + 24x - 36)$

9. $2(x^2 + 3x)(x^2 + 4)^{-4}(-x^3 - 6x^2 + 8x + 12)$

11. $3(3x + 4)^{11}(9x - 4)/x^4$

13. $20(2x - 1)^4/(2x + 1)^6$

15. $2(x + 2)(2x - 3)^2(x^2 + 1)^3(13x^3 + 7x^2 - 19x + 3)$

17. $\dfrac{3x^5 - 3x^2}{x^3 + 1}$

Exercise Set 4.6

1. $\frac{1}{2}(15x^{1/2} - 3x^{-1/2} + x^{-3/2})$

3. $\dfrac{2x}{3(x^2 + 1)^{2/3}}$

5. $-\frac{4}{3}(2x^2 - x - 5)^{-7/3}(4x - 1)$

7. $\dfrac{2(-12y + 1)}{5y^{3/5}(3y + 1)^3}$

9. $\dfrac{-2a^2x}{(x^2 + a^2)^{1/2}(x^2 - a^2)^{3/2}}$

11. $\dfrac{(t^2 - 1)^2(-19t^2 + 54t + 1)}{3(3 - t)^{2/3}}$

13. $\dfrac{(1 + x^{1/2})(1 - x^{3/2})}{x^{1/2}(1 + x^2)^{3/2}}$

15. $\dfrac{(x^3 - 4x + 2)(3x^2 - 4)}{|x^3 - 4x + 2|}$

17. $\dfrac{-5(x + 2)|x - 3|}{(x - 3)^3|x + 2|}$

19. $0, \pm\sqrt{\frac{2}{5}}$

Exercise Set 4.7

1. $12x^3 - 15x^2 + 6$, $36x^2 - 30x$, $72x - 30$; $-150, 204, -174$

3. $8(2x - 3)^3$, $48(2x - 3)^2$, $192(2x - 3)$; $64, 192, 384$

5. $\dfrac{x}{2(x^2 + 2)^{3/4}}, \dfrac{4 - x^2}{4(x^2 + 2)^{7/4}}$

7. $(3x - 4)(9x - 4)$, $54x - 48$

Exercise Set 4.8

1. $3x/4y$

3. $\dfrac{6x^{4/3}y^{2/3} + 6x^{1/3}y^{5/3} - 2y}{x - 6x^{4/3}y^{2/3} - 6x^{1/3}y^{5/3}}$

5. $2x/3y, -4/3y^3$

7. $-y^{1/2}/x^{1/2}, a^{1/2}/2x^{3/2}$

9. $y - 2x/2y - x, -18/(2y - x)^3$

11. $1, \frac{2}{3}$

Exercise Set 4.9

1. $y = 4x - 1$, $x + 4y = 13$

3. $x - 3y = -7$, $3x + y = 9$

5. $3x + 2y = -5, -2x + 3y = 12$

7. $2x + 2y = a, y = x$

9. $x = -2$

13. $a = -\frac{3}{2}, b = 7, c = -3$

15. $(2, -3), (8, 45)$

17. $3x + 2y = 8, y = -\frac{21}{46}x - \frac{676}{299}$

19. $(-1, 1), \left(\dfrac{1 \pm \sqrt{11}}{2}, \dfrac{6 \pm \sqrt{11}}{2}\right)$

Exercise Set 4.10

1. (a) $1/\sqrt{3}$ (c) undefined   (e) $-1/\sqrt{3}$

2. (a) $33.7°$ (c) $111.8°$ (e) $63.4°$

3. (a) $136.4°$ (c) $127.9°$

4. (a) $32.5°, 109.7°, 37.9°$

5. $2x + y = 1, x - 2y = -12$ **7.** $151.4°$

9. $0°, 135°$ **11.** $53.1°, 126.9°$

Exercise Set 5.1

1. max. $f(3) = 7$, min. $f(-1) = 9$

3. max. $f(3) = 49$, min. $f(\pm\sqrt{2}) = 0$

5. max. $f(-1) = \sqrt[3]{6}$, min. $f(\frac{3}{2}) = \sqrt[3]{-\frac{1}{4}}$

7. max. $f(-1) = f(3) = 0$, min. $f(1) = -2$

9. max. $f(\frac{3}{4}) = \frac{3}{4}\sqrt[3]{\frac{1}{4}}$, min. $f(3) = -3\sqrt[3]{2}$

11. max. $f(3) = \frac{9}{10}\sqrt{10}$, min. $f(0) = 0$

13. max. $f(1) = f(3) = -3$, min. $f(-1) = -9$

15. max. $f(0) = 0$, no min.

17. max. $f(3) = 5$, no min.

Exercise Set 5.2

1. $\frac{25}{4}$ **3.** $\frac{2}{3}$ **5.** $-\frac{1}{2} + \dfrac{7\sqrt{3}}{6}$

Exercise Set 5.3

1. (a) $(-\infty, -2], [1, +\infty)$ (b) $[-2, 1]$

(c) $f(-2) = 11$, a rel. max.; $f(1) = -\frac{5}{2}$, a rel. min.

3. (a) $(-\infty, -3], [3, +\infty)$ (b) $[-3, 3]$

(c) $f(-3) = 162$, a rel. max.; $f(3) = -162$, a rel. min.

5. (a) $[-1, 1]$ (b) $(+\infty, -1], [1, +\infty)$

(c) $f(1) = 2$, a rel. max.; $f(-1) = -2$, a rel. min.

7. (a) $(-\infty, \frac{3}{2}]$ (b) $[\frac{3}{2}, +\infty)$ (c) $f(\frac{3}{2}) = \sqrt[3]{\frac{9}{4}}$, a rel. max.

9. (a) $(-\infty, \frac{4}{3}]$ (b) $[\frac{4}{3}, 2]$ (c) $f(\frac{4}{3}) = \frac{4}{3}\sqrt{\frac{2}{3}}$, a rel. max.

11. (a) $[\frac{1}{2}, +\infty)$ (b) never decreasing   (c) no relative extrema

13. (a) $[-1, 0], [1, +\infty)$ (b) $[0, 1]$

(c) $f(0) = 1$, a rel. max.; $f(x) = 0$, a rel. min. where
$$x \in \{1\} \cup (-\infty, -1]$$

15. F   **17.** T   **19.** T

Exercise Set 5.4

1. (a) all points in graph   (b) never concave downward

(c) no inflection points

3. (a) where $x < 2$ (b) where $x > 2$ (c) $(2, 41)$ (d) 27

5. (a) where $1 < x < 3$ (b) where $x < 1$ or $x > 3$

(c) $(1, -10), (3, -26)$ (d) -16 at $(1, -10)$, 0 at $(3, -26)$

7. (a) where $x > -1$ (b) where $x < -1$ (c) $(-1, -6)$ (d) 5

9. (a) never concave upward   (b) where $x > -\frac{1}{2}$

(c) no inflection points

Exercise Set 5.5

1. $f(\frac{2}{3}) = \frac{2}{3}$, a rel. min.   **3.** $f(-1) = 0$, a rel. min.

5. $f(-\frac{7}{3}) = \frac{284}{27}$, a rel. max.; $f(1) = -8$, a rel. min.

7. $f(0) = 12$, a rel. max.; $f(1) = 7$ and $f(-2) = -20$ are rel. min.

9. $f(4) = \frac{5}{4}$, a rel. max.; $f(-2) = -\frac{5}{2}$, a rel. min.

11. $f(-\frac{24}{7}) = \frac{576}{49}\sqrt[3]{\frac{4}{7}}$, a rel. max.; $f(0) = 0$, a rel. min.

13. no relative extrema

Exercise Set 5.6

1. 0   **3.** doesn't exist   **5.** $-\frac{1}{3}$

7. 0   **9.** 3   **11.** $-\frac{1}{2}$

13. 0   **15.** 0   **17.** 0

Exercise Set 5.7

1. $-\infty$ **3.** $+\infty$ **5.** $+\infty$

7. $+\infty$ **11.** $+\infty$

Exercise Set 6.1

1. $f(-2) = 16$, abs. max.   **3.** $f(2) = -15$, abs. min.

5. Let $y = f(x)$. Then $f(-1) = \frac{2}{3}$, abs. max.; $f(0) = 0$, abs. min.

7. no abs. extrema

9. $f(3/\sqrt{2}) = 9$, abs. max.; $f(-3/\sqrt{2}) = -9$, abs. min.

Exercise Set 6.2

1. 10, 10   **3.** 6×6

5. $2\sqrt[3]{25} \times 2\sqrt[3]{25} \times \sqrt[3]{25}$

7. $b\sqrt{3}/6$ units above the base where $b = $ length of the base

11. $5 - \sqrt{5} = \sqrt{30} - 10\sqrt{5}$ **13.** rad. $= r/2$, alt. $= h/2$

15. rad. $= 2a\sqrt{2}/3$, alt. $= \frac{4}{3}a$

17. (a) $8 - 2\sqrt{3}$ mi.   (b) 0 mi.

19. 2 men   **21.** $h/r = 2\sqrt{2}$

Exercise Set 6.3

1. (a) 360   **3.** 49   **5.** 482,000

7. 4.58 weeks   **9.** 36¢   **11.** $t = 200$

Exercise Set 6.4

1. $0 \leq t < 2$; moving in positive direction, losing speed

   $t = 2$; at rest

   $2 < t \leq 4$; moving in negative direction, gaining speed

   $s = 0, 64$, and 0, respectively, when $t = 0, 2$, and 4

3. $1 \leq t < 3\sqrt[3]{2}$; moving in negative direction, losing speed

   $t = 3\sqrt[3]{2}$; at rest

   $t > 3\sqrt[3]{2}$; moving in positive direction, gaining speed

   $s = 56$ and $9\sqrt[3]{2}$, respectively, when $t = 1$ and $3\sqrt[3]{2}$

5. $t < -3$; moving in positive direction, losing speed

   $t = -3$; at rest

   $-3 < t < 1$; moving in negative direction, gaining speed

   $t = 1$; moving in negative direction, coasting

   $1 < t < 5$; moving in negative direction, losing speed

   $t = 5$; at rest

   $t > 5$; moving in positive direction, gaining speed

   $s = 80, -48$, and -176, respectively, when $t = -3, 1$, and 5

7. $t < -2/\sqrt{3}$; moving in positive direction, gaining speed

   $t = -2/\sqrt{3}$; moving in positive direction, coasting

   $-2/\sqrt{3} < t < 0$; moving in positive direction, losing speed

   $t = 0$; at rest

   $0 < t < 2/\sqrt{3}$; moving in negative direction, gaining speed

   $t = 2/\sqrt{3}$; moving in negative direction, coasting

   $t > 2/\sqrt{3}$; moving in negative direction, losing speed

9. (a) 128 ft/sec   (b) 582 ft   (c) 192 ft/sec   (d) $8\sqrt{582}$ ft/sec

Exercise Set 6.5

1. 512π in^3/sec

3. $2/9\pi$ ft/sec

5. $\frac{16}{15}$ ft/sec

7. $\frac{10}{3}\sqrt{\frac{2}{3\pi}}$ ft, $-3\sqrt{6\pi}/5$ ft^2/sec

9. 6 ft/sec

11. $8/\sqrt{15}$ ft/sec

13. $\sqrt{2015}/8$ ft/sec

Exercise Set 6.6

1. (a) 15.6   (c) $\frac{121}{60}$

2. (a) 0.0032   (c) $\frac{1}{7200}$

3. (a) $\frac{752}{3}$ (c) 4.9964

4. (a) $\frac{2}{81}$ (c) 1.944×10^{-5}

5. (a) $-\frac{26}{9}$ (c) $\frac{72}{7}$

7. 0.09π in.^{3}

9. $\frac{1}{2}$% increase

11. 2.5 in.^{2}

Exercise Set 6.7

1. (a) $dy = 6x^2\, dx$ (b) $dy = -1.92$, $\Delta y = -1.910416$

3. (a) $dy = \frac{1}{2}x^{-1/2}\, dx$ (b) $dy = -0.0025$, $\Delta y \approx -0.0050$

5. (a) $dy/dx = 3x^2 + 4x - 7$, $d^2y/dx^2 = 6x + 4$

   (c) $dy/dx = -6/(x-3)^2$, $d^2y/dx^2 = 12/(x-3)^3$

6. (a) $(4t^3 + 12t^2 - 8)\, dt$

Exercise Set 7.1

1. $\dfrac{3x^2}{2} + C$

3. $\dfrac{3\sqrt[3]{4}}{4}x^{4/3} + C$

5. $\dfrac{t^5}{5} - \dfrac{t^3}{3} - \dfrac{5t^2}{2} + 7t + C$

7. $\dfrac{2x^{3/2}}{3} - 6x^{1/2} - \dfrac{2}{x^{1/2}} + C$

9. $2x + \dfrac{5}{x} - \dfrac{1}{x^3} + C$

11. $\dfrac{(x^4 + 1)^{3/2}}{6} + C$

13. $\dfrac{(2s - 1)^9}{18} + C$

15. $-2\sqrt{3 - t - t^2} + C$

Exercise Set 7.2

1. $y = x^3 - 2x^2 - 2x + C$

3. $s = -\dfrac{1}{2t} + C_1 t + C_2$

5. $y = \frac{2}{3}\sqrt{x^3 + 1} - \frac{3}{2}$

7. $y = \frac{1}{10}(2x + 3)^5 + \frac{49}{10}$

9. $y = -\frac{1}{12}t^4 + 2t - 1$

11. $y = -4x^{1/2} + \frac{1}{4}x + \frac{23}{4}$

13. $f(x) = 2x^2 - 20x + 47$

15. 60 ft, $\dfrac{3 + \sqrt{15}}{2}$ sec

17. $32\sqrt{30}$ ft/sec

19. $\frac{25}{3}$ ft/sec^2

21. 3.00 mtr/sec^2

23. $8\sqrt{165}/33$ mi/sec, $40/\sqrt{33}$ mi/sec

Exercise Set 7.3

1. 55

3. $-\frac{182}{9}$

5. 960

7. 2390

9. $2^n - 1$

Exercise Set 7.4

1. (a) $\underline{S}_P = mb^2/4$, $\overline{S}_P = 3mb^2/4$

(c) $\underline{S}_P = mb^2(n-1)/2n$, $\overline{S}_P = mb^2(n+1)/2n$

2. (a) $\underline{S}_P = \frac{15}{2}$, $\overline{S}_P = \frac{117}{8}$ (c) $\underline{S}_P = \frac{2197}{216}$, $\overline{S}_P = \frac{2873}{216}$

5. (a) $k(b-a)$ (c) $\dfrac{b^4 - a^4}{4}$

Exercise Set 7.5

1. $\frac{121}{8}$

3. $\frac{295}{16}$

5. yes

7. no

9. a

11. $\frac{1}{3}$

Exercise Set 7.7

1. $x^2 - 9$

3. $-|x^{1/2} - 1|$

5. $\dfrac{x(1 - |x|)}{|x|(1 + x^2)}$

7. 13

9. $\dfrac{3 + \sqrt{33}}{2}$

11. a negative root between -2 and -1, a positive root between 0 and 1 and between 3 and 4

Exercise Set 7.8

1. $\frac{56}{3}$

3. 6

5. $\frac{1568}{5}$

7. $\frac{209}{96}$

9. $\frac{16}{5}$

11. $\frac{29}{2}$

13. -1

15. $\frac{33}{2}$

17. $\frac{9}{2}$

19. $-\frac{2}{9}$

21. $\frac{9}{14}(4\sqrt[3]{2} - 1)$

23. $\frac{2}{3}(3\sqrt{3} - 2\sqrt{2})$

25. $6(\sqrt{3} - \sqrt{2})$

Exercise Set 8.1

1. 44

3. $\frac{32}{3}$

5. $\frac{1}{6}a^2$

7. $\frac{52}{3}$

9. $\frac{16}{9}\sqrt{3}$

11. $\frac{9}{2}$

13. $\frac{148}{3}$

15. $\frac{16}{3}$

Exercise Set 8.2

1. (a) $250\pi/3$ (c) $\frac{1000}{3}$

3. $\pi h^2\left(r - \dfrac{h}{3}\right)$ 5. $a^2h/3$ 7. $512\pi/15$

9. (a) $512\pi/5$ (b) 32π (c) $32\pi/3$ (d) $160\pi/3$
11. (a) $1024\pi/7$ (b) 64π (c) $2048\pi/35$ (d) $\pi(656 - 9\sqrt[3]{4})/5$
13. $32\pi/15$ 15. $2\pi/15$
17. $2\pi^2 kr^2$ 19. $2r^3/3$

Exercise Set 8.3

1. (a) $1120\pi/3$ (b) $592\pi/3$
3. (a) $512\pi/5$ (b) 32π (c) $32\pi/3$ (d) $160\pi/3$
5. (a) $1024\pi/7$ (b) 64π (c) $2048\pi/35$ (d) $\pi(656 - 9\sqrt[3]{4})/5$
7. $27\pi/2$ 9. $2\pi^2 kr^2$

Exercise Set 8.4

1. (a) 45 in-lb (b) 135 in-lb (c) $36\sqrt{3}$ lb
3. 7 in

5. (a) $Gm_1 m_2\left(\dfrac{1}{a} - \dfrac{1}{b}\right)$ (b) $\dfrac{Gm_1 m_2}{a}$

7. 6400 ft-lb 9. $7{,}350{,}000\pi$ ft-lb
11. $108{,}400$ ft-lb 13. $8640\,\delta(3\pi - 4)$ ft-lb

Exercise Set 8.5

1. (a) $140{,}400$ lb (b) $62{,}400$ lb (c) $327{,}600$ lb
3. (a) 5200 lb (c) $(5200 + 1560\pi)$ lb
5. (a) 9072π lb (b) 6048 lb
7. $2.12 \cdot 10^6$ lb

Exercise Set 9.1

1. (a) 1.79176 (c) -0.69315 (e) 2.89037

3. $x(2 \ln 3x + 1)$ 5. $\dfrac{24[\ln(x+1)]^2}{x+1}$ 7. $\dfrac{-x}{x^4 - 1}$

9. $\dfrac{x - 1}{\sqrt{x^2 + 1}}$ 11. $\dfrac{1}{x + y + 1}$

Exercise Set 9.2

1. $3(x + 2)(8x^2 - 5x - 12)$

3. $\dfrac{30x^3 + 38x^2 + 12x + 140}{3(x^2 + 4)^{1/2}(2x + 3)^{4/3}}$

5. $\dfrac{(x + 1) \ln (x + 1) - (x - 1) \ln (x - 1)}{(x^2 - 1)[\ln (x + 1)]^2}$

7. $\frac{1}{4} \ln |4x + 3| + C$

9. $\dfrac{6x + \ln |9x + 12|}{9} + C$

11. $\frac{1}{4}(\ln x)^2 + C$

13. $\frac{1}{3} \ln 4$

15. $3 - 5 \ln \frac{4}{3}$

Exercise Set 9.3

1. no inverse function

3. $y = f^{-1}(x) = \dfrac{7 - x}{2}$

5. $y = f^{-1}(x) = \dfrac{-x}{2(x - 1)}$

7. $y = f^{-1}(x) = \dfrac{1 + x^2}{1 - x^2}$

9. $y = f^{-1}(x) = 3 + \sqrt{16 + x}$

11. no inverse function

13. no inverse function

15. no inverse function

Exercise Set 9.4

1. $\frac{1}{3}$

3. $\frac{1}{3}$

5. $\frac{1}{4}$

7. $\frac{1}{4}$

9. 1

11. $(f^{-1})'(y) = \dfrac{1}{\sqrt{1 + (f^{-1}(y))^2}}$

Exercise Set 9.5

1. (a) $1/x$ (c) $-x$ (e) $3x$ (g) $|x| \sqrt{e}$

3. $-(x - 1)e^{-(x-1)^2/2}$

5. $\dfrac{a}{2}(e^{ax} - e^{-ax})$

7. $\dfrac{-e^{1/x}}{x^2}$

9. $\dfrac{2e^{2x}(x^2 - x + 1)}{(x^2 + 1)^2}$

11. $\dfrac{-4}{e^{2x} - e^{-2x}}$

13. $\dfrac{e^{x-y} - e^x}{e^{x-y} - e^y}$

15. $-\dfrac{1}{2e^{2x}} + C$

17. $\frac{1}{3}e^{x^3} + C$

19. $\ln (e^x + 1) + C$

21. $\dfrac{e^{2x} - 4x - e^{-2x}}{2} + C$

23. $\dfrac{1}{3(1 - 3e^x)} + C$

25. $e^x - \ln (1 + e^x) + C$

Exercise Set 9.6

1. (a) 0.6990   (c) 0.3170

7. $2 \ln 10(10^{2x+3})$

9. $3^{x^2} 2^{4x+1}(2 \ln 2 + x \ln 3)$

11. $\dfrac{2^{3x}[3(x^2 + 1) \ln 2 - 2x]}{(x^2 + 1)^2}$

13. $\dfrac{x^{1/x}(1 - \ln x)}{x^2}$

15. $xe^{-x}e^{-x}\left(\dfrac{1}{x} - \ln x\right)$

17. $(x + 2)^{\ln x}\left[\dfrac{(x + 2)\ln(x + 2) + x\ln x}{x(x + 2)}\right]$

19. $\dfrac{(2x - 3)\log_{10} e}{x^2 - 3x - 5}$

21. $\dfrac{\log_{10} e}{x(x - 1)}$

23. $\dfrac{(\log_{10} e)^2}{x\log_{10} x}$

25. $-\dfrac{1}{2^{3x}3\ln 2} + C$

27. $\dfrac{-1}{10^{x^2+3x-1}\ln 10} + C$

29. $\dfrac{3^{2x} + 4x\ln 3 - 3^{-2x}}{2\ln 3} + C$

31. $\dfrac{-1}{(\ln a)^2 a^{a^x}} + C$

37. approx. $995,750,000

Exercise Set 9.7

1. $L = 3.9e^{(1.2)(10^{-5})T}$

3. (a) 11.9 in.   (b) 6 mi

5. 108

7. approx. 0.3 yr

9. (a) 5.9 million, 9.8 million   (b) 9.8 million, 26.6 million

Exercise Set 10.1

1. (a) $\sin\dfrac{4}{3}\pi = -\dfrac{\sqrt{3}}{2}$ $\cos\dfrac{4}{3}\pi = -\dfrac{1}{2}$ $\tan\dfrac{4}{3}\pi = \sqrt{3}$

$\csc\dfrac{4}{3}\pi = -\dfrac{2}{\sqrt{3}}$ $\sec\dfrac{4}{3}\pi = -2$ $\cot\dfrac{4}{3}\pi = \dfrac{1}{\sqrt{3}}$

(c) $\sin\dfrac{11}{6}\pi = -\dfrac{1}{2}$ $\cos\dfrac{11}{6}\pi = \dfrac{\sqrt{3}}{2}$ $\tan\dfrac{11}{6}\pi = -\dfrac{1}{\sqrt{3}}$

$\csc\dfrac{11}{6}\pi = -2$ $\sec\dfrac{11}{6}\pi = \dfrac{2}{\sqrt{3}}$ $\cot\dfrac{11}{6}\pi = -\sqrt{3}$

(e) $\sin\dfrac{13}{6}\pi = \dfrac{1}{2}$ $\cos\dfrac{13}{6}\pi = \dfrac{\sqrt{3}}{2}$ $\tan\dfrac{13}{6}\pi = \dfrac{1}{\sqrt{3}}$

$\csc\dfrac{13}{6}\pi = 2$ $\sec\dfrac{13}{6}\pi = \dfrac{2}{\sqrt{3}}$ $\cot\dfrac{13}{6}\pi = \sqrt{3}$

Exercise Set 10.2

1. $-(2x + 3)\sin(x^2 + 3x)$

3. $2\sin 4x$

5. $2e^{-\cos y}(1 + y\sin y)$

7. $3x^2\sin\dfrac{1}{x} - x\cos\dfrac{1}{x}$

9. $-3(1 + \cos^2 2x)^{1/2}\sin 4x$

11. $\cos^3 x\sin^2 x(3\cos^2 x - 4\sin^2 x)$

13. $\dfrac{\cos(x-y) + y\sin x}{\cos x + \cos(x-y)}$

15. -2 17. $\frac{1}{2}$ 19. $\frac{2}{3}$ 21. $\frac{2}{5}$

25. $f\left(\dfrac{\pi}{3} + 2m\pi\right) = f\left(\dfrac{4\pi}{3} + 2m\pi\right) = \dfrac{3\sqrt{3}}{16}$, a rel. max.;

$f\left(\dfrac{2\pi}{3} + 2m\pi\right) = f\left(\dfrac{5\pi}{3} + 2m\pi\right) = -\dfrac{3\sqrt{3}}{16}$, a rel. min.

27. $f(2m\pi) = 3$, a rel. max.; $f(\pi + 2m\pi) = -5$, a rel. min.

29. yes, yes

Exercise Set 10.3

1. $2x\sec^2 x^2$ 3. $-2\csc 2x$ 5. $\frac{3}{2}\tan 3x\sqrt{\sec 3x}$

7. $-\dfrac{1}{(t+1)^2}\csc\dfrac{t}{t+1}\cot\dfrac{t}{t+1}$ 9. $-\csc x(\csc^2 x + \cot^2 x)$

11. $2\sec^2 x\tan x(\sec^2 x + \tan^2 x)$ 13. $\csc x$

15. $\dfrac{\sec^2(x-y) - \sec(x+y)\tan(x+y)}{\sec^2(x-y) + \sec(x+y)\tan(x+y)}$

21. 0

23. $f\left(\dfrac{\pi}{6}\right) = \dfrac{3\sqrt{3} - 2\pi}{3}$, a rel. min.;

$f\left(\dfrac{5\pi}{6}\right) = -\dfrac{3\sqrt{3} + 10\pi}{3}$, a rel. max.

Exercise Set 10.4

1. $\frac{1}{10}$ rad/sec   3. 500 rad/hr

5. $\dfrac{60\sqrt{66}\pi(1 - \sqrt{22})}{11}$ in/sec   7. $5(1 + 2^{2/3})^{3/2}$ ft

9. $\frac{15}{7}$ mi from the point on the beach nearest A

11. 4π sec; 2 cm; $\sin 5\pi/12$ cm/sec (downward)

Exercise Set 10.5

1. $\frac{1}{2}\sin 2x + C$ 3. $\frac{1}{3}\sec 3x + C$

5. $-\frac{1}{2}\ln|\cos x^2| + C$ 7. $-\frac{1}{8}\cos^4 2y + C$

9. $\frac{1}{12}\tan^3 4x + C$ 11. $-\frac{2}{5}\csc^5\frac{1}{2}x + C$

13. $\dfrac{(2 + \sin 3x)^{4/3}}{4} + C$ 15. $-\frac{1}{2}\ln|1 + 2\cot t| + C$

17. $-\ln|\csc t + \cot t| - t + C$ 19. 1

21. $\frac{1}{4}\ln 2$ 23. $\frac{1}{3}$

25. $\frac{1}{8}$ 29. 2

Exercise Set 10.6

1. (a) $\pi/3$ (c) $-\pi/6$

2. (a) $\frac{2}{3}$ (c) $\sqrt{15}/4$

3. (a) $\pi/4$ (c) $\frac{33}{65}$

7. $-1/\sqrt{x-x^2}$

9. $\dfrac{4-x-x^2}{(4-x^2)^{3/2}}$

11. $3x\left(x\sin^{-1}\dfrac{3}{x} - \dfrac{|x|}{\sqrt{x^2-9}}\right)$

13. $\dfrac{\sqrt{1-x^2y^2}+y}{\sqrt{1-x^2y^2}-x}$

15. $\frac{1}{10}$ rad/sec

Exercise Set 10.7

1. (a) $\pi/6$ (c) $\pi/3$

2. (a) $\frac{3}{2}$ (c) $2\sqrt{2}$

3. (a) $\frac{1}{13}$

7. $-|x|/x\sqrt{1-x^2}$

9. $\dfrac{1}{t^2+1}$

11. $\dfrac{-|x|}{x(1+x^2)}$

13. $\dfrac{2(1-4x\tan^{-1}2x)}{(1+4x^2)^2}$

15. 26.5 ft

17. $-\frac{1600}{4349}$ rad/sec

Exercise Set 10.8

3. $\dfrac{1}{3}\sin^{-1}\dfrac{3x}{5} + C$

5. $\sec^{-1}2x + C$

7. $\ln|\tan^{-1}y| + C$

9. $-2\tan^{-1}(\cos x) + C$

11. $\dfrac{1}{2}\tan^{-1}\dfrac{e^x}{2} + C$

13. $\dfrac{1}{\sqrt{11}}\sec^{-1}\dfrac{x+2}{\sqrt{11}} + C$

15. $\sin^{-1}\dfrac{x+1}{3} + C$

17. $\dfrac{1}{4}\ln|2x^2-x+3| + \dfrac{5}{2\sqrt{23}}\tan^{-1}\dfrac{4x-1}{\sqrt{23}} + C$

19. $\pi/6$

21. $\pi/6\sqrt{3}$

Exercise Set 10.9

13. $\dfrac{1}{2\sqrt{x}}\sinh\sqrt{x}$

15. $2\coth 2x$

17. $\operatorname{sech} x$

19. $\frac{1}{3}\cosh 3x + C$

21. $\frac{1}{4}\sinh^2 2x + C$

23. $\frac{1}{12}\cosh^3 4s - \frac{1}{4}\cosh 4s + C$

25. $x - \tanh x + C$

Exercise Set 11.1

1. $\dfrac{x}{2}\sin 2x + \dfrac{1}{4}\cos 2x + C$ 3. $-\frac{1}{2}y^2 e^{-v^2} - \frac{1}{2}e^{-v^2} + C$

5. $x\tan^{-1}x - \frac{1}{2}\ln(1+x^2) + C$

7. $\dfrac{2x}{9}(3x+5)^{3/2} - \dfrac{4}{135}(3x+5)^{5/2} + C$

9. $x\ln x - x + C$

11. $x^2\sqrt{2x+1} - \frac{2}{3}x(2x+1)^{3/2} + \frac{2}{15}(2x+1)^{5/2} + C$

13. $-\frac{1}{2}r^2\cos 2r + \frac{1}{2}r\sin 2r + \frac{1}{4}\cos 2r + C$

15. $\dfrac{t}{2}\sin(\ln t) - \dfrac{t}{2}\cos(\ln t) + C$

17. $\frac{1}{10}e^x \sin 3x - \frac{3}{10}e^x \cos 3x + C$

21. $\dfrac{x^5}{5}\ln x - \dfrac{x^5}{25} + C$ 23. $\pi(\frac{5}{4}e^{3/2} - 2)$

Exercise Set 11.2

1. $\frac{3}{2}\sin^2 \frac{1}{3}x - \frac{3}{2}\sin^4 \frac{1}{3}x + \frac{1}{2}\sin^6 \frac{1}{3}x + C$

3. $\frac{1}{2}\sin 2t + C$ 5. $-\frac{1}{2}\cos 2x + \frac{1}{6}\cos^3 2x + C$

7. $-\frac{1}{4}\cos^4 y + \frac{1}{6}\cos^6 y + C$

9. $-\frac{2}{3}\cos^{3/2}x + \frac{2}{7}\cos^{7/2}x + C$

11. $\frac{1}{2}x - \frac{3}{4}\sin\frac{2}{3}x + C$

13. $\frac{1}{16}x - \frac{1}{64}\sin 4x + \frac{1}{48}\sin^3 2x + C$

15. $\frac{1}{16}\sin 8x + \frac{1}{4}\sin 2x + C$

17. $-\frac{1}{12}\cos 6x + \frac{1}{4}\cos 2x + C$

Exercise Set 11.3

1. $-\dfrac{\cot^2 x}{2} - \ln|\sin x| + C$ 3. $-\frac{1}{3}\cot 3x - \frac{1}{9}\cot^3 3x + C$

5. $\frac{1}{5}\sec^5 x - \frac{1}{3}\sec^3 x + C$ 7. $\frac{1}{6}\tan^3 2x + \frac{1}{10}\tan^5 2x + C$

9. $-\frac{2}{7}\csc^7 \frac{1}{2}x + \frac{2}{5}\csc^5 \frac{1}{2}x + C$ 11. $\frac{1}{2}\tan 2y - \frac{1}{2}\cot 2y - 4y + C$

13. $\frac{1}{4}\tan^4 t + \frac{1}{6}\tan^6 t + C$

15. $\frac{1}{3}\sec^3 x - \sec x + x - \tan x + \frac{1}{3}\tan^3 x + C$

17. $\frac{1}{3}$ 19. $\dfrac{20\sqrt{2} + 8}{21}$

Exercise Set 11.4

1. $\dfrac{x}{4\sqrt{4-x^2}} + C$ 3. $\ln\left(\sqrt{y^2+3} + y\right) + C$

5. $\dfrac{x}{2}\sqrt{4x^2-25} - \dfrac{25}{4}\ln\left(2x + \sqrt{4x^2-25}\right) + C$

7. $\dfrac{x}{18}\sqrt{9x^2 + 16} - \dfrac{8}{27}\ln\left(3x + \sqrt{9x^2 + 16}\right) + C$

9. $\dfrac{\sqrt{x^2 - 4}}{4x} + C$

11. $\dfrac{1}{14\sqrt{7}}\left[\tan^{-1}\dfrac{t + 2}{\sqrt{7}} + \dfrac{\sqrt{7}(t + 2)}{t^2 + 4t + 11}\right] + C$

13. $-\sqrt{4 - 3x - x^2} - \dfrac{1}{2}\sin^{-1}\dfrac{2x + 3}{5} + C$

15. $\sqrt{x^2 - 2x - 5} - \dfrac{1}{2}\ln\left(x - 1 + \sqrt{x^2 - 2x - 5}\right) + C$

17. $2\sin^{-1}\dfrac{e^x}{2} - \dfrac{e^x\sqrt{4 - e^{2x}}}{2} + C$

19. $\dfrac{-\ln t}{\sqrt{(\ln t)^2 - 1}} + C$

21. $\dfrac{x - 2}{2\sqrt{x^2 - 4}} + C$

23. $\dfrac{-\sqrt{9 - x^2}}{2x^2} - \dfrac{1}{6}\ln\left|\dfrac{3 - \sqrt{9 - x^2}}{x}\right| + C$

25. $\frac{1}{36} - \frac{1}{27}\sin^{-1}\frac{3}{5}$

27. $\dfrac{\sqrt{6} - 2}{2}$

29. $\frac{1}{10}$

35. πab

Exercise Set 11.5

1. $\dfrac{1}{6}\ln\left|\dfrac{x - 3}{x + 3}\right| + C$

3. $\ln\dfrac{|x - 1|^3}{(x + 2)^4} + C$

5. $\ln\dfrac{|2x + 1|^{1/2}}{|x - 2|^3} + C$

7. $-\dfrac{2}{x} + \ln\dfrac{(x - 3)^4}{|x|^3} + C$

9. $\ln|x - 2| - \dfrac{4}{x - 2} - \dfrac{1}{2(x - 2)^2} + C$

11. $\dfrac{3}{2}x^2 + 9x - 18\ln|x + 1| - \dfrac{9}{x + 1} + C$

13. $x - \dfrac{4}{3}\ln|x - 2| + \dfrac{2}{3}\ln|x^2 - 2x + 4| - \dfrac{4}{\sqrt{3}}\tan^{-1}\dfrac{x - 1}{\sqrt{3}} + C$

15. $\dfrac{1}{12}\ln\left|\dfrac{x - 3}{x + 3}\right| + \dfrac{1}{6}\tan^{-1}\dfrac{x}{3} + C$

17. $\ln\dfrac{(x - 1)^4}{|x^2 + x + 2|} + \dfrac{4}{\sqrt{7}}\tan^{-1}\dfrac{2x + 1}{\sqrt{7}} + C$

19. $-\ln|x - 1| - 2\ln(x^2 + 4) - \dfrac{1}{2}\tan^{-1}\dfrac{x}{2} + \dfrac{2}{x^2 + 4} + C$

Exercise Set 11.6

1. $\frac{4}{3}x^{3/4} - 2x^{1/2} + 4x^{1/4} - 4\ln\left(1 + x^{1/4}\right) + C$

3. $\frac{2}{7}(x + 2)^{7/2} - \frac{8}{5}(x + 2)^{5/2} + \frac{8}{3}(x + 2)^{3/2} + C$

5. $2 \tan^{-1} \sqrt{e^x - 1} + C$

7. $2x^{1/2} - 4x^{1/4} + 4 \ln (x^{1/4} + 1) + C$

9. $-\frac{8}{3}(2 - \sqrt{x + 1})^{3/2} + \frac{4}{5}(2 - \sqrt{x + 1})^{5/2} + C$

11. $2\sqrt{2t + 4} + 2\sqrt{t - 1} - 2\sqrt{6}\left(\tan^{-1} \sqrt{\dfrac{t + 2}{3}} + \tan^{-1} \sqrt{\dfrac{t - 1}{6}} \right) + C$

13. $-\sin^{-1} \dfrac{1 - x}{\sqrt{2x}} + C$ if $x > 0$, $\sin^{-1} \dfrac{1 - x}{\sqrt{2x}} + C$ if $x < 0$

15. $\dfrac{2}{\sqrt{3}} \tan^{-1} \left(\dfrac{2 \tan \dfrac{x}{2} - 1}{\sqrt{3}} \right) + C$

17. $\dfrac{1}{\sqrt{5}} \ln \left| \dfrac{2 \tan \dfrac{x}{2} + 1 - \sqrt{5}}{2 \tan \dfrac{x}{2} + 1 + \sqrt{5}} \right| + C$

19. $-\ln \left| 1 - \tan \dfrac{x}{2} \right| + C$ **21.** $-\frac{1}{3} \cot \frac{3}{2} x + C$

23. $\dfrac{1}{5} \ln \dfrac{\left| 3 \tan \dfrac{x}{2} + 1 \right|}{\left| \tan \dfrac{x}{2} - 3 \right|} + C$

25. $\dfrac{1}{2} \tan^{-1} \left(\tan \dfrac{x}{2} \right) - \dfrac{5}{6} \left(\dfrac{5}{3} \tan \dfrac{x}{2} + \dfrac{4}{3} \right) + C$

Exercise Set 12.1

1. $(y - 4)^2 = -12(x - 2)$ **3.** $(x - 2)^2 = 16(y + 3)$

5. $(x - 4)^2 = -8(y - 1)$ **7.** $(y - 2)^2 = -5(x - 1)$

9. $y = \frac{1}{6}x^2 + \frac{2}{3}x + \frac{11}{3}$, $x = \frac{2}{3}y^2 - 7y + 13$

11. (a) $(0, 0)$ (b) $(0, -\frac{5}{2})$ (c) $y = \frac{5}{2}$ (d) $(-5, -\frac{5}{2})$ and $(5, -\frac{5}{2})$

13. (a) $(2, 0)$ (b) $(\frac{1}{2}, 0)$ (c) $x = \frac{7}{2}$ (d) $(\frac{1}{2}, 3)$ and $(\frac{1}{2}, -3)$

15. (a) $(-3, 3)$ (b) $(-3, 5)$ (c) $y = 1$ (d) $(-7, 5)$ and $(1, 5)$

17. (a) $(-\frac{5}{2}, -5)$ (b) $(-1, -5)$ (c) $x = -4$
(d) $(-1, -2)$ and $(-1, -8)$

19. (a) $(2, 1)$ (b) $(\frac{7}{5}, 1)$ (c) $x = \frac{13}{5}$ (d) $(\frac{7}{5}, \frac{11}{5})$ and $(\frac{7}{5}, -\frac{1}{5})$

23. $x^2 + 2xy + y^2 - 16x + 16 = 0$

25. 20 ft from nearest end   **27.** $y = mx + \dfrac{p}{m}$

Exercise Set 12.2

1. $\dfrac{x^2}{36} + \dfrac{y^2}{20} = 1$ **3.** $\dfrac{(y + 1)^2}{41} + \dfrac{(x - 4)^2}{16} = 1$

5. $\dfrac{(x-1)^2}{36} + \dfrac{(y-2)^2}{20} = 1$

7. $\dfrac{(x+3)^2}{12} + \dfrac{(y+1)^2}{6} = 1$

9. $\dfrac{y^2}{20} + \dfrac{x^2}{5} = 1$

11. (a) $(\pm 3, 0)$ (b) $(\pm 5, 0)$
    (c) $(0, \pm 4)$
    (d) $(3, \pm \frac{16}{5}), (-3, \pm \frac{16}{5})$

13. (a) $(5, -2 \pm \frac{2}{3}\sqrt{5})$ (b) $(5, 0), (5, -4)$ (c) $(\frac{11}{3}, -2), (\frac{19}{3}, -2)$
    (d) $(\frac{37}{9}, -2 \pm \frac{2}{3}\sqrt{5}), (\frac{57}{9}, -2 \pm \frac{2}{3}\sqrt{5})$

15. (a) $(3, 2 \pm \frac{1}{5}\sqrt{7})$ (b) $(3, \frac{6}{5}), (3, \frac{14}{5})$ (c) $(\frac{12}{5}, 2), (\frac{18}{5}, 2)$
    (d) $(\frac{51}{20}, 2 \pm \frac{1}{5}\sqrt{7}), (\frac{69}{20}, 2 \pm \frac{1}{5}\sqrt{7})$

17. (a) $(0, 2 \pm 3\sqrt{2})$ (b) $(0, 2 \pm 3\sqrt{3})$ (c) $(-3, 2), (3, 2)$
    (d) $(-\sqrt{3}, 2 \pm 3\sqrt{2}), (\sqrt{3}, 2 \pm 3\sqrt{2})$

19. (a) $(1 \pm 2\sqrt{5}, 3)$ (b) $(-5, 3), (7, 3)$
    (c) $(1, 7), (1, -1)$ (d) $(1 \pm 2\sqrt{5}, \frac{17}{3}), (1 \pm 2\sqrt{5}, \frac{1}{3})$

21. (a) $(4 \pm \sqrt{7}, 1)$ (b) $(0, 1), (8, 1)$
    (c) $(4, -2), (4, 4)$ (d) $(4 \pm \sqrt{7}, \frac{13}{4}), (4, \pm \sqrt{7}, -\frac{5}{4})$

23. $y = \frac{2}{3}x - \frac{8}{3}$

31. $y \mp \dfrac{4}{\sqrt{29}} = \dfrac{1}{2}\left(x \pm \dfrac{50}{\sqrt{29}}\right)$

Exercise Set 12.3

1. $\dfrac{x^2}{16} - \dfrac{y^2}{20} = 1$

3. $\dfrac{(y+4)^2}{25} - \dfrac{(x-3)^2}{39} = 1$

5. $\dfrac{(y-7)^2}{15} - \dfrac{(x-1)^2}{10} = 1$

7. $\dfrac{(x+1)^2}{8} - \dfrac{(y-2)^2}{4} = 1$

9. $\dfrac{(y-2)^2}{4} - \dfrac{(x+5)^2}{5} = 1$

11. (a) $(\pm 6, 0)$ (b) $(\pm 3, 0)$ (c) $(\pm 6, 9), (\pm 6, -9)$ (d) $y = \pm\sqrt{3}x$

13. (a) $(1, 4 \pm 2\sqrt{13})$ (b) $(1, 0), (1, 8)$
    (c) $(-8, 4 \pm 2\sqrt{13}), (10, 4 \pm 2\sqrt{13})$ (d) $y - 4 = \pm\frac{2}{3}(x - 1)$

15. (a) $\left(5 \pm \dfrac{5\sqrt{85}}{42}, 4\right)$ (b) $(\frac{35}{6}, 4)(\frac{25}{6}, 4)$

    (c) $\left(5 \pm \dfrac{5\sqrt{85}}{42}, \dfrac{166}{49}\right), \left(5 \pm \dfrac{5\sqrt{85}}{45}, \dfrac{226}{49}\right)$ (d) $y - 4 = \pm\frac{6}{7}(x - 5)$

17. (a) $(\pm 2\sqrt{5}, 1)$ (b) $(\pm 4, 1)$ (c) $(\pm 2\sqrt{5}, 2), (\pm 2\sqrt{5}, 0)$
    (d) $y = 1 \pm \frac{1}{2}x$

19. (a) $(-3 \pm \sqrt{29}, 2)$ (b) $(-5, 2), (-1, 2)$
    (c) $(-3 \pm \sqrt{29}, \frac{29}{2}), (-3 \pm \sqrt{29}, -\frac{21}{2})$
    (d) $y - 2 = \pm\frac{5}{3}(x + 3)$

21. (a) $(2, 0), (2, 6)$ (b) $(2, 1), (2, 5)$
    (c) $(\frac{9}{2}, 0), (\frac{9}{2}, 6), (-\frac{1}{2}, 0), (-\frac{1}{2}, 6)$ (d) $y - 3 = \pm\dfrac{2}{\sqrt{5}}(x - 2)$

23. $y = \dfrac{\sqrt{10}}{5}(-2x + 7)$ **31.** $(9.7, 3.6)$

Exercise Set 12.4

1. (a) $(3, 4)$ (c) $(-3, 2)$ **3.** $4x'^2 - y'^2 + 10 = 0$

5. $2x'^2 + x'y' - y'^2 + 155 = 0$ **7.** $y' = x'^3 - 7x'$

9. (a) $\left(\dfrac{-2\sqrt{3} - 3}{2}, \dfrac{-2 + 3\sqrt{3}}{2} \right)$ (c) $(-2, -2\sqrt{3})$

11. $-\dfrac{5}{\sqrt{13}}x' - \dfrac{12}{\sqrt{13}}y' = 4$ **13.** $2x'^2 - 2\sqrt{2}x' + 3\sqrt{2}y' = 17$

15. $\dfrac{x'^2}{2} + \dfrac{y'^2}{22} = 1$

Exercise Set 12.5

1. $\dfrac{x'^2}{12/7} + \dfrac{y'^2}{12} = 1,\ \theta = 45°$ **3.** $\dfrac{x'^2}{2} + \dfrac{y'^2}{4} = 1,\ \theta = \cos^{-1}\dfrac{1}{\sqrt{10}}$

5. $\dfrac{(x' + 2\sqrt{10})^2}{4} - \dfrac{(y' + \sqrt{10})^2}{16} = 1,\ \theta = \cos^{-1}\dfrac{3}{\sqrt{10}}$

7. $(y' - 4\sqrt{5})^2 = -5,\ \theta = \cos^{-1}\dfrac{2}{\sqrt{5}}$

9. $(x' - 4\sqrt{3})^2 = 8(y' + 5),\ \theta = 60°$

Exercise Set 13.1

1. (a) 6.856 (c) 10.149 **3.** (a) 2.080 (c) 3.936

5. 1.328 **7.** 0.739 **9.** 1.310

Exercise Set 13.2

1. $\sin x = x - \dfrac{x^3}{3!} + \dfrac{(\cos c)x^5}{5!},\ c$ between 0 and x.

3. $x^4 = 16 + 32(x - 2) + 24(x - 2)^2 + 8(x - 2)^3 + (x - 2)^4$

5. $\ln x = (x - 1) - \dfrac{1}{2}(x - 1)^2 + \dfrac{1}{3}(x - 1)^3 - \dfrac{1}{4}(x - 1)^4 + \dfrac{1}{5}(x - 1)^5 -$

$\dfrac{1}{6c^6}(x - 1)^6,\ c$ between 1 and x.

7. $\sqrt{1 - x} = 1 - \dfrac{1}{2}x - \dfrac{1}{8}x^2 - \dfrac{1}{16}x^3 - \dfrac{5}{128}x^4 - \dfrac{7x^5}{256(1 - c)^{9/2}}$,

c between 0 and x.

9. $\tan^{-1}x = 1 - \dfrac{1}{3}x^3 + \dfrac{(c - c^3)x^4}{(1 + c^2)^4},\ c$ between 0 and x.

11. $\cosh x = 1 + \dfrac{x^2}{2!} + \dfrac{x^4}{4!} + \dfrac{(\sinh c)x^5}{5!},\ c$ between 0 and x.

13. 0.5150   **15.** 0.6164   **17.** 1.6487

19. 0.1823   **21.** 0.0415

23. (a) $|\text{error}| < \dfrac{1}{750}$ (b) $|\text{error}| < \dfrac{1}{375{,}000}$

Exercise Set 13.3

1. 5.3125   **3.** 0.7850   **5.** 1.9101

7. 0.6570   **9.** $\frac{5}{48}$ **11.** $\frac{1}{300}$

13. $\dfrac{\pi^3}{1728}$ **15.** $\frac{1}{750}$ **17.** less than $\dfrac{\sqrt{15}}{2} \cdot 10^{-3}$

Exercise Set 13.4

1. 0.7854   **3.** 1.9101   **5.** 0.6577   **7.** 1.2093

9. $\frac{1}{75,000}$ **11.** $3 \cdot 10^{-5}$ (assuming that $M = 50$)

13. less than $\sqrt{15} \cdot 10^{-2}$

Exercise Set 13.5

1. $\frac{1}{18}$ **3.** 0   **5.** $-\pi$ **7.** 0

9. 1   **11.** 0   **13.** 3   **15.** $-\frac{1}{2}$

17. $\frac{1}{3}$ **19.** 1

Exercise Set 13.6

1. 0   **3.** 0   **5.** 1   **7.** 1

9. 1   **11.** 0   **13.** 0   **15.** $-\frac{1}{2}$

17. 1   **19.** 0   **21.** 1   **23.** 1

Exercise Set 13.7

1. $\frac{2}{3}$ **3.** $\frac{1}{9}$ **5.** $-\dfrac{1}{\sqrt{2}}$ **7.** $\frac{1}{6}\ln 4$

9. doesn't exist   **11.** $2(1 - e^{-\sqrt{2}})$ **13.** $\frac{2}{9}\sqrt{2}$

15. $\frac{3}{2}$ **17.** doesn't exist   **19.** 0

21. doesn't exist   **23.** $\dfrac{s}{k^2 + s^2}$ **25.** π

29. volume $= \pi$

Exercise Set 14.1

1. 0   **3.** $\frac{27}{8}$ **5.** sequence diverges

7. 1   **9.** 1   **11.** 0

13. $\frac{1}{4}$ **15.** 1   **17.** 0

19. 1   **21.** 0   **23.** increasing, divergent

Exercise Set 14.2

    1. diverges   **3.** 0   **5.** diverges

    7. diverges   **9.** $1/(e-1)$ **11.** $\frac{1}{2}$

    13. $\displaystyle\sum_{n=1}^{+\infty} \frac{1}{(2n-1)(2n+1)}$ **15.** $\displaystyle\sum_{n=1}^{+\infty} \frac{2}{3^n}$

    17. $\displaystyle\sum_{n=1}^{+\infty} \frac{n+1-n^2}{(n+1)!}$

Exercise Set 14.3

    1. converges   **3.** diverges   **5.** converges

    7. diverges   **9.** converges   **11.** converges

    13. converges   **15.** diverges   **17.** converges

    19. diverges   **21.** converges

    23. (b) $1 + \ln 25.5 \leq \displaystyle\sum_{i=1}^{50} \frac{1}{i} \leq 1 + \ln 50$

Exercise Set 14.4

    1. converges   **3.** converges   **5.** diverges

    7. converges   **9.** converges   **11.** converges

    13. 0.632   **15.** 0.631

Exercise Set 14.5

    1. converges   **3.** converges   **5.** diverges

    7. converges   **9.** diverges   **11.** converges

    13. converges   **15.** converges   **17.** converges

    19. diverges   **21.** diverges   **23.** converges

    25. converges   **27.** converges   **29.** converges

Exercise Set 14.6

    1. $(-1, 1)$ **3.** $(-4, 0)$ **5.** $[-1, 1]$

    7. $[-1, 3)$ **9.** $[\frac{5}{3}, \frac{7}{3})$ **11.** $(-1, 3]$

    13. $(-10, 8)$ **15.** $\left(1 - \dfrac{1}{\sqrt{2}}, 1 + \dfrac{1}{\sqrt{2}}\right)$ **17.** $[-1, 1)$

    19. R

Exercise Set 14.7

    1. $x - \dfrac{x^3}{3!} + \dfrac{x^5}{5!} - + \cdots + \dfrac{(-1)^n x^{2n+1}}{(2n+1)!} + \cdots;\ R$

3. $1 - \dfrac{x^2}{2!} + \dfrac{x^4}{4!} - + \cdots + \dfrac{(-1)^n x^{2n}}{(2n)!} + \cdots; \ R$

5. $1 - \dfrac{x}{3 \cdot 1!} + \dfrac{x^2}{3^2 \cdot 2!} + \cdots + \left(-\dfrac{1}{3}\right)^n \dfrac{x^n}{n!} + \cdots; \ R$

7. $(x-1) - \dfrac{(x-1)^2}{2} + \dfrac{(x-1)^3}{3} - \dfrac{(x-1)^4}{4} + - \cdots; \ (0, 2]$

9. $2x - \dfrac{4}{3}x^3 + \dfrac{4}{15}x^5 - + \cdots + \dfrac{(-1)^n (2x)^{2n+1}}{(2n+1)!} + \cdots; \ R$

11. $\dfrac{1}{4} + \dfrac{x}{16} + \dfrac{x^2}{64} + \dfrac{x^3}{256} + \cdots; \ (-4, 4)$

Exercise Set 14.8

1. $1 + \dfrac{1}{2}x - \dfrac{1}{8}x^2 + \dfrac{1}{16}x^3 + \cdots + \dfrac{(1/2)(-1/2) \cdots ((3/2) - n)}{n!} x^n +$
$\cdots; \ [-1, 1]$

3. $1 + x + \dfrac{3}{4}x^2 + \cdots + \dfrac{(n+1)x^n}{2^n} + \cdots; \ (-2, 2)$

5. $x + \dfrac{1}{2}x^3 + \dfrac{3}{8}x^5 + \cdots + \dfrac{1 \cdot 3 \cdots (2n-1)}{2^n n!} x^{2n+1} + \cdots; \ (-1, 1)$

7. $1 - \dfrac{x^2}{3!} + \dfrac{x^4}{5!} + \cdots + \dfrac{(-1)^n x^{2n}}{(2n+1)!} + \cdots; \ R$

9. $x + \dfrac{1}{6}x^3 + \dfrac{3}{40}x^5 + \cdots + \dfrac{1 \cdot 3 \cdots (2n-1)}{2^n n!(2n+1)} x^{2n+1} + \cdots; \ [-1, 1]$

NOTE: The proof that this series represents, $\sin^{-1} x$ when $x = \pm 1$, is difficult.

11. $x^2 - \dfrac{1}{3}x^4 + \dfrac{2}{45}x^6 + \cdots + \dfrac{(-1)^{n+1} 2^{2n-1} x^{2n}}{(2n)!} + \cdots; \ R$

13. $x - \dfrac{x^3}{2 \cdot 3} + \dfrac{x^5}{4 \cdot 2! \cdot 5} + \cdots + \dfrac{(-1)^n x^{2n+1}}{2^n n!(2n+1)} + \cdots; \ R$

15. $\dfrac{2^{|m|} \, |m(m-1) \cdots (m-n)| \, |x|^{n+1}}{(n+1)!}$

19. (a) $1 + 2x + 3x^2 + 4x^3 + \cdots$

   (c) $x - \dfrac{x^2}{2} + \dfrac{x^3}{3} - \dfrac{x^4}{4} + - \cdots$

19. (e) $2x + \dfrac{2x^3}{3} + \dfrac{2x^5}{5} + \cdots + \dfrac{2x^{2n+1}}{2n+1} + \cdots$

21. (a) 0.6931

Exercise Set 15.1

1. $2\sqrt{2}; \ \frac{3}{4}\pi$ **3.** $5; \ \cos^{-1}\frac{3}{5} \approx 53.1°$

5. $\sqrt{29}; \ \pi + \cos^{-1}\dfrac{2}{\sqrt{29}} \approx 248.2°$

7. $(2, -4)$ **9.** $(9, -2)$

13. (a) $\langle 2, 10 \rangle$ (b) $\langle -8, -8 \rangle$
       (c) $2\sqrt{26}$ (d) $8\sqrt{2}$

15. (a) $\langle 3, 10 \rangle$ (b) $\langle 11, 6 \rangle$
       (c) $\sqrt{109}$ (d) $\sqrt{157}$

17. (a) $\langle 6, 8 \rangle$ (c) $\langle -25, -7 \rangle$ (e) $5\sqrt{13}$
       (g) $\sqrt{449}$ (i) $-\dfrac{5}{\sqrt{29}}, \dfrac{2}{\sqrt{9}}$

19. $m = -\frac{1}{5}, n = -\frac{8}{5}$

Exercise Set 15.2

1. -11 **3.** -10 **5.** $\frac{3}{4}\pi$

7. $\cos^{-1}\left(\dfrac{-2\sqrt{5}}{25}\right) \approx 100.3°$ **9.** (a) 6   (b) $-\frac{8}{3}$ **11.** 0

Exercise Set 15.3

1. (b) $x = \frac{3}{4}y^2$ (c) $y = \frac{1}{3}x + 1$
     (d) no horizontal tangents   (e) $(0, 0)$

3. (b) $y = \sqrt{\dfrac{8-x}{2}}$ (c) $x + 8y = 16$

3. (d) no horizontal tangents   (e) $(8, 0)$

5. (b) $(x - 3)^2 + \dfrac{(y - 2)^2}{16} = 1$ (c) $4\sqrt{3}x + y = 10 + 12\sqrt{3}$
     (d) $(3, 6), (3, -2)$ (e) $(2, 2), (4, 2)$

7. (b) $y = (1/x)$ (if $x > 0$)
     (c) $x + 4y = 4$ (d) no horizontal tangents   (e) no vertical tangents

9. (b) $y = \dfrac{x^2}{2}$ (if $|x| \leq 2$)   (c) $y = x - \dfrac{1}{2}$
     (d) $(0, 0)$ (e) no vertical tangents

11. (b) $x^2 + y^2 = 1$ (if $(x, y) \neq (0, -1)$)   (c) $4x + 3y = 5$
     (d) $(0, 1)$ (e) $(1, 0)$ and $(-1, 0)$

15. $x = t, y = 2 - \frac{1}{4}t^2$

17. $x = 1 + 4\cos t, y = -2 + 4\sin t$

Exercise Set 15.4

1. $4\sqrt{5}$ **3.** 12   **5.** $\frac{1}{27}(37\sqrt{37} - 1)$

7. $6a$ **9.** 8   **11.** π

13. $\ln(\sqrt{2} + 1)$ **15.** $2\sqrt{2} + 2\ln(1 + \sqrt{2})$

Exercise Set 15.5

1. $(-\infty, 3]$ **3.** $\{-1, 1\}$ **5.** $(1, 2]$

7. $\langle 3, \frac{27}{4}\rangle$; $\langle 2, 9\rangle$ **9.** $\langle 1, \frac{1}{2}\rangle$; $\langle 0, -\frac{1}{2}\rangle$

11. $\langle -\frac{1}{2}, \frac{1}{2}\rangle$; $\langle 0, 0\rangle$

Exercise Set 15.6

1. $\langle 0, 4\rangle$; $\langle 6, 2\rangle$; 4 **3.** $\langle \frac{5}{4}, \frac{3}{4}\rangle$; $\langle \frac{3}{4}, \frac{5}{4}\rangle$; $\frac{\sqrt{34}}{4}$

5. $\langle -2, 2\rangle$; $\langle 2, 0\rangle$ **7.** $\langle \frac{4}{3}, -2\rangle$; $\langle -\frac{16}{9}, 4\rangle$

9. $2\sqrt{2}$

11. (a) $\dfrac{v_o{}^2}{g}\sin^2\alpha$ (b) v_o (c) $\dfrac{v_o{}^2\sin^2\alpha}{2g}$

Exercise Set 16.1

1. $(-3, 0)$ **3.** $(-2, -2\sqrt{3})$ **5.** $(4, 0)$

7. $(-1, 1)$ **9.** $(2\sqrt{2}, \frac{1}{4}\pi)$ **11.** $(2, \frac{4}{3}\pi)$

13. $(4, \pi)$ **15.** $(2\sqrt{2}, \frac{11}{6}\pi)$ **17.** $x^2 + y^2 = 16$

19. $(x - \frac{3}{2})^2 + y^2 = \frac{9}{4}$ **21.** $xy = 1$

23. $(x - \frac{3}{2})^2 + (y - 1)^2 = \frac{13}{4}$

25. $(x^2 + y^2)^2 - 2ax(x^2 + y^2) - a^2y^2 = 0$ **27.** $r = 2\sec\theta$

29. $2r\cos\theta + 3r\sin\theta = 5$

31. $r^2\sin 2\theta = 2a$ **33.** $r = 2(\cos\theta - 2\sin\theta)$

37. $r\sin\theta - r_1\sin\theta_1 =$

$\dfrac{r_2\sin\theta_2 - r_1\sin\theta_1}{r_2\cos\theta_2 - r_1\cos\theta_1}(r\cos\theta - r_1\cos\theta_1)$

Exercise Set 16.2

25. $-\frac{1}{2}$

Exercise Set 16.3

9. $r = \dfrac{-9}{4 - 3\cos\theta}$ **11.** $r = \dfrac{3}{1 - 3\sin\theta}$

Exercise Set 16.4

1. $(1, \frac{\pi}{6})$, $(1, \frac{5}{6}\pi)$ **3.** $(2, 0)$, $(2, \pi)$

5. $(2 - \sqrt{2}, \frac{\pi}{4})$, $(2 + \sqrt{2}, \frac{3}{4}\pi)$, $(2 + \sqrt{2}, \frac{5}{4}\pi)$, $(2 - \sqrt{2}, \frac{7}{4}\pi)$

7. pole, $(\frac{1}{2}, \frac{7}{6}\pi)$, $(\frac{1}{2}, \frac{11}{6}\pi)$, $(0, 1)$, $(\pi, 1)$,

$\left(\dfrac{-5 + \sqrt{17}}{4}, \sin^{-1}\dfrac{-1 + \sqrt{17}}{4}\right)$, $\left(\dfrac{-5 + \sqrt{17}}{4}, \pi - \sin^{-1}\dfrac{-1 + \sqrt{17}}{4}\right)$

9. pole, $\left(\dfrac{\sqrt{2}}{2}, \dfrac{n\pi}{8}\right)$ where n is any odd integer.

Exercise Set 16.5

1. $\frac{1}{24}\pi^3$ **3.** $a^2\pi/16$ **5.** $\frac{8}{3}$

7. 4π **9.** 6π **11.** $\frac{9}{2}\pi$

13. a^2 **15.** $\frac{1}{8}\pi$ **17.** $(2\pi - 3\sqrt{3})/2$

19. $(a^2(\pi + 4))/2$ **21.** $(a^2(5\pi - 8))/4$ **23.** $(a^2(\pi - 2))/8$

25. $(2\pi - 3\sqrt{3})/4$ **27.** 16 **29.** $\frac{3}{2}\pi$

Exercise Set 17.1

1. (b) $3\sqrt{6}$ (c) $(-\frac{1}{2}, \frac{3}{2}, 1)$

3. (b) $\sqrt{89}$ (c) $(-1, \frac{1}{2}, 4)$

5. $(-3, 5, 6)$; 9

7. $\left(\frac{1}{2}, \frac{5}{2}, -3\right)$; $\dfrac{\sqrt{58}}{2}$ **9.** $3x + 4y - 5z + 3 = 0$

Exercise Set 17.2

1. (a) $\sqrt{26}$ (b) $\dfrac{1}{\sqrt{26}}, \dfrac{-5}{\sqrt{26}}, 0$ (c) $78.7°$, $168.7°$, $90°$

3. (a) $\sqrt{29}$ (b) $\dfrac{4}{\sqrt{29}}, \dfrac{2}{\sqrt{29}}, \dfrac{-3}{\sqrt{29}}$ (c) $42.0°$, $68.2°$, $123.9°$

5. $(1, 1, 1)$ **7.** $(2, 4, -3)$

9. $\langle 10, -18, 6 \rangle$ **11.** $\langle 13, -15, -36 \rangle$

13. $\langle -124, 184, -290 \rangle$ **15.** $\dfrac{3}{5\sqrt{2}}, -\dfrac{1}{\sqrt{2}}, \dfrac{4}{5\sqrt{2}}$

17. $\dfrac{-12}{\sqrt{29}}$ **19.** $139.8°$

Exercise Set 17.3

1. $\langle -6, -5, 3 \rangle$ **3.** $\langle -16, -4, 8 \rangle$ **5.** -14

7. $\langle -10, -6, 12 \rangle$ **9.** $\langle 2, -3, -1 \rangle$

11. $\dfrac{7}{3\sqrt{6}}, \dfrac{1}{3\sqrt{6}}, \dfrac{2}{3\sqrt{6}}$ **13.** $5\sqrt{13}$

15. $\frac{8}{3}$

Exercise Set 17.4

1. $x = -3 - t, \; y = 1 + 2t, \; z = 5 + 4t$

3. $x = 3 - 5t, \; y = 5 + t, \; z = 1 + 3t$

5. $x = 1 + t, \; y = 2 - t, \; z = -1 + \frac{4}{3}t$

7. $\dfrac{69}{\sqrt{509}}$ **9.** no intersection

11. $x = -3 - 13s,\ y = -1 - 5s,\ z = 2 + 4s$

13. $\dfrac{\sqrt{470}}{5}$

Exercise Set 17.5

1. $5x - 2y + z = -9$

3. $-11x + 3y + 4z = -5$

5. $-3x + 2y + 11z = 25$

7. $7x + 10y + 8z = 22$

9. $4x - 13y + 5z = -5$

11. $\left(\frac{45}{14}, \frac{3}{7}, -\frac{17}{14}\right)$

13. $x = t,\ y = -14 + 5t,\ z = 10 - 3t$

15. $x = \frac{3}{2} + 2t,\ y = \frac{7}{4} - 7t,\ z = \frac{3}{4} + 5t$

Exercise Set 17.6

13. $y^2 + z^2 = (\ln x)^2$

15. $4x^2 + y^2 + 4z^2 = 36$

17. $x^2 - 9y^2 - 9z^2 = 9$

Exercise Set 17.8

1. (a) $\mathbf{P}'(t) = \langle 2, -4, 2t \rangle;\ \mathbf{P}'(2) = \langle 2, -4, 4 \rangle$

$\mathbf{P}''(t) = \langle 0, 0, 2 \rangle;\ \mathbf{P}''(2) = \langle 0, 0, 2 \rangle$

(b) $x = 5 + 2t,\ y = -5 - 4t,\ z = 4 + 4t$

3. (a) $\mathbf{P}'(t) = e^{-2t}\langle \cos t - 2 \sin t, -\sin t - 2 \cos t, -2 \rangle$

$\mathbf{P}'(\frac{1}{2}\pi) = e^{-\pi}\langle -2, -1, -2 \rangle$

$\mathbf{P}''(t) = e^{-2t}\langle 3 \sin t - 4 \cos t, 3 \cos t + 4 \sin t, 4 \rangle$

$\mathbf{P}''(\frac{1}{2}\pi) = e^{-\pi}\langle 3, 4, 4 \rangle$

3. (b) $x = e^{-\pi} - 2e^{-\pi}t,\ y = -e^{-\pi}t,\ z = e^{-\pi} - 2e^{-\pi}t$

11. $\frac{20}{3}$

13. $2\sqrt{10}\pi$

15. $2\sqrt{3}$

17. $x = t,\ y = 2t,\ z = 4 - 5t^2$

19. $x = t,\ y = \sqrt{4 - t^2},\ z = \sqrt{4 - t^2}$

Exercise Set 17.9

1. (a) $\langle \frac{1}{3}, \frac{2}{3}, \frac{2}{3} \rangle$ (b) $\langle -\frac{2}{3}, \frac{2}{3}, -\frac{1}{3} \rangle$ (c) $\frac{1}{18}$

(d) 4 (e) 2

3. (a) $\left\langle -\dfrac{2}{\sqrt{17}}, \dfrac{3}{\sqrt{17}}, -\dfrac{2}{\sqrt{17}} \right\rangle$ (b) $\left\langle -\dfrac{4}{\sqrt{221}}, \dfrac{6}{\sqrt{221}}, \dfrac{13}{\sqrt{221}} \right\rangle$

(c) $\dfrac{2\sqrt{221}}{289}$ (d) $-\dfrac{4}{\sqrt{17}}$ (e) $\dfrac{2\sqrt{221}}{17}$

5. (a) $\langle -\frac{3}{5}, -\frac{4}{5}, 0 \rangle$ (b) $\langle \frac{4}{5}, -\frac{3}{5}, 0 \rangle$

(c) $\dfrac{24\sqrt{2}}{125}$ (d) $\dfrac{7\sqrt{2}}{10}$ (e) $\dfrac{12\sqrt{2}}{5}$

7. (a) $\dfrac{\langle e^{-\pi}, -e^{-\pi}, 1\rangle}{\sqrt{2e^{-2\pi}+1}}$ (b) $\dfrac{\langle -e^{-2\pi}, e^{-2\pi}+1, e^{-\pi}\rangle}{\sqrt{2e^{-4\pi}+3e^{-2\pi}+1}}$

(c) $\dfrac{2e^{-\pi}\sqrt{e^{-2\pi}+1}}{(e^{-2\pi}+2)^{3/2}}$ (d) $\dfrac{-2e^{-2\pi}}{\sqrt{2e^{-2\pi}+1}}$

(e) $\dfrac{2e^{-\pi}\sqrt{2e^{-4\pi}+3e^{-\pi}+1}}{2e^{-2\pi}+1}$

11. points where $t = n\pi/4$, n being any odd integer.

13. $\dfrac{4}{5\sqrt{5}}$ **15.** $\dfrac{1}{|a|}\operatorname{sech}^2 \dfrac{x}{a}$

17. $\dfrac{275\sqrt{2}}{4}$ lb

Exercise Set 17.10

1. (a) $(\sqrt{3}, 1, -4)$ (c) $\left(\dfrac{6}{5}, -\dfrac{3\sqrt{21}}{5}, 1\right)$

2. (a) $(3\sqrt{2}, \tfrac{3}{4}\pi, 5)$ (c) $(2\sqrt{5}, \pi - \tan^{-1}\tfrac{1}{2}, -3)$

3. (a) $\left(-\dfrac{\sqrt{6}}{2}, \dfrac{\sqrt{6}}{2}, 1\right)$ (c) $\left(\dfrac{2\sqrt{10}}{3}, -\dfrac{4\sqrt{2}}{3}, -1\right)$

4. (a) $(4, \tfrac{5}{6}\pi, \tfrac{5}{3}\pi)$ (c) $(5\sqrt{2}, \tfrac{1}{4}\pi, \pi + \tan^{-1}\tfrac{4}{3})$

5. (a) $\left(2\sqrt{5}, \cos^{-1}\left(-\dfrac{2}{\sqrt{5}}\right), \dfrac{1}{6}\pi\right)$ (c) $\left(5, \cos^{-1}\left(-\dfrac{4}{5}\right), \dfrac{2}{3}\pi\right)$

6. (a) $(\sqrt{3}, \tfrac{3}{4}\pi, 1)$ (c) $(\tfrac{12}{5}, \tfrac{1}{3}\pi, -\tfrac{6}{5})$

7. (a) $r = \dfrac{6+4z}{\cos\theta + 2\sin\theta}$

(b) $\rho = \dfrac{6}{\sin\phi\cos\theta + 2\sin\phi\sin\theta - 4\cos\phi}$

9. (a) $r^2 = 3z^2$ (b) $(\phi - \tfrac{\pi}{3})(\phi - \tfrac{2\pi}{3}) = 0$

11. (a) $z = 2r\cos\theta - r^2$ (b) $\rho\sin^2\phi - 2\sin\phi\cos\theta + \cos\phi = 0$

13. (a) $r^2(1 + \sin^2\theta) = 4(4 - z^2)$ (b) $\rho^2[\sin^2\phi(1 + 2\sin^2\theta) + 4\cos^2\phi] = 16$

15. (a) $x^2 + y^2 = 9$ (b) $\rho\sin\phi = 3$

17. (a) $x^2 + (y-1)^2 = 1$ (b) $\rho = \dfrac{2\sin\theta}{\sin\phi}$

19. (a) $z = 4xy$ (b) $\rho = \dfrac{\cos\phi}{2\sin^2\phi\sin 2\theta}$

21. (a) $y = -\sqrt{3}x$ (b) $\theta = \tfrac{2}{3}\pi$

23. (a) $x^2 + y^2 + (z-2)^2 = 4$ (b) $r^2 = 4z - z^2$

Exercise Set 18.1

1. (a) $-\tfrac{3}{4}$ (b) $\tfrac{9}{16}$ (c) $3/y^2$

(d) $\dfrac{3(4 + h)}{4(2 + h)^2}$

3. (a) -4 (b) $\frac{13}{5}$ (c) $\dfrac{2x + h}{z}$

(d) $\dfrac{4}{z(\sqrt{y + h} + \sqrt{y})}$ (e) $\dfrac{4}{-3 + h}$

5. (a) -36 (b) $-27 - 2h$

7. $\mathcal{D}_f = R^2; \; \mathcal{R}_f = R$

9. $\mathcal{D}_f = \{(x, y) : 9x^2 + 16y^2 \le 25\}; \; \mathcal{R}_f = [0, 5]$

11. $\mathcal{D}_f = \{(x, y) : x > 2y \text{ and } x \ne -y\}, \; \mathcal{R}_f = R$

13. $\mathcal{D}_f = \{(x, y) : x \le 0 \text{ and } x < 3y\} \cup \{(x, y) : x \ge 0 \text{ and } x > 3y\}$
$\mathcal{R}_f = [0, +\infty)$

15. $\mathcal{D}_f = \{(x, y, z) : z \ge 0 \text{ and } z > -y\} \cup \{(x, y, z) : z \le 0 \text{ and } z < -y\}, \; \mathcal{R}_f = [0, \pi]$

Exercise Set 18.2

1. $\delta = \epsilon/7$ 3. $\delta = \sqrt{\epsilon/3}$

5. $\delta = $ smaller of 1 and $\epsilon/12$

Exercise Set 18.3

1. 30   3. $\sqrt{15}$

5. doesn't exist   7. 1

9. doesn't exist   11. doesn't exist

Exercise Set 18.4

1. $\{(x, y) : x \ge 2y\}$

3. $\{(x, y) : x > y \text{ and } x < -y\} \cup \{(x, y) : x < y \text{ and } x > -y\}$

5. $\left\{(x, y) : -\dfrac{x + 2}{2} \le y < \dfrac{x}{4}\right\} \cup \left\{(x, y) : \dfrac{x}{4} < y \le -\dfrac{x + 2}{2}\right\}$

7. $\{(x, y) : y \le 2x, \; -1 \le x + y \le 1, \text{ and } x + y \ne 0\}$

9. continuous   11. continuous   13. discontinuous

15. discontinuous   17. continuous

Exercise Set 18.5

1. $6xy^4 - 5y^3; \; 12x^2y^3 - 15xy^2$

3. $-3y \sin^2 e^{-xy} \cos e^{-xy}; \; -3x \sin^2 e^{-xy} \cos e^{-xy}$

5. $\dfrac{2y}{x^2 - y^2}; \; \dfrac{-2x}{x^2 - y^2}$

7. $\dfrac{-xy}{|x|(x^2 + y^2)}; \; \dfrac{|x|}{x^2 + y^2}$

9. $-\sqrt{1 + x^3}; \ \sqrt{1 + y^3}$

11. $D_1 f(x, y, z) = 3x^2 y + 2y^2 z; \ D_2 f(x, y, z) = x^3 - 4xyz - 3z^3;$
$D_3 f(x, y, z) = 2xy^2 - 9yz^2$

13. $D_1 f(u, v, w) = v \tan^{-1} \dfrac{w}{uv} - \dfrac{uv^2 w}{u^2 v^2 + w^2};$

$D_2 f(u, v, w) = u \tan^{-1} \dfrac{w}{uv} - \dfrac{u^2 vw}{u^2 v^2 + w^2}$

$D_3 f(u, v, w) = \dfrac{u^2 v^2}{u^2 v^2 + w^2}$

15. $D_1 f(x, y, z) = \dfrac{y^2 + z^2}{(x^2 + y^2 + z^2)^{3/2}};$

$D_2 f(x, y, z) = \dfrac{-xy}{(x^2 + y^2 + z^2)^{3/2}};$

$D_3 f(x, y, z) = \dfrac{-xz}{(x^2 + y^2 + z^2)^{3/2}}$

17. $\left\langle \dfrac{x}{\sqrt{x^2 + 2y^2}}, \dfrac{2y}{\sqrt{x^2 + 2y^2}} \right\rangle; \ \left\langle \dfrac{1}{3}, -\dfrac{4}{3} \right\rangle$

19. $\langle y^2 - 2xz, 2xy + 2z^2, 4yz - x^2 \rangle; \ \langle 5, 6, 7 \rangle$

21. -4 23. $\frac{2}{9}, \frac{8}{27}$ 25. $0, 0$

Exercise Set 18.6

1. $\eta_1(h, k) = \eta_2(h, k) = h$

Exercise Set 18.7

1. (a) $2y^2 \, dx + 4xy \, dy + 2x \, (dy)^2 + 4y \, dx \, dy + 2 \, dx \, (dy)^2;$
$2y^2 \, dx + 4xy \, dy$
   (b) $-0.792218; \ -0.80$

3. (a) $e^{-xy}(e^{-x \, dy - y \, dx - dx \, dy} - 1); \ e^{-xy}(-y \, dx - x \, dy)$
   (b) $0.0052; \ 0.0049$

5. 4.998

7. $\frac{40}{3}\pi$ in.^{3}

9. 112.5 in.^{3}

Exercise Set 18.8

1. $8t + 12t^2 - 4t^3$ 3. $\frac{9}{8}t^8$

5. $\dfrac{18s - 5t}{(2s + t)(3s - 2t)}, \dfrac{2s - 6t}{(2s + t)(3s - 2t)}$

7. $0, 0, 2 \cos 2\theta$ 9. $-e^{-s}\sqrt{2}, 0$

11. $\dfrac{7\sqrt{19}}{38}$ 13. $0, 0, -1$

15. $\dfrac{y(y^2 + 2x^2)\cos t + x(x^2 + 2y^2)\sin t}{\sqrt{x^2 + y^2}}$

17. $-3,\ -2v,\ -1,\ -2v$

21. $\dfrac{66}{\sqrt{145}}$ ft/sec

23. increasing at 17.2π in.^{2}/sec

Exercise Set 18.9

1. $11\sqrt{2}$

3. $\dfrac{-9\sqrt{7} - 14}{28}$

5. $\dfrac{-1}{6\sqrt{26}}$

7. direction of $\langle -\tfrac{4}{5}, \tfrac{3}{5} \rangle$ or $\langle \tfrac{4}{5}, -\tfrac{3}{5} \rangle$

11. (a) and (b) $\dfrac{-11\sqrt{5}}{120}$ (c) $\dfrac{\sqrt{97}}{48}$ (d) $\left\langle \dfrac{4}{97}, \dfrac{-9}{\sqrt{97}} \right\rangle$

13. (a) and (b) $\dfrac{-80}{\sqrt{13}}$ (c) $4\sqrt{37}$ (d) $\left\langle \dfrac{-1}{\sqrt{37}}, \dfrac{6}{\sqrt{37}} \right\rangle$

15. 0

Exercise Set 18.10

1. $4x - 3y - z + 5 = 0;\ x = -1 + 4t,\ y = 2 - 3t,\ z = -5 - t$

3. $-3x + 2y - 4z - 15 = 0;\ x = -3 - 3t,\ y = 1 + 2t,\ z = -1 - 4t$

5. $-y + \sqrt{3}z - \dfrac{5\pi\sqrt{3}}{3} = 0;\ x = 1,\ y = -t,\ z = \dfrac{5}{3}\pi + \sqrt{3}t$

13. $\dfrac{\partial F}{\partial x}\dfrac{\partial G}{\partial x} + \dfrac{\partial F}{\partial y}\dfrac{\partial G}{\partial y} + \dfrac{\partial F}{\partial z}\dfrac{\partial G}{\partial z} = 0$

15. $x = 1 + 2t,\ y = -2 + 5t,\ z = 2 + 4t$

17. $x = -2 + 3t,\ y = 1 + 3t,\ z = 3 - t$

Exercise Set 18.11

1. $6xy^2;\ 6x^2y + 12y^3;\ 6x^2y + 12y^3;\ 2x^3 + 36xy^2 - 6y$

3. $\dfrac{4x^{10} + 12x^2y^8}{(y^8 - x^8)^{3/2}};\ \dfrac{-16x^3y^7}{(y^8 - x^8)^{3/2}};\ \dfrac{-16x^3y^7}{(y^8 - x^8)^{3/2}};\ \dfrac{4x^4(5y^8 - x^8)}{y^2(y^8 - x^8)^{3/2}}$

5. $-24xy(x^2 - y^2 + 2z^2);\ 48xz(x^2 - y^2 + 2z^2);$
$6(x^2 - y^2 + 2z^2)(-x^2 + 5y^2 - 2z^2);\ -48yz(x^2 - y^2 + 2z^2)$

7. $-z^2 \sin xz \sec yz;$

$-xz \sin xz \sec yz + yz \cos xz \sec yz \tan yz + \cos xz \sec yz;$

$y^2 \sin xz \sec^3 yz + y^2 \sin xz \sec yz \tan^2 yz +$
$2xy \cos xz \sec yz \tan yz - x^2 \sin xz \sec yz;$

$yz \sin xz \sec^3 yz + yz \sin xz \sec yz \tan^2 yz +$
$\sin xz \sec yz \tan yz + xz \cos xz \sec yz \tan yz$

9. $6y^2;\ 0;\ 0$

19. $4t^2 \dfrac{\partial^2 F}{\partial x^2} + 12t^3 \dfrac{\partial^2 F}{\partial x \, \partial y} + 9t^4 \dfrac{\partial^2 F}{\partial y^2} + 2 \dfrac{\partial F}{\partial x} + 6t \dfrac{\partial F}{\partial y}$

21. (a) $4u^2 \dfrac{\partial^2 F}{\partial x^2} + 8uv \dfrac{\partial^2 F}{\partial x \, \partial y} + 4v^2 \dfrac{\partial^2 F}{\partial y^2} + \dfrac{2 \, \partial F}{\partial x}$

(b) $-4uv \dfrac{\partial^2 F}{\partial x^2} + (4u^2 - 4v^2) \dfrac{\partial^2 F}{\partial x \, \partial y} + 4uv \dfrac{\partial^2 F}{\partial y^2} + 2 \dfrac{\partial F}{\partial y}$

(c) $4v^2 \dfrac{\partial^2 F}{\partial x^2} - 8uv \dfrac{\partial^2 F}{\partial x \, \partial y} + 4u^2 \dfrac{\partial^2 F}{\partial y^2} - 2 \dfrac{\partial F}{\partial x}$

Exercise Set 18.12

1. $f(4, -4) = -18$, absolute minimum; no absolute maximum

3. no absolute extrema

5. $f(6, 2) = 22$, absolute maximum; no absolute minimum

7. $f(3, 3) = -25$, minimum value; $f(4, 0) = 50$, maximum value

9. $f(-\frac{1}{2}, 0) = \frac{1}{2}$, minimum value; $f(\frac{1}{2}, \pm\sqrt{3}/2) = \frac{11}{2}$, maximum value

11. $N/3, N/3, N/3$

13. $\sqrt{3}$

15. $450 per television set; $50 per service policy

17. $8\sqrt{3}abc/9$

19. plane $6x + 3y + 2z = 18$

21. line $y = \frac{29}{35}x - \frac{1}{35}$

Exercise Set 19.1

1. 70   **3.** 21   **5.** $15(2 - \sqrt{2})/2$

7. $\frac{4}{3}$ **9.** $2a^4/15$

11. $\dfrac{2h(3a^2 + 2h^2)}{3(a^2 + h^2)^{3/2}}$

Exercise Set 19.2

1. 180   **3.** $\frac{2}{3}a^3$ **5.** 6

7. $\frac{27}{2}\pi$ **9.** $\frac{63}{2} \sin^{-1} \frac{2}{3} + 17\sqrt{5}$

Exercise Set 19.3

1. e^{-1} **3.** $-\frac{2}{3}$ **5.** $8(2\sqrt{2} - 1)/3$

7. 9   **9.** $\frac{61}{105}$ **11.** $(4 - \sqrt{3})/8$

13. $e - 1$

Exercise Set 19.4

1. $(\frac{7}{4}, \frac{13}{4})$ **3.** $(0, 4r/3\pi)$ **5.** on the x axis a distance $\frac{3}{8}\pi$ from the origin.

7. $(\frac{17}{5}, -2)$ **9.** $(\frac{8}{5}, \frac{6}{5})$ **11.** $(\frac{20}{9}, \frac{16}{7})$

13. $(3\pi a/16, 3\pi a/16)$ **19.** $312\sqrt{10}\pi/5$ **23.** $(0, 2a/\pi)$

Exercise Set 19.5

1. $\pi a^5/20$ **3.** $\pi(3 - \sqrt{5})$ **5.** $\frac{9}{16}\pi$

7. $5\pi a^2/4$ **9.** $2\pi a^3(8 - 3\sqrt{3})/3$ **11.** $\frac{4}{3}\pi$

13. $(\frac{5}{6}, 0)$ **15.** $(\frac{3}{10}a, \frac{3}{10}a)$

Exercise Set 19.6

1. 4   **3.** $\frac{3}{4}$ **5.** 0

7. $\frac{1}{24}a^4$ **9.** $36\sqrt{3}/35$

11. $\displaystyle\int_0^4 \int_{-1/2\sqrt{16-z^2}}^{1/2\sqrt{16-z^2}} \int_{z/2}^{\sqrt{4-x^2}} f(x, y, z)\, dy\, dx\, dz$

11. (b) $\displaystyle\int_0^2 \int_0^{2y} \int_{\sqrt{4-y^2}}^{\sqrt{4-y^2}} f(x, y, z)\, dx\, dz\, dy$

13. $\dfrac{9}{2}$ **15.** $\left(\dfrac{a}{4}, \dfrac{b}{4}, \dfrac{c}{4}\right)$ **17.** $\dfrac{88}{315}k;\ \sqrt{\dfrac{11}{21}}$

19. $(0, 0, \frac{8}{7})$ **21.** $\frac{128}{21}m$

Exercise Set 19.7

1. $(0, 0, \frac{25}{7})$ **3.** $\frac{4}{3}\pi a^3$

5. $\frac{1}{2}ma^2$ **(b)** $\frac{3}{2}ma^2$

7. on the axis a distance $\frac{2}{5}a$ from the center

9. on the axis a distance $\frac{4}{5}h$ from the vertex

11. $\dfrac{ka^4\pi(8\pi - 9\sqrt{3})}{6}$ **13.** $\left\langle 0, 0, \pi G\mu k \ln\left(\dfrac{b}{a}\right)\right\rangle$

Exercise Set 19.8

1. $\frac{4}{3}$ **3.** -1 **5.** 0

7. $\frac{32}{5}$ **9.** $k(5 - \sqrt{5})/5$ **11.** $\frac{4}{9}$

Exercise Set 19.9

1. -4 **3.** $-\frac{1}{2}\pi$ **5.** $2 - \dfrac{\sqrt{3}}{2}$

7. does not exist

11. $-\frac{1}{2}\pi$ **13.** $\ln 2/8 - e^{-1}$ **21.** -8

Exercise Set 20.2

1. $f_0(x) = 1;$

$\qquad f_1(x) = 1 + x + \dfrac{x^2}{2};$

$$f_2(x) = 1 + x + x^2 + \frac{x^3}{6};$$

$$f_3(x) = 1 + x + x^2 + \frac{x^3}{3} + \frac{x^4}{24}$$

3. $f_0(x) = 0;$

$f_1(x) = x - 1;$

$$f_2(x) = \frac{x^3}{3} - \frac{x^2}{2} + x - \frac{5}{6};$$

$$f_3(x) = \frac{x^5}{15} - \frac{x^4}{8} + \frac{x^3}{3} - \frac{5x^2}{12} + x - \frac{103}{120}$$

5. $f_0(x) = 0;$

$f_1(x) = x;$

$$f_2(x) = x + \frac{x^3}{3};$$

$$f_3(x) = x + \frac{x^3}{3} + \frac{2x^5}{15} + \frac{x^7}{63}$$

Exercise Set 20.3

1. $4\sqrt{y} = y^2 + C$

3. $\sin^{-1} x + \cot y = C$

5. $y = \ln(C - e^{-x})$

7. $x\sqrt{5 - 2y} = C$

9. $y = \dfrac{C}{1 - \cos x}$

11. $y = x - \dfrac{2}{C_1} \ln|C_1 x + 1| + C_2$

13. $\left| \dfrac{x\sqrt{3} + 2y + x}{x\sqrt{3} - 2y + x} \right|^{1/\sqrt{3}} = Cx^2$

15. $e^{-x^2/2y^2} = Cxy$

17. $2\tan^{-1}\left(\dfrac{5y - 7}{5x - 1}\right) - \ln\left(x^2 - \dfrac{2}{5}x + y^2 - \dfrac{14}{5}y + 2\right)^{1/2} = C$

19. $4\sqrt{y} = x^2$

21. $\sin^{-1} x + \cot y = \frac{1}{6}\pi + 1$

23. $y = \ln(2 - e^{-x})$

25. $5x^2 - 2x^2 y = 3$

27. $y = \dfrac{1}{1 - \cos x}$

29. $\left| \dfrac{x\sqrt{3} + 2y + x}{x\sqrt{3} - 2y + x} \right|^{1/\sqrt{3}} = x^2 \left(\dfrac{\sqrt{3} - 1}{3 + \sqrt{3}}\right)^{1/\sqrt{3}}$

31. $2e^{y^2 - 4x^2/8y^2} = xy$

33. $2\tan^{-1}\left(\dfrac{5y - 7}{5x - 1}\right) - \ln\left(x^2 - \dfrac{2}{5}x + y^2 - \dfrac{14}{5}y + 2\right)^{1/2} =$

$2\tan^{-1}\dfrac{13}{14} - \dfrac{1}{2}\ln\dfrac{73}{5}$

35. $I = \dfrac{E}{R}(1 - e^{-(R/L)t})$

37. $2x^2 + y^2 = K$

39. $(x - K)^2 + y^2 = K^2$

41. approx. 6 days

Exercise Set 20.4

1. $x^2 y = C$

3. $\frac{3}{2}x^2 - 2xy - 4x + \frac{5}{2}y^2 + y = C$

5. $\frac{1}{3}x^3 + x \ln y = C$

7. $x^2 + x \tan y - xy + \frac{1}{2}y^2 = C$

9. $\dfrac{x^2 \tan^{-1} y}{2} - \dfrac{x^2}{2} + e^y = C$

13. $y = 2x \ln x + Cx$

15. $x - \dfrac{2y}{x} + \dfrac{y^2}{2} = C$

17. $\dfrac{x}{y} + y = C$

19. $y = \frac{3}{4} + Ce^{-4x}$

21. $y = (2x^{5/2}/3) + Cx$

23. $y = \cos x \sin x + C \cos x$

25. $y = \dfrac{x}{2} - \dfrac{1}{2\sqrt{1+x^2}} \ln \left(x + \sqrt{1+x^2} \right) + \dfrac{C}{\sqrt{1+x^2}}$

27. $I = \dfrac{E_0}{L^2\omega^2 + R^2}(R \sin \omega t - \omega L \cos \omega t) +$

$\left(I_0 + \dfrac{E_0 \omega L}{L^2\omega^2 + R^2} \right) e^{-(R/L)t}$

Exercise Set 20.5

1. $y = C_1 e^{4x} + C_2 e^{-2x}$

3. $y = e^x(C_1 \cos \sqrt{3}x + C_2 \sin \sqrt{3}x)$

5. $y = (C_1 + C_2 x)e^{4x}$

7. $y = C_1 e^{\sqrt{5}x} + C_2 e^{-\sqrt{5}x}$

9. $y = C_1 + C_2 e^{-4x}$

11. $y = C_1 e^{(-1+\sqrt{13}/2)x} + C_2 e^{(-1-\sqrt{13}/2)x}$

13. $y = e^{\sqrt{5}x}(C_1 \cos x + C_2 \sin x)$

15. $y = (C_1 + C_2 x)e^{-2\sqrt{3}x}$

17. $y = -\frac{1}{2}e^{4x} - \frac{3}{2}e^{-2x}$

19. $y = e^{\sqrt{5}x}[\cos x + (2 - \sqrt{5}) \sin x]$

21. $y = (1 - 5x)e^{4x}$

Exercise Set 20.6

1. $y = C_1 e^{-2x} + C_2 e^x - \frac{3}{2}$

3. $y = e^{-3x}(C_1 + C_2 x) + \frac{1}{16}e^x$

5. $y = C_1 e^{(-1+\sqrt{13}/2)x} + C_2 e^{(-1-\sqrt{13}/2)x} + \frac{2}{9} - \frac{1}{3}x$

7. $y = e^x(C_1 \cos 2x + C_2 \sin 2x) - \frac{1}{13} \cos 3x - \frac{3}{26} \sin 3x$

9. $y = C_1 \cos 2x + C_2 \sin 2x + \frac{3}{4}x \sin 2x$

11. $y = \frac{1}{6}e^{-2x} + \frac{7}{3}e^x - \frac{3}{2}$

13. $y = e^{-3x}(-\frac{17}{16} - \frac{1}{4}x) + \frac{1}{16}e^x$

15. $y = e^x(\frac{1}{13} \cos 2x + \frac{33}{52} \sin 2x) - \frac{1}{13} \cos 3x - \frac{3}{26} \sin 3x$

17. $I = e^{-(5/2)t} \left(\dfrac{1240}{977} \cos \dfrac{5\sqrt{7}}{2}t - \dfrac{1320}{977} \sin \dfrac{5\sqrt{7}}{2}t \right) -$

$\dfrac{1240}{977} \cos 40t + \dfrac{160}{977} \sin 40t$

19. $x = \frac{7}{9} \sin 3t - \frac{1}{3}t \cos 3t$

Exercise Set 20.7

1. $f(x) = a_0(1 - 2x + 2x^2 - \frac{4}{3}x^3 + \frac{2}{3}x^4 - \frac{4}{15}x^5 + \frac{4}{45}x^6$
$- + \cdots) = a_0 e^{-2x}$

3. $f(x) = a_0\left(1 - \frac{1}{2!}x^2 + \frac{1}{4!}x^4 - + \cdots\right)$
$+ a_1\left(x - \frac{1}{3!}x^3 + \frac{1}{5!}x^5 - + \cdots\right) = a_0 \cos x + a_1 \sin x$

5. $f(x) = a_0(1 + x^2 + \frac{1}{3}x^3 + \frac{1}{4}x^4 + \frac{1}{12}x^5 + \cdots) +$
$a_1(x + \frac{1}{2}x^2 + \frac{1}{2}x^3 + \frac{5}{24}x^4 + \frac{11}{120}x^5 + \cdots) = C_1 e^{2x} + C_2 e^{-2x}$
if $C_1 + C_2 = a_0$ and $2C_1 - C_2 = a_1$

7. $f(x) = a_1\left(x - \frac{2x^2}{2} + \frac{2^2 x^3}{3 \cdot 2^2} - \frac{2^3 x^4}{4 \cdot 3^2 \cdot 2^2} + \frac{2^4 x^5}{5 \cdot 4^2 \cdot 3^2 \cdot 2^2} - + \cdots\right)$

9. $f(x) = a_1 x$

11. $f(x) = 2 + a_2\left(x^2 - \frac{x^3}{1} + \frac{x^4}{2 \cdot 1} - \frac{x^5}{3 \cdot 2 \cdot 1} + \frac{x^6}{4 \cdot 3 \cdot 2 \cdot 1} - + \cdots\right)$

Appendix

TABLE I   Trigonometric Functions

Degrees	Radians	Sin	Cos	Tan	Cot		
0	0.0000	0.0000	1.0000	0.0000		1.5708	90
1	0.0175	0.0175	0.9998	0.0175	57.290	1.5533	89
2	0.0349	0.0349	0.9994	0.0349	28.636	1.5359	88
3	0.0524	0.0523	0.9986	0.0524	19.081	1.5184	87
4	0.0698	0.0698	0.9976	0.0699	14.301	1.5010	86
5	0.0873	0.0872	0.9962	0.0875	11.430	1.4835	85
6	0.1047	0.1045	0.9945	0.1051	9.5144	1.4661	84
7	0.1222	0.1219	0.9925	0.1228	8.1443	1.4486	83
8	0.1396	0.1392	0.9903	0.1405	7.1154	1.4312	82
9	0.1571	0.1564	0.9877	0.1584	6.3138	1.4137	81
10	0.1745	0.1736	0.9848	0.1763	5.6713	1.3963	80
11	0.1920	0.1908	0.9816	0.1944	5.1446	1.3788	79
12	0.2094	0.2079	0.9781	0.2126	4.7046	1.3614	78
13	0.2269	0.2250	0.9744	0.2309	4.3315	1.3439	77
14	0.2443	0.2419	0.9703	0.2493	4.0108	1.3265	76
15	0.2618	0.2588	0.9659	0.2679	3.7321	1.3090	75
16	0.2793	0.2756	0.9613	0.2867	3.4874	1.2915	74
17	0.2967	0.2924	0.9563	0.3057	3.2709	1.2741	73
18	0.3142	0.3090	0.9511	0.3249	3.0777	1.2566	72
19	0.3316	0.3256	0.9455	0.3443	2.9042	1.2392	71
20	0.3491	0.3420	0.9397	0.3640	2.7475	1.2217	70
21	0.3665	0.3584	0.9336	0.3839	2.6051	1.2043	69
22	0.3840	0.3746	0.9272	0.4040	2.4751	1.1868	68
23	0.4014	0.3907	0.9205	0.4245	2.3559	1.1694	67
24	0.4189	0.4067	0.9135	0.4452	2.2460	1.1519	66
25	0.4363	0.4226	0.9063	0.4663	2.1445	1.1345	65
26	0.4538	0.4384	0.8988	0.4877	2.0503	1.1170	64
27	0.4712	0.4540	0.8910	0.5095	1.9626	1.0996	63
28	0.4887	0.4695	0.8829	0.5317	1.8807	1.0821	62
29	0.5061	0.4848	0.8746	0.5543	1.8040	1.0647	61
30	0.5236	0.5000	0.8660	0.5774	1.7321	1.0472	60
31	0.5411	0.5150	0.8572	0.6009	1.6643	1.0297	59
32	0.5585	0.5299	0.8480	0.6249	1.6003	1.0123	58
33	0.5760	0.5446	0.8387	0.6494	1.5399	0.9948	57
34	0.5934	0.5592	0.8290	0.6745	1.4826	0.9774	56
35	0.6109	0.5736	0.8192	0.7002	1.4281	0.9599	55
36	0.6283	0.5878	0.8090	0.7265	1.3764	0.9425	54
37	0.6458	0.6018	0.7986	0.7536	1.3270	0.9250	53
38	0.6632	0.6157	0.7880	0.7813	1.2799	0.9076	52
39	0.6807	0.6293	0.7771	0.8098	1.2349	0.8901	51
40	0.6981	0.6428	0.7660	0.8391	1.1918	0.8727	50
41	0.7156	0.6561	0.7547	0.8693	1.1504	0.8552	49
42	0.7330	0.6691	0.7431	0.9004	1.1106	0.8378	48
43	0.7505	0.6820	0.7314	0.9325	1.0724	0.8203	47
44	0.7679	0.6947	0.7193	0.9657	1.0355	0.8029	46
45	0.7854	0.7071	0.7071	1.0000	1.0000	0.7854	45
		Cos	Sin	Cot	Tan	Radians	Degrees

TABLE II Integrals

1. $\displaystyle\int u^n \, du = \frac{u^{n+1}}{n+1} + C \qquad$ if $n \neq -1$

2. $\displaystyle\int \frac{du}{u} = \ln |u| + C$

3. $\displaystyle\int e^u \, du = e^u + C$

4. $\displaystyle\int a^u \, du = \frac{a^u}{\ln a} + C$

5. $\displaystyle\int \sin u \, du = -\cos u + C$

6. $\displaystyle\int \cos u \, du = \sin u + C$

7. $\displaystyle\int \sec^2 u \, du = \tan u + C$

8. $\displaystyle\int \csc^2 u \, du = -\cot u + C$

9. $\displaystyle\int \sec u \tan u \, du = \sec u + C$

10. $\displaystyle\int \csc u \cot u \, du = -\csc u + C$

11. $\displaystyle\int \tan u \, du = -\ln |\cos u| + C$

12. $\displaystyle\int \cot u \, du = \ln |\sin u| + C$

13. $\displaystyle\int \sec u \, du = \ln |\sec u + \tan u| + C$

14. $\displaystyle\int \csc u \, du = -\ln |\csc u + \cot u| + C$

15. $\displaystyle\int \sin^2 u \, du = \tfrac{1}{2} u - \tfrac{1}{4} \sin 2u + C$

16. $\displaystyle\int \cos^2 u \, du = \tfrac{1}{2} u + \tfrac{1}{4} \cos 2u + C$

17. $\displaystyle\int \cosh u \, du = \sinh u + C$

18. $\displaystyle\int \sinh u\,du = \cosh u + C$

19. $\displaystyle\int \frac{du}{\sqrt{a^2 - u^2}} = \sin^{-1} \frac{u}{a} + C$

20. $\displaystyle\int \frac{du}{a^2 + u^2} = \frac{1}{a} \tan^{-1} \frac{u}{a} + C$

21. $\displaystyle\int \frac{du}{u\sqrt{u^2 - a^2}} = \frac{1}{a} \sec^{-1} \frac{u}{a} + C$

22. $\displaystyle\int \frac{du}{u^2 - a^2} = \frac{1}{2a} \ln \left| \frac{u - a}{u + a} \right| + C$

23. $\displaystyle\int \frac{du}{a^2 - u^2} = \frac{1}{2a} \ln \left| \frac{a + u}{a - u} \right| + C$

24. $\displaystyle\int \frac{du}{\sqrt{u^2 + a^2}} = \ln \left(u + \sqrt{u^2 + a^2} \right) + C$

25. $\displaystyle\int \frac{du}{\sqrt{u^2 - a^2}} = \ln \left| u + \sqrt{u^2 - a^2} \right| + C$

26. $\displaystyle\int \sqrt{a^2 - u^2}\,du = \frac{u}{2} \sqrt{a^2 - u^2} + \frac{a^2}{2} \sin^{-1} \frac{u}{a} + C$

27. $\displaystyle\int u^2\sqrt{a^2 - u^2}\,du = -\frac{1}{4} u\,(a^2 - u^2)^{3/2} + \frac{1}{8} a^2 u\sqrt{a^2 - u^2} + \frac{1}{8} a^4 \sin^{-1} \frac{u}{a} + C$

28. $\displaystyle\int u^3\sqrt{a^2 - u^2}\,du = \frac{1}{5}(a^2 - u^2)^{5/2} - \frac{1}{3} a^2(a^2 - u^2)^{3/2} + C$

29. $\displaystyle\int (a^2 - u^2)^{3/2}\,du = \frac{1}{4} u(a^2 - u^2)^{3/2} + \frac{3}{8} a^2 u(a^2 - u^2)^{1/2} + \frac{3}{8} a^4 \sin^{-1} \frac{u}{a} + C$

30. $\displaystyle\int \sin^4 u\,du = \frac{3}{8} u - \frac{\sin 2u}{4} + \frac{\sin 4u}{32} + C$

31. $\displaystyle\int \cos^4 u\,du = \frac{3}{8} u + \frac{\sin 2u}{4} + \frac{\sin 4u}{32} + C$

32. $\displaystyle\int \sec^3 u\,du = \frac{1}{2} \sec u \tan u + \frac{1}{2} \ln \left| \sec u + \tan u \right| + C$

33. $\displaystyle\int \csc^3 u\,du = -\frac{1}{2} \csc u \cot u - \frac{1}{2} \ln \left| \csc u + \cot u \right| + C$

34. $\displaystyle\int \sin^{-1} \frac{u}{a}\,du = u \sin^{-1} \frac{u}{a} + \sqrt{a^2 - u^2} + C$

35. $\displaystyle\int u \sin^{-1} \frac{u}{a}\, du = \left(\frac{u^2}{2} - \frac{a^2}{4}\right) \sin^{-1} \frac{u}{a} + \frac{u}{4} \sqrt{a^2 - u^2} + C$

36. $\displaystyle\int \sqrt{u^2 + a^2}\, du = \frac{u \sqrt{u^2 + a^2}}{2} + \frac{a^2}{2} \ln\left(u + \sqrt{u^2 + a^2}\right) + C$

37. $\displaystyle\int u e^{au}\, du = \frac{e^{au}}{a^2}\, (au - 1) + C$

38. $\displaystyle\int u^n e^{au}\, du = \frac{1}{a}\, u^n e^{au} - \frac{n}{a} \int u^{n-1} e^{au}\, du$

39. $\displaystyle\int \ln u\, du = u \ln u - u + C$

40. $\displaystyle\int u \ln u\, du = \frac{u^2}{2} \ln u - \frac{u^2}{4} + C$

41. $\displaystyle\int e^{au} \sin bu\, du = \frac{e^{au}}{a^2 + b^2}\, (a \sin bu - b \cos bu) + C$

42. $\displaystyle\int e^{au} \cos bu\, du = \frac{e^{au}}{a^2 + b^2}\, (a \cos bu + b \sin bu) + C$

TABLE III Natural Logarithms

N	0	1	2	3	4	5	6	7	8	9
1.0	0.0000	0100	0198	0296	0392	0488	0583	0677	0770	0862
1.1	0953	1044	1133	1222	1310	1398	1484	1570	1655	1740
1.2	1823	1906	1989	2070	2151	2231	2311	2390	2469	2546
1.3	2624	2700	2776	2852	2927	3001	3075	3148	3221	3293
1.4	3365	3436	3507	3577	3646	3716	3784	3853	3920	3988
1.5	4055	4121	4187	4253	4318	4383	4447	4511	4574	4637
1.6	4700	4762	4824	4886	4947	5008	5068	5128	5188	5247
1.7	5306	5365	5423	5481	5539	5596	5653	5710	5766	5822
1.8	5878	5933	5988	6043	6098	6152	6206	6259	6313	6366
1.9	6419	6471	6523	6575	6627	6678	6729	6780	6831	6881
2.0	6931	6981	7031	7080	7129	7178	7227	7275	7324	7372
2.1	7419	7467	7514	7561	7608	7655	7701	7747	7793	7839
2.2	7885	7930	7975	8020	8065	8109	8154	8198	8242	8286
2.3	8329	8372	8416	8459	8502	8544	8587	8629	8671	8713
2.4	8755	8796	8838	8879	8920	8961	9002	9042	9083	9123
2.5	9163	9203	9243	9282	9322	9361	9400	9439	9478	9517
2.6	9555	9594	9632	9670	9708	9746	9783	9821	9858	9895
2.7	9933	9969	°0006	°0043	°0080	°0116	°0152	°0188	°0225	°0260
2.8	1.0296	0332	0367	0403	0438	0473	0508	0543	0578	0613
2.9	0647	0682	0716	0750	0784	0818	0852	0886	0919	0953
3.0	1.0986	1019	1053	1086	1119	1151	1184	1217	1249	1282
3.1	1314	1346	1378	1410	1442	1474	1506	1537	1569	1600
3.2	1632	1663	1694	1725	1756	1787	1817	1848	1878	1909
3.3	1939	1969	2000	2030	2060	2090	2119	2149	2179	2208
3.4	2238	2267	2296	2326	2355	2384	2413	2442	2470	2499
3.5	1.2528	2556	2585	2613	2641	2669	2698	2726	2754	2782
3.6	2809	2837	2865	2892	2920	2947	2975	3002	3029	3056
3.7	3083	3110	3137	3164	3191	3218	3244	3271	3297	3324
3.8	3350	3376	3403	3429	3455	3481	3507	3533	3558	3584
3.9	3610	3635	3661	3686	3712	3737	3762	3788	3813	3838
4.0	1.3863	3888	3913	3938	3962	3987	4012	4036	4061	4085
4.1	4110	4134	4159	4183	4207	4231	4255	4279	4303	4327
4.2	4351	4375	4398	4422	4446	4469	4493	4516	4540	4563
4.3	4586	4609	4633	4656	4679	4702	4725	4748	4770	4793
4.4	4816	4839	4861	4884	4907	4929	4951	4974	4996	5019
4.5	1.5041	5063	5085	5107	5129	5151	5173	5195	5217	5239
4.6	5261	5282	5304	5326	5347	5369	5390	5412	5433	5454
4.7	5476	5497	5518	5539	5560	5581	5602	5623	5644	5665
4.8	5686	5707	5728	5748	5769	5790	5810	5831	5851	5872
4.9	5892	5913	5933	5953	5974	5994	6014	6034	6054	6074
5.0	1.6094	6114	6134	6154	6174	6194	6214	6233	6253	6273
5.1	6292	6312	6332	6351	6371	6390	6409	6429	6448	6467
5.2	6487	6506	6525	6544	6563	6582	6601	6620	6639	6658
5.3	6677	6696	6715	6734	6752	6771	6790	6808	6827	6845
5.4	6864	6882	6901	6919	6938	6956	6974	6993	7011	7029

TABLE III (Continued)

N	0	1	2	3	4	5	6	7	8	9
5.5	1.7047	7066	7084	7102	7120	7138	7156	7174	7192	7210
5.6	7228	7246	7263	7281	7299	7317	7334	7352	7370	7387
5.7	7405	7422	7440	7457	7475	7492	7509	7527	7544	7561
5.8	7579	7596	7613	7630	7647	7664	7681	7699	7716	7733
5.9	7750	7766	7783	7800	7817	7834	7851	7867	7884	7901
6.0	1.7918	7934	7951	7967	7984	8001	8017	8034	8050	8066
6.1	8083	8099	8116	8132	8148	8165	8181	8197	8213	8229
6.2	8245	8262	8278	8294	8310	8326	8342	8358	8374	8390
6.3	8405	8421	8437	8453	8469	8485	8500	8516	8532	8547
6.4	8563	8579	8594	8610	8625	8641	8656	8672	8687	8703
6.5	1.8718	8733	8749	8764	8779	8795	8810	8825	8840	8856
6.6	8871	8886	8901	8916	8931	8946	8961	8976	8991	9006
6.7	9021	9036	9051	9066	9081	9095	9110	9125	9140	9155
6.8	9169	9184	9199	9213	9228	9242	9257	9272	9286	9301
6.9	9315	9330	9344	9359	9373	9387	9402	9416	9430	9445
7.0	1.9459	9473	9488	9502	9516	9530	9544	9559	9573	9587
7.1	9601	9615	9629	9643	9657	9671	9685	9699	9713	9727
7.2	9741	9755	9769	9782	9796	9810	9824	9838	9851	9865
7.3	9879	9892	9906	9920	9933	9947	9961	9974	9988	°0001
7.4	2.0015	0028	0042	0055	0069	0082	0096	0109	0122	0136
7.5	2.0149	0162	0176	0189	0202	0215	0229	0242	0255	0268
7.6	0281	0295	0308	0321	0334	0347	0360	0373	0386	0399
7.7	0412	0425	0438	0451	0464	0477	0490	0503	0516	0528
7.8	0541	0554	0567	0580	0592	0605	0618	0631	0643	0656
7.9	0669	0681	0694	0707	0719	0732	0744	0757	0769	0782
8.0	2.0794	0807	0819	0832	0844	0857	0869	0882	0894	0906
8.1	0919	0931	0943	0956	0968	0980	0992	1005	1017	1029
8.2	1041	1054	1066	1078	1090	1102	1114	1126	1138	1150
8.3	1163	1175	1187	1199	1211	1223	1235	1247	1258	1270
8.4	1282	1294	1306	1318	1330	1342	1353	1365	1377	1389
8.5	2.1401	1412	1424	1436	1448	1459	1471	1483	1494	1506
8.6	1518	1529	1541	1552	1564	1576	1587	1599	1610	1622
8.7	1633	1645	1656	1668	1679	1691	1702	1713	1725	1736
8.8	1748	1759	1770	1782	1793	1804	1815	1827	1838	1849
8.9	1861	1872	1883	1894	1905	1917	1928	1939	1950	1961
9.0	2.1972	1983	1994	2006	2017	2028	2039	2050	2061	2072
9.1	2083	2094	2105	2116	2127	2138	2148	2159	2170	2181
9.2	2192	2203	2214	2225	2235	2246	2257	2268	2279	2289
9.3	2300	2311	2322	2332	2343	2354	2364	2375	2386	2396
9.4	2407	2418	2428	2439	2450	2460	2471	2481	2492	2502
9.5	2.2513	2523	2534	2544	2555	2565	2576	2586	2597	2607
9.6	2618	2628	2638	2649	2659	2670	2680	2690	2701	2711
9.7	2721	2732	2742	2752	2762	2773	2783	2793	2803	2814
9.8	2824	2834	2844	2854	2865	2875	2885	2895	2905	2915
9.9	2925	2935	2946	2956	2966	2976	2986	2996	3006	3016

$$\ln 10 = 2.30259 \qquad \ln (a \cdot 10^n) = \ln a + n \ln 10$$

TABLE IV Exponential Functions

x	e^x	$\text{Log}_{10}(e^x)$	e^{-x}	x	e^x	$\text{Log}_{10}(e^x)$	e^{-x}
0.00	1.0000	0.00000	1.000000	**0.50**	1.6487	0.21715	0.606531
0.01	1.0101	.00434	0.990050	0.51	1.6653	.22149	.600496
0.02	1.0202	.00869	.980199	0.52	1.6820	.22583	.594521
0.03	1.0305	.01303	.970446	0.53	1.6989	.23018	.588605
0.04	1.0408	.01737	.960789	0.54	1.7160	.23452	.582748
0.05	1.0513	0.02171	0.951229	**0.55**	1.7333	0.23886	0.576950
0.06	1.0618	.02606	.941765	0.56	1.7507	.24320	.571209
0.07	1.0725	.03040	.932394	0.57	1.7683	.24755	.565525
0.08	1.0833	.03474	.923116	0.58	1.7860	.25189	.559898
0.09	1.0942	.03909	.913931	0.59	1.8040	.25623	.554327
0.10	1.1052	0.04343	0.904837	**0.60**	1.8221	0.26058	0.548812
0.11	1.1163	.04777	.895834	0.61	1.8404	.26492	.543351
0.12	1.1275	.05212	.886920	0.62	1.8589	.26926	.537944
0.13	1.1388	.05646	.878095	0.63	1.8776	.27361	.532592
0.14	1.1503	.06080	.869358	0.64	1.8965	.27795	.527292
0.15	1.1618	0.06514	0.860708	**0.65**	1.9155	0.28229	0.522046
0.16	1.1735	.06949	.852144	0.66	1.9348	.28663	.516851
0.17	1.1853	.07383	.843665	0.67	1.9542	.29098	.511709
0.18	1.1972	.07817	.835270	0.68	1.9739	.29532	.506617
0.19	1.2092	.08252	.826959	0.69	1.9937	.29966	.501576
0.20	1.2214	0.08686	0.818731	**0.70**	2.0138	0.30401	0.496585
0.21	1.2337	.09120	.810584	0.71	2.0340	.30835	.491644
0.22	1.2461	.09554	.802519	0.72	2.0544	.31269	.486752
0.23	1.2586	.09989	.794534	0.73	2.0751	.31703	.481909
0.24	1.2712	.10423	.786628	0.74	2.0959	.32138	.477114
0.25	1.2840	0.10857	0.778801	**0.75**	2.1170	0.32572	0.472367
0.26	1.2969	.11292	.771052	0.76	2.1383	.33006	.467666
0.27	1.3100	.11726	.763379	0.77	2.1598	.33441	.463013
0.28	1.3231	.12160	.755784	0.78	2.1815	.33875	.458406
0.29	1.3364	.12595	.748264	0.79	2.2034	.34309	.453845
0.30	1.3499	0.13029	0.740818	**0.80**	2.2255	0.34744	0.449329
0.31	1.3634	.13463	.733447	0.81	2.2479	.35178	.444858
0.32	1.3771	.13897	.726149	0.82	2.2705	.39521	.440432
0.33	1.3910	.14332	.718924	0.83	2.2933	.36046	.436049
0.34	1.4049	.14766	.711770	0.84	2.3164	.36481	.431711
0.35	1.4191	0.15200	0.704688	**0.85**	2.3396	0.36915	0.427415
0.36	1.4333	.15635	.697676	0.86	2.3632	.37349	.423162
0.37	1.4477	.16069	.690734	0.87	2.3869	.37784	.418952
0.38	1.4623	.16503	.683861	0.88	2.4109	.38218	.414783
0.39	1.4770	.16937	.677057	0.89	2.4351	.38652	.410656
0.40	1.4918	0.17372	0.670320	**0.90**	2.4596	0.39087	0.406570
0.41	1.5068	.17806	.663650	0.91	2.4843	.39521	.402524
0.42	1.5220	.18240	.657047	0.92	2.5093	.39955	.398519
0.43	1.5373	.18675	.650509	0.93	2.5345	.40389	.394554
0.44	1.5527	.19109	.644036	0.94	2.5600	.40824	.390628
0.45	1.5683	0.19543	0.637628	**0.95**	2.5857	0.41258	0.386741
0.46	1.5841	.19978	.631284	0.96	2.6117	.41692	.382893
0.47	1.6000	.20412	.625002	0.97	2.6379	.42127	.379083
0.48	1.6161	.20846	.618783	0.98	2.6645	.42561	.375311
0.49	1.6323	.21280	.612626	0.99	2.6912	.42995	.371577
0.50	1.6487	0.21715	0.606531	**1.00**	2.7183	0.43429	0.367879

TABLE IV (Continued)

x	e^x	$\text{Log}_{10}(e^x)$	e^{-x}	x	e^x	$\text{Log}_{10}(e^x)$	e^{-x}
1.00	2.7183	0.43429	0.367879	**1.50**	4.4817	0.65144	0.223130
1.01	2.7456	.43864	.364219	1.51	4.5267	.65578	.220910
1.02	2.7732	.44298	.360595	1.52	4.5722	.66013	.218712
1.03	2.8011	.44732	.357007	1.53	4.6182	.66447	.216536
1.04	2.8292	.45167	.353455	1.54	4.6646	.66881	.214381
1.05	2.8577	0.45601	0.349938	**1.55**	4.7115	0.67316	0.212248
1.06	2.8864	.46035	.346456	1.56	4.7588	.67750	.210136
1.07	2.9154	.46470	.343009	1.57	4.8066	.68184	.208045
1.08	2.9447	.46904	.339596	1.58	4.8550	.68619	.205975
1.09	2.9743	.47338	.336216	1.59	4.9037	.69053	.203926
1.10	3.0042	0.47772	0.332871	**1.60**	4.9530	0.69487	0.201897
1.11	3.0344	.48207	.329559	1.61	5.0028	.69921	.199888
1.12	3.0649	.48641	.326280	1.62	5.0531	.70356	.197899
1.13	3.0957	.49075	.323033	1.63	5.1039	.70790	.195930
1.14	3.1268	.49510	.319819	1.64	5.1552	.71224	.193980
1.15	3.1582	0.49944	0.316637	**1.65**	5.2070	0.71659	0.192050
1.16	3.1899	.50378	.313486	1.66	5.2593	.72093	.190139
1.17	3.2220	.50812	.310367	1.67	5.3122	.72527	.188247
1.18	3.2544	.51247	.307279	1.68	5.3656	.72961	.186374
1.19	3.2871	.51681	.304221	1.69	5.4195	.73396	.184520
1.20	3.3201	0.52115	0.301194	**1.70**	5.4739	0.73830	0.182684
1.21	3.3535	.52550	.298197	1.71	5.5290	.74264	.180866
1.22	3.3872	.52984	.295230	1.72	5.5845	.74699	.179066
1.23	3.4212	.53418	.292293	1.73	5.6407	.75133	.177284
1.24	3.4556	.53853	.289384	1.74	5.6973	.75567	.175520
1.25	3.4903	0.54287	0.286505	**1.75**	5.7546	0.76002	0.173774
1.26	3.5254	.54721	.283654	1.76	5.8124	.76436	.172045
1.27	3.5609	.55155	.280832	1.77	5.8709	.76870	.170333
1.28	3.5966	.55590	.278037	1.78	5.9299	.77304	.168638
1.29	3.6328	.56024	.275271	1.79	5.9895	.77739	.166960
1.30	3.6693	0.56458	0.272532	**1.80**	6.0496	0.78173	0.165299
1.31	3.7062	.56893	.269820	1.81	6.1104	.78607	.163654
1.32	3.7434	.57327	.267135	1.82	6.1719	.79042	.162026
1.33	3.7810	.57761	.264477	1.83	6.2339	.79476	.160414
1.34	3.8190	.58195	.261846	1.84	6.2965	.79910	.158817
1.35	3.8574	0.58630	0.259240	**1.85**	6.3598	0.80344	0.157237
1.36	3.8962	.59064	.256661	1.86	6.4237	.80779	.155673
1.37	3.9354	.59498	.254107	1.87	6.4883	.81213	.154124
1.38	3.9749	.59933	.251579	1.88	6.5535	.81647	.152590
1.39	4.0149	.60367	.249075	1.89	6.6194	.82082	.151072
1.40	4.0552	0.60801	0.246597	**1.90**	6.6859	0.82516	0.149569
1.41	4.0960	.61236	.244143	1.91	6.7531	.82950	.148080
1.42	4.1371	.61670	.241714	1.92	6.8210	.83385	.146607
1.43	4.1787	.62104	.239309	1.93	6.8895	.83819	.145148
1.44	4.2207	.62538	.236928	1.94	6.9588	.84253	.143704
1.45	4.2631	0.62973	0.234570	**1.95**	7.0287	0.84687	0.142274
1.46	4.3060	.63407	.232236	1.96	7.0993	.85122	.140858
1.47	4.3492	.63841	.229925	1.97	7.1707	.85556	.139457
1.48	4.3929	.64276	.227638	1.98	7.2427	.85990	.138069
1.49	4.4371	.64710	.225373	1.99	7.3155	.86425	.136695
1.50	4.4817	0.65144	0.223130	**2.00**	7.3891	0.86859	0.135335

TABLE IV (Continued)

x	e^x	$\text{Log}_{10}(e^x)$	e^{-x}	x	e^x	$\text{Log}_{10}(e^x)$	e^{-x}
2.00	7.3891	0.86859	0.135335	**2.50**	12.182	1.08574	0.082085
2.01	7.4633	.87293	.133989	2.51	12.305	1.09008	.081268
2.02	7.5383	.87727	.132655	2.52	12.429	1.09442	.080460
2.03	7.6141	.88162	.131336	2.53	12.554	1.09877	.079659
2.04	7.6906	.88596	.130029	2.54	12.680	1.10311	.078866
2.05	7.7679	0.89030	0.128735	**2.55**	12.807	1.10745	0.078082
2.06	7.8460	.89465	.127454	2.56	12.936	1.11179	.077305
2.07	7.9248	.89899	.126186	2.57	13.066	1.11614	.076536
2.08	8.0045	.90333	.124930	2.58	13.197	1.12048	.075774
2.09	8.0849	.90768	.123687	2.59	13.330	1.12482	.075020
2.10	8.1662	0.91202	0.122456	**2.60**	13.464	1.12917	0.074274
2.11	8.2482	.91636	.121238	2.61	13.599	1.13351	.073535
2.12	8.3311	.92070	.120032	2.62	13.736	1.13785	.072803
2.13	8.4149	.92505	.118837	2.63	13.874	1.14219	.072078
2.14	8.4994	.92939	.117655	2.64	14.013	1.14654	.071361
2.15	8.5849	0.93373	0.116484	**2.65**	14.154	1.15088	0.070651
2.16	8.6711	.93808	.115325	2.66	14.296	1.15522	.069948
2.17	8.7583	.94242	.114178	2.67	14.440	1.15957	.069252
2.18	8.8463	.94676	.113042	2.68	14.585	1.16391	.069563
2.19	8.9352	.95110	.111917	2.69	14.732	1.16825	.067881
2.20	9.0250	0.95545	0.110803	**2.70**	14.880	1.17260	0.067206
2.21	9.1157	.95979	.109701	2.71	15.029	1.17649	.066537
2.22	9.2073	.96413	.108609	2.72	15.180	1.18128	.065875
2.23	9.2999	.96848	.107528	2.73	15.333	1.18562	.065219
2.24	9.3933	.97282	.106459	2.74	15.487	1.18997	.064570
2.25	9.4877	0.97716	0.105399	**2.75**	15.643	1.19431	0.063928
2.26	9.5831	.98151	.104350	2.76	15.800	1.19865	.063292
2.27	9.6794	.98585	.103312	2.77	15.959	1.20300	.062662
2.28	9.7767	.99019	.102284	2.78	16.119	1.20734	.062039
2.29	9.8749	.99453	.101266	2.79	16.281	1.21168	.061421
2.30	9.9742	0.99888	0.100259	**2.80**	16.445	1.21602	0.060810
2.31	10.074	1.00322	.099261	2.81	16.610	1.22037	.060205
2.32	10.176	1.00756	.098274	2.82	16.777	1.22471	.059606
2.33	10.278	1.01191	.097296	2.83	16.945	1.22905	.059013
2.34	10.381	1.01625	.096328	2.84	17.116	1.23340	.058426
2.35	10.486	1.02059	0.095369	**2.85**	17.288	1.23774	0.057844
2.36	10.591	1.02493	.094420	2.86	17.462	1.24208	.057269
2.37	10.697	1.02928	.093481	2.87	17.637	1.24643	.056699
2.38	10.805	1.03362	.092551	2.88	17.814	1.25077	.056135
2.39	10.913	1.03796	.091630	2.89	17.993	1.25511	.055576
2.40	11.023	1.04231	0.090718	**2.90**	18.174	1.25945	0.055023
2.41	11.134	1.04665	.089815	2.91	18.357	1.26380	.054476
2.42	11.246	1.05099	.088922	2.92	18.541	1.26814	.053934
2.43	11.359	1.05534	.088037	2.93	18.728	1.27248	.053397
2.44	11.473	1.05968	.087161	2.94	18.916	1.27683	.052866
2.45	11.588	1.06402	0.086294	**2.95**	19.106	1.28117	0.052340
2.46	11.705	1.06836	.085435	2.96	19.298	1.28551	.051819
2.47	11.822	1.07271	.084585	2.97	19.492	1.28985	.051303
2.48	11.941	1.07705	.083743	2.98	19.688	1.29420	.050793
2.49	12.061	1.08139	.082910	2.99	19.886	1.29854	.050287
2.50	12.182	1.08574	0.082085	**3.00**	20.086	1.30288	0.049787

TABLE IV (Continued)

x	e^x	$\text{Log}_{10}(e^x)$	e^{-x}	x	e^x	$\text{Log}_{10}(e^x)$	e^{-x}
3.00	20.086	1.30288	0.049787	**3.50**	33.115	1.52003	0.030197
3.01	20.287	1.30723	.049292	3.51	33.448	1.52437	.029897
3.02	20.491	1.31157	.048801	3.52	33.784	1.52872	.029599
3.03	20.697	1.31591	.048316	3.53	34.124	1.53306	.029305
3.04	20.905	1.32026	.047835	3.54	34.467	1.53740	.029013
3.05	21.115	1.32460	0.047359	**3.55**	34.813	1.54175	0.028725
3.06	21.328	1.32894	.046888	3.56	35.163	1.54609	.028439
3.07	21.542	1.33328	.046421	3.57	35.517	1.55043	.028156
3.08	21.758	1.33763	.045959	3.58	35.874	1.55477	.027876
3.09	21.977	1.34197	.045502	3.59	36.234	1.55912	.027598
3.10	22.198	1.34631	0.045049	**3.60**	36.598	1.56346	0.027324
3.11	22.421	1.35066	.044601	3.61	36.966	1.56780	.027052
3.12	22.646	1.35500	.044157	3.62	37.338	1.57215	.026783
3.13	22.874	1.35934	.043718	3.63	37.713	1.57649	.026516
3.14	23.104	1.36368	.043283	3.64	38.092	1.58083	.026252
3.15	23.336	1.36803	0.042852	**3.65**	38.475	1.58517	0.025991
3.16	23.571	1.37237	.042426	3.66	38.861	1.58952	.025733
3.17	23.807	1.37671	.042004	3.67	39.252	1.59386	.025476
3.18	24.047	1.38106	.041586	3.68	39.646	1.59820	.025223
3.19	24.288	1.38540	.041172	3.69	40.045	1.60255	.024972
3.20	24.533	1.38974	0.040762	**3.70**	40.447	1.60689	0.024724
3.21	24.779	1.39409	.040357	3.71	40.854	1.61123	.024478
3.22	25.028	1.39843	.039955	3.72	41.264	1.61558	.024234
3.23	25.280	1.40277	.039557	3.73	41.679	1.61992	.023993
3.24	25.534	1.40711	.039164	3.74	42.098	1.62426	.023754
3.25	25.790	1.41146	0.038774	**3.75**	42.521	1.62860	0.023518
3.26	26.050	1.41580	.038388	3.76	42.948	1.63295	.023284
3.27	26.311	1.42014	.038006	3.77	43.380	1.63729	.023052
3.28	26.576	1.42449	.037628	3.78	43.816	1.64163	.022823
3.29	26.843	1.42883	.037254	3.79	44.256	1.64598	.022596
3.30	27.113	1.43317	0.036883	**3.80**	44.701	1.65032	0.022371
3.31	27.385	1.43751	.036516	3.81	45.150	1.65466	.022148
3.32	27.660	1.44186	.036153	3.82	45.604	1.65900	.021928
3.33	27.938	1.44620	.035793	3.83	46.063	1.66335	.021710
3.34	28.219	1.45054	.035437	3.84	46.525	1.66769	.021494
3.35	28.503	1.45489	0.035084	**3.85**	46.993	1.67203	0.021280
3.36	28.789	1.45923	.034735	3.86	47.465	1.67638	.021068
3.37	29.079	1.46357	.034390	3.87	47.942	1.68072	.020858
3.38	29.371	1.46792	.034047	3.88	48.424	1.68506	.020651
3.39	29.666	1.47226	.033709	3.89	48.911	1.68941	.020445
3.40	29.964	1.47660	0.033373	**3.90**	49.402	1.69375	0.020242
3.41	30.265	1.48094	.033041	3.91	49.899	1.69809	.020041
3.42	30.569	1.48529	.032712	3.92	50.400	1.70243	.019841
3.43	30.877	1.48963	.032387	3.93	50.907	1.70678	.019644
3.44	31.187	1.49397	.032065	3.94	51.419	1.71112	.019448
3.45	31.500	1.49832	0.031746	**3.95**	51.935	1.71546	0.019255
3.46	31.817	1.50266	.031430	3.96	52.457	1.71981	.019063
3.47	32.137	1.50700	.031117	3.97	52.985	1.72415	.018873
3.48	32.460	1.51134	.030807	3.98	53.517	1.72849	.018686
3.49	32.786	1.51569	.030501	3.99	54.055	1.73283	.018500
3.50	33.115	1.52003	0.030197	**4.00**	54.598	1.73718	0.018316

TABLE IV (Continued)

x	e^x	$Log_{10}(e^x)$	e^{-x}	x	e^x	$Log_{10}(e^x)$	e^{-x}
4.00	54.598	1.73718	0.018316	**4.50**	90.017	1.95433	0.011109
4.01	55.147	1.74152	.018133	4.51	90.922	1.95867	.010998
4.02	55.701	1.74586	.017953	4.52	91.836	1.96301	.010889
4.03	56.261	1.75021	.017774	4.53	92.759	1.96735	.010781
4.04	56.826	1.75455	.017597	4.54	93.691	1.97170	.010673
4.05	57.397	1.75889	0.017422	**4.55**	94.632	1.97604	0.010567
4.06	57.974	1.76324	.017249	4.56	95.583	1.98038	.010462
4.07	58.557	1.76758	.017077	4.57	96.544	1.98473	.010358
4.08	59.145	1.77192	.016907	4.58	97.514	1.98907	.010255
4.09	59.740	1.77626	.016739	4.59	98.494	1.99341	.010153
4.10	60.340	1.78061	0.016573	**4.60**	99.484	1.99775	0.010052
4.11	60.947	1.78495	.016408	4.61	100.48	2.00210	.009952
4.12	61.559	1.78929	.016245	4.62	101.49	2.00644	.009853
4.13	62.178	1.79364	.016083	4.63	102.51	2.01078	.009755
4.14	62.803	1.79798	.015923	4.64	103.54	2.01513	.009658
4.15	63.434	1.80232	0.015764	**4.65**	104.58	2.01947	0.009562
4.16	64.072	1.80667	.015608	4.66	105.64	2.02381	.009466
4.17	64.715	1.81101	.015452	4.67	106.70	2.02816	.009372
4.18	65.366	1.81535	.015299	4.68	107.77	2.03250	.009279
4.19	66.023	1.81969	.015146	4.69	108.85	2.03684	.009187
4.20	66.686	1.82404	0.014996	**4.70**	109.95	2.04118	0.009095
4.21	67.357	1.82838	.014846	4.71	111.05	2.04553	.009005
4.22	68.033	1.83272	.014699	4.72	112.17	2.04987	.008915
4.23	68.717	1.83707	.014552	4.73	113.30	2.05421	.008826
4.24	69.408	1.84141	.014408	4.74	114.43	2.05856	.008739
4.25	70.105	1.84575	0.014264	**4.75**	115.58	2.06290	0.008652
4.26	70.810	1.85009	.014122	4.76	116.75	2.06724	.008566
4.27	71.522	1.85444	.013982	4.77	117.92	2.07158	.008480
4.28	72.240	1.85878	.013843	4.78	119.10	2.07593	.008396
4.29	72.966	1.86312	.013705	4.79	120.30	2.08027	.008312
4.30	73.700	1.86747	0.013569	**4.80**	121.51	2.08461	0.008230
4.31	74.440	1.87181	.013434	4.81	122.73	2.08896	.008148
4.32	75.189	1.87615	.013300	4.82	123.97	2.09330	.008067
4.33	75.944	1.88050	.013168	4.83	125.21	2.09764	.007987
4.34	76.708	1.88484	.013037	4.84	126.47	2.10199	.007907
4.35	77.478	1.88918	0.012907	**4.85**	127.74	2.10633	0.007828
4.36	78.257	1.89352	.012778	4.86	129.02	2.11067	.007750
4.37	79.044	1.89787	.012651	4.87	130.32	2.11501	.007673
4.38	79.838	1.90221	.012525	4.88	131.63	2.11936	.007597
4.39	80.640	1.90655	.012401	4.89	132.95	2.12370	.007521
4.40	81.451	1.91090	0.012277	**4.90**	134.29	2.12804	0.007447
4.41	82.269	1.91524	.012155	4.91	135.64	2.13239	.007372
4.42	83.096	1.91958	.012034	4.92	137.00	2.13673	.007299
4.43	83.931	1.92392	.011914	4.93	138.38	2.14107	.007227
4.44	84.775	1.92827	.011796	4.94	139.77	2.14541	.007155
4.45	85.627	1.93261	0.011679	**4.95**	141.17	2.14976	0.007083
4.46	86.488	1.93695	.011562	4.96	142.59	2.15410	.007013
4.47	87.357	1.94130	.011447	4.97	144.03	2.15844	.006943
4.48	88.235	1.94564	.011333	4.98	145.47	2.16279	.006874
4.49	89.121	1.94998	.011221	4.99	146.94	2.16713	.006806
4.50	90.107	1.95433	0.011109	**5.00**	148.41	2.17147	0.006738

TABLE IV (Continued)

x	e^x	$\text{Log}_{10}(e^x)$	e^{-x}	x	e^x	$\text{Log}_{10}(e^x)$	e^{-x}
5.00	148.41	2.17147	0.006738	**5.50**	244.69	2.38862	0.004087
5.01	149.90	2.17582	.006671	5.55	257.24	2.41033	.003888
5.02	151.41	2.18016	.006605	5.60	270.43	2.43205	.003698
5.03	152.93	2.18450	.006539	5.64	284.29	2.45376	.003518
5.04	154.47	2.18884	.006474	5.70	298.87	2.47548	.003346
5.05	156.02	2.19319	0.006409	**5.75**	314.19	2.49719	0.003183
5.06	157.59	2.19753	.006346	5.80	330.30	2.51891	.003028
5.07	159.17	2.20187	.006282	5.85	347.23	2.54062	.002880
5.08	160.77	2.20622	.006220	5.90	365.04	2.56234	.002739
5.09	162.39	2.21056	.006158	5.95	383.75	2.58405	.002606
5.10	164.02	2.21490	0.006097	**6.00**	403.43	2.60577	0.002479
5.11	165.67	2.21924	.006036	6.05	424.11	2.62748	.002358
5.12	167.34	2.22359	.005976	6.10	445.86	2.64920	.002243
5.13	169.02	2.22793	.005917	6.15	468.72	2.67091	.002134
5.14	170.72	2.23227	.005858	6.20	492.75	2.69263	.002029
5.15	172.43	2.23662	0.005799	**6.25**	518.01	2.71434	0.001931
5.16	174.16	2.24096	.005742	6.30	544.57	2.73606	.001836
5.17	175.91	2.24530	.005685	6.35	572.49	2.75777	.001747
5.18	177.68	2.24965	.005628	6.40	601.85	2.77948	.001662
5.19	179.47	2.25399	.005572	6.45	632.70	2.80120	.001581
5.20	181.27	2.25833	0.005517	**6.50**	665.14	2.82291	0.001503
5.21	183.09	2.26267	.005462	6.55	699.24	2.84463	.001430
5.22	184.93	2.26702	.005407	6.60	735.10	2.86634	.001360
5.23	186.79	2.27136	.005354	6.65	772.78	2.88806	.001294
5.24	188.67	2.27570	.005300	6.70	812.41	2.90977	.001231
5.25	190.57	2.28005	0.005248	**6.75**	854.06	2.93149	0.001171
5.26	192.48	2.28439	.005195	6.80	897.85	2.95320	.001114
5.27	194.42	2.28873	.005144	6.85	943.88	2.97492	.001060
5.28	196.37	2.29307	.005092	6.90	992.27	2.99663	.001008
5.29	198.34	2.29742	.005042	6.95	1043.1	3.01835	.000959
5.30	200.34	2.30176	0.004992	**7.00**	1096.6	3.04006	0.000912
5.31	202.35	2.30610	.004942	7.05	1152.9	3.06178	.000867
5.32	204.38	2.31045	.004893	7.10	1212.0	3.08349	.000825
5.33	206.44	2.31479	.004844	7.15	1274.1	3.10521	.000785
5.34	208.51	2.31913	.004796	7.20	1339.4	3.12692	.000747
5.35	210.61	2.32348	0.004748	**7.25**	1408.1	3.14863	0.000710
5.36	212.72	2.32782	.004701	7.30	1480.3	3.17035	.000676
5.37	214.86	2.33216	.004654	7.35	1556.2	3.19206	.000643
5.38	217.02	2.33650	.004608	7.40	1636.0	3.21378	.000611
5.39	219.20	2.34085	.004562	7.45	1719.9	3.23549	.000581
5.40	221.41	2.34519	0.004517	**7.50**	1808.0	3.25721	0.000553
5.41	223.63	2.34953	.004472	7.55	1900.7	3.27892	.000526
5.42	225.88	2.35388	.004427	7.60	1998.2	3.30064	.000501
5.43	228.15	2.35822	.004383	7.65	2100.6	3.32235	.000476
5.44	230.44	2.36256	.004339	7.70	2208.3	3.34407	.000453
5.45	232.76	2.36690	0.004296	**7.75**	2321.6	3.36578	0.000431
5.46	235.10	2.37125	.004254	7.80	2440.6	3.38750	.000410
5.47	237.46	2.37559	.004211	7.85	2565.7	3.40921	.000390
5.48	239.85	2.37993	.004169	7.90	2697.3	3.43093	.000371
5.49	242.26	2.38428	.004128	7.95	2835.6	3.45264	.000353
5.50	244.69	2.38862	0.004087	**8.00**	2981.0	3.47436	0.000336

TABLE IV (Continued)

x	e^x	$\text{Log}_{10}(e^x)$	e^{-x}	x	e^x	$\text{Log}_{10}(e^x)$	e^{-x}
8.00	2981.0	3.47436	0.000336	**9.00**	8103.1	3.90865	0.000123
8.05	3133.8	3.49607	.000319	9.05	8518.5	3.93037	.000117
8.10	3294.5	3.51779	.000304	9.10	8955.3	3.95208	.000112
8.15	3403.4	3.53950	.000289	9.15	9414.4	3.97379	.000106
8.20	3641.0	3.56121	.000275	9.20	9897.1	3.99551	.000101
8.25	3827.6	3.58293	0.000261	**9.25**	10405	4.01722	0.000096
8.30	4023.9	3.60464	.000249	9.30	10938	4.03894	.000091
8.35	4230.2	3.62636	.000236	9.35	11499	4.06065	.000087
8.40	4447.1	3.64807	.000225	9.40	12088	4.08237	.000083
8.45	4675.1	3.66979	.000214	9.45	12708	4.10408	.000079
8.50	4914.8	3.69150	0.000204	**9.50**	13360	4.12580	0.000075
8.55	5166.8	3.71322	.000194	9.55	14045	4.14751	.000071
8.60	5431.7	3.73493	.000184	9.60	14765	4.16923	.000068
8.65	5710.1	3.75665	.000175	9.65	15522	4.19094	.000064
8.70	6002.9	3.77836	.000167	9.70	16318	4.21266	.000061
8.75	6310.7	3.80008	0.000159	**9.75**	17154	4.23437	0.000058
8.80	6634.2	3.82179	.000151	9.80	18034	4.25609	.000056
8.85	6974.4	3.84351	.000143	9.85	18958	4.27780	.000053
8.90	7332.0	3.86522	.000136	9.90	19930	4.29952	.000050
8.95	7707.9	3.88694	.000130	9.95	20952	4.32123	.000048
9.00	8103.1	3.90865	0.000123	**10.00**	22026	4.34294	0.000045

TABLE V   Squares, Square Roots, Cubes, and Cube Roots

N	N^2	$\sqrt{N}$	N^3	$\sqrt[3]{N}$	N	N^2	$\sqrt{N}$	N^3	$\sqrt[3]{N}$
1	1	1.000	1	1.000	51	2,601	7.141	132,651	3.708
2	4	1.414	8	1.260	52	2,704	7.211	140,608	3.733
3	9	1.732	27	1.442	53	2,809	7.280	148,877	3.756
4	16	2.000	64	1.587	54	2,916	7.348	157,464	3.780
5	25	2.236	125	1.710	55	3,025	7.416	166,375	3.803
6	36	2.449	216	1.817	56	3,136	7.483	175,616	3.826
7	49	2.646	343	1.913	57	3,249	7.550	185,193	3.849
8	64	2.828	512	2.000	58	3,364	7.616	195,112	3.871
9	81	3.000	729	2.080	59	3,481	7.681	205,379	3.893
10	100	3.162	1,000	2.154	60	3,600	7.746	216,000	3.915
11	121	3.317	1,331	2.224	61	3,721	7.810	226,981	3.936
12	144	3.464	1,728	2.289	62	3,844	7.874	238,328	3.958
13	169	3.606	2,197	2.351	63	3,969	7.937	250,047	3.979
14	196	3.742	2,744	2.410	64	4,096	8.000	262,144	4.000
15	225	3.873	3,375	2.466	65	4,225	8.062	274,625	4.021
16	256	4.000	4,096	2.520	66	4,356	8.124	287,496	4.041
17	289	4.123	4,913	2.571	67	4,489	8.185	300,763	4.062
18	324	4.243	5,832	2,621	68	4,624	8.246	314,432	4.082
19	361	4.359	6,859	2.668	69	4,761	8.307	328,509	4.102
20	400	4.472	8,000	2.714	70	4,900	8.367	343,000	4.121
21	441	4.583	9,261	2.759	71	5,041	8.426	357,911	4.141
22	484	4.690	10,648	2.802	72	5,184	8.485	373,248	4.160
23	529	4.796	12,167	2.844	73	5,329	8.544	389,017	4.179
24	576	4.899	13,824	2.884	74	5,476	8.602	405,224	4.198
25	625	5.000	15,625	2.924	75	5,625	8.660	421,875	4.217
26	676	5.099	17,576	2.962	76	5,776	8.718	438,976	4.236
27	729	5.196	19,683	3.000	77	5,929	8.775	456,533	4.254
28	784	5.292	21,952	3.037	78	6,084	8.832	474,552	4.273
29	841	5.385	24,389	3.072	79	6,241	8.888	493,039	4.291
30	900	5.477	27,000	3.107	80	6,400	8.944	512,000	4.309
31	961	5.568	29,791	3.141	81	6,561	9.000	531,441	4.327
32	1,024	5.657	32,768	3.175	82	6,724	9.055	551,368	4.344
33	1,089	5.745	35,937	3.208	83	6,889	9.110	571,787	4.362
34	1,156	5.831	39,304	3.240	84	7,056	9.165	592,704	4.380
35	1,225	5.916	42,875	3.271	85	7,225	9.220	614,125	4.397
36	1,296	6.000	46,656	3.302	86	7,396	9.274	636,056	4.414
37	1,369	6.083	50,653	3.332	87	7,569	9.327	658,503	4.431
38	1,444	6.164	54,872	3.362	88	7,744	9.381	681,472	4.448
39	1,521	6.245	59,319	3.391	89	7,921	9.434	704,969	4.465
40	1,600	6.325	64,000	3.420	90	8,100	9.487	729,000	4.481
41	1,681	6.403	68,921	3.448	91	8,281	9.539	753,571	4.498
42	1,764	6.481	74,088	3.476	92	8,464	9.592	778,688	4.514
43	1,849	6.557	79,507	3.503	93	8,649	9.644	804,357	4.531
44	1,936	6.633	85,184	3.530	94	8,836	9.695	830,584	4.547
45	2,025	6.708	91,125	3.557	95	9,025	9.747	857,375	4.563
46	2,116	6.782	97,336	3.583	96	9,216	9.798	884,736	4.579
47	2,209	6.856	103,823	3.609	97	9,409	9.849	912,673	4.595
48	2,304	6.928	110,592	3.634	98	9,604	9.899	941,192	4.610
49	2,401	7.000	117,649	3.659	99	9,801	9.950	970,299	4.626
50	2,500	7.071	125,000	3.684	00	10,000	10.000	1,000,000	4.642

TABLE VI Common Logarithms

x	0	1	2	3	4	5	6	7	8	9
1.0	.0000	.0043	.0086	.0128	.0170	.0212	.0253	.0294	.0334	.0374
1.1	.0414	.0453	.0492	.0531	.0569	.0607	.0645	.0682	.0719	.0755
1.2	.0792	.0828	.0864	.0899	.0934	.0969	.1004	.1038	.1072	.1106
1.3	.1139	.1173	.1206	.1239	.1271	.1303	.1335	.1367	.1399	.1430
1.4	.1461	.1492	.1523	.1553	.1584	.1614	.1644	.1673	.1703	.1732
1.5	.1761	.1790	.1818	.1847	.1875	.1903	.1931	.1959	.1987	.2014
1.6	.2041	.2068	.2095	.2122	.2148	.2175	.2201	.2227	.2253	.2279
1.7	.2304	.2330	.2355	.2380	.2404	.2430	.2455	.2480	.2504	.2529
1.8	.2553	.2577	.2601	.2625	.2648	.2672	.2695	.2718	.2742	.2765
1.9	.2788	.2810	.2833	.2856	.2878	.2900	.2923	.2945	.2967	.2989
2.0	.3010	.3032	.3054	.3075	.3096	.3118	.3139	.3160	.3181	.3201
2.1	.3222	.3243	.3263	.3284	.3304	.3324	.3345	.3365	.3385	.3404
2.2	.3424	.3444	.3464	.3483	.3502	.3522	.3541	.3560	.3579	.3598
2.3	.3617	.3636	.3655	.3674	.3692	.3711	.3729	.3747	.3766	.3784
2.4	.3802	.3820	.3838	.3856	.3874	.3892	.3909	.3927	.3945	.3962
2.5	.3979	.3997	.4014	.4031	.4048	.4065	.4082	.4099	.4116	.4133
2.6	.4150	.4166	.4183	.4200	.4216	.4232	.4249	.4265	.4281	.4298
2.7	.4314	.4330	.4346	.4362	.4378	.4393	.4409	.4425	.4440	.4456
2.8	.4472	.4487	.4502	.4518	.4533	.4548	.4564	.4579	.4594	.4609
2.9	.4624	.4639	.4654	.4669	.4683	.4698	.4713	.4728	.4742	.4757
3.0	.4771	.4786	.4800	.4814	.4829	.4843	.4857	.4871	.4886	.4900
3.1	.4914	.4928	.4942	.4955	.4969	.4983	.4997	.5011	.5024	.5038
3.2	.5051	.5065	.5079	.5092	.5105	.5119	.5132	.5145	.5159	.5172
3.3	.5185	.5198	.5211	.5224	.5237	.5250	.5263	.5276	.5289	.5302
3.4	.5315	.5328	.5340	.5353	.5366	.5378	.5391	.5403	.5416	.5428
3.5	.5441	.5453	.5465	.5478	.5490	.5502	.5514	.5527	.5539	.5551
3.6	.5563	.5575	.5587	.5599	.5611	.5623	.5635	.5647	.5658	.5670
3.7	.5682	.5694	.5705	.5717	.5729	.5740	.5752	.5763	.5775	.5786
3.8	.5798	.5809	.5821	.5832	.5843	.5855	.5866	.5877	.5888	.5899
3.9	.5911	.5922	.5933	.5944	.5955	.5966	.5977	.5988	.5999	.6010
4.0	.6021	.6031	.6042	.6053	.6064	.6075	.6085	.6096	.6107	.6117
4.1	.6128	.6138	.6149	.6160	.6170	.6180	.6191	.6201	.6212	.6222
4.2	.6128	.6138	.6149	.6160	.6170	.6180	.6191	.6201	.6212	.6222
4.3	.6335	.6345	.6355	.6365	.6375	.6385	.6395	.6405	.6415	.6425
4.4	.6435	.6444	.6454	.6464	.6474	.6484	.6493	.6503	.6513	.6522
4.5	.6532	.6542	.6551	.6561	.6571	.6580	.6590	.6599	.6609	.6618
4.6	.6628	.6637	.6646	.6656	.6665	.6675	.6684	.6693	.6702	.6712
4.7	.6721	.6730	.6739	.6749	.6758	.6767	.6776	.6785	.6794	.6803
4.8	.6812	.6821	.6830	.6839	.6848	.6857	.6866	.6875	.6884	.6893
4.9	.6902	.6911	.6920	.6928	.6937	.6946	.6955	.6964	.6972	.6981
5.0	.6990	.6998	.7007	.7016	.7024	.7033	.7042	.7050	.7059	.7067
5.1	.7076	.7084	.7093	.7101	.7110	.7118	.7126	.7135	.7143	.7152
5.2	.7160	.7168	.7177	.7185	.7193	.7202	.7210	.7218	.7226	.7235
5.3	.7243	.7251	.7259	.7267	.7275	.7284	.7292	.7300	.7308	.7316
5.4	.7324	.7332	.7340	.7348	.7356	.7364	.7372	.7380	.7388	.7396

TABLE VI   (Continued)

x	0	1	2	3	4	5	6	7	8	9
5.5	.7404	.7412	.7419	.7427	.7435	.7443	.7451	.7459	.7466	.7474
5.6	.7482	.7490	.7497	.7505	.7513	.7520	.7528	.7536	.7543	.7551
5.7	.7559	.7566	.7574	.7582	.7589	.7597	.7604	.7612	.7619	.7627
5.8	.7634	.7642	.7649	.7657	.7664	.7672	.7679	.7686	.7694	.7701
5.9	.7709	.7716	.7723	.7731	.7738	.7745	.7752	.7760	.7767	.7774
6.0	.7782	.7789	.7796	.7803	.7810	.7818	.7825	.7832	.7839	.7846
6.1	.7853	.7860	.7868	.7875	.7882	.7889	.7896	.7903	.7910	.7917
6.2	.7924	.7931	.7938	.7945	.7952	.7959	.7966	.7973	.7980	.7987
6.3	.7993	.8000	.8007	.8014	.8021	.8028	.8035	.8041	.8048	.8055
6.4	.8062	.8069	.8075	.8082	.8089	.8096	.8102	.8109	.8116	.8122
6.5	.8129	.8136	.8142	.8149	.8156	.8162	.8169	.8176	.8182	.8189
6.6	.8195	.8202	.8209	.8215	.8222	.8228	.8235	.8241	.8248	.8254
6.7	.8261	.8267	.8274	.8280	.8287	.8293	.8299	.8306	.8312	.8319
6.8	.8325	.8331	.8338	.8344	.8351	.8357	.8363	.8370	.8376	.8382
6.9	.8388	.8395	.8401	.8407	.8414	.8420	.8426	.8432	.8439	.8445
7.0	.8451	.8457	.8463	.8470	.8476	.8482	.8488	.8494	.8500	.8506
7.1	.8513	.8519	.8525	.8531	.8537	.8543	.8549	.8555	.8561	.8567
7.2	.8573	.8579	.8585	.8591	.8597	.8603	.8609	.8615	.8621	.8627
7.3	.8633	.8639	.8645	.8651	.8657	.8663	.8669	.8675	.8681	.8686
7.4	.8692	.8698	.8704	.8710	.8716	.8722	.8727	.8733	.8739	.8745
7.5	.8751	.8756	.8762	.8768	.8774	.8779	.8785	.8791	.8797	.8802
7.6	.8808	.8814	.8820	.8825	.8831	.8837	.8842	.8848	.8854	.8859
7.7	.8865	.8871	.8876	.8882	.8887	.8893	.8899	.8904	.8910	.8915
7.8	.8921	.8927	.8932	.8938	.8943	.8949	.8954	.8960	.8965	.8971
7.9	.8976	.8982	.8987	.8993	.8998	.9004	.9009	.9015	.9020	.9025
8.0	.9031	.9036	.9042	.9047	.9053	.9058	.9063	.9069	.9074	.9079
8.1	.9085	.9090	.9096	.9191	.9106	.9112	.9117	.8122	.9128	.9133
8.2	.9138	.9143	.9149	.9154	.9159	.9165	.9170	.9175	.9180	.9186
8.3	.9191	.9196	.9201	.9206	.9212	.9217	.9222	.9227	.9232	.9238
8.4	.9243	.9248	.9253	.9258	.9263	.9269	.9274	.9279	.9284	.9289
8.5	.9294	.9299	.9304	.9309	.9315	.9320	.9325	.9330	.9335	.9340
8.6	.9345	.9350	.9355	.9360	.9365	.9370	.9375	.9380	.9385	.9390
8.7	.9395	.9400	.9405	.9410	.9415	.9420	.9425	.0430	.9435	.9440
8.8	.9445	.9450	.9455	.9460	.9465	.9469	.9474	.9479	.9484	.9489
8.9	.9494	.9499	.9504	.9509	.9513	.9518	.9523	.9528	.9533	.9538
9.0	.9542	.9547	.9552	.9557	.9562	.9566	.9571	.9576	.9581	.9586
9.1	.9590	.9595	.9600	.9605	.9609	.9614	.9619	.9624	.9628	.9633
9.2	.9638	.9643	.9647	.9652	.9657	.9661	.9666	.9671	.9675	.9680
9.3	.9685	.9689	.9694	.9699	.9703	.9708	.9713	.9717	.9722	.9727
9.4	.9731	.9736	.9741	.9745	.9750	.9754	.9759	.9763	.9768	.9773
9.5	.9777	.9782	.9786	.9791	.9795	.9800	.9805	.9809	.9814	.9818
9.6	.9823	.9827	.9832	.9836	.9841	.9845	.9850	.9854	.9859	.9863
9.7	.9868	.9872	.9877	.9881	.9886	.9890	.9894	.9899	.9903	.9908
9.8	.9912	.9917	.9921	.9926	.9930	.9934	.9939	.9943	.9948	.9952
9.9	.9956	.9961	.9965	.9969	.9974	.9978	.9983	.9987	.9991	.9996

Index